RESEARCH COMPANION TO CONSTRUCTION ECONOMICS

ELGAR COMPANIONS TO THE BUILT ENVIRONMENT

This important and timely series brings together critical and thought-provoking contributions on the most pressing topics and issues in built environment research. Comprising specially-commissioned chapters from leading academics these comprehensive *Companions* feature cutting-edge research, help to define the field and are written with a global readership in mind. Equally useful as reference tools or high-level introductions to specific topics, issues, methods, innovations and debates, these *Companions* will be an essential resource for academic researchers and postgraduate students in building, construction management and related disciplines.

Titles in the series include:

Research Companion to Building Information Modeling
Edited by Weisheng Lu and Chimay J. Anumba

Research Companion to Construction Economics
Edited by George Ofori

Research Companion to Construction Economics

Edited by

George Ofori

Professor, School of the Built Environment and Architecture, London South Bank University, UK

ELGAR COMPANIONS TO THE BUILT ENVIRONMENT

Cheltenham, UK • Northampton, MA, USA

Published by
Edward Elgar Publishing Limited
The Lypiatts
15 Lansdown Road
Cheltenham
Glos GL50 2JA
UK

Edward Elgar Publishing, Inc.
William Pratt House
9 Dewey Court
Northampton
Massachusetts 01060
USA

Paperback edition 2024

A catalogue record for this book
is available from the British Library

Library of Congress Control Number: 2022931153

This book is available electronically in the **Elgar**online
Economics subject collection
http://dx.doi.org/10.4337/9781839108235

ISBN 978 1 83910 822 8 (cased)
ISBN 978 1 83910 823 5 (eBook)
ISBN 978 1 0353 3734 7 (paperback)

Printed and bound by CPI Group (UK) Ltd, Croydon, CR0 4YY

This book is for my children: Abena, Amma, Akosua and Kofi.
They changed my life.

Contents

List of contributors

Abdul-Rashid Abdul-Aziz, Professor, Faculty of Built Environment, Universiti Malaysia Sarawak, Malaysia.

Afzan Binti Ahmad Zaini, Associate Professor, Faculty of Built Environment, Universiti Malaysia Sarawak, Malaysia.

Nii A. Ankrah, College of Engineering and Physical Sciences, Aston University, UK.

Chimay J. Anumba, Professor, College of Design, Construction and Planning, University of Florida, USA.

Rick Best, Associate Professor, Faculty of Society and Design, Bond University, Australia.

Jan Bröchner, Professor, Department of Technology Management and Economics, Chalmers University of Technology, Sweden.

Anita Cerić, Professor, Faculty of Civil Engineering, University of Zagreb, Croatia.

Albert P.C. Chan, Professor, Faculty of Construction and Environment, Hong Kong Polytechnic University, Hong Kong SAR.

Ezekiel Chinyio, Senior Lecturer, Faculty of Science and Engineering, University of Wolverhampton, UK.

Obas John Ebohon, Professor, School of the Built Environment and Architecture, London South Bank University, UK.

Stephen Gruneberg, Honorary Professor, Bartlett School of Sustainable Construction, Faculty of the Built Environment, University College London, UK.

Suraya Ismail, Director of Research, Khazanah Research Institute, Kuala Lumpur, Malaysia.

Hongbin Jiang, Partner, Capital Project and Infrastructure, PricewaterhouseCoopers China, Beijing, People's Republic of China.

John Kelsey, Associate Professor, Bartlett School of Sustainable Construction, Faculty of the Built Environment, University College London, UK.

Jeroen Klink, Professor, Center for Engineering, Modeling and Applied Sciences, Universidade Federal do ABC, Brazil.

Mohan Kumaraswamy, Honorary Professor, Faculty of Engineering, University of Hong Kong, Hong Kong SAR; and University of Moratuwa, Sri Lanka.

Samuel Laryea, Associate Professor, School of Construction Economics and Management, University of the Witwatersrand, South Africa.

Jorge Lopes, Professor, Department of Construction and Planning, Polytechnic Institute of Bragança, Portugal.

Weisheng (Wilson) Lu, Professor, Faculty of Architecture, University of Hong Kong, Hong Kong SAR.

Gangadhar Mahesh, Associate Professor, Faculty of Engineering, National Institute of Technology Karnataka, India.

Emmanuel Manu, Associate Professor, School of Architecture, Design and the Built Environment, Nottingham Trent University, UK.

Jim Meikle, Honorary Professor, Bartlett School of Sustainable Construction, Faculty of the Built Environment, University College London, UK.

Esther A. Obonyo, Associate Professor, College of Engineering, Pennsylvania State University, USA.

George Ofori, Professor, School of the Built Environment and Architecture, London South Bank University, UK.

Alex Opoku, Associate Professor, College of Engineering, University of Sharjah, UAE; Honorary Associate Professor, Bartlett School of Sustainable Construction, University College London, UK.

Emmanuel Kingsford Owusu, Research Assistant Professor, Faculty of Construction and Environment, Hong Kong Polytechnic University, Hong Kong SAR.

Asheem Shrestha, Senior Lecturer, Deakin University, Australia.

Sukhtaj Singh, Research Student, Faculty of Science and Engineering, University of Wolverhampton, UK.

Subashini Suresh, Reader, Faculty of Science and Engineering, University of Wolverhampton, UK.

Gerard de Valence, Senior Lecturer, School of the Built Environment, University of Technology, Sydney, Australia.

Ron Watermeyer, Director, Infrastructure Options, South Africa.

Edmundo Werna, Associate Professor, School of the Built Environment and Architecture, London South Bank University, UK.

Meng Ye, Associate Professor, School of Economics and Management, Southwest Jiaotong University, People's Republic of China.

Foreword

When I accepted my appointment as an economist in one of the largest construction firms in the United Kingdom (UK), I thought it should be possible to apply the principles of economics to some of the aspects of construction. I did not expect that the efforts I was making in undertaking both macroeconomic and microeconomic analyses from the perspective of one contractor would grow to become a field of knowledge. I am impressed by how the field of construction economics has developed since the first efforts I made with colleagues at University College London (UCL) to build the framework for the field. This framework was drawn from industrial economics, and I always thought that field would provide a permanent home for studies on the construction industry, its procedures, processes and products. The critiques of my notes and ideas by the students on the MSc in Building Economics and Management course over the initial years of the course helped me to write, and continue to develop, *Economic Theory and the Construction Industry*. The tentativeness of the title shows that I did not think a separate field would emerge.

The principles that I had developed were applied and developed further in the work I did for government agencies on performance improvement in the economy including the construction industry. I worked at the National Economic Development Office (NEDO) on the National Plan in the 1960s. When I left NEDO, I went to UCL. Working with colleagues at UCL and later at Reading University we analysed the role of construction in the economies of both developed and developing countries. I was fortunate to undertake consultancies on the construction industry, often funded by world organisations, in a number of countries, notably Egypt, Latvia, Russia and Sri Lanka.

I am happy that many governments seem to have a better understanding of the construction industry and how judicious levels of investment in it and stimulation of improvements in its performance can be used for the good of the economy and the nation.

In my early writings I observed that it was rare to find countries formulating policies to develop their construction industries even as they mapped out ambitious national development plans. Today, the *Construction Sector Deal* in the UK is an example of a conscious effort by government to prepare the construction industry to deliver on the ambitions of a national plan, in this case, the *Industrial Strategy*. This underlines the importance of construction economics: it is necessary for us to provide policy makers with the basis for decision making to ensure that an efficient and innovative construction industry enables the attainment of national aspirations. I have always said that clients get the industry they deserve, and this is more apt today than ever before. There is much scope for the government–industry initiatives such as those by the Construction Leadership Council to succeed in bringing about change in the industry for the benefit of its clients, the component organisations and practitioners, as well as society as a whole.

I am disappointed that construction economics has not yet attained recognition as a branch of economics, but I believe this will come if the rate of development of the field is at least maintained.

I think this collection is a highly valuable contribution to the field of construction economics. I like the broad sweep of topics in this collection. It is good to see that construction economics has grown to explore not only the theories and concepts of economics but also elements of the environment, society and technology. I like the exploration of the philosophy of construction economics, its application in antiquity, and its own nature and historical development. The strong theoretical foundations on which the chapters are based, and the projections for the future, are distinctive features of this book. I consider this to be truly an indispensable *Research Companion to Construction Economics*. I commend the authors of the chapters for producing such authoritative pieces. This is book will influence research and practice in construction economics for many years to come.

Patricia M. Hillebrandt
April 2021

Preface

When I was approached to put together and edit a collection of chapters on construction economics, I had no choice but to accept the invitation. I had been taught building economics in my undergraduate degree course in Building Technology at the University of Science and Technology, Kumasi, Ghana; and on the MSc course in Construction Economics and Management at University College London (UCL), I was taught by Dr Patricia Hillebrandt. In her highly interactive lecture sessions, we went through, section by section, the contents of *Economic Theory and the Construction Industry*, which Dr Hillebrandt had written a few years earlier. In writing my dissertation for the masters degree, and then doing the research for my doctoral degree, I had access to all the reports of the Building Economics Research Unit, which had been set up by Professor Duccio Turin, as well as the special tutelage of Dr Hillebrandt.

I am grateful to Dr Hillebrandt for writing the Foreword to the book. I also thank Professor Will Hughes for the great help he gave me on this book. He provided me with a great deal of incisive critical comments on the first draft of the original introduction, which served as a guide for the authors on what the book was setting out to achieve.

I thank my colleagues in construction economics research who accepted my invitation to write chapters for the book. I am happy to have received a positive response to my request from so many eminent researchers in construction economics. I appreciate the hard work all of them put in, especially the quick response to my many requests. I thank them also for being willing to review the draft chapters written by each other. Judging from the revisions made to the first drafts, the comments were useful in enabling the authors to improve the drafts.

The subject of construction economics focuses on the application of economic methods to every phase of the planning, design, construction and maintenance of built items. Over time, the field has been extended beyond the minimisation of capital cost on projects to include life cycle cost considerations, the idea of value, sustainable construction and climate change, and applications of technology. Attention has also been extended to include consideration of companies and organisations, as well as the strategic, industry-level considerations involving the economy and construction markets, government policy and regulations, and international finance and economics.

Construction economics is an important subject of research, owing to the strategic significance of the construction industry. However, the development of construction economics as a distinct scientific discipline has been very slow. In what way can this book be the turning point for greater pace, and achievement in future? What is different about this book, which makes one convinced that it is what will stimulate effective progress?

This is a major contribution to the body of knowledge on construction economics. It discusses some aspects of the subject for the first time. These include the philosophy and principles of the subject, the form it took in antiquity, and the historical development of the field. It considers contemporary issues on all the topics. It looks forward into the future and draws out a path for the development of the subject.

The book does not address every topic in construction economics. It has not settled all the questions on construction economics. The debate on some of the issues will, and should, continue. However, I believe readers will obtain great insights into the subject. I am proud to have been able to help in making this book possible. I have every confidence in saying that this is a *Research Companion to Construction Economics*.

George Ofori
April 2021

List of abbreviations

4IR	Fourth Industrial Revolution
ABC	activity based costing
ADB	Asian Development Bank
AEC	architecture, engineering and construction
AECO	architecture, engineering, construction and operation
AfDB	African Development Bank
AI	artificial intelligence
AMA	Analysis of Main Aggregates
APM	Association for Project Management
ASCE	American Society of Civil Engineers
BCA	Building and Construction Authority
BCA	Business Council of Australia
BDN	Blue Dot Network
BERU	Building Economics Research Unit
BIM	building information modelling
BLOC	basket of locally obtained commodities
BMI	Big Mac Index
BOCC	basket of construction components
BQ	bills of quantities
BREEAM	Building Research Establishment Environmental Assessment Method
BRI	Belt and Road Initiative
BRI	*Building Research and Information*
CAGR	compound annual growth rate
CBA	cost–benefit analysis
CEEC	European Council of Construction Economists
CEO	chief executive officer
CHORDS	catalyzer, hinderer, owner, regulator, disrupter, synergizer
Ci3	Construction Industry Institute India
CIB	International Council for Research and Information in Building and Construction
CIC	Construction Industry Council
CIDB	Construction Industry Development Board

CIOB	Chartered Institute of Building
CIRC	Construction Industry Review Committee
CIS	Commonwealth of Independent States
CITP	Construction Industry Transformation Programme
CLC	Construction Leadership Council
CME	*Construction Management and Economics*
CPD	continuing professional development
CPI	consumer price index
CPS	cyber-physical system
CVA	construction value added
DFC	Development Finance Corporation
DfMA	design for manufacturing and assembly
DT	digital twins
ECAM	engineering, construction and architectural management
EIA	environmental impact assessment
ENR	*Engineering News-Record*
EPC	engineering, procurement and construction
EPC+F	engineering, procurement and construction with financing
ESA	European System of Accounts
FDI	foreign direct investment
FIABCI	International Federation of Real Estate Developers
FINT	First International Network on Trust
GDP	gross domestic product
GFCF	gross fixed capital formation
GGDC	Groningen Growth and Development Centre
GICA	Global Infrastructure Connectivity Alliance (GICA)
GIP	Global Infrastructure Programme
GLF-CEM	Global Leadership Forum for Construction Engineering and Management Programs
GNI	gross national income
HDI	Human Development Index
HM Government	Her Majesty's Government
HNDA	Housing Need and Demand Assessment
HPI	house price index
ICE	Institution of Civil Engineers
ICG	Infrastructure Client Group

ICGP	Inter-organizational Committee on Principles and Guidelines for Social Impact Assessment
ICMS	International Construction Measurement Standards
ICO	investment, construction and operation
ICP	International Comparison Program
ICT	information and communication technology
ICTAD	Institute of Construction Training and Development
ILO	International Labour Office
IoT	Internet of Things
IPMS	International Property Measurement Standards
ISIC	International Standard Industrial Classification of All Economic Activities (also SIC)
ISO	International Organization for Standardization
JCDC	*Journal of Construction in Developing Countries*
KPI	key performance indicator
LCA	life cycle assessment
LCC	life cycle costing
LEED	Leadership in Energy and Environmental Design
LIC	low-income country
LMIC	lower middle-income country
M&A	mergers and acquisitions
MDG	Millennium Development Goal
MEA	Millennium Ecosystem Assessment
MIC	middle-income country
MMC	modern methods of construction
MNC	multinational corporation
MSD	musculoskeletal disorder
NASA	National Aeronautics and Space Administration
NCS	net capital stock
NEC	New Engineering Contract
NEDO	National Economic Development Office
NII	New Infrastructure Initiatives
NRM	New Rules of Measurement
NUA	New Urban Agenda
OECD	Organisation for Economic Co-operation and Development
OFT	Office of Fair Trading
OLI	ownership, location and internalization

OLI+S	ownership, location and internalization plus specialty
PCICB	Provisional Construction Industry Co-ordination Board
PFI	private finance initiative
POE	post-occupancy evaluation
PPR	post-project review
PPP	public–private partnership
PPP	purchasing power parity
PSIB	Process and System Innovation in Building and Construction (PSIB) Programme
PWT	Penn World Table
R&D	research and development
RIBA	Royal Institute of British Architects
RICS	Royal Institution of Chartered Surveyors
RIVANS	relationally integrated value networks
ROCE	return on capital employed
RoI	return on investment
RTI	repeated transaction index
SAR	Special Administrative Region
SDG	Sustainable Development Goal
SIA	social impact assessment
SME	small and medium-sized enterprises
SNA	System of National Accounts
SROI	social return on investment
TCE	transaction cost economics
TFV	transformation, flow and value generation
UCL	University College London
UMIC	upper-middle-income country
UN-Habitat	United Nations Human Settlements Programme
UNDP	United Nations Development Programme
UNTEC	National Union of Construction Economists
VDC	virtual design and construction
VfM	value for money
WEF	World Economic Forum
WLC	whole life costing
WTO	World Trade Organization

1. Introduction to the *Research Companion to Construction Economics*

George Ofori

INTRODUCTION

This *Research Companion*

A review of the literature on construction economics shows that a new, comprehensive reference book on the body of knowledge of the subject is needed. The reasons are: there is a debate on many of the aspects of the field of knowledge, including many of the topics it covers, such as its main concern, the construction industry; and there is a need for effort to systematically build up the subject in order to enable it to claim the status of a separate field of knowledge. To fill this gap, the required reference book will have to be a compendium of existing knowledge which also reviews the origins of the subject, explores new ideas, and contributes to the development of new knowledge on the overall field. This Elgar *Research Companion to Construction Economics* seeks to fulfil this need.

The *Research Companion* presents a picture of the state-of-the-art of the subject. It creates a platform for presenting, discussing and challenging the existing theories and concepts, and the existing paradigms of the field. It enables a critical examination of complex issues in the field to be undertaken. It also provides the opportunity for some new issues and underexplored ones in the field to be discussed.

The book comprises original chapters authored by specially invited globally recognised experts on the particular topics. This is an influential collection which represents a broad-ranging coverage of the field of construction economics.

Aim and Objectives of the Book

The aim of the book is to present a state-of-the-art review of construction economics and indicate possible new directions for the subject. It is intended to be an authoritative reference text on the subject. The objectives of the book are to:

1. present a comprehensive review of the literature on the main topics which comprise the body knowledge of construction economics;
2. explore the fundamental issues and underpinnings of the subject of construction economics such as its philosophy; its history; its nature, boundaries and linkages to related areas such as segments of mainstream economics; its theories and main concepts; and the methods of analysis it uses;
3. propose new topics and subtopics to be explored with the view to enabling the subject to grow;
4. draw the outline of the future of the subject of construction economics.

Producing the Book

The carefully selected authors of the chapters were provided with a guide which set out the rationale for the book; outlined its aims and objectives; and indicated the timeline for the production of the book. Each author was given a topic on which they had a solid body of work, together with a brief outline of the possible content prepared by the editor. The authors were given the flexibility to suggest possible fine-tuning of the topics or the guidance notes. The first draft of each chapter was peer reviewed by one of the other authors, as well as the editor.

Nature of the Chapters

Each chapter of the book presents a thorough scholarly analysis of the selected topic. Each chapter analyses the existing knowledge on the topic, compares the various views on it, and presents a reference point for further advanced research leading to further development of the subject. This first chapter of the book provides an introduction of the book, and presents a synopsis of the various chapters.

Each chapter of the book presents and discusses current definitions and the main concept of the topic; considers historical development and current state; discusses likely future development; and proposes aspects for further research and further development of the field of knowledge. As each chapter forms part of an important reference book intended for researchers, students, teachers and practitioners, it is based on a strong review of the literature. Each chapter presents the state of the art of the topic it covers. Each chapter is an authoritative and enduring piece on the topic. Each chapter is an important addition to the body of knowledge on construction economics.

SYNOPSES OF THE CHAPTERS

A synopsis of each of the chapters in the book is now presented.

Chapter 2 by George Ofori

Ofori writes the first substantive chapter in the book. It has the title: 'Construction economics: its origins, significance, current status and need for development'. Ofori notes that construction economics should be an essential field of knowledge and research because of the uniqueness of the construction industry, process, procedures and products. The completed items are necessary for other economic and social activities which lead to economic growth, long-term development and improvement in quality of life. Ofori observes that, despite this importance of construction, construction economics has not yet been established as a distinct field of knowledge, or branch of economics. It is not recognised as a scientific discipline in its own right. Ofori states that the aim of the chapter is to discuss responses to these key questions on the subject of construction economics. What is construction economics? How has it developed? What are some of its principles, theories and concepts? What are its techniques, which are applied in research and practice? What is its status, such as its position among the established bodies of knowledge? What is its future?

First, Ofori discusses the nature of construction economics with a brief review of the historical development of the subject, which considers some of the major milestones and driving forces. Second, he analyses the components of construction economics, at the industry, company and project levels, and outlines the main themes and techniques of these components, putting the focus on some of the major contributors and the key works. He presents examples from various countries. Finally, Ofori suggests that there is a need to develop the subject, and highlights some possible areas which such developmental efforts should cover.

Chapter 3 by Stephen Gruneberg and John Kelsey

In the chapter entitled 'The philosophy of construction economics', Gruneberg and Kelsey set out a number of issues in which they subject issues in construction economics to philosophical consideration. They note that to enable readers to understand the nature of the philosophy of construction economics, a number of preliminary steps are necessary. The first step taken by Gruneberg and Kelsey is to define philosophy. The second is to define economics; the third is to define construction economics; and the fourth is to define the philosophy of economics. They then address the main task, and take up the discussion of the philosophy of construction economics. Gruneberg and Kelsey set out the conditions for a construction industry that steers a course that is positive and inspiring. They suggest that central to this approach are six principles of construction: an industry that is competitive; an industry that is productive and embraces innovation; an industry that produces a quality output; an industry that is efficient and whose output is efficient; an industry that employs a workforce that is professional in its attitude, behaviour and skills; and an industry that has an excellent reputation and has confidence and pride in itself.

Gruneberg and Kelsey note that in a single chapter it is only possible for them to direct the reader to a number of potential areas of debate. Therefore, they note that they consider it necessary to state the key issues rather than carrying out in-depth analysis of the issues. They observe that the questions they address draw attention to other matters which may be explored in the future, and which are considered in the concluding the chapter of this book.

Chapter 4 by Gerard De Valence

In the chapter on 'The nature of construction economics', De Valence suggests that it is challenging to define construction economics because of the range of topics associated with the production and maintenance of the built environment, the economic role of the construction industry, and the porous boundaries between the subject and related disciplines such as urban and regional economics, housing economics, cost engineering and, in particular, construction management.

De Valence discusses the interrelated propositions about the definition, boundaries and historical development of construction economics. He reviews construction economics research across the segments of the field: the construction process, the construction firm and the construction industry. De Valence considers three periods: 1966 to 1989 is one of emergence and establishment, as themes and topics are investigated and the literature develops; 1990 to 2007 covers the expansion of construction economics as new approaches such as transaction cost economics are applied to construction; and 2008 to 2020 brings the story to the present. In each period, research contributions to the three areas and the topics that research has covered are discussed. This is followed by an assessment of the development of construction economics across the areas and topics in the literature, along with issues such as the availability and quality of relevant data, which has constrained empirical research on construction economics. Gruneberg concludes the chapter with the argument that the scope of construction economics is a strength because it reflects the nature of an evolving economy and a construction industry which can be defined in various ways. The openness of construction economics to ideas from economics has led to important insights into construction firms, markets, and the relationship between the industry and the wider economy.

Chapter 5 by Jan Bröchner

Bröchner addresses the topic, 'Construction economics in antiquity'. He reviews the current state of research on the economics of the construction industry in ancient Greece and Rome. Bröchner notes that the sources for construction costs and the role of construction in the economy are literary texts, inscriptions and other archaeological finds. Grain and olives for prehistoric building workers are recorded in Linear B texts on clay tablets from 13th century BC Pylos in the southwest Peloponnese. Bröchner points out that, in the 5th century BC, Herodotus provides insights into the economic aspect of major building projects. Athenian democracy was an audit society, and there are detailed inscribed data for Acropolis temples because of the need for permanent and displayed accounts related to public procurement. He mentions that in Aristotle's *Politics*, tyrants are considered to have reasons to keep the people poor and occupied through large public construction projects.

Bröchner reveals that in ancient Rome there was less concern with public scrutiny of projects, while both the Republic and the Empire were able to finance huge investments in building. The Romans developed new construction technologies, and although we lack original data on work rates for specific tasks, cost effects of innovative technologies for wall construction and concrete vaulting have been estimated. Finally, Bröchner shows that Diocletian's Price Edict of AD 301 includes maximum wages for categories of construction workers and prices for construction materials, clearly indicating a command economy with a large bureaucracy, trying to reduce inflation.

Chapter 6 by Jorge Lopes

Lopes considers the topic, 'Construction in the economy and in national development'. He notes that buildings and other construction assets constitute a significant part of a country's physical and economic infrastructure, and that construction infrastructure plays a role as a capital input into the production and wealth generation. The economic impact can be transformative, especially at low-to-middle levels of income per capita. Lopes suggests that the close association between physical capital and different measures of the national economy is one of the reasons why physical infrastructure has been considered a powerful engine of economic growth and development. However, a significant number of studies have not shared the generalised view on the positive role of construction investment, that is, on the magnitude of investment and the direction of causality between construction investment and economic growth.

Lopes reviews the main strands in the literature on the role the construction sector plays in the national economy and in economic development. He also assesses the development pattern of the construction sector in two groups of sub-Saharan African countries pertaining to the middle-income status of economic development, for the period between 2000 and 2018. Lopes concludes that the results of the study show that the share of construction (measured as a proportion of construction value added in gross domestic product) revolve around a norm that is determined by a country's level of built assets prior to the period of reference.

Chapter 7 by Nii A. Ankrah and Emmanuel Manu

Ankrah and Manu write on the topic, 'Construction project economics'. They explain that, at the project level, the body of knowledge of construction economics has largely focused on showcasing tools and techniques that support decision-making throughout the life cycle of a project. Specifically, techniques such as cost–benefit analysis, option appraisal, project budgeting, cost analysis and planning, cost modelling, estimating of building costs, and whole life costing have been the bedrock of construction project economics, and have supported the management of the economic problem of making optimal choices regarding the allocation and use of scarce resources such as land, labour, capital, equipment, time and management on construction projects. In the chapter, Ankrah and Manu discuss the main concepts underpinning these techniques, and consider their historical development and limitations, and current thinking regarding their application.

Ankrah and Manu note that some emerging global trends are driving the development for application of a new approach to construction project economics by shifting the focus from cost and financial value to increasing emphasis on the environmental implications and social value contributions of construction projects. These trends include circular thinking, carbon accounting, social accounting, natural capital accounting, new procurement methodologies, and application of smart off-site manufacturing approaches in construction. Ankrah and Manu explore some of these trends, their likely future development, their implications for the optimal utilisation of project resources, and the tools and techniques needed to embed them into standard project practice. They highlight the connections to mainstream economics thinking throughout the discussion in the chapter. Based on the current gaps in knowledge around these issues which are identified, Ankrah and Manu propose aspects for further research to support more informed and rational decision-making in relation to construction projects.

Chapter 8 by Mohan Kumaraswamy and Gangadhar Mahesh

In the chapter on 'Dynamics of construction industry development', Kumaraswamy and Mahesh note that imperatives for improving construction industry performance have triggered high-powered reports with promising recommendations in many countries in the recent past. However, the resulting improvements have rarely met expectations. They suggest that the reasons for these shortfalls have been traced to various root causes, including: a lack of system-level understanding of construction industry linkages with the environment in which it operates; institutional inertia; and a resistance to shift industry culture and mindsets from adversarial to collaborative working modalities in fragmented teams.

In the chapter, Kumaraswamy and Mahesh probe the failure of decision-makers to appreciate and leverage the linkages between the construction industry and the national economy, from the dual perspectives of: (1) construction economics; and (2) the overarching goal of national development, which itself needs a reliable and efficient construction industry. For example, lack of work continuity disrupts the development of high-performing construction organisations, personnel and technologies. Kumaraswamy and Mahesh note that to address these for the longer term, policy-makers should be appraised of an important construction industry–national economy nexus that may be better appreciated through a dual macro–micro construction economics 'bifocal' lens. Such an overview could inform policy development, for example, to justify providing alternative work opportunities in public infrastructure augmentation, and rehabilitation during troughs in demand for construction. Kumaraswamy and Mahesh also provide insights into industry issues previously identified in Australia and the Netherlands; and lessons learned when implementing landmark industry reform reports around the turn of the century, in the Hong Kong SAR, Singapore and the United Kingdom. They provide examples of other important issues and proposed remedial measures in developing construction industries: a recent development initiative in India, and previous exercises in Sri Lanka.

Chapter 9 by Abdul-Rashid Abdul-Aziz and Afzan Binti Ahmad Zaini

Abdul-Aziz and Zaini consider the topic, 'Applications of mainstream economic theories to the construction industry: transaction costs'. They note that, at its core, transaction cost economics, which was pioneered by Williamson in the 1970s, addresses the make or buy decision. It posits that the optimum governance structure (that is, the hierarchy or the market) is the one that achieves economic efficiency through minimisation of transaction costs. External transaction costs arise whenever goods or services are procured through the market as opposed to internally. They cover search and information costs, bargaining and decision costs, and policing and enforcement costs. Abdul-Aziz and Zaini explain that transaction cost economics employs concepts such as 'bounded rationality', 'asset specificity' and 'opportunism'. They note that there has been refinement since it was first introduced, notably the recognition that potential arrangements can be a market hierarchy hybrid that occupies anywhere along the continuum between the two extremes. They observe that transaction cost economics has influenced scholars in many disciplines such as politics, management, business and social sciences, and there has been increasing interest in it among construction researchers.

Abdul-Aziz and Zaini note that some construction economics scholars opine that transaction cost economics has limited explanatory power for various practices of the construction industry. The most significant reason those critics give is that so many issues are excluded from the framework. However, the limitations highlighted have not stopped other scholars in construction from still applying transaction cost economics when undertaking studies such as analysing subcontracting, supply chains, public–private partnerships, alliancing and even project delays. One way to compensate for the shortcomings of the concept is to use it in combination with other theories to obtain more credible theoretical lenses for investigation of specific construction economics phenomena.

Chapter 10 by Alex Opoku

In the chapter on 'Construction industry and the Sustainable Development Goals (SDGs)', Opoku notes that the 2030 Agenda for sustainable development, which set out 17 Sustainable Development Goals, 169 targets and 232 indicators, is a major global initiative towards socio-economic development in all countries. The construction industry has a critical role to play in achieving the SDGs; it can influence the realisation of the SDGs by formulating policies and regulatory frameworks that drive the adoption of sustainable construction practices. The SDGs provide the construction industry with a new lens through which global needs and desires can be translated into business solutions. The SDGs provide a framework for all types of construction business organisations to play a crucial role towards the realisation of the SDGs by embracing the opportunities they present.

Opoku explores the role of the construction industry in achieving the 2030 Agenda for sustainable development. He suggests that, to successfully deliver the targets of the 2030 Agenda, the industry should engage with construction businesses in aligning business strategies with the SDGs. He suggests that the companies in the construction industry should integrate the SDGs into their long-term business strategies; and companies and practitioners should collaborate with government agencies, industry peers and policy-makers to attain the SDG targets and the broader objectives of the 2030 Agenda for sustainable development.

Chapter 11 by Obas John Ebohon

Ebohon addresses the topic, 'Sustainability economics and the construction industry'. He notes that, like the concept of sustainability, sustainability economics is an evolving concept that has its roots in ecological or environmental economics. Thus, it has been approached from different perspectives. Nevertheless, these perspectives have in common the emphasis that sustainability economics places on the problem associated with economic growth and attendant pollution, and the challenges posed to sustaining economic growth given the insatiable demands on environmental resources on the one hand, and the exhaustible, non-renewable and finite nature of natural resources on the other. Ebohon notes that concerns about potential trade-offs and conflicts between equity and economic efficiency in the pursuit of economic growth remain the focus of sustainability economics.

Ebohon observes that there is intense debate about the existence of such conflicts: whether human-made capital can replace natural capital, economic growth continues and attendant pollutions can be managed and contained – the 'weak sustainability' concept. However, there are some natural capital services, such as the ecosystem, that human capital cannot replace; hence the view is that natural capital and man-made capital are complimentary but not interchangeable. Ebohon notes that, this being the case, it is necessary to decouple economic growth from natural resources, reducing the consumption intensity in order to enhance the absorptive or regenerative capacity of the natural environment. Considering the disproportionate amount of natural resources construction accounts for, the predominant share in total global solid wastes generated, and the huge volume of carbon emissions, this debate is critical to the survival of the construction industry.

Ebohon critically examines the case for adopting the strong sustainability approach to construction activities. He also analyses the indirect risks posed to the natural environment and the global ecosystems, but also the direct risks to the construction industry itself, if it continues with the 'business as usual' or weak sustainability approach in its activities.

Chapter 12 by Jim Meikle and Asheem Shrestha

On the topic, 'International construction data: a critical review', Meikle and Shrestha note that construction is a difficult economic activity to measure and record. As a result, construction data is often unreliable and analyses over time and across countries are particularly problematic, although data is an essential element of much research on construction, so researchers should understand and be able to deal with its shortcomings.

Meikle and Shrestha describe the availability and characteristics of both official statistical data and other quantitative data on international construction and discuss the more important economic indicators and ratios that emerge from it. Meikle and Shrestha focus on data that support analysis of construction in the economy, construction prices and construction productivity, but also comment on other construction-related data, including on building and infrastructure stock, housing production numbers and construction materials consumption, data that can help indicate construction volume. The final section of their chapter concludes with a review of the existing situation and makes recommendations for future work by construction researchers and others.

Chapter 13 by Rick Best

Best writes on 'Measuring and comparing construction costs in different locations: methods and data'. He reviews the issues related to measuring and comparing construction costs between countries and within countries, with emphasis on measurement at the levels of projects, types of work and industries. Best notes that while cost is often a key parameter in productivity and other studies, and cost comparisons are often attempted, there are many factors that complicate such attempts, and researchers are constantly searching for more robust methods that alleviate or eliminate at least some of the complexities.

The topics addressed by Best in the chapter include: the nature of construction, mainly the lack of homogeneity across types of items and individual projects; comparison methods – a summary of previous studies and current exercises; the link between spatial and temporal cost comparisons, the Law of One Price; sources of construction cost data and their reliability; methods for converting international costs to a common base – purchasing power parity, 'real' exchange rates; comparability and representativeness of the components of cost comparison exercises; and a summary of the strengths and weaknesses of existing methods.

Chapter 14 by Hongbin Jiang

Jiang addresses the topic, 'New trends in international construction'. He notes that international construction is in a state of transformation; the future will require innovation, collaboration, optimisation and an understanding of the trends that will shape the industry. He reviews the main concepts of international construction and analyses the current and future trends. Jiang notes that international construction has long been complementary to studies on international economics. He revisits the main analytical frameworks for international construction in the contemporary global market, and presents the current profiles in terms of growth capacity, market coverage, business modalities and sector presence. He proposes a new framework which has six dimensions: the endogenous catalysing factors (Catalyzer), the hindering factors (Hinderer), the ownership essentials (Owner), the regulatory factors (Regulator), the disruptive exogenous factors (Disruptor) and the synergizing factors (Synergizer). He uses this new CHORDS framework to analyse trends in international construction.

Jiang observes that the disruptions of technology, environmental issues, energy transition, and so on, have big implications for the long-term global trends including demographic and social change, shifts of global economic power, rapid urbanisation, climate change and resource scarcity. He identifies six major trends which he expects to emerge in international construction: (1) consolidated connectivity in the physical domain, information and financial flows; (2) continued high-tech disruption in process transformation, smart facilities, and information and communication (ICT) expansion; (3) demanding resilience in environmental, operational and financial sustainability; (4) synergized diversification in business models, sector presence and financing modality; (5) improved standardization in technical standards, procurement and legal procedures; and (6) sophisticated oligopoly of firms and nations in the midst of the frictions of the trends in both globalization and deglobalization.

Jiang discusses three major current schemes: the Belt and Road Initiatives advocated by China from 2013; the Global Infrastructure Programme launched by UK Prosperity Funds in 2018; and the Blue Dot Network established by the United States government in 2019. Jiang analyses the potential future impacts of these schemes in the international construction arena. He makes proposals for further research and further development of the field of knowledge of construction economics.

Chapter 15 by Anita Cerić

Cerić writes on the topic, 'Economics of trust in construction'. She notes that the thesis that trust is central to construction projects comes from both empirical research and theoretical foundations of related fields of social science, such as economics, sociology and psychology. However, much of the work in construction management studies trust empirically, without theoretical foundations in economics, sociology or psychology. Cerić provides an overview of the principal–agent theory (which is part of the new institutional economics) and its application to the construction management field. She notes that the theory provides a framework for the interplay of economics, sociology and psychology in research on the relationships between the project owner, the contractor and their respective project managers. According to the theory, information asymmetry occurs between project participants, which leads to communication risk. Trust minimises information asymmetry and it thus reduces the communication risk. Thus, building trust and creating a reputation for trustworthy behaviour is valuable for every project. In construction projects, the role of trust is even more important as non-contractual relationships outnumber contractual ones.

In the chapter, Cerić presents a framework for investigating trust. Trust involves project participants interacting at interpersonal and interfirm levels. Trust involved in interfirm relationships falls mainly into the domain of economics, but both sociology and psychology can contribute to the study; trust involved in intrafirm relationships relies chiefly on sociology, but economics and psychology are also relevant to its study; and trust involved in interpersonal relationships falls mainly in the domain of psychology, but both sociology and economics can also contribute to the study. Cerić suggests that research on trust in construction projects needs to focus on both empirical research and its theoretical foundation, and there should be a close relationship between researchers and practitioners which is conducive to the advancement of both theory and practice.

Chapter 16 by Edmundo Werna and Jeroen Klink

Werna and Klink address the topic: 'The builders of cities: prospects for synergy between labour and the built environment'. They analyse labour in construction in the context of urban development, stressing the importance of research and policies on the connection between the construction workers who build cities and towns, and the settlements they produce. Werna and Klink note that the rapid urbanisation throughout the world is forecast to continue; thus, the already significant number of construction workers in urban settlements is due to increase. Construction is a major provider of employment in urban areas directly or indirectly, especially for the poor. Thus, the quality and quantity of employment in construction have a significant impact on urban development and urban poverty alleviation. Trends in urban development also have an impact on the construction industry and its workers. Werna and Klink point out both complementarities and gaps between construction economics and mainstream urban economics. Whereas mainstream urban economics works with concepts such as Marshallian economies of agglomeration, there is not much research on the strategic contribution of the construction industry to the generation of positive agglomeration effects in urban centres. Werna and Klink observe that changes in labour in construction are related to the rapid changes in the industry. Also, paradigms for urban growth and blueprints have emerged including the SDGs, especially number 11 (about cities) and number 8 (about employment), the New Urban Agenda of the United Nations, and the construction and urban agendas of the World Economic Forum.

Werna and Klink observe that given the labour–construction–cities nexus, the research questions are: Do these trends combine or clash? Is their mutual impact a vicious or virtuous cycle? What is happening in the interface between labour economics, construction economics and urban economics? Many challenges need to be better understood. Cities and towns have never had as much wealth and innovation as now. However, there are huge poverty deficits and socio-economic differences, in both the Global North and the Global South. It is not clear how the panacea paradigms of 'smart cities', 'sustainable cities' or 'resilient cities' will bridge the inequalities. Cities usually include the epitome of cutting-edge construction. The construction industry is embracing the 4th industrial revolution, with cyber-technology. Yet, there is no evidence that construction using the advanced technologies adopted will trickle down to the slum areas.

How do the challenges faced by urban construction workers impact on their capacity to deliver quality products? How are they rooted in urban development? What are the main features of the construction industry that are necessary to build the cities and towns according to the new urban paradigms? What is the necessary profile of the construction workforce to serve the industry and build the cities of tomorrow? Recommendations made by Werna and Klink of initiatives for promoting decent work in construction may be significant for urban development in general, and urban poverty alleviation in particular.

Chapter 17 by Samuel Laryea

In the chapter on 'Economic principles of bidding for construction projects', Laryea discusses economic principles and considerations influencing the bidding price of a construction project; the way that certain factors influence the economic behaviour of client and contractor in the bidding process; and some of the theories and principles that have been used to explain bidding behaviour in construction. He notes that most contracts in the construction industry are awarded through a tender or bidding process used by clients to obtain an economic price from competing contractors. Bidding is a complex process of interaction and exchange between a client and the competing contractors. The bidding approach of contractors is influenced by both micro-level economic needs and macro-level economic factors. While bidding contractors aim to cover their costs, they also try to gain a competitive advantage with their pricing, balancing their pricing needs with other realities that may constrain perfect rationality in deciding the final price. The microeconomic theory of the behaviour of individual competitive markets suggests that a bidding price may depend on the market in which it takes place, and a firm's particular circumstances.

Laryea notes that a contractor's bidding process comprises: the commercial review process, which helps to make a bid/no-bid decision; estimating, where the actual project costs are considered; and the adjudication process, where the directors of a firm take a commercial view of the estimated cost in the context of the firm's particular circumstances, market conditions and risk. The directors ultimately try to pitch the bidding price between cost and value in order to win the work. Laryea highlights key economic considerations at play at each stage of the bidding process. The relationship between cost, price and value ultimately influences the final bidding price that is negotiated and agreed between the client and contractor. In closing, Laryea highlights the increasing adoption of more strategic procurement and contracting approaches by clients to engage contractors in longer-term partnering relationships to achieve better value and outcomes. Current global economic conditions will increase clients' demand for better value and improved outcomes. This requires innovative and more efficient supply models from contractors. Further studies should develop better understanding of the leadership role of the client in achieving project economy and ways to enhance project information, communication among parties, pricing data and procurement processes required to develop and achieve accurate, cost-effective and reliable bids.

Chapter 18 by Ron Watermeyer

Watermeyer writes on the economics of 'Procurement and delivery management'. He notes that the construction industry delivers its products in a project-specific environment, which on every project involves: (1) different combinations of funders, clients, professionals, site conditions, materials and technologies, general contractors, specialist contractors, skills, workforces, client requirements and stakeholders; and (2) risk events that can significantly impact on project outcomes during the protracted delivery process. Watermeyer notes that client procurement and delivery management practices (the client buying functions) are central to the performance of the construction supply chain and have a direct impact on the realisation of the client's value proposition for the project, that is, the promise of measurable benefits. He states that procurement is the process which creates, manages and fulfils contracts, while delivery management is the critical leadership role played by a knowledgeable client to plan, specify, procure and oversee the delivery of projects. Delivery management is required to translate the value proposition associated with a business case into project outcomes. Watermeyer notes that there is often a significant gap between what was planned and what was achieved. Research has found that the root causes of project failures commonly relate to a lack of governance and poor procurement and delivery management practices, all of which are under the control of the client.

Watermeyer describes the role of the client, examines the reasons for poor project outcomes, and outlines client practices aimed at consistently delivering value for money. Such practices focus on client governance and organisational ownership, the provision of chief executive officer (CEO)-level client leadership, and strategic and tactical approaches to procurement, all of which support effective implementation.

Chapter 19 by Suraya Ismail

Ismail discusses the topic, 'The economics of housing policy and construction: developing a responsive supply sector'. She notes that in their housing policies, most nations attempt to overcome the problems of severe housing shortages and lack of affordability created by prevailing market conditions. The underlying assumptions of utility-maximising households and profit-maximising firms create market inefficiencies such as speculative purchases, sub-standard housing, high vacancy rates, and rapid house price escalations leading to decreased affordability for a significant proportion of society.

Ismail observes that housing affordability is a function of both house prices/rent and income within a specific housing market area. This infers an analysis into the demand and supply of local housing conditions which is shaped by a country's social and institutional context. She notes that the policy response usually covers: (1) the demand side, for example, tenure policies, taxation, interest and mortgage rates; and (2) the supply side, such as land and property rights, spatial planning and the firms in the construction industry. Ismail provides evidence to show that enhancing the efficacy of the supply side will render the provision of housing generally more affordable to different segments of society.

Ismail proposes that the problem of inadequate supply of housing be framed from the perspectives of both the institutional arrangement and the governance of firms in the construction project coalition. This involves analysing the national business systems as manifested by the temporal clustering of firms within the procurement routes. She suggests that improvements at the micro-analytical level of construction projects will increase the general affordability, and hence enhance the production and consumption of houses.

Chapter 20 by Ezekiel Chinyio, Sukhtaj Singh and Subashini Suresh

Chinyio, Singh and Suresh discuss 'A review of stakeholder management in construction'. They note that many organisations would have different interests in the operations or projects of a particular firm. When these interests do not align with those of the firm or its projects, there would be a potential for covert or overt conflicts which could manifest as opposition, hostility, sabotage, antagonism, and so on. Chinyio, Singh and Suresh suggest that the costs and time consequences of stakeholder management impinge on the economics of an organisation and thus the industry. They note that stakeholder management initially gained prominence in business management, and subsequently attracted the attention of construction and other disciplines. They point out that many scholars attribute the modern formalisation of stakeholder management to Ronald Freeman, but its art is part of general human relations at individual and corporate levels. Stakeholder management is now entrenched in organisational developments and practised overtly or covertly in negotiations, dispute resolution, leadership functions, and so on.

Chinyio, Singh and Suresh note that a threefold theory of stakeholder management is commonly reported: the descriptive, instrumental and normative. Strategies, tactics and tools for stakeholder engagement are being deployed, and these include the use of information technology (IT). Stakeholder management continues to gain recognition in construction practice in such as in procurement and in contracts and their administration. For instance, the New Engineering Contract (NEC) promotes stakeholder management through its risks and issues management procedure. Chinyio, Singh and Suresh suggest that the stakeholder management phraseology will likely appear increasingly in contracts, procurement and other construction practices. They observe that while stakeholder management still faces some challenges, some enablers are enhancing its implementation. Chinyio, Singh and Suresh suggest areas of future research on stakeholder management, including the assessments of the time spent on it, the cost of its implementation, its tangible and intangible benefits, and the role of technology in its deployment.

Chapter 21 by Weisheng (Wilson) Lu and Meng Ye

In the chapter on 'The global construction market', Lu and Ye note that economic globalisation, trade liberalisation, advanced technology, fast transportation and convenient communication have all catalysed the globalisation of construction. Their aim in the chapter is to provide an overall picture of the global construction market. Lu and Ye start by clarifying some of the concepts and terms, such as whether construction is a sector or an industry; the term 'architecture, engineering and construction'; whether construction makes products or provides services; and market segments and geographic dispersal. They then empirically describe the historical development and status quo of the construction market from a global perspective. Lu and Ye note that global construction has developed into a massive market worth trillions of dollars, encompassing the developed world as well as hotspots in the emerging countries.

Lu and Ye observe that a considerable portion of the global market is occupied by contractors who 'follow the money' to new continents for corporate growth and to minimise the risk of vicissitudes of individual markets. They note that there is a clear tendency for these firms to expand to other continents or sectors through vigorous business practices such as mergers and acquisitions, but the outcomes are mixed. In concluding, Lu and Ye's discussion focuses on the future of the global construction market. They suggest that the market is now confronting great uncertainty triggered by the rising populism, increasing xenophobia, and distrust of globalisation, which are further exacerbated by the Covid-19 pandemic. Lu and Ye note that there must be a 'new normal', but no one knows what it will look like, and how it will affect the global construction market.

Chapter 22 by Albert P.C. Chan and Emmanuel Kingsford Owusu

Chan and Owusu discuss 'Relational impacts of corruption on the procurement process: implications for economic growth in developing countries'. They note that the economic state of a nation is influenced by both micro and macro indicators, including government spending. With the increasing global demand for infrastructure projects and housing, governments are often pushed to allocate a considerable amount of resources for such spending, which is estimated to amount to at least 30 percent of the gross domestic product (GDP) of developing countries. Chan and Owusu note that such a high budget allocation often renders public procurement processes vulnerable to corruption, which may increase the overall public expenditure significantly, either directly or indirectly. In the chapter, they examine two critical indicators of corruption, which pose a significant threat to economic growth: (1) the proneness of the procurement activities to corruption; and (2) the criticality of the causal factors of corruption and their impact on the procurement process.

Chan and Owusu use two non-probability sampling techniques (that is, purposive and snowball) to secure the participation of 62 experts involved in construction-related works in the context of developing countries. The findings from Chan and Owusu's study revealed that the leading causes of corruption were: personal greed, inadequate sanctions, inappropriate political interference and lack of rigorous supervision. Network analysis showed that the overall impact exerted by the causal factors is highly critical at the contract stage, the contract administration stage and the post-contract stage.

To Chan and Owusu, their chapter contributes to the enhancement of the understanding of the critical causal factors that instigate corruption during the different phases of the procurement process. They also note that their chapter reveals the critical activities and stages of the procurement process that are most affected by the established causal factors. Chan and Owusu believe that their findings provide an insightful discourse that will be valuable to project stakeholders and policy-makers in developing countries who seek to mitigate or eradicate corruption in all its guises.

Chapter 23 by Chimay J. Anumba and Esther A. Obonyo

Anumba and Obonyo discuss 'Economic considerations in the procurement and deployment of construction informatics applications'. They note that the construction industry has witnessed a significant increase in the uptake of construction informatics applications. From being notoriously slow in the deployment of emerging information and communication technologies, the industry is now leveraging new applications in the construction project delivery process. In the chapter, Anumba and Obonyo explore the economic considerations in the procurement and deployment of these applications. They start with a brief background on the nature of the construction industry and the adoption of these technologies in the construction industry. This is followed by an overview of current and emerging technologies that are of interest to the industry. Anumba and Obonyo then discuss at length the specific economic considerations in procuring and deploying these technologies, including in-house development, off-the-shelf solutions, costs, training and return on investment. They then draw conclusions on what construction organisations should do to ensure that they maximise the benefits of their technology investments.

Chapter 24 by George Ofori

In the final chapter of this book: 'The future: new directions of construction economics research', Ofori provides a brief summary of the book and considers some issues which can be further studied in order to contribute to the development of construction economics. He suggests that whereas construction economics is an important subject owing to the strategic role of the industry it studies and builds knowledge on, the subject has not developed significantly; it has been left behind by many other subfields of economics of a similar age. Ofori outlines the aspects of construction economics which require to be explored in order to develop the field.

After considering various possible ways, he presents a path towards a better future for construction economics. Ofori stresses that progress will require a systematic and concerted effort, that certain prerequisites should be in place, and that leadership and action by all researchers in the field will be needed. Ofori identifies some possible actors in this process of development.

Categorising the Subjects of the Chapters

Construction economics is generally divided into three areas: construction industry economics, construction company economics and construction project economics. Table 1.1 presents a categorisation of the subjects of the chapters of the book into these three areas, and another 'temporary' general category.

CONCLUSION

The preview provided by the synopses of the chapters presented in this introduction shows that, taken together, the chapters, which are authored by accomplished researchers on the particular topics, provide a comprehensive coverage of construction economics in terms of the breadth of subject and the time period. They not only present the state of the art of the field of knowledge, but also they provide a window to the future of the subject. The book shows that the subject has an illustrious history, where it was useful in guiding the allocation of scarce resources in the provision of both public and private buildings and items of infrastructure, and was also influential in governance, professionalism and enhancement of the quality of life. It also shows that construction economics has some way to go to become a recognised field of knowledge. Researchers and students of construction economics will find this book a very useful resource.

Table 1.1 Categorisation of subjects of the chapters of the book

No.	Title	Features of construction economics	Construction industry economics	Construction company economics	Construction project economics
1	Introduction	X	X	X	X
2	Construction economics: its origins, significance, current status and need for development	X			
3	The philosophy of construction economics	X			
4	The nature and development of construction economics	X			
5	Construction economics in antiquity	X			
6	Construction in the economy and in national development		X		
7	Construction project economics				X
8	Dynamics of construction industry development		X		
9	Applications of mainstream economic theories to the construction industry: transaction costs			X	
10	Construction industry and the Sustainable Development Goals (SDGs)		X	X	X
11	Sustainability economics and the construction industry		X	X	X
12	International construction data: a critical review		X	X	X
13	Measuring and comparing construction costs in different locations: methods and data		X	X	X
14	New trends in international construction		X	X	
15	Economics of trust in construction			X	X
16	The builders of cities: prospects for synergy between labour and the built environment		X		
17	Economic principles of bidding for construction projects			X	X
18	Procurement and delivery management			X	X
19	The economics of housing policy and construction: developing a responsive supply sector		X	X	
20	A review of stakeholder management in construction			X	X
21	The global construction market		X	X	
22	Relational impacts of corruption on the procurement process: implications for economic growth in developing countries			X	X

No.	Title	Features of construction economics	Construction industry economics	Construction company economics	Construction project economics
23	Economic considerations in the procurement and deployment of construction informatics applications		X	X	X
24	The future: new directions of construction economics research	X	X	X	X

2. Construction economics: its origins, significance, current status and need for development

George Ofori

INTRODUCTION OF CONSTRUCTION ECONOMICS

Construction economics should be an essential field of economics because constructed items (buildings and infrastructure) constitute a distinct and specific category of national assets providing the foundation for, and enabling, all the economic activities and long-term development of any country. For example, Rees (2016) notes that, between 1880 and 1929, industrialisation and urbanisation expanded in the United States (US) faster than ever before. Urbanisation accelerated because of new methods of building, while the concentration of people in small areas gave a boost to economic activity and increased industrial growth. Thus, industrialisation and urbanisation reinforced one another, enhancing the speed with which economic growth occurred. Fedorov (2005) presents the historical development of the construction industry in Russia from the 1930s. He notes that the huge scale of construction undertaken in the Soviet Union during the first five-year plan for the country's industrialisation (1928–1933) made the construction industry one of the most important areas in the programme for the transformation of the national economy.

More recently, Dato' Sri Mohd Najib bin Tun Haji Abdul Razak, Prime Minister of Malaysia, noted: 'The importance of the construction industry to Malaysia's economy cannot be overstated. Construction in Malaysia has traditionally been a substantial driver of growth, and looking to the future, we expect this trend to continue and expand' (CIDB, 2017, p. 4). In the United Kingdom (UK), Her Majesty's Government (HM Government, 2018) notes that: 'Construction underpins our economy and society. Few sectors have such an impact on communities across the UK or have the same potential to provide large numbers of high-skilled, well-paid jobs' (p. 1). HM Government (2018) presents this overview of the UK construction industry: economic output – £413 billion (2018), 8.6 per cent of gross domestic product (GDP); employment – 3.1 million, including in contracting (2.3 million), associated manufacturing and professional services; structure – 405 000 firms, 900 000 sole traders, 300 large contractors; and 41 per cent of the workforce are self-employed. Box 2.1 presents an account of the importance of the construction industry to Hong Kong Special Administrative Region (SAR) and the need to develop it to prepare it for its imminent tasks.

BOX 2.1 HOW CONSTRUCTION MADE HONG KONG AND HOW IT CAN BE ENABLED TO CONTINUE TO DO SO

The construction industry ... has played a key role in positioning Hong Kong as one of the most recognisable, dynamic and admired cities in the world. It has been an instrumental driver of economic growth and enabler of social development for many generations. Whilst the past has been highly productive, the Industry is facing a challenging future.

A core aspect of the challenge ahead is centred on the significant volume of predicted construction activity. Over the next 10 years, construction investment of approximately HK$2.5 trillion to HK$3 trillion is expected, a material increase over the HK$1.9 trillion recorded in the past 10 years. Whilst a healthy and growing pipeline is encouraging, questions remain as to the capabilities and resources available to deliver against this pipeline. This includes an increasingly ageing construction workforce, a tendency to lag in innovation and in the adoption of advanced technologies as well as being labelled one of the most expensive construction markets in the world.

In recent years the Industry has also witnessed a series of incidents related to certain high profile megaprojects. These incidents have included unsatisfactory cost performance, commissioning delays, site safety incidents and in a more recent case, alleged issues related to the quality of construction delivery. These events have led to heightened levels of media scrutiny, reduced levels of public confidence and challenges in recruiting the next generation of high performing talent.

To address these challenges and ensure a bright and prosperous future, the Government of the Hong Kong Special Administrative Region ... is taking the initiative to be a leading agent for change. This is presented in Construction 2.0 – an expression of the Industry changes required across three key pillars: Innovation, Professionalisation and Revitalisation.

Source: Development Bureau of Government of Hong Kong SAR and KPMG (2018, p. 2).

Despite the importance of construction, unlike related areas such as urban economics, real estate economics and housing economics, construction economics has not yet established its legitimacy as a field of knowledge, although ‘Construction’ (code no. 74) is a subject under section L, ‘Industrial Organisation’, in the classification system of the *Journal of Economic Literature*, which is described as being ‘a standard method of classifying scholarly literature in the field of economics’ (American Economic Association, 2020). It is suggested that most mainstream economists have not heard about Construction Economics (Bröchner, 2018). Many researchers continue to question whether it is a scientific discipline in its own right (Rutter, 1993) or simply an agglomeration of disparate economic theories, principles and techniques which are occasionally applied to the construction industry and its constituent organisations and their operations (Bon, 1989).

Aim of the Chapter

What are the origins of construction economics? How has it developed? What is its current nature? What are some of its theories and concepts? What are its techniques, which are applied in research and practice? What is its status, such as its position among the established bodies of knowledge? What is its future? The aim of this chapter is to discuss these key points on the subject of construction economics. It is pertinent to note that ‘construction economics’ is often used interchangeably with the narrower ‘building economics’, and is often combined with ‘construction management’, as in (‘construction management and economics’), another emerging field.

A Brief History

The subjects of building economics and engineering economics, which have existed for many decades, are focused on the project level of building or civil engineering construction activity, respectively. Some of the earlier books on these subjects include Stone (1966) and DeGarmo (1973), respectively. Duccio Turin was appointed the London Master Builders Association Chair of Building at University College London (UCL) in 1966. His seminal inaugural lecture, entitled: 'What do we mean by building' (see Turin, 1980) explained the nature of the construction industry and the construction process, and highlighted the implications of their features for efficiency on projects. Turin established the Building Economics Research Unit at UCL in the late 1960s; it focused on macro, strategic issues such as the role of construction in the economy and in national development. In another seminal work, Steven Groak, an architect who was also teaching at UCL (Groak, 1993), focused on the construction process. The books and special issues of journals dedicated to some of the giants of the field of construction economics, such as Duccio Turin and Patricia Hillebrandt, have been useful additions to the body of knowledge (see, for example, Koenigsberger and Groak, 1978; Raftery, 1994).

The book *Economic Theory and the Construction Industry* (Hillebrandt, 1974) is often considered to be the first major work on the field of construction economics (although it had been preceded by others, such as Stone, 1966, as discussed below). Patricia Hillebrandt was the first economist to be employed by a contractor in the UK (Richard Costain Ltd), before joining UCL. In the preface to the first edition of the book, Hillebrandt (1974, p. xiii) stated:

> As far as I know, this is the first book to treat the construction industry from this point of view. It is, however, a field in which many questions remain unanswered. Some of them can be considered from theoretical standpoints alone, but most require applied research in the field to test the validity of hypotheses. It is hoped that this first tentative outline of the map will tempt researchers to study selected areas in greater detail.

It is pertinent to consider the extent to which Hillebrandt's hope has been fulfilled, and an attempt to do this is made in this chapter.

Bon (1999a) suggests that building economics emerged as a distinct field in the mid-1970s, induced by the energy crises; Ranko Bon, an economist, was working at the University of Reading. However, there had been much earlier works on building economics. For example, Croizé (2009) discusses two doctoral theses published in France in 1921 (Sellier, 1921) and 1946 (Olchanski, 1946) which considered the economics of housebuilding and arrived at two different conclusions, one favouring a focus on individual houses, and the other on housing blocks. A book entitled *Methods of Building Cost Analysis* by the US Building Research Institute (1962) includes chapters on techniques for economic analysis of building designs, and case studies of life-cycle costing of thermal envelopes and building services including computer applications. P.A. Stone, who worked at the UK Building Research Station and the National Institute of Economic and Social Research, had also produced works on the economics of building (Stone, 1966, 1967), planning and development (such as Stone, 1959).

There had also been earlier works such as the comprehensive study on the structure, practices, problems and development of the British construction industry during various stages of the 20th century by Marian Bowley (1966), a lecturer in political economy at UCL. Carl W. Condit, a professor of architecture at Northwestern University, suggested that a similar analysis of the US construction industry would be beneficial (Condit, 1968). Bowley's work and the study on costs and price formation in the US by Cassimatis (1969) have been, and remain, influential in the field (Manseau and Shields, 2005). Subsequently, Bon (1989), in *Building as an Economic Process: An Introduction to Building Economics*, sought to establish theoretical foundations for building economics, using the perspective of the Austrian school of economics.

The launch of the journal *Construction Management and Economics*, in 1983, provided further impetus to the development of construction economics. *Habitat International* and, to some extent, *Building Research and Information* (which had been an international periodical for national building research stations) had been the main vehicles for the release of infrequent works on the subject. Some works on the economics of construction, such as Strassman (1970), had appeared in mainstream economics journals. The Working Commission of the International Council of Research and Innovation in Building and Construction (CIB) on Building Economics (CIB W055) was formed in 1970.

Chang (2015) presents a review of the historical development of construction economics, including that of areas of economics which are relevant to construction. He notes that Friedman (1956) applied the new area of 'decision theory and application' in economics to construction bidding strategy. The study of building cycles and determinants of fluctuations in construction demand were undertaken by authors including Nobel Prize winner Simon S. Kuznets (1958) and Abramovitz (1961). Rostow (1975) presents a useful review of works on long waves in economies. Construction was also of interest in the area of 'development economics', of which another Nobel laureate Arthur Lewis was one of the pioneers (see, for example, Lewis, 1955) as they considered the role of the construction industry in economic development. The works included that by Strassmann (1970), and the book by Hirschman (1958). Bon (1977, 1986) studied input–output analysis of construction. Interest in bidding in construction among economists has continued since Friedman's (1956) seminal work. They include Whittaker (1981), and King and Mercer (1985). Chang (2015) noted that construction issues are receiving increasing attention from mainstream economists. The works he highlighted included: Bajari et al. (2009), Levin and Tadelis (2010) and Bajari et al. (2014) which covered issues including contracts and public procurement. These papers show that mainstream economists find construction topics are intriguing for.

Industrial economists have analysed many aspects of construction. Two such studies can be considered. Marion (2015) studies the effect of horizontal subcontracting on firm bidding strategies in California highway construction auctions. Subcontractors are hired by prime contractors prior to the auction, and the subcontractor may also be a competitor in the primary auction. Marion (2015) found that horizontal subcontracting, which is intended to improve productive efficiency, actually softens the subcontractor's bid strategy, since winning the auction may entail losing subcontracting business. Barrus and Scott (2020) studied asphalt paving auctions in Kentucky, US. They note that aspects of highway procurement auctions facilitate collusive outcomes. They analysed firms' bid participation decisions, including variables affecting costs as well as competitive and strategic effects. They concluded that in geographic markets where firms face only a few rivals, county boundaries serve as a coordinating mechanism for softening competition; they influence firms' decisions on whether and how much to bid.

A paper by Robert G. Eccles (1981), an economist at Harvard University, which is frequently cited by construction economics researchers, is on the quasifirm, a stable organisational unit, which is formed when a general contractor retains the services of trade subcontractors instead of vertically integrating the trades (when conditions permit) because of the transaction cost implications of construction technology. Economists who have written on the importance of investment in infrastructure in economic growth include Lawrence Summers, the former Chief Economist of the World Bank and US Secretary of the Treasury (Summers and De Long, 1993; Summers, 2017). It is also pertinent here to note that the editors of the *Oxford Encyclopaedia of Economic History* invited contributions on construction economics, and included three entries on the subject: on a historical overview and technological change (Hughes and Hillebrandt, 2003), industrial organisation and markets (Menes, 2003a), and regulation (Menes, 2003b). The mainstream economists who study topics on construction use the construction industry as a context for testing existing theories (in the cases Chang, 2015, mentioned, these theories were auction theory, transaction cost economics and contract theory); they do not attempt to develop an integrated theoretical understanding of construction activity. In other words, the mainstream economists do not aim to contribute to the development of a body of knowledge of construction economics.

NATURE OF CONSTRUCTION ECONOMICS

What is Construction Economics?

Construction economics applies economic theory, concepts and analytical tools to the construction industry, the companies and other organisations comprising it, and the projects it undertakes. It has been recognised as a field of knowledge in some countries for some decades. For example, in the *Great Soviet Encyclopaedia* (Ionas, 2010), construction economics, which developed in the Soviet Union as an independent branch of economics, was described as: (1) dealing with the organisational forms of management and planning of capital investment and construction production; (2) conducting research on the economic efficiency of capital investment and of scientific and technological progress in construction; (3) working out the economic principles underlying construction planning, the standardisation of construction work, industrialisation of construction, and the lowering of estimated costs and capital investments per unit of output; (4) studying the resources of capital construction and effectiveness of their utilisation; and (5) substantiating economic methods of management and incentives in capital construction (Ionas, 2010).

CIB W055 on Building Economics described itself as being concerned with: characteristics of construction firms; characteristics of construction markets; applying microeconomic theory; and cost studies and design economies (Skitmore and Marston, 1999). More recently, Newton et al. (2016) noted that the main areas of attention for research of W055 include: (1) characteristics of the construction firm – strategic, managerial and production-based theories; transaction costs and contracting; mergers and acquisitions, market entry and international construction; technology uptake models and construction firms; (2) characteristics of construction markets – identifying construction firms and markets; imperfect competition in construction; game theory in construction bidding and contracting; auction markets and bidding for construction projects; (3) applying macroeconomic theory – use of input–output data for analysis of construction industry; asset prices, monetary policy and building cycles; stages of development and construction activity; (4) theoretical issues – methodology in construction economics; the property market and demand for new building; measuring construction productivity; and (5) cost studies and design economics – cost modelling; life-cycle costing and sustainability; value management.

Some factors giving impetus to further development of construction economics are occasionally highlighted. For example, CIB W055 Building Economics notes that the grand challenges of the subject include (Newton et al., 2016): (1) macroeconomics – demographic change and the ageing population; globalisation and international comparisons; industry-level structural change, infrastructure provision and the new economy; and sustainability and natural, industrial and conflict-driven disasters; and (2) microeconomics – business analytics and data; forecasting, uncertainty and risk; and industrialisation and modern methods of construction.

It is worth noting that the two words, 'construction' and 'economics', are sometimes put together to denote things other than the area of knowledge. The Chartered Institute of Building (CIOB, n.d.) attempts to explain 'construction economics' by stating that construction has 'a unique role in economic growth and acts as a key barometer of economic conditions', and that: 'A healthy construction industry is synonymous with a healthy economy'. It highlighted the need for 'policies and initiatives aimed at recognising the positive impact construction has on economies, societies and communities'. In another example, the Bachelor of Construction (Construction Economics) programme at a New Zealand university actually educates quantity surveyors, and prepares them for membership of professional surveying institutions (UNITEC, 2020).

The Construction Economists

Who are construction economists? Bon (1999b), after a survey on what construction economists did, found that most of them are either quantity surveyors (a profession in the UK and Commonwealth countries) or (mainstream) economists. The former adopted the title 'building economists' in the UK in the late 1970s. The *Construction Economist* is the official journal of the Canadian Institute of Quantity Surveyors.[1] Commission 10 on Construction Economics and Management of the International Federation of Surveyors, among other objectives, seeks to promote best practice for construction project and cost management globally; foster research appropriate to the better understanding of construction project and cost management practice around the world; and advance construction project and cost management by education and research and continuing professional development (Muse, 2020).

The French National Union of Construction Economists (UNTEC) (2020) notes that the knowledge of the construction economist is based on proven analysis methods, libraries of data and experience gained in the field. They can reconcile the design, technical choices and the budget. UNTEC (2020) notes in its aptly titled paper on its organisation, 'Construction economics: several skills, one profession', that: 'A specialist in costs for construction or renovation operations, the Economist determines, quantifies and estimates the work' (p. 1). In Scandinavia, researchers with a background in economics, business studies and civil engineering deal with construction economics. The European Council of Construction Economists (CEEC) notes that 'there are over 100,000 members of organisations representing the discipline of the construction economist in Europe, whose education, training and professional standards are supervised by professional bodies'.[2]

The CEEC notes that the profession of construction economist has developed along different lines in each European state; a survey of the competencies and functions of construction economists in several countries in Europe showed that the roles differ from one country to the next (CEEC, 2017). In general, knowledge of construction economics is required in a wide range of disciplines and professions, including construction management, project management, quantity surveying, building surveying, civil engineering, architecture, building services engineering, valuation and real estate development and asset management. It has been suggested that other built environment professionals should study construction economics in order to be equipped to address the critical problems on projects (Bon, 1989).

SEGMENTS OF CONSTRUCTION ECONOMICS

It is suggested that the subject of construction economics can be subdivided into construction project economics, construction firm economics and construction industry economics (Hillebrandt, 1974; Ofori, 1994). Recent books which also consider the three segments include Gruneberg and Francis (2019). The three established segments of Construction Economics are now considered.

Construction Economics at Industry Level

The development of the body of knowledge on construction industry economics has benefited from the conceptual foundation built by Turin (1969), who considered the role that construction activity and product, and the construction industry, play in the economy, and the nature of developments in the industry as the economy develops. National-level industry reviews in various countries follow the pattern set in Turin's work (see, for example, National Advisory Council on Construction, 2014). Paul Strassman (1970), a professor at the University of Chicago, also discussed the role of construction in economic development and made similar conclusions to those of Turin on how the contribution of construction to GDP changes as an economy develops. He unveiled the concept of the 'middle-income country bulge' (Strassman, 1970) which has been referred to as the 'Bon curves' (after a much later description of the phenomenon by Ranko Bon; Bon and Crosthwaite, 2000) by many researchers (see, for example, Ruddock and Lopes, 2006; Chia, 2011).

As noted above, the first book which applied economic theories and principles to the construction industry was Hillebrandt (1974). The second edition, Hillebrandt (1984), discussed topics including: the nature of the construction industry; the structure of the UK industry including contracting and the professions; determinants of demand and forecasting demand and output; the construction process; output in the UK and abroad; the resources of the industry; productivity and efficiency; and research and development. Ball (1988) analysed the changes in the economic organisation of the British construction industry in the 1970s and early 1980s, considering its social and economic structure and the causes of its poor record. He described how the major firms survived an economic recession in this period by substantially restructuring their operations, relationships with clients, workforces and subcontractors; and considered the influence of trade unionism and the role of other agencies in the building process. Some of the more recent works which have covered construction industry economics are the collection of chapters edited by De Valence (2011), and the works of Myers (2017) and Gruneberg and Francis (2019).

Hillebrandt (2000), in the third edition of her book, notes that the importance of construction in the economy stems from three of its characteristics: its size; that it provides predominantly investment goods; and that government is the client for a large part of its work. In essence, the issues in construction industry economics include: the nature of the construction industry and its impact on efficiency and effectiveness; relationships between construction and other sectors of the economy; how to plan and time interventions in the industry to induce desired changes in the economy; how best to utilise the backward and forward linkages of construction activity to stimulate activities in the economy; exploring the labour generation potential of construction activity; and how to develop the industry to improve performance on projects and in companies.

As early as the 1950s, Colean and Newcomb (1952) had considered the issue of stabilising the levels of activity in construction, and its impact. Lange and Mills (1979) discussed the possibility of using the construction industry as the 'balance wheel' of the economy through adjustments in government's investment. For example, in the battle against Covid-19, the UK government has adopted the slogan, 'Build back better' (HM Treasury, 2021), and Muse (2020) is among many who suggest that construction will be critical in countering the economic effects of the coronavirus by providing a stimulus to the world economy. On the post-Covid-19 future, the UK Construction Leadership Council (2020) notes that: 'Construction is uniquely placed to drive the national economic recovery. It operates throughout the UK, employs 3.1m workers, and exports billions of pounds of products and services.' The city of Brisbane in Australia introduced a range of measures to help the city to recover from the impacts of coronavirus, designed to stimulate new construction, particularly smaller-scale projects (Brisbane City Council, 2020). The planning and building economic recovery initiatives included: reductions of application fees; automatic extension of time for existing development approvals; faster assessment of requests to change development approvals; infrastructure charges incentives; and application of a green buildings incentive. Long-term fluctuations in construction activity and their impacts on the rest of the economy, and input–output analyses (Bon, 1977) have also long been applied in arguments to highlight the importance of the construction industry.

A much highlighted element of the importance of construction is its employment generation potential. There has been interest in employment generated by construction. In the US, federally aided highways have been studied every three years since 1958. Robert Ball, an economist at the Bureau of Labor Statistics, estimated that almost 24 000 workers were employed for one full year in the US for each $1 billion spent in 1980 for new construction (Ball, 1981). Over half of the jobs were created in industries that produce, sell and deliver construction materials and equipment. The fewest jobs were generated in commercial office buildings and civil works land projects (nearly 22 000 jobs per $1 billion) and the largest number were in public housing (26 000 jobs per $1 billion). A similar study was earlier done by Claiborne M. Ball, also of the bureau (Ball, 1965). Long before then, Peter Anthony Stone compared data on employment from construction spending in 1937 with that in 1936 (Stone, 1937).

The importance of the construction industry in socio-economic development generated a segment of construction industry economics dedicated to the developing countries, which has yielded some significant works. An example is a report by a team of experts who studied the construction industry in Tanzania (Ministry of Works, 1977), which argued for direct planning and management of the development of the industry. The first full books on construction in the developing countries were World Bank (1984) and Wells (1986). The segment has a global research group, CIB W107, on Construction in Developing Countries; its main publishing outlet is *Journal of Construction in Developing Countries* (launched in 2005), although works on it have been published in *Construction Management and Economics* and *Habitat International* since the early 1980s. Giang and Low (2011), Ofori (2015) and Ullal and Tombesi (2021) present useful reviews of works on construction in developing countries. The literature has facilitated policy formulation in construction industry development in many countries (see Ofori, 2012; and Ministry of Urban Development and Construction, 2012).

In recent years, as the nature and importance of the construction industry has been better understood, the contribution construction makes to the attainment of macro-level economic and social objectives are highlighted in national development plans and strategies (HM Government, 2018) and are evident in broad international declarations such as the Millennium Development Goals (Ofori, 2007), the Sustainable Development Goals (The Economist Intelligence Unit, 2017) and the New Urban Agenda (United Nations, 2017). The need to develop construction to enable it to make its due contribution to national socio-economic development is highlighted in reviews of national industries, such as in Malaysia (CIDB, 2017) and the UK (HM Government, 2018), and at the global level (Boston Consulting Group and World Economic Forum, 2016).

The reviews of national construction industries by government-appointed task groups, of which there have been many in the UK since 1934 (Bossom, 1934) (although the Simon, 1944, report is more often referred to as the first in the series; see Murray and Langford, 2003), have contributed to the literature on construction industry economics. A number of these works were published just before the new millennium. They included the Egan Report in the UK (Egan, 1998), the Construction 21 Report of Singapore (Construction 21 Steering Committee, 1999) and the Construction Industry Review Committee (2001) report in Hong Kong SAR. These reports typically present a review of the state of the industry, its achievements and weaknesses, initiatives undertaken thus far, and the future challenges. In most cases, they compared the construction industry to other sectors in the economy and its counterparts overseas. They made radical suggestions for change, as is evident in the key words in the titles of the reports: 're-thinking' in the UK (Egan, 1998); and 're-inventing' in Singapore (Construction 21 Steering Committee, 1999). Reviews of these reports have been instructive in explaining the nature of the task of performance improvement in construction (Construction Excellence, 2009; Murray and Langford, 2003).

Some recent examples of the national industry reviews are the Construction 25 Report (HM Government, 2013) and Construction Sector Deal (HM Government, 2018) of the UK, and the Construction Industry Transformation Programme of Malaysia (CIDB, 2017). The Construction 2.0 report of Hong Kong SAR (Development Bureau and KPMG, 2018) sets out three pillars: innovation, professionalism and revitalisation. On the way forward, it stresses performance measurement, and indicates areas where targets should be set, under each of the pillars. The government has set up the HK$1 billion Construction Innovation and Technology Fund 'to promote innovation and investment in construction delivery' (p. 2), set out various initiatives, and called on all stakeholders to contribute to the development process, with the message that it is 'Time to Change'. The report urges the industry to provide feedback 'as to how the industry can be developed in a productive and sustainable manner' (p. 2).

It is also pertinent to note that some authors have been sceptical of the effectiveness of such national programmes (Green, 2011). Others point out that the emphases have been put on the wrong places (Smyth, 2018). However, following such strategic policy guidelines, countries such as Singapore have made considerable progress in developing their construction industries.

Some international studies of construction industries have also been undertaken. The CIB-commissioned Revaluing Construction study found that the performance of the construction industry has been criticised for many years in many countries, and this has resulted in a series of strategic reviews and initiatives aimed at changing the nature of the industry to one that effectively delivers the objectives of its clients (Barrett, 2007). Based on experience in many countries, the study suggests these key areas for action: a holistic notion of construction; a shared vision amongst stakeholders; a balance between markets and social capital; dynamic decisions and information; evolving knowledge and attitudes; awareness of the systemic contribution of construction; and promotion of the full value delivered to society. More recently, Rwelamila and Abdul-Aziz (2021) presented studies of 13 developing and emerging countries, examining the legal and policy framework, administrative infrastructure and procedures, and implementation mechanisms of their construction industries, as well as the challenges, current activities, prospects and future plans and programmes for the development of their industries. The intention is to foster 'translative adaptation', with countries taking lessons from more successful countries as a guide for a process of self-exploration.

Also pertinent to note are the reports on construction, its performance and its challenges which in the past decade or so have been frequently published by the major consulting organisations. These include World Economic Forum and Boston Consulting Group (2015), and Barbosa et al. (2017), a report by McKinsey Global Institute.

Construction Firm Economics

Construction companies are consulting firms – architects, planners, various engineers, various surveyors and project managers – contractors, operators and facilities managers. Construction companies in the same segments of the industry come in many forms. The companies and their employees are subject to various regulations and requirements, including registration and compliance with health and safety, and environmental and social considerations. The operating environments of the companies also differ among countries, as do business practices and government procedures (Davis et al., 2000). An example of the difficulties construction faces with regard to regulation could be the restrictions against the spread of Covid-19. Irvine (2020) reports that in Australia, in August 2020, construction remained largely open. When work is being done on buildings more than three storeys in height (excluding basements), a maximum of 25 per cent of the site's normal workforce is allowed on site. However, if the building is three storeys or fewer, a maximum of five workers is allowed. On the other hand, construction on schools and hospitals continued as normal, regardless of height. Architects, consultants and other professionals are ordered to work from home, unless they are conducting 'research where Australia has a competitive advantage and which cannot be shut down and requires on-site attendance'.

Many other factors make construction companies worth studying. They generally have little control over demand; they are mostly price takers. Moreover, construction firms should be able to deal with periodic recessions even as they work on projects with long periods of gestation. They obtain inputs from other sectors and they need to manage relationships among a variety of business partners for each project. The companies face high levels of competition in many segments owing to the low barriers to entry; they have relatively low bargaining power compared with clients; and encounter high levels of risks. The margins in construction are relatively small, and the industry has a high occurrence of losses and high levels of corporate mortality. Box 2.2 presents information on the financial performance of UK construction contractors in 2019.

BOX 2.2 CONSTRUCTION CONTRACTING COMPANIES IN THE UK, 2019

In the UK, the total turnover of the Top 100 construction companies was £72 billion in 2018, which is 4 percent higher than the previous year at the same companies. The industry's 10 biggest contractors generated revenue of £33.1bn was 46 percent of the total revenue for the top 100. The 100 companies generated pre-tax profits of over £1.1bn, also 4 percent up on the previous year but there were major issues. Turnover fell at three of the top 10 contractors; and 34 of the top 100 companies and many of the firms made losses. Amey lost £428m in 2018 from a disastrous highways deal in Birmingham and problems in the waste and utilities sector; and two other top 10 contractors, Interserve and Laing O'Rourke, also made significant losses. Some 13 companies in the Top 100 traded at a pre-tax loss in financial year 2018; 12 of those companies also traded at a loss in the previous year. Of the top 100, 48 companies either traded in the red or suffered from reduced profitability at a pre-tax level. Many of the larger firms were issuing profit warnings. The industry was losing companies. Costain had collapsed the previous year. Dawnus, a Welsh firm, was ranked 76th by turnover in the 2018 list but went bankrupt soon after, as did Birmingham-based Shaylor, ranked 97th in 2018.

Many familiar regional names outside the top 100 also went bankrupt in 2018; they included Bolton-based Forrest and Scottish firms Havelock Europa and Crummock. Britain's oldest contractor, West Country firm R Durtnell – which could trace its roots back to 1591 – collapsed in June 2018. Between 2010 and the end of 2018, the construction industry of England and Wales lost 28 000 companies according to data from the Insolvency Service. Worse, with Brexit uncertainty in the economy, some clients were withholding investment and main contractors with large workforces are coming under pressure. Many construction firms had left the stock market in recent years, including Amco, Metnor, Tolent, Styles & Wood and in 2019, Interserve, when the group's lenders took control.

Source: Construction Index (2019).

Arguably, the first major work on construction economics which focused on the construction firm was the pair of related books by Hillebrandt and Cannon (1989) and Cannon and Hillebrandt (1990). A more recent one is Gruneberg and Ive (2000). In essence, these works considered: factors which influence the growth and development of firms; the firm in different market conditions, especially during recessions; and the different corporate participants and how their behaviour influences others in the value chain.

The international dimension of construction business has also attracted interest from researchers. Owing to the particular features of the construction process, of the item to be built, and of the inputs required, all things being equal, the local firm has all advantages over any international player. So, why have some firms been so successful outside their own home markets? Marc Linder (1994), a professor at the University of Iowa, wrote a seminal book, *Projecting Capitalism: A History of the Internationalization of the Construction Industry* in the series, Contributions in Economics and Economic History. Linder (1994) discusses how, from the 1840s to the 20th century, large Western European and US construction companies undertook major infrastructure projects around the world. He sought to disprove the belief that construction is a localised industry of many small firms in perfect competition, sheltered from world markets, by showing that the American and European firms were able to transcend local and national markets and effectively incorporate the countries in Africa, Asia and South America into the world's construction and finance markets. International construction has given rise to discussions on the notion of competition in construction (Flanagan et al., 2007). Other concepts discussed include entry mode considerations (Chen, 2008), corporate growth decisions (Utama et al., 2018), bid or no bid decisions (Han and Diekman, 2001) and risk analysis (Viswanathan et al., 2020). Studies have involved the application of concepts from mainstream economics such as Porter's 'diamond framework' and Dunning's 'eclectic paradigm' (Low et al., 2004; Han et al., 2010), and efforts to build an analytical framework for international construction (Ofori, 2003).

Construction Project Economics

At the project level, the construction economics body of knowledge has yielded techniques and tools which have been applied in practice in undertaking project evaluation, feasibility studies, project budgeting, estimating of building costs, cost analysis and planning, cost modelling and life-cycle costing. Until recently, the focus in construction project economics had been on the minimisation of capital costs. For example, the *Great Soviet Encyclopedia* (1979) notes that an important function of construction economics is to develop a system of interrelated cost and physical standards for the technological and economic regulation and management of construction production in all phases of the investment process (Ionas, 2010). The early books on construction project economics included Ferry and Brandon's well-known textbook which was first published in 1964 with Douglas Ferry as the original author, Peter Brandon as a later additional author, and which is now published as Kirkham (2014); and the books of Stone (1966, 1967), Edmeades (1971) and Seeley (1972). Other works have been Smith and Jaggar (2007) and Ashworth and Perera (2015).

Much of what is covered in construction project economics relates to the design variables of buildings (such as the shape, height, choice of materials and techniques) and how their judicious selection and combination can lead to reduction of costs. Tan (1999) found that cost variation with building height is not only affected by technology, but also that building design, demand and institutional factors play important roles. He provided a model for estimating construction cost variation with building height. Picken and Ilozor (2003) also studied the cost–height relationship of tall buildings in Hong Kong, and concluded that, contrary to the convention in construction economics that construction cost per square metre would increase as buildings become higher, a different set of criteria should be applied in judging how height affects cost, depending on the context and commonality of tall buildings in the location under consideration. Lowe et al. (2007) studied the relationships between total construction cost and design-related variables. The cost comparison of alternative methods of construction is also a popular subject for research, as shown in Jaillon and Poon (2010). The accuracy of cost estimates has also been a popular topic for several years (Skitmore, 1988; Fellows and Liu, 2000; Ling and Boo, 2001; Akintoye, 2000).

Ahlfeldt and McMillen (2018) considered a topic at the nexus of construction project economics and real estate economics by finding the influence of land value on development decisions, including consideration of the height of tall buildings and construction cost elasticities. Comparison of building cost with revenue has been given some attention in a sub-theme of development economics, as in the US work by Ruegg and Marshall (1990). Cost–benefit analysis, which is rooted in welfare economics, has also been covered in the works on project economics (Stone, 1967; Seeley, 1972).

The possible influence of the procurement approach on project performance, such as on cost and time, has also been studied, most notably in strategic industry reviews such as Latham (1994), Construction 21 of Singapore (which supported design and build), and Malaysia's Construction Industry Masterplan 2006–2015 (CIDB, 2007) (which proposed project partnering). For these reasons, novel approaches in project procurement such as design and build, and partnering, have been influential during various periods. Frequently cited works on the choice of procurement approach include Love et al. (1998).

The economic aspects of the attainment of the various performance parameters, such as time, safety and health, and environmental performance, have been among the most commonly studied subjects. Among the most cited of these works in the literature on construction economics are Kaming et al. (1997) and Koushki et al. (2005). More recent books on construction project economics have covered a comprehensive set of topics. For example, Ashworth and Perera (2015) includes chapters on: development appraisal, whole life costing, value management, risk analysis and management, and sustainability in construction. The advent of public–private partnerships saw the consideration of new concepts to establish pricing methods on the sides of the parties, 'willingness to pay' of the users, and the viability gap (Yescombe, 2007; World Bank Group, 2014).

Newcombe's (2003) paper on stakeholder mapping is one of the most cited in construction management and economics. He argues that 'the concept of client, which has prevailed throughout the twentieth century, is now obsolete and is being replaced by the reality of project stakeholders' (p. 841). He shows that it is key to analyse the power, predictability and interest of key project stakeholders. The study by Olander (2007) on stakeholder impact analysis is also well cited.

The economics of value, which had been considered by authors such as Stone (1976), has been given impetus by the policy concern with social value in public procurement, such as in the UK, where regulations require such considerations. For example, the Social Value Act requires public authorities 'to have regard to economic, social and environmental well-being in connection with public services contracts; and for connected purposes' (HM Government, 2012, p. 1). Value has been highlighted in some recent UK reports on the construction industry (Bentley, 2018). This wider perspective of value will further strengthen the established concepts of value engineering and value management, which were adopted from manufacturing but are narrowly client-focused (Kelly et al., 2014; Churcher, 2017). The boundaries of construction project economics continue to be expanded. For example, a current six-year research project examining 1800 homes built using modern methods of construction identified 16 major themes that it will monitor as it tests different types of the modern methods, to provide long-term data to inform decisions about emerging construction technologies (CIAT News, 2021). The themes are: pace of build, cost of build, labour productivity, planning issues, pre-manufactured value, levels of construction wastage, construction logistics, delivery performance 'quality rating', energy efficiency performance, sales performance, life cycle, economic rationale, social value and well-being.

FURTHER DEVELOPMENT OF CONSTRUCTION ECONOMICS

Restating the Nature of Construction Economics from First Principles

Construction economics as a concept predates the current attempts to define and conceptualise it. Those who commission, design, engineer, construct, manage, operate and own constructed items have always been faced with the same challenges which affect decision-making on the economics of the items (Ruegg and Marshall, 1990). The main considerations can be outlined from the perspectives of the major stakeholders in construction. Clients wish to lower costs or increase the profits or benefits generated from the projects. To realise this objective, the items must be located, designed, engineered, constructed, managed and operated, considering the economic consequences of these decisions. Architects need to consider the capital, owning and operating costs of items built through the use of alternative designs and materials. Engineers should also consider the safety and reliability of the item, and the economy of alternative designs and sizes of building and mechanical systems and components over their lifetime. Construction companies need to select cost-effective materials, equipment and techniques and manage their human and financial resources effectively. Managers and operators of the facilities should establish cost-effective maintenance, repair, and replacement and renovation programmes and systems. The community is concerned with the service, utility and economic impact of the built item, minimisation of inconvenience during construction, and durability in its use.

From the considerations in the last paragraphs, although it should be recognised that the level of awareness and knowledge of each of the stakeholders is being increasingly enhanced over time, construction economics, as an idea, has existed since transactions to build have been entered into. For example, reference is often made to passages in old religious texts to the need to consider the budget before undertaking construction projects.[3]

Developing the Field of Construction Economics

There have been attempts to realise progress in the endeavour to build a body of knowledge of construction economics. For example, CIB W055 launched the Foundations of Building Economics Programme with the view to organising existing knowledge of the field by assembling already published seminal papers on the subject in one place (Skitmore and Marston, 1999). While Bon (1989) lauded this effort, it did not involve the creation of new knowledge. Bon (1999a) suggests that building economics is still in its infancy despite the importance of buildings in a country's capital stock and annual expenditures in the economy. He notes that the field has been long in emerging because it still lacks theoretical foundations.

The field of construction economics requires further development. Many of the seminal works on the field which should have provided a foundation for establishing the building blocks of the field have not been well known. These include Turin's inaugural lecture (see Turin, 1980) which explained the nature of the construction industry and the construction process, and highlighted the implications of their features for efficiency on projects; the chapters in the collection edited by Turin (1975) which considered concepts on capacity, productivity and information on construction; the state-of-the-art review by Kafandaris (1980); Bon's (1989) monograph; and the entries on construction economics in the *Oxford Encyclopaedia of Economic History*, especially Hughes and Hillebrandt (2003). These works should have continued to be influential.

There are disagreements on many aspects of construction economics. For example, authors argue about the boundaries of the construction industry (see Hillebrandt, 2000; Gruneberg and Francis, 2019). This makes it difficult to establish the size of the industry and its importance to the economy and national development. In the national accounts statistics, whereas construction is the only part of the economy which appears in two major groups of data – value added and capital formation (Hillebrandt, 2000) – a narrow definition of the industry is adopted, following the Standard Industrial Classification (SIC). A broader consideration would enable the capacity, capability, needs and performance of construction to be studied. In turn, this would make it easier to make effective arguments for the importance of the construction industry, and the need for its development.

There has also been a long debate on the appropriate conceptualisation of the very nature and composition of the construction industry. Winch (2003) classifies construction as a 'complex-product industry' where projects and project teams are largely bespoke, and small changes in one aspect of the work may lead to large impacts elsewhere in the process. Some authors disagree with the common view of a single construction industry and argue that it is more correct to envisage a construction sector comprising a number of distinct industries (Carassus, 2004; Meikle, 2019). Groak (1994) suggests that it would be more fruitful be to regard construction as organised agglomerations of projects, rather than as a discrete industry or a fixed constellation of firms, and that analyses of construction should pay greater attention to its external linkages. Finally, Carassus et al. (2006) suggested that as the construction industry no longer provides single buildings or items infrastructure, but a variety of services and improvements to the human environment, it is more appropriate to adopt as the framework for analysis, 'the entire construction and property sector – the "built environment cluster"' (p. 169). Turin (1969) presented an early matrix of the construction industry which was further developed by Ofori (1989, 1993), which should have been a useful point of departure for discussions on the structure of the industry. The argument over whether construction is a production or service industry (Hillebrandt, 2000) is also relevant here.

There is also the question of whether there is a scientific discipline which can be referred to as 'construction economics'. For example, Rutter (1993) considers whether there is 'such a thing' as 'construction economics'. He discusses the nature of business decisions, incremental analysis, the nature of construction economics, contract pricing theory and operational decisions. Ofori (1994) considered whether the subject of construction economics can be considered to be a *bona fide* field of knowledge. He notes that, despite the importance of construction in national economies and in socio-economic development, the field is fledgling. Its foundations are weak and there is confusion about many of its concepts and terms. He concluded that construction economics lacks a conceptual structure, a key attribute of a discipline. He highlighted main areas where further study is required, and suggested that the development of construction economics should be managed. Some authors, including Myers (2003) and Chang (2015), have taken this discussion further and reached similar conclusions: that the field requires further development in order to become a recognised segment of economics. For example, the study of Myers (2003) found that construction economics continues to lack any coherent conceptual structure, but concluded that, in future, the sustainability agenda could provide the impetus for researchers to agree on a common purpose and conceptual approach.

It can be suggested that construction economics has not taken some of the opportunities which have come its way to develop. These include the failure to build economic theories, principles and techniques around public–private partnership (PPP) projects and their variant forms. A well-cited paper, Cui et al. (2018), presents a review of studies on PPP infrastructure projects; and Chan et al. (2010) presents critical success factors for such projects. The World Bank has led the compilation of knowledge on PPP projects. There is controversy over the performance of such projects in terms of long-term cost, overall value and equity (Sherratt et al., 2020), and construction economics researchers have the opportunity to investigate viable options and approaches.

Another topic which requires to be studied is the cost impact of regulation on construction (the 'regulation burden'). Mayer and Somerville (2000) study the relationship between land use regulation and residential construction using data from 44 US metropolitan areas between 1985 and 1996. They consider regulations as adding explicit costs, uncertainty or delays to the development process. They found that land use regulation reduces the level of the steady state of new construction; there can be up to 45 per cent fewer starts, and price elasticities more than 20 per cent lower than those in less-regulated markets. Another study by the US National Association of Home Builders (Emrath, 2016) found that in 2016, on average, regulations at all levels (local, state and federal) accounted for 24.3 per cent of the final price of a new single-family home built for sale. Slowey (2016) highlighted additional forthcoming regulations in the US: the Department of Labor's new stipulations on exposure of workers to silica; that department's overtime rules, which raise the exempt threshold for workers from overtime payment from $23 660 to $47 476 per year; the Davis–Bacon Act requirement that contractors pay prevailing wages on public projects (a rate often higher than firms would normally pay); and residential construction regulations, whose costs include permits, utility connection fees, impact fees and the cost to comply with new building codes and standards.

There are many other questions, including: in what way has construction economics helped in the development of appropriate government policies for the development of the construction industry, and of corporate practices for competitiveness, success and profitability in construction? Finally, how should the field of construction economics change?

CONCLUSION

Whereas there is no agreement on the origins of construction economics, there is a consensus on its components. Construction economics brings together ideas and techniques from mainstream economics and some other existing areas of knowledge, and has been developing foundations, such as design economics, procurement, bidding and contract analyses, life-cycle costing and project evaluation (in the project segment); on resource efficiency, corporate growth strategies and internationalisation (in the company segment); and on market analyses, short-term economic impact, long-term socio-economic impact, input–output and linkage effect analyses, and implications of legislation and policy (in the industry segment). There are also the techniques for rent or buy, refurbish or rebuild and maintenance strategies (which are applicable to the product segment). Construction economics also concerns the constructed items as assets. It should also be about the individual building and infrastructure as well as portfolios of items, and the national stock of these items over their entire lives. For this reason, it would be appropriate to suggest another segment: construction product economics.

Moreover, it is evident that construction economics can have a dimension at the level of the individual person, whether as an employee, employer (given the preponderance of small companies), or one of the many other stakeholders, such as a member of the community on which a project has impact.

Construction economics covers an important subject of national significance. It should have the most sophisticated approaches and techniques to apply to the complex tasks it faces. Construction economics will not emerge as an important field of study until it has developed its own theoretical foundations and unique analytical techniques and approaches which are also of interest to other fields of knowledge. Developing construction economics to this level is a critical task which its researchers should be committed to.

NOTES

1. http://www.ciqs.org/english/the-construction-economist.
2. https://www.ceecorg.eu/who-we-are/.
3. Such as the Bible, Luke 24: 25–33: 'Which of you, wishing to build a tower, does not first sit down and count the cost, whether he has enough to complete it?'

REFERENCES

Abramovitz, M. (1961) The nature and significance of Kuznets cycles. *Economic Development and Cultural Change*, 9, 225–248.

Ahlfeldt, G.M. and McMillen, D.P. (2018) Tall buildings and land values: height and construction cost elasticities in Chicago, 1870–2010. *Review of Economics and Statistics*, 100:5, 861–875.

Akintoye, Akintola (2000) Analysis of factors influencing project cost estimating practice. *Construction Management and Economics*, 18:1, 77–89, DOI: 10.1080/014461900370979.

American Economic Association (2020) JEL Classification System / EconLit Subject Descriptors. https://www.aeaweb.org/econlit/jelCodes.php?view=jel.

Ashworth, A. and Perera, S. (2015) *Cost Studies of Buildings*, 6th edition. Routledge, Abingdon.

Bajari, P., Houghton, S. and Tadelis, S. (2014) Bidding for incomplete contracts: an empirical analysis of adaptation costs. *American Economic Review*, 104, 1288–1319.

Bajari, P., McMillan, R. and Tadelis, S. (2009) Auctions versus negotiations in procurement: an empirical analysis. *Journal of Law, Economics, and Organization*, 25, 372–399.

Ball, C.M. (1965) Employment effects of construction expenditures. *Monthly Labor Review*, February, pp. 154–158.

Ball, M. (1988) *Rebuilding Construction: Economic Change in the British Construction Industry*. Routledge, London.

Ball, R. (1981) Employment created by construction spending. *Monthly Labor Review*, December, pp. 38–44.

Barbosa, F., Woetzel, J., Mischke, J., Ribeirinho, M.J., Sridhar, M., et al. (2017) *Reinventing Construction through a Productivity Revolution*. McKinsey Global Institute, https://www.mckinsey.com/industries/capital-projects-and-infrastructure/our-insights/reinventing-construction-through-a-productivity-revolution.

Barrett, P. (2007) *Revaluing Construction*. Wiley-Blackwell, Chichester.

Barrus, D. and Scott, F. (2020) Single bidders and tacit collusion in highway procurement auctions. *Journal of Industrial Economics*, 68:3, 483–522, https://doi.org/10.1111/joie.12233.

Bentley, A. (2018) *Procuring for Value*. Construction Leadership Council, London.

Bon, R. (1977) Some conditions of macroeconomic stability of multiregional input–output models. *Economic Analysis*, 16:1, 23.

Bon, R. (1986) Comparative stability analysis of demand-side and supply-side input–output models. *International Journal of Forecasting*, 2:2, 231–235.
Bon, R. (1989) *Building as an Economic Process: An Introduction* to *Building Economics*. Prentice Hall, Englewood Cliffs, NJ.
Bon, R. (1999a) The future of building economics: a note. http://www.reading.ac.uk/kqFINCH/wkc1/source/bon/Future.pdf.
Bon, R. (1999b) What do the building economists do? Some results of an international survey. http://www.reading.ac.uk/kqFINCH/wkc1/source/bon/2000/Bon.pdf.
Bon, R. and Crosthwaite, D. (2000) *The Future of International Construction*. Thomas Telford, London.
Bossom, A.C. (1934) *Building to the Skies: The Romance of the Skyscraper*. The Studio, New York.
Boston Consulting Group and World Economic Forum (2016) *Shaping the Future of Construction – A Landscape in Transformation: An Introduction*. World Economic Forum, Geneva.
Bowley, M. (1966) *The British Building Industry: Four Studies in Response and Resistance to Change*. Cambridge University Press, Cambridge.
Brisbane City Council (2020) Planning and building economic recovery initiatives. https://www.brisbane.qld.gov.au/planning-and-building/planning-and-building-economic-recovery-initiatives.
Bröchner, J. (2018) Construction economics and economics journals. *Construction Management and Economics*, 36:3, 175–180.
Building Research Institute (1962) *Methods of Building Cost Analysis*, Publication No. 1002. Washington, DC.
Cannon, J. and Hillebrandt, P.M. (1990) *The Modern Construction Firm*. Macmillan, London.
Carassus, J. (2004) From the construction industry to the construction sector system. In Carassus, J. (ed.) *The Construction Sector System Approach: An International Framework*. Report by CIB W055-W065 Construction Industry Comparative Analysis Project Group. CIB, Rotterdam, pp. 5–18.
Carassus, J., Andersson, N., Kaklauskas, A., Lopes, J., Manseau, A., Ruddock, L. and Valence, G. de (2006) Moving from production to services: a built environment cluster framework. *International Journal of Strategic Property Management*, 10(3), 169–184.
Cassimatis, P.J. (1969) *Economics of the Construction Industry*. National Industrial Conference Board, New York.
Chan, A.P.C., Lam, P.T.I., Chan, D.W.M., Cheung, E. and Ke, Y. (2010) Critical success factors for PPPs in infrastructure developments: Chinese perspective. *Journal of Construction Engineering and Management*, 136:5, 484–494.
Chang, C-Y. (2015) A festschrift for Graham Ive. *Construction Management and Economics*, 33:2, 91–105, DOI: 10.1080/01446193.2015.1039044.
Chartered Institute of Building (CIOB) (n.d.) Policy: construction economics. https://www.ciob.org/policy/construction-economics.
Chen, C. (2008) Entry mode selection for international construction markets: the influence of host country related factors. *Construction Management and Economics*, 26:3, 303–314, DOI: 10.1080/01446190701882382.
Chia, F.C. (2011) Revisiting the 'Bon curve'. *Construction Management and Economics*, 29:7, 695–712, DOI: 10.1080/01446193.2011.578959.
Churcher, D. (2017) *Value Management and Value Engineering – RICS Professional Standards and Guidance, UK*, 1st edition. RICS, London, https://www.rics.org/globalassets/rics-website/media/upholding-professional-standards/sector-standards/construction/black-book/value-management-and-value-engineering-1st-edition-rics.pdf.
CIAT News (2021) Homes England MMC study to focus on cost, safety and wastage. 10 March, https://architecturaltechnology.com/resource/homes-england-mmc-study-to-focus-on-cost-safety-and-wastage.html.
Colean, M.L. and Newcomb, R. (1952) *Stabilising Construction: The Record and Potential*. McGraw-Hill, New York.
Condit, C.W. (1968) Review: the British building industry: four studies in response and resistance to change by Marian Bowley. *Journal of the Society of Architectural Historians*, 7:4, 303, https://online.ucpress.edu/jsah/article-abstract/27/4/303/56301/Review-The-British-Building-Industry-Four-Studies?redirectedFrom=fulltext.

Constructing Excellence (2009) *Never Waste a Good Crisis: A Review of Progress since Rethinking Construction and Thoughts for Our Future*. Constructing Excellence, UK.
Construction 21 Steering Committee (1999) *Re-inventing Construction*. Ministry of Manpower and Ministry of National Development, Singapore.
Construction Index (2019) Top 100 Construction Companies 2019. https://www.theconstructionindex.co.uk/news/view/top-100-construction-companies-2019.
Construction Industry Development Board (CIDB) (2007) *Construction Industry Master Plan Malaysia 2006–2015*. Kuala Lumpur.
Construction Industry Development Board (CIDB) (2017) *Construction Industry Transformation Programme*. Kuala Lumpur.
Construction Industry Review Committee (2001) *Construct for Excellence*. Hong Kong SAR Government, Hong Kong.
Construction Leadership Council (2020) *Roadmap to Recovery: An Industry Recovery Plan for the UK Construction Sector*. https://www.constructionleadershipcouncil.co.uk/wp-content/uploads/2020/06/CLC-Roadmap-to-Recovery-01.06.20.pdf.
Croizé, J-C. (2009) Academic views on the economics of construction: French variations (1920–1970). In *Proceedings of the Third International Congress on Construction History*, Cottbus, May.
Cui, C., Liu, Y., Hope, A. and Wang, J. (2018) Review of studies on the public–private partnerships (PPP) for infrastructure projects. *International Journal of Project Management*, 36:5, 773–794.
Davis, Langdon & Everest (2000) *Spon's European Construction Costs Handbook*, 3rd edition. Spon Press, London.
DeGarmo, E.P. (1973) *Engineering Economy*. Macmillan, New York.
De Valence, G. (2011) *Modern Construction Economics: Theory and Applications*. Spon, London.
Development Bureau of Government of Hong Kong SAR and KPMG (2018) *Construction 2.0: Time to Change*. Development Bureau, Hong Kong SAR, https://www.psgo.gov.hk/assets/pdf/Construction-2-0-en.pdf.
Eccles, R.G. (1981) The quasifirm in the construction industry. *Journal of Economic Behavior and Organization*, 2:4, 335–357.
Edmeades, D.H. (1971) *Building Economics and Measurement (metric): Part 1*. Estates Gazette, London.
Egan, J. (1998) *Re-thinking Construction: Report of the Construction Task Force*. Department of Trade and Industry, London.
Emrath, P. (2016) *Government Regulation in the Price of a New Home Special Study for Housing Economics*. HousingEconomics.com, NAHB, https://www.nahbclassic.org/fileUpload_details.aspx?contentTypeID=3&contentID=250611&subContentID=670247&channelID=311.
European Council of Construction Economists (CEEC) (2017) *QS/EC Competencies in Europe: The Position of the QS/EC*. https://www.ceecorg.eu/wp-content/uploads/2018/04/CEEC-QS-EC-Competencies-presentation-Rotterdam-2017-v3-.pdf.
Fedorov, S. (2005). Construction history in the Soviet Union – Russia: 1930–2005 emergence, development and disappearance of a technical discipline. *Construction History*, 21, 81–97, retrieved 8 November 2020, from http://www.jstor.org/stable/41613897.
Fellows, R. and Liu, A.M.M. (2000) Human dimensions in modelling prices of building projects. *Engineering Construction and Architectural Management*, 7:4, 362–372.
Flanagan, R., Lu, W., Shen, L. and Jewell, C. (2007) Competitiveness in construction: a critical review of research. *Construction Management and Economics*, 25:9, 989–1000, DOI: 10.1080/01446190701258039.
Friedman, L. (1956) A competitive bidding strategy. *Operational Research*, 4, 104–112.
Giang, D.T.H. and Low, S.P. (2011) Role of construction in economic development: review of key concepts in the past 40 years. *Habitat International*, 35:1, 118–125, https://doi.org/10.1016/j.habitatint.2010.06.003.
Green, S. (2011) *Making Sense of Construction Improvement*. Wiley-Blackwell, Chichester.
Groak, S. (1993) *The Idea of Building: Thought and Action in the Design and Production of Buildings*. Taylor & Francis, London.
Groak, S. (1994) Is construction an industry? *Construction Management and Economics*, 12:4, 287–293, DOI: 10.1080/01446199400000038.
Gruneberg, S. and Francis, N. (2019) *The Economics of Construction*. Agenda Publishing, Newcastle.

Gruneberg, S. and Ive, G.J. (2000) *The Economics of the Modern Construction Firm*. Palgrave, Basingstoke.
Han, S.H. and Diekmann, J.E. (2001) Making a risk-based bid decision for overseas construction projects. *Construction Management and Economics*, 19:8, 765–776, DOI: 10.1080/01446190110072860.
Han, S.H., Kim, D., Jang, H-S. and Choi, S. (2010) Strategies for contractors to sustain growth in the global construction market. *Habitat International*, 34, 1–10, 10.1016/j.habitatint.2009.04.003.
Hillebrandt, P.M. (1974) *Analysis of the British Construction Industry*. Macmillan, Basingstoke
Hillebrandt, P.M. (1984) *Analysis of the British Construction Industry*, 2nd edition. Macmillan, Basingstoke.
Hillebrandt, P.M. (2000) *Economic Theory and the Construction Industry*, 3rd edition. Macmillan, Basingstoke.
Hillebrandt, P.M. and Cannon, J. (1989) *The Management of Construction Firms: Aspects of Theory*. Macmillan, London.
Hirschman, A.O. (1958) *The Strategy of Economic Development*. Yale University Press, New Haven.
HM Government (2012) *Public Services (Social Value) Act 2012*, Chapter 3, https://www.legislation.gov.uk/ukpga/2012/3/enacted/data.pdf.
HM Government (2013) *Construction 2025: Industrial Strategy: Government and Industry in Partnership*. London.
HM Government (2018) *Industrial Strategy: Construction Sector Deal*. London.
HM Treasury (2021) *Build Back Better: Our Plan for Growth*. London.
Hughes, W. and Hillebrandt, P.M. (2003) Construction industry: historical overview and technological change. In Mokyr, J. (ed.), *Oxford Encyclopaedia of Economic History*, Vol. 1. Oxford University Press, Oxford, pp. 504–512.
Ionas, B.A. (2010) *The Great Soviet Encyclopedia*, 3rd edition (1970–1979). Gale Group, Farmington Hills, MI.
Irvine, J. (2020) From brothels to knackeries, the intricacies of lockdown astound. *Sydney Morning Herald*, 5 August, https://www.smh.com.au/business/the-economy/from-brothels-to-knackeries-the-intricacies-of-lockdown-astound-20200805-p55iqt.html.
Jaillon, L. and Poon, C-S. (2010) Design issues of using prefabrication in Hong Kong building construction. *Construction Management and Economics*, 28:10, 1025–1042, DOI: 10.1080/01446193.2010.498481.
Kafandaris, S. (1980) The building industry in the context of development, *Habitat International*, 5:3–4, 289–322.
Kaming, P.F., Olomolaiye, P.O., Holt, G.D. and Harris, F.C. (1997) Factors influencing construction time and cost overruns on high-rise projects in Indonesia. *Construction Management and Economics*, 15:1, 83–94, DOI: 10.1080/014461997373132.
Kelly, J., Male, S. and Graham, D. (2014) *Value Management of Construction Projects*, 2nd edition. Wiley Blackwell, Chichester.
King, M. and Mercer A. (1985) Problems in determining bidding strategies. *Journal of the Operational Research Society*, 36, 915–923.
Kirkham, R. (2014) *Ferry and Brandon's Cost Planning of Buildings*, 9th edition. Wiley-Blackwell, Chichester.
Koenigsberger, O.H. and Groak, S. (1978) *Essays in Memory of Duccio Turin*. Pergamon, Oxford.
Koushki, P.A., Al-Rashid, K. and Kartam, N. (2005) Delays and cost increases in the construction of private residential projects in Kuwait. *Construction Management and Economics*, 23:3, 285–294, DOI: 10.1080/0144619042000326710.
Kuznets, S.S. (1958) Long swings in the growth of population and in related economic variables. *Proceedings of the American Philosophical Society*, 102: 25–52.
Lange, J.E. and Mills, D.Q. (eds) (1979) *The Construction Industry: Balance Wheel of the Economy*. Lexington Books, Lexington, MA.
Latham (1994) *Constructing the Team*. HMSO, London.
Levin, J. and Tadelis, S. (2010) Contracting for government services: theory and evidence from US cities. *Journal of Industrial Economics*, 58:3, 507–541.
Lewis, W.A. (1955) *The Theory of Economic Growth*. Irwin, Homewood, IL.

Linder, M. (1994) *Projecting Capitalism: A History* of the *Internationalization* of the *Construction Industry*, 1st edition (Contributions in Economics and Economic History). Greenwood, Westport, CT.
Ling, T.Y. and Boo, J.H.S. (2001) Improving the accuracy estimates of building of approximate projects. *Building Research and Information*, 29:4, 312–318.
Love, P.E.D, Skitmore, M. and Earl, G. (1998) Selecting a suitable procurement method for a building project. *Construction Management and Economics*, 16:2, 221–233, DOI: 10.1080/014461998372501.
Low, S.P., Jiang, H. and H.Y. Leong, C.H.Y. (2004) A comparative study of top British and Chinese international contractors in the global market. *Construction Management and Economics*, 22:7, 717–731, DOI: 10.1080/0144619042000202780.
Lowe, D.J., Emsley, M.W. and Harding, A. (2007) Relationships between total construction cost and design related variables. *Journal of Financial Management of Property and Construction*, 12:1, pp. 11–24.
Manseau, A. and Shields, R. (2005) *Building Tomorrow: Innovation in Construction and Engineering*. Routledge, Abingdon.
Marion, J. (2015) Sourcing from the enemy: horizontal subcontracting in highway procurement. *Journal of Industrial Economics*, 63:1, 100–128, https://doi.org/10.1111/joie.12065.
Mayer, C.J. and Somerville, C.T. (2000) Land use regulation and new construction. *Regional Science and Urban Economics*, 30:6, 639–662.
Meikle, J. (2019) A response to George Ofori's special note. *Journal of Construction in Developing Countries*, 24:2, 207–208, https://doi.org/10.21315/jcdc2019.24.1.10.
Menes, R. (2003a) Industrial organization and markets. In Mokyr, J. (ed.), *Oxford Encyclopaedia of Economic History*. Oxford University Press, Oxford, pp. 512–516.
Menes, R. (2003b) Regulation. In Mokyr, J. (ed.), *Oxford Encyclopaedia of Economic History*. Oxford University Press, Oxford, pp. 516–518.
Ministry of Works (1977) *Local Construction Industry Study*. Dar-es-Salaam.
Ministry of Urban Development and Construction (2012) *Construction Industry Policy*. Addis Ababa.
Murray, M. and Langford, D. (eds) (2003) *Construction Reports 1944–98*. Blackwell Science, Oxford.
Muse, A. (2020) Construction economics is fundamental in responding to coronavirus: how improved project and cost management of construction is critical to recovering economies. *FIG Working Week*, Amsterdam, May 10–14, https://www.fig.net/fig2020/articles/FIG2020_10_Construction_Economics_and_Management.htm.
Myers, D. (2003) The future of construction economics as an academic discipline. *Construction Management and Economics*, 21:2, 103–106, DOI: 10.1080/0144619032000056117.
Myers, D. (2017) *Construction Economics: A New Approach*. Routledge, New York.
National Advisory Council on Construction (2014) *National Policy on Construction – Formulated under the Provisions of the Construction Industry Development Act No.33 of 2014*. Ministry of Housing and Construction, Colombo.
National Union of Construction Economists (UNTEC) (2020) *Construction Economics: Several Skills, One Profession*. Paris.
Newcombe, R. (2003) From client to project stakeholders: a stakeholder mapping approach. *Construction Management and Economics*, 21:8, 841–848, DOI: 10.1080/0144619032000072137.
Newton, S., Ruddock, L. and Gruneberg, S. (2016) *W55 Construction Industry Economics Research Roadmap*. https://www.cibworld.nl/app/attach/tLFKJEfc/20152487/9e3dab6fb020f23bfe51e13bd07309c6/Les_Ruddock.pdf.
Ofori, G. (1989) A matrix for the construction industries of developing countries. *Habitat International*, 13:3, 111–123.
Ofori, G. (1993) *Managing Construction Industry Development: Lessons from Singapore's Experience*. Singapore University Press, Singapore.
Ofori, G. (1994) Establishing construction economics as an academic discipline. *Construction Management and Economics*, 12:4, 295–306.
Ofori, G. (2003) Frameworks for analysing international construction. *Construction Management and Economics*, 21:4, 379–391, DOI: 10.1080/0144619032000049746.
Ofori, G. (2007) Millennium Development Goals and construction: a research agenda. In *CME 2007 Conference – Construction Management and Economics: Past, Present and Future*, pp. 1–13.

Ofori, G. (2012) *New Perspectives in Construction in Developing Countries*. Spon Press, Abingdon, https://doi.org/10.4324/9780203847343.

Ofori, G. (2015) Nature of the construction industry, its needs and its development: a review of four decades of research. *Journal of Construction in Developing Countries*, 20:2, 115–135.

Olander, S. (2007) Stakeholder impact analysis in construction project management. *Construction Management and Economics*, 25:3, 277–287, DOI: 10.1080/01446190600879125.

Olchanski, C. (1946) Le Logement des Travailleurs Français. Thesis, Faculté de Droit, Paris, 1945. Librairie Générale de Droit et de Jurisprudence, Paris.

Picken, D.H. and Ilozor, B.D. (2003) Height and construction costs of buildings in Hong Kong. *Construction Management and Economics*, 21:2, 107–111, DOI: 10.1080/0144619032000079671.

Raftery, J. (1994) Festschrift: in honour of Patricia Hillebrandt. *Construction Management and Econom ics*, 12:4, 283–286, DOI: 10.1080/01446199400000037.

Rees, J. (2016) *Industrialization and Urbanization in the United States, 1880–1929*. Oxford Research Encyclopedias: American History. Oxford University Press, New York, https://doi.org/10.1093/acrefore/9780199329175.013.327.

Rostow, W.W. (1975) Kondratieff, Schumpeter, and Kuznets: trend periods revisited. *Journal of Economic History*, 35:4, 719–753, DOI: https://doi.org/10.1017/S0022050700073745.

Ruddock, L. and Lopes, J. (2006) The construction sector and economic development: the 'Bon curve'. *Construction Management and Economics*, 24:7, 717–723, DOI: 10.1080/01446190500435218.

Ruegg, R.T., and Marshall, H.E. (1990) *Building Economics: Theory and Practice*. Springer, Boston, MA.

Rutter, D.K. (1993) Construction Economics: Is there such a thing? Construction Papers, No 18, Chartered Institute of Building, Ascot.

Rwelamila, P.D. and Abdul-Aziz, A.R. (eds) (2021) Improving the Performance of Construction Industries for Developing Countries: Programmes, *Initiatives, Achievements* and *Challenges*. Routledge, Abingdon.

Seeley, I.H. (1972) *Building Economics: Appraisal and Control of Building Design Cost and Efficiency*, 1st edition. Macmillan, Basingstoke.

Sellier, H. (1921) La Crise du Logement et l'Intervention Publique en Matière d'Habitation Populaire dans la Région Parisienne. Thesis, Faculté de Droit, Paris. Office Public d'Habitations à Bon Marché du Département de la Seine, Paris.

Sherratt, F., Sherratt, S. and Ivory, C. (2020) Challenging complacency in construction management research: the case of PPPs. *Construction Management and Economics*, 38:12, 1086–1100, DOI: 10.1080/01446193.2020.1744674.

Simon, E. (1944) *The Placing and Management of Contracts*. HMSO, London.

Skitmore, R. Martin (1988) Factors affecting estimating accuracy. *Cost Engineering*, 30:12, 12–17.

Skitmore, M. and Marston, V. (eds) (1999) *Cost Modelling*. Spon, London.

Slowey, K. (2016) Regulation roundup: 5 crucial issues impacting the construction industry. *Construction Dive*, https://www.constructiondive.com/news/regulation-roundup-5-crucial-issues-impacting-the-construction-industry/420165/.

Smith, J. and Jaggar, D. (2007) *Building Cost Planning for the Design Team*, 2nd edition. Elsevier, Oxford.

Smyth, H. (2018) *Castles in the Air? The Evolution of British Main Contractors*, Bartlett School of Construction and Project Management, University College London.

Stone, P.A. (1937) *Construction Expenditures and Employment: 1937 Compared with 1936*. Work Projects Administration, Division of Research, Statistics and Records, USA.

Stone, P.A. (1959) The economics of housing and urban development. *Journal of the Royal Statistical Society*, Series A, 122:4, 417–483.

Stone, P.A. (1966) *Building Economy: Design, Production and Organisation: A Synoptic View,* 1st edition. Pergamon, Oxford.

Stone, P.A. (1967) *Building Design Evaluation: Costs-In-Use*. E & FN Spon, London.

Stone, P.A. (1976) *Building Economy*. Pergamon, Oxford.

Strassmann, W.P. (1970) The construction sector in economic development. *Scottish Journal of Political Economy*, 17:3, 391–409.

Summers, L. (2017) Public infrastructure investment in the national interest. http://larrysummers.com/2017/01/16/public-infrastructure-investment-in-the-national-interest/.

Summers, L. and DeLong, B. (1993) How strongly do developing economies benefit from equipment investment? *Journal of Monetary Economics,* 32:3, 395–415.

Tan, W. (1999) Construction cost and building height. *Construction Management and Economics*, 17:2, 129–132, DOI: 10.1080/014461999371628.

The Economist Intelligence Unit (2017) *The Critical Role of Infrastructure for the Sustainable Development Goals*. London.

Turin, D.A. (1969) The construction industry: Its economic significance and its role in development. University College Environmental Research Group (UCERG), London.

Turin, D.A. (Ed.) (1975) *Aspects of the Economics of Construction*. Goodwin, London.

Turin, D.A. (1980) What do we mean by building? Inaugural lecture as London Master Builders Professor, University College London, 14 February 1966. *Habitat International*, 5:3–4, 271–288.

UNITEC (2020) Bachelor of Construction (Construction Economics) Programme overview. https://www.unitec.ac.nz/career-and-study-options/quantity-surveying/bachelor-of-construction-construction-economics.

United Nations (2017) *The New Urban Agenda*. UN, Nairobi.

Utama, W.P., Chan, A.P.C., Gao, R. and Zahoor, H. (2018) Making international expansion decision for construction enterprises with multiple criteria: a literature review approach. *International Journal of Construction Management*, 18:3, 221–231, DOI: 10.1080/15623599.2017.1315527.

Viswanathan, S.K., Tripathi, K.K. and Jha, N.K. (2020) Influence of risk mitigation measures on international construction project success criteria – a survey of Indian experiences. *Construction Management and Economics*, 38:3, 207–222, DOI: 10.1080/01446193.2019.1577987.

Wells, J. (1986) *The Construction Industry in Developing Countries: Alternative Strategies for Development*. Croom Helm, London.

Whittaker, J. (1981) General bidding model. *Journal of Operational Research*, 32, 11–17.

Winch, G. (2003) Models of manufacturing and the construction process: the genesis of re-engineering construction. *Building Research and Information*, 31:2, 107–118. DOI: https://doi.org/10.1080/09613210301995.

World Bank (1984) *The Construction Industry: Issues and Strategies in Developing Countries*. World Bank, Washington DC.

World Bank Group (2014) *A Checklist for Public-Private Partnership Projects*. Washington, DC, https://ppp.worldbank.org/public-private-partnership/sites/ppp.worldbank.org/files/documents/global_checklist_ppp_g20_investmentinfrastructure_en_2014.pdf.

World Economic Forum and Boston Consulting Group (2015) *Shaping the Future of Construction*. https://www.weforum.org/global-challenges/projects/future-of-construction.

Yescombe, E.R. (2007) *Public-Private Partnerships: Principles of Policy and Finance*. Elsevier, Burlington, MA.

3. The philosophy of construction economics

Stephen Gruneberg and John Kelsey

INTRODUCTION

Philosophy

Philosophy can be defined as 'the use of reason in understanding such things as the nature of the real world and existence, the use and limits of knowledge, and the principles of moral judgement' (Cambridge 2020). To unpack this definition, it is necessary to ask two sets of questions:

1. To what extent can anyone claim that certain types of entities exist and that we have knowledge about their properties and interrelationships? By what means do we receive, develop, modify and communicate such knowledge, whether interpersonally or intertemporally?
2. To the extent that human entities are capable of autonomous judgement, intentional decision-making and action, on what basis can anyone claim that some actions or entities are to be preferred to others?

Economics

While the potential scope of philosophy is wide, the range of economics is somewhat narrower, although highly contested. Smith (1776/1977) entitled his treatise as *An Inquiry into the Nature and Causes of the Wealth of Nations*. Marshall (1920/1961, pp. 4–5) defined economics as 'A study of men [*sic*] as they live and move and think in the ordinary business of life. It examines that part of individual and social action which is most closely connected with the attainment and with the use of the material requisites of well-being.' This was a criticism of Smith's omission of the human element in his definition.

Robbins (1932, p. 15) defined the scope of the subject as follows: 'Economics is the science that studies human behavior as a relationship between ends and scarce resources which have alternative uses'. This was a direct criticism of Marshall's inclusion of welfare, since Robbins thought that this introduced an unscientific element based on value judgements. However, Robinson (1962, p. 1) had a somewhat more sceptical view: 'Economics itself … has always been partly a vehicle for the ruling ideology of each period as well as partly a method of scientific investigation.' That was evidently intended as a criticism of Robbins.

What can be seen in the philosophy questions posed above is that in looking at the way society maintains itself in existence there should be knowledge both about what is and what ought to be the case. Economics used to be classified in terms of 'positive' and 'normative' to reflect this division, although it is now recognized that economic 'science' is not necessarily value-free (Hausman et al. 2017, pp. 240–250)

Economics has its origins in 'political economy', which could be defined as 'economics in the real world'. This is implied in the titles of the works of major early political economists such as Smith's (see above), Ricardo's (1817) *The Principles of Political Economy and Taxation* and Marx's (1867/1990) *Capital: A Critique of Political Economy*. These books set economics in the context of dynamic industrialization, unfree trade, imperial expansion, slavery, strikes, oppressive taxation, landlords, corrupt businessmen, war and revolution. Some claim that too much contemporary economics has forgotten these origins (and indeed the real world) by omitting essential areas such as the study of history (Hodgson 2001) and power or social hierarchy (Ozanne 2016).

The Economics of Construction

The 'construction industry' can be subdivided by type into various broad markets: housing, commercial property, civil infrastructure (Cabinet Office 2011, p. 5), together with a significant market for repairs and maintenance. However, the downstream and upstream activities of developers, major clients, designers and supply chains also have to be considered as part of a much wider construction sector (Chartered Institute of Building 2020, pp. 6–10). Therefore, the economics of construction encompasses a wide range of considerations particularly at the level of the firm and different markets, as well as interactions with other industries and the whole economy. The subject matter extends well beyond 'building economics' as taught at a number of universities, and which is about cost control, tendering, life-cycle building costs, and so on, with the emphasis on the project as the unit of analysis (Seeley 1996).

Philosophy of Economics

The philosophy of economics is a wide subject area and the reader should consult works such as Reiss (2013), Hausman (2008) or Hausman et al. (2017) for comprehensive overviews of the domain. For now, the main concepts of philosophy can be applied to economics as questions about the methodology of economic enquiry and the ethical basis for economic action. An overview of the philosophy of economics can be provided by expanding the two philosophy questions set out above and applying them to economics, which produces the following questions:

1. What problems do economic analyses pose in the philosophy of science in terms of both measurement and method?
2. In accounting for relationships between economic phenomena and economic agents, how do we account for causation, causal tendencies and causal mechanisms?
3. How much realism is needed in economic model building?
4. To what extent does observed behaviour of production agents conflict with standard economic models of competitive and non-cooperative agents?
5. If economies are populated by rational, self-interested and maximizing consumers, what is it that they are trying to maximize?
6. Can collective economic phenomena be analysed without an explanation derived from the behaviour of their constituent parts?
7. What basis is there for economic agents to act ethically, and for 'society' in the form of government to intervene in economic transactions and institutions?
8. What elements have been ignored in mainstream economic analysis?

PHILOSOPHY OF THE ECONOMICS OF CONSTRUCTION

The eight questions above can now be examined in turn with respect to significant issues in the economics of construction.

Observation and Measurement

The extent to which accurate data can be obtained for conducting economic research was questioned by Morgenstern (1950). The basis for the use of such data for forecasting has been challenged by Streissler (1970), although more sympathetically treated by Kennedy (1998). Even then the available data, whether from business or government sources, has been largely collected and classified according to pre-existing theoretical constructs from Pacioli (1494), Keynes (1936) and Leontief (1941/1951). That still leaves the problem of adjusting data for intertemporal and international comparison (Balk 2008; Best and Meikle 2015).

Obtaining accurate statistics for the industry is an ongoing struggle, with considerable margins of error. The existing Standard Industrial Classification then fragments the elements of construction such that some of these get amalgamated with professional services or manufacturing (Chartered Institute of Building 2020, pp. 6–10). If the United Kingdom (UK) statistics are problematic, then they are even more so in many other countries where data collection is less well organized. Such problems are further explored in Best and Meikle (2015, 2019). Constructed assets form a large proportion of fixed capital formation added each year to the capital stock (Gruneberg and Folwell 2013). This poses another measurement and conceptual problem. There has been an acrimonious debate among economists for at least 50 years about the extent to which it is legitimate to even talk of an aggregate capital stock (Felipe and McCombie 2013), let alone to use such a concept in macroeconomic (mainly growth) modelling.

Causation

Causation has proved to be one of the most contested areas in philosophy. There are questions to be asked about what causation is, as well as what causes and causal mechanisms are, particularly in the area of the social sciences. These are questions which go back to Aristotle and even before his time. The mantra 'correlation is not causation' goes back to the work of Hume (1739, p. 75), although he uses the word 'contiguity' instead of correlation. He could not see anything other than the correlation between X (the cause or independent variable) and Y (the thing caused by X or the dependent variable), the temporal precedence of X over Y together with the spatio-temporal abutment of both variables. However, Hume claims that the 'necessary connection' between the variables is not actually observable. On the other hand, Kant (1787/1999, pp. 432–451) provided several counter-examples to show that Hume's reductionist account was flawed.

The problem is that the form of Hume's statement of cause is itself too simple. Causal claims instead come in the form of 'X causes Y under Z', where Z represents some set of conditions. Typically, in economics, it is claimed that an increase in X causes an increase or decrease in Y 'other things being equal' (*ceteris paribus*). In physical science, experiments can be conducted in a *ceteris paribus* context. However, for nearly all causal claims in social science this is not possible. Therefore, in order to test a hypothesis, it is assumed that, for instance,

$$Y = a + bX + e$$

where: a is some constant,
b is a coefficient representing the quantitative effect of X upon Y, and
e represents the deviation from the relationship caused by extraneous factors which change simultaneously with the variables under investigation.

This is very much the approach taken in econometrics.

However, an issue arises with possible mis-specification of the model such that, in fact, X and Y are some function of a third variable W, in the absence of which it only appears that Y is caused by X. Further problems can arise where model data is aggregated for groups with dissimilar behaviour.

Layered Causation and Critical Realism

The subject of layered causation and critical realism follows arguments set out in Smyth et al. (2007). This paper was originally written to apply philosophy to project management, but can equally be applied elsewhere. A possible approach which extends multiple causal conditions is to be found in the development of critical realism (Gorski 2013). Here, the causal relationships are examined within a layered structure of reality. There is the set of actual observable events, within the second layer of all possible events. Underlying these is a layer of deep structures and fundamental causal mechanisms, some of which may not be available for direct observation.

There are causal powers such as the power of human beings to act in certain ways. Whether those powers are activated is contingent upon susceptibilities to other influences. How they are activated is part of a causal or generative mechanism. The activation may be modified by other causal powers activated at the same time. However, there is still divided opinion about the extent to which causal powers can be directly ascribed to abstract concepts such as social structures (Kaidesoja 2007). However, individual combinations of the same causal mechanisms and causal powers may give rise to different observable events (Franck 2003). Therefore, formulating hypotheses and interpreting tests is a rather complex affair which may well be suitable for complex social events. Building construction as a socio-technical process (Lowe et al. 2018) is subject to invariant physical causal mechanisms, together with both individual and institutional human causal powers exercised in a social context and subject to error.

Causation examples

Consider the question: ‘What causes structural failure in reinforced concrete frame buildings over 15m high?’

So, in the first two layers might be included ‘all actual/possible reinforced concrete framed buildings’. The third layer would include: (1) gravity; (2) the strength and other properties of the materials of which the building is comprised; and (3) the geological forces governing seismic activity. However, human agency must be added. One must consider the causal power of human beings to competently design, construct, and use and maintain buildings within their design limits. The way in which this occurs constitutes generative mechanisms from which variable outcomes may occur.

Consider the following events: the collapse of the Sampoong Store in June 1995 in Seoul, South Korea; and the collapses of multi-storey reinforced frame buildings following the earthquake in Northern Pakistan in October 2005.

It has been shown in Gardner et al. (2002) that the store collapse in Korea was caused by the unauthorized addition of an additional floor to the building, together with the transfer of plant from one part of the roof to another. There was also evidence of corrupt behaviour by the client, designer and regulatory officials. It has also been shown in Maqsood and Schwartz (2008) that the collapses of buildings in Pakistan were caused by inadequate design and the use of inferior materials in construction. There is evidence of failure to follow prescribed codes for the design of the government buildings, and corrupt regulatory failure.

In the first case, we can cite the causal mechanisms of gravity and material properties. In the second, underlying seismic risk should be added. However, there should also be added the misdirection of the causal powers to design and build, together with the misdirection of the causal powers of others to regulate what is built. In these two cases, the apparent causes may be technical (material failure), but the underlying cause can be traced back to actions which are really a combination of error and possibly ethical failures motivated by economic gain. So what initially appears as a technical problem (the proximate cause) can actually be traced back to economic motive (the ultimate cause). This may have philosophical implications, particularly in legal cases (Conway-Jones 2002).

Multifactor Causation

In practical terms, construction performance events – particularly failure – are often analysed using Ishikawa analysis or 'root cause' analysis (Ishikawa 1968; Wilson et al. 1993). These methods also recognize a complex causal reality behind each such event whose structure they seek to expose. Such events can also be thought of more simply as caused by a cumulative set of conditions which are individually unnecessary and individually insufficient, but collectively sufficient to cause the event (Mackie 1980/2002). The practical application of this can be seen in the 'Swiss Cheese' model of accident analysis (Reason 2000). In analysing, for instance, why transportation projects have cost overruns, Flyvbjerg et al. (2004) cite the example of the Channel Tunnel, whose technical risks, both during and after construction (Kirkland 1995, pp. 3–12; Kirkland 2002), exposed false optimism in the forecasts (Anguera 2006). Breaking down the standard financial appraisal model into its component parts of revenue, capital cost and finance cost, one can analyse each forecast item and the individual causes of deviation of actual variables against those forecast:

1. Revenue reduction and delay.
   a. Overoptimistic traffic forecast.
      i. Competition from low-cost airlines.
      ii. New technology ferries.
      iii. Duty-free discrimination in favour of ferries.
      iv. Deliberate misrepresentation.
   b. Overoptimistic construction schedule.
      i. Controllable factors behind delay.
      ii. Uncontrollable factors behind delay.
   c. Period of reduced traffic revenue following 1996/2008 fires.

2. Construction cost increase.
   a. Overoptimistic construction cost/schedule.
      i. Controllable factors behind cost increase and delay.
      ii. Uncontrollable factors behind cost increase and delay.
   b. Repair cost of post-completion fires in 1996/2008.
3. Finance cost increase.
   a. Overoptimistic finance requirement forecast.
      i. Overoptimistic construction cost/schedule leading to greater capitalized interest.
      ii. Interest rate increases.

There are a cumulative set of factors with interlocking effects. One can construct a much more complex analysis by examining the assumptions made in constructing the forecasts, but the above is enough to make the point.

Predictive Causality

In spite of the insufficiency of correlation alone, some economists have nonetheless argued that regularities in correlation can, under certain circumstances, be used as a substitute predictor even in the face of a lack of a theoretically adequate model. In particular (following Occam's razor), the fewer the variables involved, the better (Friedman 1953). The problem is that good predictability over a period of time may hide underlying complex causal networks. If some causal factor or mechanism changes, then the predictive model may fail without any means of explaining the failure.

A more sophisticated argument of this sort can be found in Granger (1969), although it is acknowledged that spurious correlations can arise (Granger and Newbold 1974). Even this model has come under fire for similar reasons (Maziarz 2015). However, there is usefulness in network models of what are most easily observed as probability-based events (Pearl 2000, pp. 133–172) to predict, for example, critical path-based models of construction or project costs through means such as Monte Carlo analysis (Špačková and Straub 2012).

PARADIGMS AND MODELS

Kuhn (1962) showed that while much 'normal' scientific activity was involved with experimentation and hypothesis testing within a specific framework of assumptions or beliefs, there was another key activity involved in challenging and changing the framework or paradigm within which 'normal' science was carried out. Rather than adding new data, this involved a reinterpretation of existing data. Therefore, the models and underlying assumptions of economics are a highly significant and contested area.

Homo Economicus

In constructing models, economists use the hypothetical entity of *Homo economicus*, who is a rational decision-maker seeking to maximize personal, household or corporate wealth on a basis of 'self-interest with guile'. This model was challenged by Simon (1957, p. 198) as well as Kahneman and Tversky (2000) and Thaler (2015) as part of a competing body of theory under the general heading of behavioural economics. Here, more realistic models of human behaviour are proposed. Much of this work has been based on decision experiments, but more recently some has been based on models in cognitive psychology (Wilhelms and Reyna 2015) and observed brain activity (Charron et al. 2008).

While economics can be classified as a social science, the boundary between it and other social sciences is contested territory and some such as Fine (1999) claim that economics is attempting to colonize other social sciences. One example he gives is the replacement of the term 'labour' by 'human capital', as if the human input into production can be considered conceptually identical with that of plant and machinery. The lack of understanding of the reality of industrial labour in particular is criticized by Braverman (1974). One can also see it as an attempt to undermine the conceptual basis of Marxian political economy in which a line is drawn between those who own the means of production and those who are employed to produce. Now, it is evident that there are those few whose specific abilities put them in a special earnings class, such as top actors, musicians and sports practitioners, but they are even further removed from machine-like operation. The machine analogy does not work in considering site-based construction labour, which has its own peculiar organization based on specialization, process complexity and subcontracting (Eccles 1981).

Information in Competitive Markets

One of the early key assumptions for neoclassical competitive markets was perfect information. Asymmetry of information has been the source of much economic theory, including economics of contracts, auctions and transaction costs (Williamson 1975, p. 40). Additionally, much was made of the ability of the price system itself to communicate information (Hayek 1945). The problem is that it may communicate incorrect or ambiguous information. Suppose there is a rise in house prices. This may mean that there is a shortage relative to demand. It may mean that demand has increased due to a higher wage level. It could reflect a change in tax or subsidy treatment of housing. It could also reflect a reduction in borrowing costs or a decline in other income-producing assets. It could reflect the high point of a speculative bubble which is then reversed with a later fall in prices.

A consumer who wants a tin of baked beans knows what is involved in buying such a product, where and when it can be obtained, an expected price, and the high probability that there will be no quality problems. Consumers who want a building are in a very different position. They need professional help even to articulate what it is they want in terms that can be translated into a design brief. Initially, at least, building clients do not know what they want, what it will cost, when it will be available, and whether there will be quality problems either in the standard of construction or in the suitability of the design to work for its intended purpose. Information is therefore accumulated over the life of the project, but it may arrive too late for critical decisions to be changed. However, there are costs in ever-greater project definition, which then gives rise to an information–time/cost trade-off and attempts to optimize this such as the technique of real options analysis (Garvin and Nord 2012).

The one-off nature of construction purchase decisions also requires modification of market assumptions. If the consumer of baked beans does not like the taste of the beans they can go back to the store and complain to the manager. They will likely do so if the can contains a foreign object. However, many will take the much easier step of buying a different brand or buying from another store ('exit'). Some may only have the choice of complaint ('voice') if there is only one local store which may only sell one brand of beans ('loyalty'; enforced in this case) (Hirschman 1970). 'Exit' is not available to the one-off construction client. They are stuck with the building they have. Their only route of redress is through the contract for design and/or construction together with any relevant warranties.

Cooperative versus Non-Cooperative Behaviour

The consumer of baked beans does not normally experience a change in bargaining power in relation to the baked beans seller. Construction has been notorious for adversarial behaviour rooted in the short-term nature of the client–contractor or architect–contractor or contractor–subcontractor relationship (Latham 1994). However, during the course of a construction project there may be significant changes (such as in the design) made or proposed by the various parties which may affect behaviour. Risk is increased due to the dynamic nature of bargaining power during the life of a construction project when contractors can increase their adversarial behaviour during the period of their greatest bargaining power (Chang and Ive 2007).

Information asymmetry and bargaining power also play a part, as does the fragmentation of the construction supply chain. One can, indeed, find interdepartmental adversarial behaviour in large organizations, but at least the two departments cannot normally sue one another. Observed adversarial behaviour supports the model of self-interest with guile, and economists have sought to explore this further in the form of non-cooperative game theory with adversaries regarding hard negotiation as a zero-sum game (Von Neumann and Morgenstern 1944; Osborne and Rubinstein 1994). However, it has been observed in the field of construction that it might be much more productive and beneficial to all parties if a more cooperative approach was taken, based on longer-term relationships. In longer-term cooperative behaviour, trust is an important component. A key reason for the low level of trust in construction is because it is a project-orientated industry. As a result, continuity is often lost between the contractors and their clients on completion of projects. Consequently, there is little incentive to invest in a longer-term relationship (Latham 1994; Mosey 2019).

In recent years there has been considerable improvement in the ability to exploit multi-contract relationships and non-adversarial approaches to behaviour (Mosey 2019, pp. 1–38). However, economics has had less to say in this area, although there is now an increasing volume of literature on cooperative game theory (Gintis 2009). One further aspect which must now be taken into account is the extent to which emerging technology will shape organizational form and behaviour (Woodward 1965/1981). The emergence of such technologies as building information modelling, robotics and three-dimensional (3D) printing may force the construction sector into less fragmented and more collaborative forms of organization (Mosey 2019, pp. 9–10). Having said that, the ability of the construction sector to resist change should not be underestimated (Bowley 1966).

The Ultimate Object of Consumer Decisions: Maximizing Value

In construction project management and economics, the idea of value management has become important (Green and Sergeeva 2019). This involves exploring with the client and main delivery stakeholders both the priorities of what the client sees as benefits and the most economical way of delivering them. However, what is value in owner-occupied housing? The

UK market is dominated by secondhand prices, and most new stock is built by speculative developers who have to interpret through market research what value means to house buyers. How are consumers actually supposed to judge the value of the houses? This is especially difficult given the fluctuations in house prices and the generally poor quality of new homes in the UK (Ryan-Collins et al. 2017, pp. 86–101).

Structure and Agency

In other social sciences such as sociology, it is understood that structure and agency must be considered as interdependent (Giddens 1984). Economists, uncomfortable with independent aggregate variables, have tried to find micro-foundations for macroeconomic variables (Clower 1967). However, it is clear that the macroeconomy, individual markets and other 'collective' variables, such as culture, law and media, constrain individual perceptions and behaviour. Individuals may have powers of autonomous decision-making but only within certain limits. Macroeconomic variables might indeed be theoretically conceived of as a consequence of millions of individual economic decisions by individuals. However, that ignores the extent to which those decisions were themselves constrained by macro or structural variables. This is a chicken and egg problem. The reality requires the recognition of bi-directional causality between both micro and macro observations. A good in-depth discussion of this problem can be found in Hoover (2009).

Construction and Macroeconomic Crisis

The majority of the housing stock in the UK is owner-occupied, the choice of this mode of occupation being the result of a deliberate policy of favourable tax treatment which, together with restrictive planning, has fuelled a long-run rise in house prices, and the conversion of housing from a pure consumer product into what is seen as a relatively low-risk asset class whose tax-free capital gains can be extracted in later life to supplement dwindling and uncertain pensions (for middle-income groups at least). Alternatively, value can be extracted from property as collateral for loans used to increase present consumption at the expense of the future. This (together with regulatory problems in the banking sector) resulted in unsustainable levels of private debt which was one key factor in the Global Financial Crisis of 2007–2009. Housing demand is also vulnerable to macroeconomic variables such as interest rates and fluctuations in output and employment (Piazzesi and Schneider 2016).

Elsewhere, one can point to excessive borrowing to undertake high-risk property speculation in Thailand and elsewhere as a prime cause of the Asian currency crisis of 1997, which had knock-on effects in the Russian Federation and Latin America, particularly Argentina (Kim and Haque 2002). Here again one can see the bi-directional causal effect between individual decision-makers and economic aggregates.

ETHICS AND INTERVENTION

Ethics can be defined as a system of accepted beliefs that control behaviour, especially such a system based on morals' (Cambridge 2020). However, ethics is not about prescribing for individuals or societies how to act in certain circumstances. Rather, it is about considering the different bases on which such matters are decided and the role that competing considerations play in decision-making (Hausman et al. 2017, pp. 1–2). Having said that, it is necessary to examine this through the lens of real issues in construction economics.

Kant (1785/2012, pp. 33, 38, 392) defined the 'categorical imperative' as actions can only be justified that can be done by all under the same conditions and circumstances. If an individual carries out an action, which cannot be generally permitted to others in the same circumstances, then that action cannot be justified on moral grounds. For example, if everyone behaved in the same way, then collusive tendering would restrict the number of firms competing for the work and clients would be deterred from developing projects that were no longer viable.

Uff (2003) argues that an ethical approach may be used to help solve a number of problems in the construction industry. Resolving some of the ethical issues in the construction sector offers one approach to possibly creating a fairer and more prosperous future for constructors and their clients. A unified code might also be helpful in this respect (Mason 2009).

Ethics and Distribution

The philosophical principles of utilitarianism (Bentham 1780/2009) included the idea of 'the greatest good for greatest number', which was a potentially dangerous and revolutionary idea in the highly unequal economies of the 19th century. To counter this, and the even more radical ideas of Marx, economists accepted the idea of Pareto optimality whereby no one can be made better off without making someone else worse off (Mornati 2013). The problem of this idea is apparent, since it attempts to by-pass the idea of reallocation of some otherwise arbitrarily given distribution of wealth. Economists attempted to justify this distribution as a natural and just outcome of the marginal products of labour and capital in competitive markets (Clark 1891).

In the 20th century, while still using the concept of utility, the idea of welfare economics was introduced in terms of both individual and aggregate welfare (Bergson 1938; Samuelson 1947; Arrow 1951/1963; Sen 1970), which coincided with the New Deal in the United States in the 1930s and the Welfare State in the 1940s in the UK. In particular, controversy arose around the idea of interpersonal comparisons of utility and the extent to which governments might intervene in the distribution of income and wealth. A famous thought experiment was proposed whereby people in an 'original state' should choose a social and economic system prior to them knowing their own place within it (Rawls 1971). This led to the idea of 'justice as fairness' and an argument for a more egalitarian distribution of income and wealth. This was opposed by the thought experiment of Nozick (1975), who argued from a state of anarchy that would result in a minimalist state in which defence against personal violence and violation of property rights would be the only legitimate state functions. Nozick's underlying principle is that the rights of individuals cannot be interfered with even for the purpose of attempting to create a fairer society.

Finally, in the 21st century the concept of well-being has again become a focus for economic argument (Stiglitz et al. 2009), while at the same time the growing economic inequality of the last 30 years has come under criticism (Piketty 2014).

Government, Distribution and Construction

How society should be governed is a large topic ranging in modern times from a pessimistic view of human society (Hobbes 1651; Schmitt 1921/2014) leading to more authoritarian or oligarchical forms of government, to a more optimistic view based on the protection of individual liberties (Locke 1689/1966; Nozick 1975) or the general will of society and its citizens (Rousseau 1762/1997; Taylor and Nanz 2020).

Whatever the political outlook of governments, a substantial part of construction industry demand comes from government, although the mode of procurement and distributional effects may vary widely according to political inclinations. One form of government intervention controls what can be built and where. One of the bases for this is the doctrine that the state should control and redistribute development rights for the benefit of the community. There is much criticism aimed at the UK planning system by developers, because of its speed and restrictive approach. However, the system which constrains them also constrains other developers and to an extent is a relative preserver of property values through refusing permission for locally incompatible forms of development. So each planning decision poses a potential ethical dilemma (Campbell 2012). At the time of writing in August 2020 there are proposals to relax a number of these controls.

There is a need to justify planning decisions when a proposed building benefits some and harms the interests of others. The interests of all parties can be weighted, and if the benefits to the winners are greater than the costs to the losers the project can be said to be justified. Unfortunately, offsetting the benefits to some against the costs to others is not straightforward. For example, there is the problem of attributing the value of individual costs and benefits, when people value items differently. The value of a benefit varies between people. A poor person evaluates costs differently compared to a wealthy individual. How should a benefit to a developer worth £1 million be compared to a cost of £1 to 1 million people? (Boardman et al. 2018, pp. 27–51).

Infrastructure rests on the right of the government to carry out works for the benefit of the community such as constructing roads, railways, urban drainage, water and power utilities, as well as law-based and social infrastructure. Whether this is funded through user charges, taxation or borrowing, it does, at the very least, increase social productivity and therefore represents a form of redistributive activity (Haughwout 2000). Sometimes this also requires actual forcible removal of individual property rights. The owner will be entitled to compensation, but the exchange is clearly not voluntary (Ryskamp 2007; Miceli 2011). Government support for subsidised housing and planning conditions requiring the construction of affordable housing represents redistribution, since the market will generally not offer such housing except through compromising on space or build quality standards. Since poor housing is a causal variable in health and educational attainment, improving this is a constituent of a policy of promoting social well-being (Tunstall 2020, pp. 16, 197–199).

Normative Assumptions

There are two schools of thought about the nature of man in society, which are embodied in the thinking of John Locke and Thomas Hobbes. Locke thought that man came together freely in the Social Contract to collaborate and benefit from mutual help (Locke 1690/1980, *Second Treatise* sections 14, 123, 163, 171, 194), whereas Hobbes thought that life was 'nasty, brutish and short' (Hobbes 1651, p. 78) and people only came together because they needed to, in order to survive (Hobbes 1651, p. 80). Some interpret the Hobbesian point of view in construction as a dog-eat-dog battle for survival, with fierce, narrow-margin competition among firms which thrives in oligopoly and too often tips players over the line into corrupt acts (CIOB 2013, pp. 1, 23).

What would be the features, operating environment and market conditions for a construction industry which would steer a course that is not a constant battle for survival, but positive and inspiring, moral and decent? From the discussion, six principles of construction can be highlighted:

- an industry that is competitive, while it comprises firms which are conscious of their social responsibility;
- an industry that is productive and embraces innovation;
- an industry that produces a quality output and seeks to provide optimum value to clients and society;
- an industry that is efficient and whose output is efficient;
- an industry that employs a workforce that is professional in its attitude, behaviour and skills; and
- an industry that has an excellent reputation, has confidence and pride in itself, and is valued and trusted by society.

A major professional institution in UK construction has the general ethical code now outlined (CIOB 2018):

1. Members shall at all times act with integrity, honesty and trustworthiness.
2. Members shall treat others with respect, fairness and equality at all times.
3. Members shall discharge their duties with complete fidelity and probity. In particular they shall:
   3.1 not divulge any information of a confidential nature relating to business activities;
   3.2 avoid conflicts of interest and any actions or situations that are inconsistent with their obligations as a member of the Institute;
   3.3 at no time improperly offer, solicit or accept gifts or favours with a view to obtaining preferential treatment or gaining an unfair advantage.
4. Members shall not discriminate on the grounds of gender, race or ethnic origin, sexual or gender orientation, creed, nationality, disability or age.

Among more specific items in the code are the following:

5. Members shall not undertake work for which they knowingly lack sufficient professional or technical competence, or the adequate resources to meet their obligations.
6. Members shall ensure that all work undertaken is in accordance with good practice and current standards.
7. Members shall act in the best interests of the client at all times and deliver good customer care and service. [...]
10. Members shall keep themselves informed of current thinking and developments appropriate to the type and level of their responsibility. They should be able to provide evidence that they have undertaken sufficient study and personal development to fulfil their professional obligations in accordance with Institute's policy for Continuing Professional Development (CPD). [...]
16. Company Members shall ensure that staff engaged in the construction process have achieved, or are working towards, appropriate qualifications and/or training.

The code of the CIOB implies not just ethical but also competent, professional and efficient behaviour which benefits the professionals themselves and/or the client. Condition 10 also introduces a dynamic element which implies a commitment to continuous improvement and innovation. A further implication is that the organizations (and indeed the whole industry) that employ members must allow them to act appropriately.

Two more clauses from another code are also relevant (Institution of Civil Engineers 2017)

3. All members shall have full regard for the public interest, particularly in relation to matters of health and safety, and in relation to the well-being of future generations.
4. All members shall show due regard for the environment and for the sustainable management of natural resources.

This states that professionals and, by implication, the construction industry have wider obligations beyond the client relationship to societies both present and future.

SOME MORAL DILEMMAS IN CONSTRUCTION

Employer–Employee Power Relationship

Earlier, the interrelationship of construction with socio-economic power structures was discussed. Construction, like many industries, is faced with the problem of principled behaviour being replaced by that based on power. Employers have power over their employees, who can often be easily replaced at short notice. For example, piece rate workers or those on zero hours contracts can be dismissed instantly, and day rate workers at a day's notice, depending on the conditions of employment. These arrangements can range from monthly salaries to day rates and even piece rates, which mean that workers are paid for the actual items they make. The longer the period of payment, the more security and bargaining power is given to the employee. Indeed, the terms of employment enable employers to shift risk on to employees. In extreme cases, conditions are imposed which make the relationship close to a modern form of slavery (Da Graça Prado 2020). Similar considerations may apply to relationships between large contractors and small subcontractors, particularly with respect to delays in payment.

Unethical or Illegal Behaviour

Vee and Skitmore (2003) found that construction contractors had a reputation for unethical behaviour and there were many disputes between clients and their builders. Notwithstanding this, they also found that 90 per cent of survey respondents subscribed to a professional code of ethics, 45 per cent had an ethical code in their employing organizations, 84 per cent considered good ethical practice to be an important organizational goal, and 93 per cent thought 'business ethics should be driven or governed by personal ethics'. No respondents were aware of any cases of employers attempting to force their employees to initiate or participate in unethical conduct. Nevertheless, all respondents had witnessed or experienced some form of unethical conduct. The percentage of respondents reporting unethical behaviour is given in brackets: unfair conduct (81 per cent), negligence (67 per cent), conflict of interest (48 per cent), collusive tendering (44 per cent), fraud (35 per cent), breaches of confidentiality (32 per cent), bribery (26 per cent) and violation of environmental ethics (20 per cent). A similar study by Doran (2004) reports on a survey of the views of 270 American architects, engineers, construction managers, general contractors and subcontractors on the ethical state of the construction industry in the US. As many as 84 per cent of the sample had encountered or observed examples of unethical behaviour in the previous year, with 34 per cent saying this had occurred many times.

Conflicts of interest arise when, for example, the aim of a consultant is to satisfy two contradictory interests such as those of a client and those of a local authority, where both the client and the local authority have hired the same consultant. This situation can arise where the consultant is using their position for personal financial gain. A similar situation can arise where clients award contracts to companies in which they hold an interest, or to former employees and friends. Conflicts of interest can also arise where there is a need to maintain impartiality, while representing a client. Vee and Skitmore (2003) also found unethical behaviour extended to knowingly damaging the environment by contaminating the soil, degrading vegetation and eroding soil. In recent years, environmental issues have moved up the moral and ethical agenda in general, and the construction industry in particular. The broader agendas of sustainability and conservation are also highly topical.

Infringements of confidentiality can occur through revealing tendering or product information to competitors. Consultants may betray their clients by discussing client details. There is not always agreement about unethical conduct. For example, while some might argue in agreement with Vee and Skitmore (2003), that underbidding is unethical in order to win work in the tendering process, it could equally be argued that although it may be seen as sharp practice, it is nevertheless the responsibility of the appointing party to ensure all bids in the tendering process can be delivered, as in the principle of *caveat emptor* (buyer beware).

Regarding professionalism, integrity and client focus, Zhang et al. (2006) reported that employees often wish to take an active role in developing themselves to the highest potential and strive for personal and professional excellence. Workers often also valued personal integrity, self-respect and dignity, and that all people should be treated with courtesy and respect. High ethical and moral standards were closely linked to honesty, loyalty and responsibility. At the same time, their client focus implied that every effort should be made to satisfy or exceed their internal and external customers' needs, to build a long-term relationship with their clients. Zhang et al. (2006) also pointed out that key issues for employees were the lack of two-way communication, the importance of teamwork, and the difficulty of maintaining a work–life balance. Although not necessarily directly moral issues in themselves, unethical behaviour often impacts on those aspects of work as a consequence.

Discrimination based on gender, race or sexual orientation exists in most parts of modern society. Unfortunately, sexism, racism and homophobia seem to be more prevalent in construction than elsewhere (Naoum et al. 2020; Caplan et al. 2009; Ramchum 2015). This raises questions of both ethics and efficiency. It cannot be efficient for the industry as a whole to be perceived as a poor in terms of discriminatory behaviour.

An example of widespread bid rigging was discovered in the UK. It involved as many as 103 construction firms and the activities covered 199 tenders of projects over a six-year period (Office of Fair Trading 2009). The firms were fined a total of £129 million. However, some see the common occurrence of such practice as arising in opposition to competitive tendering with wafer-thin margins, and the so-called 'winner's curse' of those who have underbid their competitors and then have to deliver with a very small margin for error. An ethical question might be posed as to whether contractors end up getting a fair reward for the risks they take.

The relatively large sums involved in construction transactions make the industry particularly vulnerable to bribery, which economists view as a form of rent-seeking behaviour (Lambsdorf 2002). Additionally, in some countries construction is also a target for money-laundering and penetration by those involved in other forms of illegal activity. It is clear that such problems need to be addressed at a state level (although in some countries it is the state that is the problem, as is evident in delays in payments due to contractors of public projects, and failure to enforce existing regulations and guidelines). Many governments have been instituting increasingly stringent laws, such as the Bribery Act 2010, and forming stronger institutions such as Australia's Commonwealth Integrity Commission (a federal anti-corruption agency).

MISSING ELEMENTS FROM CONSTRUCTION ECONOMICS

What Construction? To Build Now and in the Future: Time, Aggregation and Social Cost

In health and educational investment decisions, the net social benefits at the macroeconomic level exceed the expected private individual benefits. Private decisions, if taken individually, would result in an aggregate outcome which nobody wanted.

In the standard investment appraisal model, forecast net receipts are discounted into the future. That is generally not a problem since it is to be expected that, apart from discounting for time preference, any investment now will eventually encounter diminishing net benefits. The problem arises when a project has distant negative revenues, such as land remediation after mining or nuclear power station decommissioning costs. There is a temptation that when the time comes a company or a country short of funds may simply abandon a site rather than pay for its remediation. Now this could be solved by annual contributions to an escrow sinking fund to meet such costs when they fall due. Policing this, however, might be a tricky problem.

What is much more problematic, however, is when projects by individual firms are appraised as profitable with individually small environmental damage, but appraised collectively are liable to cause catastrophic and irreversible change in the environment or the climate. While the problem here might be stated in terms of intergenerational equity it could more simply be stated by assigning a future period value of minus infinity. It does not matter how far this occurs in the future since the value of minus infinity cannot be discounted. Similarly, even if there is only a 10 per cent probability of the occurrence the use of expected monetary values still yields minus infinity. The problem here is that such matters can only be resolved by collective appraisal at whole industry or governmental and intergovernmental levels.

Construction Output and Socio-Economic Power Structures

Marxian political economy takes into account the way the whole of society maintains itself in existence, and in particular the radical change represented by the change from a feudal to a capitalist mode of production, involving the struggle between employers and employees in the distribution of income, the persistence of economic crises and the accumulation of a significant portion of total wealth by an elite group, giving them disproportionate economic and social power (Marx 1867/1990).

Critical theory, drawing from not only the later Marx but also the early Marxian idea of alienation, argues for interdisciplinary social research (with both moral and practical objectives) into the restrictive life led by most people, in order to show how human beings can be emancipated from what is seen as domination and oppression in a modern industrial society (Marcuse 1964/2002).

Green (2001) argues that university research in construction is dominated by an agenda set by management elites which plays down, in particular, the well-being of those who actually work in the industry. This is further developed in Green (2011), which is a critique of both Egan (1998) and more generally the promotion of 'best practice' which has a narrow focus on technical efficiency and management fashions rather than a wider view of construction in the context of social and political discourse.

Finally, postmodernism in architecture is eclectic in style and employs a self-conscious and sometimes playful remix of styles which lends itself well to designing iconic buildings. These draw attention to themselves as distinctive from, rather than blending in with, the surrounding environment, which makes them ideal for offices for corporations wishing to promote their brand of finance, insurance or management consultancy. This is part of a wider trend which is also open to the criticism that it primarily serves the interest of an elite group (Sklair 2017).

CONCLUSION AND FURTHER QUESTIONS FOR CONSIDERATION

In this chapter the surface has barely been scratched in examining the way philosophy can be applied to construction economics. The authors hope that readers will have been stimulated to see 'common' problems in a different light and to further explore the subject themselves. To this end, some further questions are set out below.

To what extent should construction companies press government for better statistics, and themselves be prepared to disclose more information to assist this process? To the extent that statistical inaccuracies have to be accepted, what caution should be exercised in interpreting econometric studies?

While construction economics is not a separate discipline, how do the peculiarities of the construction industry and markets challenge the methodology of mainstream economics? How might a greater use of behavioural economics and game theory inform understanding of both cooperative and non-cooperative conduct among economic actors in construction?

In the last three years (2017–2020) two of the largest contractors in the UK have failed and some other large firms are struggling. Is the business model of contracting still viable? How might construction economics be applied to suggest and analyse alternatives, particularly in the context of emerging technologies? To what extent should industry actors take on ethical responsibilities both in their own companies and in their supply chains; particularly in relation to labour practices, pollution and climate change?

Professionals are taught to regard the client's wishes as paramount. If a contractor is approached to build a camp whose main purpose is the incarceration and possible death of people from minority communities, should they accept – maybe on the grounds that if they did not build it someone else would? Are there contracts from which a professional should walk away?

REFERENCES

Anguera, R. (2006) The Channel Tunnel: an ex-post economic evaluation. *Transportation Research Part A Policy and Practice* 40(4), 291–315.

Arrow, K.J. (1951/1963) *Social Choice and Individual Values* (2nd edition). New Haven, CT, Yale University Press.

Balk, B.M. (2008) *Price and Quantity Index Numbers: Models for Measuring Aggregate Change and Difference*. Cambridge, CUP.

Bentham, J. (1780/2009) *An Introduction to the Principles of Morals and Legislation*. Dover Philosophical Classics. Mineola, NY, Dover Publications.

Bergson, A. (1938) A reformulation of certain aspects of welfare economics. *Quarterly Journal of Economics* 52(2), 310–334

Best, R., and Meikle, J. (eds) (2015) *Measuring Construction: Prices, Output and Productivity*. Abingdon, Routledge.

Best, R., and Meikle, J. (eds) (2019) *Accounting for Construction: Frameworks, Productivity, Cost and Performance*. Abingdon, Routledge.

Boardman, A.E., Greenberg, D.H., Vining, A.R., and Weimer, D.L. (2018) *Cost Benefit Analysis: Concepts and Practice* (5th edition). Cambridge, CUP.

Bowley, M. (1966) *The British Building Industry: Four Studies in Response and Resistance to Change*. Cambridge, CUP.

Braverman, H. (1974) *Labor and Monopoly Capital*. New York, Free Press.

Cabinet Office (2011) *Government Construction Strategy*. London, Cabinet Office.

Cambridge (2020) *English Dictionary*. https://dictionary.cambridge.org/dictionary/english/ (accessed 30 September 2020).

Campbell, H. (2012) Planning ethics and rediscovering the idea of planning. *Planning Theory* 11(4), 379–399.

Caplan, A., Aujla, A., Prosser, S., and Jackson, J. (2009) Race discrimination in the construction industry: a thematic review. Research Report 23. Manchester, Equality and Human Rights Commission.

Chang, C.Y., and Ive, G.J. (2007) Reversal of bargaining power in construction projects: meaning, existence and implications. *Construction Management and Economics* 25(8), 845–855.

Charron, S., Fuchs, A., and Oullier, O. (2008) Exploring brain activity in microeconomics. *Revue D'Économie Politique* 118, 97–124.

Chartered Institute of Building (CIOB) (2013) *Corruption in the UK Construction Industry*. Chartered Institute of Building, Ascot.

Chartered Institute of Building (CIOB) (2018) *Rules and Regulations of Professional Competence and Conduct*. Bracknell, CIOB.

Chartered Institute of Building (CIOB) (2020) *The Real Face of Construction 2020*. Bracknell, CIOB.

Clark, J.B. (1891) Distribution as Determined by a Law of Rent. *Quarterly Journal of Economics* 5(3) 289–318.

Clower, Robert W. (1967) A reconsideration of the microfoundations of monetary theory. *Western Economic Journal* 6, 1–9.

Conway-Jones, D. (2002) Factual causation in toxic tort litigation: a philosophical view of proof and certainty in uncertain disciplines. *University of Richmond Law Review* 35(4), 875–941.

Da Graça Prado, M. (2020) *Changing the Game: A Critical Analysis of Labour Exploitation in Mega Sport Event Infrastructure*. London, Engineers Against Poverty.

Doran, D. (2004) FMI/CMAA *Survey of Construction Industry Ethical Practices*. Mclean, VA, Construction Management Association of America.

Eccles, R. (1981) Bureaucratic versus craft administration: the relationship of market structure to the construction firm. *Administrative Science Quarterly* 26(3), 449–469.

Egan, J. (1998) *Rethinking Construction*. London, Department of Trade and Industry.

Felipe, J., and McCombie, J.S.L. (2013) *The Aggregate Production Function and the Measurement of Technical Change: 'Not Even Wrong'*. Cheltenham, UK and Northampton, MA, USA, Edward Elgar Publishing.

Fine, B. (1999) A question of economics: is it colonizing the social sciences? *Economy and Society* 28(3), 403–425.

Flyvbjerg, B., Skamris Holm, M.K., and Buhl, S.L. (2004) What causes cost overrun in transport infrastructure projects? *Transport Reviews* 24(1), 3–18.
Franck R. (2003) Causal analysis, systems analysis, and multilevel analysis: philosophy and epistemology. In Courgeau, D. (ed.), *Methodology and Epistemology of Multilevel Analysis*, Methodos Series, Vol. 2. Dordrecht, Springer, pp. 175–198.
Friedman, M. (1953) The methodology of positive economics. In Friedman, M., *Essays in Positive Economics*. Chicago, IL, University of Chicago Press.
Gardner, N., Huh, J.S., and Chung, L. (2002) Lessons from the Sampoong department store collapse. *Cement and Concrete Composites* 24, 523–529.
Garvin, M.J., and Nord, D.N. (2012) Real options in infrastructure projects: theory, practice and prospects. *Engineering Project Organization Journal* (March–June) 2, 97–108.
Giddens, A. (1984) *The Constitution of Society: Outline of a Theory of Structuration*. Cambridge, Polity.
Gintis, H. (2009) *Game Theory Evolving: A Problem-Centered Introduction to Modeling Strategic Interaction*. Princeton NJ, Princeton University Press.
Gorski, P.S. (2013) What is critical realism? And why should you care? *Contemporary Sociology: A Journal of Reviews* 42(5), 658–670.
Granger, C.W.J. (1969) Investigating causal relations by econometric models and cross-spectral methods. *Econometrica* 37(3), 424–438.
Granger, C.W.J., and Newbold, P. (1974) Spurious regressions in econometrics. *Journal of Econometrics* 2, 111–130.
Green, S.D. (2001) Towards a critical research agenda in construction management. CIB World Building Congress, Wellington, New Zealand.
Green, S.D. (2011) *Making Sense of Construction Improvement: A Critical Review*. Chichester, John Wiley.
Green, S.D., and Sergeeva, N. (2019) Value creation in projects: towards a narrative perspective. *International Journal of Project Management* 37(5), 636–651.
Gruneberg, S., and Folwell, K. (2013) The use of gross fixed capital formation as a measure of construction output. *Construction Management and Economics* 31(4), 359–368.
Haughwout, A.F. (2000) *Public Infrastructure Investments, Productivity and Welfare in Fixed Geographic Areas*. New York, Federal Reserve Bank of New York.
Hausman, D. (ed.) (2008) *The Philosophy of Economics: An Anthology* (3rd edition). Cambridge, CUP.
Hausman, D., McPherson, M., and Satz, D. (2017) *Economic Analysis, Moral Philosophy and Public Policy* (3rd edition). Cambridge, CUP.
Hayek, F. (1945) The use of knowledge in society. *American Economic Review* 35(4), 519–530.
Hirschman, A.O. (1970) *Exit, Voice, and Loyalty: Responses to Decline in Firms, Organizations, and States*. Cambridge, MA, Harvard University Press.
Hobbes, T. (1651) *Leviathan*. London, Andrew Crooke.
Hobbes, T. (2017) *Leviathan*. Harmondsworth, Penguin.
Hodgson, G. (2001) *How Economics Forgot History*. London, Routledge.
Hoover, K. (2009) Microfoundations and the ontology of macroeconomics. In Kincaid, H., and Ross, D. (eds), *The Oxford Handbook of Philosophy of Economics*. Oxford, OUP, pp. 386–409.
Hume, D. (1739) *A Treatise of Human Nature*. London, John Noon.
Institution of Civil Engineers (2017) *Code of Professional Conduct*. London, ICE.
Ishikawa, Kaoru (1968) *Guide to Quality Control*. Tokyo, JUSE.
Kahneman, D., Tversky, A. (eds) (2000) *Choices, Values and Frames*. New York, Cambridge University Press.
Kaidesoja, T. (2007) Exploring the concept of causal power in a critical realist tradition. *Journal for the Theory of Social Behavior* 37(1), 63–87.
Kant, I. (1785/2012) *Groundwork of the Metaphysics of Morals* [original in German]. Cambridge, CUP.
Kant, I. (1787/1999) *Critique of Pure Reason* [original in German]. Cambridge, CUP.
Kennedy, P. (1998) *A Guide to Econometrics* (4th edition). Oxford, Blackwell.
Keynes, J.M. (1936) *The General Theory of Employment, Interest and Money*. London, Macmillan.
Kim, S.H., and Haque, M. (2002) The Asian financial crisis of 1997: causes and policy responses. *Multinational Business Review* 10(1), 37–44.
Kirkland, C.J. (ed.) (1995) *Engineering the Channel Tunnel*. London, Chapman & Hall.

Kirkland, C.J. (2002) The fire in the Channel Tunnel. *Tunnelling and Underground Space Technology* 17(2), 129–132.
Kuhn, T. (1962) *The Structure of Scientific Revolutions*. Chicago, IL, University of Chicago Press.
Lambsdorf, J.G. (2002) Corruption and rent-seeking. *Public Choice* 113, 97–125.
Latham, M. (1994) *Constructing the Team*. London: HMSO.
Leontief, W. (1941/1951) *Structure of the American Economy 1919–1929: An Empirical Application of Equilibrium Analysis* (2nd edition). New York, OUP.
Locke, J. (1689/1966), *The Second Treatise of Government*. Oxford, Blackwell.
Locke, J. (1690/1980) *The Second Treatise of Government*. Cambridge, Hackett.
Lowe, R., Chiu, L.F., and Oreszczyn, T. (2018) Socio-technical case study method in building performance evaluation. *Building Research and Information* 46(5), 469–484.
Mackie, J.L. (1980/2002) *The Cement of the Universe: A Study of Causation*. Oxford, OUP.
Maqsood, T., and Schwarz, J. (2008) Analysis of building damage during the 8 October 2005 earthquake. *Pakistan Seismological Research Letters*, 79, 163–177.
Marcuse (1964/2002) *One Dimensional Man: Studies in the Ideology of Advanced Industrial Society*. Abingdon, Routledge.
Marshall, A. (1920/1961) *Principles of Economics*, edited with annotations by C.W. Guillebaud (two volumes). London: Macmillan & Co.
Marx, K. (1867/1990) *Capital: a Critique of Political Economy* [original in German]. London, Penguin.
Mason, J. (2009) Ethics in the construction industry: the prospects for a single professional code. *International Journal of Law in the Built Environment* 1(3), 194–205.
Maziarz, M. (2015) A review of the Granger-causality fallacy. *Journal of Philosophical Economics: Reflections on Economic and Social Issues* 8(2), 86–105.
Miceli, T.J. (2011) *The Economic Theory of Eminent Domain*. New York, Cambridge University Press.
Morgenstern, O. (1950) *On the Accuracy of Economic Observations*. Princeton, NJ, Princeton University Press.
Mornati, F. (2013) Pareto Optimality in the work of Pareto. *European Journal of Social Sciences* 51(2), 65–82.
Mosey, D. (ed.) (2019) *Collaborative Construction Procurement and Improved Value*. Chichester, Wiley-Blackwell.
Naoum, S.G., Harris, J., Rizzuto, J., and Egbu, C. (2020) Gender in the construction industry: literature review and comparative survey of men's and women's perceptions in UK construction consultancies. *Journal of Management in Engineering* 36(2), 04019042.
Nozick, R. (1975) *Anarchy, State and Utopia*. New York, John Wiley.
Office of Fair Trading (2009) *Bid Rigging in the Construction Industry in England*. London, OFT.
Osborne, M.J., and Rubinstein, A. (1994) *A Course in Game Theory*. Cambridge, MA, MIT Press.
Ozanne, A. (2016) Why power matters for economics. In Ozanne, A., *Power and Neoclassical Economics: A Return to Political Economy in the Teaching of Economics*. Palgrave Pivot, London, pp. 13–18.
Pacioli, L. (1494) *Rules of Double Entry Bookkeeping* (from *Summary of arithmetic, geometry, proportions and proportionality*) [Original in Italian] IICPA Publications.
Pearl, J. (2000) *Causality: Models, Reasoning, and Inference*. Cambridge, CUP.
Piazzesi and Schneider, M. (2016) Housing and macroeconomics. In Taylor, J.B., and Uhlig, H. (eds), *Handbook of Macroeconomics Vol. 2B*. Amsterdam, Elsevier, pp. 1547–1640.
Piketty, T. (2014) *Capital in the Twenty-First Century*. Cambridge, MA, Belknap Press.
Ramchum, R. (2015) Survey results: homophobia remains rife in construction industry. *Architects' Journal*, 24 August.
Rawls, J.B. (1971) *A Theory of Justice*. Cambridge, MA, Belknap Press of Harvard University Press.
Reason, J. (2000) Human error: models and management. *British Medical Journal* 320(7237), 768–770.
Reiss, J. (2013) *The Philosophy of Economics: A Contemporary Introduction*. New York, Routledge.
Ricardo (1817) *The Principles of Political Economy and Taxation*. London, John Murray.
Robbins, L. (1932) *Essay on the Nature and Significance of Economic Science*. London, Macmillan.
Robinson, J. (1962) *Economic Philosophy*. Harmondsworth, Penguin.
Rousseau (1762/1997) *On the Social Contract and Other Later Political Writings*. Cambridge, CUP [original in French].

Ryan-Collins, J., Lloyd, T., and Macfarlane, L. (2017) *Rethinking the Economics of Land and Housing*. London, Zed Books.
Ryskamp, J. (2007) *The Eminent Domain Revolt: Changing Perceptions in a New Constitutional Epoch*. New York, Algora Publishing.
Samuelson, P. (1947) *Foundations of Economic Analysis*. Harvard, MA, Harvard University Press.
Schmitt, C. (1921/2014) *Dictatorship: From the Origin of the Modern Concept of Sovereignty to Proletarian Class Struggle* [original in German]. Cambridge, Polity Press.
Seeley, I.H. (1996) *Building Economics: Appraisal and Control of Building Design, Cost and Efficiency* (4th edition). Basingstoke, Macmillan.
Sen, A. (1970) *Collective Choice and Social Welfare*. San Francisco, CA, Holden-Day.
Simon, H. (1957) *Models of Man. Social and Rational. Mathematical Essays on Rational Human Behavior in a Social Setting*. New York, John Wiley.
Sklair, L. (2017) *The Icon Project: Architecture, Cities and Capitalist Globalization*. New York, OUP.
Smith (1776/1977) *An Inquiry into the Nature and Causes of the Wealth of Nations*. Chicago. IL, University of Chicago Press.
Smyth, H.J., Morris, P.W.G., and Kelsey, J. (2007) Critical realism and the management of projects: epistemology for understanding value creation in the face of uncertainty. Paris, EURAM, 16–19 May.
Špačková, O., and Straub, D. (2012) Dynamic Bayesian network for probabilistic modeling of tunnel excavation processes. *Computer Aided Civil and Infrastructure Engineering* 28(1), 1–21.
Stiglitz, J.E., Sen, A.K., and Fitoussi, J.-P. (2009) *Measuring Economic Performance and Social Progress*. Paris, Commission on the Measurement of Economic Performance and Social Progress.
Streissler, W.E. (1970) *On the Pitfalls of Econometric Forecasting*. London, IEA.
Taylor, C., and Nanz, P. (2020) *Reconstructing Democracy. How Citizens are Building from the Ground Up*. Boston, MA, Harvard University Press.
Thaler, Richard H. (2015) *Misbehaving: The Making of Behavioral Economics*. New York, W.W. Norton & Company.
Tunstall, R. (2020) *The Fall and Rise of Social Housing*. Bristol, Policy Press.
Uff, J. (2003) Duties at the legal fringe: ethics in construction law. Centre for Construction Law and Management, The Michael Brown Foundation 4th Public Lecture, King's College London, 19 June.
Vee, C., and Skitmore, R.M. (2003) Professional ethics in the construction industry. *Engineering Construction and Architectural Management* 10(2), 117–127.
Von Neumann, J., and Morgenstern, O. (1944) *Theory of Games and Economic Behaviour*. Princeton , NJ, Princeton University Press.
Wilhelms, E.A., and Reyna, V.F. (2015) *Neuroeconomics, Judgement and Decision Making*. New York, Psychology Press.
Williamson, O.E. (1975) *Markets and Hierarchies: Analysis and Antitrust Implications*. New York, Free Press.
Wilson, P.F., Dell, L.D., and Anderson, G.F. (1993) *Root Cause Analysis: A Tool for Total Quality Management*. Milwaukee, WI, ASQ Quality Press.
Woodward, J. (1965/1981) *Industrial Organisation: Theory and Practice* (2nd edition). Oxford, OUP.
Zhang, X., Austin, S.A., and Glass, J. (2006) Linking individual and organisational values: a case study in UK construction. In Boyd, D. (ed.), *Proceedings of the 22nd Annual ARCOM Conference*, Association of Researchers in Construction Management, 4–6 September, Birmingham, UK, Vol. 2. pp. 833–842.

4. The nature and development of construction economics

Gerard de Valence

INTRODUCTION

In the opening paragraph of *Economic Theory and the Construction Industry*, Hillebrandt (1974) writes that construction economics is 'the application of the techniques and expertise of economics to the construction industry', a definition that has been broadly endorsed for nearly five decades. The book was, and is, an important milestone in the development of construction economics. It was the first time construction economics had been presented as a distinct form of industry economics, a branch of economics that had a burst of development in the 1940s and 1950s, and it linked an extensive, well-developed body of work on the macroeconomic role of the construction industry with the prevailing economic paradigm of the time, the neo-classical synthesis. The analysis of the demand and supply sides of the construction industry explicitly incorporated four characteristics: the physical nature of the product; the structure of the industry and organization of the construction processes; the characteristics of demand; and the method of price determination, either by tendering or by some form of negotiation. Hillebrandt (1974, p. 9) concludes: 'In view of these unique characteristics of construction, there is a need for the development of a new theoretical economic analysis, or at least for adaptation of existing theory, to assist in the understanding of the workings of the construction process, the construction industry and the construction firm.' This chapter is an assessment of how these challenges have been met in the following decades.

The chapter mainly focuses on theoretical developments in construction economics in the treatment of processes, the industry and firms. The period since the 1960s is divided into three and, in each period, research contributions to the three broad areas Hillebrandt identified of firm, industry and process are discussed. The first period from 1966 to 1989 is one of emergence and establishment, as topics are investigated and the literature develops. The second period from 1990 to 2007 covers the expansion and development of construction economics as new approaches such as transaction cost economics (TCE) and lean construction are applied. The third period from 2008 to 2020 is one of consolidation, and brings the story to the present.

Although a discussion on the nature of a field of inquiry implies some delineation of scope, as in the topics for investigation, the contents of this and other recent books are in themselves evidence of the scope of construction economics. Many of the topics investigated by construction economics are in this *Research Companion*, where they are well documented; more can be found in the other publications discussed. The more limited discussion here focuses on the development of theoretical understanding of the workings of the construction process, the construction industry and the construction firm. The conclusion argues that construction economics is in the process of establishing a distinctive research agenda and that the scope of construction economics is a strength rather than a weakness, because this reflects the nature of an evolving economy and a construction industry which can be defined in various ways. The openness of construction economics to ideas from economics has led to important insights into construction firms and markets, and the relationship between the industry and the wider economy.

A note on presentation and handling of sources is required. The chapter is an overview of research, not a compendium, and tells the story with broad brush strokes with some prior knowledge of basic concepts assumed. The approach followed uses representative publications from each period for the topics discussed. Individual papers are cited, and contributions to this volume are recommended for the overview they provide of the sources and material found in construction economics. Most of the works cited in this chapter are used as a shorthand way of providing links to more extensive research on specific topics. Particularly useful are the relevant review papers and journal special issues that have been used where possible.[1] Some points of interest and supporting material are in the endnotes.

In 2007 a conference on 25 years of publications in *Construction Management and Economics* provided the opportunity for an assessment of developments in construction economics research, and identified the areas and topics that had been addressed in the literature. The issues raised in those reviews of construction economics form part of the discussion of the third period, along with the availability and quality of relevant data (see Meikle and Shrestha, Chapter 12), which has been a significant constraint on empirical research in construction economics. Nevertheless, the scope of construction economics has been broad, and an economic perspective has been turned onto many aspects of construction and its relationship with other industries, the economy and society. As the contributions to this *Companion* demonstrate, this research overlaps topics studied by other branches of economics such as macroeconomics (see Lopes, Chapter 6), development (see Kumaraswamy and Mahesh, Chapter 8), urban economics (see Werna and Klink, Chapter 16) and housing economics (see Ismail, Chapter 19). There is also an overlap with construction management research, as the contributions here on bidding (see Laryea, Chapter 17), procurement (see Watermeyer, Chapter 18) and stakeholder management (see Chinyio et al., Chapter 20) demonstrate. That the boundaries and therefore the profile of construction economics are indistinct is discussed.

Hillebrandt's book (there were eventually three editions, in 1974, 1985, 2000) provided a template for later texts (such as Briscoe 1988; Raftery 1991; Runeson 2000; Myers 2004 [2017]). These books, and many others, have approached the adaptation of current economic theory to building and construction in a way that both reflects the industry and remains accessible to students in construction and built environment courses. This context is important. The origins of construction economics are not in the economics departments of universities, but in courses with diverse students who might take one or two economics subjects, such as Hillebrandt's postgraduate classes. These courses typically sit in schools of the built environment, construction, engineering or project management.

Also in the 1960s, many undergraduate quantity surveying courses in the United Kingdom (UK) and other countries such as those in the Commonwealth where the profession is established, became building economics courses, where key topics such as cost estimating and control, measurement methods and building technology were taught. Accreditation for these courses required an introductory economics subject. As a result, some of the most useful work in construction economics, including that published in journals, employs fairly basic economic concepts, but requires detailed knowledge of the construction industry and its characteristics. This is a distinguishing feature of construction economics, one shared with other branches of industry economics such as health, transport and urban economics, where industry-specific knowledge is combined with economic theory to analyse and understand structure and dynamics under specific conditions. Economic theory, in turn, is based on models of markets, supply and demand, and so on.

Models take many different forms (Morgan 2012), depending on discipline, but share the purpose of explaining and the method of testing by prediction (Page 2018). They can be formal (that is, mathematical) or informal, and are based on evidence and data. Models of the construction industry, and theories about how and why the industry functions as it does, have progressed through a series of stages of increasing sophistication. The contribution of construction economics research has been fundamental to that evolution. This chapter focuses on development of models of firms, industry and processes for two reasons. First, because they are shared with other branches of economics that are focused on specific industries, they are the bridge between them and construction economics. If construction economics has developed distinctive models of firms, industry and processes, that argues for a distinctive identity for construction economics. Second, these models fundamentally underpin researchers' approach to their work, but are not much discussed because their basic outlines are widely known and agreed upon. As the following discussion shows, much of construction economics research is based on construction statistics and cost data, and is empirical rather than theoretical, but that does not mean such models are not important and are not incorporated into that research. (Note that 'construction statistics' is used strictly here, referring only to the data produced by national statistical agencies. Construction data includes other public and private sector sources.)

PERIOD 1: EMERGENCE AND ESTABLISHMENT 1966–1989

The 1960s is taken as the starting point for construction economics research. On the one hand, prior research into the economics of construction was concerned with the macroeconomic role of the industry; on the other hand was building economics, called 'project economics' by Ofori (1994). In the decades leading up to the 1970s, construction had attracted the attention of economists because of its role in investment and its macroeconomic importance. When Kuznets (1930) compiled the first set of long-run statistics on United States (US) output, from 1869, he found a 15–25 year pattern in construction that came to be known as the Kuznets cycle, beginning an enduring interest by macroeconomists in the relationship between construction activity and the business cycle (Riggleman 1933; Duca et al. 2010). The business cycle is short-run fluctuations in the rate of economic growth, and construction is the component of investment expenditure that is most sensitive to economic conditions. Thus, in many countries, the level of construction work done makes it the most volatile sector of the economy as it goes through these periods of expansion and contraction (Barras 2010).

The macroeconomic role of the industry was the subject of Lange and Mills (1979), whose book coined the phrase 'the balance wheel of the economy' to describe the counter-cyclical effect of construction, particularly housing, in response to a lowering of interest rates in a recession. This enduring idea is heard today in calls for increased infrastructure expenditure to counter 'secular stagnation', as the slowdown in gross domestic product (GDP) growth rates after the financial crisis that started in 2007 is known (Summers 2016).

Also, during the 1960s a profile of building and construction emerged from economic histories of the industry in the UK (Bowley 1966), US (Fitch 1973) and Australia (Hutton 1967):

the industry had many small firms, typically subcontractors, a small number of medium-sized firms, and a small number of large contractors; demand was uncertain and unstable, so firms minimized fixed costs and capital spending; incremental innovation and problem-solving were strengths, but new technology spread slowly (Bowley 1960); and productivity grew slowly, if at all (Dacy 1965; Cassimatis 1970). Among the topics addressed by economic histories of construction were building materials and labour costs, prices and price indexes, industry structure and subcontracting, housing policies, training and productivity. As an economics of construction began to emerge, these topics were carried across the divide between history and the present to become part of the research agenda found in construction economics.

The issues this profile raised were addressed in the earliest research on the organization of construction. Stinchcombe (1959) contrasted bureaucratic and craft systems of work administration: manufacturing has mass production with economies of scale through standardization of tasks, but construction uses standardized products and parts. In craft production, work administration and control is given to workers and foremen, but they do not make decisions on product type, design and price, which are made by others, variously referred to as administrators, bureaucrats, clients and employers. Stinchcombe argued that bureaucratic administration requires long production runs and predictable work-flow, while uncertainty and variability in work-flow will make subcontracting and the craft system more efficient. Eccles (1981a) discussed industry structure with extensive subcontracting, and in (Eccles 1981b) argued the relationship between contractors and subcontractors can be characterized as a 'quasifirm'.

By the 1960s, national accounting had developed and more data was becoming available (Vanoli 2005). Attention shifted to the role of construction in long-run economic growth as the relationship between infrastructure investment, capital formation and GDP growth rates was established, and became a major focus of economic research after Aschauer (1989). In industrialized economies, construction was typically around half of gross capital formation (GCF) and around 10 per cent of GDP. However, rapidly industrializing economies in Asia and elsewhere could have construction shares of GDP of 20 or 30 per cent, or more (Strassmann 1970). This is a pattern found repeatedly by the World Bank (2013): for countries to take off into growth rates above 10 per cent a year, very high levels of GCF may be required (Girardi and Mura 2014).

Construction's role in economic development was the focus of a report to the United Nations by Turin (1966), on *The Place of Construction in Economic Development*, another founding text in construction economics.[2] By putting construction at the centre of the analysis of the macroeconomy in development research, Turin (1966) and Strassman (1970) provided the foundations for the incorporation of the characteristics of construction into the analysis of construction firms and markets. That project was initially pursued at the Building Economics Research Unit (BERU), established by Turin at University College London in the late 1960s after his appointment as the first Professor in Building in 1966. The BERU began a series of studies on the building process, the functions of its participants, how the industry responds to demand, and the possibilities of 'industrializing' the industry. By the mid-1970s that research had led to a Building Economics and Management MSc course, the first edition of Hillebrandt's *Economic Theory and the Construction Industry* and Turin's (1975) *Aspects of the Economics of Construction*.[3] The treatment of processes, industry and firms in those two books provide a reference point for the first phase of development of construction economics to the end of the 1980s.

Firms

The textbook neoclassical model of a firm is sparse and lacks detail. Based on cost and demand functions a firm maximizes profit as a 'black box' that mysteriously but efficiently turns inputs into outputs as its contribution to total industry output. Firms are optimizers of scarce resources constrained by technological capacity. In introductory texts this is presented graphically as the set of choices a firm faces to minimize marginal cost and maximize profit based on marginal revenue. Firms here are price takers, they are small relative to the market, do not have market power, and products are homogeneous. Hillebrandt (1974) used this model in chapters on costs, demand and equilibrium for a firm. However, Hillebrandt ranged widely, with alternative views on the objectives of the firm (growth, revenue and managers' incentives), costs for project-based firms, revenue curves and mark-ups, product differentiation, and the effects of different types of markets included. That discussion incorporated the characteristics of construction, based on Hillebrandt's familiarity with the British industry, and came to two key conclusions. The first was perfect competition due to ease of entry. Even with a limited, selected number of tenderers there is 'effective competition', with the same outcome as perfect competition. The short discussion of imperfect competition (pp. 136–138) is conventional and construction is not mentioned. The second conclusion was that marginal analysis is appropriate for project-based firms. This model of the firm can also be found in Briscoe (1988) who, similarly to Hillebrandt and Cannon (1990), attempted to reconcile neoclassical economics with construction industry characteristics.

At this time, at the end of the 1980s, three books that took significantly different approaches to the industry and the firm appeared. Ball (1988) gives a Marxist analysis of British construction, thus focusing on social relations and the contracting system, and claims: 'One theoretical avenue which seems of little use in studying the industry is to apply neoclassical economic theories of the firm' (p. 19). The dual role given to firms as producers (of buildings) and merchants (purchasing inputs) foreshadowed the trade credit/cash farming literature to come, with Ball arguing that the merchanting role predominates at the expense of wages and productivity.

By contrast, Bon's (1989) *Building as an Economic Process* applied Austrian economics to construction. This branch of economics, a polar opposite to Marxism, is relevant to construction because it emphasizes capital and the capital stock, the pivotal role of investment in capital formation, the explicit role of the time taken for investment decisions to be fulfilled, the cyclic nature of economic and building activity, and the possibility of production plans or projects failing. Firms here are vehicles for investment, preparing plans and sourcing the capital required, and the building process is a series of decisions on the use of capital. Bon strongly linked construction (supply) to the property market (demand) through capital flows and the need for ongoing repair and maintenance, and suggested that this relationship should be the basis of a research agenda in construction economics. However, much of that research is now in the atheoretical area of life-cycle costs, and Bon's book remains the only use of Austrian economics in construction economics.[4] Despite the appeal of the Austrian approach, it had its heyday in the 1980s and has since faded because it is discursive, lacking formal models and empirical methods (Blaug 1992).

In their book, Hillebrandt and Cannon (1989, p. 6) argue that it is 'inherently difficult to relate the economic structure, behaviour and performance of contracting firms to theoretical models'. Their alternative is managerial economics (on firm decision-making) and management theory (on business strategy, organization theory and human resources), explored in seven of the eight other contributions to the book. This illustrates the large grey area between construction management and construction economics, where topics such as these are of mutual interest and cannot be considered from an economic viewpoint without reference to industry custom and practice.

Thus, at the end of the 1980s, there were four distinctly different concepts of construction firms, supported by detailed analysis of the industry in Britain, where these researchers worked. These were: a neoclassical concept of firms in competitive contracting markets; a Marxist view of firms acting as producers and merchants; an Austrian model form of firms circulating capital between construction and property; and managerial firms organized for construction.

Industry

One way in which construction economics differentiates itself from construction management is by emphasizing the management of firms rather than projects, although this border is an open one in both directions. A clearer boundary is the analysis by construction economics of the industry, as the context for firms, which is not found in construction management with its focus on projects. In the 1980s, industry analysis followed the structure–conduct–performance framework associated with Bain (1959) and the Harvard School, with the challenges from TCE, empirical industrial organization studies and game theoretic approaches still developing. As noted above, the analysis of the conduct (management) of construction firms in construction economics was deeply embedded in an understanding of the structure of the industry, which is different from the manufacturing industries studied in industrial organization.

In the 1980s journals started publishing construction economics research, with *Construction Management and Economics* (*CME*) establishing itself as the lead journal:

> [*CME*] was founded in 1982, and the *Journal of the Construction Division of ASCE* was replaced by the *Journal of Construction Engineering and Management* in 1983. These events helped to separate construction economics from urban, property and housing economics, and to facilitate the decision-economics of construction by conjoining construction economics with construction management. (Ive and Chang 2007, p. 1591)

CME claimed that 'construction economics includes: design economics, cost planning, estimating and cost control, the economic functioning of firms within the construction sector, and the relationship of the sector to national and international economies'. This agenda also broadly became that of the International Council for Research and Information in Building and Construction (CIB) Working Commission on Building Economics. The first six volumes of *CME* had papers on tendering and cost estimating, within the project economics stream. The first construction economics paper was on productivity (Bowlby and Schriver 1986).

A major topic of empirical research in *CME* and elsewhere in the 1980s was the linkages between construction and other industries, using the input–output tables from national statistical agencies. Initiated by Bon, a series of papers with collaborators compared cross-industry flows in Organisation for Economic Co-operation and Development (OECD) countries, and demonstrated the fundamental role of the industry in capital formation (collected in Bon 2000). These input–output studies on the relationship of construction to other industries confirmed the importance of site work as a driver of output across the economy, but did not provide an industry model.

Processes

The foundational paper on the construction process is another from Turin (2003[5]). In 'Building as a process' he defined this as 'building as an activity concerned with the best possible use of inputs to produce a desired output' (Turin 2003, p. 181), and argued that traditional building of one-off projects with temporary teams was less efficient than alternatives that better integrated design, manufacturing and assembly. The effects of fragmentation/integration have been a theme in many industry reports where, following Turin, the differing roles of users, clients, professionals, contractors and manufacturers under different contractual arrangements, and the importance of information flows, or lack of them, are the basis of the discussion. His process has eight stages and four different versions. The versions were: a traditional 'maximum fragmentation'; a manufacturing-orientated 'component' version; a 'model' version controlled by the contractor; and a 'process' version that would allocate control of a stage to the appropriate participant. This results in a matrix with horizontal time (stages) and vertical participant axes, and reflected Turin's view that industrialization was necessary to improve industry performance. It became the reference model of the industry used by researchers for the next couple of decades (see Groak 1993), and the influence of this approach can be seen in the reports from the inquiries into the UK industry reviewed in Murray and Langford (2002). As they document, government policies for performance improvement in the UK often focused on contractual relations, an approach widely followed elsewhere.

Although neither its purpose nor its intention, Turin's model predicted the evolution of the industry over the following decades, as contractors moved to take a central role in managing the process, as a comparison of the traditional one-off version with his model version shows. With the emergence of global multidisciplinary firms that combine consulting, contracting and design, the international industry looks like a combination of Turin's model and process versions, with contractors and the professions sharing responsibility for product design.

PERIOD 2: DEVELOPMENT AND EXPANSION 1990–2007

Processes

Turin's stages model was product-based, and appropriate for an industry that was rapidly industrializing with standardization of components and increasing use of prefabricated concrete. However, by the end of the 1980s, the issues facing the industry had changed and the focus had shifted to industry performance and processes. A core issue on the construction statistics side was (and is) the low rate of productivity growth of, more or less, 0 per cent a year across the OECD. On the construction project side there was increasing pressure to improve value for clients and users by reducing costs and increasing quality. The process-based model of the industry that emerged focused on the functions of participants in construction projects. It extended the construction economics synthesis from the level of the project to that of the industry, an important step, and broadened the range of participants. It also clearly assigned participants, or actors, to either the demand side or the supply side for construction projects.

In 2003, there was a special issue of *Building Research and Information* (*BRI*) on Re-valuing Construction. The papers considered how the industry operates and the organization of the building process, production strategies, standardization and supply chain management, and thus provide an opportunity to assess developments since Turin's earlier conception of the building process. Turin focused on variations in process under different contractual conditions. In 2003 this was no longer the main issue: the focus is on management of a 'value stream' and re-engineering processes to create more value for owners and users. Three of the eight *BRI* papers were from a lean construction perspective, from leading advocates Koskela, Ballard, Tommelein and others, and three were on off-site manufacturing in different forms, from Winch, Gann, Gibbs and others. Processes were unpicked at a much greater level of detail, and the outline Turin drew had been filled in with ideas from business management such as lean production, business process re-engineering and value management. Construction processes were more industrialized, although this was unevenly spread across both countries and the industry, and many new management techniques had been introduced. Importantly, there was now a conceptualization of industrialized construction as a production process, rather than a series of stages, and of projects as a form of production, bringing together on-site work and off-site fabrication.

The founding text for lean construction was Koskela (1992), who emphasized the importance of a theory of production that reflected construction characteristics that he called the transformation, flow and value generation (TFV) framework. Over the next decade his ideas were taken up and developed by a global group of collaborators in the International Group for Lean Construction. They argued that project management attempts to manage by scheduling, cost and output measures, but lean construction attempts to manage the value created by all the work processes used between project conception and delivery. One of the core ideas in lean construction is a system of production control called 'the last planner', usually a front-line supervisor such as a construction foreman or a design leader, and lean also brought together the product design, engineering, fabrication and logistics aspects of construction projects. A pair of book chapters summarized lean construction theory (Koskela et al. 2002) and the lean project delivery system (Ballard et al. 2002), which was a stage model like Turin's. In contrast to Hillebrandt and Cannon's (1989) view of managerial economics as primarily about business strategy, lean argues that managerial actions affect the design, operation and improvement of a production system and, for construction, the three major factors are the one-of-a-kind nature of projects, site production and temporary organization; three features that are embedded in the construction management research literature.

Koskela's TFV theory of production was an explicit attempt to introduce an alternative to what he called the 'economic view' of production. He edited a special issue of *BRI* in 2008 on developing theories of the built environment that shows how difficult the terrain traversed by construction economics can be. His editorial asked:

> has this reactive discussion that has taken natural science as a reference point, or a straw man, been somewhat misdirected? Could all the energy used for criticizing or undermining the use of the natural science-oriented world view in research have been more productively used in the proactive establishment of different kinds of sciences more suitable for the tasks at hand? (Koskela 2008, p. 215)[6]

The six other papers in the special issue discuss in depth and at length challenges, issues and approaches to the built environment, three of them from a design perspective. However, in contrast to the earlier *BRI* special issue on revaluing construction there is no data, and while there is much theorizing and many propositions, there are no predictions. Fellows and Liu (2020, p. 586) described this as a debate about 'management/economics of construction or management/economics in construction'.

Industry

The initial industry model was framed by the development of the International Standard Industrial Classification of All Economic Activities (ISIC) in the 1960s and Turin's stages model in the 1970s. This bridged the gap between 'statistical construction' defined by the output of industries with the SIC codes for construction (de Valence 2019a), and 'project construction', by recognizing inputs from all participants while accepting the SIC categories that defined them. It has been the baseline model of the industry for five decades, and could be described as the construction economics synthesis, combining economists' work on national accounting with built environment researchers' work on the role of the industry in the economy. Examples of the widespread use of this model are found whenever participants from outside the construction SIC are included in research into construction processes, such as supply chain research. While this seems intuitive and obvious, the need to do so is a result of the industry boundaries drawn by the definitions in the SIC. The manufacturing industry includes product design within its SIC, but architectural and engineering design firms are not included in the Construction SIC. Ive and Gruneberg (2000) called the collection of industries contributing to the production of the built environment the 'broad construction industry'.

Ive and Gruneberg (2000) was a comprehensive analysis of the industry. They started by defining 'the construction sector as all production activities contributing to the production of the built environment' (p. 5). As with their book on construction firms, this took a heterodox approach. There is discussion on national accounts and financial accounts, measurement of output and business cycles, multiplier effects, social relations of production and ownership of land. They include formal models, examples are labour costs, stocks of capital, a production function and producers' surplus, and there is an analysis of UK construction statistics. Again, the contrast with Hillebrandt (1974) is striking, with the range of topics broadened to include the macroeconomic role of construction, the financing and capital structure of firms, productivity and the role of subcontractors as producers in their own right.

Around the turn of the millennium the first estimates were produced of the total value of production of the built environment. Using value-added and employment data from the Australian and UK national statistics agencies at the industry level, these estimates showed that construction directly contributes around half the total value of production of buildings and structures (Ive and Gruneberg 2000; de Valence 2001; Pearce 2003). This strand of research has continued, recently in Francis (2019) and a special 2019 issue of *Engineering, Construction and Architectural Management* (*ECAM*) on 'The true value of construction and the built environment to the economy'. The editorial starts:

> There has been a long-standing view that the construction industry as a whole is under-achieving and the industry, in many other countries, is seen as one of the least productive sectors in the economy. A prerequisite for discussion and analysis of the sector's value has to be the production of appropriate information to enable the development of models for the sector. (Ruddock et al. 2019, p. 738)

The eight contributions ranged from using the term 'built environment sector' to describe the broad industry, to sustainability and capital stock issues, productivity, and project economics. There is a common theme around assessing the importance of construction and the built environment to social welfare and, highlighting the empirical nature of construction economics, six of the eight papers are on aspects of measurement of value.

A new mapping of participants in the construction industry along these lines was developed by Gann and his colleagues in 1992 (Ando et al., 1992). Instead of stages, they grouped participants into supply, site-based and demand categories within an institutional framework. A version of the diagram is in Figure 4.1, which found its way into many official reports and has been used extensively by researchers since. This was part of the development of the complex product systems approach by Hobday (2000), Gann (2001) and others, which argued that construction was a service industry that requires coordination of multidisciplinary teams to organize production, which generally becomes more complex as projects get larger. This is a product-based industry model, with construction projects as the products and contractors (the project-based firms) in the centre.

Firms

Issues with the model of firms with perfect information, constant technology and no market power had created a broad research agenda in economics. Two streams in that agenda appeared in construction economics papers in the 1990s. The first was on the boundaries of the firm and TCE based on the work of Williamson (1975), who described firms as a 'nexus of contracts'. First applied to construction by Winch (1989), this became an active research stream (see Abdul-Aziz and Ahmad-Zaini, Chapter 9). Firms in TCE minimize transaction costs by choosing internal production or external supply, the make–buy decision. This added another reason

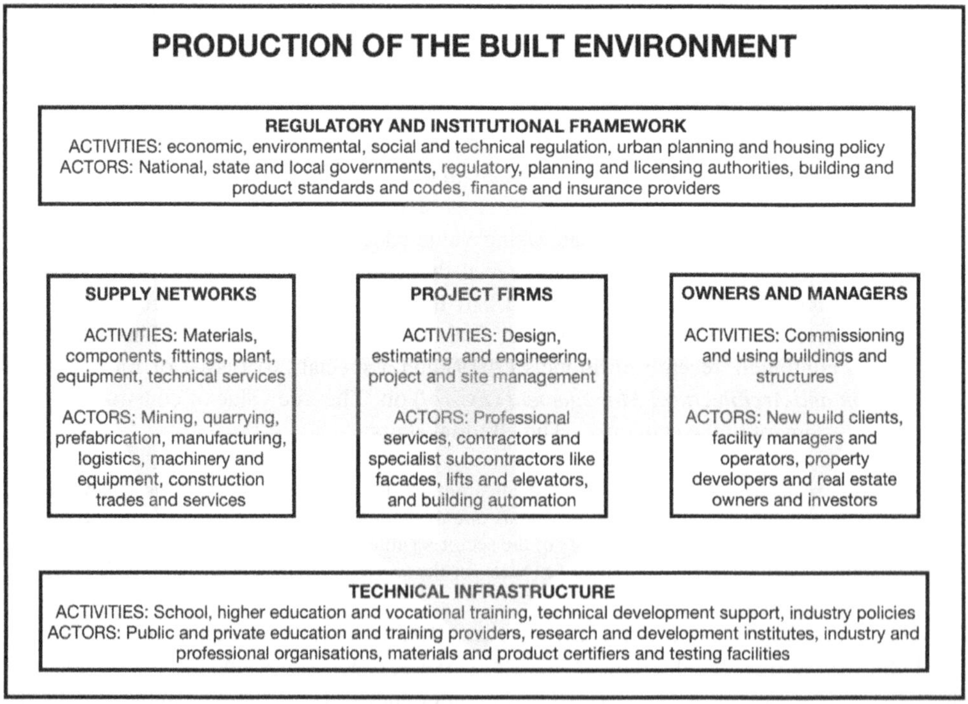

PRODUCTION OF THE BUILT ENVIRONMENT

REGULATORY AND INSTITUTIONAL FRAMEWORK
ACTIVITIES: economic, environmental, social and technical regulation, urban planning and housing policy
ACTORS: National, state and local governments, regulatory, planning and licensing authorities, building and product standards and codes, finance and insurance providers

SUPPLY NETWORKS
ACTIVITIES: Materials, components, fittings, plant, equipment, technical services
ACTORS: Mining, quarrying, prefabrication, manufacturing, logistics, machinery and equipment, construction trades and services

PROJECT FIRMS
ACTIVITIES: Design, estimating and engineering, project and site management
ACTORS: Professional services, contractors and specialist subcontractors like facades, lifts and elevators, and building automation

OWNERS AND MANAGERS
ACTIVITIES: Commissioning and using buildings and structures
ACTORS: New build clients, facility managers and operators, property developers and real estate owners and investors

TECHNICAL INFRASTRUCTURE
ACTIVITIES: School, higher education and vocational training, technical development support, industry policies
ACTORS: Public and private education and training providers, research and development institutes, industry and professional organisations, materials and product certifiers and testing facilities

Source: Based on Gann and Salter (2000) and Gann (2001).

Figure 4.1 *Projects as products*

for subcontracting to the flexibility and minimizing fixed cost explanation already established; specialization by subcontractors results in lower-cost supply under specific contract conditions. The extension of TCE to construction introduced issues such as the hold-up problem on required investment, incentives and contracts, and reframed information asymmetry between participants as a principal–agent problem (see Cerić, Chapter 15).

In their book on construction firms, Gruneberg and Ive (2000) included a chapter on TCE, and much else. Added to the Hillebrandt and Cannon (1989) model of the firm were mix-price and contestable markets, industry capacity, productivity, decision-making under uncertainty and models of pricing, cost and investment. Construction firms were differentiated by Gruneberg and Ive by specialization, size and growth rates. Firms manage portfolios of projects in markets that have barriers to entry and can become concentrated. This is a considerably more complex model of construction firms. Their prologue concludes: 'The aim of this book is to give a clear understanding of some of the economic issues directly confronting construction firms in their operations and provide the economic basis for planning and decision making' (Gruneberg and Ive 2000, p. xvii). However, while they follow Hillebrandt in seeing construction economics as being about management decisions, they also include discussion on capital circuits and the social structure of accumulation, which builds on Ball (1988).

Economic Development and Construction

The relationship between economic development and investment in construction is generally understood to follow an inverted U-shaped curve. In developing countries, the level of construction output as a share of GDP rises as the economy grows, reflecting the investment required to generate that growth. Typically, the rate of growth of construction output is higher than the rate of growth of GDP at this stage. As countries become middle-income and the stock of built assets accumulates, construction's share of GDP levels out. Construction's share of GDP starts to fall in high-income countries as the stock of built assets becomes more productive, and repair and maintenance becomes more significant. A high level of capital investment and construction of infrastructure has long been recognized as a characteristic of industrializing countries, and is clearly related to the stage of economic development of a country (Lopes 2008). This inverted U was found by Strassman (1970), who called it the 'middle-income country bulge', but it became known as the 'Bon curve' much later after Bon (1992) drew it as a simplified, stylised curve, as shown in Figure 4.2.

It was difficult to get data sets for this research. Turin's (1966, 1978) pioneering work used 46 and 78 countries, respectively, and the latter found an S-shaped curve for developing countries as the rate of increase of construction's share was rapid at first but levelled out and stabilized over time. Bon (1992) had only six countries in his data set (Finland, Ireland, Italy, Japan, UK, US). Many of the subsequent studies supporting Bon also do not include much data. Most of them are descriptive, typically grouping countries into four categories based on per capita income and then calculating an average of construction's share of GDP in each group. There are many reasons why these average values are likely to be biased, such as non-stationarity of the data, changes in composition of groups over time, omitted variables and outliers. Lewis (2008) notes how variable data is across countries grouped by income levels. Using a data set of 205 countries Choy (2011) found only qualified support for the Bon Curve. Giradi and Mura (2014) tested their 'construction development curve' using 148 countries, finding that the curve fits better if economic development is measured by alternative indicators instead of

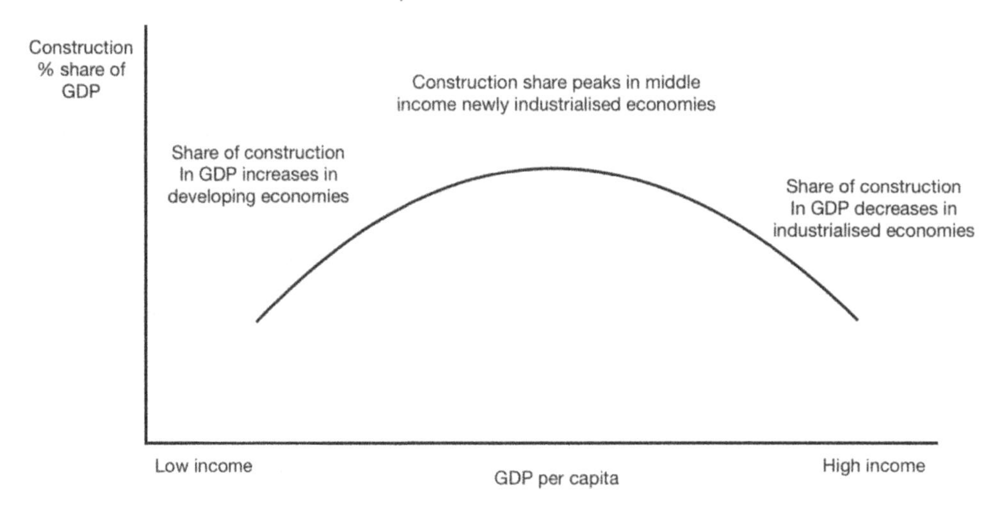

Source: Based on Strassman (1970), Bon (1992), Lopes (2008).

Figure 4.2 *Construction and industrialization*

per capita GDP, such as life expectancy and a broad economic development index. Population density, demographic growth and credit expansion did not explain cross-country variation in the share of construction in output in their model.

PROGRESS IN CONSTRUCTION ECONOMICS IN PERIODS 1 AND 2

The proposition that building economics was not established as an academic discipline was first made in Bon's *Building as an Economic Process,* where the 'objective of this book is to assemble in one place those concepts that may contribute to the development of building economics as a distinct discipline' (Bon 1989, p. 25). In Bon (2001) the future of building economics was seen to lie in fields such as corporate real estate and facilities management, topics connected to building use and reuse decisions and building life cycles. Ofori (1994) argued that construction economics had not yet developed to the point where it could be recognized as a distinct part of general economics, due to a lack of consensus on the 'main concerns and contents' and a lack of a coherent theory (p. 304). Ofori also argued for the term 'construction economics' as preferable to 'building economics' because of its wider scope. Myers (2003) saw the future of construction economics in sustainability: it could provide both a common purpose and a conceptual approach, thus solving the two major problems identified by Bon and Ofori. While sustainability became a large and growing field of study that construction economics has contributed to, and does provide common purpose, it has not been the basis for a theory because it is based on empirical studies of building, materials and process performance. As noted above, there is a largely one-way flow of ideas from theory to empirical research.

The contribution to this debate from Ive and Chang (2007) was in a paper that addressed the relationship between construction economics, economics and management. Their concern was the extent of progress towards recognition of construction economics as a subdiscipline of economics, measured by citations and authorship across the journals of the main discipline and the 'putative subdiscipline'. Papers published in *Construction Management and Economics* (*CME*) between 2000 and 2006 were examined and classed as 'economics of construction', 'construction management' or 'building economics'. Ive and Chang found a substantial body of papers categorized as 'building economics', because of their lack of reference to recognized economics, and identified a largely one-way traffic in ideas from economics to construction economics. They were pessimistic about the prospect of construction economics finding an area where a breakthrough to subdiscipline status could be possible:

> The economics of construction should 'ideally' face two ways: back towards the sources of its ideas (which should include the economics profession), to whom it can report on applications of theory, and forward towards the users of its normative work, to whom it can make recommendations. Meanwhile it also needs to look 'sideways' at itself, developing positive analysis whose value lies in adding to our understanding of why construction is organized as it is – something of critical importance for the development of CE [construction economics], but which is not perhaps a main concern either to mainstream economists or to construction 'users'. (Ive and Chang 2007: 1597)

Ive and Chang concluded that, without a theoretical breakthrough recognized by mainstream economics, the best that construction economics can aspire to is applying propositions from economics to the understanding of behaviour and explanation of institutions within construction. This point is important, and is revisited below. That conclusion was later supported by a citation analysis of construction economics and economics journals by Bröchner (2018, p. 179): 'While CME authors often cite articles published in top economics journals, citations in the opposite direction simply do not exist, although the construction industry does figure in mainstream articles, but then with few exceptions only in the periphery of article topics'.

So, from one perspective it could be taken that construction economics has in some way failed through lack of external recognition. Another perspective is to ask what progress has been made in construction economics in terms of the scope of topics addressed and development of industry models. In their review, Ive and Chang (2007, p. 1595) found 63 papers in ten broad two-digit JEL classes (codes derived from the *Journal of Economic Literature*), of which:

> 36 papers represent pre-established fields within the earlier economics of construction that would be found in a similar analysis of the contents of CME in its first decade (bidding strategy; input–output; building cycles; multinational firms; market structure; firm performance, size and scope; economic development). (Mainly) new fields are: organizational behaviour/transaction costs/property rights; futures pricing/decision making under risk and uncertainty; and innovation (27 papers).

The expanding scope of construction economics can be seen in this broadening of the research agenda, developing from building economics in the 1960s to Ofori's (1990) widely adopted distinction between construction project economics and construction industry economics. Over the next two decades, as the range of topics in *CME* and the other journals on economic aspects of construction increased, these became four categories in de Valence (2006) and finally five in de Valence (2011):

1. Building economics, or construction project economics.
2. Construction economics, or construction industry economics.
3. Facility sustainability, or environmental economics applied to buildings and structures.
4. Theories of industrial organization applied to building and construction.
5. Macroeconomic theories applied to building and construction.

Period 2 ends in 2007 with the conference to mark 25 years of publication of *CME*.[7] There were 161 papers across a large and varied range of topics, around a third of which are recognizably construction economics. This accords with the Pietroforte et al. (2007) paper presented at the conference, which found for 1983–2006, CME papers divided into three perspectives: firms 38 per cent, clients 29 per cent, and industry 33 per cent. A similar analysis of the *Journal of Construction Engineering and Management* showed overwhelming interest in the operations of construction firms (more than 65% of cases) and significantly less emphasis on project or industry level of analysis, 19% and 15% respectively (Pietroforte and Stefani 2004). The conference provided the opportunity for Bröchner, Chang, Hughes, Ive, Pietroforte, and others, to reflect on the state of construction economics research based on the journal's papers, and the significant role the journal had played in the development of construction economics. However, after 2007 a number of books focused construction economics on issues around the measurement and performance of construction, as a distinctive research agenda developed.

PERIOD 3: CONSOLIDATION 2008–2020

Issues around the measurement, structure and performance of the building and construction industry, and its relationship with the manufacturing, professional services and materials industries, have become the focus of construction economics in six books published since 2008.[8] The contributions developed topics identified within the scope of construction economics in Periods 1 and 2, but they ranged widely, and again consolidated the boundaries of construction economics while continuing to introduce ideas from elsewhere in economics. For example, the six books have contributions on the activities of large, international contractors that dominate the global construction industry, a topic that has been, and is, of continuing interest (see Jiang, Chapter 14 and Lu and Ye, Chapter 21). However, the global perspective in these books, while not new, marked another expansion of the topics and issues addressed, to include developments in market analysis, contractor strategies and, in particular, international cost comparisons and construction data. The following summary of the topics covered in the first five of these edited books illustrates the scope of construction economics research and current areas of interest.

The first two books ranged across practical, empirical and theoretical topics. In *Economics for the Modern Built Environment* (Ruddock 2008) seven contributions were on macroeconomic topics such as the economic effects of capital formation and investment, using construction statistics. There were five studies of markets and contractors, with three contributors emphasizing the increasing divergence between global firms and local markets and two country case studies. The book brought a great deal of data together, and updated previous work in empirical construction economics on measuring construction activity and the broad construction industry. *Modern Construction Economics: Theory and Application* (de Valence 2011) took an industry economics/industrial organization approach with contributions on market structure and competition, auctions and innovation. There were two contributions on production theory and three on methodology and experimental methods. Three of the contributions attacked the model of perfectly competitive markets with price-taking firms, arguing that construction markets have significant barriers to entry, and thus can be concentrated and oligopolistic.

Between them the two books covered many of the topics and techniques established in Period 2, and they carried on earlier debates over production theory and methodology. They included global and national research using macroeconomics, research based on industry economics, and case studies with managerial economics. Importantly, they consolidated the expansion of the focus of construction economics from the SIC construction industry and its activity and management, and made the case for construction economics being about the economics of the built environment. Bridging the gap between the urban scale of the built environment, and new building and construction projects, which typically only deliver a small fraction of the total stock each year, has always been the fundamental challenge for construction economics; Bröchner's (2008, p. 27) view was that 'the lack of access to large amounts of data ... remains a barrier between highly aggregated urban modelling and smaller case studies'. (This issue is discussed further below.)

In *Measuring Construction: Prices, Output and Productivity*, Best and Meikle (2015) put the focus on data quality and international comparisons of construction costs, raising issues in the collection and use of construction data. As their introduction makes clear, 'there are standard methods for measurement of physical building work, but the same cannot be said for the characteristics of the construction industry' (p. 1). The 12 contributions covered measurement of construction work, productivity and prices at the global, national, industry and project levels. Their conclusion was:

> there is no 'correct' answer to any of the questions this book explores ... It is perhaps only by applying a variety of techniques to the various problems and comparing the results that we obtain that we will know if we are getting closer to developing an acceptable set of tools and methods. (p. 256)

It is argued below that a multiple models approach is required to tackle the 'various problems' with construction data.

In *Accounting for Construction: Frameworks, Productivity, Cost and Performance* (Best and Meikle 2019), the dozen contributions considered different ways of measuring construction. With chapters on construction statistics, productivity, costs and data, the book both reviewed and extended previous studies. An 'important thread' in the book was 'the lack of consistency in the way construction industry data is collected and how it is aggregated and/or disaggregated' (p. xiii). This thread became the focus of the next book, *Global Construction Data* (Gruneberg 2019). The ten contributions included three on construction statistics, while four used cost data, and the other three covered innovation, architectural services and international contractors' make–buy decisions. The reliability and quality of construction statistics is a well-known issue, going back to the 1960s, and the shortcomings of the System of National Accounts (SNA) and ISIC have not been overcome in the revisions since then. Those shortcomings were the focus of earlier attention by Cannon (1994) and Briscoe (2006), among others, and are also a major theme in the books edited by Best and Meikle.

The reason for this renewed focus on data quality and reliability is the fundamental role that data is playing in the current topics construction economics is pursuing. Measuring progress on the United Nations Sustainable Development Goals (see Opoku, Chapter 10), or in moving to a low carbon economy (see Ebohon, Chapter 11), are examples where data are required on how construction contributes to emissions, and how production and maintenance of the built environment functions. While new data sources will undoubtedly become available, it is important to understand the content and coverage of current data, especially data from national statistical agencies and the international data aggregators such as the United Nations, OECD and World Bank (see Meikle and Shrestha, Chapter 12).

Construction statistics are varied in their quality and extent. Therefore, identifying what is being measured as 'construction' is a major task (see Best, Chapter 13). The supply chain for building and construction projects brings suppliers from many industries together. Construction output, measured as value added, is around half the total output of the broad construction industry. Measurement of the broad construction industry, by adding industry-level data together, has been one research topic uniquely pursued by construction economics (reviewed in de Valence 2019b). The broad construction industry is the supply side of a wider economic sector, where it meets demand from private and public sector clients. Bringing the demand and supply side together creates a model of the built environment sector, the collection of industries responsible for the production and maintenance of the stock of buildings and structures (that is, the human built world). This research is a significant attempt to define and measure construction as an industry at differing scales – from components to buildings, structures and the urban fabric, from projects to firms – and to incorporate both the industry and its clients. This is a far more expansive idea of an 'industry' and its processes than was possible in the 1960s.

Industry processes have also been subject to increased interest and measurement. Several different methods for measuring productivity have been proposed. An *ECAM* special issue in 2019 on 'The true value of construction and the built environment to the economy' had contributions on national wealth and investment, productivity and the capital stock, and measuring the built environment sector (Ruddock et al. 2019). Performance measures for projects have been proposed; there have also been suggestions on how model buildings or types can be used for cost comparisons. How costs and processes vary across countries has been a particular area of research (see Best, Chapter 13).

After their analysis of 25 years of publications in *CME*, Ive and Chang (2007) suggested that the number of construction economics papers would support a 'slim quarterly journal'. The content of these edited books and the *ECAM* special issue, spread over more than a decade since, suggests that might be about right. While these journals are important, overall the proportion of construction economics papers in them is low, compared to the number of project and construction management orientated papers published. Although an inevitable outcome in built environment journals and construction conference proceedings, as a specialized subfield this is one reason why construction economics has struggled to establish an identity alongside the many more construction management and project papers published. These books addressed that problem by providing a visible collection of construction economics research from a diverse group of researchers that is clearly differentiated from the management literature, a necessary condition for construction economics as an academic discipline.

The sixth book was Gruneberg and Francis's (2019) *The Economics of Construction*. Gruneberg and Francis provide 'a game theory account of the behaviour of firms', the approach typically taken in other branches of industry economics. They discuss aspects of firms' business models, financing, contractual disputes and power relations at greater length than Hillebrandt, building on the research of Periods 1 and 2. Another difference is the use of case studies of the collapse of the UK contractor Carilion in 2018, Grenfell Tower, construction for the London Olympics, and manufactured housing. These illustrate how the business environment that a construction firm faces has become significantly more complex over the decades. Hillebrandt's turnover and profit-maximizing firm has evolved into one primarily concerned with growth and survival. While that may be a matter of degree, it is not insignificant. Gruneberg and Francis argue that contracting markets compete profits down to the point where firms cannot invest in productivity improvements. In Hillebrandt, prices, costs and profits for a project were determined by a conventional marginal analysis, producing an equilibrium result. In Gruneberg and Francis, the last two chapters point to an emerging field of research on the economics of construction projects, combining project financial and feasibility studies with procurement strategies. Although they do not develop the link, this field also draws on research applying TCE to construction.

Construction firms operate in an industry which Gruneberg and Francis describe as 'a highly fragmented project-based industry, with very low profit margins and a high risk of failure for the many firms operating in a very complex supply chain'. This is a description that Hillebrandt would have agreed with, but others would challenge as being too simplistic (Laryea and Hughes 2011), and it does not take into account the coexistence of an industry with a majority of small firms, and the substantial market power of relatively few large multinational corporations in the construction supply chain (see de Valence 2011; McCloughan 2004).

As the first and last of the books that have been discussed, the differences and similarities in the topics and treatment between *Economic Theory and the Construction Industry* and *The Economics of Construction* provide a partial snapshot of where and how construction economics has been developing. The content of both these and the fourth edition of Myers' (2004 [2017]) *Construction Economics: A New Approach* are outlined in Table 4.1. There is much in common, starting with the fundamentals of neoclassical theory on firms and markets, and their analysis of competition, demand, tendering, costs and prices is similar. The books include finance and the sort of managerial economics that Ive and Chang (2007) referred to as 'decision making economics'. However, Gruneberg and Francis take these topics further than Hillebrandt, and introduce different ones such as game theory, innovation and productivity. Myers emphasizes environmental issues and sustainability. Even without detailed descriptions of the chapter contents, the comparison is an indicator of developments in construction economics over four decades, and the extent of the range of the current knowledge base in construction economics.

Table 4.1 is a partial snapshot because these are three among many contributions to the development of construction economics, and reflect those researchers' interests. The scope of construction economics is wider than the topics they cover, primarily because these books are intended to be used as textbooks for construction economics subjects; and wider than this discussion focused on models and data has allowed. For example, an important area with its own history is bidding theory (see Laryea, Chapter 17) and associated research on auctions (Drew 2011). Other important topics not included in these books are TCE, financing/trade credit research using corporate data (Ive and Murray 2015), and industry analysis using input–output tables (Bon 2000). Also, Goh (2008) discusses the range of quantitative methods used in

Table 4.1 *Example of developments in construction economics*

Chapter	Hillebrandt (1974)	Myers (2004 [2017])	Gruneberg and Francis (2019)
1	The nature of construction economics	Introduction to basic concepts	Getting to grips with construction statistics
2	Relation of construction to the economy	Economic systems for economic allocation	Economic theory of markets and construction
3	Basic concepts in economics	The market mechanism	Running a construction firm
4	Demand for housing	The theory of demand	The firm and economies of growth
5	Demand for industrial and commercial building	The theory of supply	Productivity and the construction market
6	Demand for social-type construction	Clients and contractors	The game of construction
7	How demand is put to the industry	Costs of the construction firm	The underlying causes of conflict in construction
8	The firm and its objectives	Types of market structure in construction	Construction and cyclicality
9	Costs of the construction firm	Markets for green buildings and infrastructure	Projects
10	Market supply curve	Market failure and government intervention	The economics of construction project management
11	Equilibrium in market situations	Environmental economics	
12	Demand curves facing the firm	Managing the macroeconomy	
13	Price determination for a project	The economy and construction: measurement and manipulation	
14	Conclusions on costs, revenue and equilibrium of the contracting firm	The business case: inflation and expectations	
15		Sustainable construction	

construction economics, and there may be potential in developing construction econometrics if data reliability issues can be addressed.

CONCLUDING COMMENTS

The review of construction economics research since 2008 shows the development of a distinctive research agenda focused on two key topics: the data that is used in construction economics research, which is typically applied and empirical; and the measurement and performance of construction processes, industry and firms. It draws on three traditional areas of economics: industry economics/industrial organization; development and economic growth; and macroeconomics and public finance. To these are added three industry-specific areas: cost estimating and project economics; definition of units of output (such as components, buildings, value added); and measurement and comparison of costs and output across projects and countries. This combination means that construction economics research generally employs basic economic concepts, but requires detailed knowledge of the industry, technology and institutions.

What does this short history of developments in construction economics show about its nature? Are there distinctive characteristics of construction economics research that distinguish it? Since the 1970s, construction economics research has ranged across productivity and value for money, environmental performance and sustainability, the delivery process and strategies, the financing, viability and competitiveness of construction firms, optimization of the roles of participants and processes, technological and institutional development, construction statistics and measurement, input–output data, and the economic role and structure of the industry. Underpinning that research are models of the industry and its firms and processes; models that have become richer and more refined. From the fragmented perfectly competitive industry of the 1970s, by 2020 the model of the industry is a hierarchical supply chain managed by contractors. Larger firms in two layers of monopolistic competition and oligopoly compete among themselves, above a deep layer of small firms in perfectly competitive markets. The industry maintains key elements of hierarchical governance though contracts between tiers of subcontractors and suppliers, optimizing supply chains through standardization of parts and products. Some firms act variously as contractors, subcontractors, designers or consultants, depending on circumstances.

One reason why it is challenging to define the nature and scope of construction economics is the wide range of issues and topics associated with the production and maintenance of the built environment, which is the economic role of the construction industry in its broader sense. This wide range makes for porous boundaries between construction economics and related disciplines such as transport, urban and regional economics, housing economics, cost engineering and, in particular, construction management. Whether this is a strength or a weakness, an opportunity or a threat, has been subject to debate by construction economics researchers.

This review of construction economics research has focused on theoretical and empirical development in the three areas identified by Hillebrandt (1974): the construction process, the construction firm and the construction industry. The development of construction economics has been divided into three periods. The first from 1966 to 1989 is one of emergence and establishment, as themes and topics are investigated and the literature develops; and the second from 1990 to 2007 covers the expansion of construction economics as new approaches were applied to construction. In the third period of consolidation from 2008 to 2020, a number of publications are reviewed.

Despite the innovative nature of her work, it would be fair to say that Hillebrandt (1974) did not develop a new theoretical analysis. Although the last part of her Chapter 13 on price determination is on bidding theory, her book lacks a conclusion with some theoretical principles of construction economics outlined. Given the difficulties she and Turin had with reconciling the complexity of construction of the built environment with the stylized models of economics, that is not surprising. Nevertheless, while the 'peculiarities of construction firms, processes and markets' have continued to perplex us, our models of firms, processes and industry have developed, becoming more complex and multifaceted as the range of economic theories applied has expanded. The combining of economic theory and techniques with industry-specific knowledge is the first distinctive characteristic of construction economics research.

The basic tools of economics are models such as those for supply and demand, marginal costs, utility, consumer preferences and market structure. The application of those models to an industry is an empirical exercise, based on the characteristics of a specific industry. This has generally seen ideas travel from economics to industry studies and, as Fellows and Liu (2020) note, careful borrowing of theories from relevant disciplines is advantageous. This openness of construction economics to ideas from mainstream economics has led to important insights into construction firms, processes and the relationship between the industry and the wider economy; and the scope of construction economics is a strength because it reflects the nature of an evolving economy and a construction industry which can be defined in various ways. The research on defining and measuring construction at many different scales is the second distinctive characteristic of construction economics, as this is not done elsewhere. These scales range from projects to firms and the broader construction industry, which includes all participants in the supply chain on the supply side, and production and maintenance of the built environment on the demand side.

There is, however, limited feedback from industry economics to mainstream economics because industry and organizational research has not led to rethinking of the relevant economic models. An example of this is the use of bid rent curves and the monocentric city model in urban economics; a useful economic model, but not one that will affect mainstream economics. Similarly, construction project cost curves and a contractor-centric industry model does not change economic principles. So, like other branches of industry economics, construction economics research works within current economic theory, applying tools and techniques from a wide range of fields in economics, and has not produced results that challenge mainstream economic theory. However, for construction economics there is a key difference. For many industries, data is plentiful; for example, health and transport economists have access to hospital records and traffic volumes, and those have become recognized subdisciplines in economics. The development of discrete choice models to analyse transport decisions led to a Nobel Prize in Economic Sciences for Daniel McFadden in 2000, and his model of supply (capacity) and demand (distance travelled per person) is widely used.

Some industries provide the data required for detailed economic modelling to researchers, either directly or through regulators. Examples are supermarkets (bar codes), banking (transactions and lending), and airlines (passengers and distance). By comparison, construction economics has a more limited set of data available, largely restricted to construction statistics, cost data and corporate financial records, supplemented by surveys of various sorts. Generally, firms involved in construction do not share data because it is, or may be, a source of competitive advantage, or is confidential due to commercial and contractual reasons. Government agencies at differing levels hold specific data, which may or may not be available and compatible with other data. Therefore, the third distinguishing feature of construction economics research is the limited range of data available to work with, and the variable quality of that data. This may be one reason for the unidirectional flow of citations between construction economics and economics found by Bröchner (2018). On the other hand, the example of the Bon curve shows that when construction economics has a proposition that can be tested with data, crossover from construction economics to other domains in economics can occur.

It is not that economists are not interested in construction, and the peculiarities of construction firms, contracting markets and construction projects, with construction used as an example or representative industry, and topics such as auctions and contracts also of interest to construction economics. However, the application of economics to construction by economists was done by those within the built environment, such as Hillebrandt, Ball, Ive, Gruneberg and Runeson. In a fourth distinguishing feature of construction economics, many of the other researchers included here were not economists, but found uses for economic theory; examples are Turin, Bon, Meikle, Winch, Gann and Koskela. The development of construction economics has been multidisciplinary and multifaceted, which is appropriate because a multiple models approach is required to disentangle and analyse the variety of and range of issues associated with the construction industry.

The problem of low visibility in the midst of construction management and project management publications is also a distinctive feature of construction economics, and was not one encountered in the emergence of other fields in industry economics such as health and transport. These emerged from economics departments as their home base before branching out, whereas construction economics came from construction and built environment departments or schools. This meant that much of the potential readership of construction economics research were not very interested or particularly well informed on economics, which takes a different approach to core issues such as the relationship between data, theory and testing. Over the last decade, as construction management research has become more 'anthropological' (Chan 2020), this divide between construction economics and construction management, where research is increasingly based on sociology, has widened. Construction economics has succeeded in developing a distinctive research agenda on the economics of production, maintenance and management of the built environment. It is now developing the tools and techniques needed to investigate those topics under a wide range of conditions in many different countries.

NOTES

1. Most of the papers cited here are from *Building Research and Information*, *Construction Management and Economics* and *Engineering, Construction and Architectural Management*.
2. This report was reprinted, reissued, updated and revised over the next ten years, until Duccio Turin's untimely death in 1976. His Festschrift in *Habitat International* (Turin 1978) reprints 'Construction and development' from 1973, with papers from the contributors on construction statistics, costs and developing countries, reflecting the range and influence of his work. On his work at University College London (UCL) the editors' memoir notes: 'His UK research programmes initially were concerned with the available statistics on the UK construction industry ... He introduced his ideas [to the BERU] about the building process, about housing and construction, and about the particular issues in developing countries' (Groák and Koenigsberger, 1978, p. 18).
3. Turin and Patricia Hillebrandt were colleagues at UCL when Turin was Director of the BERU. Hillebrandt and John Andrews founded the Building Economics and Management MSc in 1974, which runs today as the Construction Economics and Management MSc at UCL.
4. After retiring in 2003 from Reading University, and as a founding editor of *CME*, Ranko Bon became an artist. There is a 2016 addendum to his 1987 entry on his blog residua.org, called 'Building economics be damned', where he says the book:

   > went over the head of the entire field – students, teachers, and the research community. To this day, only a few among my former colleagues appreciate it for what it is worth. The only remaining hope is that someone somewhere will pick it up one fine day and bring it back to life, as it were. Not the book itself, but the ideas it offers. Alas, that hope is also thinning at a clip.
5. Reprinted in a 2003 special issue of *BRI* but first published in the *Proceedings of the Bartlett Society* in 1967. In his introduction to the paper, Graham Winch says that it was translated into French, German and Italian, and 'Turin argues for the importance of viewing construction as a process, but he does not lose sight of the relevance of the contractual arrangements between the parties' (Winch 2003, p. 179).
6. Runeson and de Valence (2015) provided their answer to this question in a *CME* paper 'A critique of the methodology of building economics: trust the theories' which argued that predictions and theory are inseparable.
7. The full proceedings can be downloaded from http://centaur.reading.ac.uk/31329/.
8. The books discussed are: Ruddock (2008) *Economics for the Modern Built Environment*; de Valence (2011) *Modern Construction Economics: Theory and Application*; Best and Meikle (2015) *Measuring Construction: Prices, Output and Productivity*; Best and Meikle (2019) *Accounting for Construction: Frameworks, Productivity, Cost and Performance*; Gruneberg (2019) *Global Construction Data*; and Gruneberg and Francis (2019) *The Economics of Construction*.

The work continues. Both with this *Research Companion to Construction Economics*, and with Best and Meikle (forthcoming), *Describing Construction: Industries, Projects and Firms*, due in 2022.

REFERENCES

Ando, F.Y., Funo, S., Furusaka, S., Matsudome, S., Matsumura, S., et al. (1992). A comparative study on technological change in the house building industries in Japan and the United Kingdom (1). *Housing Research Foundation Annual Report*, 18, 379–392.

Aschauer, D.A. (1989). Is public expenditure productive? *Journal of Monetary Economics*, 23(2), 177–200.

Bain, J.S. (1959). *Industrial Organisation*, Wiley, New York.

Ball, M. (1988). *Rebuilding Construction: Economic Change and the British Construction Industry*, London: Routledge.

Ballard, G., Tommelein, I., Koskela, L. and Howell, G. (2002). Lean construction tools and techniques. In Best, R. and de Valence, G. (eds), *Building in Value: Design and Construction*, Oxford: Butterworth-Heinemann, pp. 227–255.

Barras, R. (2010). *Building Cycles: Growth and Instability*, Chichester: Wiley-Blackwell.

Best, R. and Meikle, J. (eds) (2019). *Accounting for Construction: Frameworks, Productivity, Cost and Performance,* London: Taylor & Francis.

Best, R. and Meikle, J. (eds.) (2015). *Measuring Construction: Prices, Output and Productivity*, Abingdon: Routledge.

Best, R. and Meikle, J. (eds) (forthcoming). *Describing Construction: Industries, Projects and Firms*, London: Taylor & Francis.

Blaug, M. (1992). *The Methodology of Economics: Or, How Economists Explain*, Cambridge: Cambridge University Press.

Bon, R. (1989). *Building as an Economic Process: An Introduction to Building Economics*, Englewood Cliffs, NJ: Prentice Hall.

Bon, R. (1992). The future of international construction secular patterns of growth and decline. *Habitat International*, 16, 119–128.

Bon, R. (2000). *Economic Structure and Maturity: Collected Papers in Input–Output Modelling and Applications*, Aldershot: Ashgate.

Bon, R. (2001). The future of building economics: a note. *Construction Management and Economics*, 19, 255–258.

Bowlby, R. and Schriver, W. (1986). Observations on productivity and composition of building output in the United States, 1972–82. *Construction Management and Economics*, 4(1), 1–18.

Bowley, M. (1960). *Innovation in Building Materials*, London: Gerald Duckworth.

Bowley, M. (1966). *The British Building Industry: Four Studies in Response and Resistance to Change*, Cambridge: Cambridge University Press.

Briscoe, G. (1988). *The Economics of the Construction Industry*, London: Mitchell.

Briscoe, G. (2006). How useful and reliable are construction statistics? *Building Research and Information*, 34(3), 220–229.

Bröchner, J. (2008). Developing construction economics as industry economics. In Ruddock, L. (ed.), *Economics for the Modern Built Environment*, London: Taylor & Francis, pp. 14–28.

Bröchner, J. (2018). Construction economics and economics journals. *Construction Management and Economics*, 36(3), 175–180.

Cannon, J. (1994). Lies and construction statistics. *Construction Management and Economics*, 12(4), 307–312.

Cassimatis, P. (1970). Economics of the construction industry. Conference Board, Studies in Business Economics No. 111.

Chan, P.W. (2020). Revisiting basics: theoretically-grounded interesting research that addresses challenges that matter. *Construction Management and Economics*, 38(1), 1–10.

Choy, C. F. (2011). Revisiting the 'Bon curve'. *Construction Management and Economics*, 29(7), 695–712.

Dacy, D.C. (1965). Productivity and price trends in construction since 1947. *RES*, 47(4), 406–414.
de Valence, G. (2001). Defining an industry: what is the size and scope of the Australian building and construction industry? *Australian Journal of Construction Economics and Building*, 1(1), 53–65.
de Valence, G. (2006). Future development of construction economics. *Construction Management and Economics*, 24, 661–668.
de Valence, G. (ed.) (2011). *Modern Construction Economics: Theory and Application,* London: Taylor & Francis.
de Valence, G. (2019a). Comparing construction in national industrial classification systems. In Best, R. and Meikle, J. (eds), *Accounting for Construction: Frameworks, Productivity, Cost and Performance,* London: Taylor & Francis, pp. 31–45.
de Valence, G. (2019b). Reframing construction within the built environment sector. *Engineering, Construction and Architectural Management*, 26(5), 740–745.
Drew, D. (2011). Competing in construction auctions: a theoretical perspective. In de Valence, G. (ed.), *Modern Construction Economics,* London: Taylor & Francis, pp. 63–79.
Duca, J.V., Muellbauer, J. and Murphy, A. (2010). Housing markets and the financial crisis of 2007–09: lessons for the future. *Journal of Financial Stability* 6(4), 203–217.
Eccles, R. (1981a). Bureaucratic versus craft administration: the relationship of market structure to the construction firm. *Administrative Science Quarterly*, 26(3), 449–469.
Eccles, R.G. (1981b). The quasifirm in the construction industry. *Journal of Economic Behavior and Organization*, 2, 335–357.
Fellows, R. and Liu, A. (2020). Borrowing theories: contextual and empirical considerations. *Construction Management and Economics*, 38(7), 581–588.
Fitch, J.M. (1973). *American Building: The Historical Forces that Shaped It*, New York: Shocken Books.
Francis, N. (2019). Measuring construction activity in the UK. In Gruneberg, S. (ed.), *Global Construction Data*, London: Taylor & Francis, pp. 3–17.
Girardi, D. and Mura, A. (2014). The construction–development curve: evidence from a new international data set. *IUP Journal of Applied Economics*, 13(3), 7–26.
Gann, D. (2001). Putting academic ideas into practice: technological progress and the absorptive capacity of construction organizations. *Construction Management and Economics, 19*(3), 321–330.
Gann, D. and Salter, A.J. (2000). Innovation in project-based service-enhanced: the construction of complex products and systems. *Research Policy*, 29, 955–972.
Goh, B.H. (2008). The state of applications of quantitative analysis techniques to construction economics and management (1983 to 2006). *Construction Management and Economics*, 26(5), 485–497.
Groak, S. (1993). *The Idea of Building: Thought and Action in the Design and Production of Buildings*, London: Taylor & Francis.
Groák, S. and Koenigsberger, O. (1978). Introduction. *Habitat International*, 3, 1–2.
Gruneberg, S. (ed.) (2019). *Global Construction Data*, London: Taylor & Francis.
Gruneberg, S. and Francis, N. (2019). *The Economics of Construction*, London: Agenda Publishing.
Gruneberg, S and Ive, G. (2000). *The Economics of the Modern Construction Firm*, London: Macmillan.
Hillebrandt, P. (1974). *Economic Theory and the Construction Industry*, Basingstoke: Macmillan.
Hillebrandt, P.M. and Cannon, J. (1989). *The Management of Construction Firms: Aspects of Theory*, London: Macmillan.
Hillebrandt, P.M. and Cannon, J. (1990). *The Modern Construction Firm*, London: Macmillan.
Hobday, M. (2000). The project-based organisation: an ideal form for managing complex products and systems? *Research Policy*, 29, 871–893.
Hutton, J. (1967). *Building and Construction in Australia*, Melbourne: F.W. Cheshire.
Ive, G.J. and Chang, C.-Y. (2007). Have the Economics of Construction got Closer to Becoming a Recognised Sub-discipline of Economics? Can it do so? *Construction Management and Economics 25th Anniversary Conference,* Reading, p. 1585–1605.
Ive, G. and Gruneberg, S. (2000). *The Economics of the Modern Construction Sector*, London: Macmillan.
Ive, G. and Murray, A. (2015). Trade Credit in the UK Construction Industry: An empirical analysis of construction contractor financial positioning and performance. Dept. of Business, Innovation and Skills, London.

Koskela, L. (1992). Application of the new production philosophy to construction. Technical Report No. 72, Center for Integrated Facilities Engineering, Dept. of Civil Engineering, Stanford University, Stanford, CA.

Koskela, L. (2008). Is a theory of the built environment needed? *Building Research and Information*, 36(3), 211–215.

Koskela, L., Howell, G., Ballard, G. and Tommelein, I. (2002). The foundations of lean construction. In Best, R. and de Valence, G. (eds), *Building in Value: Design and Construction*, Oxford: Butterworth-Heinemann, pp. 211–225.

Kuznets, S. (1930). *Secular Movements in Production and Prices: Their Nature and their Bearing upon Cyclical Fluctuations*, New York: Houghton Mifflin.

Lange, J.E. and Mills, D. (eds) (1979). *The Construction Industry: The Balance Wheel of the Economy*, Lexington, MD: Lexington Books / D.C. Heath & Co.

Laryea, S. and Hughes, W. (2011). Risk and price in the bidding process of contractors. *Journal of Construction Engineering and Management*, 137(4), 248–258.

Lewis, T. (2008). Quantifying the GDP-construction relationship. In Ruddock, L. (ed.), *Economics for the Modern Built Environment*, London: Taylor & Francis, pp. 34–59.

Lopes, J. (2008). Investment in construction and economic growth: a long-term perspective. In Ruddock, L. (ed.), *Economics for the Modern Built Environment*, London: Taylor & Francis, pp. 94–112.

McCloughan, P. (2004). Construction sector concentration: evidence from Britain. *Construction Management and Economics*, 22, 979–990.

Morgan, M. (2012). *The World in the Model: How Economists Work and Think*, Cambridge: Cambridge University Press.

Murray, M. and Langford, D. (2002). *Construction Reports 1944–98*, Oxford: Wiley-Blackwell.

Myers, D. (2003). The future of construction economics as an academic discipline. *Construction Management and Economics*, 21, 103–106.

Myers, D. (2004 [2017]). *Construction Economics: A New Approach*, 4th edn, Abingdon: Routledge.

Ofori, G. (1990). *The Construction Industry: Aspects of its Economics and Management,* Singapore: Singapore University Press.

Ofori, G. (1994). Establishing construction economics as an academic discipline. *Construction Management and Economics*, 12, 295–306.

Page, S. (2018). *The Model Thinker*, New York: Basic Books.

Pearce, D. (2003). *The Social and Economic Value of Construction*, London: nCRISP.

Pietroforte, R., Costantino, N., Falagario, M. and Gould, J.S. (2007). A review of the abstracts of *Construction Management and Economics*, 1983–2006. *Construction Management and Economics 25th Anniversary Conference,* Reading, pp. 1523–1532.

Pietroforte, R. and Stefani, T. (2004). *Journal of Construction Engineering and Management*: a review of the years 1983–2000. *Journal of Construction Engineering and Management*, 130, 440–448.

Raftery, J. (1991). *Principles of Building Economics*, Oxford: BSP Professional Books.

Riggleman, J.R. (1933). Building cycles in the United States, 1875–1932. *Journal of the American Statistical Association*, 28(182), 174–183.

Ruddock, L. (ed.) (2008). *Economics for the Modern Built Environment*, London: Taylor & Francis.

Ruddock, L., Gruneberg, S. and Ruddock, S. (2019). Assessing the true value of construction and the built environment to the economy. *Engineering, Construction and Architectural Management*, 26(5), 738–739.

Runeson, G. (2000). *Building Economics*, Deakin: Deakin University Press.

Runeson, G. and de Valence, G. (2015). A critique of the methodology of building economics: trust the theories. *Construction Management and Economics*, 33(2), 117–125.

Stinchcombe, A.L. (1959). Bureaucratic and craft administration of production: a comparative study. *Administrative Science Quarterly*, 4, 168–187.

Strassmann, W.P. (1970). The construction sector in economic development. *Scottish Journal of Political Economy*, 17(3), 391–409.

Summers, L. (2016). The age of secular stagnation: what it is and what to do about it. *Foreign Affairs*, February, 487–495.

Turin, D. (1966). *The Place of Construction in Economic Development*, London: BERU.

Turin, D.A. (1969). *The Construction Industry: Its Economic Significance and its Role in Development*. Building Economics Research Unit, University College London.
Turin, D. (ed.) (1975). *Aspects of the Economics of Construction*, London: Goodwin.
Turin, D. (1978). Construction and development. *Habitat International*, 3(1–2), 33–45.
Turin, D. (2003). Building as a process. *Building Research and Information*, 31(2), 180–187.
Vanoli, A. (2005). *A History of National Accounting*, Amsterdam: IOS Press.
Williamson, O. (1975). *Markets and Hierarchies: Analysis and Antitrust Implications: A Study in the Economics of Internal Organization*, New York: Free Press.
Winch, G. (1989). The construction firm and the construction project: a transaction cost approach. *Construction Management and Economics*, 7, 331–345.
Winch, W. (2003). Introduction to Duccio Turin's 'Building as a process'. *Building Research and Information*, 31(2), 179.
World Bank (2013). *Measuring the Real Size of the World Economy: The Framework, Methodology and Results of the International Comparison Program – ICP*, Washington DC: World Bank.

5. Construction economics in antiquity

Jan Bröchner

INTRODUCTION

This chapter is about the economics of construction in ancient Greece and Rome, covering a time span of more than two millennia. What and why do we know about construction economics in ancient Greece and Rome? We do have project accounts for many public buildings, knowledge about construction labour and insights into project management, estimating, procurement and corruption. The overview presented here begins with contemporary inscriptions, followed by literary sources and then by studies where archaeology has been the dominant input. The order of presentation is roughly chronological for each category of sources. Finally, there are reflections on what bearing all this has on current studies of construction economics.

The dawn of construction economics is in Mesopotamia: the Garšana archives from the Sumerian city-state of Ur, about 2030 BC, include data on rations issued to unskilled female construction workers, engaged to carry bricks (Heimpel, 2009); the Code of Hammurabi, dated about 1750 BC, has a price control regulation (228) which states that builders will receive 2 shekels per area unit (equivalent to about 26 m^2) of a completed house (Richardson, 2000, p. 108f.).

EPIGRAPHIC SOURCES

Epigraphy, the study of inscriptions, has supported in recent years a shift in studies of the ancient world. In particular, as Bresson (2012) highlights, new institutional economics, with an added focus on transaction costs, has thrown light on the impact of the cost of the institutional framework itself, on production in general and construction in particular.

Prehistory: The Mycenaean Palace Culture

What is believed to be the palace of Nestor at Pylos in the southwest Peloponnese was destroyed by fire about 1200 BC, but the conflagration also ensured that the temporary administrative records, inscribed in Linear B on more than 1000 unbaked clay tablets, were preserved for posterity. From the viewpoint of construction economics, the important tablet is Fn 7, which lists allotments of grain to a group of 20 *toikhodomoi* (wall builders), 5 *priētēres* (sawyers) and a single *pantektōn* (general building project manager); wall builders and sawyers received each 1.2 litres of grain daily and the *pantektōn* 3.2 litres (Palaima, 2015). Although the Mycenaeans invested heavily in monumental architecture (palaces, fortifications and tombs) and public works (roads, bridges and dams), Montecchi (2013) observes that the records are selective, and do not cover all the main economic interests of the ruling elites, which explains why they mention builders only occasionally.

Greek Public Projects: Accounts on Marble

While we do have remains of palatial archives from the Mycenaean era, it is inscriptions that we depend on for construction cost data from the Greek cities and Hellenistic kingdoms. These records on marble concern only public building projects, whereas no accounts have been left from private projects. On the other hand, the public projects had considerable effects on society and the economy.

Greek democracy was closely tied to accountability and suspicion of officials being corrupted (Fröhlich, 2004). If the people were to be able to exercise control over officials, accessible archival documentation was required. This reflects the evolution in Athens of democratic institutions during the first half of the fifth century BC. Epstein (2013) has considered three groups of building accounts, from Athens and Eleusis, and pointed out that the inscribed documents served as symbols and as means of attaining both transparency and accountability. Athenians could look at the *stele* with the Parthenon accounts and understand that the officials had been subject to their annual audit, when any citizen could be present and challenge any official.

A systematic analysis of the categories of craftsmen who were employed in large construction projects in Greek sanctuaries has been provided by Feyel (2006). This is based on inscriptional evidence from the late fifth century to the second century BC. There were no general contractors; instead, projects were undertaken by many small contractors supervised by the clients. In principle, payment was in the form of daily wages or by piecework: daily wages were for non-specialized workers, piecework payments for those who were specialist craftsmen. New craftsmen were recruited, often from other regions. Lowest price was usually the criterion for being awarded a contract. It is a complex picture that emerges, and the craftsmen appear as anything but a homogeneous group.

Athens

In his overview of public building in Athens under Pericles during the fifth century BC, Shear (2016, pp. 57–69) discusses the expense accounts of the builders, which were summarized in inscriptions by the committee of overseers. These were elected annually from the people, working under the supervision of the Council; each year they presented their accounts for audit, and eventually published the inscription. A remaining example of an expense item in the Parthenon building accounts for 434/3 BC is: '16 392 drachmas to the sculptors of the pediment sculptures, wages' (Osborne and Rhodes, 2017, pp. 258–263). When Stanier (1953) attempted to estimate the total cost of the Parthenon, data from the fragmentarily preserved inscriptions had to be supplemented by a number of other sources, in particular expenditure on similar buildings. He chose the temple of Asklepios at Epidaurus (see below), where detailed inscriptions remain, recalculating costs according to the dimensions of the Parthenon and the difference in resources needed for transport of marble and limestone. The total cost was estimated at 469 talents. One Attic talent of silver amounts to about 26 kg of silver and corresponds to 6000 drachmas.

Close to the Parthenon on the Acropolis, there remain 30 fragments of the accounts for the Erechtheum project, dating from 409/8 to 405/4 BC (Randall, 1953); the accounts for 409/8 and 408/7 have been edited by Osborne and Rhodes (2017, pp. 482–499). This project is the only one where the accounts indicate how many workers did what. The status of skilled and unskilled workers is recorded; about 19 per cent were slaves, the rest being free citizens or metics ('resident aliens'). Among the unskilled workers, metics were predominant; slaves were found in the categories of skilled masons and carpenters. This pattern of status and skills has been explained by Epstein (2008) by referring to the occasional nature of major building projects; slave owners had to ensure continuous employment for their slaves. The client function was organized as a building commission with five citizens, one of whom was the overseeing architect. This architect received an annual salary corresponding to 1 drachma per day, while the workers were paid wages by the day (usually 1 drachma) or for piecework, as masons were paid by the block and by measure for finishing walls. A few prices of materials are also recorded, for example, 1 drachma for a fifth of a talent of lead, about 5 kg. We also find that the commission had to buy boards and costly papyrus sheets for writing up the accounts, later to be inscribed on marble.

Slightly later we have the rebuilding of the Piraeus walls in 394–391 BC (Rhodes and Osborne, 2003, pp. 46–49). The accounts include payments for yoke-teams bringing stones, 160 dr., for iron tools, 53 dr., related to daily-paid work. There is also piecework for a defined stretch of the wall, 790 dr., including the actual bringing-up of the stones.

Epidaurus

The total cost for building a Greek temple is known only in one case, the temple of Asklepios in Epidaurus, constructed about 370 BC and amounting to between 23 and 24 talents of silver, as estimated by Alison Burford (1969) in her pioneering study of the sanctuary construction projects. The cult of Asklepios in Epidaurus as hero and god of healing became famous during the fifth century BC, possibly influenced by the plague of Athens described by Thucydides. The Asklepios temple belongs to the first decade of the fourth century BC, with a project duration of four years and eight months. Work on the spectacular tholos circular building was begun about 380 BC, but it took over 30 years to complete (Prignitz, 2014). Initially, it can be seen that bondsmen were required for all work over 1000 dr. and most contractors received advance payments. Only a few contractors were paid only on completion of their tasks, usually those who made deliveries. Penalties paid by contractors for delayed delivery of supplies are listed in later inscriptions. A typical item in the accounts is: 'Protagoras took up the contract for treating the surface of the beam below the ceiling and the waved moulding, for 767 dr.; his guarantor was Sotimos' (Burford, 1969, p. 214).

Three categories of building costs can be identified: for materials supply, for transport of materials to the site, and for construction. Materials supplied are basically stone (limestone, marble), timber (silver fir, cypress, elm), metal (lead, iron, bronze, gold), ceramic roof tiles and glue. Transport usually refers to carts with pairs of oxen, or by sea. Carting more expensive and easily damaged marble the 7 miles from the harbour of Epidaurus cost 1¾ dr. per ton per mile, while the corresponding unit cost of transporting limestone at least 11 miles by land and 25 miles by sea was only about 6 per cent of that rate (Burford, 1969, p. 191). Reanalysing data from the inscriptions, Salmon (2001) has estimated a global work rate of 0.134 m^2 per man/day for the stone parts of the whole Asklepios temple; and as for timber for the ceiling, the man/day rate would be lower, 0.027 m^2. A special analysis of timber in the temple commissioners' accounts has been made by Meiggs (1982, pp. 423–430).

Delphi

In Delphi, the Apollo temple rebuilt in the sixth century collapsed in 373/2 BC. Detailed accounts for re-rebuilding the temple have survived (Bousquet, 1988; Davies, 2001). There was reuse of materials from what remained of the archaic temple. By now, the demands for public accountability had grown, and the techniques of accounting for the building process were well developed. There were several components in the Delphi system of public procurement:

1. Commissioning a detailed specification for the rebuild, displaying it publicly, and requiring that variations from it should be recorded.
2. Creating a board of officials (*naopoioi*, temple builders) as a building committee, with an architect permanently on site.
3. Breaking down the project into work packages, to be contracted for separately.
4. Requiring contractors to provide performance bondsmen.
5. Inspecting the completed work, paying the final one-tenth retained of the contract sum.
6. Employing secretaries to keep detailed records and accounts, to archive records in chests or boxes, and to publish accounts on stone.

This can be said to have been common practice in fourth century BC Greek cities. Just as the sources of finance for this project were many, so the specialist workers engaged came from many parts of Greece. Accounts of the Delphi *naopoioi* for 345/4–343/2 BC are given by Rhodes and Osborne (2003, pp. 328–337). Examples of items are ceiling beams for the colonnade: 'from the tenth we gave 1400 dr'. To Xenodorus the architect, stipends corresponding to 30 dr. or 60 dr. per month. For three benches on which the *naopoioi* sit was paid 9 dr. How the *naopoioi* engaged in the procurement of materials is illustrated by their expedition in 335 BC to buy cypress timbers for the temple roof structure (Meiggs, 1982, p. 430f.; Bousquet, 1988, p. 88). Accompanied by their architect, they sailed across the Corinthian Gulf to Sicyon and there selected 17 timbers, which they bought in six transactions. The total cost was high and amounted to about 3000 dr. Additionally, the cost of transport was not negligible, 500 dr. The heavy cost of transporting stone to the site led to the contractors receiving advance payments.

Other Greek City States

The overseers in Eleusis supervised all building activities in the sanctuary, publishing specifications and paying for construction (Shear, 2016, p. 165). Expense accounts from about 330 BC cover new walls and gates, a new entrance and a new portico. Purchases, with prices per piece, and uses of timber have been tabulated and analysed also for Eleusis by Meiggs (1982, pp. 433–440).

Penalties are found in a mid-fourth century standard form of contract from the city of Tegea in the Peloponnese (Rhodes and Osborne, 2003, pp. 286–297). This unique standard form reveals concern with competition, with penalties for creating consortia of more than two contractors (each is fined 50 dr., while any whistleblower is to receive 25 dr.), and unless express permission has been given, no contractor may have contracts for more than two pieces of work: the fine is 50 dr. per month for each supernumerary contract.

A public–private contract with the city of Eretria on the island of Euboea is still extant. The contractor/concessionaire will pay the costs of drying out the swamp or lake of Ptechai, but the financial details are not to be seen in the contract. The inscription is dated by Knoepfler (2001) shortly after 318 BC.

From the 314–166 BC period, when the island of Delos was independent, there are many inscriptions (Davis, 1937). There is a payment schedule for construction contracts dated 297 BC: one-tenth retained until completion of the work; of the remaining 90 per cent, half the sum would be paid when the contractor had established surety, one-quarter when one-third of the work had been done, and one-quarter when two-thirds had been done. Fines and penalties are recorded in these inscriptions; the contractor could be fined if he kept less than four skilled workers in his regular force, if requirements in the specifications were not met, and if the contractual time limit was not respected (the contract imposed fines per day). Among the transaction costs that faced a contractor was that he had to pay for exhibiting the contract, including specifications, inscribed on a marble stele, and even a minimum size for the letters in the inscription was regulated. As with Eleusis, Meiggs (1982, pp. 441–457) has made a detailed analysis of references to timber in the Delian inscriptions.

Lebadeia in Boeotia had an important federal sanctuary of Zeus Basileus, built before 220 BC (Pitt, 2014). The inscribed contracts and accounts must have given a monumental impression, displayed on a wall which may have been more than 20 metres long. The principle for payment schedules was basically identical to the one at Delos. Imposing quality penalties on contractors was a frequent task for the building commission: for poor workmanship, low quality of materials and other failures to follow the precise letter of the specifications. They also fined contractors who set blocks in place without first allowing inspection of their work.

Roman Inscriptions

The Romans never published detailed accounts for public construction projects, but Roman law recognized a need for efficient publicizing of calls for tenders (Ulpian, *Digesta* 14.3.11.3–4). Rosillo (2003) has collected information on the many types of procurement fraud recognized. There exists only one completely preserved inscription with a construction contract, dated 105 BC, including the price (1500 sesterces), for a wall with a gateway in front of the sanctuary of Serapis in Puteoli, modern Pozzuoli north of Naples (Anderson, 1997, p. 74f.). (One sesterce then equalled about 0.2 drachma; note that exchange rates developed over time.)

What is preserved otherwise is total costs for a range of projects. Inscriptions from Italy are dominated by prices of public baths and road projects (Duncan-Jones, 1982, pp. 124ff.). Here, the costliest project where there is a figure, 2 million sesterces and more, is the Baths of Neptune in Ostia, dated AD 139. Four inscriptions concerning road projects indicate that the average cost was about 22 sesterces per foot of road, with little variation. Total costs for a wider range of projects, in more than 60 inscriptions, are known from the provinces in Africa (Duncan-Jones, 1982, pp. 75ff.). Here, the highest cost is 600 000 sesterces for a temple in Lambaesis (in what is now eastern Algeria), constructed around AD 200. Building costs are given for other temples, theatres, public baths, and also for rebuilding and restoration of structures. Barresi (2003) has studied imperial public construction in what is now Turkey, relying primarily on dedicatory inscriptions on buildings and having a focus on marble costs.

Correspondence from the Mons Claudianus quarry in Egypt's Eastern Desert gives documentary evidence, dated roughly to the late second century AD, of how imperial officials ordered marble columns (Russell, 2013, pp. 211–214). Although no costs are included in these letters written on broken pottery, they illustrate how the Romans combined two strategies of quarry prefabrication: columns made according to detailed specifications or selected from quarry stock. While Greek practice had been to rely more on local quarries, the Romans had marble transported to Italy over considerable distances, which also led to a challenging task of coordinating quarry activities with site production. It appears that site contractors would sometimes supervise quarry work, and that finishing of column capitals and bases could take place within the quarries.

From late antiquity we have inscriptions in more than 40 locations which together allow us to read Diocletian's Prices Edict of AD 301 (Corcoran, 2000, pp. 205–233). The Edict was an imperial attempt to curb inflation by fixing maximum prices and wages for a total of about 1500 items. Wages for categories of building labour are given in denarii per day, to which should be added maintenance, perhaps equivalent to 11 denarii per day (1 denarius = 4 sesterces) (Bernard, 2017). The lowest wages, 50 denarii, were received by stonemasons, joiners, woodworkers, lime burners, tesserae makers, cartwrights and iron workers. Higher wages reflecting a skill premium were allowed for marble workers and mosaicists (60), wall painters/plasterers (75) and picture painters (150). An *architectus magister* (master architect) would receive 100 denarii per month and student as a maximum wage.

A contemporary observer, Lactantius, noted that the edict led to bloodshed and that people were afraid to offer goods for sale; the edict was withdrawn. Immediately following this passage in his *On the Deaths of the Persecutors* (7.6–10), Lactantius complains of Diocletian's infinite passion for building: basilicas, a circus, a mint, a weapons factory, a house for his wife and one for his daughter. A large part of the city of Nicomedia, today's İzmit, was demolished to make space for all this; when the buildings were finished, the emperor would say that they were 'not right and should be built in another way'.

The Edict lists 19 types of marbles and their maximum prices. Long (2017) has been able to analyse the marble prices with multiple regression, where the three explanatory variables are marble hardness, marble colour and distance from quarry. The regression coefficient for distance indicates that each additional mile overland cost 0.34 denarii. Also, relying on legal texts and archaeological studies of quarries in the region of Aphrodisias (west of Denizli in Turkey) and operating in a framework of institutional economics, she has reconstructed how the exchange of local building stone took place in a competitive market with local landowners, municipalities and artisans as suppliers.

LITERARY SOURCES

Ancient historians and biographers reveal information on construction projects. What they report is often anecdotal, but they serve to illustrate attitudes to the economic aspects of monumental construction projects, while we sometimes have to doubt whether the cost data we find in the manuscript tradition are dependable. Original texts by the ancient authors, also translated into English, can be found easily on the web today and are not included in the list of references.

Authors in the Greek World

The first historical narrative we have is that of Herodotus, writing in the fifth century BC. His interest in economic aspects is manifest (Spengler, 1955), but there is less on construction costs specifically. After a disastrous fire in 548/47 BC, the great Apollo temple in Delphi was to be replaced, and the amphictyony responsible for the sanctuary had procured a new temple for 300 talents (*Histories* 2.180). The Delphians themselves were required to provide a quarter of this sum, and they went around to the cities; the Greeks living in Egypt gave one-third of a talent, while pharaoh Amasis donated 1000 talents of alum (about 26 tonnes of alum) towards the project (the earliest Egyptian coinage appears only after this time). We are also told that the Alcmaeonids from Athens took over the execution of the project, and since they were an old, reputable and very wealthy family, they increased the specified quality (*Histories* 5.62). Thus they built the east front of the temple using Parian marble instead of coarse limestone, finishing the work which would remain until 373/72 BC in about 510 BC. As to megaprojects, Herodotus criticized the huge investments for the Cheops pyramid (*Histories* 2.124f.). When visiting Giza, he was told that hieroglyphs on the pyramid listed the enormous volumes of purgative, onion and garlic that were issued to the workforce. The total value for just these vegetables was given as 1600 silver talents, but Lloyd (1995) doubts whether Herodotus understood this right.

Pericles was criticized for spending thousands of talents paid by the allies of Athens on building costly temples in the 440s and 430s (Plutarch, *Pericles* 12.1–5), and replied that these projects gave rise to all sorts of craftsmanship and a variety of demands, which would stimulate every art and move every hand, bringing almost the whole city into paid occupation, so that it is at the same time beautified and maintained. Those not in active military service would not receive state benefits as unemployed; instead, he proposed major construction projects that would need many specialists and workers. The speed of construction is also emphasized by Plutarch. Kallet-Marx (1989) warns that Plutarch's claim that tribute from the allies was the dominant source of finance for the temples does not fit the inscribed accounts. Shear (2016), in his overview of building under Pericles, suspects that Plutarch, writing many centuries afterwards, was influenced by enemies of Pericles; the inscriptions for the Acropolis temples reveal that many specialist craftsmen were recruited from other cities, and that a shortage of skills was actually a major obstacle to the projects. Once the Parthenon had been dedicated in 438 BC, work started on the monumental entrance to the Acropolis. The project lasted five years (Plutarch, *Pericles* 13.7) and Harpocration reports that the total cost was 2000 talents (*Lexicon*, under *Propylaia tauta*).

About 100 years later, Aristotle questioned why tyrants engage in major projects (*Politics* 1313b): they wish to make their subjects poor and employed during the day so that they will not conspire against the ruler. Among Aristotle's examples of projects that kept the population poor and busy, there are the pyramids in Egypt and also the projects in the island of Samos launched in the sixth century by Polycrates; Shipley (1987, p. 93), however, interprets Polycrates' purpose as not keeping 'the people poor and too busy to think, but to keep them prosperous and loyal'.

Fast-tracking of a Sicilian megaproject occurs in about 400 BC, when Dionysius I as ruler of Syracuse extended and reinforced the city fortifications, the largest in the Greek world (Diodorus Siculus, *Library* 14.18.2–8). Recruiting 60 000 workers, to which must be added many working in the quarry and having 6000 pairs of oxen, the project was completed in 20 days, heavily incentivized and with Dionysius himself not only managing but also setting an example by participating in the heavy work.

After the major earthquake in Rhodes 227/6 BC, which brought down the city walls and toppled the harbour Colossus, various Hellenistic states offered generous financing for rebuilding the city by donating money and contributing other resources (Polybius, *Histories* 5.88–90). Ptolemy III sent 100 master builders with 350 assistant masons from Egypt, together with 14 silver talents annually for their pay; also, among much else, 3000 talents of bronze for restoring the Colossus, which the Rhodians actually failed to do. Bringmann (2001) takes this as a striking example of royal liberality among the Hellenistic kingdoms.

The Roman Republic

The first Roman work on agriculture was written in about 160 BC by Cato the Elder. Despite his reputation for austerity, which included that the walls of his houses were not plastered (Plutarch, *Cato Maior* 4.4), he would behave as a good client who conscientiously paid building contractors (*On agriculture* 14.3–4). Thus he would add 25 per cent to the price for building in an unhealthy location where work cannot be done in summer. Cato also states prices per ordinary roof tile (*tegula*), 1 sesterce, and higher prices for gutter and joint tiles.

Rome invested heavily in water supply and road infrastructure; government expenditure on construction projects dominated public finance (Polybius, *Histories* 6.13.3). The long Aqua Marcia aqueduct, completed in 140 BC, cost according to Frontinus in his *On the aqueducts of Rome* (7.4) 180 million sesterces, about 2000 sesterces/m. Walker and Dart (2011) point to Frontinus as an early writer on project management, and Leveau (2001) reviews what is known about aqueduct costs and finance. The extensive reliance on public procurement and leasing in the Republic was considered by Polybius (*Histories* 6.17) to be an important mechanism for integrating public and private interests during the first half of the second century BC. Contracts throughout Italy for the construction and repair of public buildings, leasing of rivers, harbours, gardens, mines and land, meant that 'almost everyone' was involved in these projects, some as having been awarded contracts by the censors, others as the contractors' partners or as their performance bondsmen. Contractual issues were a matter for the senate, and people were thus at the mercy of the senate; this contributed to the stability of the Roman political system, according to Polybius. Cicero tells a horror story of corrupt procurement of temple renovation (*Against Verres* 2.1.127–154). A conspectus of how public building was organized during the Late Republic and the Early Empire has been provided by Strong (1968).

Vitruvius: On Architecture

In the Ten Books on Architecture, Vitruvius's manual from the last years of the first century BC, there are three passages dealing with economic aspects of construction. First, he explains that an architect should learn arithmetic to estimate the cost of buildings (*On Architecture* 1.1.4). Later, he says that *distributio* is the suitable management of resources and site, as well as a thrifty consideration of project costs (1.2.8). The concept of *distributio* has earlier been translated as 'economy', while Rowland and Howe (Vitruvius, 1999) have chosen 'allocation', but it is still difficult to find a precise English term for what appears to mean disposition/design/administration. A third passage (*On Architecture* preface 10.1) describes the payment scheme applied by the city of Ephesus to public procurement. The contractor/architect had to submit an estimate of project cost, and as a performance security his property was mortgaged. If the project cost was as estimated, he would be honoured. If there was a cost overrun of no more than a quarter of his estimate, the city would pay the total cost, but if the excess cost was higher, the difference would be taken from his assets. Considering innovative construction materials, Vitruvius was suspicious of concrete, Roman concrete walls being mortarized rubble. As Rihll (2013) notes, he stated (*On Architecture* 2.8.8) that property valuators assumed that concrete walls depreciated linearly over 80 years.

The Roman Empire

The emperor Claudius is associated with infrastructure investment around AD 50. Two aqueducts, Aqua Claudia and Anio Novus with a total length of 156 km, serving all the hills of Rome, are believed to have cost 350 million sesterces, but the figure (Pliny the Elder, *Natural History* 36.122) is uncertain here. Claudius is also the protagonist of an infrastructure cost denial story:

> Understanding this [i.e., the need for a harbour close to the mouth of the Tiber to ensure food supply], [Claudius] undertook to construct a harbour and was not deterred, although when he asked how much the cost would be, the architects told him that 'you do not want to do this' because they thought that he would balk at the size of the expenses if he learned this in advance; but he had in mind a project worthy of the spirit and greatness of Rome, and he brought it to completion. (Dio Cassius, *Roman History* 60.11.3)

Quantity surveying is recognized in an agricultural treatise from this time: Columella (*On Agriculture* 5.1.3) identifies the role of quantity surveyors (*mensores*) in construction, stating that architects think that there is another profession which measures what has been built and calculates the cost of the completed project.

Construction innovation was not welcomed by Vespasian as emperor around AD 75: 'To a mechanical engineer, who promised to transport heavy columns to the Capitol at small expense, he gave no mean reward for the invention, but declined to use it, saying that he must be allowed to feed the common people' (Suetonius, *Divus Vespasianus* 18). Vespasian is nevertheless one of the few emperors who Suetonius appreciates, and the anecdote illustrates his concern for the *plebs*. Brunt (1980) takes this text as his starting point for an analysis of the free/unfree status of construction workers in Rome. Unfree labour had to be kept continuously employed but was combined with unskilled and casual self-employed workers. The importance of the lack of continuity in building programmes has been echoed by Temin (2013, p. 130). Bernard (2017) asks what organizational structures allowed Rome's labour supply to meet demand for monumental architecture, and again what the role was of slave labour, unskilled casual labour, forced labour and contractual labour, given that building activities were seasonal and specialized.

Award criteria in public procurement receive a special twist by Plutarch in a fragmentarily preserved dialogue, 'Whether vice can be sufficient to cause unhappiness': 'Cities, certainly, when they announce the procurement of temples or colossal statues, listen to the competing craftsmen, who provide estimates and models; then they select the one who will do this at lowest cost as well as better and more quickly' (*Moralia* 498E). Plutarch continues by imagining the procurement of 'making a life unhappy', where Fortune and Vice submit tenders.

Corruption in public projects gives rise to concern c. AD 110 in exchanges of letters between Pliny the Younger and the Emperor Trajan. First, having just arrived in Prusa (modern Bursa in Turkey), Pliny is looking into the city finances and finds illegitimate expenses, which makes him ask Trajan for a quantity surveyor to be able to recover money from contractors (*Letters* 10.17–18). The second occasion was a question of accounts for a public project overseen by Dio Chrysostom, the sophist, and Trajan replies that Pliny should examine the accounts (10.81–82). Dio's own version of the controversies surrounding his colonnade project, nearly completed in AD 105–106, is found in his three speeches 40, 45 and 47 (Salmeri, 2000). In another letter (10.37) to Trajan, Pliny describes how the citizens of Nicomedia had expended 3 329 000 sesterces on an aqueduct, left unfinished and then demolished; again, they had decided on 200 000 sesterces for another aqueduct, which was then abandoned. Pliny asked the Emperor to send an expert to avoid a repetition. Trajan's response (10.38) was to ask Pliny to investigate whose fault this was, suspecting local corruption.

Cost overrun in public projects is also recognized. Herodes Atticus, wealthy sophist and benefactor, wrote to the emperor Hadrian in AD 135 and asked for 3 million drachmae (then about 12 million sesterces) to build an aqueduct for Alexandria Troas, close to Troy, known from the Iliad (Philostratus, *Lives of the Sophists,* 548). He was appointed to take charge of the water supply. When outlay had reached 7 million dr., and provincial administrators wrote to the Emperor and complained, the Emperor expressed his disapproval to Atticus, who responded haughtily that the Emperor should not let himself be irritated by small matters, and what had been spent in excess of the 3 million would be given to the city by his son. A few decades later, when Aulus Gellius wishes to give an example of how the word *praeterpropter* (more or less) is used, expectations of cost overrun is the occasion:

> Several builders, who had been called upon to construct new baths, were standing round and exhibited various plans for baths on sheets of parchment. When [Fronto] had selected one he asked what the whole cost of completing the project would be. And when the architect had said that about 300 000 sesterces would be needed, one of Fronto's friends said 'And *praeterpropter*, another 50 000.' (*Attic Nights* 19.10.3–4)

THE ARCHAEOLOGICAL APPROACH

In recent decades, archaeologists investigating the prehistory of the Americas, such as the Maya ruins in Central America, have developed what is called architectural energetics (Abrams and Bolland, 1999). This is a modelling procedure that translates the material remains of buildings into cost estimates expressed in person-days of labour, sometimes also in terms of animal labour. Each activity in the building process, including quarrying, transporting of materials, construction on site and decorating, is analysed together with the volume of materials needed. Erasmus (1965) studied Mexican peasants at work and measured their output in order to understand the construction of a Maya ceremonial centre. Work rate data are given in the Chinese *Yingzao Fashi* manual from about AD 1100 (Low, 2007), but as Turner (2018) warns in his comparative study of earthmoving work, there are challenges in transferring work rate data from one cultural context to another. Since the 1980s, French archaeologists had performed a number of studies developing the related concept of a *chaîne opératoire* (Sellet, 1993), which refers to the succession of technical transformations, also including the mental processes required, from procurement of raw materials to the end result in a project, originally applied mostly to the production of Palaeolithic stone tools. This concept was borrowed from ethnology and other fields of social sciences.

A Paradigm: Estimating Costs of the Baths of Caracalla

The gigantic Baths of Caracalla, the second-largest thermae in Imperial Rome, were built between AD 211 and 216/17. Janet DeLaine (1997) analysed and reconstructed all the processes involved in building the baths, thus allowing an estimate of actual costs. She was finally able to estimate that the project employed at least 9000 men per year, costing the equivalent of 12 million rations of wheat. Unlike earlier architectural energetics researchers, her focus was on the process of building rather than the social and cultural processes. Sufficient was preserved of the building complex for a reconstruction of its volume and materials to be undertaken. It was then possible to estimate the minimum amount of labour required to produce the materials and construct the building, including the transportation required to move the materials to the site. A schedule of the processes needed to produce the materials was used, and the labour inputs of different types for the actual erection of the structure were calculated, based on the properties of materials and the physical limitations of what a worker can carry out in a given time, which was derived from data in nineteenth century manuals such as those of Hurst (1865) and especially Pegoretti (1863–1864). Finally, the related prices and wages as given in the Prices Edict of Diocletian were applied to the manpower and materials.

Twenty years after this pioneering work, DeLaine (2017) noted that she had acquired many followers. She then pointed out issues raised by the method. First, construction is site-specific, and it is essential to have an archaeologically correct reconstruction, although the geometry can be simplified for computational purposes. The whole process of construction, the *chaîne opératoire*, should be covered, which includes temporary structures such as scaffolding, baskets, tools and other supplies, as well as services from smiths. Historical and experimental data have to be applied consistently. If an obsolete technology is involved with no later data available, it will require careful argument and justification when labour assumptions are made. A combination is often needed of actual prices of materials, piecework or completed objects such as column capitals, and the cost of labour for more or less specialized workers, but there are problems of data compatibility.

Prehistoric Greece

The Cretan Neopalatial architecture (seventeenth to fifteenth century BC) has been considered using the energetic approach to estimate the total number of work-hours necessary for erecting a building, including the collecting, transporting and manufacturing of materials (Devolder, 2013, 2017). This provides an understanding of the builders' behaviour, especially regarding the selection and use of the different materials, and the levelling of the terrain prior to construction. A broad sample of structures shows how the inhabitants were engaged in constructing their own houses, in contrast with the large size and partly specialized manpower involved in elaborate building projects.

There have been many studies of Mycenaean citadels, tholoi graves, waterworks, roads and bridges, but it is only in recent years that the question has been addressed of how the monumental building programmes contributed to socio-economic and political changes in the period until 1100 BC. The Leiden University SETinSTONE project (Brysbaert, 2017) has investigated the human and natural resources of several sites, combining *chaîne opératoire* and cross-craft interaction approaches to capture the practical building processes and inherent social practices. As shown in a study of the citadel of Tiryns (Brysbaert, 2013), stone from quarries and wood from the forests were brought to the centres by means of men and oxen, requiring considerable effort. Skilled and unskilled builders and artisans added value to these raw materials by turning them into finished composite multifunctional and impressive structures.

Archaic and Classical Greece

Patay-Horváth (2019) has looked at seven Doric temples (sixth to fourth century BC) in the Western Peloponnese, from a small one with an estimated construction cost of 8 talents to the 300–400 talents for the temple of Zeus at Olympia. Matching literary sources and the ruins, he emphasizes the variety of political and socio-economic conditions that shaped the funding of these projects. A similar ambition of widening the understanding of the financial background of projects was already evident when Martin (1973) looked at temples in Sicily (Selinus, Akragas) and Poseidonia (Roman Paestum, in southwestern Italy). Martin applied data from Burford's 1969 study of the accounts from Epidaurus. There are now also two DeLaine-inspired studies of the Greek colonies in western Sicily. Fitzjohn (2013) investigated archaic construction in Megara Hyblaea, north of Syracuse. Lancaster (2019) applied architectural energetics to an archaic building in Syracuse itself and the mid-sixth century fortifications of the Kamarina settlement in the same region. Work rates were applied separately for stone, mudbrick, wood and supervision (10 per cent) and transport.

Cheaper public projects might use more timber than marble. Pakkanen (2013) has analysed the economics of fourth century BC shipsheds for the Athenian navy at the Zea harbour in Piraeus, relying on Hurst's Handbook for work rates and applying architectural energetics along with the *chaîne opératoire*. His conclusion is that the shipsheds cost no more than 4 talents per year to build and on average engaged fewer than 100 workers each building season.

The Roman Republic

Combining archaeology with textual sources, Bernard (2018) has investigated public construction in Mid-Republican Rome, which means the 396–168 BC period. His study is an example of new institutional economics, emphasizing the role of social structures and ideology for the economy. Focus is on a cost analysis of Rome's 11 km of circuit walls, applying the method developed by DeLaine, which leads to an estimate of no less than 6.8 million person-days of skilled and unskilled labour by slaves and free citizens. After this extreme effort, there was a change from coercive labour systems to a market economy; coinage and public procurement of construction appear as parallel phenomena around 300 BC. Bernard also identifies the spread of innovative construction technologies during the period: reliance on a greater range of stone types for different structural needs and changes in lifting technology.

Imperial Rome

After her pioneering 1997 work on the Baths of Caracalla, an early follower of which was Lancaster's (1998) study on building Trajan's markets in Rome, DeLaine has concentrated on Ostia, the harbour city of Rome, and on rational economic choice when new construction technologies appeared. Assuming a model wall 12 m long by 3 m high by 0.6 m wide, she found that the size of pieces in core and facing is as important in establishing labour requirements for concrete as is the type of facing (*opus incertum* with small uncut tuff blocks, gradually over time replaced by *opus reticulatum* with diamond-shaped small tuff blocks), that reuse of precisely cut rectangular or square blocks of stone is essential in understanding the economics of brick and reticulate, and that using blocks of stone was much more labour-intensive than concrete (DeLaine, 2001). By 'concrete' is meant strongly mortared rubble construction rather than present-day concrete. An innovative facing material, brick, appears as an economical choice, once the supply of reusable stone blocks has been exhausted. Reticulate has high material production cost, while brick is costly to transport. The results from this study can be seen in a broader perspective of the economic effects of Roman construction innovation, as outlined by Wilson (2006), who notes the little interest traditionally shown by economic historians in the construction industry, despite the massive building activities.

The *opus reticulatum*, where small blocks are squared to a uniform size and laid diagonally, implies modularity and standardized size of facing elements, just as with bricks, which enabled the employment of less skilled workers, and also that contractors now could rely on a greater diversity of suppliers. Later, DeLaine (2018) exploited further opportunities provided by Ostia to examine economic choice in construction. Three case studies – a mausoleum, a warehouse with an auction room, and a third case with colonnaded porticoes of three monuments – reveal a not unexpected tension between strategies for minimizing construction expenses and requirements of the patron's self-representation.

The economic ability of emperors to finance major building projects grew over time, and they increasingly had ways of acquiring funds unavailable during the Republic. As shown by Lancaster (2005, pp. 10ff.), the engagement in large and technologically advanced building projects, and having the financial and material resources available, meant that the innovative Roman technology of concrete vaulted construction had time to develop. Well-known examples of how this technology was exploited in Rome are the Pantheon and the Baths of Caracalla and of Diocletian.

Late Antiquity

After the principal imperial residence in the West was moved to Ravenna in AD 402, new circuit walls were built there and also around Classe, its satellite town. The inputs of materials and labour have been estimated by Snyder (2020), basically following DeLaine's approach. This approach has been combined with agent-based modelling when exploring the building of a large Byzantine structure, the sixth or seventh century AD Cistern at Hebdomon outside Constantinople (Snyder et al., 2018). Starting from the remains of the cistern walls, their thickness, extent and height, there are three classes of agents in this model: people, sites and ox carts. People are divided into workers and managers. Clay, sand and limestone are quarried; while the resulting types of material are clay, sand, facing stone, rubble, lime, bricks, crushed bricks and mortar. Nothing is known about wages paid to craftsmen in the Byzantine period, unlike during classical antiquity (Bouras, 2002), and the rate at which a given task can be carried out had to be derived from later manuals.

SUSTAINABLE CONSTRUCTION?

Sustainability is not a main theme in ancient Greece and Rome. Denunciation of luxury building was nevertheless a recurring literary topic, and Horace laments that developments encroach on agriculture in Italy (*Odes* 2.15): land for cultivating grain, vines and olives will soon disappear under the palaces of the rich, he wrote in this 23 BC prophecy, roughly contemporary with Vitruvius. Plato (*Republic* 381A6–9) thought that houses well constructed and in good condition are the least affected by time and other external loads. Permanence of structures expressed the architectural virtue of *firmitas* (Vitruvius, *On Architecture* 1.3.2).

Reuse of marble and other costly building materials is a practice that recurs throughout antiquity. Already the fifth century BC Parthenon project engaged in extensive reuse of stone from the unfinished Older Parthenon, earlier razed by the Persians (Shear, 2016, pp. 79–82). Archaeologists discovered in the 1990s that the fifth century BC temple of Athena in Pallene had been dismantled and removed in Roman times, then carefully reassembled as the temple of Ares in the Athenian Agora (Shear, 2016, pp. 250ff.). In late antiquity, there was more extensive reuse of primarily marble from older buildings and ruins, but also renovation (Ng and Swetnam-Burland, 2018; Underwood, 2019), partly to be explained in labour cost terms as in the case of Ostia (DeLaine, 2001), and sometimes also because an object or a building was understood as a repository of historical identity. Several temples were recirculated as churches in late antiquity, often in the late fourth century, and many of these remain, although they are seldom used for religious services today. One exception is the current cathedral of Syracuse, where the Sicilians converted the fifth century BC temple of Athena into a church by piercing the *cella* walls and thus forming arcades between the nave and the aisles, as well as walling in the spaces between the Doric columns in the outer colonnade (Van Ooijen, 2019). There would then be indoor space sufficient for the whole congregation. After Syracuse fell to the Arabs in 878, the church was converted into a mosque, and then back again by the Normans, who took the city in 1085.

Social sustainability is a concern that does emerge occasionally. Aristotle held dim views on the social effects of major projects, as we have seen. Plutarch's interpretation of the Periclean megaprojects in Athens belongs here, and Vespasian's engagement for the urban poor was illustrated by the anecdote quoted earlier where he rejects an innovative technology which could have raised construction efficiency.

IMPLICATIONS FOR CONSTRUCTION ECONOMICS

Economic historians tend to regret that so much was spent on prestige construction projects during the Roman Empire. Kehoe (2007, p. 550) thought that resources could have been used for 'more directly productive purposes', and Long (2017) saw it as 'a drain on capital investment that could have otherwise supported other, more generative pursuits'. On the other hand, basic elements of urban, water, sanitation, transportation and communication systems were well developed during the late Roman empire (Goldsmith, 2015, p. 84). This was a question not only of engineering skills but also of an institutional framework for contracting and operating infrastructure.

Does the study of construction economics in the ancient world deliver insights that can be applied to the developing countries of today? The answer is complex. Carugati et al. (2019)

argue that Athens in the fourth century BC can already be viewed as a developed society, despite the lack of features such as industrialization, rapid technological change, and the existence of centralized and liberal government. From a construction viewpoint, the role of unfree labour in Athenian society is particularly interesting; the expense accounts from major public projects indicate that at least some slaves, both private and public, enjoyed forms of freedom and economic independence, while free citizens employed as unskilled workers on a daily basis might have been in a more precarious situation. The reason why the ancient Greek accounts can be read today is the high degree of transparency that was associated with public procurement in the Athenian democracy, discouraging officials from corruption.

Why should the study of ancient Greek and Roman construction be relevant to the future of construction economics? First, it should be noted that standard practice among most researchers appearing in this chapter is to frame construction as the built environment sector (de Valence, 2019), analysing supply of materials as well as on-site production in an economic perspective. Despite the development of technology over millennia, not least the emergence of modern information and communication technologies, fundamental challenges of construction remain: the engagement of skilled versus unskilled workers when major projects come and go, locations shift and there are seasonal influences; choosing the best degree of prefabrication, here illustrated by the relation between quarry and site production, including considerations of logistics. We can identify a continuity of industry custom and practice in estimating, contracting and managing projects. This is an argument for looking more closely at how the Greco-Roman world approached problems that we still grapple with, such as competition and corruption in public procurement.

REFERENCES

Abrams, E.M. and Bolland, T.W. (1999). Architectural energetics, ancient monuments, and operations management. *Journal of Archaeological Method and Theory*, *6*(4), 263–291.

Anderson, J.C. (1997). *Roman architecture and society*. Baltimore, MD: Johns Hopkins University Press.

Barresi, P. (2003). *Province dell'Asia Minore: costo dei marmi, architettura pubblica e committenza*. Roma: 'L'Erma' di Bretschneider.

Bernard, S.G. (2017). Workers in the Roman imperial building industry. In K. Verboven and C. Laes (eds), *Work, labour, and professions in the Roman world* (pp. 62–86). Leiden: Brill.

Bernard, S. (2018). *Building Mid-Republican Rome: labor, architecture, and the urban economy*. Oxford: Oxford University Press.

Bouras, C. (2002). Master craftsmen, craftsmen, and building activities in Byzantium. In A.E. Laiou (ed.), *The economic history of Byzantium: from the seventh through the fifteenth century* (Vol. 2, pp. 539–554). Washington, DC: Dumbarton Oaks Research Library and Collection.

Bousquet, J. (1988). *Étude sur les comptes de Delphes*. Athènes: École française d'Athènes.

Bresson, A. (2012). Greek epigraphy and ancient economics. In J. Davies and J. Wilkes (eds), *Epigraphy and the historical sciences* (pp. 223–247). Oxford: Oxford University Press.

Bringmann, K. (2001). Grain, timber and money: Hellenistic kings, finance, buildings and foundations in Greek cities. In Z. Archibald, J. Davies, V. Gabrielsen and G.J. Oliver (eds), *Hellenistic economies* (pp. 205–214). London: Routledge.

Brunt, P.A. (1980). Free labour and public works in Rome. *Journal of Roman Studies*, *70*, 81–100.

Brysbaert, A. (2013). Set in Stone? Socio-economic reflections on human and animal resources in monumental architecture of Late Bronze Age Tiryns in the Argos plain, Greece. *Arctos – Acta Philologica Fennica*, *47*, 49–96.

Brysbaert, A.N. (2017). SETinSTONE: an impact assessment of the human and environmental resource requirements in Late Bronze Age Mycenaean monumental architecture. *Antiquity, 91*(358), 1–7.
Burford, A. (1969). *The Greek temple builders at Epidauros: a social and economic study of building in the Asklepian sanctuary, during the fourth and early third centuries BC.* Liverpool: Liverpool University Press.
Carugati, F., Ober, J. and Weingast, B.R. (2019). Is development uniquely modern? Ancient Athens on the doorstep. *Public Choice, 181*(1–2), 29–47.
Corcoran, S. (2000). *The empire of the tetrarchs: imperial pronouncements and government AD 284–324* (rev. edn). Oxford: Oxford University Press.
Davies, J.K. (2001). Rebuilding a temple: the economic effects of piety. In D.J. Mattingly and J. Salmon (eds), *Economies beyond agriculture in the classical world* (pp. 209–229). London: Routledge.
Davis, P.H. (1937). The Delian building contracts. *Bulletin de Correspondance Hellénique, 61*, 109–135.
DeLaine, J. (1997). *The Baths of Caracalla: A Study in the Design, Construction, and Economics of Large-Scale Building Projects in Imperial Rome* (Journal of Roman Archaeology, Suppl. 25). Portsmouth, RI: Journal of Roman Archaeology.
DeLaine, J. (2001). Bricks and mortar: exploring the economics of building techniques in Rome and Ostia. In D.J. Mattingly and J. Salmon (eds), *Economies beyond agriculture in the classical world* (pp. 230–268). London: Routledge.
DeLaine, J. (2017). Quantifying manpower and the cost of construction in Roman building projects: research perspectives. *Archeologia dell'architettura, 22*, 13–19.
DeLaine, J. (2018). Economic choice in Roman construction: case studies from Ostia. In A. Brysbaert, V. Klinkenberg, A. Gutiérrez Garcia-M. and I. Vikatou (eds), *Constructing monuments, perceiving monumentality and the economics of building: theoretical and methodological approaches to the built environment* (pp. 243–270). Leiden: Sidestone Press.
de Valence, G. (2019). Reframing construction within the built environment sector. *Engineering, Construction and Architectural Management, 26*(5), 740–745.
Devolder, M. (2013). *Construire en Crète minoenne: Une approche énergétique de l'architecture néo-palatiale.* Leuven/Liège: Peeters.
Devolder, M. (2017). Architectural energetics and Late Bronze Age Cretan architecture: measuring the scale of Minoan building projects. In Q. Letesson and C. Knappett (eds), *Minoan architecture and urbanism: new perspectives on an ancient built environment* (pp. 57–79). Oxford: Oxford University Press.
Duncan-Jones, R. (1982). *The economy of the Roman Empire: quantitative studies* (2nd edn). Cambridge: Cambridge University Press.
Epstein, S. (2008). Why did Attic building projects employ free laborers rather than slaves? *Zeitschrift für Papyrologie und Epigraphik, 166*, 108–112.
Epstein, S. (2013). Attic building accounts from euthynae to stelae. In M. Faraguna (ed.), *Archives and archival documents in ancient societies: legal documents in ancient societies IV, Trieste 30 September – 1 October 2011* (pp. 127–141). Trieste: EUT Edizioni Università di Trieste.
Erasmus, C.J. (1965). Monument building: some field experiments. *Southwestern Journal of Anthropology, 21*(4), 277–301.
Feyel, C. (2006). *Les artisans dans les sanctuaires grecs à travers la documentation financière en Grèce.* Athènes: École française d'Athènes.
Fitzjohn, M. (2013). Bricks and mortar, grain and water: tracing tasks and temporality in Archaic Sicily. *World Archaeology, 45*(4), 624–641.
Fröhlich, P. (2004). *Les cités grecques et le contrôle des magistrats: IVe–Ier siècle avant J.-C.* Genève: Droz.
Goldsmith, H. (2015). Actions and innovations in the evolution of infrastructure services. In A. Picot, M. Florio, N. Grove and J. Kranz (eds), *The economics of infrastructure provisioning: the changing role of the state* (pp. 23–91). Cambridge, MA: MIT Press.
Heimpel, W. (2009). *Workers and construction work at Garšana* (CUSAS 5). Bethesda, MD: CDL Press.
Hurst, J.T. (1865). *A hand-book of formulæ, tables, and memoranda for architectural surveyors, and others engaged in building.* London: E. & F.N. Spon.
Kallet-Marx, L. (1989). Did tribute fund the Parthenon? *Classical Antiquity, 8*(2), 252–266.

Kehoe, D.P. (2007). The early Roman Empire: production. In W. Scheidel, I. Morris and R. Saller (eds), *The Cambridge economic history of the Greco-Roman world* (pp. 543–569). Cambridge: Cambridge University Press.
Knoepfler, D. (2001). Le contrat d'Éretrie en Eubée pour le drainage de l'étang de Ptéchai. In P. Briant (ed.), *Irrigation et drainage dans l'Antiquité, qanâts et canalisations souterraines en Iran, en Égypte et en Grèce* (pp. 41–79). Paris: Thotm éditions.
Lancaster, J. (2019). To house and to defend: the application of architectural energetics to south-east Archaic Greek Sicily. In L. McCurdy and E.M. Abrams (eds), *Architectural energetics in archaeology: analytical expansions and global explorations* (pp. 95–113). London: Routledge.
Lancaster, L. (1998). Building Trajan's Markets. *American Journal of Archaeology*, *102*, 283–308.
Lancaster, L.C. (2005). *Concrete vaulted construction in Imperial Rome: innovations in context.* Cambridge: Cambridge University Press.
Leveau, P. (2001). Aqueduct building: financing and costs. In D.R. Blackman and A.T. Hodge (eds), *Frontinus' legacy: essays on Frontinus' de aquis urbis Romae (pp. 84–108).* Ann Arbor, MI: University of Michigan Press.
Lloyd, A.B. (1995). Herodotus on Egyptian buildings: a test case. In A. Powell (ed.), *The Greek World* (pp. 273–300). London: Routledge.
Long, L.E. (2017). Extracting economics from Roman marble quarries. *Economic History Review*, *70*(1), 52–78.
Low, S.P. (2007). Managing building projects in Ancient China: a comparison with modern day project management principles and practices. *Journal of Management History*, *13*(2), 192–210.
Martin, R. (1973). Aspects financiers et sociaux des programmes de construction dans les villes grecques de Grande-Grèce et de Sicile. In P. Romanelli (ed.), *Economia e società nella Magna Grecia: Atti del XII Convegno di studi nella Magna Grecia, Taranto, 8–14 ottobre 1972* (pp. 185–205). Naples.
Meiggs, R. (1982). *Trees and timber in the ancient Mediterranean world.* Oxford: Oxford University Press.
Montecchi, B. (2013). *Luoghi per lavorare, pregare, morire: Edifici e maestranze edili negli interessi delle élites micenee.* Firenze: Firenze University Press.
Ng, D.Y. and Swetnam-Burland, M. (2018). Introduction: reuse, renovation, reiteration. In D.Y. Ng and M. Swetnam-Burland (eds), *Reuse and renovation in Roman material culture: functions, aesthetics, interpretations* (pp. 1–23). Cambridge: Cambridge University Press.
Osborne, R. and Rhodes, P.J. (eds) (2017). *Greek historical inscriptions 478–404 BC.* Oxford: Oxford University Press.
Pakkanen, J. (2013). The economics of shipshed complexes: Zea, a case study. In D. Blackman, B. Rankov, K. Baika, H. Gerding and J. Pakkanen, *Shipsheds of the Ancient Mediterranean* (pp. 55–75). Cambridge: Cambridge University Press.
Palaima, T. (2015). The Mycenaean mobilization of labor in agriculture and building projects: Institutions, individuals, compensation, and status in the Linear B tablets. In P. Steinkeller and M. Hudson (eds), *Labor in the Ancient World* (Vol. 5, pp. 617–648). Dresden: Islet.
Patay-Horváth, A. (2019). Greek temple building from an economic perspective: case studies from the Western Peloponnesos. In P. Sapirstein and D. Scahill (eds), *New directions and paradigms for the study of Greek architecture* (pp. 168–177). Leiden: Brill.
Pegoretti, G. (1863–1864). *Manuale pratico per l'estimazione dei lavori architettonici, stradali, idraulici e di fortificazione per uso degli ingegneri ed architetti* (2 vols, 2nd edn). Milano: Domenico Salvi.
Pitt, R.K. (2014). Just as it has been written: inscribing building contracts at Lebadeia. In N. Papazarkadas (ed.), *The epigraphy and history of Boeotia: new finds, new prospects* (pp. 373–394). Leiden: Brill.
Prignitz, S. (2014). *Bauurkunden und Bauprogramm von Epidauros (400–350): Asklepiostempel, Tholos, Kultbild, Brunnenhaus.* München: C.H. Beck.
Randall, R.H. (1953). The Erechtheum workmen. *American Journal of Archaeology*, *57*(3), 199–210.
Rhodes, P.J. and Osborne, R. (eds) (2003). *Greek historical inscriptions 404–323 BC.* Oxford: Oxford University Press.
Richardson, M.E.J. (2000). *Hammurabi's laws: text, translation and glossary.* London: T&T Clark International.
Rihll, T.E. (2013). Depreciation in Vitruvius. *Classical Quarterly*, *63*(2), 893–897.

Rosillo, C. (2003). Fraude et contrôle des contrats publics à Rome. In J.-J. Aubert (ed.), *Tâches publiques et entreprise privée dans le monde romain* (pp. 57–94). Genève: Université de Neuchâtel.
Russell, B. (2013). *The economics of the Roman stone trade.* Oxford: Oxford University Press.
Salmeri, G. (2000). Dio, Rome, and the civic life of Asia Minor. In S. Swain (ed.), *Dio Chrysostom: politics, letters, and philosophy* (pp. 53–92). Oxford: Oxford University Press.
Salmon, J. (2001). Temples the measures of men: public building in the Greek economy. In D.J. Mattingly and J. Salmon (eds), *Economies beyond agriculture in the Classical World* (pp. 195–208). London: Routledge.
Sellet, F. (1993). Chaîne opératoire: the concept and its applications. *Lithic Technology*, *18*(1–2), 106–112.
Shear, T.L., Jr (2016). *Trophies of victory: public building in Periklean Athens.* Princeton, NJ: Department of Art and Archaeology, Princeton University.
Shipley, G. (1987). *A history of Samos 800–188 BC.* Oxford: Oxford University Press.
Snyder, J.R. (2020). Defending a new capital: Ravenna, Classe, and the revival of the construction industry in late antiquity. In E. Intagliato, S.J. Barker and C. Courault (eds), *City walls in late antiquity: an empire-wide perspective* (pp. 87–100). Oxford: Oxbow.
Snyder, J.R., Dilaver, Ö., Stephenson, L.C., Mackie, J.E. and Smith, S.D. (2018). Agent-based modelling and construction: reconstructing antiquity's largest infrastructure project. *Construction Management and Economics*, *36*(6), 313–327.
Spengler, J.J. (1955). Herodotus on the subject matter of economics. *Scientific Monthly*, *81*(6), 276–285.
Stanier, R.S. (1953). The cost of the Parthenon. *Journal of Hellenic Studies*, *73*, 68–76.
Strong, D.E. (1968). The administration of public building in Rome during the Late Republic and Early Empire. *Bulletin of the Institute of Classical Studies*, *15*, 97–109.
Temin, P. (2013). *The Roman market economy.* Princeton, NJ: Princeton University Press.
Turner, D.R. (2018). Comparative labour rates in cross-cultural contexts. In A. Brysbaert, V. Klinkenberg, A. Gutiérrez Garcia-M. and I. Vikatou (eds), *Constructing monuments, perceiving monumentality and the economics of building: theoretical and methodological approaches to the built environment* (pp. 195–218). Leiden: Sidestone Press.
Underwood, D. (2019). *(Re)using ruins: public building in the cities of the Late Antique West, A.D. 300–600.* Leiden: Brill.
Van Ooijen, J.A. (2019). Resilient matters: the cathedral of Syracuse as an architectural palimpsest. *Architectural Histories*, *7*(1), 26, 1–12.
Vitruvius (1999). *Ten books on architecture.* I.D. Rowland and T.N. Howe (eds). Cambridge: Cambridge University Press.
Walker, D. and Dart, C.J. (2011). Frontinus – a project manager from the Roman Empire era. *Project Management Journal*, *42*(5), 4–16.
Wilson, A.I. (2006). The economic impact of technological advances in the Roman construction industry. In E. Lo Cascio (ed.), *Innovazione tecnica e progresso economico nel mondo romano* (pp. 225–236). Bari: Edipuglia.

6. Construction in the economy and in national development

Jorge Lopes

INTRODUCTION

One of the main areas in the construction economics literature deals with the relationship between the construction industry and economic development. According to this line of thought, the role of construction is associated with changes in the economic development process due to changes in the structure of the construction industry and variations in levels of physical investment during the development process. Construction infrastructure plays a role as a capital input into production and wealth generation. The economic impact can be transformative, especially at low to middle levels of income per capita (OECD, 2013). As pointed out by Maddison (1987), the close association between physical capital and different measures of the national economy is one of the reasons why physical infrastructure has been considered a powerful engine of economic growth and development.

Early seminal works on the role of the construction industry in the process of development (Strassman, 1970; Turin, 1973; World Bank, 1984; Wells, 1986) found a positive correlation between several measures of the construction output and the level of national income per capita. However, the dominant paradigm that has emerged is that the share of construction in gross domestic product (GDP) tends to increase in the first stages of development, to stabilize in the middle-income range and to decrease in the latter stages of development (Strassman, 1970; Bon, 1992). The main argument behind this proposition, as explained by Bon (1992), is the special historical link between the construction industry and the process of urbanization and industrialization, particularly with the manufacturing industry, the construction sector's main partner in the development process. As pointed out by Rodrik (2015), it was the industrial revolution that enabled sustained economic growth in Europe and in North America for the first time. It was also industrialization that permitted catch-up and convergence with the West by a relatively smaller number of non-Western nations: Japan in the late 19th century and the Republic of Korea, Taiwan and a few East Asian countries after the 1960s. The earlier stages of development are characterized by intense processes of urbanization, demographic growth, creation of basic infrastructure and construction of industrial plants (Bon, 1992; Girardi and Mura, 2014). Thus, the construction sector tends to grow faster than the rest of the economy during this phase, increasing its share in national output. In later stages, as these processes reach maturity and start slowing down, growth in construction investment tends to slow down with respect to the overall economy. It should be noted that a significant number of studies (such as de Long and Summers, 1991; Ganesan, 2000; Lopes et al., 2002; Yiu et al., 2004) have not shared the generalized view on the positive role of construction investment; namely, on the magnitude of investment and the direction of causality between construction and economic growth.

However, the positive association between infrastructure and economic growth has been the subject of the debate on the part played among several writers on growth accounting (see for example, Caselli, 2005; Calderon et al., 2015, for a detailed review) and international organizations such as the World Bank in its Structural Adjustment Programmes for Africa in the late 1980s to early 1990s. Calderon et al. (2015) used an infrastructure-augmented production function approach to consider the contribution of infrastructure capital to aggregate productivity and output. Their panel data set consisted of 88 developed and developing countries, over the period 1960 to 2000. They found that marginal product of infrastructure was higher when the (relative) infrastructure stock was lower, but then diminished at higher levels. Banerjee et al. (2020) analysed the effect of access to transportation networks on regional economic outcomes over a 20 year period (1986–2006) of rapid income growth in China. They concluded that proximity to transportation networks has a moderately sized positive causal effect on per capita GDP levels across sectors, but no effect on per capita GDP growth. Similarly, Ansar et al. (2016) examined the relationship between infrastructure investment and economic growth in China over the period 1984–2008. They reported that overinvestment in underperforming projects in China did lead to economic fragility. The study also suggests that a massive programme of infrastructure investment is not a viable development programme for other developing countries, which may look to China as a model for development.

Kodongo and Ojah (2016) analysed the relationship between infrastructure and economic growth, for a panel of 45 sub-Saharan African countries, over the period 2000–2011. They found that it is the spending on infrastructure and increments in the access to infrastructure that influence economic growth and development in sub-Saharan Africa; and these significant associations, particularly infrastructure spending, are more pronounced in the lesser-developed countries than the relatively more developed countries of the region. The World Bank's *World Development Report 1994* (World Bank, 1994) documents substantial cross-country differences in the efficiency with which public infrastructure is used. According to this line of thought, it could be argued that the productivity of public capital would be improved through adequate maintenance and upgrading of existing infrastructure stock, and by prioritizing investments that modernize production and enhance international competitiveness (Lopes, 2012).

A particular feature related to the role of infrastructure in the process of development is a trend of deindustrialization in the majority of developing countries, particularly in Latin America and sub-Saharan Africa since the 1980s. This phenomenon is known as 'premature deindustrialization'. Using data drawn from the Groningen Growth and Development Centre (GGDC) 10-Sector Database, Rodrik (2015) found that developing countries – with some exceptions, confined largely to Asia – have experienced falling shares in both employment and real value added of the manufacturing sector, especially since the 1980s. Moreover, since 1990, developing countries have reached peak manufacturing employment at incomes per capita that are around a third of the levels experienced before 1990. For manufacturing value added at constant prices, the corresponding ratio is less than a half.

The mixed results on the relationship between infrastructure (construction) spending and economic growth point to the importance of the level of capital stock on the future development pattern of developing countries, and that of the construction industry in these countries. In highly developed countries, the construction sector is no longer focused on large-scale production but on the services provided by the built environment (Carassus et al., 2006). Being a dynamic process, modelling of the construction sector and economic development relationship has to deal with such changes (Ruddock and Lopes, 2006), and with the changes in the industrial structure of developing countries. Empirical tests of the Strassman (1970) and Bon (1992) model have provided mixed results. Choy (2011) concluded that the inverted U-shaped curve holds within most developed countries over time, but not across countries at a given time.

Girardi and Mura (2014) employed panel data techniques to analyse the economic development and construction relationship. The sample covered the majority of developed and developing countries, over the period 2000–2011. They found that the construction–development curve is asymmetric with respect to its maximum; that is, the share of construction in GDP decreases at a slowing pace after industrialization, approaching some kind of plateau in mature economies. The peak in construction shares is reached at per capita income levels closer to the origin, a pattern similar to that of manufacturing in Rodrik's (2015) study. Sun et al. (2013) studied panel data on construction and other economic indicators in advanced and emerging (in 2011) countries of Europe, over the period 2000–2011. The empirical results indicate that a country's geography, demographics and economic conditions are key determinants of a norm around which construction's shares revolve. The results also showed that, in many European countries, construction's shares overshot relative to their norms before the 2008 global crisis, but they had fallen significantly since the crisis. Lopes (2012) analysed a dataset covering 45 sub-Saharan countries, over the period 1990–2008. He found that in the middle-income economies the share of construction in GDP remained practically stagnant from 2000 onwards, in spite of a rapid growth in GDP per capita in the same period.

As pointed out by Stern (1991), research on growth accounting has been stimulated by, and has stimulated, the documentation and analysis of the empirical growth process by economic historians and statisticians. Important contributions have come from a particularly valuable set of data, which has provided re-computations of national income and physical capital on the basis of purchasing power parity (PPP). For over four decades, the Penn World Table (PWT) has been a standard source of data on national economies (Ruddock and Ruddock, 2019). Making use of prices collected across countries in benchmark years by the International Comparison Program (ICP) and using these prices to construct purchasing power parity exchange rates, PWT converts national economic data to a common currency (USD) allowing intercountry comparisons. The latest version of the PWT (version 9.1) (Feenstra et al., 2015) is a database with information on relative levels of income, output, capital and productivity, covering 182 countries between 1950 and 2017. The gross fixed capital formation (GFCF) estimates include the value of structures, machinery, equipment and other assets. The capital detail file includes information on investment at constant national prices, the investment deflator, the current-cost net capital stock and the capital stock deflator. Intuition and descriptive overviews of dynamics observed in the middle-income countries suggest that the level of capital stock is an additional variable for modelling the construction and economic development relationship in developing countries.

It appears that the level of a country's net capital stock (particularly built capital stock) captures some dimensions of the level of urbanization and, to some extent, of the industrial structure. Thus, this study represents a novel approach in two interrelated ways: it uses the most comprehensive dataset that allows comparable prices of consumption and investment goods across countries and across time; and to overcome some of the heterogeneity problems in growth accounting exercises, the sample analysed here comprises countries of the same African subregion and with the same economic development status (middle-income countries).

The rest of this chapter is organized as follows: it first discusses the concepts of developed and developing economies. The dataset used for the analyses and methodological issues concerning the indicators of economic activity are discussed in the next section. A brief literature review on the macroeconomic level of the construction sector is presented next. The following section presents an overview of the long-run relationship between GDP and construction-related indicators for the world, economic groupings, sub-Saharan Africa and two economically dominant powers of the world: the United States (US) and China. The chapter then presents a model of the development pattern of the construction industry in the middle-income economies in sub-Saharan Africa over the period between 2000 and 2018. Two groups of countries are established: one in which the capital–output ratio was equal or greater than 4 in the period 1990–1999 (average for the period); and the other in which the capital–output ratio was less than 4 in the period 1990–1999 (average for the period). Some concluding remarks are presented in the final section.

DEVELOPED AND DEVELOPING ECONOMIES

The classification of countries according to their development status does not have a definitive basis. There is no established convention for the designation of 'developed' and 'developing' countries in the United Nations system (Ruddock and Lopes, 2006). In common practice, the G7 members (France, Germany, Italy and the United Kingdom in Europe; Japan in Asia; and Canada and the United States in North America) and Australia and New Zealand in Oceania are considered developed economies. However, China, despite being the world's leading merchandise trader and exporter and accounting for 32 per cent of world exports in office and telecommunications equipment in 2018 (United Nations, 2019b), is considered a developing economy in the United Nations system. Besides the broad division of the world by regions and their subregions, the *World Economic Situation and Prospects* of the United Nations (UN, 2020a) classifies, for analytical purposes, all countries of the world into three broad categories: developed economies, economies in transition and developing economies. Countries characterized as developed economies are those of North America, Europe – except Albania, Bosnia and Herzegovina, Georgia, Montenegro, North Macedonia and Serbia, and the European countries that are part of the Commonwealth of Independent States (CIS) – Japan in Asia, and Australia and New Zealand in Oceania. Developing economies are those in Africa, Asia except Japan and countries in Central Asia that are part of the CIS, Latin America and the Caribbean and Oceania, except Australia and New Zealand. Economies in transition are those that are part of the CIS, Albania, Bosnia and Herzegovina, Georgia, Montenegro, North Macedonia and Serbia. The World Bank, through the World Development Indicators, is more straightforward. It classifies the economies into the following categories, according to their level of gross national income (GNI) per capita: low-income countries (LICs), middle-income countries (MICs) subdivided into lower-middle-income countries (LMICs) and upper-middle-income countries (UMICs), and high-income countries (HICs).

It is worth noting that there is an extensive literature pointing to the fact that GNI and GNI per capita are rather poor and mono-dimensional measures of economic development, and worth stressing the need for broader and more comprehensive indicators (Stiglitz et al., 2010; Girardi and Mura, 2014). GNI and GNI per capita do not, for example, include negative processes associated with economic activities such as the degradation and pollution of the environment, or indicate how national income is distributed amongst a country's population. Since 1990, The United Nations Development Programme (UNDP) publishes annually the Human Development Report (UNDP, 1990–). The UNDP constructs a Human Development Index (HDI), which considers per capita income, mean and expected years of schooling and life expectancy at birth. Countries are ranked by the index from lowest (0) to highest (1), and compared with the ranking by GNI per capita. In the low-to-middle-income range, some countries rank low by GNI per capita and high (relatively) by HDI, and vice versa. However, the general feature is that a high (low) ranking by the HDI tends to correspond a high (low) ranking by GNI per capita. In their study on the relationship between construction and economic development, Girardi and Mura (2014) found that there is no significant difference in the results whether GNI per capita or HDI are chosen for the measure of economic development. Thus, for the sake of simplicity and the fact that GNI per capita is a component of any composite index of a country's welfare, GNI per capita is chosen to be the indicator of economic development.

The figures for the benchmark year 2010 are GNI per capita of US$1005 or less for LICs, GNI per capita higher than US$1006 and less than US$3975 for LMICs, GNI per capita higher than US$3976 but less than US$12 275 for UMICs, and GNI per capita of US$12 276 or higher for HICs. Thus, all the sub-Saharan African countries analysed in this chapter belonged to the middle-income status in 2010 except Equatorial Guinea, which was by then considered a HIC. It is worth noting all the countries in SSA that belonged to the MIC status in 2010 were also in the same income development status in 2019. Former low-income economies in sub-Saharan Africa that jumped to the next stage of economic development in 2019 are Comoros, Tanzania (mainland) and Zimbabwe.

DATASET AND METHODOLOGICAL ISSUES

The main indicators of economic activities used for the analysis are construction value added (CVA), GDP per capita and gross fixed capital formation (GFCF). In the production approach, the estimates of GDP are obtained as the summation of the value added of all economic activities. CVA is calculated the same way as in any other productive sector, but includes the activities of the construction activity proper only. For example, it excludes the output of the building materials industry, which is accounted for in the manufacturing sector. The choice of CVA as the main indicator of the construction industry activity rather than GFCF in construction is that the production approach has generally been utilized by the national statistics offices in sub-Saharan Africa as the main approach to compile the estimates of GDP (de Vries et al., 2014). In order to measure the importance of construction in a country's economy, and for cross-country comparisons, it is calculated as a share of GDP. GDP per capita introduces the variable population into the aggregate domestic product of a nation. In this sense, it appears to be a better indicator of a country's welfare than GDP (Ruddock and Lopes, 2006).

Building and other construction products are also a component of a country's GFCF as described in the System of National Accounts (SNA) 2008 of the United Nations (UN, 2009). According to the SNA 2008, GFCF consists of the purchase of goods (and services) that are used in production for more than one year. This publication classifies capital stock statistics according to: type of assets; institutional sectors; and economic sectors as described in the International Standard Classification of All Economic Activities (ISIC revision 4) (United

Nations, 2008). In terms of type of assets, the built capital stock is comprised of dwellings, and other buildings and structures (including land improvements). Other productive fixed assets that are recognized in both the European System of Accounts (ESA) 2010 (Eurostat, 2013) and SNA 2008 are: transport equipment; machinery and other equipment; cultivated biological assets; cost of ownership transfer on non-produced assets; and intellectual property products. Within the context of national accounting, there are two types of capital measures, each reflecting a different role of capital (OECD, 2009, 2013). The first type of measure looks at capital in its function as a provider of services in production. The second type of capital measure captures its role as a store of wealth. Its aggregate is the net capital stock that captures the market value of capital goods. Unlike gross capital stock (which does not take depreciation of assets into account), net capital stock (NCS) is part of an economy's balance sheet in the context of income and wealth accounting (Lopes et al., 2019). Most modern works, at both national and international levels, that publish capital stock data (see for instance Feenstra et al., 2015; Eurostat, various years; EU KLEMS, 2017; Derbyshire et al., 2011) are based on the perpetual inventory methodology outlined in the OECD Manual *Measuring Capital* (OECD, 2009).

The statistical sources used for the analysis are the Analysis of the Main Aggregates and Main Aggregates and Detailed Tables, both from the United Nations (UN, 2020b, 2020c), Penn World Tables – Version 9.1 (Feenstra et al., 2015) and the World Bank's World Development Indicators 2012 (World Bank, 2012). The United Nations database presents various sets of economics series detailing the evolution of GDP and its components (production, expenditure and income approaches), in different statistical formats, at current and at 2015 constant prices, in national currencies and in US dollars, over the period 1970–2018. The Penn World Tables – version 9.1 (PWT 9.1) present GDP and NCS data at current PPP (2011 USD) and at constant 2011 prices (2011 USD), going back, for some countries, as far as 1950.

As the figures in constant prices are based on national accounts data, it is possible to compare the capital–output ratio across time and across countries. Calculations from the data presented in PWT 9.1 show essentially no systematic pattern of cross-country variation in the capital–output ratio, particularly in the middle to high income levels of economic development. However, for highly developed countries, a slight trend of decline is observed with increasing per capita income across time, as implied by the Solow model (Solow, 1956). It is worth noting that the depreciation rates in PWT 9.1 vary across countries and across time, reflecting assets composition for a given country and at a given time. Calculations of the capital–output ratio (from PWT 9.1 data) also indicate that the majority of European countries were in a range of 3 to 5 in the period 1980–2000, so a value of 4 for the capital–output ratio seems reasonably constructed for the analysis. It is worth noting that for developing countries, the capital–output ratios constructed from the PWT 9.1 (national accounts-based constant 2011 prices) are generally higher than those measured at national currency prices.

As pointed out by Feenstra et al. (2015), the price level of the capital stock tends to be similar to that of consumption for both low- and high-income economies. Due to its lower rate of depreciation, buildings and other constructed facilities form a large part of the price index of the capital stock, and since they are non-traded, the price level for structures tends to be lower for poorer countries. Therefore, the price level of the capital stock is usually lower than that of GDP. Assuming that structures investment represents about 80 per cent of a country's NCS (Inklaar and Timmer, 2013), that figure corresponds to a built stock–GDP ratio of

3.2. As noted above, data on GNI per capita for the year 2010 are obtained from the World Development Indicators 2012 (World Bank, 2012).

The composition of the groups of countries analysed in this chapter is as follows:

- Group 1: corresponding to the group of countries in which the capital–output ratio was equal to or greater than 4 in the period 1990–1999 (average for the period). This group comprises: Angola, Cabo Verde, Eswatini, Lesotho, Mauritius, Sao Tome and Principe, Seychelles, South Africa and Zambia.
- Group 2: corresponding to the group of countries in which the capital–output ratio was less than 4 in the period 1990–1999 (average for the period). This group comprises: Botswana, Cameroon, Congo Republic, Côte d'Ivoire; Djibouti, Gabon, Ghana, Equatorial Guinea, Mauritania, Namibia, Nigeria and Senegal.

THE MACRO LEVEL OF THE CONSTRUCTION SECTOR: A BRIEF REVIEW

There are three main strands in the literature at the macroeconomic level of the construction industry. The first one deals with the relationship between construction and economic growth and development (Strassman, 1970; Turin, 1973; Bon, 1992; Ruddock and Lopes, 2006). The second tries to assess whether construction investment leads GDP growth, or vice versa (De Long and Summers, 1991; Dasgupta et al., 2014). The third strand employs input–output analysis to study the pull and push effects of the construction sector within the national economy (Bon and Pietroforte, 1990; Pietroforte and Gregori, 2003; Gloser et al., 2017). Early influential papers investigating the role of construction in economic development (Turin, 1973; Strassman, 1970; Drewer, 1980; Edmond and Miles, 1984) were based on Keynesian economic philosophy, particularly on the feature related to the role of capital formation in the process of economic growth and development.

In this line of thought, construction plays a unique role in economic growth and is often seen as a key barometer of economic conditions. Construction increases a country's physical infrastructure (including the housing stock) which is a critical factor for long-term growth (OECD, 2013). The construction industry has also historically been linked with the process of industrialization and urbanization, particularly since the advent of the industrial revolution. Railway systems and canals played an important role in the development of different regions of Europe, North America and some parts of Latin America (Rostow, 1963; Donaldson and Hornbeck, 2016).

The most influential of these early works was that of Turin (1973). Turin's study consisted of 87 countries in all regions of the world, representing all stages of economic development, and it considered data over the period 1955–1965. Two of the observations made by Turin can be presented as follows: (1) capital formation in construction was 6–9 per cent of GDP in developing countries and 10–15 in industrialized countries; and (2) using cross-country comparisons, there exists a direct relationship between the level of income per capita and the level of construction activity measured as a share of total value added. These findings have been supported by many other studies in the literature (World Bank, 1984; Edmond and Miles, 1984; Wells, 1986; Ofori, 1988). However, as pointed out by Chiang and Low (2011), some of these early studies noted that the role of construction in the economy would decline when countries reach the middle-income development stage (Strassman, 1970); that more construction activity does not necessarily mean more economic growth when resources are misallocated (Drewer, 1980); and that the required level of construction value added for sustained growth in the economy would be 5 per cent of GDP (Edmonds, 1979).

Bon (1990, 1992), analysing a sample of countries representing all stages of economic development, presented a development pattern of the construction industry based on stages of development. The main point from Bon's (1990, 1992) model is that the share of the construction industry in GDP follows an inverted U-shaped pattern of development: that is, it increases in the first stages of development, stagnates in the middle-income range and starts to decrease in the latter stages of development. Moreover, in the long run, in the most advanced industrialized countries, construction will decline not only relatively but also absolutely. Bon (1992) seemed to follow Maddison's (1987) observation that as an economy becomes more developed and more fully realizes its economic potential, the economy becomes less dependent on any single sector to stimulate economic growth and development. However, Ruddock and Lopes (2006) and Carassus et al. (2006) noted that there was no evidence of the decline in absolute terms in the most economically advanced countries.

With the availability of long and more reliable time-series data and the development of econometric methodology related to the study of economic relationships between variables, a new strand of studies has emerged. Some of these studies have applied econometric analysis within Granger's 1969 framework (Granger, 1969) to test the causality link between construction investment and different measures of the national aggregate. One view in this strand is that construction spending causes growth in the national aggregate as it creates physical facilities that are needed in the development of other productive activities (Anaman and Osei-Amponsah, 2007; Lean, 2001). The contrasting view holds that GDP causes construction growth (Tse and Ganesan, 1997; Yiu et al. 2004; Khan, 2008; Lopes et al., 2011). Other studies suggested a bidirectional relationship between different segments of the construction industry's activity and the national economy (Ozkan et al., 2012; Green, 1997; Chen and Zhu, 2008).

Yiu et al. (2004) found that, for Hong Kong, the real growth of the aggregate economy leads the real growth of the construction output and not vice versa, at least in the short term. On the other hand, Wong et al. (2008), using more recent data covering a longer period of Hong Kong's high-income status, concluded that the direction of the causality is from the construction sector, particularly the civil engineering subsector, to GDP. Along the same lines, Anaman and Osei-Amponsah (2007) analysed the relationship between the construction industry and the macroeconomy in Ghana, based on time series data from 1968 to 2004, and found that the construction industry leads economic growth in Ghana. Khan (2008) studied data on the construction industry and other indicators of economic activity for Pakistan for the period 1950–2000, and also reported that the direction of causality is from construction to GDP. Chen and Zhu (2008) analysed provincial data on housing investment in three main regions of China, and found that there was a bi-directional Granger causality between GDP and housing investment for the country as a whole, while the impact of housing investment on GDP behaved differently in the three regions. In an earlier study, Green (1997) examined whether residential and non-residential investments Granger cause GDP, and whether GDP Granger causes each of these types of construction investments. Using quarterly national income and products data for the period 1959 to 1992, he found that housing investment causes GDP, but is not caused by it; while non-residential investment does not cause GDP, but is caused by it. Thus, he concluded, housing investment leads and other types of construction investment lag the business cycle.

Sahoo and Dash (2009) analysed infrastructure development and economic growth in India, based on data from the period between 1970 and 2006. The variables in the study included data on the energy, telecommunications and aviation industries. Overall, Sahoo and Dash concluded that infrastructure stocks and total infrastructure investment played a major role in economic growth in India. Specifically, they found that infrastructure development had a significant positive contribution towards economic growth, and that the direction of causality was from infrastructure investment to GDP. Wilhelmsson and Wigren (2011) investigated the causal relationships between construction investment flows and economic growth in 14 Western European countries, using panel data techniques. They found that residential construction Granger causes GDP in the short and long run, and this effect was not true when it comes to infrastructural and other building construction. Erol and Unal (2015) examined the causal relationship between construction investment and economic growth in Turkey from 1998 to 2014. The results of the study indicate that, for the entire sample period, GDP growth in Turkey preceded construction industry growth with two to four quarter lags, but not vice versa. They concluded that the construction industry is not a driver of GDP growth, but a follower of fluctuations in the macroeconomy. In a similar vein, the results of a more recent study on the relationship between construction investment and GDP in Nigeria (Ogunbiyi et al., 2017) showed that a unidirectional causality exists between construction investment and economic growth in Nigeria, with the causality running from the latter to the former.

The review in this section suggests that the causal relationships between construction (infrastructure) spending and aggregate output growth are mixed, with some relationships being unidirectional, bidirectional or neither, creating ambiguity for policy developments. It goes without saying that the term 'Granger causality', as Wong et al. (2008) put it, is not a true causality concept but a statistical tool which, in principle, concerns only the predictability of time-series variables.

AN OVERVIEW ON THE LONG-RUN RELATIONSHIP BETWEEN GDP AND CONSTRUCTION INDUSTRY-RELATED INDICATORS

Figure 6.1 presents the evolution of GDP (constant 2015 USD) in the world and major regions of the world for the period 1970–2018. The first remarkable feature of note is that the evolution pattern of GDP in all major regions of the world is one of an increasing trend throughout the period of analysis. The world GDP increased from US$18.1 trillion in 1970 to US$81.9 trillion in 2018, an annual average rate of increase 3.15 per cent, compared to an annual average rate of increase of the world population of 1.53 per cent (UN, 2019b). The data also show that there was a spectacular catching-up process on the part of Asia, which became the largest economic bloc in 2008. Asia's GDP increased by 960 per cent in the period 1970 to 2018, compared to 394 per cent, 315 per cent, 311 percent, 275 per cent and 180 per cent for Africa, Latin America and the Caribbean, Oceania, Northern America and Europe, respectively, during the same period (UN, 2020c). China, which is the most populous country in the world and has experienced a remarkable growth in GDP from 1990 onwards, had a significant impact in this growth process.

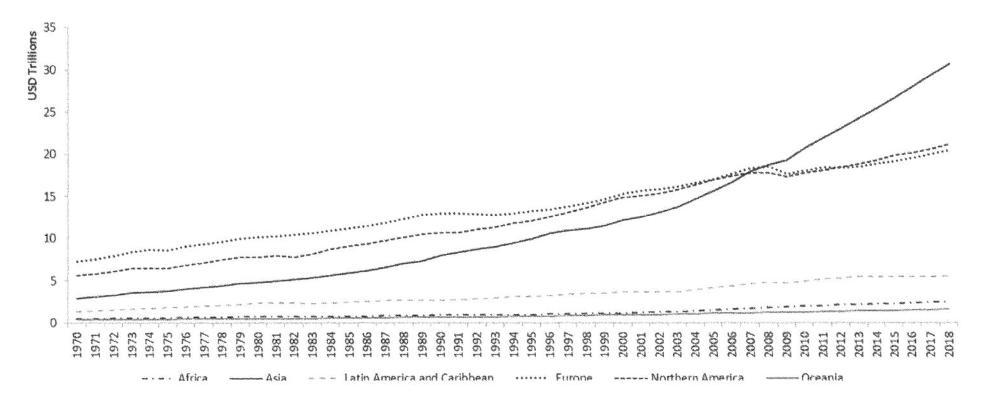

Note: NPISH: non-profit institutions serving households.
Source: Adapted from United Nations (2020c).

Figure 6.1 *Gross domestic product by region and subregion (constant 2015 prices)*

Table 6.1 presents the weights and rates of growth for GDP, CVA and GFCF for major economic groupings, sub-Saharan Africa (SSA) and the two economically dominant powers of the world, the US and China, in the period 1970 to 2018. It is shown that in 2015 the developed economies represented 58 per cent of the world output, developing economies 39.3 per cent, and transition economies 2.3 per cent. SSA, which represented about 67 per cent of the African economy in 2015, accounted for 2.1 per cent of the world GDP despite representing about 13 per cent of the world population (UN, 2019a). Combined together, the US (24.4 per cent) and China (14.7 per cent) represented 39.1 per cent of the world GDP in 2015. The evolution of GFCF shows a remarkable uniformity across the period of analysis, which is in keeping with the findings of previous works (Lopes, 2009; Ruddock and Lopes, 2006). From a close examination of the data in Table 6.1(a) and (b), the evolution of GDP is similar to that of GFCF in all economic groupings, as is clearly illustrated in Figure 6.2. The proportion of GFCF was in the range of 22–23 per cent of GDP in the period 1970 to 2003, and from then on it experienced a slightly increasing trend, reaching 25.2 per cent of GDP in 2015. Again, the economically dominant countries are the US and China, but the ranking order changes. China, which experienced a spectacular growth in this indicator from 1980 onwards, accounted for 26 of the total for the world in 2015.

Table 6.1(c) presents the growth rates of CVA in the major economic groupings – SSA, US and China – in the period 1970 to 2018. It also presents the economic groupings and the countries' weights in the total for the world in 2015. The world CVA represented 5.6 per cent of GDP (more precisely, of total value added) in 2015. The world total value added was US$71.4 trillion in 2015 (2015 constant prices), which corresponds to a CVA of US$4.03 trillion. In 2014, the proportion of CVA in total GFCF in construction was, on average, 41 per cent in the European Union (Gloser et al., 2017). Considering that minor maintenance and repairs are not included in GFCF in construction, and assuming as a rough measure that CVA represents 45 per cent of GFCF in construction, the construction expenditure in the world amounted to US$8.96 trillion in 2015. This is higher than the GDP of any country except the US and China. As in the case of GFCF, the developing economies' weight for CVA in the total for the world is higher than that for GDP. The share of construction in total value added, in 2015, was 4.9

Table 6.1 *Weights and real growth rates of (a) gross domestic product, (b) gross fixed capital formation and (c) construction value added*

Major area, subregion and country	Base year weights as percentage of total for world in 2015	Base year share as percentage of total GDP in 2015	Average rates of growth (%)				Annual rate of change (%)					
			1970–1979	1980–1989	1990–1999	2000–2010	2013	2014	2015	2016	2017	2018
(a) Weights and real growth rates of gross domestic product												
World	100.0		4.1	3.1	2.8	2.8	2.7	3.0	3.0	2.7	3.3	3.1
Developed economies	58.0		3.6	3.1	2.5	1.4	1.3	2.0	2.4	1.7	2.4	2.3
Developing economies	39.3		5.8	3.2	5.0	5.8	5.0	4.7	4.3	4.2	4.6	4.3
Transition economies	2.7		4.9	3.9	-5.4	5.3	2.5	1.1	-1.4	1.1	2.3	2.8
Sub-Saharan Africa	2.1		3.4	1.5	1.7	5.8	5.4	5.2	3.0	1.2	2.7	3.0
US	24.4		3.5	3.5	3.4	1.6	1.8	2.5	2.9	1.6	2.4	2.9
China	14.7		6.0	9.9	10.6	10.5	7.8	7.3	6.9	6.7	6.8	6.6
(b) Weights and real growth rates of gross fixed capital formation												
World	100.0	25.2	4.2	2.8	2.7	3.4	3.7	3.8	3.2	2.7	3.9	3.4
Developed economies	47.9	20.8	3.2	3.6	2.8	0.2	1.6	3.5	2.9	1.9	3.5	2.8
Developing economies	49.8	32.0	8.4	0.9	6.3	8.7	5.8	4.4	4.0	3.5	4.2	4.0
Transition economies	2.3	21.5	3.5	1.5	-16.2	9.3	2.6	-0.9	-6.9	1.6	6.0	4.6
Sub-Saharan Africa	1.8	21.6	5.7	-4.7	3.0	5.7	4.1	6.5	2.6	-1.8	1.1	-0.4
US	19.6	20.4	4.4	3.9	5.4	-0.2	3.6	5.1	3.2	1.9	3.7	4.1
China	26.0	44.0	7.5	6.8	14.7	14.2	9.3	6.8	6.7	6.8	4.4	4.8
(c) Weights and real growth rates of construction value added												
World	100.0	5.6	2.1	1.9	0.6	1.5	3.6	3.5	2.9	3.0	3.3	3.0
Developed economies	49.9	4.9	1.0	1.8	0.4	-1.3	1.4	2.0	2.4	1.9	3.3	2.6
Developing economies	46.8	6.7	7.5	1.0	4.2	7.2	6.5	5.6	3.7	4.3	3.4	3.3
Transition economies	3.2	7.1	6.3	5.6	-13.0	7.8	0.5	-0.9	-0.5	1.1	1.5	3.6
Sub-Saharan Africa	2.1	5.7	3.0	-1.7	2.5	10.4	9.7	5.7	4.6	3.9	4.7	3.4
US	17.3	3.8	0.3	2.1	2.2	-3.3	2.5	1.9	4.7	3.4	2.3	2.2
China	18.6	6.8	4.4	9.6	10.9	12.6	9.7	9.1	6.8	7.2	3.5	4.5

Source: United Nations (2020b).

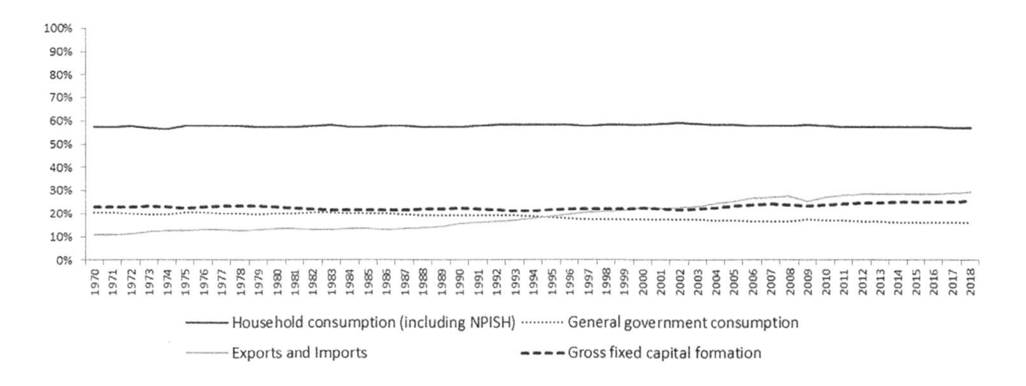

Source: Adapted from United Nations (2020c).

Figure 6.2 Share in nominal world GDP by type of expenditure

per cent for the developed economies, 6.7 percent for the developing economies and 7.1 per cent for the transition economies.

A particular feature of note from Table 6.1(b) and (c) are the figures for the share of CVA in total value added and the share of GFCF in GDP, in 2015, for China and the developing economies as a whole. Whereas for the former indicator the figures for China (6.8 per cent) and developing economies as a whole (6.7 per cent) are practically the same, the figures for the latter are markedly different: 44 per cent and 32 per cent for, respectively, China and the developing economies as a whole. Considering that the proportion of CVA in GFCF is about 22.5 per cent (world average, for the period 1970–2018), it appears that the figures for CVA might be underreported in China's national accounts statistics. This issue warrants a fresh look in further research. Table 6.1(c) also indicates that the growth rates of CVA of the world as a whole as well as of the economic groupings were, in general, lower than those of GDP and GFCF throughout the period of analysis. Data constructed from UN (2020c) show that the world CVA increased in volume by 133 per cent (2015 constant prices) in the period 1970 to 2018, compared to 402 per cent and 303 per cent for GFCF and GDP, respectively. This increase in construction volume also holds generally for individual countries and for all the economic groupings. In contrast, as can be constructed from the figures presented in Table 6.1, there was a markedly decreasing trend of the share of CVA in total value added throughout the period of analysis.

For the world as a whole, the proportion of CVA in total value added (measured at 2015 constant prices) was 10.4 per cent in 1970, 8.7 per cent in 1980, 7.8 per cent in 1990, 6.4 per cent in 2000 and, on average, 5.5 per cent in the period 2010–2018. These figures are in line with the proposition (Strassman, 1970; Bon, 1992) that the share of construction in GDP tends to increase in the first stages of development, to stabilize in the middle-income range, and to decrease in advanced industrial countries; and this pattern also holds for individual countries across time as well. It is worth noting that in Bon's (1992) terminology, the less developed countries (LDCs), newly industrialized countries (NICs) and advanced industrial countries (AICs) correspond, respectively, to the World Bank's World Development Indicators: LICs and LMICs, UMICs and HICs. However, the figures on CVA presented here do not corroborate Bon's (1992) assumption that the proportion of construction will ultimately decrease not only relatively but also absolutely, a fact that has been noted by several writers (Carassus et al., 2006; Ruddock and Lopes, 2006; Choy, 2011).

A particular feature worthy of note, as suggested in the figures presented in Table 6.1, is the role played by China in the evolution pattern of construction and construction-related industries, particularly since the 1990s. Data drawn from UN (2020c) indicate that China has ranked first in the world since 2014 (constant 2015 prices) when the construction output is measured using the production approach (valued added). In terms of GFCF in construction, China ranks first in the world, at least, since 2011, according to the data presented in the International Comparison Program (ICP) 2017 (World Bank, 2020). Measured in US dollars (exchange rate-based), GFCF in construction in China reached US$2178.4 billion in 2011, compared to US$1338.4 billion for the US in the same year. China's influence has also been felt at the international level, particularly in terms of construction services, export of building materials-related products and funding of investment projects. Export of construction services from China reached US$12 250.5 million in 2017, or 14.6 per cent of the total for the world (UN, 2019b). Regarding the export of building materials-related products and construction plant and equipment, China's performance in the world market has been spectacular. Table 6.2 presents the value of export of building materials-related products and civil engineering equipment from China in 2018.

Table 6.2 Value of export of construction-related products from China, 2018

Products	US$ million	World share (%)
Floor coverings	2 997.5	18.1
Lime, cement and fabricated construction materials (except glass and clay)	6 243.2	21.0
Clay construction materials and refractory construction materials	7 248.6	26.1
Iron and steel bars, rods, angles, shapes and sections	11 008.9	13.5
Manufactures of base metal, not elsewhere specified (nes)	32 531.4	18.0
Civil engineering and contractors' plant and equipment; parts thereof	13 545.2	13.2
Sanitary, plumbing and heating fixtures and fittings, nes	6 196.7	30.1

Source: UN (2019b).

With regard to sub-Saharan Africa, the share of construction in total value added of 5.7 per cent might signal room for further construction investment, when it is compared to that of developing economies as a whole (6.7 per cent). This observation is in keeping with the results of a recent study on the infrastructure investment needs for African countries (Global Infrastructure Hub, 2018) which showed that Africa's investment gap (investment needs minus current investment trends) for the period up to 2040 is 28 per cent of investment needs. This investment gap is forecast to widen further to 43 per cent, if investment needs include the Sustainable Development Goals. A question that merits further investigation is whether the low-income economies in sub-Saharan Africa will account for the great majority of this additional construction investment. Or whether a specific segment (such as affordable housing) of the construction market in certain middle-income countries in sub-Saharan Africa merits special attention from their governments.

MODELLING THE RELATIONSHIP BETWEEN CONSTRUCTION AND GDP PER CAPITA IN THE MIDDLE-INCOME COUNTRIES OF SSA

Data

Tables 6.3a and 6.3b present GNI per capita for the benchmark year (2010), and GDP per capita and the share of construction in GDP for the years 2000, 2010 and 2018 for Group 1 and Group 2, respectively. It is evident that all the countries analysed had middle-income category status in 2010 except Equatorial Guinea, which was by then considered a high-income economy.

Table 6.3a GNI per capita, GDP per capita and CVA as a share of GDP in Group 1

Country	GNI p.c. in 2010 (USD)	GDP p.c. (USD 2015 constant prices)			CVA as a share of GDP (%)		
	2010	2000	2010	2018	2000	2010	2018
Angola	3 960	2 445	3 989	3 625	0.072	0.104	0.135
Cabo Verde	3 160	2 084	3 009	3 349	0.10	0.101	0.091
Eswatini	2 600	2 476	3 310	3 791	0.041	0.032	0.031
Lesotho	1 080	688	1 022	1 214	0.085	0.053	0.033
Mauritius	7 740	5 542	7 833	10 322	0.05	0.062	0.041
S. Tome & P.	1 270	1 108	1 384	1 623	0.062	0.059	0.062
Seychelles	9 480	10 914	11 982	16 683	0.035	0.051	0.029
S. Africa	6 100	4 503	5 558	5 636	0.023	0.036	0.036
Zambia	1 070	759	1 192	1 344	0.078	0.104	0.113

Table 6.3b GNI per capita, GDP per capita and CVA as a share of GDP in Group 2

Country	GNI p.c.(curr. USD)	GDP p.c. (USD 2015 constant prices)			CVA as a share of GDP		
	2010	2000	2010	2018	2000	2010	2018
Botswana	6 980	4 654	5 747	7 173	0.048	0.062	0.066
Cameroon	1 160	1 055	1 184	1 382	0.013	0.044	0.054
Congo R.	3 280	1 942	2 521	2 012	0.046	0.138	0.138
Cote d'Ivoire	1 650	1 322	1 182	1 651	0.022	0.031	0.041
Djibuti	1 280	1 102	1 745	3 027	0.070	0.080	0.041
E. Guinea	14 680	5 270	13 828	8 488	0.039	0.10	0.041
Gabon	7 760	8 120	6 858	7 044	0.022	0.047	0.04
Ghana	1 600	1 044	1 411	1 964	0.045	0.074	0.082
Mauritania	1 060	1 049	1 350	1 614	0.015	0.040	0.057
Namibia	4 650	3 149	4 179	4 767	0.017	0.026	0.025
Nigeria	1 180	1 474	2 442	2 553	0.024	0.027	0.035
Senegal	1 050	997	1 151	1 363	0.015	0.018	0.026

Source: World Bank (2012) for GNI p.c.; United Nations (2020c) for GDP p.c.; author's calculations based on data from United Nations (2020c) for CVA.

It can also be seen that in general the countries, irrespective of their grouping, experienced sustained economic growth (measured by GDP per capita at 2015 constant prices) in the period

2000–2018. The exceptions from this development pattern, and for the period 2010–2018, are Angola (in Group 1); and Congo Republic and Equatorial Guinea (in Group 2), which are, with Nigeria, the top oil-exporting countries in sub-Saharan Africa. The high volatility in the market price of oil tends to affect the economic performance of oil income-reliant countries, particularly those with weak industrial structure. Nonetheless, the general economic performance of the countries in this period contrasted with the development pattern in most countries in sub-Saharan Africa in the 1980s and early 1990s, as can be observed in Table 6.1(a).

The Model

The model is tested using a statistical method for the equality of two correlations (Hogg et al., 2015).

Let $i = 1, 2$ corresponding to the two groups whose correlations are being compared. Sample correlations R_{ij} were observed based on n_{ij} observations for $i = 1, 2$ and $j = 1, \ldots, n_i, n_1$ is 9 and n_2 is 12 so R_{1_1} ..... R_{1_9} and $R_{2_1} \ldots R_{2_{12}}$ can be observed, and n_{ij} is 19 (that is, time series data for the period 2000–2018).

For each sample correlation it is necessary to evaluate:

$$W_{ij} = \frac{1}{2} \ln \frac{1 + R_{ij}}{1 - R_{ij}} \tag{6.1}$$

And let:

$$W_i = \frac{1}{n_i} \sum_{j=1}^{n_i} W_{ij} \tag{6.2}$$

be the averages of W_{ij} in the two groups.

Also, define:

$$v = \frac{1}{n_1^2} \times \sum_{j=1}^{n_1} \frac{1}{n_{1j-3}} + \frac{1}{n_2^2} \times \sum_{j=1}^{n_2} \frac{1}{n_{2j-3}} \tag{6.3}$$

Then, a $100_{\alpha\%}$ test of the null hypothesis $H_0 : \rho_1 = \rho_2$ against the alternative hypothesis $H_1 : \rho_1 \angle \rho_2$ is obtained by comparing the test statistic:

$$z = \frac{W_1 - W_2}{\sqrt{v}} \tag{6.4}$$

with the lower N (0, 1) critical value - z_α e.g. for a 5% test, $z_\alpha = 1.645$.

Analysis

In this analysis, the hypothesis $H_0 : \rho_1 = \rho_2$ is tested against the alternative hypothesis $H_1 : \rho_1 \angle \rho_2$ at an α = 0.05 significance level. The groups corresponding to these correlation coefficients (Groups 1 and 2) are defined above. The variables used in this model are $ICVA_{2015}$ (the share of construction value added in GDP at 2015 constant prices) and $GDPpc_{2015}$ (GDP per capita also measured in US$, at 2015 constant prices).

The sample correlations R_{1j} and R_{2j} are based on n_{1j} and n_{2j} observations (time series data for the period 2000–2018) for Groups 1 and 2, respectively. As referred to n_1 is 9 and n_2 is 12 and n_{1j} and n_{2j} are 19.

Thus, for Group 1:

$$W_{1j} = \frac{1}{2}\ln\frac{1+R_{1j}}{1-R_{1j}} \qquad W_1 = \frac{1}{n_1}\sum_{j=1}^{n_1} W_{1j}$$

Angola: $R_{1_1} = 0.7899$

Cape Verde: $R_{1_2} = -0.21948$

Eswatini: $R_{1_3} = -0.74539$

Lesotho: $R_4 = -0.26211$

Mauritius: $R_{1_5} = -0.55201$

São Tome & Principe: $R_{1_6} = 0.3464$

Zambia: $R_7 = 0.11945$

Seychelles: $R_{1_8} = 0.4195$

South Africa: $R_{1_9} = 0.9605$

$W_1 = 0.08272$

For Group 2:

$$W_{2j} = \frac{1}{2}\ln\frac{1+R_{2j}}{1-R_{2j}} \qquad W_2 = \frac{1}{n_2}\sum_{j=1}^{n_2} W_{2j}$$

Botswana: $R_{2_1} = 0.58621$

Cameroon: $R_{2_2} = 0.84249$

Congo: $R_{2_3} = 0.78046$

Côte d'Ivoire: $R_{25} = 0.4214$

Equatorial Guinea: $R_{2_6} = 0.67502$

Gabon: $R_{2_7} = 0.73498$

Ghana: $R_{2_8} = 0.90549$

Mauritania: $R_{2_9} = 0.90719$

Nigeria: $R_{2_{11}} = 0.72073$

Senegal: $R_{2_{12}} = 0.94992$

$W_2 = 1.156823$

under $H_0 : \rho_1 = \rho_2$, the null hypothesis,

$$Z = \frac{W_1 - W_2}{\sqrt{v}} \sim N(0,1)$$

and at an α= 0.05 significance level,

$$Z = \frac{0.08272 - 1.156823}{0.113948} = -9.426 \angle -1.645.$$

Therefore, the null hypothesis H_0 is rejected in favour of the alternative hypothesis H_1.

The evolution of the construction industry (measured as a share of CVA in GDP) illustrated in Figure 6.3 is in line with the results of the above statistical test that shows that the pattern of the construction industry in the two groups presents distinct developments from the period between 2000 and 2018. Figure 6.3 indicates that in the countries in which the capital-output ratio was equal or greater than four in the period 1990–1999 (Group 1), the share of CVA in GDP (both indicators measured at 2015 constant prices) remained, for the group average, practically constant during the period 2000–2018 (about 6.3 per cent of GDP). However, a slight trend of decline is observed in the late years of the period, falling to 6.3 per cent of GDP in 2018 from 6.9 per cent in 2009. The construction volume increased, in general, absolutely but not relatively. This pattern also holds generally for individual countries, disregarding annual fluctuations that characterize the construction industry activity.

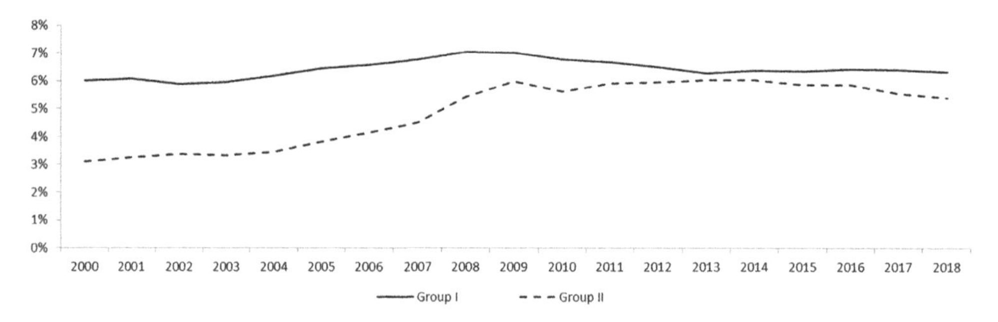

Source: Author's calculations based on data from United Nations (2020c).

Figure 6.3 Evolution of the Share of CVA in GDP in the middle-income countries of SSA

The other side of the picture shows that in the group of countries in which capital–output ratio was less than four in the period 1990–1999 (Group 2), the share of CVA in GDP (2015 ICVA) increased remarkably during the period of analysis. It increased from about 3.1 per cent of GDP in 2000 to 6.0 per cent in 2013. Again, a trend of decline is observed in the latter years of the period, reaching 5.5 per cent in 2017. The construction volume increased relatively, not only absolutely, accompanying the evolution of the general economy.

The results of this study are in keeping with the findings of a previous work (Lopes et al., 2002) that there is a minimum required level of construction investment (a share of CVA in GDP of around 5 per cent) for an efficient functioning of the construction industry in sub-Saharan Africa. Regarding this critical value, many studies of the economic history (see, for example, Kuznets, 1961; Maddison, 1987) of today's advanced industrial countries and some upper-middle-income countries, looking at as far back as 1870, have referred to a value of around 20 per cent as the share of capital formation in a country's domestic expenditure.

More recently (after the Second World War), when international bodies, particularly the United Nations, started publishing data on the components of GDP, there has been a remarkable uniformity across countries, with 20–25 per cent being the average contribution of capital formation to a country's domestic expenditure. As the proportion of construction value added in total GFCF was, on average, 22.5 per cent in the period between 2000 and 2018, it appears that a value around 5 per cent as the contribution of CVA to the national aggregate is reasonably constructed. This figure for CVA depends on the year chosen as the basis, and whether this indicator is compounded at current or constant prices. It is also worth noting that the 5 percent value is consistent with Syrquin and Chenery's (1989) value pertaining to the construction sector, in their study of norms for structural change and per capita income.

CONCLUSIONS AND FURTHER DEVELOPMENT

This chapter has presented a review of the concepts and features related to the relationship between infrastructure (construction) investment and economic growth and development. It has also discussed the methodological issues related to the data and indicators used for analysing the construction and economic growth nexus.

The findings from the panel data approach used in this analysis suggest that the development patterns of the construction industry in the middle-income economies in sub-Saharan Africa follow rather than lead economic growth. These findings lend credibility to the proposition that as a country reaches an advanced stage, construction output tends to decrease relatively but not absolutely. The type of activity changes as countries develop, and countries with an established building stock find that construction activity becomes more oriented towards repair and maintenance of their infrastructure and built and urban environment (Lopes et al., 2019). Built assets are a major component of a country's produced capital, particularly in developing countries. The total wealth of a country also comprises natural capital and intangible capital, the latter consisting of human and social capital (Lange et al., 2018). Depending on the circumstances, the relative importance of each type of capital needs to be considered in any developing country's development strategy. Accordingly, when addressing the link between the construction industry and the economic and social aspects of the sustainable development goals, developing countries should focus on the planning and development of investment projects that have real impact and will be sustainable over the long term.

The statistical test for the equality of two correlations developed in the study is highly significant: the share of construction in GDP in the two groups of middle-income countries in sub-Saharan Africa during the period 2000 to 2018 are markedly different. However, it is well known that statistical association does not reveal causation. What is offered here is not a prediction but a prospect of the development pattern of the construction industry of the middle-income countries in sub-Saharan Africa in the near future. Nonetheless, the results of this study suggest that the size of the construction sector is not just a function of per capita output, but is also related to broader socio-economic trends, namely urbanization, industrialization and creation of basic infrastructure. Moreover, the level of the capital stock captures some dimensions of these development dynamics.

Some suggestions for further research are now put forward:

- Studies should use the level of net capital stock as a control variable in the construction–economic growth regression models.
- It would be useful to expand the sample of the study by including the middle-income countries of other regions of the world. This would shed more light on the current comprehension of the role the construction sector plays in the process of economic growth and development.
- There is a need for studies on the built capital stock of individual countries in less developed countries. These accounting exercises based on national accounts data would be a useful complement of those derived from the Penn World Tables.

REFERENCES

Anaman, K.A., and Osei-Amponsah, C. 2007. Analysis of Causality Links between the Growth of the Construction Industry and the Growth of the Macroeconomy in Ghana. *Construction Management and Economics*, 25, 951–961.

Ansar, A., Flyvbjerg, B., Budzier, A., and Lunn, D. 2016. Does Infrastructure Lead to Economic Growth or Economic Fragility? Evidence from China. *Oxford Review of Economic Policy*, 32(3), 360–390. Doi: 10.10.93/oxrep/grw022.

Banerjee, A., Duflo, E., and Qian, N. 2020. On the Road: Access to Transportation Infrastructure and Economic Growth in China. *Journal of Development Economics*, 145, 102442. https://doi.org/10.1016/j.jdeveco.2020.102442.

Bon, R. 1990. The World Building Market, 1970–1985. *Proceedings of the CIB W65 International Symposium of Building Economics and Construction Management*, Vol. 1, pp. 16–47. Sydney, 14–21 March.

Bon, R. 1992. The Future of International Construction: Secular Patterns of Growth and Decline. *Habitat International*, 16(3), 119–128.

Bon, R., and Pietroforte, R. 1990. Historical Comparison of Construction Sectors in the United States, Japan, Italy and Finland Using Input–Output Tables. *Construction Management and Economics*, 8, 233–247.

Calderon, C., Moral-Benito, E., and Servén, L. 2015. Is infrastructure Capital Productive? A Dynamic Heterogeneous Approach. *Journal of Applied Econometrics*, 30(2), 177–198.

Carassus, J., Andersson, N., Kaklauskas, A., Lopes, J., Manseau, A., et al. 2006. Moving from Products to Services: A Built Environment Cluster Framework. *International Journal of Strategic Property Management*, 10(3), 169–181. doi: 10.1081/64875CX.2006.9637551.

Caselli, F. 2005. Accounting for Cross-Country Income Differences. In Philippe Aghion and Steven Durlauf (ed.), *Handbook of Economic Growth*, Vol. 1, pp. 679–741. Elsevier, Amsterdam.

Chen, J., and Zhu, A. 2008. The Relationship between Housing Investment and Economic Growth in China. A Panel Analysis Using Quarterly Provincial Data. Working Paper 2008: 17, Department of Economics, Uppsala University.

Chiang, D T.H., and Low, S.P. 2011. Role of Construction in Economic Development. Review of the Key Concepts of the Part 40 Years. *Habitat International*, 35(1), 118–125. doi: 10.1016/j.habitatint.2010.06.

Choy, C.F. 2011. Revisiting the 'Bon Curve'. *Construction Management and Economics*, 29(7), 695–712. doi: 10.1081/01446193.2011.578959.

Dasgupta, B., Lall, S.V., and Lozano-Gracia, N. 2014. *Urbanization and Housing Investment*. Social, Urban, Rural and Resilience Global Practice Group, World Bank.

de Long, J., and Summers, L. 1991. Equipment Investment and Economic Growth. *Quarterly Journal of Economics*, 106, 445–502.

Derbyshire, J., Gardiner, B., and Waights, S. 2011. Estimating the Capital Stock for the NUTS 2 Regions of the EU-27. Working Paper No. 1/2011, European Commission, Brussels.

de Vries, G., Timmer, M., and de Vries, K. 2014. Structural Transformation in Africa: Static Gains and Dynamic Losses. IARIW 33rd General Conference, Rotterdam, 24–30 August.
Drewer, S. 1980. Construction and Development: A New Perspective. *Habitat International*, 5(3–4), 395–428.
Donaldson, D., and Hornbeck, R. 2016. Railroads and American Economic Growth: 'A Market Access' Approach. *Quarterly Journal of Economics*, 131(May), 799–858. https://doi.org/10.1093/qje/qjw002.
Edmonds, G.A. 1979. The Construction Industry in Developing Countries. *International Labour Review*, 118(3), 355–369.
Edmonds, G.A., and Miles, D.W.J. 1984. *Foundations for Change: Aspects of the Construction Industry in Developing Countries*. Intermediate Technology Publications, London.
Erol, I., and Unal, U. 2015. Role of the Construction Sector in Economic Growth: New Evidence from Turkey. MRPA Paper Nº 68263. https://mrpa.ub.uni-munchen.de/68263.
EU-KLEMS. 2017. *Growth and Productivity Accounts 2017 Release: Descriptions of the Methodologies and General Notes*, September (Revised, July 2018). EU-KLEMS.
Eurostat. 2013. *European System of Accounts 2010*. https://ec.europa.eu/eurostat/documents/3859598/5925693/KS-02-13-269-EN.PDF/44cd9d01-bc64-40e5-bd40-d17df0c69334 (accessed 22 June 2020).
Eurostat. Various years. *Annual National Accounts*. http://ec.europa.eu/eurostat/data/database (accessed 22 June 2020).
Feenstra, R., Inklaar, R., and Timmer, M. 2015. The Next Generation of Penn World Tables, *American Economic Review*, 105(10), 3150–3182.
Ganesan, S. 2000. *Employment Technology and Construction Development, With Cases Studies in Asia and China*. Routledge, London.
Girardi, D. and Mura, A. 2014. The Construction Development Curve: Evidence from an International Dataset. *IUP Journal of Applied Economics*, 13(3), 7–26.
Global Infrastructure Hub. 2018. *Global Infrastructure Outlook: Investment Needs in the Compact with Africa Countries*. A G20 Initiative, Global Infrastructure Hub – Oxford Economics.
Gloser, J., Baker, P., Giustozzi, L., Hanz-Weiss, D., Merkus, E., et al. 2017. The European Construction Value Chain. Performance, Challenges and Role in GVC. WIIW Research Report, May, Vienna Institute for Economic Studies.
Granger, C. 1969. Investigating Causal Relations by Economic Methods and Cross-spectral Methods. *Econometrica*, 37(3), 424–438.
Green, R.K. 1997. Follow the Leader: How Changes in Residential and Non-residential Investment Predict Changes in GDP. *Real Estate Economics*, 25(2), 253–270.
Hogg, V., Tanis E., and Zimmerman, D. 2015. *Probability and Statistical Inferences* (9th edition). Pearson, Hoboken, NJ.
Inklaar, R., and Timmer, M. 2013. Capital, Labour and TFP in PWT8.0. Groningen Growth and Development, Centre, July.
Khan, R.A. 2008. Role of Construction in Economic Growth: Empirical Evidence from Pakistan Economy. *First International Conference on Construction in Developing Countries ICCIDC–I)*, Karachi, Pakistan, 4–5 August, pp. 279–290.
Kodongo, O., and Ojah, K. 2016. Does Infrastructure Really Explain Economic Growth in Sub-Saharan Africa? ERSA Working Paper 653, Economic Research Southern Africa.
Kuznets, S. 1961. *Capital in the American Economy: Its Formation and Financing*. Princeton University Press. http://papers.nber.org/books/kuzn61-1.
Lange, GM., Wodon, Q., and Carey, K. (eds). 2018. *The Changing Wealth of Nations 2018: Building a Sustainable Future*. Washington, DC: World Bank. https:/doi:10.1596/978-1-4648-1046-6.
Lean, S.C. 2001. Empirical Tests to Discern Linkages between Construction and other Economic Sectors in Singapore. *Construction Management and Economics*, 13, 253–262.
Lopes, J. 2009. Investment in Construction and Economic Growth: A Long-term Perspective. In L. Ruddock (ed.), *Economics for the Built Environment*. Taylor & Francis, London and New York, pp. 94–112.
Lopes, J. 2012. Construction in the Economy and its Role in Socio-economic Development. In G. Ofori (ed.), *New Perspectives in Construction in Developing Countries*. Spon Press, London and New York, pp. 40–71.

Lopes, J., Nunes, A., and Balsa, C. 2011. The Long-Run Relationship between the Construction Sector and the National Economy in Cape Verde. *International Journal of Strategic Property Management*, 15(1), 48–59.

Lopes, J., Oliveira, R., and Abreu, M.I. 2019. Estimating the Built Environment Stock in Cape Verde. *Engineering, Construction and Architectural Management*. http://doi.org/10.1108/ECAM-07-2018-0290.

Lopes, J., Ruddock, L., and Ribeiro, F.L. 2002. Investment in Construction and Economic Growth in Developing Countries. *Building Research and Information*, 30(3), 152–159.

Maddison, A. 1987. Growth and Slowdown in Advanced Capitalist Economies. *Journal of Economic Literature*, 25, 649–698.

OECD (2009). *Measuring Capital – OECD Manual* (2nd edn). OECD, Paris.

OECD (2013). Understanding the Value of Transport Infrastructure – Guidelines for Macro-level Measurement of Spending and Assets. International Transport, Forum, OECD.

Ofori, G. 1988. Construction Industry and Economic Growth in Singapore. *Construction Management and Economics*, 6(1), 57–70.

Ogunbiyi, M., Olawale, S.O., and Oyaromade, R. 2017. The Relationship Between Construction Sector and Economic Growth in Nigeria: 1981–2013. Department of Civil Engineering, College of Science, Engineering and Technology, Osun State University, Osogbo, Osun State, Nigeria.

Ozkan, F., Ozkan, O., and Gunduz, M., 2012. Causal Relationship between Investment Policy and Economic Growth in Turkey. *Technological Forecasting and Social Change*, 79, 362–370.

Pietroforte, R., and Gregory, T. 2003. An Input–Output Analysis of the Construction Sector in Highly Developed Economies. *Construction Management and Economics*, 21, 319–327.

Rodrik, D. 2015. Premature Deindustrialization. Economics Working Papers, No. 107, School of Social Science, Institute for Advanced Study, Princeton, NJ.

Rostow, W.W. 1963. The Leading Sectors and the Take-off. In W.W. Rostow (ed.), *The Economics of Take-off into Sustained Growth. Proceedings of a Conference Held by the International Economic Association*. Macmillan Press, London.

Ruddock, L., and Lopes, J. 2006. The Construction Sector and Economic development: The 'Bon Curve'. *Construction Management and Economics*, 24(7), 717–723.

Ruddock, L., and Ruddock, S. 2019. Wealth Measurement and the Role of Built Assets: An Empirical Comparison. *Engineering Construction and Architectural Management*. http://doi.org/10.1108/ECAM-05-2018-0190.

Sahoo, P., and Dash, R.K. 2009. Infrastructure Investment and Economic Growth in India. *Journal of the Asia Pacific Economy*, 14(4), 351–365. https://doi.org/10.1080/13547860903168340.

Solow, A. 1956. A Contribution to the Theory of Economic Growth. *Quarterly Journal of Economics*, 70, 65–94.

Stern, N. 1991. The Determinants of Growth. *Economic Journal*, 101(404), 122–133. http://www.jstor.org/stable/2233847.

Stiglitz, J., Sen, A., and Fitoussi, J.P. 2010. *Mismeasuring Our Lives: Why the GDP Doesn't Add Up*. New Press, New York.

Strassman, P. 1970. The Construction Sector in Economic Development. *Scottish Journal of Political Economy*, 17(3), 390–410.

Sun, Y., Mitra, P., and Simone, A. 2013. The Driving Force between Boom and Bust in Construction Sector in Europe. IMF Working Paper, WP/13/81.

Syrquin, M., and Chenery, H. 1989. Patterns of Development, 1950–1980. World Bank Discussion Paper No. 41.

Tse, R.Y.C., and Ganesan, S. 1997. Causal relationship between Construction Flows and GDP: Evidence from Hong Kong, *Construction Management and Economics*, 15(4), 371–376.

Turin, D. 1973. *The Construction Industry: Its Economic Significance and its Role Development*. UCERG, London.

United Nations (UN). 2008. *International Standard Classification of All Economic Activities (ISIC), Rev. 4*. Department of Economic and Social Affairs, Statistical Division, New York.

United Nations (UN). 2009. *The System of National Accounts (SNS) 2008*. United Nations Statistical Division, New York.

United Nations (UN). 2019a. *World Population Prospects 2019, Online Edition. Rev. 1*, Department of Economic and Social Affairs, Population Division New York.
United Nations (UN). 2019b. *2018 International Trade Statistics Yearbook, Vol. 2. Trade by Product.* United Nations Statistical Division, New York.
United Nations (UN). 2020a. *World Economic Situation and Prospects.* United Nations Statistical Division, New York.
United Nations (UN). 2020b. *Analysis of the Main Aggregates.* United Nations Statistical Division, New York.
United Nations (UN). 2020c. *Main Aggregates and Detailed Tables.* United Nations Statistical Division, New York.
United Nations Development Programme. 1990–. *Human Development Report.* United Nations Development Programme, New York.
Wells, J. 1986. *The Construction Industry in Developing Countries: Alternative Strategies for Development*. Croom Helm, London.
Wilhelmsson, M., and Wigren, R. (2011). The Robustness of the Causal and Economic Relationship between Construction Flows and Economic Growth: Evidence from Western Europe. *Applied Economics*, 43(7), 891–900, DOI: 10.1080/00036840802600020.
Wong, J., Chiang, Y., and Nge, T. 2008. Construction and economic Development. The Case of Hong Kong. *Construction Management and Economics*, 26, 815–826.
World Bank. 1984. *The Construction Industry: Issues and Strategies in Developing Countries.* IBRD, World Bank, Washington, DC.
World Bank. 1994. *World Development Report 1994: Infrastructure for Development.* Oxford University Press, World Bank, Washington, DC.
World Bank. 2012. *World Development Indicators 2012.* World Bank, Washington, DC.
World Bank. 2020. Purchasing Power Parities and the Size of World Economies: Results from the 2017 International Comparison Program. Washington, DC, The World Bank. databank.worldbank.org/source/icp-2017.
Yiu, C.Y., Lu, X.H., Leung, M.Y., and Jin, W.X. 2004. A Longitudinal Analysis on the Relationship between Construction output and GDP in Hong Kong. *Construction Management and Economics*, 22(4), 339–345.

7. Construction project economics

Nii A. Ankrah and Emmanuel Manu

INTRODUCTION

The built environment is realised from the consumption of natural and man-made resources of limited availability. Therefore, efficiency in the consumption of these resources is critical. In addition, the use of these resources for one project results in denial to other projects, or purposes for which there is an opportunity cost (Hillebrandt, 2000). Thus, this creates competition between alternative desirable uses of the resources. In construction, to facilitate decision-making with regard to the best use of the limited means to achieve the most desirable ends, economic analysis is required in order to establish priorities among competing projects (ICE, 1976). This is fundamentally what construction project economics is about, and this chapter provides an overview of the main principles and techniques underpinning such economic analysis.

The chapter proceeds as follows. It first discusses the main concepts of economic analysis at the level of the construction project, their historical development, and current thinking regarding their application. It then interrogates the research in construction project economics, using a bibliometric analysis, to identify the emerging trends. This is followed by an exploration of the emerging trends in detail, and their implications for project economics. The chapter then considers further research on, and development of, the field of knowledge to support more informed and rational decision-making in relation to construction projects, as well as wider implications for mainstream economics.

ECONOMICS AT THE PROJECT LEVEL

Construction economics covers a broad spectrum of themes that broadly fit into the three categories of the industry level (including its interaction with the wider economy), the firm level, and project-level economics (Hillebrandt, 2000). At the industry level, the concern is with the construction market, and the underpinning research and resulting literature have shone a light on the demand and supply issues within the construction market, and how the industry fits into, or interacts with, the wider economy (Ofori, 1994). Research at this level has addressed labour markets, productivity, output, housing, transport and energy infrastructure, industrialisation and government policy (Bröchner, 2018). At the firm level, the focus has been on business cycles and organisational behaviour in the light of limited resources and the uncertainty of demand that is particularly prevalent in the construction industry. These themes align closely with mainstream economics thinking in the areas of transaction cost economics, behavioural economics, finance, development and industrial organisation as reflected in the economics journals (see for instance Bröchner, 2018).

The third, and arguably most important, component in this triad is project-level economics. The project level is the genesis of the entire construction market (Myers, 2017). This is most important because demand for construction always crystallises in a project which the supply side, represented by a complex web of firms in a supply chain, must respond to. The supply of construction and consultancy services to deliver a project is against a background of limited resources, and this is where firm-level economics comes to the fore to secure the efficient delivery of the project. Optimisation of resources at the project level, among other factors, can influence efficiency and profitability at firm level, which aggregates to industry-level productivity and performance in relation to other set parameters.

With the exception of housing, demand for construction is derived, meaning that the item to be constructed is not required for its own sake, but only to enable the production of goods and services required in the wider economy (Myers, 2017). This includes construction outputs such as factories, hotels, garages, shops and warehouses. The WEF (2016) report affirms this point as it describes the construction industry as a 'horizontal' industry with considerable interaction with other sectors providing the infrastructure through which value creation within the economy occurs. This implies that when demand for goods and services declines, as is often the case during periods of economic uncertainty, there can be a disproportionate fall in the demand for construction (Hillebrandt, 2000; Ribeirinho et al., 2020). An illustration of this volatility is the Spanish construction market, which as cited by WEF (2016) went into decline, losing over 50 per cent of its output, following the financial crisis between mid-2007 and early 2009. In contrast, the services sector stagnated but saw very little decline over the same period according to data from Trading Economics (2020). Such uncertainty elicits a range of responses by firms to secure flexibility and efficiency, and limit their risk exposure to the fluctuations in demand which fundamentally affects the composition and structure of the industry. WEF (2016) argues that firms with strong processes in place and the ability to adapt their business models to new market opportunities survive the turbulence, whilst those that are inflexible disappear. This shows the nature of the complex project, firm and industry-level interactions that exist within the context of the wider economy.

At the project level, the economics of construction relates to an understanding of factors driving construction cost and value. The efficiency of projects depends on both design choices and construction processes. Therefore, economics at this level has focused on design morphology and cost geometry, architectural and engineering design variables affecting the cost of projects. This has led to economic and functional analysis of the project as a whole or elements of a project, and an interrogation of alternative specifications and their cost implications. In addition to these cost drivers, various other organisational and macroeconomic factors also have an impact on cost, including interest rates, inflation, procurement practices, organisational overheads and location variables that influence resource costs (see for example Windapo et al., 2017). In a study of the South African construction market, Windapo et al. (2017) catalogued several determinants of these costs, categorised into resource, project, stakeholder requirement and macroeconomic factors.

On the other hand, value is a measure of the benefit derived from an intervention. Typically, this is quantified in monetary terms to allow comparison with costs as a basis for evaluating the business case for a project. Value, or more accurately 'market value', can be viewed as the comparison between a product and what someone is willing to give up in order to obtain it (Hendriks, 2005). Lockwood and Rutherford (1996), in a study of the determinants of value in industrial buildings in Texas, argued that market value is dependent on the 'physical characteristics of the property, national market conditions, regional market conditions, interest rates and location of the property' (p. 262). This means that market value is a price that will occur under certain conditions contingent upon the interaction of the forces of demand and supply (French, 2004; Hendriks, 2005). However, value is more than just monetary, and construction projects offer value in various ways, not all of which are reflected in the market value of the built items. This is particularly true for non-market developments such as churches or community centres. Therefore, economic (or perhaps more accurately, financial) analysis of projects has, for the most part, been found unreliable. This unreliability results not only from the definition and measurement of value, but also from the numerous assumptions made to facilitate the financial analysis of cost and value in an industry that is beset with uncertainty and risk. Evidence of this can be found in the large number of projects around the world that have cost more than expected, or failed to realise their projected benefits (Ribeirinho et al., 2020; Judy, 2015; Flyvbjerg et al., 2014; Ahiaga-Dagbui and Smith, 2014).

The tendency of construction projects to exceed forecast cost has been a long-standing challenge of the industry. Flyvbjerg et al. (2003, 2016) attribute the phenomenon to optimism of human judgement under conditions of insufficient information leading to underestimation of the costs, completion time and risks of planned actions, with a corresponding overestimation of the benefits of those same actions. Ahiaga-Dagbui and Smith (2014) present a broader range of systemic factors to explain the unreliability in project forecasting, in their quest to develop a system dynamics and causal mapping approach that clarifies the complex interactions which compromise the reliability of project appraisal. Factors that they identify (some of which are in common with Flyvbjerg et al., 2003) include: optimism bias, prospect theory, strategic misrepresentation and the Dunning–Kruger effect, which collectively affect the quality of forecasting; and risk and uncertainty, scope creep, managerial and technical difficulties, which collectively cause overruns of forecasts. The solutions offered by these authors for improving the economic evaluation of projects are reference class forecasting (Flyvbjerg et al., 2003) and artificial neural networks and data mining (Ahiaga-Dagbui and Smith, 2014). In Hong Kong, the Development Bureau set up the Project Cost Management Office in 2016 to drive industry-wide improvements in relation to the cost and project management of capital works projects.

From a project life cycle approach, costs begin to accrue even before the decision to build, or the selection of the site. Under this approach, ahead of land acquisition and project development, there is a need for analysis of public or private developers' investment perspectives, and a reliable appraisal of the viability of the project considering the return on investment (RoI) from a private sector perspective, or value for money (VfM) from a public sector proposition of the proposed project. Therefore, methodologies for undertaking such appraisals become critical considerations.

The body of knowledge supporting the requirement to ensure economic efficiency of construction projects has largely focused on showcasing tools and techniques that support decision-making throughout the life cycle of a project. A comparison by Goh (2008) of 41 construction economics articles published between 1983 and 2006 in the two top-ranking construction management journals, *Construction Management and Economics* and American Society of Civil Engineers' (ASCE) *Journal of Construction Engineering and Management*, identified a reliance on conventional techniques, which is emblematic of an industry that is reputed to be slow to change. Whilst such techniques may not respond fully to the appraisal needs of modern projects, it is positive to see the industry embracing methods forged in mainstream economics. Specifically, techniques such as cost–benefit analysis (CBA), development appraisal, project budgeting, cost analysis and planning, cost modelling, and whole-life costing have been the bedrock of construction project economics. They have supported the management of the economic problem of making optimal choices regarding the allocation and use of scarce resources such as land, labour, capital, equipment and time for construction projects. Whilst Goh (2008) concludes that there is a need to develop tools that incorporate intelligence in order to respond to increasingly complex problems, the importance of the traditional tools and techniques to the industry cannot be overstated. Therefore, there is a need to review their theoretical bases as well as any current thinking regarding their application. Ashworth and Perera (2015) provide a historical review of building economics that traces the advent of some of these approaches. Some of this history is covered in the discussion below.

Cost–Benefit Analysis

In the context of construction projects, CBA concerns the investment problem. It is a long-standing technique for determining the net benefits of a proposal relative to the alternative, including the status quo (Boardman et al., 2018). From a historical perspective, it is reported by Persky (2001) that this technique was pioneered in the United States (US) by the Army Corps of Engineers in the 1920s. As an economic appraisal technique, its roots lie in welfare economics (Persky, 2001; Nyborg, 2012; Broughel, 2020). Historically, its main application has been in the assessment of the impact of government's social intervention policy (see, for instance, Goldfarb, 1975; Persky, 2001). Today, its application is not only in this arena, but also in the environmental economics domain, which is where it offers great value to construction project economics by providing a lens through which to evaluate the environmental impacts of construction activities. In theory, CBA relies on the analysis of all the costs incurred, including costs to society, of an investment decision, versus the total benefits, including social benefits, accruing from the decision.

CBA typically finds application in evaluation of interventions such as policies, programmes, projects and regulations. It is about applying efficiency considerations to expenditure decisions (Musgrave, 1969). It responds to the RoI or VfM question. However, it is not always objective due to the intangibility of some of the benefits and externalities which, consequently, it fails to adequately account for (Weiss, 2000; Broughel, 2020). Indeed, it has been argued further that it suffers even more serious shortcomings beyond just the above, by attributing characteristics of individuals to society and wiping out the future with a social discount rate (Weiss, 2000; Broughel, 2020). The theory and practice of CBA has been widely covered in many authoritative sources such as Drèze and Stern (1987) and more recently Boardman et al. (2018). For example, Drèze and Stern (1987) provide a detailed theoretical account of the way CBA should proceed, deriving several rules to aid the appraisal of projects by governments. On the other hand, Boardman et al. (2018) offer one of several textbooks on CBA which seek to explain how to conduct CBA, or interpret and utilise the results of CBA. It offers practical guidance to CBA, over the more formal mathematical approaches available in sources such as Drèze and Stern (1987). In construction economics research, this technique has found application in the evaluation of various interventions (see, for example, Ikpe et al., 2012; Pitonak and Pepucha, 2016). No theoretical perspectives or rules were found in this review of the literature highlighting contributions of construction researchers to the refinement of the technique of CBA. Much of the research on this technique is in the economics literature. This is evidence of the application of mainstream economics ideas, concepts and tools to tackle construction problems. The challenges of the technique noted above by authors such as Musgrave (1969) and Drèze and Stern (1987) still remain to this day.

Development Appraisal

Development appraisal is designed to aid assessment of the financial viability of projects. The Royal Institution of Chartered Surveyors (RICS, 2019) defines it as a financial appraisal of a development that assesses the residual site value or the development profit. At the heart of development appraisal is the question of value and how this should be assessed. Value is the result of the comparison between a product and what someone is willing to give up in order to obtain it (Hendriks, 2005). The current practice of spatial development valuation is said to trace its roots to the works of classical economists such as David Ricardo and Adam Smith. Ricardo's thesis, which appears to stem from critical reading of *The Wealth of Nations*, is that 'value in use is utility' (Hollander, 1904). Together with scarcity and necessary expenditure of labour, this confers value in exchange (Hollander, 1904). These economists identified land as an important factor of production with an attendant initial acquisition cost and annual rent driven by open market value determined by an interaction between demand and supply which, in turn, also was determined by various factors such as location, land use planning, ancillary infrastructure and population. Thus, development appraisal uses a range of valuation techniques to measure outflows of a development and the inflows or market value of the completed project. A typical example is the residual approach/development method.

Traditionally, this approach to spatial development valuation is used for determining the value of development properties, that is, obsolete properties in city centres earmarked for refurbishment or redevelopment, or greenfield sites in the countryside earmarked for development (Pagourtzi et al., 2003; French, 2004; Shapiro et al., 2013). The rationale is to determine the latent value of the redevelopable properties or land. Since it is difficult to obtain evidence of sales or rents of similar properties, the method estimates the value of developable properties as the surplus of the market value for the proposed development after making allowance for all the cost items and the developer's profit. The estimate of the value is based on the highest and best use of land for which planning permission has been obtained. 'Highest and best use' is a key concept supporting land use and value decisions (Geltner et al., 2018; Dotzour et al., 1990). It underlies all considerations of the development potential of any piece of land and, consequently, the economic use of the land.

Another technique is the cost method, also known as the contractor's test, replacement cost and depreciated replacement cost methods (see Pagourtzi et al., 2003; French, 2004; Shapiro et al., 2013). It is based on the economic principle of substitution. The principle proposes that a prudent purchaser of a spatial development will not expend more than the cost of obtaining a development that provides equivalent utility through either purchase or construction (see Park and Park, 2004; Shapiro et al., 2013). The method is usually used for the valuation of spatial developments which rarely come to the market, that is, specialised properties such as church premises, town halls and community centres, schools and police stations, for which there is no evidence of sales to inform the determination of market value (Pagourtzi et al., 2003). The method can also be used for other developments such as residential, factory and warehouse properties, which do not earn income.

Some development appraisal approaches involve the application of discounted cash flows and net present value calculations to determine project viability.

Cost Planning

Cost planning is a technique that helps the designer to control total cost and make the best use of the money available (Browning, 1961). The Royal Institute of British Architects (RIBA, 2020) describes cost planning as the process that cost managers (commonly referred to as quantity surveyors or commercial managers) use to develop costs relating to various elements within a project at various stages through its planning phase. Ashworth and Perera (2015) describe cost planning as a budget distribution technique that is implemented during the design stages of a building project. It allows a balanced distribution of expenditure, implying more rational design. The level of precision and detail captured in the cost planning process depends on the stage or level within the project cycle when it is being used, and the amount of design information that is available to work with (Kirkham, 2014).

Formal cost planning was arguably the most significant development in cost management for about half a century since it emerged in the 1950s and 1960s. It was introduced by the United Kingdom (UK) Ministry of Education to standardise the approach to managing costs during the procurement of an extensive programme of school projects. Fundamentally this was to ensure comparable outcomes in terms of cost per place, and some degree of parity in the allocation of funding across school projects. The introduction of cost planning has been described by Ashworth and Perera (2015) as 'a milestone in practice' at that time. They suggest that cost planning revolutionised cost management practice by defining a new set of ideas and procedures that were yet to be exploited and tested in practice. It originated from professional practice, with efforts made towards understanding building costs and the development of cost centres for analysis (Ashworth and Perera, 2015). It was not until publication of the first edition of the RICS *New Rules of Measurement: Order of Cost Estimating and Cost Planning for Capital Building Works* (NRM1) in 2009, that there was any significant development of this technique. Ashworth and Perera (2015) provide a comprehensive review of cost planning from a historical perspective, signposting many of the seminal publications on this technique, the early ones of which included Nisbet (1961), Browning (1961) and Ferry (1964). The review described the process and principles of cost planning, and demonstrated the application of the technique.

The main evolution of this technique in recent times have been precipitated by the publication of an updated Plan of Work by the Royal Institute of British Architects (RIBA) which specified clear stages in the project development cycle where different formal cost plans with varying degrees of detail must be produced, and the NRM1 by RICS which provided rules of measurement for building works to standardise the cost planning process and deliver better forecasting and organisation of building information.

Whole-Life Costing

Prior to the development of whole-life costing (WLC), a widely used technique to measure the costs of ownership of a building was life cycle costing (LCC) (see, for example, Baker, 1978). The RICS guide to LCC defines it as a tool to assist in assessing the cost performance of construction work (RICS, 2016). Similar to other project cost techniques previously discussed, LCC's primary function is to facilitate choices where alternative solutions to a client's needs are being considered. It takes into account the initial capital cost and all relevant costs over a defined period of time. BS ISO 15686-5:2008: Buildings and Constructed Assets – Service

Life Planning – Part 5: Lifecycle Costing also defines LCC as a 'methodology for systematic economic evaluation of the life cycle costs over a period of analysis, as defined in the agreed scoping'. Its development represented a departure from the traditional reliance on economic evaluation of projects based only on initial cost estimates or construction bids (Baker, 1978).

WLC is a broader concept and can include costs (and revenues) associated with the provision of the construction works but not included in the client's costs (RICS, 2016). Robinson and Kosky (2000) and Constructing Excellence (2004) defined it as 'The systematic consideration of all relevant costs and revenues associated with the acquisition and ownership of an asset'. RICS (2014) defined it as 'a structured approach addressing all costs in connection with a building or facility (including construction, maintenance, renewals, operation, occupancy, environmental and end of life)'. The reason why these techniques have become important in project economics is the recognition that the cost of operating and maintaining a building often far outweighs the initial capital cost. Studies have shown that for every £1 of capital cost there is £5 operational expenditure over the life of the building (O'Brien, 2015).

The anticipated cost profile of an asset over its planned life is generated with a discounted cash flow method used to calculate a single cost figure. This enables the project team to analyse the impact of the capital cost decisions. A higher initial capital expenditure can often be justified by taking into account the impact this will have in terms of maintenance, servicing and other forms of operational costs associated with managing the building. It is used as part of the business case:

- to determine affordability of a project;
- to appraise options to decide on the most economically advantageous solution over the life of an asset;
- to control the design development within the running cost and the capital cost budgets; and
- to provide a set of instructions and a budget for the facilities manager.

ISO 15686 and BS 8544-2013 provide guidance on standard structures for WLC. ISO 15686, specifically Part 5, provides requirements and guidelines for performing LCC analyses of buildings and constructed assets and their parts, whether new or existing. BS 8544-2013 establishes rules and a methodology for LCC of maintenance during the in-use phases of buildings. Two key concerns in the application of this technique are the period of analysis and the discount rate. Much research has been undertaken on the application of this technique in the construction literature; some recent examples are Alqahtani and Whyte (2016) and Olubodun et al. (2010).

Cost Modelling

Cost modelling has also evolved over time to embrace better modelling techniques, better and more information, as well as the improved understanding of project risks and the role of the modeller (for example, Hardcastle, 1992; Fellows and Liu, 2000). Cost modelling is a system of creating a predicted cost using design information for a product that has yet to be built. It emerged from the inadequacy of traditional cost data (Dean, 1993), which was often a cause of company failure (Kadiri, 2015). Harris et al. (2013) suggest that there are two broad types of cost modelling in use: single cost and multiple cost modelling. The chosen method will depend on the level of information and time available (Kadiri, 2015).

Cost models are anchored in activity-based costing (ABC) (see Kaplan and Cooper, 1988), which assigns costs to products or services based on the amount of resources they use. However, Dean (1993), and later Robson et al. (2016), argue that ABC is not relevant as there is more involved in working out costs than simply costing the activity. On the other hand, competent estimators will combine activity costs and market condition judgement. The deficiencies of cost models led to the emergence of newer methods such as Monte Carlo simulation which attempts to present a range of outcomes, and reference class forecasting in cost modelling (see Flyvbjerg et al., 2016). Reference class forecasting is a method of predicting the future by looking at similar past situations and their outcomes. It relies on data rather than on assumptions. It is reported (Flyvberg et al., 2016) that the theories behind reference class forecasting were developed by Daniel Kahneman and Amos Tversky, and that this helped Kahneman to win the Nobel Prize in Economic Sciences. This is another example of how the principles of project appraisals in construction are well grounded in mainstream economics theories. Kahneman (2011, p. 251, in Flyvbjerg et al., 2016) is reported to have projected reference class forecasting in construction as 'the single most important piece of advice regarding how to increase accuracy in forecasting through improved methods'.

Value Management

Finding that the business case for an investment decision is weak, based on some of the above evaluation methods, does not necessarily lead to a 'no go' decision. Value management is a method of identifying opportunities to manage solutions and increase value, thereby enhancing the business case for the project. It encompasses the concepts of value engineering and value analysis. Value management usually incorporates a series of workshops, interviews and reviews, through which the project requirements are evaluated against the means of achieving them (Kelly et al., 2004; Constructing Excellence, 2004). Value management emerged from the demands of the US manufacturing industry during the Second World War. Shortages of skilled labour, raw materials and component parts occasioned the need to look for acceptable substitutes that reduced costs, improved the product or, better still, both.

The concept of 'value analysis' was developed by Lawrence Miles, an electrical engineer with the General Electric Company, and Harry Erlicher, a general purchasing agent with the same firm (Alazemi, 2011; Miles, 1968), who adopted a functional approach to the purchasing requirements of the company. This involved the analysis of a component part of a product in terms of the function it performed, and the search for an alternative solution to provide that function (rather than the product) at lower cost. Throughout the 1940s and 1950s the use of the concept further developed and expanded within the US, becoming a procedure which could be used during the design or engineering stages. In 1954, the term 'value engineering' was introduced by the US military (Alazemi, 2011). During the 1960s value engineering spread to the United Kingdom (UK) manufacturing industry and in the US it was introduced to the construction industry.

The concept of 'value management' was first used within the UK construction industry in the 1980s. While value techniques are now used by manufacturers on a global scale, their application within the construction sector is mainly found in the US, UK, Australia and Hong Kong. The position of value management is now established by legislation in both the US and New South Wales, Australia. In the UK, several bodies have issued documents providing recommendations and guidelines; however, the government has stopped short of any mandatory requirement. The principal tools and processes involved in value management are discussed in detail in Kelly and Male (2016).

Implications for Project Decision-Making

It follows from the above discussions that cost minimisation and financial value optimisation have been the main drivers of project decisions and the main focus of much of the extant literature in this area. Given the uncertainties inherent in these methods, to manage the risk of project failure, contractual mechanisms are used to shift risk down the supply chain. The track record of project overruns – including the Channel Tunnel (80 per cent cost overrun with less than half the projected revenues), Denver International Airport (200 per cent cost overrun with only half the projected traffic), the City of Boston's Central Artery project (470 per cent over budget), and more recent examples such as Crossrail in London (26 per cent over budget at the time of writing) (Flyvbjerg et al., 2003; Ahiaga-Dagbui et al., 2017; Georgiadis, 2020) – raises fundamental questions about the ability of conventional approaches to enable the effective economic evaluation of projects in order to anticipate and adequately capture the risk factors that drive cost overruns, and the efficacy of strategies to manage the forecast costs and financial benefits.

With the evidence from many large projects as discussed, conventional economic analysis and financial forecasts do not always provide a sound basis for decision-making. Another case study from Hong Kong is illustrative. In the 2011–12 Policy Agenda, the development of a new broadcasting house of Radio Television Hong Kong was announced. The construction cost had been estimated at HK$1.6 billion in 2009. However, a revised cost estimate in 2013 put the cost at HK$6.1 billion, ultimately leading to a rejection of the funding application by the Public Works Subcommittee due to concerns regarding the scope of the project and the associated cost estimate (Development Bureau, 2018).

What is further apparent from the above review is that researchers (such as Flyvbjerg et al., 2003; Ahiaga-Dagbui and Smith, 2014) are developing enhanced modelling techniques and applying artificial intelligence to the process. To identify the state of the art on these matters and the implications for construction project economics, a search of five widely used databases in construction research (Scopus, Zetoc, Emerald, Science Direct and Ethos) was undertaken using a range of keywords. A summary of the results is presented in Table 7.1. Many of the articles returned were in fields unconnected with construction or engineering. Similarly, some of the databases returned articles published in non-peer-reviewed magazines such as *Engineering News-Record* (*ENR*). These articles were excluded from the review. It is noteworthy that the Boolean search term 'construction project economics' returned only one article in Scopus, demonstrating that it is not a term of art, and is not yet recognised as a subject in its own right.

PROJECT ECONOMICS RESEARCH

To find the trends in the research related to construction project economics, the articles above were reviewed to identify networks that would signpost the significant areas of research in this field. The software VOSviewer was used to create these networks based on bibliometric data to help visualise the areas of research. Figures 7.1 to 7.3 show the results of this biblio-

Table 7.1 *Past construction project economics related research*

Keywords used	Number of articles returned Scopus	Zetoc	Emerald	E-thos	Science Direct
'construction project economics'	1	-	-	-	-
Construction AND 'project economics'	34	5	30	8	-
'construction economics'	190	160	710	25	-
'building economics'	95	-	270	-	248
Total	320	165	1010	33	248
Exclusion of non-construction and non-academic papers	99	23	30	6	38

metric analysis. Themes are shown as nodes. The bigger the node, the bigger the number of citations the word has. The distance between two nodes indicates their correlation (Van Eck and Waltman, 2014).

From Figure 7.1, it can be seen that there are six broad themes in the extant literature. These are: cost control; cost estimation; industrial economics; economic and social effects; risk; and economic analysis.

It is evident that cost lies at the heart of the network, reinforcing the point made previously that much of the focus in construction project economics is on cost. It manifests itself in themes such as cost estimation and cost control, typified by scrutiny by researchers of issues

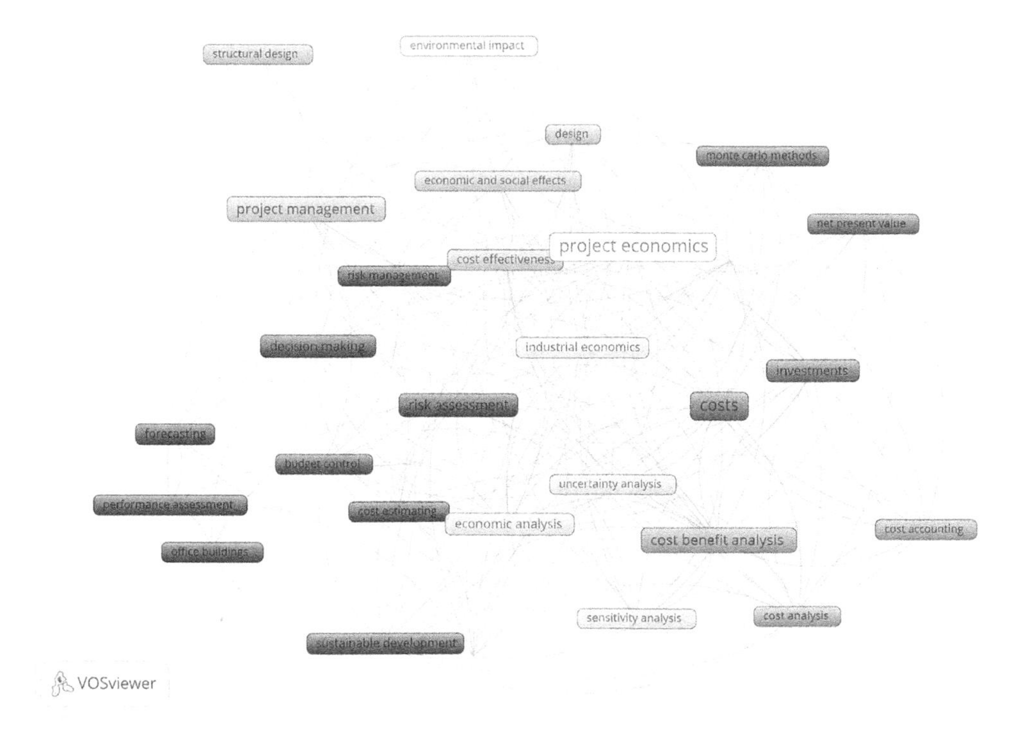

Figure 7.1 *Bibliometric network visualisation of past research on building economics and construction project economics*

around design economics, estimating practice, and estimating accuracy with the point being made several times that there is a high degree of inaccuracy in the estimation of project costs. An example of studies that highlights this problem in construction project economics is Enshassi et al. (2005), which identifies some of the principal risks that threaten estimating accuracy. Another study that addresses the issue of construction cost is Watts et al. (2007), which explored the economics of individual super-tall tower developments with respect to construction costs. It discussed, among other things, the development, design and construction challenges that have an impact on cost. Li et al. (2012) focus on a much-neglected part of project appraisal, which is the transaction cost associated with projects.

Similarly, Ive and Chang (2007) examined, through a transaction cost economics prism, the economic implications of construction procurement systems, highlighting the trade-offs clients make in selecting procurement systems. Procurement systems define the governance regime and contractual mechanisms of projects. They are varied and can be complex owing to the long-term and uncertain nature of many projects, and the need for transaction-specific investments, and this can add significantly to project costs. Transaction cost economics was developed by Williamson (1979), who argues that it is necessary to match governance structures with transactional attributes in a discriminating way. This is central to the study of economics and definitely central to the efficient organisation of construction project delivery. This is another example of how mainstream economic thinking provides essential perspectives for sense-making in construction project economics. Other studies such as Dandan et al. (2019) have also explored factors driving estimating accuracy, identifying issues such as client experience and project team experience as two significant factors.

The network also showcases research where project economics overlaps the economics of the firm. Khosrowshahi and Kaka (2007) present a cashflow management model looking to rebalance the traditional dominance of 'project economics' in shaping 'corporate economics' which consequently places the corporate strategy at the mercy of the projects. They refer to the high risk of insolvency within the industry as evidence of a lack of attention to corporate strategy. They suggest that the conventional cashflow forecasting approach that focuses largely on individual projects, compounded by the practice of running several projects in parallel, creates a situation that places severe strain on corporate resources to an extent that is sometimes 'detrimental to the success and indeed survival of the company'. These arguments are compelling and accord with previous arguments pointing to the causal interaction between the economics of projects and firms (see, for example, Myers, 2017; WEF, 2016; Hillebrandt, 2000).

A fourth theme brings together research addressing design economics, decision-making, and economic and social effects of projects. Such research shows a willingness to embrace not just the input cost of projects in economic analysis, but also the impacts of projects, both economic and social. A particular interest of researchers has been the economics of tall towers. Although it is now dated, Jaafari (1988) is an exemplar of such studies. Worked examples are used to illustrate how design concepts can be investigated, first in terms of feasibility of construction considering such matters as the proposed construction method, the craneage requirements, cycle time and cycle programme evaluation; and second, in terms of potential cost competitiveness and the evaluation of the overall project economics (Jaafari, 1988).

The fifth theme covers studies that begin to explore alternative approaches to managing the uncertainties and imperfections in the market (e.g., Piekarski, 1984). This theme also includes environmental impacts and the broader sustainable development requirements of construction and engineering projects. Although it is widely acknowledged that environmental concerns must be considered in economic calculations on projects, for a long time the complexity of doing so adequately confounded the consideration of the environmental dimension by practising project economists. Weiss (2000) reviewed the methodological challenges of an environmental approach to project economics, highlighting the key considerations for project appraisal from this perspective. The main concerns relate to the perspective (national or international) to be adopted given the transnational nature of project impacts, monetising environmental effects and the problem of time. More recently, Stasiak-Betlejewska and Potkány (2015) presented a comparison of the construction cost analysis for an energy-efficient timber house that is designed to meet sustainable aspects and traditional construction. They also assess the annual savings in energy costs resulting from the more sustainable design.

In response to the lack of precision and uncertainties in the construction market, there has also been extensive use of Monte Carlo simulation techniques in the evaluation of project viability. This technique facilitates the comprehensive analysis of the impact of risk in the absence of complete data. For instance, Hileman (2003) demonstrated that when evaluating project economics, plant owners which use a combination of three methods will gain improved understanding of the capital project and make better decisions whether to invest in a given project. The three methods include benchmarking, Monte Carlo risk analysis and a historical RoI correlation. Sachs et al. (2007) also developed a novel fuzzy set approach for quantifying qualitative information on risks (QQIR) which addresses some of the weaknesses in risk assessment associated with the economic appraisal of projects.

In an exploration of the role of inflation and how it affects the economy and the construction industry, Musarat et al. (2020) concluded that the inflation rate is neglected in most of construction project economics and budgeting, leading to project cost overruns as building materials prices, labour wages and machinery hire rates change annually. Musarat et al. (2020) subsequently proposed a framework that highlights the strong relationship between the inflation rate and the construction industry to be used for future budget estimation to eliminate project cost overruns that occur due to the inflation rate. It has been noted that the unreliability of forecasts and estimates is due to a range of complex interacting factors (see Flyvbjerg et al., 2014; Ahiaga-Dagbui and Smith, 2014).

The final theme draws together a variety of techniques used for the economic or financial appraisal of projects. An example of research that has contributed to this theme is by Berge et al. (2009), who developed an economic model to evaluate the impact of various operational and construction strategies on the project economics of bioreactor landfill infrastructure. A review of the application of quantitative techniques in construction economics research was undertaken by Goh (2008). The review revealed increasing application of conventional techniques in construction economics research, as opposed to construction management research where there was greater integration of artificial intelligence (AI) techniques. Goh (2008) recommends that researchers should better enable themselves to build tools that incorporate intelligence as innovative solutions for increasingly complex problems. Therefore, AI features as one of the innovations in cost modelling.

In terms of innovations in this area, a number of articles have also focused on estimating in a building information modelling (BIM) environment (examples are Olatunji and Sher, 2015; Babatunde et al., 2019). From a project management and estimating perspective, Olatunji and Sher (2015, pp. 25) defined BIM as 'a digital system for facilitating a data-rich, object-oriented, intelligent and parametric representation of a construction project, from which views and data appropriate to various users' needs can be extracted and analysed to generate information and enhance decision-making on project economics, and improve project delivery processes'. This definition is consistent with the benefits anticipated from BIM adoption in construction project delivery. Georgiadou (2019) discussed the benefits of BIM as comprising improvements in cost efficiency, quality assurance and on-time delivery, improvements in collaboration and communication amongst stakeholders (client, design teams and contractor), design optimisation and an integration of life cycle thinking and sustainability into design decision-making. The implications for construction project economics is that BIM and its related plug-in tools can be used to assess and optimise designs through clash detection and avoidance (which eliminates rework), and choice of more economical geometries and orientations; assess and optimise energy performance and carbon contributions by simulating and tweaking the designs; facilitate extraction of design parameters for estimating and costing purposes; and optimise decisions relating to the on-site construction process. Also, because the ambition is to continuously capture and integrate project life cycle performance data onto BIM models, the resulting BIM meta-data that is captured throughout the life cycle of an asset can be fed back to enhance project appraisal and decision-making on future projects.

It is noteworthy that a significant proportion of articles on the evaluation of projects have been from the oil and gas or, more generally, the energy sector (for example, Hileman, 2003; Trinh and Moghanloo, 2020). It is clear from the review above that a variety of approaches have been adopted, including both financial and economic valuation approaches. To identify further trends, a chronological network was extracted from VOSviewer as shown in Figure 7.2.

Figure 7.2 confirms a historical focus on financial costs and revenues, and a progressive shift towards non-financial value with the consideration of environmental and sustainability impacts in the last two decades leading to greater emphasis on CBA, life cycle costing and

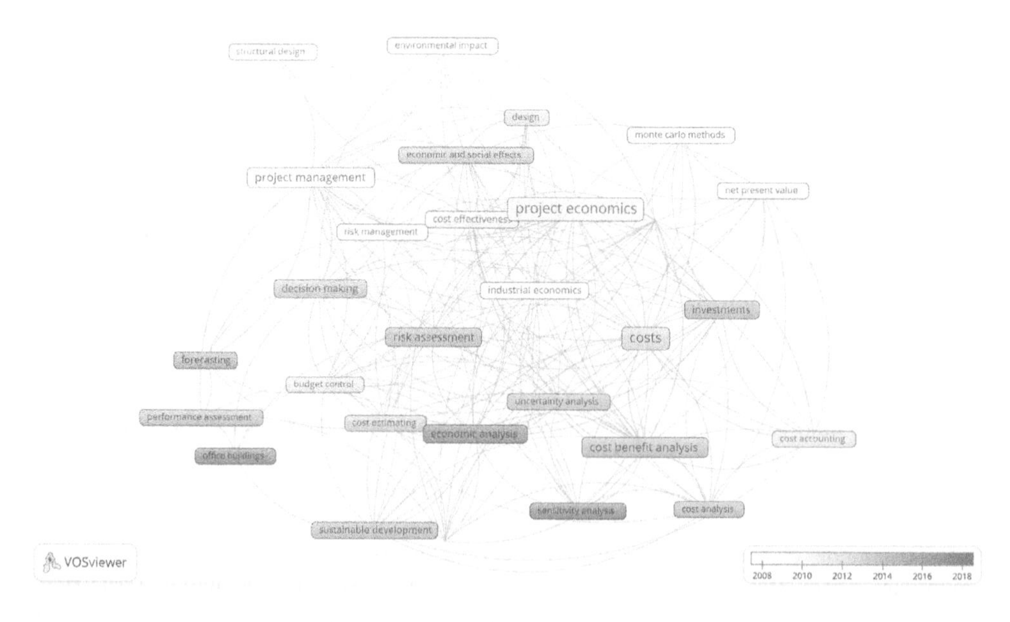

Figure 7.2 Chronological overlay visualisation of past research on building and project economics

whole-life value, risk assessment and innovation directly connected to economic analysis and AI. As shown in Figure 7.3, much of this research is from the US, UK, China, Canada and Germany. It is noticeable also that the later research is from countries such as Malaysia, Taiwan, Switzerland, the Netherlands, Russia, the United Arab Emirates (UAE) and Saudi Arabia.

The next section develops these concepts and reviews the techniques underpinning them.

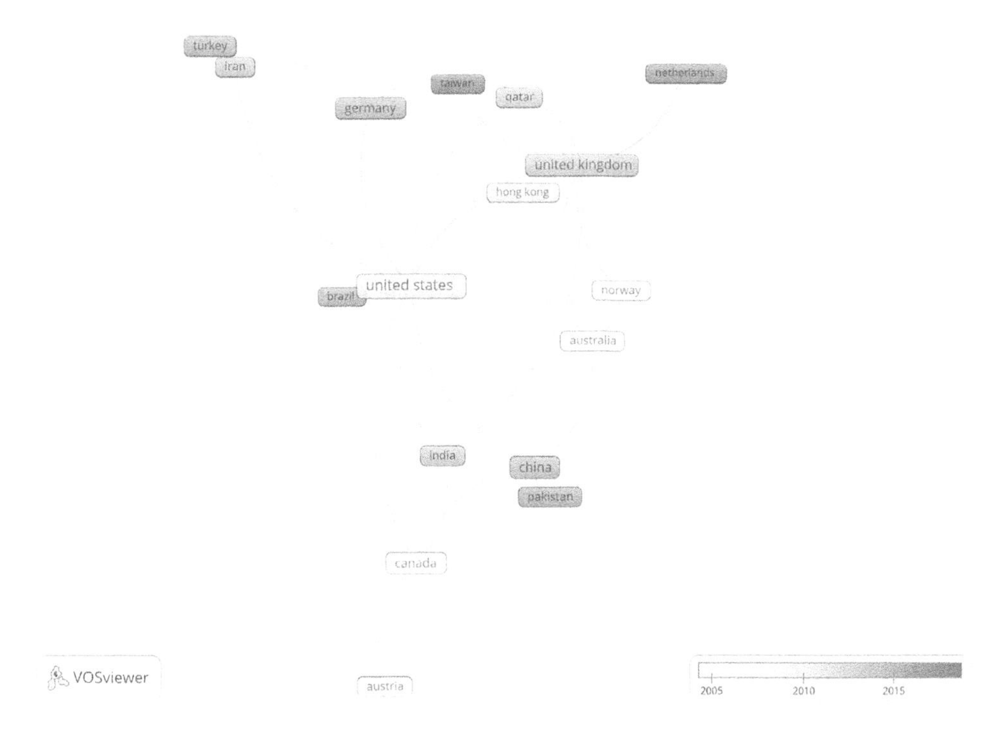

Figure 7.3 *Geographical distribution of past research on building and project economics*

DEVELOPMENTS IN PROJECT ECONOMICS

It is clear from the above review that there are emerging trends driving a new approach to construction project economics. Whilst economics at the project level requires the optimisation of resources to achieve the 'triple bottom line' of financial, social and environmental value, most of the earlier techniques and approaches for appraising the value of projects have focused predominantly on notions of financial value (for example, cost of resources and revenues generated from projects) and how these can be optimised. However, there is now an increasing emphasis on the environmental and social value contributions from the construction industry, which has translated into a greater need to optimise resources at the project level to achieve a harmonious balance between financial, social and environmental value (see, for example, Watson et al., 2016; Weiss, 2000). Some of these emerging trends include circular thinking, carbon accounting, social accounting, natural capital accounting, new procurement methodologies, and smart off-site manufacturing approaches (see, for example, Celik et al., 2017;

Jowitt et al., 2012). These developments are driving a redefinition of the concept of value to embrace the broader impacts of projects. Another significant shift appears to be the emergence of a more holistic approach to project evaluation in the form of growth in the use of system dynamics and reference class forecasting supported by the application of big data and AI to aid risk assessment (see, for example, Sawhney and Muse, 2019; An et al., 2007).

Project Economics and Social Value

Construction projects can generate significant unintended adverse consequences for neighbouring communities (Celik et al., 2017). Apeldoorn (2013) demonstrates, through case studies, that these adverse impacts are greater in densely populated areas as compared to areas with lower density. During the appraisal and decision-making phase, it is common practice to evaluate the social impact of construction projects. Social impact assessments (SIAs) have been widely used in project appraisal to assess and minimise the social impacts which have been defined by the Inter-organizational Committee on Principles and Guidelines for Social Impact Assessment (ICGP, 2003, p. 231) as: 'the consequences to human populations of any public or private actions that alter the ways in which people live, work, play, relate to one another, organize to meet their needs, and generally cope as members of society'. The ICGP proposed six principles that should be used to conduct SIA: achieve extensive understanding of local and regional settings that will be affected by the action or policy; focus on key elements of the human environment; identify research methods, assumptions and significance; provide quality information for use in decision-making; ensure that any environmental justice issues are fully described and analysed; and undertake evaluation, monitoring and mitigation.

From these ICGP principles, the traditional focus of SIA is to minimise negative consequences that communities will experience due to the proposed construction project. This can only be achieved by assessing the social costs arising from the adverse social impacts. Social costs have been defined in economics as the overall impact of economic activity on the welfare of society (Celik et al., 2017). These social costs owing to construction activities can take the form of disruption to common life patterns of society around the construction zones (Apeldoorn, 2013), on issues such as traffic congestion, pollution, economic activity, and ecology, health and well-being (Allouche et al., 2000). The scope of SIA has increasingly evolved to embrace a corresponding focus on maximising the benefits that projects bring to the affected communities (Vanclay et al., 2015), as a form of social value benefit.

The concept of social value initially grew out of early ideas on corporate social responsibility, ethical economics and the social enterprise (Emerson, 2003; Richmond et al., 2003). This social value focus has continued to diffuse and influence the procurement and delivery of construction projects. For example, in the UK, this trend has now been driven further by the Public Services (Social Value) Act (2012), which mandates that public sector entities should consider requirements for achieving social value within their procurement contracts. To comply with the Social Value Act, public sector clients will have to move beyond the traditional emphasis on costs when appraising projects, by considering wider aspects of value to society over the project's lifetime, extending the evaluation of projects beyond profit-driven motivations. The social value benefits of projects can be optimised through avenues such an increase in local content (that is, jobs for local people and local procurement), removal of barriers that make it possible for local enterprises to supply goods and services, and through the provision of training and support to local people (Vanclay et al., 2015).

Social return on investment (SROI) has been used as a social impact valuation approach for evaluating the social value contributions of projects. SROI allows for monetary values to be assigned to social returns using financial proxies. The social returns can then be compared against the initial investment by computing an SROI ratio of costs to social outcomes, much like the cost–benefit ratio in cost–benefit analysis. For example, SROI has been applied by Watson et al. (2016) to evaluate the social value of designs for cancer support centres from the perspective of the building users. Social value will continue to be an important consideration in the economic evaluation of projects during the project appraisal and decision-making process. This will continue to drive the growth in use of social accounting methods that focus on establishing a comparative market value for non-market social outputs that are derived from projects.

Project Economics and Environmental Value

Another important trend is the increased emphasis on minimising the negative impact of construction projects on the environment (eco-efficiency) or pushing the boundaries further by ensuring that projects, in fact, add to environmental value through their positive and regenerative effects (eco-effectiveness). This growing environmental focus as a project-level objective dates back to Ofori (1992), who questioned whether in addition to cost, time and quality, the environment would become a fourth project objective for construction clients. The climate change agenda has also become another powerful and urgent force for change and requires a response, particularly at the front end of conceptualising and defining the scope of projects for construction clients (Morris, 2017).

To minimise the negative impacts of construction projects on the environment, environmental impact assessments (EIAs) are often conducted as part of project appraisals, with the necessary mitigation requirements having to be satisfied before licences or permits to build are issued. Yang (2017, pp. 21) defined an EIA as: 'a set of comprehensive methods and system to analyse, predict, and assess the potential environmental impact that might be incurred by the project planning and construction, and to raise solutions and measures to prevent or mitigate the adverse environmental impact and to follow up and monitor afterwards'. During an EIA study, the focus has predominantly been on alterations that the proposed construction project can make to the biophysical environment, although Celik et al. (2017) acknowledged that this process can also be extended to incorporate the cultural impact on the neighbouring society, alongside other socio-economic impacts.

In addition to this traditional conduct of EIAs as part of project appraisals, a growing trend in project appraisal is the focus on carbon accounting and the minimisation of whole-life carbon of construction projects, in alignment with the climate change situation addressed by Morris (2017). This is due to the potentially significant contributions of construction projects to carbon emissions throughout their life cycle. Therefore, the emissions impact of construction projects are evaluated and measured using carbon accounting techniques. These carbon accounts can be used for benchmarking and/or optimising project designs through the implementation of appropriate carbon reduction measures. As part of project-level economic analysis, whole-life carbon appraisals measure both anticipated operational emissions and the embodied emissions of proposed designs, with results at the project level ultimately informing a measure of the built environment's total carbon impact. In line with this development, RICS (2017) has proposed an approach and guidelines for whole-life carbon assessment for the built environment, based on the EN 15978:2011 which sets out a life cycle assessment (LCA) method for assessing the environmental impacts of built projects.

Carbon costs can also be computed and integrated into cost–benefit analyses by applying prevailing carbon prices, although the effectiveness of this approach will depend largely on carbon pricing policy. Jowitt et al. (2012) expressed concerns about translating carbon emissions into carbon costs as part of the financial appraisal of infrastructure projects, because these costs could be significantly discounted due to low carbon prices, despite the effects of the carbon emission on the climate remaining high. This integration of carbon accounting techniques into mainstream project appraisal and decision-making processes will continue to define a shift from conventional assessments of whole-life cost towards whole-life carbon appraisal (Jowitt et al., 2012).

Beyond the assessment of environmental and carbon emissions impact, the contribution of construction projects to natural ecosystems is likely to receive growing interest. This stems from the perspective that rather than the focus on minimising negative environmental impacts (eco-efficiency), construction projects can be used more ambitiously to contribute positively to natural ecosystems (eco-effectiveness). This eco-effectiveness ambition can be achieved by designing and constructing built assets to purify the surrounding air, sequester carbon, fix nitrogen, distil water, accrue solar energy as fuel, build up soil, create a microclimate and provide a habitat for thousands of species (Ankrah et al., 2013). For projects to contribute towards longer-term sustainability, this eco-effectiveness ambition will continue to grow, resulting in the integration of techniques such as natural capital accounting to the economic appraisals of construction projects.

Natural capital accounting is an ecosystem accounting approach for valuing the stocks of natural capital and flows of services they generate for people (ONS, 2017). Ecosystems comprise the living and non-living natural resources in the air, water and on land, with natural capital having a much broader coverage to include minerals and fossil fuels that are below the ground (Mace, 2019). Changes in ecosystem asset accounts can only be evaluated by understanding, as a baseline, the extent and condition of the different land covers or habitats. Ecosystems' features on land can comprise woodlands, wetlands, water or soil, all of which will support different kinds of biodiversity. These ecosystems provide services to society such as amenity, air quality, recreation, climate regulation, water purification and flood protection. Natural capital accounting values these ecosystems as assets in terms of their stock and the flow of services they provide to society (Hein et al., 2016), assessed in terms of their monetary values.

Construction projects can add to, or reduce, such stocks of natural capital and the quantity or quality of services they provide. For example, biophilic and regenerative project designs can contribute to natural ecosystems through integration of features such as green spaces, lakes, ponds and constructed wetlands (Dias, 2015; Kambo et al., 2019). Construction projects can also destroy ecosystem assets, reducing both the natural capital stocks and the services they provide. For example, the Hong Kong–Zhuhai–Macao Bridge megaproject runs over a conservation area for the Indo-Pacific humpbacked dolphin (Liu et al., 2018). The integration of natural capital accounting techniques into economic appraisal of projects, and even as part of EIA, will continue to advance.

Another area of growing interest related to the increased focus on optimising environmental value of construction projects, particularly from the eco-effectiveness perspective, is the ambition to transition from a linear (take–make–dispose) economy towards a circular economy in the built environment. This has led to the application of circular thinking at the project level. Circular thinking can be considered an upgrade of sustainability thinking that is grounded in the vision of eco-effectiveness as opposed to eco-efficiency (Braungart et al., 2007). Within the circular economy, built assets are meant to function as material banks that can later be mined for resources at the end of their service life (Luscuere, 2017; Copeland and Bilec, 2020). As shown in Figure 7.4, this development will potentially translate into significantly higher residual values for built assets as well as realising the desired lower carbon emissions (Cheshire, 2016).

This will have implications for life cycle assessments as the value of projects will need to reflect circularity considerations, with analysis having to be cradle-to-cradle rather than cradle-to-grave orientation. There will also be a focus on measuring the circularity of projects using circularity indices and indicators that are unique to building and infrastructure projects. These circularity indicators and indices will add to the already growing list of assessment methods, standards and benchmarks for evidencing the environmental impact and sustainability credentials of projects, such as the Building Research Establishment Environmental Assessment Method (BREEAM) Standards, and the US Green Building Council's Leadership in Energy and Environmental Design (LEED) Standards.

This emphasis on environmental economics in construction, which has now become topical, resonates strongly with similar developments in mainstream economics thinking and research (see, for example, Bröchner, 2018). Given its significant environmental impacts, one would expect that the construction project would be a natural source of data to drive mainstream environmental economics, as the examples used in Weiss (2000) illustrate. Therefore, construction

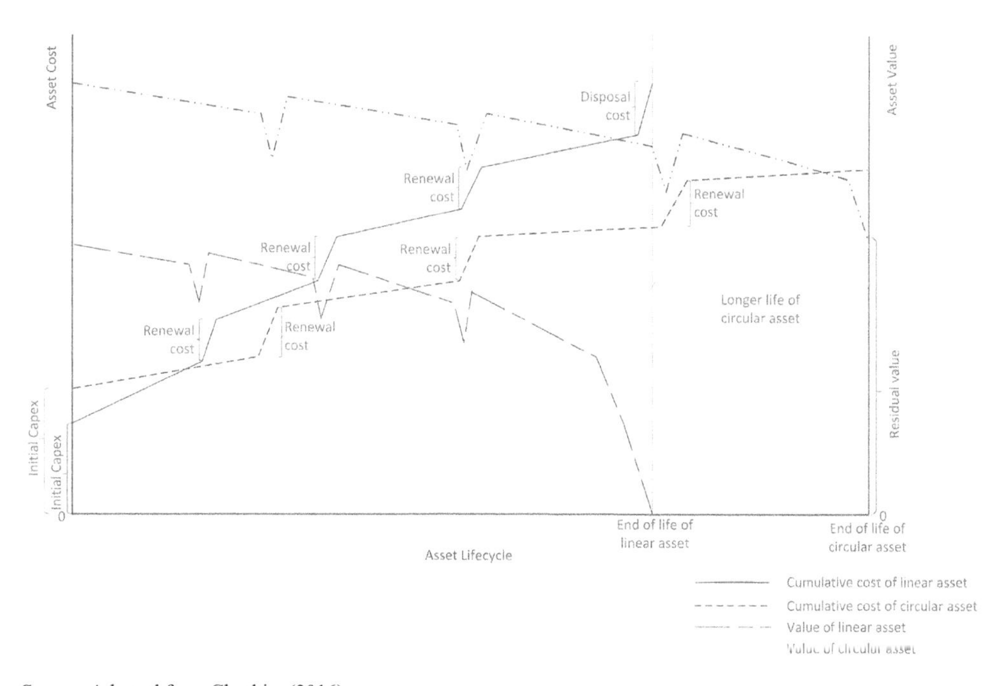

Source: Adapted from Cheshire (2016).

Figure 7.4 *Comparison between a circular building and a traditional building*

has the opportunity to lead thinking and research in this area and provide insights to inform the response to the global climate emergency.

It can be noticed from the above that there is a gradual shift from the traditional focus on financial value (cost efficiency) during project appraisals, towards approaches that account for non-financial value (environmental and social value). This trend will continue to apply to both public and private sector projects. This trend will require the development of more integrated project appraisal methods that account for financial, environmental and social value that accrue to the multiple stakeholders of the projects.

Impact of Technological Advances on Project Economics

Despite the progress that has been made in optimising the financial value of construction projects, technological advances in the delivery of projects will continue to offer further possibilities to improve cost efficiencies. Digitalisation and other innovative technologies such as augmented reality, drones, three-dimensional (3D) scanning and printing, BIM and autonomous equipment will boost productivity and efficiency in the delivery of construction projects (WEF, 2016; Ribeirinho et al., 2020). The *Construction 2.0: Time to Change* report for Hong Kong (Development Bureau, 2018) also highlights the same tools and technologies and the opportunities they offer. By embracing these modern methods of construction (MMCs), a significant proportion of construction activities will be produced in stable and more controlled environments, followed by on-site assembly using modern construction techniques. This practice will enhance the accuracy with which project costs (resource inputs) are predicted and optimised. Indeed, in North America, the market share of MMCs of new real-estate projects grew by 50 per cent from 2015 to 2018, with research and development spend in this area increasing by 70 per cent since 2013 (Ribeirinho et al., 2020).

Rapid adoption of MMCs through a platform approach will also deliver significant improvements in productivity, efficiency and quality of projects (Cabinet Office, 2020). A platform approach will enable the reuse of processes, technical solutions or assets that are repeated with a stable supply chain (Jansson et al., 2014; Hall et al., 2020; Mosca et al., 2020), and procurement processes and transactions will be streamlined and simplified for clients. Projects will be procured as a product package via product platforms as a collection of assets shared by a set of products (Robertson and Ulrich 1998). As the platform approach evolves in the construction industry, platform ecosystems will emerge as a network that operates around central focus or point of control (Thomas et al., 2014) without intermediaries. The deployment of blockchain technology and smart contracts as a decentralised network governed by peer-to-peer consensus algorithms, without the need for any central administrator (Perera et al., 2020), will radically change the way projects are procured. Rather than a complex web of discrete transactions where risks are passed down to a long and fragmented supply chain (Bröchner, 2018), these developments will drive a single-point transaction for the client on an ecosystem platform comprising of interconnected and fully integrated and blockchain-enabled supply chains. This will be potentially game-changing for economics at the project level, as transaction costs of projects will be minimised significantly.

This trend will be backed by the use of more holistic approaches to project cost evaluation such as systems dynamics and reference class forecasting supported by the application of big data and AI to aid risk assessment. Interest in these approaches is already growing due to the unreliability of project cost estimates using traditional cost estimating techniques.

System dynamic techniques, which are useful for modelling highly dynamic systems consisting of multiple interdependencies and non-linear relationships (Boateng et al., 2015), are likely to more accurately capture risks and uncertainties during economic appraisals of projects. Zhang et al. (2014) have applied system dynamics techniques to the assessment of sustainable development value of construction projects over their life cycle; a trend which is likely to become mainstream. These advancements in smart digital technologies that can capture project life cycle data (for example, BIM, digital twinning and cyber-physical systems) will continue to provide possibilities to exploit existing project datasets through data mining, big data analytics, and application of complex AI techniques for cost modelling and project cost forecasting. The use of AI techniques such as fuzzy logic models, artificial neural networks, regression models, case-based reasoning, hybrid models, diction tree, random forest, supportive vector machine, AdaBoost, scalable boosting trees and evolutionary computing techniques such as genetic algorithm (see Elmousalami, 2020) for cost modelling will increase. The idea here is to leverage the large volumes of connected data from digital twins and machine learning to predict and prioritise high-risk issues, in order to extract more accurate assessments and monetisation of the whole-life costs and value propositions of proposed projects, to optimise decision-making. This will result in greater accuracy and reliability of economic appraisals of projects (An et al., 2007; Sawhney and Muse, 2019).

IMPLICATIONS FOR THE DEVELOPMENT OF CONSTRUCTION PROJECT ECONOMICS

The growing emphasis on value rather than cost during economic appraisals of projects is already manifest in the UK construction industry. Recent UK government policy on procurement for central government bodies puts emphasis on value (Cabinet Office, 2020) and is backed by work on the Value Toolkit by the Construction Innovation Hub (see, for example, Construction Innovation Hub, 2020). The value toolkit has been developed for use across public projects to generate a value profile setting out value drivers for a client, for each project in a specific location. Value profiles for a project are assigned to five capital areas (natural, social, human, manufactured and financial capital), all with their value categories, before conversion into investment value indices (IVI) for decision-making and benchmarking (Construction Innovation Hub, 2020).

Further research implications for construction project economics as a result of these developments will involve a better understanding of the environmental and social costs and benefits of delivering construction projects using MMCs and smart digital technologies. This will enable much deeper insights to be gained into any displacement impacts arising, for example, from the social, economic and environmental costs and benefits of shifting from conventional labour-oriented construction approaches towards MMCs. Also, more research will be needed on the productivity gains and impact on project cost resulting from the rapid deployment and adoption of smart, automated, digital and other innovative technologies, as data on these developments become common. Such research should seek to quantify the impact of techniques such as automation, on productivity and cost at the project level. The increasing availability of big data on the construction industry and, more specifically, on project-level transactions, will further drive project-level economic analysis towards methods of economic analysis that have long been applied in mainstream economics. Bröchner (2018) alluded to something similar by suggesting that access to big data on construction transactions could make construction economics a subdiscipline of interest within mainstream economics.

There will also be policy implications in terms of increasingly stringent building regulations, with increasingly ambitious energy consumption requirements, and regulations that limit the use of virgin resources and promote a circular culture at the project level. These developments will further internalise the cost of any negative externalities associated with projects. This will also impact on the future cost of resources used in construction.

Based on the foregoing discussion, there is a potential opportunity for construction economics to influence mainstream economics by offering fresh perspectives on the economic theory of transaction costs, from a project-based industry that has transformed its processes and delivery models into a product-based industry. This should be viewed against the backdrop of the implications of the technological advancements that are reshaping the project procurement domain and hence transaction costs.

CONCLUSION

This chapter has reviewed the main concepts underpinning the techniques used in the economic evaluation of construction projects to aid decision-making in relation to the economic problem of making optimal choices regarding the allocation and use of scarce resources such as land, labour, capital, equipment, time and management on construction projects. As a primer for the project economics researcher, some of the conventional techniques such as cost–benefit analysis, development appraisal, cost analysis and planning, and whole-life costing, which have been the bedrock of construction project economics, have been briefly introduced. It is evident from this review that the historic focus of the body of knowledge has been on financial analysis of project costs and revenues. A fundamental flaw of this approach is that projects are frequently beset with uncertainty and risk, and also have impacts that cannot always be monetised. There is also a human dimension that manifests itself in optimism bias. The corollary is that economic evaluation of projects has been unreliable, resulting in suboptimal decision-making regarding which projects to pursue, which to abandon and which to rescope.

The chapter has highlighted some emerging trends that are driving a new approach to construction project economics by shifting the focus from cost to value, and from financial value to non-financial value. These trends include circular thinking, carbon accounting, natural capital accounting, new procurement methodologies involving smart off-site manufacturing approaches, big data analytics and AI. These developments have the potential to deliver more reliable economic evaluation of projects. However, the consideration of these concepts and application of the tools and platforms within the industry is patchy. Based on the current gaps in knowledge around these concepts, techniques and platforms, and their practical implementation, there is a need for further research to build universal standards and databases, and adapt these platforms, particularly AI, for mainstream construction application. Such research will support the diffusion of innovation across industry and among researchers. Finally, the environmental economics aspects highlighted in this review, which the construction industry is leading on in relation to the integration of renewable energy and circularity, have the potential to permeate mainstream economics.

REFERENCES

Ahiaga-Dagbui, D., Love, P.E.D., Smith, S.D. and Ackermann, F. (2017) Toward a systemic view to cost overrun causation in infrastructure projects: a review and implications for research. *Project Management Journal*, 48 (2), 88–98. http://www.pmi.org/learning/library/systemic-view-cost-overrun-causation-infrastructure-projects-10703.

Ahiaga-Dagbui, D.D. and Smith, S.D. (2014) Rethinking construction cost overruns: cognition, learning and estimation. *Journal of Financial Management of Property and Construction*, 19 (1), 38–54.

Alazemi, T. (2011) On the integration of value engineering in the procurement of public housing in the state of Kuwait. PhD thesis, University of Manchester.

Alqahtani, A. and Whyte, A. (2016) Evaluation of non-cost factors affecting the life cycle cost: an exploratory study. *Journal of Engineering, Design and Technology*, 14 (4), 818–834. https://doi.org/10.1108/JEDT-02-2015-0005.

Allouche, E.N., Ariaratnam, S.T. and Abourizk, S.M. (2000) Multi-dimensional utility model for selection of a trenchless construction method. *Construction Congress VI: Building Together for a Better Tomorrow in an Increasingly Complex World. American Society of Civil Engineers, Reston, VA*, pp. 543–553 http://dx.doi.org/10.1061/40475(278)59.

An, S.-H., Kim, G.-H. and Kang, K.-I. (2007) A case-based reasoning cost estimating model using experience by analytic hierarchy process. *Building and Environment*, 42 (7), 2573–2579.

Ankrah, N.A., Manu, E., Hammond, F.N., Awuah, K.G.B. and Tannahill, K. (2013) Beyond sustainable buildings: eco-efficiency to eco-effectiveness through cradle-to-cradle design. Sustainable Building Conference 2013, 3–5 July, Coventry, UK.

Apeldoorn, S. (2013) Comparing the costs – trenchless versus traditional methods. *MIESA*, 38 (4), 55–57.

Ashworth, A. and Perera, S. (2015) *Cost Studies of Buildings* (6th edn). Routledge, London.

Babatunde, S.O., Perera, S., Ekundayo, D. and Adeleye, T.E. (2019) An investigation into BIM-based detailed cost estimating and drivers to the adoption of BIM in quantity surveying practices. *Journal of Financial Management of Property and Construction*, 25 (1), 61–81. https://doi.org/10.1108/JFMPC-05-2019-0042.

Baker, W.R. (1978) Life-cycle costing. *Plant Engineering*, 32 (11), 171–176.

Berge, N.D., Reinhart D.R. and Batarseh E.S. (2009) An assessment of bioreactor landfill costs and benefits. *Waste Management*, 29 (5), 1558–1567.

Boardman, A.E., Greenberg, D.H., Vining, A.R. and Weimer, D.L. (2018) *Cost–Benefit Analysis: Concepts and Practice* (4th edn). Cambridge: Cambridge University Press.

Boateng, P., Ahiaga-Dagbui, D., Chen, Z. and Ogunlana, S. (2015) Modelling economic risks in megaproject construction: a systemic approach. In: Raidén, A.B. and Aboagye-Nimo, E. (eds), *Procs 31st Annual ARCOM Conference*, 7–9 September, Lincoln, UK, Association of Researchers in Construction Management, pp. 115–124.

Braungart, M., McDonough, W. and Bollinger, A. (2007) Cradle-to-cradle design: creating healthy emissions – a strategy for eco-effective product and system design. *Journal of Cleaner Production*, 15 (13–14), 1337–1348.

Broughel, J. (2020) Cost–benefit analysis as a failure to learn from the past. *Journal of Private Enterprise*, 35 (1), 105–113.

Browning, C.D. (1961) *Building Economics and Cost Planning*. Batsford, London.

Bröchner, J. (2018) Construction economics and economics journals. *Construction Management and Economics*, 36 (3), 175–180.

BS 8544: 2013 Guide for life cycle costing of maintenance during the in use phases of buildings. https://shop.bsigroup.com/ProductDetail?pid=000000000030218914.

BS/ISO 15686-5: 2017 Buildings and constructed assets. Service life planning. Life cycle costing. https://www.iso.org/standard/61148.html.

Cabinet Office (2020) *Construction Playbook: Government Guidance on Sourcing and Contracting Public Works, Projects and Programmes*. HM Government. https://www.gov.uk/government/publications/the-construction-playbook.

Celik, T., Kamali, S. and Arayici, Y. (2017) Social cost in construction projects. *Environmental Impact Assessment Review*, 64, 77–86.

Cheshire, D. (2016) *Building Revolutions: Applying the Circular Economy to the Built Environment.* RIBA Publishing, London.
Construction Innovation Hub (2020) An introduction to the Value Toolkit. https://constructioninnovationhub.org.uk/wp-content/uploads/2020/07/20200715_BR_09_ValueFrameworkReport_Digital_Pages.pdf.
Constructing Excellence (2004) Value management, constructing excellence. https://constructingexcellence.org.uk/wp-content/uploads/2015/03/value.pdf (accessed 26 November 2021).
Copeland, S. and Bilec, M. (2020) Buildings as material banks using RFID and building information modeling in a circular economy. *Procedia CIRP*, 90, 143–147.
Dandan, T.H., Sweis, G., Sukkari, L.S. and Sweis, R.J. (2019) Factors affecting the accuracy of cost estimate during various design stages. *Journal of Engineering, Design and Technology*, 18 (4), 787–819. https://doi.org/10.1108/JEDT-08-2019-0202.
Dean, E.B. (1993) Why does it cost how much? *Proceedings of AIAA 1993 Aerospace Design Conference*, February, Irvine, CA.
Development Bureau (2018) *Construction 2.0: Time to Change*. Development Bureau, Hong Kong. https://www.psgo.gov.hk/assets/pdf/Construction-2-0-en.pdf.
Dias, B.D. (2015) Beyond sustainability – biophilic and regenerative design in architecture. *European Scientific Journal, ESJ*, 11 (9). https://doi.org/10.19044/esj.2015.v11n9p%p.
Dotzour, M., Grissom, T., Liu, C. and Pearson, T. (1990) Highest and best use: the evolving paradigm. *Journal of Real Estate Research*, 5 (1), 17–32.
Drèze, Jean and Stern, Nicholas (1987) The theory of cost–benefit analysis. In: Auerbach, A.J. and Feldstein, M. (eds), *Handbook of Public Economics*, Vol. 2. Elsevier, Amsterdam, pp. 909–989.
Elmousalami, H.H. (2020) Artificial intelligence and parametric construction cost estimate modelling: state-of-the-art review. *Journal of Construction Engineering and Management* 146 (1). 03119008-1-03119008-30.
Emerson, J. (2003) The blended value proposition: integrating social and financial returns. *California Management Review*, 45 (4), 35–51.
Enshassi, A., Mohamed, S. and Madi, I. (2005) Factors affecting accuracy of cost estimation of building contracts in the Gaza Strip. *Journal of Financial Management of Property and Construction*, 10 (2), 115–125. https://doi.org/10.1108/13664380580001069.
Fellows, R. and Liu, A.M.M. (2000), Human dimensions in modelling prices of building projects. *Engineering, Construction and Architectural Management*, 7 (4), 362–372. https://doi.org/10.1108/eb021159.
Ferry (1964) *Cost Planning of Buildings* (1st edn). Crosby Lockwood & Sons, London.
Flyvbjerg B., Bruzelius, N. and Rothengatter W. (2003) *Megaprojects and Risk*. Cambridge University Press, Cambridge.
Flyvbjerg, B., Garbuio, M. and Lovallo, D. (2014) Better forecasting for large capital projects. McKinsey on Finance Number 52, Autumn.
Flyvbjerg, B., Hon, C.-K. and Fok, W.H. (2016) Reference class forecasting for Hong Kong's major roadworks projects. *Proceedings of the Institution of Civil Engineers – Civil Engineering*, 169 (6), 17–24.
French, N. (2004). The valuation of specialised property: a review of valuation methods. *Journal of Property Investment and Finance*, 22 (6), 533–541.
Geltner, D., Kumar, A. and Van de Minne, A.M. (2018) Riskiness of real estate development: a perspective from urban economics and option value theory. ASSA Annual Meeting, Philadelphia, PA, 5–7 January.
Georgiadis, P. (2020) Crossrail costs soar to £19bn and opening is pushed back to mid-2022. *Financial Times*. https://www.ft.com/content/d00bfb9e-a242-4a44-b0c2-aa289ce1eb52 (accessed 15 January 2021).
Georgiadou, M.C. (2019) An overview of benefits and challenges of building information modelling (BIM) adoption in UK residential projects, *Construction Innovation*, 19 (3), 298–320.
Goh, B.H. (2008) The state of applications of quantitative analysis techniques to construction economics and management (1983 to 2006), *Construction Management and Economics,* 26 (5), 485–497.
Goldfarb, R.S. (1975) Learning in government programs and the usefulness of cost–benefit analysis: lessons from manpower and urban renewal history. *Policy Sciences*, 6 (3), 281–299.

Hall, D.M., Whyte, J.K. and Lessing, J. (2020) Mirror breaking strategies to enable digital manufacturing. *Construction Management and Economics*, 38 (4), 322–339.
Hardcastle, C. (1992) An information model of the construction cost estimating process. PhD Thesis, Heriot Watt University, Department of Building, Edinburgh.
Harris, F., McCaffer, R. and Edum-Fotwe, F. (2013) *Modern Construction Management* (7th edn). Oxford, Wiley-Blackwell.
Hein, L., Bagstad K., Edens B., Obst C., de Jong R. and Lesschen J.P. (2016) Defining ecosystem assets for natural capital accounting. *PLoS ONE*, 11 (11), e0164460. DOI:10.1371/journal.pone.0164460.
Hendriks, D. (2005) Apportionment in property valuation: should we separate the inseparable? *Journal of Property Investment and Finance*, 23 (5), 455–470.
Hileman, M.J. (2003) Processing: combination approach improves investment decisions for capital projects. *Oil and Gas Journal*, 101 (47), 52–56.
Hillebrandt, P. (2000) *Economic Theory and the Construction Industry* (3rd edn). Palgrave Macmillan, London.
Hollander, Jacob H. (1904) The development of Ricardo's theory of value. *Quarterly Journal of Economics*, 18 (4), 455–491.
ICE (1976) *An introduction to engineering economics*. Institution of Civil Engineers.
ICGP (2003) Guidelines and Principles for Social Impact Assessment in the USA. Interorganizational Committee on Principles and Guidelines for Social Impact Assessment, *Impact Assessment and Project Appraisal*, 21 (3), 231–250, DOI: 10.3152/147154603781766293.
Ikpe, E., Hammond, F. and Oloke, D. (2012) Cost–benefit analysis for accident prevention in construction projects. *Journal of Construction Engineering and Management*, 138 (8), 991–998.
Ive, G. and Chang, C.Y. (2007) The principle of inconsistent trinity in the selection of procurement systems. *Construction Management and Economics*, 25 (7), 677–690.
Jaafari, A. (1988) Cost and performance analysis of tall structures. *Journal of Structural Engineering (United States)*, 114 (11), 2594–2611.
Jansson G., Johnsson, H. and Engström, D. (2014) Platform use in systems building. *Construction Management and Economics*, 32 (1–2), 70–82.
Jowitt, P., Johnson, A., Moir, S. and Grenfell, R. (2012) A protocol for carbon emissions accounting in infrastructure decisions. *Proceedings of the Institution of Civil Engineers, Civil Engineering*, 165 (2), 89–95.
Judy, S. (2015) Construction delays stressing Vogtle project's economics. *ENR (Engineering News-Record)*, 274 (25).
Kadiri, D.S. (2015) Construction cost models for high-rise office buildings in Nigeria. *Ethiopian Journal of Environmental Studies and Management*, 8 (Supp. 2), 874–880.
Kambo, A., Drogemuller, R. and Yarlagadda, P.K.D.V. (2019) Assessing biophilic design elements for ecosystem service attributes – a sub-tropical Australian case. *Ecosystem Services*, 39, 1–9. https://doi.org/10.1016/j.ecoser.2019.100977.
Kaplan, R.S. and Cooper R. (1988) *Cost and Effect*. Harvard Business School Press, Boston, MA.
Kelly, J. and Male, S. (2016) *Value Management in Design and Construction*. Taylor & Francis Group, London.
Kelly, J., Male, S. and Graham, D. (2004) *Value Management of Construction Projects*. Blackwell Science, Oxford. https://doi.org/10.1002/9780470773642.
Khosrowshahi, F. and Kaka, A.P. (2007) A decision support model for construction cash flow management. *Computer-Aided Civil and Infrastructure Engineering*, 22 (7), 527–539.
Kirkham, R. (2014) *Ferry and Brandon's Cost Planning of Buildings* (9th edn). Wiley-Blackwell, Hoboken, NJ.
Li, H., Arditi, D. and Wang, Z. (2012) Transaction-related issues and construction project performance. *Construction Management and Economics*, 30 (2), 151–164.
Liu, Z., Wang, L., Sheng, Z. and Gao, X. (2018) Social responsibility in infrastructure mega-projects: a case study of ecological compensation for Sousa chinensis during the construction of the Hong Kong–Zhuhai–Macao Bridge. *Frontiers of Engineering Management*, 5 (1), 98–108.
Lockwood, L.J. and Rutherford, R.C. (1996) Determinants of industrial property value. *Real Estate Economics*, 24 (2), 257–272.

Luscuere, L.M. (2017) Materials passports: optimising value recovery from materials, *Proceedings of the Institution of Civil Engineers – Waste and Resource Management*, 170 (1), 25–28, https://doi.org/10.1680/jwarm.16.00016.

Mace, G.M. (2019) The ecology of natural capital accounting. *Oxford Review of Economic Policy*, 35 (1), 54–67.

Miles, L.D. (1968) Editorial – rules of the game. *Value Engineering*, 1 (1), 5–6.

Morris, P.W.G. (2017) Climate change and what the project management profession should be doing about it – a UK perspective,.Association for Project Management, Princes Risborough, UK. https://www.apm.org.uk/media/7496/climate-change-report.pdf.

Mosca, L., Jones, K., Davies, A., Whyte, J. and Glass, J. (2020) Platform thinking for construction, transforming construction. Network Plus, Digest Series, No. 2.

Musarat, M.A., Alaloul, W.S. and Liew, M.S. (2020) Impact of inflation rate on construction projects budget: a review. *Ain Shams Engineering Journal*. 10.1016/j.asej.2020.04.009.

Musgrave, R. (1969) Cost–benefit analysis and the theory of public finance. *Journal of Economic Literature*, 7 (3), 797–806. www.jstor.org/stable/2720229 (accessed 7 July 2020).

Myers, D. (2017) *Construction Economics: A New Approach*. Routledge, New York.

Nisbet, J. (1961) *Estimating and Cost Control*. Batsford, London.

Nyborg, K. (2012) *The Ethics and Politics of Environmental Cost–Benefit Analysis*. Taylor & Francis Group, London.

O'Brien, J. (2015) How to reduce your building and estate operating costs? Constructing Excellence. https://constructingexcellence.org.uk/wp-content/uploads/2015/04/How-to-reduce-building-and-estate-operating-costs-FM-Show-March-2015.pdf (accessed 22 August 2020).

Ofori, G. (1992) The environment: the fourth construction project objective? *Construction Management and Economics*, 10 (5), 369–395.

Ofori, G. (1994) Establishing construction economics as an academic discipline. *Construction Management and Economics*, 12(4), 295.

Olatunji, O.A. and Sher, W. (2015) Estimating in geometric 3D CAD. *Journal of Financial Management of Property and Construction*, 20 (1), 24–49. https://doi.org/10.1108/JFMPC-07-2014-0011.

Olubodun, F., Kangwa, J., Oladapo, A. and Thompson, J. (2010) An appraisal of the level of application of life cycle costing within the construction industry in the UK. *Structural Survey*, 28 (4), 254–265. https://doi.org/10.1108/02630801011070966.

ONS (2017) *Principles of Natural Capital Accounting*. UK Office for National Statistics, London.

Pagourtzi, E., Assimakopoulos, V., Hatzichristos, T. and French, N. (2003) Real estate appraisal: a review of valuation methods. *Journal of Property Investment and Finance*, 21(4), 383–401.

Park, Y. and Park, G. (2004) A new method for technology valuation in monetary value: procedure and application. *Technovation*, 24, 387–394.

Perera, S., Nanayakkara, S., Rodrigo, M.N.N., Senaratne, S., and Wein, R. (2020) Blockchain technology: Is it hype or real in the construction industry? *Journal of Industrial Information Integration*, 17, 100125, https://doi.org/10.1016/j.jii.2020.100125.

Persky, J. (2001) Cost–benefit analysis and the classical creed. *Journal of Economic Perspectives*, 15 (4), 199–208. DOI: 10.1257/jep.15.4.199.

Piekarski, J.A. (1984) Simplified risk analysis in project economics. *Transactions of the American Association of Cost Engineers*, pp. D.5.1–D.5.4.

Pitonak, M. and Pepucha, L. (2016) Economic indicators to assess the bridges of rehabilitation, maintenance, monitoring, safety, risk and resilience of bridges and bridge networks. *Proceedings of the 8th International Conference on Bridge Maintenance, Safety and Management, IABMAS 2016*. CRC Press/Balkema, pp. 1735–1738.

RIBA (2020) *RIBA Plan of Work 2020 Overview*. Royal Institute of British Architects (RIBA), London.

Ribeirinho, M.J., Mischke, J., Strube, G., Sjödin, E., Blanco, J.L., Palter, R., et al. (2020) The next normal in construction: how disruption is reshaping the world's largest ecosystem. McKinsey & Company. https://www.mckinsey.com/business-functions/operations/our-insights/the-next-normal-in-construction-how-disruption-is-reshaping-the-worlds-largest-ecosystem (accessed 19 January 2021).

Richmond, B.J., Mook, L. and Quarter, J. (2003) Social accounting for non-profits: two models. *Non-profit Management Leadership*, 13 (4), 308–324.

RICS (2009) *New Rules of Measurement: Order of Cost Estimating and Cost Planning for Capital Building Works (NRM1)*. Royal Institution of Chartered Surveyors (RICS), Coventry, UK.
RICS (2014) *New Rules of Measurement (NRM) 3: Order of Cost Estimating and Cost Planning for Building Maintenance Works*. Royal Institution of Chartered Surveyors (RICS), London.
RICS (2016) *Life Cycle Costing* (1st edn). Royal Institution of Chartered Surveyors (RICS), London.
RICS (2017) *Whole Life Carbon Assessment for the Built Environment, RICS Professional Statement* (1st edn). Royal Institution of Chartered Surveyors (RICS), London.
RICS (2019) *Valuation of Development Property* (1st edn). Royal Institution of Chartered Surveyors (RICS), London.
Robertson, D. and Ulrich, K. (1998) Planning for product platforms. *Sloan Management Review*, 39 (4), 19–31.
Robinson, G.D. and Kosky, M. (2000) Financial barriers and recommendations to the successful use of whole life cycle costing in property and construction. Construction Research and Innovation Strategy Panel (CRISP).
Robson, A., Boyd, D. and Thurairajah, N. (2016) Studying 'cost as information' to account for construction improvements. *Construction Management and Economics*, 34 (6), 418–431, DOI: 10.1080/01446193.2016.1200734.
Sachs, T., Tiong, R. and Wagner, D. (2007) Political risk quantification using fuzzy set approach. *Journal of Financial Management of Property and Construction*, 12 (2), 107–126. https://doi.org/10.1108/13664380780001098.
Sawhney, A. and Muse, A. (2019) Can AI transform the way we estimate construction projects? https://www.pbctoday.co.uk/news/bim-news/ai-construction-estimates/59477/ (accessed 7 August 2020).
Shapiro, E., Davies, K. and Mackmin, D. (2013) *Modern Methods of Valuation* (11th edn). Estates Gazette, London.
Stasiak-Betlejewska, Renata and Potkány, Marek (2015) Construction costs analysis and its importance to the economy. *Procedia Economics and Finance*, 34, 35–42.
Thomas, L.D.W., Autio, E. and Gann, D.M. (2014) Architectural leverage: putting platforms in context. *Academy of Management Perspectives*, 28 (2), 98–219.
Trading Economics (2020) Spain GDP from services. https://tradingeconomics.com/spain/gdp-from-services (accessed 15 January 2021).
Trinh, K.V. and Moghanloo, R.G.G. (2020) Economic feasibility study of several utilization alternatives for a stranded offshore gas reservoir. *Proceedings of the Annual Offshore Technology Conference, 2020*, OTC 2020.
Van Eck, N.J. and Waltman, L. (2014) Visualizing bibliometric networks. In: Ding, Y., Rousseau, R. and Wolfram, D. (eds), *Measuring Scholarly Impact: Methods and Practice*. Springer, Berlin, pp. 285–320.
Vanclay, F., Esteves, A.M., Aucamp, I. and Franks, D. (2015) *Social Impact Assessment: Guidance for Assessing and Managing the Social Impacts of Projects*. International Association for Impact Assessment, Fargo, ND.
Watson, K.J., Evans, J., Karvonen, A. and Whitley, T. (2016) Capturing the social value of buildings: the promise of social return on investment (SROI). *Building and Environment*, 103, 289–301.
Watts, S., Kalita, N. and Maclean, M. (2007) The economics of super-tall towers. *Structural Design of Tall and Special Buildings*, 16 (4), 457–470.
WEF (2016) *Shaping the Future of Construction: A Breakthrough in Mindset and Technology*. World Economic Forum and Boston Consulting Group.
Weiss, J. (2000) Some reflections on project economics and environmental issues: a development economist's perspective. *Journal of Economic Studies*, 27 (1–2), 126–134.
Williamson, O.E. (1979) Transaction-cost economics: the governance of contractual relations. *Journal of Law and Economics*, 22 (2), 233–262.
Windapo, A., Odediran, S., Moghayedi, A., Adediran, A. and Oliphant, D. (2017) Determinants of Building Construction Costs in South Africa. *Journal of Construction Business and Management*, 1 (1), 8–13.
Yang, J. (2017) Environmental impact assessment. In: Yang, J. (ed.), *Environmental Management in Mega Construction Projects*. Springer, Singapore, pp. 21–26. https://doi.org/10.1007/978-981-10-3605-7_4.

Zhang, X., Wu, Y., Shen, L. and Skitmore, M. (2014) A prototype system dynamic model for assessing the sustainability of construction projects. *International Journal of Project Management*, 32 (1), 66–76.

8. Dynamics of construction industry development

Mohan Kumaraswamy and Gangadhar Mahesh

INTRODUCTION

Construction industries in developed or developing economies, have responded to periodic demands for construction industry reforms with sporadic improvement initiatives based on recommendations from government-appointed committees. Such demands may have arisen from a groundswell of pressures from construction clients and the broader community, triggered by disappointing construction industry performance levels such as in substantial schedule and budget overruns; and shortfalls in quality, health, safety, environmental and ethical safeguards. For example, work disruptions, accidents, rework and prolonged disputes, as well as low productivity, become even more conspicuous in comparison to other industries such as the aerospace or automotive industries while bid-rigging and/or overpayment scandals have led to industry investigations and reform recommendations (examples have been in Australia, Hong Kong, the Netherlands and the United Kingdom).

For example, in the United Kingdom (UK), 12 industry reform reports from 1944 to 1998 were revisited in a book edited by Murray and Langford (2003). Although the recommendations in each report were expected to cure identified industry ills, the following phrases from the Foreword and the Preface of the book highlight the failure to make progress over time, to 'radically change its structure and culture ... to improve its performance and deliver better value to its end-user clients'; adding: 'raised the same concerns ... same warnings'; 'industry continued to adopt a fragmented construction process that involved sequential procurement in spite of the recommendations given in report after report'; 'the message is strikingly similar: construction project teams must work together in true partnership and not as groups of disparate professions. Thus the call for early involvement of subcontractors'. However, such shortfalls prevailed even after the 1990s, despite the development of protocols for collaborative working following the *Constructing the Team* report (Latham, 1994) and demonstration projects on integrated teamworking with 'lean' approaches following the *Rethinking Construction* report (Egan, 1998): (1) 'partnering': first 'non-contractual partnering' based on non-binding 'partnering charters', then 'contractual partnering' through new contract forms such as the New Engineering Contract (which is used in the UK, Australia, Hong Kong and USA); and (2) 'alliancing' (for example, in the UK, Australia); and (3) 'integrated project delivery' (in the United States).

While incremental improvements have arisen from the above initiatives, it has been observed that collaborative working arrangements often degenerate after a substantial dispute arises, for example on claims for extra money or time, given traditionally adverse industry cultures. Therefore, calls for reintegrating fragmented project teams continue to this date. On the other hand, collaborative working arrangements need to be introduced and monitored with care, lest they be abused; for example, to gain access to confidential information to secure competitive advantage in future tenders. For example, while closer formal/hard relationships are expected in public–private partnerships (PPPs), Kumaraswamy et al. (2015) advocated: reinforcing these with informal/soft relationships targeting more efficient and resilient 'relationally integrated' teams; embedding in the teams suitable representatives of the 'People' who are expected to benefit from, or be affected by, the built infrastructure item; and also arranging checks and balances and effective 'relationship management' to avoid abuse of relationships that could even degenerate into collusion and corrupt practices. Wong (2014) identified the net benefits of mobilising 'social capital' (as with the end-user 'People' embedded in project teams, in the above example), targeting a more comprehensive consideration of societal needs in built infrastructure projects. This broader-based and longer-term perspective feeds well into the thrusts of this chapter, in connecting approaches to public infrastructure megaprojects to construction industry development, which in turn contributes to, and benefits from, national economic development.

Appropriately factoring in this perceived mutually reinforcing construction industry–national economy nexus may also unveil hitherto hidden force-fields that may in turn shed further light on why many construction industry reform initiatives have not yielded planned benefits. Considering one simple scenario, in the absence of proactive support mechanisms during national economic downturns, for example, to provide alternative work opportunities for deploying of core personnel, some construction organisations (both large and small) may not survive the periodic troughs in the fluctuating industry demand (Kumaraswamy et al., 2004b).

The need for unravelling such linkages had been recognised by ground-breaking researchers in this field from the 1970s, with examples of milestone findings by: Turin (1973), who probed the economic significance of the construction industry and its role in national development; Hillebrandt (1985), who applied economic theory to the construction industry; Ofori (1988), who compared the construction industry and economic growth in Singapore; and Ofori (1994) who, starting with the importance of construction in national economies and socio-economic development, identified the then state of construction economics in terms of its potential as an academic discipline. More recently, Ofori (2016) reviewed the state of knowledge on construction industries in developing countries, with a view to identifying new ways to boost industry capacities to deliver built infrastructure that improves the quality of life in such countries.

The next section of this chapter compares other recent approaches and findings and suggestions to identify and leverage connections between construction industry development and national economic development through a macro–micro construction economics dual perspective (see, for example, Ofori, 1994, 2016; El-adaway et al., 2020; Gerrard, 2020a). It drills deeper into how accelerating construction industry development could produce better-built infrastructure more reliably and efficiently, thereby helping to propel overall national development itself (e.g. Kumaraswamy, 1998, 2008, 2013; Smyth, 2020). Thus, the next section provides useful insights from a construction economics perspective on how better appreciation, operationalisation and utilisation of the critical construction industry–national economy nexus could be mutually reinforcing. The third section first provides examples of other possible reasons for slow industry development, such as misplaced priorities, missing links, inadequate strategic support; and proceeds to scan relevant thrusts and outcomes of some recent construction industry reform initiatives, with a view to help explain the slow uptake of industry reform recommendations. In the concluding observations, an attempt is made to draw upon a cross-section of the possible explanations and associated root causes for the shortfalls identified and probed in previous sections, and then provide suggestions on how some of these may be addressed with exercises to map, articulate and disseminate the linkages that can enlighten and inform 'macro' policy-makers (at economy and industry leadership levels) and 'micro' decision-makers (at organisational and project levels), to assist them in formulating win–win strategies and protocols for mutually reinforcing construction industry–national economy development, and more successful projects over the long term.

CONSTRUCTION INDUSTRY–NATIONAL ECONOMY NEXUS

Balanced and Holistic Development

This chapter aims to unveil the interplay and appropriate balance needed between the drivers, constraints and priorities of construction industry development and national economic development. For example, while a healthy, reliable and productive construction industry can be a valuable contributor to national economic development, on the other hand, a conducive economic environment is needed to fuel construction industry development. This raises 'chicken or egg' questions. While national development requires more considerations, drivers and ingredients beyond the scope of this chapter, this chapter also provides 'lay viewpoint' pointers to potential mutually reinforcing interactions that can lead to iterative win–win developments, boosting the performance of construction projects, organisations, the industry and the national economy.

An appropriate, if not optimal balance is needed for overall national development, in balancing economic and social advances with environmental impacts. The sustainability imperative and the need to combat climate change are increasingly accepted. Devastating death tolls and massive disruptions of both national and international systems by COVID-19, at the time of writing this chapter in early 2020, provide unprecedented challenges, as well as imperatives to revisit and radically revise system paradigms, including those of unbridled development. Given the increasing retreat from extensive globalisation, as countries realise the importance of self-reliance in such an international crisis, each country should make long-term socio-economic decisions which may be influenced by political agendas, rather than a longer-term vision. National economies should rise from this global mega-catastrophe. A report (Constructing Excellence, 2009) that reviewed the UK's limited progress in implementing construction industry reforms since the previous major review (Egan, 1998), quoted Egan in the foreword: 'every crisis is an opportunity'; referring at that point to an impending industry crisis following the global financial and economic crisis. This statement is amplified by the COVID-19 mega-crisis, requiring major paradigm shifts in the development of the construction industry in harmony with holistic economic and national redevelopment.

Some groundwork on developing policies and strategies for relevant reforms had been done in the past decade or more, in many countries, enabling a 'running start' adaptation to the new realities and translation to practice. For example, Manewa et al. (2016) described how some policies and related legislation in the UK that arose from the 2008 report, *Strategy for Sustainable Construction* (HM Government and Strategic Forum for Construction, 2008) and the *Construction 2025* report (HM Government, 2013) discouraged wasteful and environmentally detrimental demolition and major refurbishments, while encouraging owners and developers to seek options for adaptive reuse of buildings; with specific attention being also drawn to low embodied and operational carbon contents. Another report (Farmer, 2016) probed complementary needs and recommended parallel strategies for improving industry human resources and productivity. Thus, the knowledge base and enabling strategies have been growing, to better prepare construction industry stakeholders to face the COVID-19 imposed 'new normal' in important areas such as the 'triple bottom line' addressed by sustainability in terms of people, planet and profit.

Reverting to the broader holistic development scenario, the call for papers in 2007 for a Special Issue of the *Journal of Construction in Developing Countries* on 'Knowledge Flows for Balanced Development' articulated the intent to capture 'issues of knowledge growth, organisational development, human resource development, construction industry development and national development'. Knowledge flows were said to include 'one-way transfers and two-way flows (exchanges) of both hard and soft knowledge components between JV partners, consultants, contractors and sub-contractors etc. These also include flows between foreign and local companies, as well as between local construction organisations themselves' (Kumaraswamy (2008). The overall aim of 'balanced development' was described as:

> striking appropriate balances both within and between the development of the various stakeholders, including construction personnel, public institutions and private companies, the construction industry and the country itself. Such balancing usually targets healthy and sustainable growth of each stakeholder as well as the whole system in the long term, while achieving the desired outcomes in the short term.

For example, Kumaraswamy (2006) demonstrated how core developmental components, such as technological development, could be catalysed and harnessed to contribute to balanced economic and social growth amidst environmental harmony. Balanced development incorporating such advances could produce better-built infrastructure more reliably and efficiently by, in this example, upgrading technology levels with appropriate wider technology exchange mechanisms. Such technology exchange, say between an international contractor and a local contractor in a joint venture, is based on each party having strengths that can compensate for the shortfalls of the other (Kumaraswamy and Shrestha, 2002). Experience indicates that one-way technology transfer is rarely achieved in reality, even if mandated in a contract. However, acknowledging upfront and working towards the potential for two-way technology exchange could well justify a joint venture (Kumaraswamy, 2006). Similar collaborative mindsets and strategies could help in other construction industry developmental components and dimensions of national development.

Identifying Construction Industry–National Economy Linkages

An initial issue arises in defining the frame of reference, so that the relationships, 'numbers' (magnitudes) and their potential significance can be realistically assessed. Despite early efforts (such as Turin, 1973; Hillebrandt, 1985) in demarcating the scope of the construction industry and of construction economics as well, a snapshot by Ofori (1994) highlighted the lack of an accepted definition of the construction industry. Different approaches were said to range from some limiting it to site activity, others including planning and design, yet others covering manufacturing and supply of materials and components. Sadly, such different interpretations continue to distort the true contribution of this industry to the economy. It is hoped that this chapter and indeed the whole book that it is a part of, will bring clarity by addressing such ambiguities.

In a recent example from the UK, Gerrard (2020a), citing the Chartered Institute of Building Report on *The Real Face of Construction 2020* (CIOB, 2020) that provides a socio-economic analysis of the true value of the Built Environment, highlighted that 'Construction's economic influence in the UK is almost double that of officially recorded figures'. He observed that the Office for National Statistics (ONS) assessment of construction accounting for 6 per cent of UK economic output and employing 2.3 million people is based on a too narrow definition of the industry, since it ignored the contributions of architects, engineers and quantity surveyors, manufacturers, and other companies such as builders' merchants and plant hire providers. Adding these other aspects of the design and construction process, construction gross domestic product (GDP) was estimated (CIOB, 2020) to be close to double in size, while the Chairman of the Confederation of British Industry Construction Council, Sir James Wates, noted how the industry was 'an economic multiplier'.

More specifically, in a follow-up paper, Gerrard (2020b) added that 'every £1 spent on construction creates £2.92 of value to the UK economy'. He highlighted how the report urged addressing some critical issues, such as those in risk allocation and procurement processes (for example, moving to a two-stage system where risks are priced and allocated in the first stage before a competitive second stage), reducing ingrained adversarial behaviours, while aiming for 'industry-wide change that can deliver greater number of infrastructure projects on time and on budget'. It was argued that fundamental improvements to the business model between clients and contractors, with investments in skills, technologies and innovations, could boost productivity; while it was estimated that a productivity increase of 2 per cent in the industry could add £30 billion to the UK economy by 2030. Indeed the direct contributions of a more productive UK construction industry to more competitive manufacturing and service sectors were well articulated over 85 years ago, by a British Member of Parliament (MP) who, having worked as an architect on skyscrapers in the United States of America (USA) at the beginning of the last century, had noted much lower costs and project durations there. On returning to the UK, Bossom (1934) made a strong case for improving and streamlining design and construction practices, that is, even before the stream of UK construction industry reports from 1944 mentioned in the introduction to this chapter (Murray and Langford, 2003). For example, he claimed that bad layouts add at least 15 per cent to the production cost of the cotton industry, while also asserting that houses and factories were costing much more and taking much longer to build than they should (Bossom, 1934).

In an example from the USA, El-adaway et al. (2020) sought to study how the economic performance of the USA (as indicated by GDP) 'is impacted by the performance of the construction industry and its key players', and also how this performance could help in forecasting US GDP. For this, they modelled the relationships between GDP, total construction spending, the Standard & Poor's 500 (S&P 500) index, and the stocks of major publicly traded construction companies. They tested whether their model could have predicted the 2007–2008 financial crisis. They then investigated how governments formulated strategies to leverage their construction industries to generate economic growth. However, they could not reach a conclusion on the exact role that construction plays in overall development, because of the many variables buffeting the industry, along with its vulnerabilities to market sensitivities. Similarly, their literature review unearthed a mix of contentions, on whether construction spending stimulates economic output, or vice versa; while a third set of studies did not find any significant relationship between construction spending and economic output.

In this context, it is worth narrowing down the significant variables affecting both the construction industry and the national economy in each country separately, since contexts would introduce additional 'noise' and variabilities that may contribute to the 'ambiguities' clouding the analysis, as indicated by El-adaway et al. (2020). Deeper well-structured probes may then be formulated to help unveil specific relationships and any significant leading or lagging indicators based on the identified linkages between a specific country's construction industry and national economy. It will be important to delineate causal relationships from less significant, albeit informative (as possible indicators or proxies) correlations.

Leveraging Construction Industry–National Economy Linkages

Critical barriers to productivity linked to unrealistic risk allocation and inefficient procurement processes (Gerrard, 2020a; CIOB, 2020) are still being highlighted, although many of these have been diagnosed (along with prescribed remedies) in industry reports over many decades (Murray and Langford, 2003). Such high-powered recommendations have been complemented by further insights and proposed models and frameworks by researchers, also from the last century. For example, based on findings from a set of related studies, Kumaraswamy (1998) proposed integrated frameworks and guidelines to improve and integrate: (1) creative project packaging; (2) participant selection; and (3) operational management subsystems to enable more performance-oriented procurement, and thereby boost industry development. Industry development was then taken to include improved project performance levels and, in the longer term, development of local organisational capabilities (including capacities, competencies and cultures), and further improvements in project and construction industry performance levels. Follow-up studies by a group in Hong Kong followed the outputs of Kumaraswamy (1998). For example, findings on topics relevant to construction industry development included: (1) institutional improvements (Mahesh et al., 2010); (2) technology and knowledge exchange (Kumaraswamy and Shrestha, 2002); (3) relational contracting approaches (Rahman and Kumaraswamy, 2004) and joint risk management (Rahman and Kumaraswamy, 2005) with incentives (Rahman and Kumaraswamy, 2008), and relationally integrated value networks (Anvuur et al., 2011). Kumaraswamy (2006) showed how these and other contributory findings converged to illustrate how construction industry development would feed into more efficient and reliable infrastructure development, and thence national development (Kumaraswamy, 2013) and then developed further, as in Figure 8.1.

More recently in the UK, in the midst of the COVID-19 crisis, a paper entitled 'Construction needed intensive care before the coronavirus' (Smyth, 2020) drew attention to increasing evidence of a 'failed business model' that is underpinned by low expenditure and minimal investment. Smyth notes that, rather than focusing on return on capital investment (RoI), organisations have relied on return on capital employed (ROCE), by smart cash flow management through creative leveraging of trade credit, delaying payments to suppliers and subcontractors, and so on. The collapse of Carillion plc, a large UK contractor, and the particular disarray caused by COVID-19 in the construction industry, are seen by Smyth to have 'starkly exposed the raw edges of an unsustainable contractor business model'. Clients may tacitly condone these practices in their quest to drive prices down. However, this reflects a myopic project-focused model that is not sustainable for an organisation (whose supply chain cannot

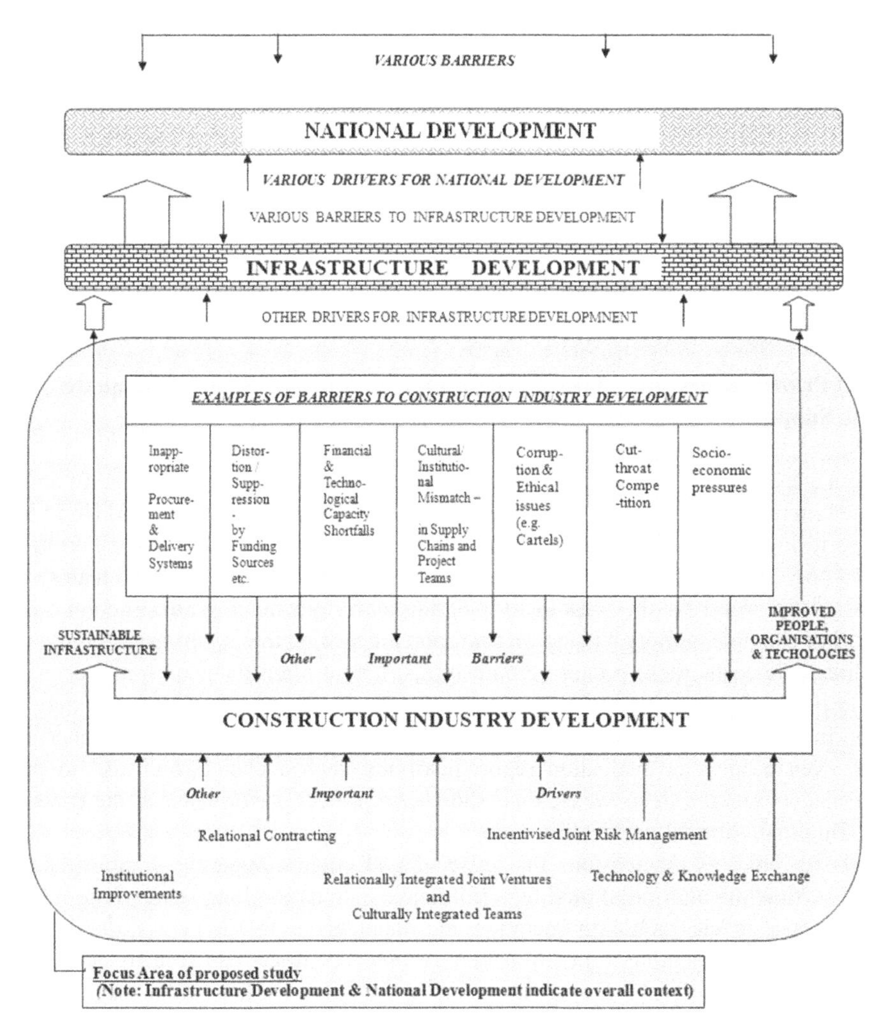

Sources: Based on Kumaraswamy (2006, 2013).

Figure 8.1 *Conceptual framework of sample drivers, barriers and linkages between construction industry development, infrastructure development and national development*

be nurtured to perform better under such a model), let alone for the industry. Smyth (2020) sees these practices as having persisted 'since the modern contractor emerged in the Second World War'. Incidentally, it was about that time that one of the major UK construction industry reform reports (Simon, 1944) was commissioned.

So, although industry ills have been receiving attention for decades, it seems that the symptoms, rather than the root causes, have been addressed. Smyth's (2020) observation that it needed intensive care before COVID-19 may turn into a dire warning unless the industry gets its act together. While also pointing to wider social implications of the resistance to change, including on health and safety, Smyth (2020) sees some positives in research through the

'Transforming Construction Network Plus', in bringing together technologies such as digitisation with organisational systems and behaviour to help transform the construction business model, also drawing inspiration from other industries such as aviation. Noting that 'social or human capital appreciates with use', he also called for investment to 'integrate across silos with better programme management within and between projects', and suggested that investing in health and safety could make industry more productive and innovative.

WHAT HAVE WE MISSED IN CONSTRUCTION INDUSTRY REFORMS?

Distorted Priorities, Missing Links, Underappreciated Realities and Inadequate Strategic Support

Why has a steady stream of construction industry reviews in the UK (Murray and Langford, 2003), and more recently in other jurisdictions such as Australia, Singapore and Hong Kong, continued to highlight the same or similar malaises, for example, those arising from fragmented supply chains and lowest bid-price selection when assembling project teams? Have 'we' missed the wood for the trees in formulating and implementing construction industry reforms? For example, although many recent reports identified the importance of more attention to life cycle costs, those procuring built infrastructure often focus on initial costs. This applies to public sector managers constrained by regulations and procedures that may deter deviation from lowest tender price, for example by suspecting any award to a higher-priced tenderer, even despite an evaluation report justifying such a choice; and also to private sector managers seeking to showcase their short-term project performance at the expense of longer-term organisational net benefits.

The above indicates the tensions that arise, if and when comparing or prioritising the project cost, timeline and initial quality/performance; versus the long-term whole-life costs and performance levels, including operation and maintenance (O&M) costs, durability and sustainability of the resulting built infrastructure. Such conflicts can lead to inappropriate decisions during planning and design (for example, designing for low initial cost, rather than sustainability) and contractor or supplier selection (for example, choosing the apparently 'cheaper' option, that may cost more in the longer term, when factoring in higher O&M including energy consumption costs) during the operational life. This suggests a need to bridge this divide by building strong links between the project management supply chains and the facilities and asset management teams. This would, for example, enable and promote a potentially useful exchange of knowledge, so that: (1) suitably detailed user requirements and priorities in advance, as well as later feedback, from the O&M teams, can better inform planners and designers to develop more usable and sustainable designs (for example, in terms of spatial planning, energy efficiency, maintainability and durability); while (2) feedforward from planners, designers and constructors can add value to as-built drawings, user manuals and building information modelling (BIM) models, in enabling facilities managers to better plan their operations and maintenance schedules by, for example, optimising the sustainable utilisation of all the design capacities and capabilities of the various parts of the built asset; by themselves, as well as a 'whole'.

The *Built Environment Project and Asset Management* journal was launched to help address the above imperative, by providing a common forum for both project management and asset management researchers and practitioners to learn from and help improve each other through two-way knowledge flows, in a common mission to deliver and sustain better-built infrastructure worldwide (Kumaraswamy, 2011). Furthermore, a research exercise in Hong Kong with parallel studies in Singapore, the UK and Sri Lanka probed the potential for providing these missing links between project management supply chains and asset management teams. Wong et al. (2014), Ling et al. (2014), Smyth et al. (2017) and De Silva et al. (2018) provided recommendations from these studies, including approaches to establishing and sustaining useful links between the supply chains, for example with 'early operator involvement' and specific knowledge exchange protocols and relationship-building (Anvuur et al., 2011) to the whole life cycle of the asset.

Given that the scope and output of the construction industry includes a sizeable segment of repairs, maintenance, refurbishments and adaptive reuse (CICD, 2011b, pp. 4–5), the facilities/asset management subsector is a significant player. The above initiatives to bridge the hitherto segregated new projects and asset management subsectors, and synergise their inputs and outputs, is a significant step forward from aiming to integrate teams within the project management subsector alone. Indeed this ties in with current trends, as in the recently reported case study on how 'The University of Birmingham is digitising its entire estate to streamline capital project delivery and control vast amounts of asset information spanning the university's premises' (BIMPLUS, 2020). The university has reportedly 'deployed a "common data environment" (CDE) to enhance management and control of digital information, and drive bottom-line improvements across every stage of a project's delivery and an asset's lifecycle', designed to 'underpin the university's capital investment programme over the following five to 10 years' and to improve decision-making. The CDE is expected to enable 'single, secure access to all documents, drawings, 3D Models and data regarding the university's circa 250 buildings'. Another important feature is in enabling the supply chain to find relevant information easily, while also protecting the information. The importance of such protection was brought home after the recent collapse of a big UK-based contractor company with a scramble to recover information held by it. Indeed, such scenarios may proliferate after the COVD-19 fallout.

The University of Birmingham case also ties in to Dame Judith Hackitt's 'Golden Thread', designed to mitigate risk, improve building quality and ensure information can be easily accessible to those who need it. This came out of the *Building a Safer Future* report commissioned by the UK Government after the Grenfell Tower fire in London (Hackitt, 2018). Illustrating the scale of the problem, it was noted (BIMPLUS, 2020), that 'up to 200,000 contractor documents had to be located for the second phase of the Grenfell inquiry'. Yet another underappreciated reality is that of the strong linkages between the construction industry and the national economy, as discussed above (and not only through more efficiently built and user-friendly infrastructure, but also through the economic multiplier effects, and so on). How these strong (but less visible) linkages could be proactively leveraged for the greater good of the public is also illustrated above. However, short-term political targets frequently displace long-term socio-economic and overall sustainability goals from national development agendas. Therefore, there is a lack of both well-positioned 'champions' and broader-based strategic support to drive through long-term construction industry development agendas. For example, although many reports highlighted the adverse effects on organisational and human resource development of peaks and troughs in construction demand – that is, of the lack of a reasonably assured continuous workload – recommendations for providing alternative work opportunities in public infrastructure augmentation, rehabilitation or adaptive reuse during demand troughs have not been actively pursued.

Identifying and Addressing Common Construction Industry Shortfalls in Developing Countries

National procurement policies, and common procurement practices of large private clients, have been found to substantially influence construction industry growth trajectories, as well as the capacity development of construction organisations in many countries. For example, Aziz and Ofori (1996) showed how government procurement policies had impacted substantially on the development of domestic contractors in Malaysia. Similarly, Gounden (1996) described a Procurement Reform Initiative in South Africa to increase participant capacities, output and competitiveness; while Kumaraswamy (1998) attributed distortions of a historically strong Sri Lankan construction industry, to inappropriate and inconsistent procurement policies that, for example, discouraged prequalifying local companies for large projects. In one measure to address this distortion, Sri Lanka Cabinet guidelines (ICTAD, 1988) helped to enforce the then guidelines of the World Bank (World Bank, 1992) and Asian Development Bank (ADB) that incentivised deploying domestic contractors in developing economies.

The World Bank sponsored a study on measures to improve the 'Domestic Construction Industry in the Roads Sub-Sector' in Sri Lanka (Kumaraswamy, 1998). This study also highlighted some generic problems in foreign-funded (whether with bilateral or multilateral funding) megaprojects in developing countries; for example, in a paradox of not prequalifying domestic contractors as tenderers, because they have inadequate experience and capacity, while these contractors are thereby denied any work opportunities to develop their experience and capacity. Thus, Sri Lanka paid the price of higher project costs that covered the mobilisation, overheads and profits of foreign consultants and contractors even for components of work that could have been handled locally and were, in any case, frequently subcontracted to local firms at much lower rates. The action plan in the ensuing report (ICTAD, 1992) included many recommendations related to improved procurement systems:

1. Rationalisation of the previous system of registration, grading and prequalification of contractors.
2. Increasing opportunities for Local Competitive Bidding (whereas most projects had been previously restricted to International Competitive Bidding or Limited International Bidding) by reducing package sizes, such as by 'slicing', either vertically (dividing the length of road) or horizontally (in terms of operations such as earthworks and surfacing).
3. Extracting the funding component envisaged for equipment from a series of projects under a forthcoming ADB programme, and using it to finance and maintain a centralised specialised heavy equipment pool, including asphalt plants to serve a group of domestic contractors.
4. Enforcing the then prevalent World Bank (1992) procurement guidelines and similar ADB guidelines for a 7.5 per cent preference margin for domestic contractors in developing economies, along with such safeguards against abuse by nominal joint ventures purely set up to gain a preferential advantage.

5. Consideration of alternative Conditions of Contract, including the New Engineering Contract.

Other recommendations included centralised information systems, import duty and tax concessions for construction equipment, and specific training programmes in technical and managerial skills. A cross-ministry Co-ordination Committee was established to implement the first four recommendations.

While a number of domestic roadworks contractors developed their capacities based on the above implementation, the contract management of many projects, whether handled by local or foreign contractors, still left much to be desired, leading to cost and time overruns (Kumaraswamy and Raza, 2014). Most of the above procurement issues that were identified in 1992 and the residual procurement issues, together with the sources of time and cost overruns that were noted in 2014, relate to construction economics, and can benefit from the integrated overview recommended in this chapter, that connects the short-term and 'micro' project performance objectives to the longer-term and 'macro' industry and national economic development goals.

Figure 8.2, developed from that by Kumaraswamy (1998), illustrates the envisaged contributors to enhancing project performance levels in the short term through improved and integrated procurement (including work packaging, contract type and team selection subsystems) and management support systems (including planning, quality, safety and monitoring subsystems); and in the longer term (including technological systems) through organisational and industry development. Contributions from improved technological systems were targeted by both: (1) encouraging technological innovations through performance-oriented procurement and longer-term mindsets with incentives for organisational research and development (R&D); and (2) two-way technology exchange with local and overseas partners, with each bringing different strengths to a project, enabling mutual learning and equipping their organisations better for forthcoming projects (Kumaraswamy et al., 2004a).

A more recent construction industry improvement initiative was launched in India; it was led by academics and with strong participation from leading construction industry representatives (Loganathan et al., 2017). It was launched from IIT Madras with a Regional Building Clients Roundtable in 2015, bringing together top-tier clients for improving industry practices and performance levels. Three more Roundtables were arranged in 2016: with Clients in Mumbai; with Consultants and Project Managers; and a Clients–Consultants Consolidation Roundtable in Chennai (Ci3, 2016). Action Teams were set up after the Mumbai Roundtable to address 19 critical issues of the Indian construction industry, as identified at the first two construction clients Roundtables. The following Action Teams were formed, each with members from academia and industry:

- Action Team 1: Identification and Formulation of Key Performance Indicators (KPIs).
- Action Team 2: Construction Project Time and Cost Reduction Strategies.
- Action Team 3: (3A) Design Processes; (3B) Technology Adoption.
- Action Team 4: Design Codes and Standards. (Note: mobilisation was deferred, given the longer-term horizon.)
- Action Team 5: Human Capital (including Labour, Technical, and Managerial, and Skills Development) and Productivity.
- Action Team 6: Construction Clients' Charter.

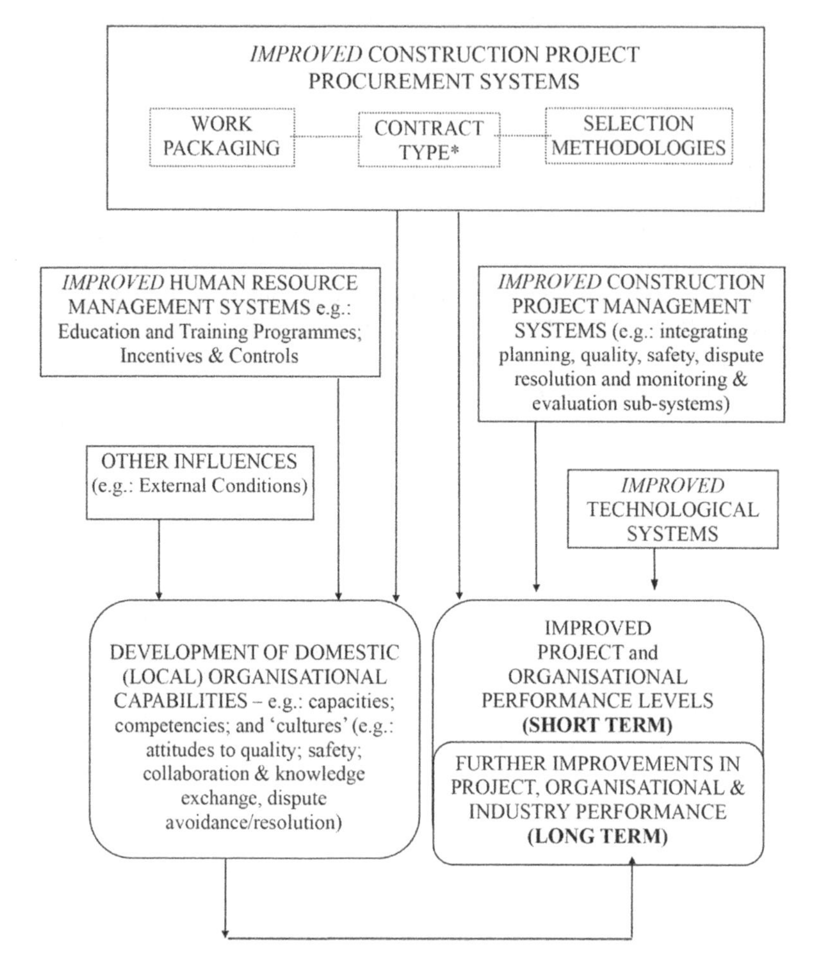

Note: * 'Contract type' includes functional grouping of financing, project management, design, construction, operation; and payment modalities and contract conditions.
Source: Developed from Kumaraswamy (1998, 2013).

Figure 8.2 *Core contributors through procurement, management, human resource and technology improvements to enhance project and organisational performance levels and industry development*

- Action Team 7: Institutional Platform for Construction Industry Improvement Initiative India. (Later renamed as Construction Industry Institute India, Ci3.)

The enthusiasm generated and the outputs from the Roundtables and some related events (Ci3, 2016) and the Action Team interactions, led to a proposal to set up a Construction Industry Institute India (Ci3) with a similar basic rationale and goals (Ci3, 2017a) to the Construction Industry Institute USA, and the European Construction Industry Institute. However, the actual setting up of Ci3 is not proceeding to plan, despite the enthusiasm and perceived net benefits. This is attributable to the difficulty of mobilising suitable champions from academia and industry who can commit to steer it through the initial years. Still, the interim outputs of the Action Teams are available as Action Team Draft White Papers (Ci3, 2017b) and an exercise is under way to consolidate these into a Ci3 White Paper.

Specific Action Team outputs have been well received by industry and academia, for example: Design Process standardization for India, from Action Team 3 (Joe et al., 2017); Key Performance Indicators (KPIs) from Action Team 1; and a Clients' Charter from Team 6 (Kumaraswamy et al., 2017a). A Clients' Charter was seen as necessary to secure the commitment of construction clients, starting with major clients who were already involved with the Ci3 initiative and enlisting others. A long-term development perspective required initial commitments to pursue better value rather than lowest cost (for example, in selecting contractors and suppliers), and to invest more in safety and environmental protection. Thus, enlisting a critical mass of clients to sign and abide by the Charter is needed to raise the bar in property developments.

Kumaraswamy et al. (2017a) described the rationale and inputs and outputs on the suggested KPIs for building clients in India. The outputs provide for organisations to choose from this sample suite of KPIs (Ci3, 2017b): (1) from three different groups, that is, design phase, construction phase, or business outcomes; and (2) at three different levels, that is, project or organization level, benchmarking club level, and industry level. They could use the chosen KPIs to benchmark internally over time or across their projects and/or with other organisations. They could also modify these if using only internally, or agree to modify jointly if benchmarking. Two more columns of the KPIs template will be populated in the next stage of development: (1) to design realistic weighting indicators that will help allow for special conditions by adjusting a typical KPI value accordingly; and (2) to collect data for, and determine typical value ranges under 'normal' conditions.

In the context of the industry–economy nexus, there is another issue that not only drives up prices, but also distorts national development in many countries, albeit to different degrees: corruption. For example, commissions that may be demanded by politicians and other decision-makers on megaprojects in some jurisdictions would need to be factored in to bid prices, while price-fixing by colluding cartels of bidders would push prices upwards. Such unethical and usually illegal practices also negate construction economics principles of lean/efficient competition-beating pricing. National economies also suffer in other ways: for example, where less beneficial (to the public) megaprojects are promoted by corrupt decision-makers and intermediate 'fixers' who envision bigger commissions in such locations; or project types where the competition is limited. In such scenarios, honest contract administrators may also encounter resistance in choosing the most appropriate procurement type, being pressurised to choose one where it is easier to reduce competition, hide inflated costs and extract windfall profits. Apart from violating many basic principles and theories, this again defies modelling through mainstream construction economics, requiring industry-specific adjustment mechanisms.

Lessons from Construction Industry Improvement Exercises in Developed Countries

The unethical practices referred to in the last paragraph are not the preserve of developing countries. For example, a recent survey by the Competition and Markets Authority (CMA) in the UK (Gerrard, 2020c) found that 'nearly a third (29%) of construction companies think it is OK to meet up with competitors to agree prices'. The CMA had observed that many recent cases involving anti-competitive practices such as price-fixing, bid-rigging and dividing markets or customers between competitors had come from the construction industry. In an example from another jurisdiction, the first case brought before the Competition Tribunal following the 2015 Hong Kong Competition Ordinance involved bid-rigging; while the first market study commissioned by the Hong Kong Competition Commission identified bid manipulation practices in the residential building renovation and maintenance market (Hickin and Ha, 2018). Grey areas persist despite various guidelines on tendering, bid-rigging and model non-collusion clauses. For example, Hickin and Ha (2018) showed in the Hong Kong context how to differentiate the rather unusual terms 'serious bid-rigging' (illegal), 'non-serious bid-rigging' (allowed, if exempted, as for some small and medium-sized enterprises) and 'joint tendering' (allowed, for example, where known in advance to the client).

In another example, the Dutch construction industry, having had to face a parliamentary inquiry following allegations of cartels among big contractors, was then subject to a comprehensive review targeting a culture change and innovation (PSIB, 2003). The objectives of that review are still relevant to many other countries today, where the word 'profitability' may be replaced with something more long term (for example, 'viability and sustainability') in the third objective: (1) improving socio-economic benefits from construction; (2) increase of added value to clients and other stakeholders; (3) increase of profitability of the construction industry; (4) creation of a competitive environment that stimulates innovation, thus enhancing the image and reliability of the industry; and (5) accumulation and rapid utilisation of knowledge.

In terms of such broader industry improvement initiatives in developed countries, in Australia a National Building and Construction Committee developed an Action Agenda for the industry in 1999 (ISR, 1999) in a report entitled *Building for Growth*. This title aptly captured the industry–economy nexus that is explored in this chapter. This report was followed by other exercises in Australia, for example, as in the 'Scope for Improvement' series of reports from 2006 to 2014 that, with support from the Australian Constructors Association, identified obstacles and pressure points faced in working to deliver successful infrastructure projects (Ashurst, 2014); and the 'Public Infrastructure' series of reports by the Australian Government Productivity Council (2014), based on an 'urgent need to comprehensively overhaul processes for assessing and developing public infrastructure projects'.

Government-backed construction industry development endeavours can be traced back to 1934 in the UK (Bossom Report), with high-profile reports published over six decades such as those by the Simon Committee in 1944, Phillips in 1948–50, Emmerson in 1962, Banwell in 1964, Tavistock Buildings in 1965, Wood in 1975, NEDO Faster Building for Industry in 1983, NEDO Faster Building for Commerce in 1988, Latham in 1994, and Egan in 1998. Some issues identified in earlier reports still re-emerged in subsequent reports. Examples include wasted resources and avoidable disputes, which arise from a fragmented industry. Industry-wide innovation, learning and change are also difficult to attain because of this fragmentation (Kumaraswamy et al., 2017a).

Some developing countries face some of these same issues, for example, those whose colonial legacies include adversarial construction project delivery systems based on separated functions of design, supervision and construction. Despite perceived merits, including safeguards against collusion, these separated functions add supervisory layers, conflicts and disputes that some of the above reforms aimed to reduce by moving towards more collaborative and joint performance-oriented procurement protocols, with less wasteful 'lean' approaches, faster adoption of suitable technologies and efficient supply chain management, in modalities based on a win–win partnering philosophy, for example in the UK (Latham, 1994; Egan, 1998) and Hong Kong (CIRC, 2001). This also aligns with the relational contracting (Rahman and Kumaraswamy, 2004) and joint risk management (Rahman and Kumaraswamy, 2005) approaches discussed above. However, why they could not be achieved in practice had been attributed to the resistance to change, the adversarial culture, and the lack of agreement and focus on common overall value (Anvuur et al., 2011).

To drill deeper to unearth other possible reasons – apart from fragmented and non-cooperating teams with multiple agendas, hence a reactive and non-conducive industry culture – for the low uptake of industry reform recommendations, a comparative three-country study helped to assess the outcomes about ten years after landmark industry reform reports around the turn of the century, in the UK (Egan, 1998), Singapore (C21, 1999) and Hong Kong (CIRC, 2001). The objectives included: assessing the roles of government agencies and the private sector in the implementation of the advocated reforms; comparing the institutional characteristics of the construction industries in Hong Kong, Singapore and the UK, and the extent to which these characteristics influenced the implementation of the above reforms; and drawing lessons from these three implementation programmes.

Kumaraswamy et al. (2010) unveiled some of the institutional constraints and enablers that may have retarded or could have had an impact on the planned reforms (CICID, 2011a, 2011b). The Hong Kong study sought: (1) to evaluate the effectiveness of the industry improvement programmes against their original objectives, while discounting, or allowing for, industry development trends that may have continued even without reform inputs, and changing aspirations, priorities and concerns; (2) to unravel reasons for presently perceived shortfalls in the above implementation and to unearth any root causes, with particular reference to relevant public agencies and private organisations involved; and (3) to unveil lessons to improve future industry improvement agendas and help realign current development trajectories.

Notably, culture is not directly highlighted as a variable in the above study, despite being identified before, because it has eluded efforts to effectively change it. Looking for causal relationships in the context of what one can 'get a handle on' to change, industry culture is seen as reflected in institutionally embedded practices (Powell and DiMaggio, 1991) shaped by the broader institutional context, including those of key organisations, legal, regulatory and contractual frameworks, and established communication channels and mindsets; that is, in the institutional characteristics of the industry. CIRC (2001) noted the difficulties in achieving the radical culture change propagated in many industry reports and for a; for example, in changing the prevalent adversarial culture. It was thus suggested to first target a 'behaviour change', which can in turn generate culture change. For this, it was proposed to enable, incentivise and in some cases mandate behaviour change through institutional improvements, for example in contractual conditions (including relational contracting principles and joint risk management practices), proactive team selection criteria (say, by rewarding desired attributes), and new regulations and codes (say, for accelerating overdue payments).

In Hong Kong, the *Construct for Excellence* report (CIRC, 2001) issued 109 recommendations under the following themes or strategic thrusts: (1) fostering a quality culture; (2) achieving value in construction procurement; (3) nurturing a professional workforce; (4) developing an efficient, innovative and productive industry; (5) improving safety and environmental performance; (6) devising a new institutional framework to drive the industry; and (7) implementation of a change programme for the industry. The government set up a Provisional

Construction Industry Co-ordination Board (PCICB) to commence implementation, pending the establishment of the statutory industry coordinating agency, now set up as the Construction Industry Council (CIC). The PCICB submitted periodic progress reports on the implementation (PCICB, 2005), while the final report to the Legislative Council (ETWB, 2007) summarised progress on each recommendation, for example as 'actions completed' and many 'to be further followed up by CIC' or 'to be further followed up by Government'. This final report conveyed an official summary of only 50 'actions substantially completed', whereas 29 were reportedly implemented in public works and had to 'be considered by CIC for rolling out to the private sector', with 34 recommendations (some possibly overlapping with the above 50 or 29) having 'to be followed up by CIC' (15) or 'by the Government' (19). This is in the context of many recommendations being originally linked to short time-frames in January 2001, such as 'within 2 years' with only a few 'within 5 years' or 'to start within 2 years', and just a few 'as soon as possible' and a couple, including for setting up the industry coordinating body, 'as soon as possible subject to enactment of the necessary legislation'. However, delays in the legislation and the consequent implementation machinery contributed to delays in implementation of many of the recommendations. The bill to set up the CIC was only first introduced into the Legislative Council in February 2004, and the CIC Ordinance emerged in May 2006 with the CIC being established in February 2007.

The main functions of the CIC are 'to forge consensus on long-term strategic issues, convey the industry's needs and aspirations to Government, as well as provide a communication channel for Government to solicit advice on all construction-related matters'. To propagate improvements across the industry, the CIC is empowered to 'formulate codes of conduct, administer registration and rating schemes, steer forward research and manpower development, facilitate adoption of construction standards, promote good practices and compile performance indicators'. However, it appears that the CIC still needs to rely on existing institutions to propagate its proposed reforms – for example, codes of conduct and guidelines – in order to improve industry practice and enhance overall performance.

Comparing the thrusts and implementation exercises, with the UK and Singapore, first, in the UK, Egan (1998) aimed at radically transforming the construction industry, with five key change drivers: committed leadership, focus on the customer, integrated processes and teams, quality-driven agenda, and commitment to people; as well as to achieve envisaged productivity gains of up to 30 per cent, price and programme certainty, vastly improved safety, and so on. The Strategic Forum, formed in 2001 to oversee these reforms, set revised targets to be achieved by the end of 2007 (Strategic Forum, 2002), and later extended this horizon to 2012 (Strategic Forum, 2007).

Second, Singapore also targeted a radical transformation, aiming to build a world-class construction industry, for which Dulaimi et al. (2001) highlighted the motivators and enablers; and to be a 'word class builder in the knowledge age' (C21, 1999; Ofori, 2002). The strategic thrusts were (C21, 1999): (1) enhancing the professionalism of the industry; (2) raising the skills level; (3) improving industry practices and techniques; (4) adopting an integrated approach to construction; (5) developing an external wing; and (6) a collective championing effort for the construction industry.

An overview reveals that it is not just that the targets keep moving and unexpected barriers spring up, as with COVID-19, but even the original targets were not achieved. For example, in the UK, the *Never Waste a Good Crisis* report comments on UK's progress in implementing the Egan (1998) recommendations (Constructing Excellence, 2009, p. 4) start with: 'there has been some progress, but nowhere near enough. Few of Egan's targets have been met in full, while most have fallen considerably short. While improvement has been achieved, too often the commitment to Egan's principles has been skin-deep'. In Hong Kong, the Construction Industry Performance Reports are published annually by the CIC. The first report in 2013 provided a performance overview in terms of productivity, health and safety and manpower over 2001 to 2011. The second report in 2014 added a fourth item: dispute resolution. The 2015 report (Construction Industry Council, 2015) also added 'environment KPIs'. However, the currently published KPIs do not include the dispute resolution KPIs (Construction Industry Council, 2020).

In comparison, the UK has a more comprehensive suite of construction KPIs, compiled on the basis of data from across the industry (Constructing Excellence, 2018). Previous versions also included a 'KPIzone' suite of products under the Constructing Excellence umbrella that helped construction organisations to measure and benchmark performance against national data. Such benchmarking at national, inter-organisational and intra-organisational (across projects) levels was envisaged in the proposal for KPIs in India (Kumaraswamy et al., 2017a; Ci3, 2017b). Realistic and reliable suites of KPIs can be powerful aids when planning, implementing and monitoring construction industry development initiatives, whether in developed or developing countries. This is another area where mainstream construction economics principles and practices, including on realistic developmental indicators, may be drawn upon to assist in evaluating construction industry development, both by itself and as a contributor to the national economy.

CONCLUDING OBSERVATIONS ON THE WAYS FORWARD

'We can't solve problems by using the same kind of thinking we used when we created them' are words attributed to Albert Einstein. He conveyed that idea in the context of the need for a new ethical perspective that could transcend nationalistic drivers and militaristic agendas, for mankind to survive in a world where nations possess nuclear and biological weapons. The need for such new thinking is needed in solving the problems in the construction industries and their linkages with national economies, given that effective solutions have not yet emerged from efforts to implement recommendations from many industry reform reports. For example, are the 'right' KPIs being used to measure performance, or should more measures be formulated that focus on evaluating the inputs, apart from the outputs? Some may argue that a performance specification should focus on outputs, rather than be concerned about the means to that end. However, since the persisting shortfalls in targeted outputs seem to arise from hitherto hidden root causes, should a few of the KPIs be designed to evaluate the status of the identified root causes, and also a few critical inputs, thereby obtaining some advance information of 'feedforward' from inputs and their processing, rather than only feedback from outputs? Much more work is needed to extend and complete both national-level exercises such as that started by Kumaraswamy et al. (2017a) in India, and international-level exercises on KPIs, such as that started by the Global Leadership Forum for Construction Engineering and Management Programs (GLF-CEM) (Hastak et al., 2017; GLF-CEM, 2017).

Highlighting and expanding on some issues described above in the section on 'Distorted Priorities, Missing Links, Underappreciated Realities and Inadequate Strategic Support', bridging critical gaps between the project management and built asset management supply chains, is important to develop a whole-life overview of not just life cycle costs, but the overall value of planned built infrastructure. This would inform and enable decisions by policy-makers on public megaprojects, and clients on smaller projects. This should be linked to other relevant emerging imperatives: addressing long-term societal needs including sustainability objectives in the planning and delivery of public built infrastructure (Wong, 2014); for this purpose, mobilising all relevant stakeholder groups and focusing on their overall best value (Kumaraswamy et al., 2017b); and injecting procurement safeguards including transparency, to ensure ethical practices and avoiding collusion in selecting first the 'right' projects and then the 'right' supply chain members.

Another critical issue worth highlighting was described above: Smyth (2020), while highlighting construction industry's long-standing need for 'intensive care' of its business model, saw possible breakthroughs from new approaches in 'Transforming Construction Network Plus', in bringing together technologies, such as digitisation, with organisational systems and behavior. Moving forward on that basis, the Infrastructure Client Group (ICG, 2020) aims to bring together the 'UK's most progressive economic infrastructure clients in partnership with government and industry', and to drive 'improvement in the development and delivery of the UK's economic infrastructure for the benefit of the end users of the assets – society'. For example, PROJECT 13 (ICE, 2018), is an industry-led initiative in the UK to rethink and reinvent infrastructure delivery models that are said to 'have failed … clients, their suppliers, operators and users of infrastructure systems and networks'. The new business model is said to be 'based on an enterprise, not on traditional transactional arrangements – to boost certainty and productivity in delivery, improve whole life outcomes in operation and support a more sustainable, innovative, highly skilled industry', as well as to promote efficiency and high performance. Potential contributions through a construction economics prism, drawing on lessons learned in other industries as well, could also inspire new thinking and pilot-testing in this regard.

If not, the procurement and delivery systems deployed on most projects are unlikely to change much for many years, from the present mix, with a large proportion of traditional/adversarial, and a smaller proportion of 'collaborative' contracts, that is, based on partnering, alliancing or 'integrated project delivery' (Kumaraswamy et al., 2017b). The move from traditional/adversarial to collaborative contracts has been slow, having started in the early 1990s (Humphreys et al., 2003). A recommendation to accelerate this move has been to inject specific selection criteria that add preferential marks in supply chain selection processes, to applicant consultants, contractors and subcontractors, for their track record (and/or potential) in collaborative contracts (Kumaraswamy and Matthews, 2000). This also targets accelerating the elusive broad-based culture change sought by various industry reform initiatives, through innovative procurement processes that used such proactive selection criteria to assemble potentially collaborative teams (including clients' representatives and supply chains) which may be expected to develop the desired project culture (Kumaraswamy et al., 2002).

More background and specifics on the criticality of procurement and management/delivery systems to project success, as well as to organisational and industry development, are discussed below, for example in providing viable strategies and useful mechanisms to move forward from fragmented supply chains, and lowest tender/bid-price selection that has often led to poor performance, project disruptions or worse. Going further, longer-term relationally integrated value networks (RIVANS) are expected to lead to closer integration based on their common value focus of thereby 'turbo-charged' supply chains (Anvuur et al., 2011; Wong et al., 2014; Ling et al., 2014; Smyth et al., 2017; De Silva et al., 2018). However, the paradigm shift envisaged by PROJECT 13 (ICE, 2018), the Infrastructure Client Group (ICG, 2020) and Smyth (2020) could, if successful, bring about a far more fundamental transformation that may have a better chance of launching the long-awaited giant leap forward, although it will need to be pilot-tested, proven and disseminated. Furthermore, the direct benefits of performance-oriented procurement are supplemented by indirect benefits such as the organisational development of suitably selected longer-term supply chain members, which in turn feeds back into greater efficiencies, since contractors, subcontractors and suppliers would be expected to build up their human and equipment resources and develop technologies that enable economies and innovations that would reduce their costs, and hence presumably their pricing to clients; subject to clients maintaining competitive mechanisms and continually benchmarking against others, for which the principles of construction economics can provide useful insights, bolstered by lessons from other industries.

As another industry-level example, to help core industry players weather troughs in construction demand during economic downturns, governments could launch some infrastructure projects. Government–industry engagement could encourage construction organisations, being reasonably confident in workload continuity, to invest in R&D, innovations and personnel upskilling. Tax incentives or seed funding could help these further. Such policy decisions, amidst limited resources, would be informed by broad, interdisciplinary investigations of the interplay between economic development and construction industry development in general, as well as in particular countries. In terms of organisational development, it is expected that larger consultants and contractors (and indeed clients), taking a longer-term view supported by being confident of work from repeat clients, would invest in R&D on breakthrough technologies such as in off-site work, including modular constructions, drones, three-dimensional (3D) printing, simulations with BIM, 'digital twins'; while small and medium-sized enterprises (SMEs) in the supply chain would also upgrade their information and knowledge management systems, training programmes, and so on. Kumaraswamy et al. (2004b) showed how such top-down and bottom-up initiatives could be synergised; Kumaraswamy et al. (2006) provided examples from specific outputs from R&D projects with industry in: (1) developing procurement and management support systems for large clients; (2) developing information and knowledge management systems for small and medium-sized contractors (which form the backbone of the industry); and (3) indicating how such initiatives could and should be integrated with synergised inputs and outputs for holistic industry development.

This chapter identifies some weaknesses in the construction industry structures and relationships, and hence overall dynamics, as contributing to shortfalls in expected outcomes from many much-heralded high-powered industry reform initiatives in many countries over recent decades. This panoramic overview also cautions as to how socio-economic drivers may deflect planned trajectories of construction industry development; for example, by injecting short-term political targets into development programmes, and distorting true priorities of long-term sustainability and societal goals. Three recent reform initiatives in Hong Kong, Singapore and Malaysia should be mentioned before closing this chapter, since these are ongoing live examples of envisaged remedial strategies. However, they may be recalibrated and revamped following the COVID-19 pandemic.

First, *Construction 2.0: Time to Change*, a report of the Hong Kong SAR Government (HKDB, 2008) on the future of the local construction industry, identifies the principal challenges and recommends remedial actions being taken and proposed following the government policy address in 2018. These are grouped under the three pillars of innovation, professionalisation and revitalisation. An example of innovative approaches is a specific thrust on off-site or industrialised construction, with pilot projects on modular integrated construction, also known as prefabricated prefinished volumetric construction (PPVC) in Singapore, or simply modular construction in many developed countries, where large three-dimensional units rather than two-dimensional panels are pre-engineered, precast and mostly prefinished in factories, then transported and installed on site (Pan and Hon, 2018; Wuni and Shen, 2020).

Second, in Singapore, the Building and Construction Authority (BCA) is driving an initiative under the banner of the Construction Industry Transformation Map based on the 'collective effort of the Industry, Institutes of Higher Learning, Union and the Government', envisioning an 'Advanced and Integrated Sector' with 'Progressive and Collaborative Firms' and 'Good Jobs for Singaporeans'. The key areas identified to transform the sector are: design for manufacturing and assembly (DfMA); green buildings; and integrated digital delivery, which includes the wider use of building information modelling (BIM). The first key area, DfMA, includes what is called modular construction, described as MiC in Hong Kong and PPVC in Singapore, as described above. The need for controlled working environments after lessons learned from COVID-19 disruptions and uncertainties will surely accelerate demands for such factory-based production and reduced on-site activities.

Third, Malaysia's Construction Industry Development Board (CIDB) has launched the Construction Industry Transformation Programme (2016–2020), being a 'national agenda to transform the construction industry to be highly productive, environmentally sustainable, with globally competitive players while focused on safety and quality standards'. 'Strategic thrusts' identified were: quality, safety and professionalism; environmental sustainability; productivity; and internationalisation and competitiveness. Eight focus areas and 11 KPIs were identified.

In summary, the panoramic overview in this chapter cautions as to how socio-economic drivers may deflect planned trajectories of construction industry development; for example, by injecting short-term political targets into development programmes, and distorting true priorities of long-term sustainability and societal goals. Therefore, from a broad brush construction economics perspective, this chapter also highlights a need to clearly identify and articulate the construction industry–national economy nexus in each country, to jointly motivate both public policy-makers and industry leaders in a given country, so that their interdependence can be mobilised into long-term win–win strategies to upgrade the performance levels of construction projects and organisations, as well as the overall industry and the national economy.

REFERENCES

Anvuur, A.M., Kumaraswamy, M.M. and Mahesh, G. (2011) Building 'relationally integrated value networks' (RIVANS). *Engineering Construction and Architectural Management*, 0969-9988, Vol. 18, No. 1, pp. 102–120.

Ashurst (2014) Scope for improvement 2015 – construction and infrastructure. 24 June. https://www.ashurst.com/en/news-and-insights/legal-updates/scope-for-improvement-2015-construction-and-infrastructure/ (accessed 4 May 2020).

Australian Government Productivity Council (2014) Public infrastructure, a series of reports. https://www.pc.gov.au/inquiries/completed/infrastructure/report (accessed 4 May 2020).

Aziz, A.A. and Ofori, G. (1996) Developing world-beating contractors through procurement policies: the case of Malaysia. *CIB W92 'North meets South' Procurement Systems Symposium Proceedings*, Durban, South Africa, ed. Taylor, R.G., pp. 1–10.

BIMPLUS (2020) Case study: how the University of Birmingham digitised its estate, based on observations of Richard Draper, BIM and digital assets manager. Estates Office, University of Birmingham, 26 April, CIOB. http://www.bimplus.co.uk/projects/case-study-how-university-birmingham-digitised-its/.

Bossom, A.C. (1934) *Building to the Skies: The Romance of the Skyscraper*, 1st edition. New York: Studio Publications.

Construction 21 Steering Committee (C21) (1999) *Reinventing Construction*. Construction 21 Review Committee, Ministry of Manpower and Ministry of National Development, Singapore.

Ci3 (2016) Completed events. Construction Industry Institute India. http://www.ci3.in/completed_events.html (accessed 2 May 2020)

Ci3 (2017a) Home page. Construction Industry Institute India. http://ci3.in/.

Ci3 (2017b) Action Team Draft White Papers. Draft White Papers, Construction Industry Institute India. http://www.ci3.in/draft_whitepapers.html (accessed 2 May 2020).

CICID (2011a) *Summary Report, Symposium: Construction Industry Development – Comparison and Acceleration*. 18 January, Centre for Innovation in Construction and Infrastructure Development, University of Hong Kong. http://www.civil.hku.hk/cicid/3_events/102/102_summary_report.pdf (accessed 27 April 2020).

CICID (2011b) *Summary Report, Focus Group Meeting II: Construction Industry Development – Comparison and Acceleration – Validation of Research Findings*. 14 October, Centre for Innovation in Construction and Infrastructure Development, University of Hong Kong. http://www.civil.hku.hk/cicid/3_events/112/112_summary.pdf (accessed 27 April 2020).

CIOB (2020) *The Real Face of Construction 2020*. Chartered Institute of Building (CIOB).

CIRC (2001) *Construct for Excellence*. Construction Industry Review Committee (CIRC), HKSAR Government, Hong Kong.

Constructing Excellence (2009) *Never Waste a Good Crisis*. Constructing Excellence, UK.

Constructing Excellence (2018) *UK Industry Performance Report 2017 – Based on the UK Construction Industry Key Performance Indicators*. https://constructingexcellence.org.uk/wp-content/uploads/2018/11/UK-Industry-Performance-Report-2017.pdf (accessed 4 May 2020).

Construction Industry Council (2015) *Hong Kong Construction Industry Performance Report for 2013*. Hong Kong. https://www.cic.hk/cic_data/pdf/research_and_data_analytics/industry_performance_report/eng/KPI%20Report%20for%202013%20(English).pdf (accessed 4 May 2020).

Construction Industry Council (2020) *Construction Industry KPI Dashboards*. http://www.cic.hk/common/KPIDashboard/dashboard.aspx?lang=en-US (accessed 4 May 2020).

De Silva, N., Weerasinghe, N., Madhusanka, H.W.N. and Kumaraswamy, M. (2018) Enablers of relational integrated value networks (RIVANS) for total facilities management (TFM). *Journal of Financial Management of Property and Construction*, Vol. 23, Issue 2, pp. 170–184. https://doi.org/10.1108/JFMPC-09-2016-0041 (accessed 26 April 2020).

Dulaimi, M.F., Ling, F.Y.Y. and Ofori, G. (2001) Building a world class construction industry – motivators and enablers. National University of Singapore, Singapore.

Egan (1998) *Rethinking Construction*. DETR, HMSO, UK.

El-adaway, I.H., Ali, G.G., Abotaleb, I.S. and Barber, H.M. (2020) Studying the relationship between stock prices of publicly traded US construction companies and gross domestic product: preliminary step toward construction–economy nexus. *Journal of Construction Engineering and Management*, Vol. 146, No. 1, 04019087-1 to 10.

ETWB (2007) Overall review of implementation of Construction Industry Review Committee recommendations. Environment Transport & Works Bureau (ETWB),

Farmer, M. (2016) *Modernise or Die: Time to decide the industry's future – The Farmer Review of the UK Construction Labour Model*. Construction Leadership Council, London.

Gerrard, N. (2020a) Construction economic influence 'almost double' previous estimate. News, Chartered Institute of Building (CIOB), 24 February. https://www.constructionmanagermagazine.com/news/ciob-constructions-economic-influence-higher-previ/ (accessed 29 April 2020).

Gerrard, N. (2020b) Every £1 spent on UK construction creates £2.92, says CBI. News, Chartered Institute of Building (CIOB), 28 February. https://www.constructionmanagermagazine.com/news/cbi-every-1-spent-uk-construction-creates-292/ (accessed 29 April 2020).

Gerrard, N. (2020c) Third of construction firms think agreeing prices is acceptable. CIOB Construction Manager, News, Chartered Institute of Building (CIOB), 26 February. https://www.constructionmanagermagazine.com/news/third-construction-firms-think-agreeing-prices-acc/ (accessed 2 May 2020).

GLF-CEM (2017) An up-date of the key performance indicators initiative. *GLF-CEM Newsletter*, Global Leadership Forum for Construction Engineering and Management Programs (GLF-CEM), Vol. 2, Issue 2, p. 4. https://www.cms.ualberta.ca/https://cloudfront.ualberta.ca/-/media/engineering/research/construction-engineering-management/construction-simulation/glf/newsletters/glf-newsletter-201712.pdf.

Gounden, S. (1996) Reconstruction and development in South Africa – the construction industry and related procurement reform. *Journal of Construction Procurement*, Vol. 2, No. 1, pp. 3–10.

Hackitt, J. (2018) *Building a Safer Future, Independent Review of Building Regulations and Fire Safety: Final Report, Cm 9607*. Her Majesty's Stationary Office, London.

Hastak, M., Ng. T., Kumaraswamy, M., Wium, J., Kagioglou, M. and Shen, G. (2017) The GLF-CEM Key Performance Indicators Initiative. White Paper, Global Leadership Forum for Construction Engineering and Management Programs (GLF-CEM). https://cloudfront.ualberta.ca/-/media/engineering/research/construction-engineering-management/construction-simulation/glf/hastak-et-al-kpi-initiatives--glfcem-newsletter-may-2017-21-58.pdf.

Hickin, J.M. and Ha, C.L. (2018) Bid-rigging and joint tendering – when are they prohibited under Hong Kong's competition ordinance? Mayer-Brown, 15 October. https://www.mayerbrown.com/en/perspectives-events/publications/2018/10/bidrigging-and-joint-tendering--when-are-they-proh (accessed 29 April 2020).

Hillebrandt, P.M. (1985) *Economic Theory and the Construction Industry*, 2nd edition. Macmillan, London.

HKDB (2018) *Construction 2.0, Time to Change*. Hong Kong Development Bureau (HKDB), Hong Kong.

HM Government (2013) *Construction 2025: Industrial Strategy: Government and Industry in Partnership*. https://www.gov.uk/government/publications/construction-2025-strategy (accessed 14 February 2021).

HM Government and Strategic Forum for Construction (2008) *Strategy for Sustainable Construction*. https://webarchive.nationalarchives.gov.uk/ukgwa/20121212135622/http://www.bis.gov.uk/files/file46535.pdf (accessed 5 May 2020).

Humphreys, P., Matthews, J. and Kumaraswamy, M.M. (2003) Pre-construction project partnering: from adversarial to collaborative relationships. *Supply Chain Management: An International Journal*, Vol. 8, No. 2, pp. 166–178.

ICE (2018) PROJECT 13, Institution of Civil Engineers (ICE). http://www.p13.org.uk/ (accessed 04 May 2020).

ICG (2020) Infrastructure Client Group (ICG), Institution of Civil Engineers. https://www.ice.org.uk/about-ice/infrastructure-client-group (accessed 5 May 2020).

ICTAD (1988) *Development of Domestic Construction Contractors - Cabinet Paper 116 of 10/8/88*. Publication ICTAD/ID/03, December, Institute for Construction, Training and Development (ICTAD) renamed Construction Industry Development Authority (CIDA), Sri Lanka.

ICTAD (1992) Consultancy reports on the 'Domestic construction industry in the roads sub-sector', Inception Report in February 1992, the Interim Report in May 1992, the Draft Final Report in July and the Final Report in September 1992. World Bank-sponsored study led by Team Leader (Technical): M.M. Kumaraswamy, Institute for Construction Training and Development (ICTAD), renamed Construction Industry Development Authority (CIDA), Sri Lanka.

ISR (1999) *Building for Growth – An Analysis of Australian Building and Construction Industries*, Department of Industry, Science and Resources (ISR), Commonwealth of Australia.

Joe, M., Sahadevan, V. and Varghese, K. (2017) Design process standardisation for building projects in India. *Conference Proceedings: 6th World Construction Symposium 2017: 'What's New and What's Next in the Built Environment Sustainability Agenda?'*, 30 June – 2 July, Colombo, Sri Lanka pp. 161–168.

Kumaraswamy, M.M. (1998) Industry development through creative project packaging and integrated management. *Journal of Engineering, Construction and Architectural Management,* Vol. 5, No. 3, pp. 228–238.
Kumaraswamy, M.M. (2006) Accelerating construction industry development. *Journal of Construction in Developing Countries*, Vol. 11, No. 1, pp. 73–96.
Kumaraswamy, M.M. (2008) Editorial. *Journal of Construction in Developing Counties* (JCDC), Special Issue on 'Knowledge Flows for Balanced Development', Vol. 13, No. 2, pp. v–vii. http://web.usm.my/jcdc/vol13_2_2008.html (accessed 30 April 2020).
Kumaraswamy, M.M. (2011) Inaugural Editorial: Integrating 'infrastructure project management' with its 'built asset management'. *Built Environment Project and Asset Management*, Vol. 1, No. 1, pp. 5–13, https://www.emerald.com/insight/content/doi/10.1108/20441241111143740/full/html (accessed 28 April 2020).
Kumaraswamy, M.M. (2013) Construction industry development. DSc thesis, Loughborough University, UK.
Kumaraswamy, M.M., van Egmond, E.L.C., Rahman, M.M. and Ugwu, J. (2004a) Technology exchange through relationally integrated joint venture teams. CIB W92 International Symposium on Procurement Systems, 'Project Procurement for Infrastructure Construction', 7–10 January, Chennai, India, pp. 326–334.
Kumaraswamy, M., Mahesh, G., Mahalingam, A. Loganathan, S. and Kalidindi S.N. (2017a) Developing a clients' charter and construction project KPIs to direct and drive industry improvements. *Built Environment Project and Asset Management Journal*, Special Issue on 'Securing Clients' Objectives Throughout Construction Project Life Cycles', Ramanayaka, C. and Sutrisna, M. (eds), Vol. 7, No. 3, pp. 253–270.
Kumaraswamy, M.M. and Matthews, J. (2000) Improved subcontractor selection employing partnering principles. *ASCE Journal of Management in Engineering*, Vol. 16, No. 3, pp. 47–57.
Kumarawamy, M., Ofori, G., Mahesh, G., Teo, E., Tjandra, I. and Wong, K. (2010) Construction industry improvement initiatives: are we really translating rhetoric into reality? *International Research Conference on Sustainability in Built Environment*, Colombo, Sri Lanka, 18–19 June, Rameezdeen, R., Senaratne, S. and Sandanayake, Y.G. (eds), 978-955-9027-33-1, BEMRU, University of Moratuwa, Sri Lanka, pp. 116–125.
Kumaraswamy, M.M., Ng, S.T., Palaneeswaran, E., Rahman, M.M., Ugwu, O.O. and Sze, E.K.K. (2004b) Top-down and bottom-up construction industry development. CIB 2004 Working Commission W107 on *'Construction in Developing Economies' International Symposium on Globalization and Construction Meeting the Challenges, Reaping the Benefits*, 17–19 November, Bangkok, pp. 497–508.
Kumaraswamy, M.M., Palaneeswaran, E., Rahman, M.M., Ugwu, O.O. and Ng, S.T. (2006) Synergising R&D initiatives for e-enhancing Management Support Systems. *Automation in Construction*, Special Issue on 'Knowledge Enabled Information Systems Applications in Construction', Vol. 15, No. 6, pp. 681–692, https://www.sciencedirect.com/science/article/abs/pii/S0926580505001457.
Kumaraswamy, M.M. and Raza, H. (2014) Pilot study on sources of Time and Cost over-runs in Roadworks projects. Report to the World Bank Colombo, Sri Lanka, July 2014, Kumaraswamy, M.M. Published by World Bank as: 'Source of time and cost over-runs in road works projects: pilot study' (2015) World Bank Group, Washington, DC. http://documents.worldbank.org/curated/en/2015/10/25127140/source-time-cost-over-runs-road-works-projects-pilot-study (accessed 30 April 2020).
Kumaraswamy, M.M., Rowlinson, S.M. and Phua, F. (2002) Accelerating cultural changes through innovative procurement processes: a Hong Kong perspective. *Journal of Construction Procurement*, Vol. 8, No. 1, pp. 3–16, May 2002.
Kumaraswamy, M.M. and Shrestha, G.B. (2002) Targeting 'technology exchange' for faster organisational and industry development. *Building Research and Information Journal*, Vol. 30, No. 3, pp. 183–195.
Kumaraswamy, M.M., Wong, K.K.W. and Chung, J.K.H. (2017b) Focusing megaproject strategies on sustainable best value of stakeholders. *Built Environment Project and Asset Management Journal*, Special Issue on 'Emerging Issues in the Built Environment Sustainability Agenda', Ramachandra, T. and Karunasena, G. (eds), Vol. 7, No. 4, pp. 441–455.

Kumaraswamy, M.M., Zou, W.W. and Zhang, J.Q. (2015) Reinforcing relationships for resilience – by embedding end-user people' in public private partnerships. *Journal of Civil Engineering and Environmental Systems*, Special Issue on 'Resilience', Milke, M. (ed.), Vol. 32, No. 1–2, pp. 119–129.
Latham (1994) *Constructing the Team*. London: HMSO.
Ling, F.Y.Y., Toh, B., Kumaraswamy, M.M. and Wong, K.K.W. (2014) Strategies for integrating design & construction and operations & maintenance supply chains in Singapore. *Structural Survey: Journal of Building Pathology and Refurbishment*, Vol. 32, No. 2, pp. 158–182.
Loganathan, S., Srinath, P., Kumaraswamy, M., Kalidindi, S. and Varghese, K. (2017) Identifying and addressing critical issues in the Indian construction industry. *Journal of Construction in Developing Countries*, Special Issue on 'AEC Sector Role to Urbanise India', Ahuja, V. and Suresh, S. (eds), Vol. 22 (Supp. 1), pp. 121–144. https://doi.org/10.21315/ jcdc2017.22.supp1.7.
Mahesh, G., Pu, S. and Kumaraswamy, M.M. (2010) Report on focus group meeting: 'Construction Industry Development Reforms: Institutional Framework, Drivers and Challenges'. 30 March, Centre for Infrastructure & Construction Industry Development, Hong Kong, http://www.civil.hku.hk/cicid/3_events/94/94_summary.pdf (accessed 25 April 2020).
Manewa, A., Siriwardena, M., Ross, A. and Madanayake, U. (2016) 'Adaptable buildings for sustainable built environment', *Built Environment Project and Asset Management*, Vol. 6, No. 2. https://doi.org/10.1108/BEPAM-10-2014-0053.
Murray, M. and Langford, D. (2003) *Construction Reports 1944–98*. Oxford: Blackwell Science.
Ofori, G. (1988) Construction industry and economic growth in Singapore. *Construction Management and Economics*, Vol. 3, pp. 33–42.
Ofori, G. (1994) Establishing construction economics as an academic discipline. *Construction Management and Economics*, Vol. 12, pp. 295–306.
Ofori, G. (2002) Singapore construction: moving towards a knowledge-based industry. *Building Research and Information*, Vol. 30, No. 6, pp. 401–412.
Ofori, G. (2016) Construction in developing countries: current imperatives and potential. *CIB World Building Congress 2016 on Intelligent Built Environment for Life*, Kahkonen, K. and Keinanen, M. (eds), Tampere, 30 May–3 July, Tampere University of Technology.
Pan, W. and Hon, C.K. (2018) Modular integrated construction for high-rise buildings. *Proceedings of The Institute of Civil Engineers – Municipal Engineer*, Vol. 1, pp. 1–5. http://dx.doi.org/10.1680/jmuen.18.00028.
PCICB (2005) Progress Report on Implementation of Recommendations of the Construction Industry Review Committee. May, HKSAR Government.
Powell, W.W. and DiMaggio, P. (eds) (1991) *The New Institutionalism in Organizational Analysis*. University of Chicago Press, Chicago, IL.
PSIB (2003) *Inventory of International Reforms in Building and Construction*. Process and System Innovation in Building and Construction (PSIB) Programme, Government Building Agency, Netherlands Government publication no.: PSIB017_S_04_2341, the Hague.
Rahman, M.M. and Kumaraswamy, M.M. (2004) Potential for implementing relational contracting and joint risk management. *ASCE Journal of Management in Engineering*, Vol. 20, No. 4, pp. 178–189.
Rahman, M.M. and Kumaraswamy, M.M. (2005) Assembling integrated project teams for joint risk management. *Journal of Construction Management and Economics* (0144-6193), Vol. 23, No. 4, pp. 365–375.
Rahman, M.M. and Kumaraswamy, M.M. (2008) Relational contracting and teambuilding: assessing potential contractual and non-contractual incentives. *Journal of Management in Engineering, ASCE*, Vol. 24, No. 1, pp. 48–63.
Simon, E. (1944) *The Placing and Management of Contracts*. HMSO, London.
Smyth, H. (2020) Construction needed intensive care before the coronavirus. *Construction Manager*, Construction News, 6 April, http://www.constructionmanagermagazine.com/news/construction-needed-intensive-care-coronavirus/ (accessed 26 April 2020).
Smyth, H., Anvuur, A. and Kusuma, I. (2017) Integrated solutions for total asset management through 'RIVANS'. *Built Environment Project and Asset Management*, Vol. 7, No. 1, pp. 5–18.
Strategic Forum (2002) *Accelerating Change, Rethinking Construction*. London.
Strategic Forum (2007) *Profiting from Integration*. Final Report, November, London.

Turin, D.A. (1973) *The Construction Industry: Its Economic Significance and Its Role in Development*. University College Environmental Research Group, London, UK.

Wong, K.K.W. (2014) Integrative approach to addressing societal needs in infrastructure development projects. PhD Thesis, University of Hong Kong.

Wong, K.K.W., Kumaraswamy, M.M., Mahesh, G. and Ling, F.Y.Y. (2014) Building integrated project and asset management teams for sustainable built infrastructure development. *Journal of Facilities Management*, Vol. 12, No. 3, pp. 187–210.

World Bank (1992) *Guidelines – Procurement under IBRD Loans and IDA Credits*, World Bank, Washington, DC.

Wuni, I.Y. and Shen, G.Q. (2020) Stakeholder management in prefabricated prefinished volumetric construction projects: benchmarking the key result areas. *Built Environment Project and Asset Management*, Vol. 10, No. 3, pp. 407–421. https://doi.org/10.1108/BEPAM-02-2020-0025.

9. Applications of mainstream economic theories to the construction industry: transaction costs

Abdul-Rashid Abdul-Aziz and Afzan Binti Ahmad Zaini

INTRODUCTION

The origins of transaction cost as an economic concept can be traced to Ronald Coase's (1937) paper entitled 'The nature of the firm' (and his subsequent work, 'The problem of social cost' which came out in 1960) in which he posed the question, 'Why does a firm emerge at all in a specialised exchanged economy' (Coase, 1937, p. 40)? In trying to answer this puzzle, he raised doubts about the prevailing assumption among economists at the time that there are no cost implications accompanying exchange relationships in markets, a conundrum which Oliver E. Williamson took it upon himself to resolve. Williamson's radical approach was to depart from the then-prevailing economic orthodoxy by integrating economics, law and organisation theory; although economics was first among equals (Williamson, 1998). Its gestation period stretched from the 1920s to the 1970s (Williamson, 2008). At the same time, he infused into the transaction construct his own original ideas regarding vertical integration and the theory of the firm (Mahoney, 2005). Following Commons (1924, 1934), he reinterpreted the basic unit of analysis as the transaction (Williamson, 1985). Reducing transaction cost then becomes the engine of analysis (Williamson, 2008).

Also, rather than characterise the firm as a production function, he considered viewing it as a governance structure to be more useful (Williamson, 1985). Governance structure was defined as the 'institutional matrix within which the integrity of a transaction is decided' (Williamson, 1993a, p. 102). The market and firm–market hybrid are the other two important economic institutions. The main hypothesis of Williamson's works is: 'align transactions, which differ in their attributes, with governance structures, which differ in their costs and competencies, in a discriminating (mainly, transaction cost economising) way' (Williamson, 1991b, p. 79). His ideas are expressed mainly in his three books *Market and Hierarchies* (1975), *The Economic Institutions of Capitalism* (1985) and *The Mechanisms of Governance* (1996a), although the condensed versions are available in several journal articles and book chapters.

The concept of the 'transaction' has spread out from its economic roots into marketing, accounting, finance, organisational theory, international business, and even law and political science (Macher and Richman, 2008). Carroll and Teece (1999) asserted that transaction cost economics (TCE) 'is perhaps the single most influential theory found in social science' (p. 3). Williamson was awarded the 2009 Nobel Memorial Prize in Economic Sciences for his seminal work. Williamson was a key figure in introducing a new strain of economic discipline, the new institution economics, that put the institutions at the heart of any economic activity (Williamson, 1993a).

This chapter begins by presenting a short summary of TCE. Construction economics scholars need to fully understand TCE before using it for their own work in order to avoid dissonance between what Williamson had in mind and the scholars' interpretations. Understanding TCE is challenging as some aspects of the theory evolved over time (such as uncertainty), were introduced but later dropped (such as ease of measurement), or were even not properly clarified (such as complexity). Moreover, the language that Williamson used does not make his work easy to read. The chapter then highlights certain outstanding conceptual and measurement issues which researchers wishing to apply the TCE schema should be mindful of. The application of TCE in construction economics studies follows next. It shows that in many such works the limitations of TCE were not recognised. There have also been instances where the theory was not even fully understood by some construction economics scholars. Before concluding, this chapter points the way forward for construction economics researchers to make a meaningful impact on TCE research.

WHAT IS TRANSACTION COST ECONOMICS?

The premise of TCE is that the main purpose and effect of economic organisation is economising on transaction costs (Williamson, 1996a). A transaction occurs 'when a good or service is traded across a technologically separable interface' (Williamson, 1985, p. 1). Put simply, according to TCE, there are cost implications whenever economic exchanges take place between two independent agents. The idea that transaction cost is not zero goes against the grain of the widely held economic assumption in the 1970s that it is so (Williamson, 2008). Transaction cost is the 'equivalent of friction in physical systems' (Williamson, 1985, p. 19).

TCE assumes that production cost is fixed and does not vary between governance modes. This is not to say that Williamson regarded production as irrelevant. As he explained, 'Holding technology constant is a pedagogical device for motivating the argument' (Williamson, 1988, p. 355), so that 'the choice between alternative internal modes for organising successive stages of production turns mainly on transaction cost rather than technology considerations' (Williamson, 1980, p. 8). For evaluation of transaction cost, *ex ante* and *ex post* transaction cost, being interdependent, must be addressed simultaneously, not sequentially at the outset (Williamson, 1985, p. 21). *Ex ante* cost encompasses 'costs of drafting, negotiating, and safeguarding an agreement' (Williamson, 1985, p. 20); whereas *ex post* cost includes 'maladaptation costs incurred when transactions drift out of alignment', 'haggling costs incurred if bilateral efforts are made to correct ex post misalignments', and 'bundling costs of effecting secure commitments' (Williamson, 1985, p. 21). Williamson (1975, 1985) posited that the optimal choice of governance depends on the relative costs of alternative institutional arrangements, which in turn depend on the characteristics of the transaction at stake. To underpin TCE, he put forward a few key concepts relating to economic actors' behavioural attributes on the one side, and transaction attributes on the other (Williamson, 1975, 1985). For him, transactions relate to economic actors, so behavioural attributes matter.

Behavioural Attributes

Bounded rationality

Williamson (1985) borrows Simon's (1957) notion that the economic decision-maker is inherently constrained by limitations of both knowledge and mental capacity. Simon challenges the neoclassical concept of the 'economic man' as acting rationally and as having all the knowledge necessary to make rational decisions. The ramification of human actors as having bounded rationality is that acquiring information is not costless.

Opportunism

Opportunism is a central concept in Williamson's TCE logic. Opportunism – the seeking of self-interest with guile – is the ultimate cause for the failure of markets and for the existence of organisations (Williamson, 1975, 1993b). Williamson (2010) regards opportunism as 'defection from the spirit of cooperation' (p. 14). For Williamson, opportunism carries attitude and behaviour connotations. He saw the opportunistic attitude as one of the rudimentary attributes of human nature (Williamson, 1975). He also regarded lying, stealing and cheating (Williamson, 1975, 1985), and 'calculated effort to mislead, distort, disguise, obfuscate, or otherwise confuse' (1985, p. 47), as opportunism. Following criticisms of his opportunism concept, Williamson (2000a, p. 39) went on the offensive with his opening statement in a paper: 'The general effect of presuming the absence of opportunism is that we enter the world of what Frank [Manuel] and Fritzie Manuel describe as "utopian fantasies" (1979: 1).'

Transaction Attributes

Uncertainty and complexity

Williamson (1975) initially made no distinction between uncertainty and complexity, but later suggested otherwise (Williamson, 1985), although he never fully explained complexity. Uncertainty is equivalent to disturbances (Williamson, 1985, 1998). The future is full of uncertainties, one of which is behavioural uncertainty. The other, which is innocent or 'non-strategic', applies to 'non-disclosure, disguise, or distortion of information' (Williamson, 1985, p. 57). Uncertainty is confined to the relations between the exchange agents. Uncertainty can be amplified through two means: more disturbances occurring and/or disturbances becoming more consequential (Williamson, 1991a). Uncertainty calls for *ex ante* costly adaptations. Transactions that take place over a longer period experience higher degrees of uncertainty as compared to 'spot markets' where the good or service is generic, and the identities of the buyers and sellers are immaterial to the transaction.

Small numbers

Opportunistic tendencies are curtailed whenever there are many competitors (Williamson, 1985). Many transactions involve a large number of bidders *ex ante* contract execution. During contract execution and at the contract renewal stage, the number shrinks substantially, resulting in contractual hazards of bilateral monopoly.

When the Market Breaks Down

The following discussion shows how the various concepts interplay, leading to market failure. If, in a particular market exchange, bounded rationality and environmental uncertainty prevail, the parties then face problems of 'information compactness'. Information compactness stems from idiosyncratic or technical information that may not be easy for third parties to decipher (Williamson, 1985). Information compactness renders complete contingency contracts (that is, contracts that anticipate all possible eventualities beforehand) too costly or infeasible to produce. For Williamson (2005), all complex contracts are unavoidably incomplete.

Incomplete contracts per se do not necessarily lead to market inefficiencies if all exchange partners share the spirit of cooperation. The situation turns for the worse if the hazard of opportunism paired with a small number of competitors also come into play.

The outcome is high transaction cost, and the remedy is for the firm to internalise the transaction (Williamson, 1996a). When the transaction is internalised, it is governed by organisational processes. Internally, organisations can economise on bounded rationality and uncertainty through adaptive, sequential decision-making, and mitigate the hazards of opportunism through internal incentives and controls that are much more extensive and refined than what the market can offer (Williamson, 1975, 1999). In the hierarchy, 'the future is permitted to unfold' (Williamson, 1975, p. 9). A small number of competitors in exchange relations is replaced by convergence in expectations by staff. The behavioural and environmental factors may change over time, and therefore the efficiency of one governance structure over another may also change (Williamson, 1975). For Williamson, pairing of bounded rationality and uncertainty and complexity, and opportunism and small numbers, are crucial.

Comparative Governance Framework

The general comparative governance framework that is central to TCE can now be considered. Williamson (1985) predicted that aligning organisational structures with exchange attributes in a discriminating way improves performance. The three exchange attributes are: (1) the degree to which the assets involved are specific to a transaction; (2) uncertainty, or the extent to which disturbance or change may affect the transaction; and (3) how frequently the transaction recurs (Williamson, 2005). A fourth dimension – ease of measurement – was included in one of Williamson's publications (Williamson, 1991b) but was later dropped without explanation. Because Williamson did not explicitly retract this fourth dimension, scholars such as Poppo and Zenger (2002) and Brown and Potoski (2005) used it in their works. The modes of governance are the market and the hierarchy which are the polar cases, and the hybrid which is a blend of the two. The three exchange attributes are briefly discussed below.

Asset specificity

TCE owes much of its predictive content to the asset specificity concept (Williamson, 1985). This concept refers to 'durable investments that are undertaken in support of particular transactions' (Williamson, 1985, p. 55). Asset specificity measures asset deployability (Williamson, 1995). The higher the specificity, the lower is the value outside of the focal transactions. Williamson initially identified four types of asset specificity (Williamson, 1983) but subsequently added another two (Williamson, 1985):

- Site specificity: immobile assets are placed nearby to minimise transportation and inventory expense, or to benefit from complementary advantages.
- Physical asset specificity: equipment and machinery involving design features specific to the transaction.
- Human assets specificity: from learning by doing, human capital becomes more efficient at producing a certain good or service than others.
- Dedicated assets: investments in assets dedicated to a particular customer.
- Brand name capital: intangible capital which can have relationship-specific attributes.

- Temporal specificity: a type of site specificity in which timely responsiveness by on-site human assets is vital.

Asset specificity inflates the transaction costs of all forms of governance and locks the parties into a mutual dependency leading to contracting hazards (Williamson, 1985, 1991a). Without going into the complicated calculus, suffice it to say that with higher asset specificity, the cost-effective choice of governance shifts from market to hybrid and finally hierarchy.

Uncertainty and complexity

The concepts of uncertainty and complexity were discussed earlier. In one publication, Williamson (1996a) postulated that uncertainty arises whenever incomplete contracting and asset specificities intersect. In another (Williamson, 2010), he equated uncertainty with contractual disturbances. Uncertainty can manifest itself in two forms: probability distribution of disturbances remains unchanged but more disturbance occurs; or disturbances become more consequential (Williamson, 1991b). For Williamson (1975), uncertainty, regardless of its source, leads to one outcome: exactness being replaced by approximation in reaching a decision. Organisations have the advantage over markets in that they can deal with uncertainty and complexity in an adaptive, sequential fashion unencumbered by opportunism hazards that afflict the market governance. Convergent expectations also help to mitigate uncertainty inside organisations.

Frequency

Frequency refers to the regularity with which transactions occur. Transactions that happen rarely are best governed by the market (Williamson, 1985). Transactions that take place more often should be integrated vertically.

Interaction of all three dimensions

How do these three dimensions – asset specificity, frequency, uncertainty/complexity and frequency – affect the choice of governance? To get round the difficulty of measuring each of them directly (Ménard, 2005), Williamson (1985, Chapter 3) came up with the 'discrete alignment principle': that is, calculative agents operating in a competitive environment select the governance structure that fits comparatively better with the attributes of the transaction on hand. Williamson (2010) explained that comparative analysis obviates the need to take absolute measures of transaction costs.

The market government mode is efficient for services that are simple, are easy to specify and measure, do not demand asset-specific investments, and involve infrequent exchange. If transactions become complex, are difficult to specify and to measure, require transaction-specific investment and involve frequent exchanges, transaction costs start to escalate due to the considerable resources needed to search for a provider/purchaser, the comprehensive contract that needs to be written, the transaction that needs to be safeguarded, monitoring of the transaction, and eventually enforcing of the contract. High transaction cost will render alternative modes of government relatively more attractive.

The alternative to incurring high transaction cost from the market mechanism is to internalise production through vertical integration. TCE predicts that the hierarchy is relied upon when: (1) asset specificity is high, uncertainty is at least of intermediate level, and transactions are recurrent; or (2) asset specificity is of intermediate degree, uncertainty is sufficiently high,

Table 9.1 *Attributes of leading generic modes of governance*

	Governance modes		
Governance attributes	Market	Hybrid	Hierarchy
Instruments			
Incentives	High-powered	Less high-powered	Low-powered
Administrative support by bureaucracy	Nil	Some	Much
Performance attributes			
Adaptation (autonomous)	Strong	Semi-strong	Weak
Adaptation (coordination)	Weak	Semi-strong	Strong
Contract law regime	Legalistic	Contract as framework	Firm as own court of ultimate appeal (fiat)

Source: Adapted from Williamson (1991a, p. 281) and Williamson (2005, p. 49).

and transactions are recurrent. Internalising the transaction when interactions are recurrent obviates the need to negotiate and write contracts each time. The hybrid cannot enjoy the incomplete contracts that the hierarchy does. It is the most susceptible to frequent disturbances, because adaptation requires mutual adaptation. It also suffers the greatest from hazards of opportunism. Overall, hybrid arrangements are adopted when asset specificity can be spread over independent partners, while uncertainties are consequential enough to require tighter coordination than what markets can provide (Williamson, 1991a).

Governance Structures

Williamson (2010, p. 15) distills the main differences between the three governance modes according to: (1) incentive intensity; (2) administrative control; (3) adaptation; and (4) contract law regime (see Table 9.1). The different governance structures have different capacities in their incentives to motivate agents. Firms have access to formal authority (fiat) that markets do not. Williamson (1991a) considers adaptation as being of two forms: autonomous and coordination. Both forms of adaptations must be analysed together to give TCE its predictive power (Williamson, 2010). Governance structures are conceived of as implicit or explicit contractual arrangements (Williamson, 1979, 1993a).

Market governance

Williamson (1975, p. 20) regarded markets as the original state of affairs: 'In the beginning there were markets'. The price mechanism coordinates transactions for the market governance, the spot market being the stereotypical market relationship in which anonymous providers and sellers converge, agree on prices and depart. The market enjoys high-powered incentives in that agents can cash directly the result of their efforts (Williamson, 1996a). Market exchange is a hands-off control mechanism. Autonomous adaptations, which manifest themselves in the market, are triggered by changes in relative prices (Williamson, 2010, p. 14). Disputes in markets are treated in a legalistic way and rely on court orders. Once the dispute is settled in court, the parties to the exchange go their separate ways.

Hierarchical governance

Hierarchical governance or the firm referred to in TCE is not the conventional firm-as-production function, but firm-as-governance structure (Williamson, 1988). The latter portrays the firm as a hierarchical system with authority as the key role. Williamson regarded internal organisation as a last resort in the face of contracting hazards and associated transaction costs. Internal incentives in hierarchies are low-powered in that performance changes by workers have little or no immediate impact on compensation (Williamson, 1985, 1988, 1996a). Internal transactions and allocation of factors are coordinated by administrative authority (or fiat) instead of prices. Command constitutes a central feature of the hierarchy structure; and with command comes control. Control permits instruction issuance, performance monitoring, disputes settlement and auditing to take place internally. Control also checks unwanted side-effects. Cooperation can limit the costs of control (Williamson, 1975, 1985). 'Attitudinal interactions' avert conflict in favour of amicable settlements. Integration harmonises interests and permits an efficient (adaptive, sequential) decision-making process to be utilised. Having a collection of people as employees allows efficiency gains and economics of scale in processing information. The hierarchy adopts consciously coordinated adaptations (Williamson, 2010). The hierarchy values continuity of relationship; it is governed by a forbearance regime whereby the firm becomes its own court of ultimate appeal (Williamson, 1991a, 1995).

All forms of governance have cost implications; bureaucracy and sub-goal pursuit are just two examples of intra-firm transfer costs (Williamson, 1985, 1993b). Low-powered incentives may also promote below-optimum worker effort.

Hybrid governance

The hybrid form of governance was introduced belatedly after Williamson had established the other two modes of governance structures (Williamson, 1991a). It was in reaction to the criticism of the simple market–hierarchy dichotomy. Hybrid governance includes franchising, joint ventures, (Williamson, 1985, p. 83), long-term contracts (Williamson, 1979), alliances (Oxley, 1999), co-operatives (Cook, 1995), networks (Thorelli, 1986), and many more. It adjusts with little influence from the price system and shares or exchanges technologies, capital, products and services. The hybrid preserves ownership autonomy while enabling semi-strong incentives (Williamson, 1991a). The hybrid mode displays adaptive capacities of both sorts – autonomous and coordination – in an intermediate degree. As with the hierarchical types of governance, continuity of relationship is valued in the hybrid form. The latter adopts the more elastic concept of 'contract as framework' (Williamson, 1991a, 2005). Credible commitment supports are drafted (these include penalties against premature termination, and specialised dispute settlement mechanisms).

CONCEPTUAL AND EMPIRICAL ISSUES

Williamson (1999) proclaimed: 'I have no hesitation, however, in declaring that transaction cost economics is an empirical success' (p. 1092). He pointed to the exponential growth of empirical applications of TCE since the 1980s, and drew attention to some of the works that show remarkable congruity between theory and evidence: Shelanski and Klein (1995), Lyons (1996), Crocker and Masten (1996) and Rindfleisch and Heide (1997). Williamson and Masten (1999) listed empirical studies which used various methods that corroborate with TCE theory. Joskow (2005) supported Williamson's proclamation: 'Indeed, it is hard to find many other areas in industrial organisation where there is such an abundance of empirical work supporting a theory of firm or market structure' (p. 344). Geyskens et al. (2006) also concurred that TCE 'stands on a remarkably broad empirical foundation' (p. 531). Macher and Richman (2008) also concurred after conducting a comprehensive review of the empirical literature on TCE across multiple fields: 'the volume of our findings lend considerable support overall in these articles for the main predictions of TCE' (p. 43).

Even so, Macher and Richman (2008) observed 'a number of lingering theoretical and empirical issues that need to be addressed' (p. 1). Williamson (2000b) himself accepted that there is still much work to be done despite TCE's accomplishments. TCE has its fair share of critics, so much so that 'criticising TCE remains a thriving industry' (Foss and Klein, 2010, p. 263). Due to space limitations in this chapter, only a few of the issues which are often highlighted are considered.

In terms of its apparent conceptual flaws, it is perhaps proper to begin with Coase (1937), whose work provided the springboard for Williamson's TCE. In writing about Coase's lifetime contribution, Skyuta (2010a) noted that he 'has always been suspect of the reliance of TCE on the concept of asset specificity and the behavioural assumptions of opportunistic holdup' (p. 44). Another critic is Herbert Simon, whose idea Williamson also borrowed. In writing about Herbert Simon, Sarasvathy (2010, p. 86) noted that there had been a series of published and unpublished arguments towards the end of the twentieth century between Simon and Williamson over the appropriateness of opportunism as an assumption within the TCE framework. Williamson (1995) himself noted that the phrase 'transaction cost' requires precise definition: 'There is nonetheless a grave problem with broad, elastic and plausible concepts – of which "transaction cost" is one and "power" is another ... Concepts that explain everything explain nothing' (p. 33).

Indeed, even the basic definition of the firm, let alone derivative issues such as the boundaries of the firms, the nature of 'hybrids' and the 'make-or-buy' decision, become 'hopelessly clouded by terminological confusion' (Hodgson, 2010). In a chapter in a book devoted to examining the persuasiveness of past empirical evidence, Sykuta (2010b) concluded that some of TCE's core concepts are difficult to define. Without clarity, concepts cannot be appropriately measured. Expressing the same concern, Klein (2005) added that there has been confusion about the definitions of those key variables; for example, asset specificity may be confused with market power. Still on the theme of multiple and at times competing definitions, Macher and Richman (2008) pointed out that TCE studies have yet to fully work out how exactly the two types of uncertainty – environmental uncertainty and behavioural uncertainty – influence organisational choice. Transaction frequency has received far less attention than asset specificity and uncertainty (Geyskens et al., 2006).

Then there are incongruities which some scholars detected in the TCE schema. Dow (1987) and Hodgson (1993) pointed to the contradiction of TCE's reliance on the bounded rationality concept while at the same time assuming that agents have the omniscience to sort between organisational forms in favour of the efficient ones. Slater and Spencer (2000) too noted that the agent must know the following in advance: what disturbances might occur, what adaptations may be required, what hazards may arise, what opportunistic behaviour might appear, and what incentives and safeguards need to be factored in. Klein (2010) questioned whether the contracting parties can trust the courts to fill in the gaps; why bother to write out every contingency?

Ménard (2005) questioned whether all contracts share the same fundamental properties, or whether they differentiate along the governance structures. Assuming production cost as a given ignores the influence of technologies on transaction modes and structures of governance (Milgrom and Roberts, 1992). Argyes (2010) pointed out that while in 'many, many cases' using the transaction as the unit of analysis is appropriate, there are cases where it is not, such as when behavioural considerations dominate, or when the governance of linked transactions cannot be separated. Macher and Richman (2008) noted that, given its theoretical centrality, it is somewhat surprising that the empirical literature is largely devoid of efforts to measure opportunism or examine its prevalence in different contexts. Williamson's own writings do not help in this regard; in one paper he indicated that, 'most economic agents are engaged in business-as-usual, with little or no thought to opportunism, most of the time' (Williamson, 1993b, p. 98), yet a few pages later he stated, 'the troublesome behaviour [i.e. opportunism] is not an arcane economic condition but is familiar and pervasive' (1993b, p. 101). Douglas (1990) found it odd that Williamson recognised firms as being varied, but human actors' productive capabilities and amenabilities as static. With regard to Williamson's assumption that all human actors are identical, his analogies of a hiker choosing travelling companions with dissimilar cooperative behaviours (Williamson, 1996b, p. 54) and the London banker's deep knowledge of the Norwegian ship-owner (Williamson, 1993c, p. 470) reflect an opposite view, thereby clouding his true standpoint. TCE ignores intrinsic forms of incentive to motivate staff (Fehr and Gächter, 2000).

There have also been criticisms from the empirical perspective. Macher and Richman (2008) noted the wide-ranging quality of past empirical works, and of the strengths of their conclusions. Two review papers are particularly striking, the first by David and Han (2004) who, having reviewed 63 TCE articles which used statistical tests, concluded that their results were mixed. As independent variables, findings pertaining to asset specificity fared best (60 per cent supported), while results for uncertainty were less convincing (24 per cent supported). Results on tests for interaction effects between asset specificity and uncertainty were also mixed (16 per cent). Frequency and performance, which are the other important variables, received considerably less scrutiny. While only a small number of studies focused on the frequency of transactions, those studies recorded higher support (69 per cent) than for asset specificity. Despite evidence that asset specificity leads to governance choice, no research had determined whether the choice is somehow 'efficient', which troubled David and Han (2004) given the central premise of comparative governance assessment within TCE. They concluded that the empirical work has provided a limited picture on TCE.

The second paper, by Carter and Hodgson (2006) who reviewed 27 of the most prominent TCE articles between 1992 and 2002 (based on citations), found none which tested all three transaction dimensions (asset specificity, uncertainty and transaction frequency) which produced results that align with predicted outcomes. Five studies were partly consistent with the framework, five were partly consistent and partly inconsistent, and one was inconclusive. It is noteworthy that Carter and Hodgson (2006) found that many of the empirical studies steered away from testing the predictions of the framework to focus mainly on developing the details of the relationship characteristics.

Several scholars pointed to the problems with measurement and use of proxies. Ménard (2005, p. 285) observed that all three variables – asset specificity, uncertainty and frequency – are 'notoriously difficult to measure', and almost all empirical literature avoided attempts at measuring transaction costs directly. Masten et al. (1991) also noted the difficulties in observing and measuring transaction costs, and hence used proxies in their own works. However, proxies have produced mixed results (Ménard and Shirley, 2005). The proxies used may be inappropriate: for example, using capital intensity or fixed costs for asset specificity as proxies is questionable (Klein, 2005). Proxies make interpreting empirical results difficult; positive results may come from TCE, but they may also come from other confounding factors (Macher and Richman, 2008). Overall, TCE still lacks a rigorous mathematical foundation (Macher and Richman, 2008).

Other authors warn against making wrongful inferences from applications of TCE. Many studies did not explicitly compare TCE with rival theories which usually posit mutually exclusive outcomes (Klein, 2005). Hodgson (2010) noted that as even positive empirical results favouring TCE can be plausible for alternative interpretations, there is a need for joint testing involving rival (or possibly complementary) interpretations. Nickerson (1997) cautioned against treating transactions as independent if interaction effects between them are significant. Correlation between transactional attributes and governance structures have been mistaken for causality (Klein, 2005). Indeed, few of the empirical studies made any effort to test the underlying causal mechanism (Bromiley and Harris, 2006). Moderating influences were often ignored (Sykuta, 2010b).

TCE AND THE CONSTRUCTION INDUSTRY

Beginning with Eccles (1981), various construction economics scholars have used TCE as the theoretical lens for their own works. The topics to which TCE has been more commonly applied include: subcontracting (Eccles, 1981; Lai, 2000; Gonzáles-Diaz et al., 2000; Kamann et al., 2006; Nyström 2019), governance structure (Gunnarson and Levitt, 1982; Reve and Levitt, 1984; Turner and Keegan, 2001), procurement system (Bajari and Tadelis, 2001; Chang and Ive, 2007; Ive and Chang, 2007; Rajeh et al., 2014), contractor selection (Lingard et al., 1998) and alliances (Love et al., 2002; Langfield-Smith, 2008; Chen et al., 2012, 2015), partnering (Greenwood and Yates, 2006), public–private partnership (PPP) projects (Jin, 2009; Ho et al., 2015), contract (Turner and Simister, 2001; Turner, 2004) and project (Piertoforte, 1997; Winch, 2001; Miller et al., 2002; Lu et al., 2015; Yan and Zhang, 2020).

The rigour with which TCE was exercised has varied among scholars. Sorell (2003) and Yik et al. (2006) glossed over the TCE schema in their works, unlike Winch (1989), Gonzáles-Diaz et al. (2000), and Constantino and Pietroforte (2002). There were scholars who treated TCE predictions as facts rather than testable suppositions. For example, Lee et al. (2009) accepted the prediction that as asset specificity increases, the modes of governance shift from market to hybrid.

Various research approaches were adopted by construction economics scholars in their application of TCE. Piertoforte (1997), Love et al. (2002), Cox and Thompson (1997), Lingard et al. (1998), Lai (2000), Winch (2001), Bremer and Kok (2000), Turner (2004) and Nyström (2019) were among those who resorted to pure theoretical discourse. Sha (2011) relied on mathematical reasoning, Lee et al. (2009) on mathematical modelling, and Gonzáles-Diaz et al. (2000) on panel data. Some, such as Chang and Ive (2007), used a combination of both mathematical and logical reasoning. The majority of construction economics researchers (for example Kamann et al., 2006; Li et al., 2013, 2014; Chen et al., 2015) used questionnaire-based surveys and quantitative data analysis. Clark and Ball (1991) collected qualitative data, and Miller et al. (2002), and Greenwood and Yates (2006) opted for the case study approach.

The same measurement difficulties faced by TCE researchers mentioned before were noted by construction economics researchers such as Constantino et al. (2001), Lee et al. (2009) and Li et al. (2013), despite the caution by Hughes et al. (2006):

> While theorists bemoan the lack of empirical evidence for transaction economics, it turns out that the main reason for this lack of data is the impossibility of collecting such data at the level of projects or companies. No company monitors cost data in this kind of detail, so there is no basis for a detailed cost–benefit analysis of different ways of working. (p. 101)

Various methods have been devised to get around the difficulty of getting hard cost data. Li et al. (2014), Guo et al. (2016) and Abdel-Galil et al. (2020) requested their respondents to estimate pre-contract and post-contract transaction costs as per contract value. Lee et al. (2009) too focused on relative values instead of absolute values. On the other hand, Rajeh et al. (2015) requested their respondents to evaluate the time spent daily in information and procurement activities relative to other project activities. Other scholars such as Gonzáles-Diaz et al. (2000) used proxies to avoid having to depend on personal evaluation by respondents who have different judgements. Similarly, Chen et al. (2015) adopted the monetary value of alliancing project as a proxy for frequency. A number of construction economics researchers, including Li et al. (2012, 2013), Lu et al. (2015), Chen et al. (2015), You et al. (2018) and Lu et al. (2020), requested their respondents to give Likert-based scores on the degree of consistency for prepared statements.

Early construction economics scholars have shed light on TCE's loss of the predictive power when applied to construction. Buckley and Enderwick (1989) found that TCE logic suggests that the construction industry should move away from subcontracting and all required manpower be internalised by construction companies. This is contrary to the reality in the industry in most countries. When commenting on Buckley and Enderwick's findings, Hughes et al. (2001) noted that TCE ignores important characteristics of the construction industry which outweigh the embedded theoretical reasons. Winch (1989) too found that TCE logic dictates that contracting work should be internalised and not governed by the market. To salvage the erroneous prediction, Winch offered his own explanation for the mismatch between theory and practice. Later, Gonzáles-Diaz et al. (2000) revisited TCE as a theoretical lens for analysing subcontracting and concluded that their results regarding uncertainty were 'not so conclusive' (p. 182). Like Winch (1989), they too produced possible explanations for the anomaly. Reflecting on past studies, Hughes et al. (2006) summed up their views:

> on a practical level, there are so many more important issues than those dealt with by transaction cost theory. So, while the cost of transaction is an important issue, it is not the deciding issue in terms of how construction work is organised and procured. (p. 101)

Indeed, Williamson (1998) had pointed out that:

> Although transaction cost economics has a broad reach – any issue that arises as or can be reformulated as a contracting problem is usefully examined through the lens of transaction cost economising – it does not tell you everything. (p. 23)

Construction economics scholars who subsequently chose to ignore the viewpoint of Hughes et al. (2006) and the caution expressed by Williamson (1988) were confronted with the same theory–practice conundrum. When TCE's prediction that the most appropriate structure for alliances does not align with reality (that is, hybrid), Langfield-Smith (2008) concluded that, 'Thus, TCE prescriptions need to be interpreted in the light of the broader control mechanisms which contribute to the overall control package' (p. 362). She added later: 'other aspects also drive the choice' (p. 362). Likewise, when studying the economics of project alliances, Chen et al. (2015) concluded that their findings 'generally support' TCE's predictions, but they also detected 'some puzzles' for which they had to come up with possible reconciliatory explanations, in particular for enforcement costs that do not increase along with asset specificity and project value, and the perceived enforcement costs that are relatively lower than set-up costs and monitoring costs.

Turner and Simister's (2001) use of TCE to predict the appropriate type of contract under varying degrees of uncertainty produced an outcome which coheres with conventional wisdom but does not align with practice. Using TCE as the interpretative framework to better understand partnering, Greenwood and Yates (2006) found that evidence for the reduction of contractual incompleteness was mixed. When using TCE to explain consortia in process engineering contracting, Clark and Ball (1991) concluded that, 'the neoclassical characteristics embodied in the transaction economics approach do not translate into empirical observation' (p. 355). Like Chang and Ive's (2007) work which was non-empirical, Nyström (2019) reasoned that asset specificity in TCE should be discarded while new ones (such as risk aversion, gains of specialization and high capital costs) be added when explaining the increase in subcontracting in the construction industry.

Some construction economics researchers chose to be pragmatic rather than be theoretically dogmatic, by taking note of the divergence between manufacturing and construction (Gonzáles-Diaz et al., 1998). In one extreme, Gebken and Gibson (2006) specifically mentioned that they used the phrase 'transaction cost', but not TCE's interpretation. Cox and Thompson (1997) used some of the TCE concepts without making reference to the theory, nor to Williamson. Others disaggregated TCE for their own studies rather than adopt the entire schema, ignoring Williamson's (1999) warning that predictions would be incorrect 'if interaction effects are missed or holistic consequences are glossed over' (p. 1102).

The constituent elements of TCE are actually inseparable. Sorell (2003) related barriers to energy efficiency in the construction of non-domestic buildings only to bounded rationality, opportunism and transaction cost. In studying the relationship between the client and the agent, Greenwood and Yates (2006) applied the contractual incompleteness and opportunism concepts to arrive at a prima facie validity of TCE as an interpretative framework for analysing partnering. To explain the causes of subcontracting, Gonzáles et al. (2000) deployed uncertainty and asset specificity. Chang and Ive (2007) relied on the notion of bargaining power and asset specificity for the selection of an appropriate procurement system. In analysing consortia in process plant contracting, Clark and Ball (1991) ignored the frequency dimension. Li et al. (2012) related transaction cost and uncertainty concepts to project performance. Rahman and Kumuraswamy (2002) and Wu et al. (2011) applied opportunism and bounded rationality, whereas Turner and Simister (2001), Sha (2011) and Rajeh et al. (2014) focused only on uncertainties for their respective works. Lu et al. (2015) also concentrated only on opportunism.

Some of the findings from the application of TCE by construction economics researchers are thought-provoking and therefore require further investigation. For example, Ho et al. (2015) proposed that PPPs are not hybrids as one might expect, but a special case of market governance. The analysis of Lu et al. (2015) also shows that opportunism has no bearing on project performance. Although they are enlightening, all these outcomes of piecemeal research should be treated with circumspection. Pragmatism carried too far may lead to conceptual contortions far removed from core TCE precepts. Here, three concepts – trust, uncertainty and project management – may be highlighted. In their argument that construction transactions are a hybrid structure, Reve and Levitt (1984) evoked the concept of trust. Langfield-Smith (2008), Ho et al. (2015) and Yan and Zhang (2020) also married the trust concept with TCE for their studies of alliances, PPPs and project performance, respectively. Such works ignored Williamson's rejection of trust in the TCE framework. He felt so strongly about trust that he devoted an entire paper to it (Williamson, 1993c, p. 463): 'It is redundant at best and can be misleading to use the term "trust" to describe commercial exchange for which cost-effective safeguards have been devised in support of more efficient exchange. Calculated trust is a contradiction in terms.'

He later noted, 'Indeed, I maintain that trust is irrelevant to commercial exchange and that reference to trust in this connection promotes confusion' (Williamson, 1993c, p. 469). Various management and organisational studies scholars have also incorporated trust in the TCE argument, but their efforts have 'had little or no influence on TCE research itself' (Bromiley and Harris, 2006, p. 126). Ali et al. (2018) noted that uncertainty can have positive implications (that is, leading to opportunity) in addition to negative (that is, leading to risk). While that may be so, the former idea does not resonate well with Williamson's writings which give uncertainty a negative connotation only. Linking project management to TCE also does not resonate well with Williamson's TCE logic. TCE views the firm as governance structure, not as production function (Williamson, 1988). Hence, by extension, stating that 'project management on behalf of a client is entirely a transaction cost' (Walker and Wing, 1999, p. 169) or analysing construction delays through the TCE lens (Sambasivan et al., 2017) are inappropriate.

Some construction economics scholars worked on providing clarity for key concepts of TCE in the context of construction. Lingard et al. (1998), Hughes et al. (2006), Li et al. (2014), Rajeh et al. (2015), Jin et al. (2017), Raji et al. (2019) and Abdel-Galil et al. (2020) identified the constituents or factors of transaction cost for the different phases of the project life cycle. Following Winch (1989), Guo et al. (2016) and Ali et al. (2018) concentrated on defining uncertainty, a task which Williamson somehow neglected (Slater and Spencer, 2000). Winch (1989) categorised uncertainty into task uncertainty, natural uncertainty, organisational uncertainty and contracting uncertainty. You et al. (2018) added environmental uncertainty to Williamson's behavioral uncertainty (Williamson, 1985). Winch (1989) associated complexity to the nature of the project (as noted above, complexity was never explained by Williamson). You et al. (2018) defined contractual complexity as 'the degree of detail of the provisions in construction contracts' (p. 3). Claiming that all the categories of asset specificity spelled out by Williamson are not applicable in the construction context, Chang and Ive (2007) came up with 'process specificity'. Lu et al. (2020) coined the term 'owner's asset specificity' without explaining what it covers. Some of these viewpoints as to what the key concepts are differ from one another, and some are even in conflict. For example, Abdel-Galil et al. (2020) classified conflict management, incomplete design and the contractor–subcontractor relationship as factors affecting transaction cost, whereas Ali et al. (2018) regarded these same three items as uncertainty. While Nyström (2019) ignored complexity, other researchers regard it as essential in their analyses (Winch, 1989). Disagreement over asset specificity, as pointed out above, can also be found in Greenwood and Yates (2006), Chen et al. (2015), Nyström (2019) and Yao et al. (2019).

Despite the advice of early construction economics scholars that TCE is ill-suited as an explanatory tool for the construction industry, it has not deterred many from still engaging it, with inevitable outcomes. The review suggests that some scholars did not fully grasp the TCE concept, as typified by treating the concept as already proven rather than being testable. Equally profound is that the valid concerns of TCE critics regarding the conceptual deficiencies of the theory were not heeded. It is pertinent to note that some researchers have been creative in getting around the measurement difficulties. However, disaggregating the TCE schema and using only certain selections of its elements flies in the face of its intended use. The validity of such piecemeal results, which incidentally can be confusing and at odds with one another, must be treated with circumspection.

THE WAY FORWARD

Milgrom and Roberts (1988, p. 450) predicted that new theories will emerge that can rival and complement TCE. Coase passed away in 2013 before he could fulfill his dream 'to construct a theory which will enable us to analyse the determinants of the institutional structure of production' (Coase, 1991, p. 73). While waiting for a new theory to emerge that is superior to TCE, Williamson made a recommendation which should be taken seriously. Williamson (2000b) invited researchers to accept pluralism, as it 'holds promise for overcoming our own ignorance' (p. 595). Indeed, whether due to this advice or to sheer serendipity, some construction economics scholars arrived at the same conclusion. After reviewing publications examining the construction industry through the theoretical lenses of TCE, supply chain management and industrial network individually, Håkansson and Jahre (2005) reflected that researchers are better served if they combine different theoretical perspectives for their studies. Having identified the poor predictive power of TCE, Carter and Hodgson (2006) stated that: 'A prominent conclusion in these studies is that an integration of TCE and competence-based explanations represents perhaps the most productive area for development' (p. 474). Hodgson (2010, p. 303) observed that TCE should also include the development of evolutionary approaches.

Others, most probably intuitively, acted out those invitations. Bridge and Tisdell (2004) integrated concepts from TCE and the resource-based view to define the boundary between the main contractor and subcontractor. Jin and Doloi (2008) too combined TCE with the resource-based view theory, to understand the practice of risk allocation in PPP projects. Rahman and Kumuraswamy (2002) extracted selected concepts of TCE to combine with relational contracting principles to come up with the 'transactionally efficient relational contracting' mechanism. Henisz et al. (2012) produced an interdisciplinary governance framework combining TCE, sociology and psychology as a guide for future research on governance of relational contracting on engineering projects. Kamann et al. (2006), Lee et al. (2009), Dolla and Laishram (2019) and Wang et al. (2019) also acted in similar fashion. Although the various theoretical strands have complementary features, they may also have contradictory assumptions. For that reason, Bygballe et al. (2013) recommended that these various strands be treated as 'arguments' in a broader debate.

CONCLUSION

Williamson passed away (in September 2020) as the finishing touches of this chapter were being added. Williamson's contribution in advancing further economic thinking has been profound and far-reaching. His legacy is not only in coming up with TCE, but also in initiating a new strand of economic thinking. His influence will continue to reverberate for years to come. Construction economics researchers intending to engage with TCE must fully grasp what Williamson had in mind, which is not easy as his work was developed and articulated over the course of a few decades. As highlighted by his critics, some key concepts were never properly explained (these include complexity, firm boundary and asset specificity), thereby creating confusion. A few conceptual incongruities remain (such as bounded rationality juxtaposition with omniscience, and firms treated as varied but human actors as static). Highlighting these lingering challenges is not to denigrate Williamson's magnificent work; rather, it is to help future construction economics researchers to navigate safely through the TCE schema. Early construction economics scholars found that TCE loses its predictive power when applied to construction, a lesson which their successors ignored, leading to the same outcome. Piecemeal research has produced a variety of theoretical contortions (such as TCE conjoined with the trust concept, and uncertainty as a positive connotation). With the amount of research material available, it is time for construction economics researchers to examine which ideas can be carried forward and which ones should be discarded, in the spirit shown by Constantino and Pietroforte (2002) when they revisited Eccles's findings 20 years later.

The most productive way for construction economics scholars to make contributions to the advancement of TCE is to incorporate multi-theoretical integration which includes TCE in their own research endeavours. A possible starting point is to re-engage with the early debate of why subcontracting is so prevalent in the construction industry, followed by organisation and governance in construction, procurement choice and contractor selection. Other areas such as green procurement, corruption, technology transfer, contractual disputes and risk management would also benefit from a TCE-grounded multi-theoretical lens approach. Some topical issues which could be studied include digital construction, big data applications in construction, and smart construction contracts. There is no better way to end this chapter than by quoting Williamson (2000b, p. 611):

> The upshot is that, its many accomplishments notwithstanding, there is a vast amount of unfinished business – refinements, extensions, new applications, more good ideas, more empirical testing, more fully formal theory. I conclude that the new institutional economics is the little engine that could. Its best days lie ahead. Who could ask for more?

REFERENCES

Abdel-Galil, E., Ibrahim, A.H., and Alborkan, A. (2020). Assessment of transaction costs for construction projects. *International Journal of Construction Management*, https://doi: 10.1080/15623599.2020.1738204.

Ali, Z., Zhu, F., and Hussain, S. (2018). Identification and assessment of uncertainty factors that influence the transaction cost in public sector construction projects in Pakistan. *Buildings*, *8*(11), 157, https://doi.org/10.3390/buildings8110157.

Argyes, N. (2010). The transaction cost as the unit of analysis. In P.G. Klein and M.E. Sykuta (eds), *The Elgar companion to transaction cost economics* (pp. 127–132). Cheltenham, UK and Northampton, MA, USA: Edward Elgar Publishing.
Bajari, P., and Tadelis, S. (2001). Incentives versus transaction costs: a theory of procurement contracts. *RAND Journal of Economics*, *32*(3), 387–407.
Bremer, W., and Kok, K. (2000). The Dutch construction industry: a combination of competition and corporatism. *Building Research and Information*, *28*(2), 98–108.
Bridge, A.J., and Tisdell, C. (2004). The determinants of the vertical boundaries of the construction firm. *Construction Management and Economics*, *22*(8), 807–825.
Bromiley, P., and Harris, J. (2006). Trust, transaction cost economics, and mechanism. In R. Bachmann and A. Zaheer (eds), *Handbook of trust research* (pp. 124–143). Cheltenham, UK and Northampton, MA, USA: Edward Elgar Publishing.
Brown, T.L., and Potoski, M. (2005) Transaction costs and contracting: the practitioner perspective. *Public Performance and Management Review*, *28*(3), 326–351.
Buckley, P.J., and Enderwick, P. (1989). Manpower management. In P. Hillerbrandt and J. Cannon (eds), *The management of construction firms: aspects of theory* (pp. 108–128). London: Palgrave Macmillan.
Bygballe, L.E., Håkansson, H., and Jahre, M. (2013). A critical discussion of models for conceptualizing gthe economic logic of construction. *Construction Management and Economics*, *31*(2), 104–118.
Carroll, G.R., and Teece, D.J. (1999). Firms, markets, and hierarchies: introduction and overview. In G.R. Carroll and D.J. Teece (eds), *Firms, markets, and hierarchies: the transaction cost economics perspective* (pp. 3–13). London: Oxford University Press.
Carter, R., and Hodgson, G.M. (2006). The impact of empirical tests of transaction costs economics on the debate on the nature of the firm. *Strategic Management Journal, 27*, 461–476.
Chang, C.-Y., and Ive, G. (2007). Reversal of bargaining power in construction projects: meaning, existence and implications. *Construction Management and Economics*, *25*, 845–855.
Chen, G., Zhang, G., and Xie, Y.M. (2012). Overview of alliancing in research and practice in the construction industry. *Architectural Engineering and Design Management*, *8*(2), 103–119.
Chen, G., Zhang, G., and Xie, Y.M. (2015). Impact of transaction attributes on transaction costs in project alliances: disaggregated analysis. *Journal of Management in Engineering, 31*(4), 04014054.
Clark, I., and Ball, D. (1991). Transaction cost economics applied? Consortia within process plant contracting. *International Review of Applied Economics*, *5*(3), 341–357.
Coase, R.H. (1937). The nature of the firm, *Economica,* 4, 386–405.
Coase, R.H. (1960). The problem of social cost, *Journal of Law and Economics*, *54*, 194–197.
Coase, R.H. (1991). The nature of the firm: origin, meaning and influence. In O.E. Williamson and S. Winter (eds), *The nature of the firm: origins, evolution and development* (pp. 34–74). Oxford: Oxford University Press.
Commons, J.R. (1924). *Legal foundations of capitalism*. New York: Macmillan.
Commons, J.R. (1934). *Institutional economics*. Madison, WI: University of Wisconsin Press.
Constantino, N., and Pietroforte, R. (2002). Subcontracting practices in USA homebuilding – an empirical verification of Eccles's findings 20 years later. *European Journal of Purchasing and Supply Management*, *8*(1), 15–24.
Constantino, N., Pietroforte, R., and Hamill, P. (2001). Subcontracting in commercial and residential construction: an empirical investigation. *Construction Management and Economics*, *19*, 439–447.
Cook, M.L. (1995). The future of US agricultural cooperatives: a neo-institutional approach. *American Journal of Agricultural Economics*, *77*, 1153–1159.
Cox, A., and Thompson, I. (1997). 'Fit for purpose' contractual relations: determining a theoretical framework for construction projects. *European Journal of Purchasing and Supply Management*, *3*(3), 127–135.
Crocker, K., and Masten, S. (1996). Regulation and administered contracts revisited: lessons from transaction-cost economics for public utility regulation. *Journal of Regulatory Economics*, *8*, 5–39.
David, R.J., and Han, S.-K. (2004). A systematic assessment of the empirical support for transaction cost economics. *Strategic Management Journal*, *25*(1), 39–58.

Dolla, T., and Laishram, B. (2019). Bundling in public-private partnership projects – a conceptual framework. *International Journal of Productivity and Performance Management*, https://doi 10.1108/IJPPM-02-2019-0086.
Douglas, M.T. (1990). Converging on autonomy: anthropology and institutional economics. In O.E. Williamson (ed.), *Organisation theory: from Chester Barnard to the present and beyond* (pp. 98–115). Oxford: Oxford University Press.
Dow, G.K. (1987). The function of authority in transaction cost economics. *Journal of Economic Behaviour and Organisation*, *8*, 13–38.
Eccles, R. (1981). The quasifirm in the construction industry. *Journal of Economic Behaviour and Organisation*, *2*, 335–357.
Fehr, E., and Gächter, S. (2000). Fairness and retaliation: the economics of reciprocity. *Journal of Economic Perspectives*, *14*(3), 159–182.
Foss, N.J., and Klein, P.G. (2010). Critiques of transaction cost economics: an overview. In P.G. Klein and M.E. Sykuta (eds), *The Elgar companion to transaction cost economics* (pp. 263–272). Cheltenham, UK and Northampton, MA, USA: Edward Elgar Publishing.
Gebken II, R.J., and Gibson, G.E. (2006). Quantification of costs for dispute resolution procedures in the construction industry. *Journal of Professional Issues in Engineering Education and Practice*, *132*(3), 264–271.
Geyskens, I., Steenkamp, J.-B.E.M., and Kumar, N. (2006). Make, buy, or ally: a meta-analysis of transaction cost theory. *Academy of Management Journal, 49*(3), 519–543.
Gonzáles, M., Arrunñada, B., and Fernández, A. (1998). Regulation as a cause of firm fragmentation: The case of the Spanish construction industry. *International Review of Law and Economics*, *18*, 433–450.
Gonzáles-Diaz, M., Arrunñada, B., and Fernández, A. (2000). Causes of subcontracting: evidence from panel data on construction firms. *Journal of Economic Behaviour and Organisation*, *42*, 167–187.
Greenwood, D., and Yates, D.J. (2006). The determinants of successful partnering: a transaction cost perspective. *Journal of Construction Procurement*, *12*(1), 4–22.
Gunnarson, S., and Levitt, R.E. (1982). Is a building construction project a hierarchy or a market? In J.O. Riis (ed.), *Proceedings of the 7th World Congress on Project Management Internet.* Copenhagen.
Guo, L., Li, H., Li, P., and Zhang, C. (2016). Transaction costs in construction projects under uncertainty. *Kybenetes*, *45*(6), 886–883.
Håkansson, H., and Jahre, M. (2005). Economic logics in the construction industry. In F. Khosrowshahi (ed.), *Proceedings of the 21st Annual ARCOM Conference* (pp. 1063–1073). SOAS, London..
Henisz, W., Levitt, R.E., and Scott, R. (2012). Toward a unified theory of project governance: economic, sociological and psychological supports for relational contracting. *Engineering Project Organisation Journal*, *2*(1–2), 37–55.
Ho, S.P., Levitt, R., Tsui, C.-W., and Hsu, T. (2015). Opportunism-focused transaction cost analysis of public-private partnerships. *Journal of Management in Engineering*, *31*(6), 04015007-3.
Hodgson, G.M. (1993). Transaction costs and the evolution of the firm. In C.N. Pitelis (ed.), *Transactions costs, markets and hierarchies* (pp. 77–100). Oxford: Basil Blackwell.
Hodgson, G.M. (2010). Limits of transaction costs analysis. In P.G. Klein and M.E. Sykuta (eds), *The Elgar companion to transaction cost economics* (pp. 297–306). Cheltenham, UK and Northampton, MA, USA: Edward Elgar Publishing.
Hughes, W., Hillebrandt, P., Greenwood, D., and Kwawu, W. (2006). *Procurement in the construction industry.* New York: Taylor & Francis.
Hughes, W., Hillebrandt, P., Lingard, H., and Greenwood, D. (2001). The impact of market and supply configurations on the costs of tendering in the construction industry. Paper presented at the CIB World Building Congress, 2–6 April, Wellington.
Ive, G., and Chang, C.Y. (2007). The principle of inconsistent trinity in the selection of procurement systems. *Construction Management and Economics*, *25*(7), 677–690.
Jin, X.-H. (2009). Allocating risks in public–private partnerships using transaction cost economics approach: a case study. *Australian Journal of Construction Economics and Building*, *9*(1), 19–26.
Jin, X.-H., and Doloi, H. (2008). Interpreting risk allocation mechanism in public–private projects: an empirical study in a transaction cost economics perspective. *Construction Management and Economics*, *26*(7), 707–721.

Jin, X. H., Zhang, G., Ke, Y., and Xia, B. (2017). Factors influencing transaction costs in construction projects: a critical review. In Y. Wu, S. Zheng, J. Luo, W. Wang, Z. Mo and L. Shan (eds), *Proceedings of the 20th International Symposium on Advancement of Construction Management and Real Estate* (pp. 949–958). Singapore: Springer.

Joskow, P. (2005). Vertical integration. In C. Ménard and M.M. Shirley (eds), *Handbook of new institutional economics* (pp. 319–348). Dordrecht: Springer.

Kamann, D.-J.F. Snijders, C., Tazalaar, F., and Welling, D.T. (2006). The ties that bind: buyer-supplier relations in the construction industry. *Journal of Purchasing and Supply Management, 12*, 28–38.

Klein, P.G. (2005). The make-or-buy decisions: lessons from empirical studies. In C. Ménard and M. M. Shirley (eds), *Handbook of new institutional economics* (pp. 435–464). Dordrecht: Springer.

Klein, P.G. (2010). Transaction cost economics and the new institutional economics. In P.G. Klein and M. E. Sykuta (eds), *The Elgar companion to transaction cost economics* (pp. 27–38). Cheltenham, UK and Northampton, MA, USA: Edward Elgar Publishing.

Lai, L.W.C. (2000). The Coasian market–firm dichotomy and subcontracting in the construction industry. *Construction Management and Economics, 18*, 355–362.

Langfield-Smith, K. (2008). The relations between transactional characteristics, trust and risk in the start-up phase of a collaborative alliance. *Management Accounting Research, 19*, 344–364.

Lee, H.-S., Seo, J.-O, Park, M., Ryu, H.-G., and Kwon, S.-S. (2009). Transaction-cost-based selection of appropriate general contractor–subcontractor relationship type. *Journal of Construction Engineering and Management, 135*(11), 1232–1240.

Li, H., Arditi, D., and Wang, Z. (2012). Transaction-related issued and construction project performance. *Construction Management and Economics, 30*(2), 151–164.

Li, H., Arditi, D., and Wang, Z. (2013). Factors that affect transaction costs in construction projects. *Journal of Construction Engineering and Management, 139*(1), 60–68.

Li, H., Arditi, D., and Wang, Z. (2014). Transaction costs incurred by construction owners. *Engineering, Construction and Architectural Management, 21(4)*, 444–458.

Lingard, H., Hughes, W., and Chinyio, E. (1998). The impact of contractor selection method on transaction costs: a review. *Journal of Construction Procurement, 4*(2), 89–102.

Love, P.E.D., Tse, R.Y.C., Holt, G.D., and Proverbs, D.G. (2002). Transaction costs, learning and alliances. *Journal of Construction Research, 3*(2), 193–207.

Lu, P., Guo, S., Qian, L., He, P., and Xu, X. (2015). The effectiveness of contractual and relational governance in construction projects in China. *International Journal of Project Management, 33*(1), 212–222.

Lu, W., Guo, W., and Zhu, Q. (2020). Effect of justice on contractor's relational behavior: moderating role of owner's asset specificity. *Journal of Construction Engineering and Management, 146*(4), 04020020.

Lyons, B. (1996). Empirical relevance of efficient contract theory: inter-firm contracts. *Oxford Review of Economic Policy, 12*, 27–52.

Macher, J.T., and Richman, B.D. (2008). Transaction cost economics: an assessment of empirical research in the social sciences. *Business and Politics, 10*(1), 1–63.

Mahoney, J.T. (2005). *Economic foundations of strategy*. Thousand Oaks, CA: SAGE Publications.

Manuel, F.E., and Manuel, F.P. (1979). *Utopian thought in the western world*. Cambridge, MA: Harvard University Press.

Masten, S.E., Meehan, J.W., Snyder, E.A. (1991). The costs of organization. *Journal of Law, Economics and Organisation, 7*(1), 1–25.

Ménard, C. (2005). A new institutional approach to organizations. In C. Ménard and M.M. Shirley (eds), *Handbook of new institutional economics* (pp. 281–318). Dordrecht: Springer.

Ménard, S.E., and Shirley, M.M. (2005). Introduction. In C. Ménard and M.M. Shirley (eds), *Handbook of new institutional economics* (pp. 1–20). Dordrecht: Springer.

Milgrom, P.J., and Roberts, J.D. (1992). *Economics, organisation and management*. Englewood Cliffs, NJ: Prentice Hall.

Milgrom, P.J., and Roberts, J.D. (1988). Economic theories of the firm: past, present, and future. *Canadian Journal of Economics, 21*(3), 444–458.

Miller, C.J.M., Packham, G.A., and Thomas, B.C. (2002). Harmonization between main contractors and subcontractors: a prerequisite for lean construction? *Journal of Construction Research, 3*(1), 67–82.

Nickerson, J. (1997) Towards an economising theory of strategy. Olin Working Paper 97-107.
Nyström, J. (2019). Updating and cleaning out: the 'make or buy' decision in construction revisited. *Emerald Reach Proceedings Series*, *2*, 3–8.
Oxley, Joanne (1999). Institutional environment and the mechanism of governance: the impact of intellectual property protection on the structure of inter-firm alliances. *Journal of Economic Behavior and Organisation*, *38*, 283–309.
Piertoforte, R. (1997). Communication and governance in the building process. *Construction Management and Economics*, *15*, 71–82.
Poppo, L., and Zenger, T. (2002). Do formal contracts and relational governance function as substitutes or complements? *Strategic Management Journal*, *23*(8), 707–725.
Rahman, M.M., and Kumaraswamy, M.M. (2002). Joint risk management through transactionally efficient relational contracting. *Construction Management and Economics*, *20*, 45–54.
Rajeh, M., Tookey, J.E., and Rotimi, J.O.B. (2014). Procurement selection model: development of a conceptual model based on transaction costs. *Australasian Journal of Construction Economics and Building-Conference Series*, *2*(2), 56–63.
Rajeh, M., Tookey, J.E., and Rotimi, J.O.B. (2015). Estimating transaction costs in New Zealand construction procurement. *Engineering, Construction, and Architectural Management*, *22*(2), 242–267.
Raji, A.U., Sarkile, K.A., and Jekadafari, M.U. (2019). Post-contract transaction costs: a waste minimisation perspective of the Nigerian construction industry. *FUTY Journal of the Environment*, *13*(1), 15–22.
Reve, T., and Levitt, R.E. (1984). Organisation and governance in construction. *International Journal of Project Management*, *2*, 17–25.
Rindfleisch, A., and Heide, J. (1997). Transaction cost analysis: past, present, and future applications. *Journal of Marketing*, *61*, 30–54.
Sambasivan, M., Deepak, T.J., Salim, A.N., and Ponniah, V. (2017). Transaction cost economics (TCE) and structural equation modeling (SEM) approach. *Engineering, Construction and Architectural Management*, *24*(2), 308–325.
Sarasvathy, S. (2010). Herbet Simon. In P.G. Klein and M.E. Sykuta (eds), *The Elgar companion to transaction cost economics* (pp. 85–91). Cheltenham, UK and Northampton, MA, USA: Edward Elgar Publishing.
Sha, K. (2011). Vertical governance of construction projects: an information cost perspective. *Construction Management and Economics*, *29*(11), 1137–1147.
Shelanski, H., and Klein, P. (1995). Empirical research in transaction cost economics: a review and assessment. *Journal of Law, Economics, and Organization*, *11*, 335–361.
Simon, H. (1957). *Administrative behaviour.* New York: Macmillan.
Slater, G., and Spencer, D.A. (2000). The uncertain foundations of transaction cost economics. *Journal of Economic Issues*, *34*(1), 61–87.
Sorrell, S. (2003) Making the link: the climate policy and the reform of the UK construction industry. *Energy Policy*, *31*, 865–878.
Sykuta, M.E. (2010a). Ronald H. Coase. In P.G. Klein and M.E. Sykuta (eds), *The Elgar companion to transaction cost economics* (pp. 39–48). Cheltenham, UK and Northampton, MA, USA: Edward Elgar Publishing.
Sykuta, M.E. (2010b). Empirical methods in transaction cost economics. In P.G. Klein and M.E. Sykuta (eds), *The Elgar companion to transaction cost economics* (pp. 152–164). Cheltenham, UK and Northampton, MA, USA: Edward Elgar Publishing.
Thorelli, H.B. (1986). Networks: between markets and hierarchies. *Strategic Management Journal*, *7*(1), 37–51.
Turner, J.R. (2004). Farsighted project contract management: incomplete in its entirety. *Construction Management and Economics*, *22*(1), 75–84.
Turner, J.R., and Keegan, A.E. (2001). Mechanisms of governance in the project-based organisation. *European Management Journal*, *19*(3), 254–267.
Turner, J.R., and Simister, S.J. (2001). Project contract management and a theory of organisation. *International Journal of Project Management*, *19*, 457–464.

Walker, A., and Wing, C.K. (1999). The relationship between construction project management theory and transaction cost economics. *Engineering, Construction, and Architectural Management*, *6*(2), 166–176.
Wang, D., Fang, S., and Fu, H. (2019). The effectiveness of evolutionary governance in mega construction projects: a moderated mediation model of relational contract and transaction cost. *Journal of Civil Engineering and Management*, *25*(4), 340–352.
Williamson, O.E. (1975). *Markets and hierarchies: analysis and antitrust implications*. New York: Free Press.
Williamson, O.E. (1979). Transaction-cost economics: the governance of contractual relations. *Journal of Law and Economics*, *22*, 233–261.
Williamson, O.E. (1980). The organisation of work. *Journal of Economic Behaviour and Organisation*, *1*, 6–31.
Williamson, O.E. (1983). Credible commitments: using hostages to support exchange. *American Economic Review*, *73*, 519–540.
Williamson, O.E. (1985). *The economic institutions of capitalism – firms, markets, relational contracting*. New York: Free Press.
Williamson, O.E. (1988). Technology and transaction costs: a reply. *Journal of Economic Behavior and Organisation*, *10*(3), 355–363.
Williamson, O.E. (1991a). Comparative economic organisation: the analysis of discrete structural alternatives. *Administrative Science Quarterly*, *36*(2), 269–296.
Williamson, O.E. (1991b). Strategising, economising, and economic organization. *Strategic Management Journal*, *12*, 75–94.
Williamson, O.E. (1993a). Transaction cost economics and organization theory. *Industrial and Corporate Change*, *2*(2), 107–156.
Williamson, O.E. (1993b). Opportunism and its critics. *Managerial and Decision Economics*, *14*, 97–107.
Williamson, O.E. (1993c). Calculativeness, trust, and economic organization. *Journal of Law and Economics*, *36*(1), Part 2, 453–486.
Williamson, O.E. (1995). Hierarchies, markets and power in the economy: an economic perspective. *Industrial and Corporate Change*, *4*(1), 21–49.
Williamson, O.E. (1996a). *The mechanisms of governance*. Oxford: Oxford University Press.
Williamson, O.E. (1996b). Economic organisation: the case for candor. *Academy of Management*, *21*(1), 48–57.
Williamson, O.E. (1998). Transaction cost economics: how it works; where it is headed. *De Economist*, *146*(1), 23–58.
Williamson, O.E. (1999). Strategy research: governance and competence perspectives. *Strategic Management Journal*, *20*, 1087–1108.
Williamson, O.E. (2000a). Strategy research: competence and government perspectives. In N. Foss and V. Mahnke (eds), *Advances in economic strategy research* (pp. 21–55). Oxford: Oxford University Press.
Williamson, O.E. (2000b). The new institutional economics: taking stock, looking ahead. *Journal of Economic Literature*, *38*(3), 595–613.
Williamson, O.E. (2005). Transaction cost economics. In C. Ménard and M.M. Shirley (eds), *Handbook of new institutional economics* (pp. 41–65). Dordrecht: Springer.
Williamson, O.E. (2008). Transaction cost economics, the precursors. *Economic Affairs*, *28*(3), 7–14.
Williamson, O.E. (2010). Transaction cost economics and the new institutional economics. P.F. Klein and E. Sykuta (eds), *The Elgar companion to transaction cost economics* (pp. 8–26). Cheltenham, UK and Northampton, MA, USA: Edward Elgar Publishing.
Williamson, O.E., and Masten, S. (1999). *The economics of transaction costs*. Cheltenham, UK and Northampton, MA, USA: Edward Elgar Publishing.
Winch, G.M. (1989). The construction firm and the construction project: a transaction cost approach. *Construction Management and Economics*, *7*, 331–345.
Winch, G.M. (2001). Governing the project process: a conceptual framework. *Construction Management and Economics*, *19*, 799–808.

Wu, J., Kumuraswamy, M.M., and Soo, G. (2011). Regulative measures addressing payment problems in the construction industry: a calculative understanding of their potential outcomes based on gametric models. *Journal of Construction Engineering and Management*, *137*(8), 566–573.
Yan, L., and Zhang, L. (2020) Interplay of contractual governance and trust in improving construction project performance: dynamic perspective. *Journal of Management Engineering*, *36*(4), 04020029
Yao, H., Chen, Y., Chen, Y., and Zhu, X. (2019). Mediating role of risk perception of trust and contract enforcement in the construction industry. *Journal of Construction Engineering and Management*, *145*(2), 04018130.
Yik, F.W.H., Lai, J.H.K., Chan, K.T., and Yiu, E.C.Y. (2006). Problems with specialist subcontracting in the construction industry. *Building Services Engineering Research and Technology*, *27*(3), 183–193.
You, J., Chen, Y., Wang, W., and Shi, C. (2018). Uncertainty, opportunistic behavior, and governance in construction projects: the efficacy of contracts. *International Journal of Project Management*, *36*(5), 795–807.

10. Construction industry and the Sustainable Development Goals (SDGs)

Alex Opoku

INTRODUCTION

The sustainable development agenda is becoming ever more important as the socio-economic and environmental conditions worsen globally. The challenge of achieving sustainable development is the ability to combine environmental conservation, social equity and economic development simultaneously (Maxwell, 2005). Even though sustainable development may have been defined differently by authors in different sectors and organisations, the concept can be integrated across different sectors and fields (Robinson, 2004). The Brundtland Commission report provides the most commonly used definition of sustainable development as: 'Meeting the needs of the present without compromising the ability of future generations to meet their own needs' (Brundtland, 1987, p. 43). However, Opoku and Ahmed (2013, p. 141) consider the involvement of humans by defining sustainable development as: 'The adjustment of human behaviour to address the needs of the present, without compromising the ability of future generations to meet their own needs'.

Sustainable development is aimed at reducing the negative impacts of human activities on the environment while ensuring socio-economic development. Environmental sustainability includes: biodiversity conversation, efficient land use and physical planning; while social sustainability involves access to decent work, quality education, good health, and respect for the rule of law and human rights. Economic sustainability involves the reduction of the negative impact of economic activities on the environment (Mensah, 2019). According to Hopwood et al. (2005), sustainable development is human-centred and the transformational view of sustainable development allows trade-offs between environmental and social dimensions with strong commitment to social issues, such as access to good health. The society depends on the environment, while the economy depends on society.

The sustainable development framework is guiding the transformation of the paradigm of socio-economic development globally from the ten specific policy reforms of the Washington Consensus to the eight Millennium Development Goals (MDGs) through to the 17 Sustainable Development Goals (SDGs) (Shi et al., 2019). According to Williamson (1994), the MDGs provide an excellent framework for political organisations and governments for achieving development. On the other hand, the Washington Consensus's ten policy reforms for developing countries were developed by the International Monetary Fund, World Bank and the United States Treasury. It was a requirement for the then George W. Bush administration's Millennium Challenge Account to support the low-income countries (Rodrik, 2002; Williamson, 2004), and it is important to note that the original ten-point Washington Consensus was mainly about macroeconomic and financial management (Williamson, 1994). The Washington Consensus provided justifications for the policy framework for addressing developmental challenges (Gore, 2000).

In order to achieve a sustainable society, sustainable development should be seen as a process and not an end-state position. It should be used as the new lens for addressing the global challenges (Robinson, 2004). Sustainable development is about the interaction between the economic and biophysical environment, and the economic aspect of sustainable development goes beyond the neoclassical approach, since the latter does not provide a complete picture of the policies and strategies required to achieve sustainable development (Mulder and Van Den Bergh, 2001). Harris (2003) argues about the validity of the economic dimension of sustainability, since neoclassical economic theory describes sustainability through the lens of human welfare growth. There has been a debate on the merits of adopting strong or weak sustainable development. However, Shi et al. (2019) express the opinion that strong sustainability should be adopted for sustainable development, as the concept is integrated into government and business strategies. Bossel (1999) believes that a sustainable society should withstand changes in values: cultural, environmental and technological.

THE SUSTAINABLE DEVELOPMENT GOALS

The United Nations (UN) SDGs were adopted by all the 193 member states of the organisation in September 2015 to address the global development challenges to end poverty, protect the planet and ensure prosperity for all (United Nations, 2015). The SDGs, which are also called the Global Goals, are applicable to both developed and developing countries although the challenges of development differ from one country to another (Corporate Citizenship, 2016). The attainment of the 17 universal goals is guided by 169 targets and measured by a set of 231 unique indicators, as indicated in Table 10.1. Although the SDGs combine the environmental, social and economic dimensions of sustainable development, Dickens et al. (2020) question whether the 231 unique SDG indicators are fit for purpose in the pursuit of sustainable development.

The SDGs replace the MDGs which covered eight goals and 21 targets, and were implemented during the period of 2000 to 2015. The SDGs are to be implemented in the subsequent 15-year period, from 2015 to 2030. While the MDGs were mainly aimed at governments in developing countries, the SDGs are global and aimed at governments, as well as business and non-governmental organisations in both developing and developed countries (Sustainable Development Solutions Network, 2016). The SDGs are guided by the principle of universality, which means that all countries and citizens have a role to play towards their achievement. The global goals provide an opportunity for all nations to protect the environment, eradicate poverty, combat the effects of climate change, reduce inequalities in society and improve human well-being (UKSSD, 2018).

To successfully achieve the SDGs by 2030, all sectors of society and all levels of government should embed the goals into all relevant regional, national and local policy processes. Government policies that capture the inclusiveness, universality and equity spirits of the SDGs are essential in meeting one of the major principles of the SDGs of leaving no one behind. Localisation of the SDG is the process whereby relevant stakeholders adapt, implement and monitor the SDG targets at different levels of operations or scenarios within and between sectors of the economy and society. Adams (2017) believes that the realisation of the SDGs will be challenging without collaboration among governments, private and public sector and civil society organisations. Thus, the attainment of the universal set of goals, targets and

Table 10.1 *The UN Sustainable Development Goals (SDGs), targets and indicators*

Goals	Description	No. of Targets	No. of Indicators
Goal 1	No Poverty: End poverty in all its forms everywhere	7	13
Goal 2	Zero Hunger: End hunger, achieve food security and improved nutrition, and promote sustainable agriculture	8	14
Goal 3	Good Health and Well-Being: Ensure healthy lives and promote well-being for all at all ages	13	28
Goal 4	Quality Education: Ensure inclusive and equitable quality education and promote life-long learning opportunities for all	10	12
Goal 5	Gender Equality: Achieve gender equality and empower all women and girls	9	14
Goal 6	Clean Water and Sanitation: Ensure availability and sustainable management of water and sanitation for all	8	11
Goal 7	Affordable and Clean Energy: Ensure access to affordable, reliable, sustainable and modern energy for all	5	6
Goal 8	Decent Work and Economic Growth: Promote sustained, inclusive and sustainable economic growth, full and productive employment and decent work for all	12	16
Goal 9	Industry, Innovation and Infrastructure: Build resilient infrastructure, promote inclusive and sustainable industrialisation and foster innovation	8	12
Goal 10	Reduced Inequalities: Reduce inequality within and among countries	10	14
Goal 11	Sustainable Cities and Communities: Make cities and human settlements inclusive, safe, resilient and sustainable	10	14*
Goal 12	Responsible Consumption and Production: Ensure sustainable consumption and production patterns	11	13
Goal 13	Climate Action: Take urgent action to combat climate change and its impacts	5	8
Goal 14	Life below Water (Oceans): Conserve and sustainably use the oceans, seas and marine resources for sustainable development	10	10
Goal 15	Life on Land (Biodiversity): Protect, restore and promote sustainable use of terrestrial ecosystems, sustainably manage forests, combat desertification, and halt and reverse land degradation and halt biodiversity loss	12	14
Goal 16	Peace, Justice and Strong Institution: Promote peaceful and inclusive societies for sustainable development, provide access to justice for all and build effective, accountable and inclusive institutions at all levels	12	24
Goal 17	Partnership for the Goals: Strengthen the means of implementation and revitalise the global partnership for sustainable development	19	24
		169	247**

Notes:
* The global statistical community is working to develop an indicator for target 11.c that could be proposed for the 2025 comprehensive review.
**The total number of indicators listed in the global indicator framework of SDG indicators is 247. However, 12 indicators are repeated under two or three different targets, resulting in 231 unique indicators.
Source: Adapted from UN (2015).

indicators of the SDGs by 2030 requires the involvement of businesses, governments and civil society as partners (SDSN, 2016).

The principle of 'leaving no one behind' underpinning the SDGs means that progress towards the achievement of the goals should be measured with reference to how well the poorest individuals and groups in society are improving in their socio-economic development. Adherence to the principle will ensure that development is equally distributed within and between nations and demographic groups (SDSN, 2016). The SDGs are transformational and are aimed at developing sustainable, inclusive and sustained economic growth cross the world. The 17 SDGs can be grouped into the five 'P areas' of People, Planet, Prosperity, Peace and Partnership, as illustrated in Figure 10.1. It is argued that these five Ps of the SDGs require the involvement of different stakeholders, such as governments, institutions and businesses, to be effectively implemented. The construction industry has the capacity to play a transformational role in the realisation of the SDGs, considering its contribution to the economy and the wider society. To play its part, the construction industry has to develop approaches that align its business strategies with the SDGs.

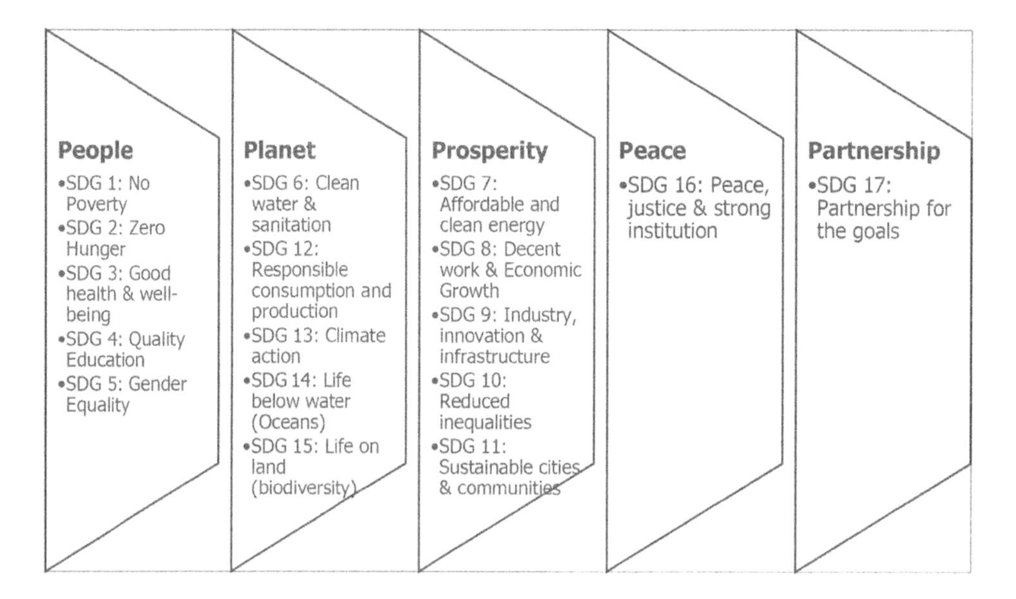

Figure 10.1 The five 'P areas' of the Sustainable Development Goals

According to Sachs et al. (2019), the SDGs can be achieved through six major societal transformations: (1) education and skills; (2) health and well-being; (3) clean energy and industry; (4) sustainable land use; (5) sustainable cities; and (6) digital technologies. The six transformations, which are guided by two key principles of 'leave no one behind' and 'ensure circularity and decoupling', are aimed at providing tangible strategies towards the realisation of the SDGs to government ministries and departments, the international community of businesses, and civil society.

The SDGs are not independent of each other. Although they are formulated and presented as individual goals, they are interlinked and intertwined in a unified framework offering the opportunity for nations and businesses to formulate strategic policy interventions and solutions that address multiple goals simultaneously (Waage et al., 2015; Nilsson et al., 2016). Owing to the integrated nature of the SDGs and the targets, any progress towards one goal or target is also linked through multifaceted feedbacks to other goals and targets, and any interactions and trade-offs between them can produce bad outcomes if the related policies and actions are not pre-designed properly (Bhaduri et al., 2016; Allen et al., 2019). According to a study by the University of Cambridge Institute for Sustainability Leadership (2017), the SDGs are deeply interconnected and none can be delivered in isolation. The institute adds that the built environment, planning, architecture and design interact with every goal.

Nations and organisations adopting the SDGs should identify the strong and positive linkages between targets and indicators and allow their prioritisation and the allocation of scarce resources to leverage efficiency, with policies and programmes targeting the goals with the greatest potential to have a positive impact and produce sustainable development outcomes (United Nations Statistical Commission, 2019). Accordingly, the United Nations Statistical Commission (2019) argues that defining and identifying linkages between the SDGs is complex because their interactions can be examined and interpreted in a variety of ways. For example, this can be done across the three dimensions of sustainable development (economic, social and environmental), or across the five thematic areas of the SDGs (people, planet, prosperity, peace and partnerships) (UNSC, 2019). Kroll et al. (2019) confirmed the synergies and trade-offs within and across the SDGs. For example, SDG 1 (No Poverty) is interlinked with many of the other goals, while SDG 12 (Responsible Consumption and Production) relates to trade-offs regarding the goals which will lead to economic progress. The study by the International Council for Science (2017) on the nature and type of interlinkages across the SDGs revealed that the SDGs can influence each other either positively or negatively. For example, there is a positive interaction between SDG 13 and SDGs 6, 7, 9, 11 and 16 (Kroll et al., 2019). Mensah (2019) adds that strategies towards the realisation of the SDGs will require trade-offs between the goals and targets.

According to Nilsson et al. (2016) the SDGs form an 'indivisible whole', which means that the goals and targets relate to, and depend on each other (Griggs et al., 2013; Nilsson et al., 2016). For example, SDG 3 (Good Health and Well-Being) cannot be achieved in isolation unless SDG 2 (Zero Hunger) has been achieved, and both goals are interlinked with SDG 10 (Reduced Inequalities) and SDG 8 (Decent Work and Economic Growth). However SDG 4 (Quality Education) has strong positive links with SDG 3 (Health and Well-Being), SDG 5 (Gender Equality), SDG 8 (Decent Work and Economic Growth) and SDG 1 (Poverty Reduction). On the other hand, SDG 1 (No Poverty), SDG 5 (Gender Equality) and SDG 6 (Clean Water) are all linked with 13 other goals. Again, SDG 11 (Sustainable Cities and Communities), which is critical for the sustainable development of the planet, has a link with all the other 16 goals; while SDG 4 (Quality Education) is linked with 14 goals (UNSC, 2019).

THE BUILT ENVIRONMENT AND THE SDGs

The built environment is literally a hub of economic activities, individuals and families, as well as society's cultural heritage. The built environment protects life and health, and the psychological and social welfare of its inhabitants; and sustains aesthetic and cultural values (Holm, 2003, as cited in Opoku, 2015). However, the built environment consumes a great deal of energy and resources while generating large amounts of waste. It is estimated that greenhouse gas emissions will rise by about 37 per cent by 2030 compared to 2005 levels (with construction contributing almost 50 per cent of this), affecting the natural and the built environment, if appropriate and efficient policies are not implemented (OECD, 2008). The solution to this problem is the planning, design and construction of improved built assets supported by appropriate management tools and regulatory frameworks that address the issues of sustainable development (Grierson, 2009, as cited in Opoku, 2015).

The connections between components of the built environment, such as individual buildings, transport systems, urban landscapes and other infrastructure, should aim at reducing environmental impacts in terms of energy, carbon emissions, waste and water. With more than half of humanity living in cities and the number of urban residents growing by nearly 73 million every year, it is estimated that urban areas account for 70 per cent of the world's gross domestic product (GDP), and the construction industry can play a crucial role in achieving SDG 11 (UN, 2018, as cited in Opoku and Akotia, 2020). Goal 11 highlights the central role of urbanisation in sustainable development, describing the need for inclusive, safe, resilient and sustainable cities and communities through relevant public policies (SDSN, 2016). Sustainable regeneration projects can also significantly contribute to social well-being (Wilson, 2015), achieving a sustainable built environment that impacts on the society's ability towards the realisation of the SDGs.

A recent UN report estimates that about 3 billion people are living in urban slum conditions due to the problem of rapid urbanisation and population growth, and would require adequate and affordable housing. The pace of constructing new affordable housing is slow compared with the global population growth, and this would require countries to develop and implement national urban plans to address this challenge (UNDESA, 2019). It is suggested that urbanisation should be at the heart of the 2030 Agenda for sustainable development. For example, Teferi and Newman (2018) believe that regenerating urban slums in the developing world could greatly contribute to the realisation of many of the SDGs including SDG 1 (No Poverty), SDG 6 (Clean Water and Sanitation), SDG 7 (Affordable and Clean Energy) and SDG 11 (Sustainable Cities and Communities). The paradigm shift towards a low-carbon society cannot be achieved without sustainable, efficient and resilient infrastructure that can withstand the effects of climate change. Maintaining sustainable and resilient infrastructure systems that can withstand floods, retain road systems and ensure the reliability of building structures is crucial for guaranteeing essential services such as energy and water supplies (Boyle et al., 2013).

The built environment has a long-term impact on the quality of life, prosperity, health, well-being and happiness of people and communities in terms of how the planning, design, management and maintenance of the built asset are undertaken (House of Lords, 2016, as cited in Opoku, 2016). The built infrastructure needs to be flexible and adaptable for future uses, and it should also be resilient with respect to the effects of climate change. The built environment should lead the fight against climate change towards a low-carbon and sustainable future. The United Kingdom government's construction strategy target of reducing carbon emissions by 50 per cent by 2025 based on a 2 degrees centigrade temperature rise target needs to be reconsidered if the built environment is to contribute significantly to the 1.5 degree centigrade target set by the Paris Agreement in the future. A sustainable built environment with energy-efficient infrastructure and services can contribute meaningfully to the effort to attain this target by reducing the demand for energy and eventually reducing the impact of climate change. The conservation of historic cities is an effective strategy in reducing carbon dioxide (CO_2) emissions when compared to their replacement by new buildings. New construction projects have more damaging impacts in the short to mid-term; however, conserving an existing building saves original energy and CO_2 investment (Lewis, 2012, as cited in Opoku, 2015). Retrofitting existing infrastructure can also increases urban resilience to disaster and ensures sustainable cities and communities (SDG 11) (Opoku, 2019a).

THE CONSTRUCTION INDUSTRY AND THE SDGs

The SDGs offer the construction industry a new opportunity to expand its focus away from the environmental dimension of sustainability (Goubran, 2019). Construction project delivery and its management could be considered to be sustainable if social, economic and environmental elements are integrated into the project delivery processes, standards and practices (Silvius, 2017). Although the construction industry has considerable social, economic and environmental impacts during the design and build process, it can play a critical role towards the achievement of the SDGs as the industry builds tomorrow's world (BDG, 2019). For example, construction and demolition activities generate large volumes of waste, and this requires the adoption of practices that will reduce the waste generated and maximise reuse with the aim of improving efficient resource use and reduce negative impact on the environment (Gálvez-Martos et al., 2018). However, Mahpour (2018) argues that the circular economy is a sustainable concept that could be adopted for the efficient exploitation of resources and the management of construction and demolition waste.

The construction industry is a key partner in the global effort towards the realisation of sustainable development by 2030 through the delivery of sustainable projects. According to Ofori (2016), the 17 SDGs can be classified, from the perspective of construction, into: (1) basic human and national needs – Goals 1, 2, 3, 4, 5 and 8; (2) what construction must do – Goals 9 and 11; (3) some of construction's results – Goals 6 and 7; and (4) inputs and methods of construction industry – Goals 12, 13, 14 and 15. Through land development, resource use, waste generation and labour practices, the construction industry has an impact on the SDGs throughout its project life cycle (RICS and UNGC, 2018). While the construction industry offers economic growth and employment opportunities, the industry has a reputation of adopting poor practices such as lack of attention to health, safety and welfare, while there are even occurrences of modern slavery affecting human and labour rights, and these can have an impact on the realisation of SDGs 1, 8, 10 and 16 (RICS and UNGC, 2018). Corrupt practices in the construction industry which are evident through planning, procurement and payments will affect SDG 16 socially, politically and economically.

The built environment sector (envisaged as the construction and property sectors) is connected to every one of the 17 SDGs, since people live in the settlements it creates (BDG, 2019). Opoku (2016) argues that the construction industry can impact highly on the realisation of a number of SDGs, including SDG 2 (End Hunger), SDG 3 (Good Health and Well-Being), SDG 4 (Quality Education), SDG 6 (Clean Water and Sanitation), SDG7 (Affordable and Clean Energy), SDG 8 (Decent Work and Economic Growth), SDG 9 (Industry, Innovation and Infrastructure), SDG 10 (Reduced Inequalities), SDG 11 (Sustainable Cities and Communities) and SDG 13 (Climate Action).

Sustainable development has been described as the new project management paradigm (Gareis et al., 2013), and the SDGs should be embedded into the early project goal-setting, the business case, project benefits and success criteria, the specifications and the design of the project outcomes. During the planning, design and construction stages as outlined in the Royal Institute of British Architects (RIBA) Plan of Works, poor practices and materials use are major causes of greenhouse gas emissions, and these impact on SDG 7, 11, 12 and 13. However, construction and the effects on the environment and communities through waste generated through end-of-life is relevant for SDGs 3, 6, 11, 12 and 15. Gareis et al. (2010) suggest that the integration of sustainability principles into project management could improve the overall project delivery by reducing the likelihood of project interruption or cancellation, manage the project complexity, and create economic benefits and social value.

The construction industry has a particularly vital role to play in preserving biodiversity (SDG 15) by leading the agenda towards the preservation of biodiversity at the heart of sustainable development (Opoku, 2019b). Biodiversity should be incorporated into the built environment by providing green urban spaces such as green roofs that can contribute to urban biodiversity conservation (Lepczyk et al., 2017). New construction projects should integrate biodiversity schemes such as the creation of habitats for wildlife by providing nest boxes, living roofs and landscapes, and schemes incorporating these elements should not add excessive cost to the overall budget of the construction projects. Integrating biodiversity into the planning, management and the regulatory process of cities and their key infrastructure is essential towards the attainment of solutions to promote biodiversity (SIDA, 2016, as cited in Opoku, 2019b).

The Role of Construction Business Towards the Realisation of the SDGs

> Business is a vital partner in achieving the Sustainable Development Goals. Companies can contribute through their core activities, and we ask companies everywhere to assess their impact, set ambitious goals and communicate transparently about the results. (Ban Ki-moon, United Nations Secretary-General, UN, 2016)

The SDGs provide business organisations with a new lens through which global needs and desires can be translated into business solutions. The SDGs provide a framework for business organisations of all sizes, sectors and geographical locations to play a crucial role towards the realisation of the SDGs by embracing the opportunities they present. Rosati and Faria (2019) noted that the business community can play a critical role in the achievement of the SDGs by planning, implementing, measuring and communicating their SDG-related activities and action though business reporting. However, business leaders are believed to be struggling with how to measure, track, communicate and demonstrate their impact on the SDGs and metrics of business success (UNGC and Accenture, 2016). The business community can contribute to several SDGs, including the creation of well-paid employment and decent work, and the promotion of inclusive and sustainable economic growth (SDG 8). Again, businesses can contribute to the development of resilient infrastructure and the creation of sustainable industries (SDG 9) with access to affordable, reliable and sustainable energy supply (SDG 7) (Economist Intelligence Unit, 2017).

A study by the UN Global Compact and Accenture reveals that 87 per cent of chief executive officers globally believe the SDGs can act as an engine of economic growth and employment and provide an opportunity to rethink approaches to sustainable development, through innovation and technological development (UNGC and Accenture, 2016). In another study by Scott and McGill (2018) involving 729 companies from 21 countries and territories across six broad industries on the adoption of the SDGs by business, 72 per cent of companies mentioned the SDGs in their annual corporate or sustainability reports, with 50 per cent identifying priority SDGs for their businesses. The research also revealed the top five priority SDGs for the business community as Goal 8 – Decent Work and Economic Growth (79 per cent), Goal 13 – Climate Action (76 per cent), Goal 12 – Responsible Consumption and Production (66 per cent), Goal 3 – Good Health and Well-Being (57 per cent), and Goal 9 – Industry, Innovation and Infrastructure (55 per cent). The SDGs will allow businesses, including construction organisations, to promote sustainable development by demonstrating how business can minimise the negative impacts and/or maximise the positive impacts on people, planet and profit. Businesses adopting the SDGs can take advantage of the benefits this brings such as the identification of future business opportunities. The SDGs can be used by businesses as the underpinning framework for shaping, steering, communicating and reporting their strategies, goals and activities (GRI et al., 2015).

The construction companies should conserve and preserve biodiversity, which is crucial for the realisation of the SDGs during the delivery of construction projects (Opoku, 2019b). The ability of construction organisations to cope with the journey towards sustainable change requires organisational learning. It is essential that construction organisations engage in organisational learning methods that incorporate sustainability and innovation. Existing organisational learning methods in the construction industry such as post-project reviews (PPRs) and post-occupancy evaluations (POEs) mostly focus on technical issues instead of the social, economic and environmental impact of the completed project (Opoku and Fortune, 2011). Not only will such reviews need to be undertaken in a routine manner on all projects, but also they should be widened to include the non-technical aspects which relate to the SDGs. Construction businesses can align business strategies with the SDGs by identifying, aligning, integrating, partnering and measuring the impact of business on the SDGs, as illustrated in Figure 10.2.

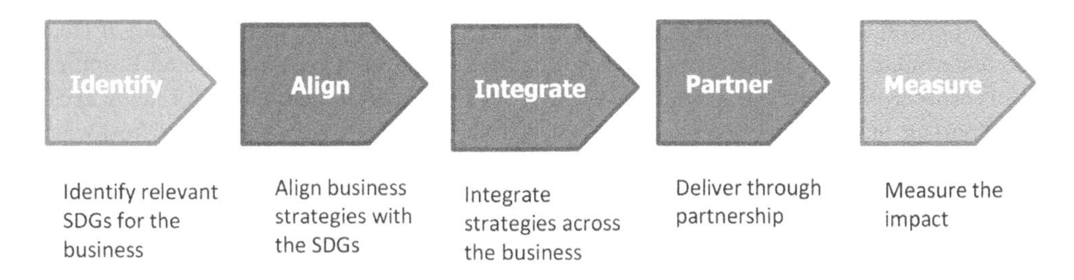

Figure 10.2 *Integrating the SDGs into business strategies*

Construction Industry's Contribution to the Achievement of the SDGs

The UN in January 2020 launched the Decade of Action (2020–2030) campaign, calling on all nations and the relevant stakeholders to accelerate the realisation of sustainable solutions needed for socio-economic transformation and the realisation of the SDGs by 2030 (UNDESA, 2019).

Reports show that progress towards the achievement of the SDGs has been slow (Globescan, 2017), and there are calls on all stakeholders to play their part towards the realisation of the goals. The construction industry can contribute to the attainment of each of the 17 SDGs, since planning, architecture and design interact with almost every goal. SDG 1, which consists of seven targets and 13 indicators, is aimed at ending all forms of poverty globally by 2030. The construction industry should implement industry-wide strategies to ensure fair wages for employees in all communities where they operate. The construction industry should formulate policies that help to provide a sustainable built environment for the poor by delivering accommodation and basic facilities for the homeless. SDG 2 is about ending hunger and achieving food security and improved nutrition, and promotes sustainable agriculture by 2030, and this goal is made up of eight targets and 14 measurable indicators. The construction industry can contribute to the development of infrastructure to enhance agricultural productivity such as green energy for agriculture, green roofs in food processing plants, and innovative food storage facilities.

SDG 3 is on good health and well-being; it consists of 13 targets and 28 indicators which are aimed at ensuring healthy lives and promoting well-being for all. The construction industry is labour-intensive, with poor working conditions in most cases; however, the industry can contribute to good health and well-being by promoting good work–life balance through job share and flexible working arrangements. Goal 4 ensures inclusive and equitable quality education that promotes life-long learning opportunities for all. This SDG is made up of ten targets and 12 indicators. The construction industry can take advantage of its expertise to research, develop and deliver educational facilities that showcase sustainable development. Construction organisations can also invest in educational programmes for communities as part of their corporate social responsibility.

Gender equality is SDG 5, made up of nine targets and 14 indicators; it is aimed at achieving gender equality and empowering all women and girls by 2030. Generally, women are disadvantaged in their career journeys due to gender bias in construction organisations; this affects women's career progression into leadership positions in the construction sector (Opoku and Williams, 2019). The promotion of gender equality in the construction industry is essential in attracting and retaining talents which can improve its productivity. The provision of clean water and sanitation is considered in SDG 6; the goal consisting of eight targets and 11 indicators to ensure availability and sustainable management of water and sanitation for all by 2030. It is important that construction organisations take the necessary steps to protect freshwater bodies close to construction sites from possible contamination.

The provision of affordable and clean energy is presented in SDG 7 to ensure access to affordable, reliable, sustainable and modern energy for all; the goal is to be achieved through five targets and six measurable indicators. Electricity usage on construction sites is one way the industry can contribute to SDG 7; the construction industry could make it a policy to source electricity for all temporary construction processes and operations from renewable energy sources. The SDGs are intended to promote decent work and attain economic growth through sustained, inclusive and sustainable economic growth, full and productive employment for all by 2030. This is captured in SDG 8 which is supported by 12 targets and underpinned by 16 indicators. The provision of employment and decent work for all promotes sustainable economic growth which is at the heart of the 2030 Agenda for sustainable development. Promoting sustainable procurement practices that protect labour rights and prevent modern slavery and child labour on construction sites should be encouraged (Chiarini et al., 2017). This should be implemented across the whole supply chain, especially when main contractors are dealing with subcontractors.

Goal 9 of the SDGs is on industry, innovation and infrastructure, and is supported by eight targets and 12 indicators to build resilient infrastructure, promote inclusive and sustainable industrialisation, and foster innovation. The construction industry should develop and align its strategies with the SDGs at both the organisational and the project levels. This could ensure that projects delivered (new build or refurbishment) demonstrate sustainable development. Goal 10 on reduced inequalities, which consists of ten targets underpinned by 14 indicators, is important for achieving prosperity and reducing inequality within and among countries by 2030. Construction organisations should develop and implement recruitment policies that provide opportunity for people in the local communities of operation and also attract and retain qualified staff from diverse underrepresented groups in the construction industry such as the ethnic minorities in various groups in all countries. SDG 11 on sustainable cities and communities is highly impacted upon by construction activities, and consists of ten targets and 14 measurable indicators to make cities and human settlements inclusive, safe, resilient and sustainable. The construction industry is crucial in developing sustainable cities and communities through the delivery of a sustainable built environment. The industry contributes to the development of communities including the provision of sustainable and affordable housing.

SDG 12 on responsible consumption and production, which is supported by 11 targets and underpinned by 13 indicators, is aimed at ensuring sustainable consumption and production patterns by 2030. Murtagh et al. (2020) add that the construction industry is one of the most environmentally damaging industries due to the amount of resources consumed during project delivery and the energy required for building operation. Therefore, construction companies should align their business strategies for construction and demolition waste recycling with the SDGs. The construction industry can play a major role towards the efficient use of natural resources and waste reduction through recycling and reuse of building materials. This will contribute to efforts toward the realisation of SDG 12, targets 12.2 and 12.5. The construction industry consumes a large amount of natural resources, and produces a significant amount of waste from its operations. Goal 13 is concerned with taking urgent action to combat climate change and its impact on society; this goal is supported by five targets and measured through eight indicators. Achieving a quality and sustainable built environment can contribute to the realisation of the SDGs through the implementation of relevant environmental protection strategies during urban development (Fang et al., 2020). Conserving and sustainably using the oceans, seas and marine resources for sustainable development is the goal of SDG 14 on life below water, with ten targets and indicators. The construction industry should adopt the use of reclamation technology to protect low-lying areas from rising sea levels and also prevent construction waste from being dumped into the seas and water bodies.

SDG 15 on biodiversity, which is supported by 12 targets and underpinned by 14 indicators, is aimed at protecting, restoring and promoting sustainable use of terrestrial ecosystems and sustainable management of forests to combat desertification and halt or reverse land degradation and biodiversity loss by 2030. The construction industry has a negative impact on biodiversity, and Opoku (2019b) argues that the industry should conserve and preserve biodiversity which is crucial for the realisation of SDG 15 during the delivery of projects. Peace, justice and strong institutions are essential for socio-economic development, and SDG 16, which consists of 12 targets and is underpinned by 24 indicators, aims to promote peaceful and inclusive societies for sustainable development by providing access to justice for all and building effective, accountable and inclusive institutions at all levels of society by 2030. The construction industry should adopt strategies towards reduction corruption and reforming its practices that create conflicts and litigation in the industry, and in their place, promote transparency in procurement.

Finally, the successful delivery of the SDGs is underpinned by partnership to strengthen the implementation of the global sustainable development. This goal is made up of 19 targets and 24 indicators. The construction industry should embrace the opportunity for exchanging experts between project, companies and possibly across countries. A study by Goubran (2019) on the role of construction activity in achieving the SDGs shows that about 17 per cent of the SDG targets are directly affected by construction activities, while 27 per cent of the targets are impacted upon indirectly. Moreover, these targets relate to all of the 17 SDGs, and therefore Goubran argues that the construction industry can expand its focus through the SDG framework.

SUMMARY AND CONCLUDING REMARKS

The SDGs represent a new direction for the global community, integrating social, economic and environmental sustainability into all policies and strategies with the view of eliminating poverty and inequality to achieve a more prosperous society. The construction industry can significantly contribute to the socio-economic development and well-being of society through the provision of employment and decent work that promotes sustainable economic growth. The construction industry is central to the delivery of many of the government's policies for sustainable development through the provision of buildings and infrastructural development. The industry can also provide solutions to previous mistakes, such as the rehabilitation of contaminated land. The construction industry can contribute to the achievement of the SDGs by developing smart cities and sustainable communities, adopting sustainable procurement, design and construction practices, and the provision of renewable energy technology such as solar on built assets.

The construction industry is a key partner in the global effort to achieve sustainable development by 2030 through the development of sustainable infrastructural projects, especially in developing countries where there is an infrastructure deficit. The achievement of the SDGs will require new investments in sustainable infrastructure globally, especially in the developing countries. The construction industry provides employment opportunities for the communities where built asset projects are undertaken. Thus, it can help to eradicate poverty and improve the standard of living of the citizenry. Governments across the globe should use the construction industry as the champion in developing the right policies and regulations. Construction businesses should collaborate with government agencies and policy-makers, and other industry peers to integrate the SDGs into long-term business strategies, and work towards the realisation of 'the world we want', another of the slogans of the SDGs. The implementation of the SDGs should be guided by the lessons learnt from the successes and failures of the MDGs, including improving partnership among the stakeholders. The future of the SDGs lies in moving away from effecting trade-offs to seeking to optimise synergies among the related SDGs during implementation. The adoption of the SDGs in 2015 changed the focus of the developmental agenda, from addressing economic and social challenges, to the inclusion of environmental and economic issues.

By stressing the commonality of the concern with development among countries at all levels of development, the SDGs stress the need for solidarity and collaboration in developing appropriate planning, accounting, assessment and monitoring techniques. These activities indicate possible themes for the development of construction economics.

REFERENCES

Adams, A.C (2017). *The Sustainable Development Goals, Integrated Thinking and the Integrated Report*. International Integrated Reporting Council (IIRC), London.

Allen, C., Metternicht, G. and Wiedmann, T. (2019). Prioritising SDG targets: assessing baselines, gaps and interlinkages. *Sustainability Science*, 14, 421–438.

BDG (2019). *Build a Better Future with the Sustainable Development Goals: A Practical Guide for Construction and Property Companies*. Bioregional Development Group (BDG), Oxford.

Bhaduri, A., Bogardi, J., Siddiqi, A., Voigt, H., Vörösmarty, C., Pahl-Wostl, C., et al. (2016). Achieving Sustainable Development Goals from a water perspective. *Frontiers of Environmental Science* 4(64). doi: 10.3389/fenvs.2016.00064.

Bossel, H. (1999). *Indicators for Sustainable Development: Theory, Method, Applications, A Report to the Balaton Group*. International Institute for Sustainable Development (IISD), Winnipeg, Canada.

Boyle, J., Cunningham, M. and Dekens, J. (2013). *Climate Change Adaptation and Canadian Infrastructures*. Report for International Institute for Sustainable Development (IISD), Winnipeg, Canada.

Brundtland, G.H. (1987). *Our Common Future: Report of the world Commission in Environment and Development*. Oxford University Press, Oxford.

Chiarini, A., Opoku, A. and Vagnoni, E. (2017). Public healthcare practices and criteria for a sustainable procurement: a comparative study between UK and Italy. *Journal of Cleaner Production*, 162, 391–339, https://doi.org/10.1016/j.jclepro.2017.06.027.

Corporate Citizenship (2016). *Advancing the Sustainable Development Goals: Business Action and Millennials' Views*. Corporate Citizenship, London.

Cambridge Institute for Sustainability Leadership (CISL) (2017). *Towards a Sustainable Economy: The Commercial Imperative for Business to Deliver the UN Sustainable Development Goals*. University of Cambridge Institute for Sustainability Leadership (CISL), Cambridge.

Dickens, C., McCartney, M., Tickner, D., Harrison, I.J., Pacheco, P. and Ndhlovu, B. (2020). Evaluating the global state of ecosystems and natural resources: within and beyond the SDGs. *Sustainability*, 12(18), 7381, https://doi.org/10.3390/su12187381.

Economist Intelligence Unit (EIU) (2017). *Meeting the SDGs: A Global Movement Gains Momentum*. A report from The Economist Intelligence Unit, London.

Fang, X., Shi, X., and Gao, W. (2020). Measuring urban sustainability from the quality of the built environment and pressure on the natural environment in China: a case study of the Shandong Peninsula region. *Journal of Cleaner Production*, 125145, https://doi.org/10.1016/j.jclepro.2020.125145.

Gálvez-Martos, J.-L., David Styles, D., Schoenberger, H. and Zeschmar-Lahl, D. (2018). Construction and demolition waste best management practice in Europe. *Resources, Conservation and Recycling*, 136, 166–178.

Gareis, R., Heumann, M. and Martinuzzi, A. (2010). Relating sustainable development and project management: a conceptual model. *PMI Research and Education Conference*. Washington, DC.

Gareis, R., Huemann, M. and Martinuzzi, A. (2013). *Project Management and Sustainable Development Principles*. Project Management Institute, Pennsylvania, PA.

Globescan (2017). *Evaluating Progress towards the Sustainable Development Goals*. Globescan Sustainability Survey. Globescan, Toronto, Canada.

Gore, C. (2000). The rise and fall of the Washington Consensus as a paradigm for developing countries. *World Development*, 28(5), 789–804.

Goubran, S. (2019). On the role of construction in achieving the SDGs. *Journal of Sustainable Research*, 1, e190020, https://doi.org/10.20900/jsr20190020.

Griggs, D., Stafford-Smith, M., Gaffney, O., Rockstrom, J., Öhman, M.C., Shyamsundar, P., et al. (2013), Sustainable development goals for people and planet. *Nature*, 495, 305–307, https://doi.org/10.1038/495305a.

GRI, UN Global Compact and WBCSD (2015). SDG Compass: the guide for business action on the SDGs, https://sdgcompass.org/wp-content/uploads/2016/05/019104_SDG_Compass_Guide_2015_v29.pdf.

Harris, J. (2003). *International Society for Ecological Economics; Internet Encyclopaedia of Ecological Economics: Sustainability and Sustainable Development*, http://isecoeco.org/pdf/susdev.pdf (accessed on 13 January 2021).
Holm, F.H. (2003). Towards a sustainable built environment prepared for climate change? A paper presented at the Global Policy Summit on the Role of Performance-Based Building Regulations in Addressing Societal Expectations. International Policy, and Local Needs, National Academy of Sciences, Washington, DC, 3–5 November.
Hopwood, B., Mellor, M. and O'Brien, G. (2005). Sustainable development: mapping different approaches. *Sustainable Development*, 13(1), 38–52.
House of Lords (2016). *Select Committee on National Policy for the Built Environment: Building Better Places*. Report of Session 2015–16. Stationery Office, London, http://www.publications.parliament.uk/pa/ld201516/ldselect/ldbuilt/100/10002.htm.
International Council for Science (2017). *A Guide to SDG Interactions: from Science to Implementation* (eds: D.J. Griggs, M. Nilsson, A. Stevance, D. McCollum), International Council for Science, Paris.
Kroll, C., Warchold, A. and Pradhan, P. (2019). Sustainable Development Goals (SDGs): are we successful in turning trade-offs into synergies? *Palgrave Communication*, 5(140), https://doi.org/10.1057/s41599-019-0335-5.
Lepczyk, C.A., Aronson, M.F.J., Evans, K.L., Goddard, M.A., Lerman, SB. and MacIvor, J.S. (2017). Biodiversity in the city: fundamental questions for understanding the ecology of urban green spaces for biodiversity conservation. *BioScience*, 67(9), 799–807.
Lewis, J.O. (2012). Renovate or demolish/rebuild: what are the drivers? EuroACE, Brussels, Renovate Europe Day, 11 October, http://www.renovateeurope.eu/uploads/REDay2012%20ppts/REDay2012_Workshop_Owen_Lewis.pdf.
Mahpour, A. (2018). Prioritizing barriers to adopt circular economy in construction and demolition waste management. *Resources, Conservation and Recycling*, 134, 216–227.
Maxwell, S. (2005). The Washington Consensus is dead! Long live the meta-narrative! Working Paper 243, Overseas Development Institute (ODI).
Mensah, J. (2019). Sustainable development: meaning, history, principles, pillars, and implications for human action: literature review. *Cogent Social Sciences*, 5(1), 1653531, https://doi.org/10.1080/23311886.2019.1653531.
Mulder, P. and Van Den Bergh, J.C.J.M. (2001). Evolutionary economic theories of sustainable development. *Growth and Change*, 32(1), 110–134.
Murtagh, N., Scott, L. and Fan, J. (2020). Sustainable and resilient construction: current status and future challenges. *Journal of Cleaner Production*, 268, 122264, https://doi.org/10.1016/j.jclepro.2020.122264.
Nilsson, M., Griggs, D. and Visbeck, M. (2016). Map the interactions between Sustainable Development Goals. *Nature*, 534, 320–322.
OECD (2008). *OECD Environmental Outlook to 2030*. Organisation for Economic Co-operation and Development. OECD Publishing, Paris.
Ofori, G. (2016). Construction in developing countries: current imperatives and potential. *Proceedings of the CIB World Building Congress 2016*, ed. Kähkönen, K. and Keinänen, M., compiled by Department of Civil Engineering, Construction Management and Economics, Tampere University of Technology, Vol. 1, 39–52. Tampere, Finland, 30 May to 3 June.
Opoku, A. (2015). The role of culture in a sustainable built environment. In: A. Chiarini (ed.), *Sustainable Operations Management: Advances in Strategy and Methodology*. Springer International Publishing, Cham, pp. 37–52.
Opoku, A. (2016). SDG2030: A sustainable built environment's role in achieving the post-2015 United Nations Sustainable Development Goals. In: P.W. Chan and C.J. Neilson (eds), *Proceedings of the 32nd Annual ARCOM Conference*, 5–7 September, Manchester, UK, Association of Researchers in Construction Management, Vol. 2, 1149–1158.
Opoku, A. (2019a). Sustainable development, adaptation and maintenance of infrastructure. *International Journal of Building Pathology and Adaptation*, 37(1), 2–5, https://doi.org/10.1108/IJBPA-02-2019-074.
Opoku, A. (2019b). Biodiversity and the built environment: implications for the Sustainable Development Goals (SDGs). *Resources, Conservation and Recycling*, 141(1) 1–7.

Opoku, A. and Ahmed, V. (2013). Understanding sustainability: a view from intra-organizational leadership within UK construction organizations. *International Journal of Architecture, Engineering and Construction*, 2(2), 133–143.
Opoku, A. and Akotia, J. (2020). Guest Editorial: Urban regeneration for sustainable development. *Construction Economics and Building*, 20(2), 1–5.
Opoku, A. and Fortune, C. (2011). Organizational learning and sustainability in the construction industry. *Built and Human Environment Review*, 4(1), 98–107.
Opoku, A. and Williams, N. (2019). Second-generation gender bias: an exploratory study of the women's leadership gap in a UK construction organisation. *International Journal of Ethics and Systems*, 35(1), 2–23, https://doi.org/10.1108/IJOES-05-2018-0079.
RICS and UNGC (2018). *Advancing Responsible Business in Land, Construction and Real Estate Use and Investment – Making the Sustainable Development Goals a Reality*. Royal Institution of Chartered Surveyors (RICS) and United Nations Global Compact (UNGC), London.
Robinson, J. (2004). Squaring the circle? Some thoughts on the idea of sustainable development. *Ecological Economics*, 48 (4), 369–384, https://doi.org/10.1016/j.ecolecon.2003.10.017.
Rodrik, D. (2002). After neoliberalism, What? Remarks at the BNDES Seminar on 'New Paths of Development', Rio de Janeiro, 12–13 September.
Rosati, F. and Faria, L.G.D. (2019). Business contribution to the Sustainable Development Agenda: organizational factors related to early adoption of SDG reporting. *Corporate Social Responsibility and Environmental Management*, 26(3), 588–597, https://doi.org/10.1002/csr.1705.
Sachs, J., Schmidt-Traub, G., Mazzucato, M., Messner, D., Nakicenovic, N. and Rockström, J. (2019). Six transformations to achieve the Sustainable Development Goals. *Nature Sustainability*, 2(9), 805–814, https://doi.org/10.1038/s41893-019-0352.
Scott, L. and Mcgill, A. (2018). From promise to reality: does business really care about the SDGs? SDG Reporting Challenge 2018. PricewaterhouseCoopers LLP, London.
Shi, L., Han, L., Yang, F. and Gao, L. (2019). The evolution of sustainable development theory: types, goals, and research prospects. *Sustainability*, 11, 7158, https://doi.org/10.3390/su11247158.
SIDA (2016). *Urban Development: Biodiversity and Ecosystems*. Swedish International Development Cooperation Agency (SIDA), Stockholm.
Silvius, G. (2017). Sustainability as a new school of thought in project management. *Journal of Cleaner Production*, 166, 1479–1493.
Sustainable Development Solutions Network (SDSN) (2016). *Getting Started with the SDGs in Cities: A Guide for Stakeholders*. UN Sustainable Development Solutions Network (SDSN). New York.
Teferi, Z.A. and Newman, P. (2018). Slum upgrading: Can the 1.5°C carbon reduction work with SDGs in these settlements? *Urban Planning*, 3(2), 52–56.
UKSSD (2018). Measuring up: how the UK is performing on the UN Sustainable Development Goals. UK Stakeholders for Sustainable Development (UKSSD), London.
United Nations (UN) (2015). Transforming our world: the 2030 Agenda for Sustainable Development, Resolution adopted by the General Assembly. Seventieth session on 25 September, A/RES/70/1.
UNDESA (2019). *The Sustainable Development Goals Report 2019*. Department of Economic and Social Affairs (DESA). United Nations Publications, New York.
UNGC and Accenture (2016). *The UN Global Compact and Accenture Strategy CEO Study, Agenda 2030: A Window of Opportunity*, https://www.accenture.com/t20161216t041642z__w__/us-en/_acnmedia/accenture/next-gen-2/insight-ungc-ceo-study-page/accenture-un-global-compact-accenture-strategy-ceo-study-2016.pdf.
United Nations (UN) (2016). World of business must play part in achieving new Sustainable Development Goals. Speech delivered by the UN Secretary-General Ban Ki-moon at the Global Economic Forum, Global Compact event on UN–Business Collaboration in Davos, Switzerland, https://news.un.org/en/story/2016/01/520492-world-business-must-play-part-achieving-new-sustainable-development-goals-un#.VqJMIPkrLcs.
United Nations (UN) (2018). Sustainable cities and human settlements. United Nations Sustainable development knowledge platform, https://sustainabledevelopment.un.org/topics/sustainablecities.
United Nations Statistical Commission (UNSC) (2019). Interlinkages of the 2030 Agenda for Sustainable Development. Prepared by the Interlinkages Working Group of the Inter-Agency and Expert Group on Sustainable Development Goal Indicators (IAEG-SDGs) United Nations Statistical Commission

(UNSC), https://unstats.un.org/unsd/statcom/50th-session/documents/BG-Item3a-Interlinkages-2030-Agenda-for-Sustainable-Development-E.pdf.
Waage, J., Yap, C., Bell S., Levy, C., Mace, G., Pegram, T., et al. (2015). Governing the UN sustainable development goals: interactions, infrastructures, and institutions. *Lancet Global Health* 3, e251–e252, https://doi.org/10.1016/S2214-109X(15)70112-9.
Williamson, J. (1994). *The Political Economy of Reform*. Institute for International Economics, Washington, DC.
Williamson, J. (2004). The Washington Consensus as policy prescription for development. A lecture in the series 'Practitioners of Development' delivered at the World Bank on 13 January. Institute for International Economics.
Wilson, A. (2015). Sustainable Development Goal 11: Make cities safe and sustainable. Retrieved from https://www.greenbiz.com/article/sustainable-development-goal-11-make-cities-and-human-settlements-inclusive-safe-and.

11. Sustainability economics and the construction industry

Obas John Ebohon

INTRODUCTION

Sustainability economics, like the concept of sustainability, is an evolving concept that has its roots in ecological, natural resource and environmental economics (Ayres et al., 2001; Baumgartner and Quaas, 2010). As a result, different perspectives on sustainability economics have emerged without precise definition (Baumgartner and Quaas, 2010). Nevertheless, a common strand running through the perspectives on sustainability economics is in the emphasis placed on the costs associated with economic growth, including externalities such as pollution (Ayres, 2008).

On a critical note, problems of externalities implicit in economic growth and development on the one hand, and the cumulative effects on the market on the other hand, are articulated comprehensively in environmental and natural resource economics (Ayres et al., 2001; Klenow and Rodríguez-Clare, 2005; Rezai et al., 2012). It is confronting intra- and intergenerational equity issues in natural resource use that affords the normative background to the notion of sustainability economics (Golub et al., 2013). While it can be argued that distributional justice is the cornerstone of ecological economics emphasising intergenerational and intrageneration equity implications of economic growth and development (Munda, 1997), physiocentric ethics which emphasise the justice between humans and nature – the rights of all living species (Krebs, 1997) – is a unique attribute of sustainability economics. In other words, there are potential conflicts and trade-offs between equity and economic efficiency related to economic growth, and this remains the focus of sustainability economics (Baumgärtner and Quaas, 2009).

The conflicts and trade-offs induced by economic growth are debated around the extent to which 'human-made capital' is a substitute for 'natural capital' – the 'weak' versus 'strong' sustainability debates (Mancebo, 2013; Neumayer, 2003; Ayres et al., 2001). If 'weak' sustainability holds, it simply means that man-made capital and natural capital are substitutes, and this has huge implications for sustainability, as it would mean that there are no limits to growth. Similarly, were 'strong' sustainability to hold, it would mean there are ecosystems services that human-made capital cannot substitute for, but complement. Where this is the scenario, there is a strong case for decoupling economic growth from natural resource consumption intensity in order to enhance the absorptive or regenerative capacity of the natural environment (Fedrigo-Fazio et al., 2016). Apart from the clear environmental implications of the industry's insatiable consumption of natural resources, unless the rapid rates of depletion aligns with the regenerative capacity of the natural environment, the survival of construction is at considerable risk.

This chapter presents a critical review of the literature on sustainability economics, drawing on the distinction it has with other branches of economics while ascertaining its significance for the practice and conduct of construction business. Analyses are undertaken of the indirect risks posed, not just to the natural environment and the global ecosystems, but also to the construction industry itself, if it continues with on the 'business as usual' trajectory. Finally, actions which have been taken in construction to address the issues of sustainability are discussed, with some examples of good practice, and proposals for the future outlined.

SUSTAINABILITY ECONOMICS

Sustainability economics is distinct from both environmental and natural resource economics that collectively define the economics of sustainability. These branches of economics evolved because of inherent weaknesses in the market mechanism with regard to accommodating or internalising the externalities (social costs and social benefits) which are normally experienced by third parties outside of the activities or transactions occurring (Helm and Pearce, 1990). It is not feasible to affix a price to these costs and benefits that may encourage or discourage their approaches to production. Concerted efforts to tackle externalities within environmental economics dates back to the pioneering works of Pigou, the Pigovian analysis (Medema, 2020), and the critical debates that ensued, such as the 'Coase theorem' that sought to extend the theoretical propositions of augmenting or correcting the market (Coase, 1960). Pigou argues that negative externalities such as pollution could be resolved through the instrumentations of taxation and subsidies, which allow the internalisation of social costs; the 'polluter pays principle' (Gaines, 1991). Introducing a tax increases the private cost of consumption or production, and when it is passed on to consumers in the form of higher prices, this should lead to reductions in demand, and hence in the production of the goods and services originating the externality, forcing the polluter to adopt the most cost-effective abatement mechanisms (Fullerton et al., 2010).

While the Pigovian analysis highlighted instances of market failures and offered a 'pragmatic approach' to externalities, several inherent weaknesses have been identified. Helm and Pearce (1990) argued that making the polluter pay will not deter pollution, particularly where the polluter can absorb the cost and pass it on to consumers in higher prices. This is particularly the case where the actual costs of externalities may not be fully ascertained, given the paucity of data and poor information, and that makes it difficult to decide optimum levels of pollution taxation to impose. Above all, the Pigovian solution to externality takes no account of other social objectives such as equity, rights and distributions (Helm and Pearce, 1990). These were the issues taken up by Coase (1960) who adopted a free market approach to the issue of externality – the 'Coase theorem' – to argue that the absence of markets and property rights explains externalities. As with the Pigovian polluter pays principle, not only does a market have to be in to existence, but also it has to be efficient for the Coasean theory to be effective. For a market and the attendant property rights to exist for any commodity or service, excludability must be possible to prevent free-riders from consuming or owning products or services not paid for and owned (Dionisio and Gordo, 2006). Despite the shortcomings of these economic approaches to externalities, the polluter pays principle and assignments of property rights remain popular, allowing the exploitation of natural resources for economic growth and development to continue, with huge consequences for the natural environment.

Sustainability economics, according to Baumgartner and Quaas (2010), has justice at its core while at the same time appealing to efficiency. While recognising the importance of growth and development, the process of generating it is critical to the sustainability debate. Efficient use of natural resources in pursuit of growth and development is critical to enhancing the regenerative capacity of the natural environment, guaranteeing its sustainability well into the future. Although economic efficiency in this sense highlights environmental benefits of reducing the insatiable demands on the natural environment through waste reductions, it also confers distributive efficiency accompanying lower prices deriving from efficiency in natural resource use. This facilitates affordability and contributes to the attainment of wider social sustainability objectives of access and well-being when benefits of waste reductions are passed on to consumers in form of lower prices (McGilivray and Clarke, 2006).

Justice is a strong distinguishing feature of sustainability economics and at the core of the normative guidelines informing the sustainable use and conservation of global resources (Baumgartner and Quaas, 2010; Glotzbach and Baumgartner, 2012). As understood in the literature, households depend on a range of ecosystem services of economic, environmental, social and cultural significance (Glotzbach and Baumgartner, 2012). The Millennium Ecosystem Assessment (MEA) defines ecosystem services as the overall benefits people or households derive from the ecosystem (MEA, 2003). There are four categories of ecosystem services:

- Life-supporting services (nutrient cycle, chlorophyll photosynthesis, soil formation and primary production).
- Provisioning services (production of food, drinking water, natural materials such as minerals, wood and combustibles).
- Regulating services (climate and tidal regulations, hydrogeological order, water purification, waste recycling, pollination and barriers to the diffusion of diseases).
- Cultural values (aesthetic, spiritual, educational and recreational values).

Justice related to human–nature interactions occurs at three basic levels (Baumgartner and Quaas, 2010). The first encounter relates to intergenerational justice or equity that is implicit in natural resource consumption. The concerns here relate to different generations, including those yet unborn, who have equal ownership rights to the natural resources deployed in effecting growth and development for satisfying current lifestyles (Barry, 1997). Similar arguments apply to those within the same generation where older and younger households have the same equal rights and access to global resources. The implications are that had previous generations exhausted or depleted the natural capital stocks, present generations would struggle to fulfil their needs. In other words, the present generation are mere custodians of the natural environment, and as guardians they are morally obliged to ensure that the way and manner of engagement with the natural environment is such that its regenerative capacity remains intact as found, or better still is left in a much better condition than it was found (Barry, 1997).

The World Commission on Environment and Development (WCED) reflects this understanding in its definition of the concept of sustainable development as a development that allows the present generation to fulfil their needs without compromising or undermining the ability of future generations to fulfil their own needs (Brundtland, 1987). According to Sen (1999), there is inherent injustice in the current use of natural resources for economic growth and development, and these are the limitations placed on the freedom, capabilities and possibilities of future generations to fulfil their own needs. This is brought about by the huge and insatiable demand for natural resources and the attendant process of converting them into products and services, generating wastes and effluence undermining the carrying capacity of the natural environment. Given the intergenerational and intragenerational equity implications of natural resource use, there is a moral obligation to conserve the natural environment, and this invites the notion of custodians of the natural environment.

The domination and subjugation of nature by humankind is the final subject of justice in sustainability economics. This is a recognition that other living species exist that have the same rights to the natural environment as humans. There is an abundance of flora and fauna that depend on the ecosystem for survival, just as humankind. This requires obligations to nature and other living species. Dominating and subjugating nature undermines the normal cycles that allow nature to recuperate and regenerate, because of the huge and insatiable demand for natural resources to fuel growth and development (Bertinellia et al., 2008). The human domination treatise suggests that economic growth should not take priority over the intrinsic values of nature, given that these values transcend natural resource exploitation for growth and development (Barry, 1997; Costanza et al., 2017).

The contemporary worldview of humans having dominion over Earth and all it contains is at variance with the rights of other living species to the natural environment for existence. Humankind is endangering nature and the ecosystem in general by rapid and intense exploitation. Estimates show that between 2015 and 2020, 10 million hectares of land were deforested annually (FAO and UNEP, 2020). Although this is less than the 16 million hectare average annual depletion in the 1990s, total primary global forests lost since 1990 stand at 80 million hectares. This holds huge significance given that a forest transcends trees: it hosts many different species of plants and animals that live in the soil, understorey and canopy (FAO and UNEP, 2020). Indeed, Mora et al. (2011) estimates that there are between 3 million and 100 million species on Earth, out of which the 7.6 billion human population is a mere 0.01 per cent. Owing to human activities, 83 per cent of all wild mammals and half of plants are already lost (Bar-On et al., 2018), and assuming the low estimates of the number of species are accurate, it is estimated that between 200 and 2000 extinctions occur annually. This increases to between 10 000 and 100 000 annual species extinctions with a higher estimate of the number of species (WWF, 2019).

According to the Intergovernmental Science-Policy Platform on Biodiversity and Ecosystem Services (IPBES, 2019), there are five ways in which human activities affect global biodiversity. Agricultural and industrial expansion is the first impact: over 85 per cent of wetlands, 75 per cent of land surface and 66 per cent of ocean areas suffer annual alterations. The second source of major impact on the ecosystem is the harvesting of plants and animals for consumption. The third impact is the effluence and solid wastes disposed of, polluting the environment, that can result in habitat loss for some species. The fourth source of biodiversity losses arises with the introduction of non-indigenous species that often drive indigenous species to extinction. The fifth and final link between human activities and biodiversity is climate change, which manifests itself in various forms including global warming, desertification, floods, bleaching of coral reefs and forest fires, amongst others, which exacerbate nature loss and undermines its resilience (IPCC, 2018). This underscores the importance of maintaining and enhancing the regenerative integrity of the natural environment irrespective of humankind's socio-economic pursuits. It goes to show that the natural environment performs a range of functions, which are first for itself, and then for humankind, and for which there is considerable debate on whether these functions can be substituted for by man-made capital (Dietz and Neumayer, 2007). This debate has been occurring within the weak and strong sustainability schools of thought (Neumayer, 2003; Dietz and Neumayer, 2007).

WEAK AND STRONG SUSTAINABILITY

The basic assumption underpinning weak sustainability is predicated upon the notion that a very high degree of substitutability exists between human-made capital and natural capital (Neumayer, 2007). This being the case, Ekins et al. (2003) argue that, essentially, the well-being generated from both natural capital and human-made capitals is indistinguishable. The view is that provided the proceeds from depleted natural resources are invested in reproducible capital to yield income streams to support future generations, it makes no difference when items of natural capital are liquidated (Solow, 1993). This view, that the total value of the aggregate stock of capital is what matters, is also supported by Neumayer (2003), who argues that as long as the stock of capital produced by the current generation is sufficient to compensate for depleted non-renewable resources and attendant pollution, concerns about the ability of future generations to fulfil their own needs is superfluous. Indeed, Hartwick (1978) articulated the weak sustainability position in what became known as 'Hartwick's rule': that accrued rents from exploiting natural capital stock by the present generation should be reinvested in reproducible capital to serve as inheritance for future generations. The inheritance transfer should be at a sufficient level to guarantee non-declining real consumption (well-being) through time. This being the case, human capital development is perfectly substitutable for natural capital (Ayres et al., 2001). This is particularly the case where technological knowledge and innovations, it is assumed, will continue to mediate the challenges associated with natural resource depletion (Koontz and Thomas, 2006).

Thus, if human-made capital can substitute for natural capital, it is possible to manage economic growth and attendant pollution successfully through technological knowledge and innovation. Buttressing this further, Japan and the Netherlands are used as illustrative examples of developed countries that have achieved growth and development and industrialised despite natural resource depletion and ecosystem damage (Ayres et al., 2001). In a similar manner, Gowdy and McDaniel (1999) referenced the small island country of Nauru as an example of weak sustainability that suffered sustained mining of phosphate since 1900 and sustained irreversible ecosystem degradation, but remained a country with one of the highest per capita incomes for several decades. The argument is that as long as income flows from liquidating natural capital are invested in reproducible capital, the equity concerns about future generations not able to fulfil their needs do not arise. This, according to Gowdy and McDaniel (1999), is the interdependence between markets, ecosystems and human well-being that reinforces the complexities of assessing the strengths and weaknesses of the weak sustainability concept.

Aside from the substitutability debate, the scope offered by advances in technological knowledge and innovations that allows economic growth to be decoupled from natural resource consumption and lowers the environmental impact of growth and development (Koontz and Thomas, 2006) should be considered. The decoupling process, the technical efficiency of the repeated use of natural resources through 'learning by doing', will have a lessening effect on the demand pressures brought to bear on the natural environment. This being the case, focus needs to shift from concerns about natural resource depletion and attendant pollution to seeking efficiency in resource use to minimise the adverse impacts of growth across all sectors of the economy (Koontz and Thomas, 2006). Generally, proponents of weak sustainability hold the view that affixing a price to nature is the best way to preserve critical natural capital (Constanza et al., 1997).

However, proponents of strong sustainability dismiss the assumptions underpinning weak sustainability, arguing that the ecosystem is made up of a stock of natural capital of biotic and abiotic environments (Brekke, 1997). Nature provides services not only for the biotic but also for the abiotic environment. These environments combine to provide the range of functions and services to all living species, including humankind, and most of these functions and services are not substitutable (Elkins et al., 2003; Brand, 2009). Ecosystem functions and services are indispensable to the survival of humankind and other species, and these functions and services cannot fully be substituted for by human-made capital (Costanza et al., 1996; Costanza et al., 1997; Daly and Farley, 2010).

Ekins et al. (2003) offer some distinctive features of natural capital, observing that unlike human-made capital, natural capital can be irreversibly consumed or depleted. Thousands of species of flora and fauna are depleted to extinction annually, and this exerts adverse pressure on the ecosystem (Bar-On et al., 2018). In particular, the elements of natural capital that are critical to well-being, how they combine to influence well-being, and the process of interactions necessary to deliver well-being, are too poorly understood to begin to affix a price or value to any components of nature. This is why Ackerman and Heinzerling (2004) hold the view that, in the absence of full knowledge of the critical elements of nature and their interactive influence on well-being, collating and converting these diverse elements into monetary values is difficult to accomplish. Thus, the ethical choices regarding trade-offs between ecological resilience, well-being, species extinctions, inter- and intragenerational justice and equity become difficult to ascertain, making optimal solutions elusive (Ackerman and Heinzerling, 2004).

Given the complexities of nature and its diverse elements and interconnections with well-being, Turner (1993) dismisses the conjecture that human-made capital and natural capital are infinitely substitutable. Although economic growth and development are necessary to enhance well-being and alleviate poverty, they must not undermine the regenerative capacity of the natural environment. For proponents of strong sustainability, the scale of human activities should be dictated by the actual carrying capacity of the ecosystem. This requires that issues of demand and efficiency be considered along with implementing ceilings, so that instead of exerting continuous pressure on the ecosystem, efficiency in resource use would be effected through technological, process and managerial innovations (Schulze et al., 2020). It also demands that non-renewable natural capital should be sustainably managed by ensuring that the rate of consumption does not exceed the regeneration capacity of the ecosystem.

SUSTAINABILITY ECONOMICS AND THE CONSTRUCTION INDUSTRY

Sustainability economics, as already argued, transcends economic sustainability, which seeks efficiency in natural resource consumption to embrace justice and equity in human–nature interactions (Baumgärtner and Quaas, 2010). Mainstreaming sustainability economics into the construction process requires a critical look at the relationship between nature and the construction industry, and the tackling of emerging questions about the extent to which the industry responds to efficiency, justice and equity, not just in its engagement with nature, but also in construction processes. The construction industry and materials manufacturing that underpin it are natural resource-intensive (Liu et al., 2018). A critical look at the various stages of the construction and post-construction process shows that the construction industry absorbs a considerable amount of natural resources and discharges significant quantities of solid wastes and effluents (Imbabi et al., 2012; Hong et al., 2016; Guo et al., 2019). The first interface between the construction industry and the natural environment occurs with irreversible site clearance for buildings and extraction of raw materials to supply to manufacturing plants (Ofori and Chan, 1998). During this process, existing ecosystems and flora and fauna habitats are disrupted (Hugh et al., 2017).

The natural resource implications of the construction process continue during construction, post-construction use of the construction output, to its eventual demolition when wastes and effluents are discharged into the environmental media (Hong et al., 2016; Huang et al., 2018; Guo et al., 2019). For example, it is estimated that buildings account for 45–50 per cent of total global energy consumption, 50 per cent of total global drinking water, 60 per cent of bulk of raw materials, and 60 per cent of timber products, of which 90 per cent is hardwood, which takes longer to grow (Lenzen and Treloar, 2002; McCormack et al., 2007). Similarly, 80 per cent of global agricultural land is lost to buildings, while accounting for 50 per cent of destruction caused to rainforests.

Apart from these direct impacts, the building construction sector also accounts for a disproportionate share in global pollution, where it produces about 23 per cent of total air pollution in cities and 50 per cent of greenhouse gas emissions (Suzuki and Oka, 1998). In a similar vein, the construction sector is answerable for 50 per cent of global landfill wastes (Arif et al., 2012; Oyedele et al., 2013), and in general it is estimated that 75 per cent of all known factors responsible for global environmental degradation are traceable in one form or another to the built environment. With the global population and rate of urbanisation rising, and the demand for housing and other built environment infrastructure and services increasing, this will intensify the demand on the natural environment, and it has the real potential of undermining the regenerative capacity of the natural environment. Under weak sustainability, this will not be seen as an issue, because of the belief that the impact of the built environment can be ameliorated through technological development and innovation.

The built environment sector, while a major source of global environmental degradation, presents itself as most ideal sector affording the best opportunity for rolling back global environmental degradation and attending to some of the critical environmental challenges. For example, according to Layke et al. (2016), a more than US$2 trillion saving on fossil fuel consumption is possible by 2030 with an 1.5–3 per cent annual rate of global energy productivity improvements. At the same time, this will create significant employment and income-generating opportunities. They further argued that residential and commercial buildings afford approximately 34 per cent of the global energy productivity improvement opportunities. Hence, when compared to other sectors, the construction industry has the largest unrealised potential for cost-effective energy and emissions savings (Layke et al., 2016).

Sustainability is mutually beneficial to both the built and the natural environments. Thus, it is necessary to slow the rate of natural resource depletion, which poses major threats to the construction industry itself. This is so because the effects of rapid depletion of natural resources are already being felt in construction (Tilton and Lagos, 2007); this will result in scarcity, leading to price rises of critical construction materials (Boyd and Caporale, 1996). This would feed into project delays, overall construction cost increases and adverse effects on affordability, and bring forth the normative issues of economic justice of equity and access to construction products.

Critical materials used to enhance the functional, structural and aesthetic integrity of construction output include zinc, tin, iron, aluminium, lead, cement, tiles, bricks, steel, glass and paint, all of which are not only natural resource-intensive to manufacture, but also produce huge amounts of effluents which are discharged into the environmental media (González and Navarro, 2006; Lawson, 2006). Observably, concrete, steel, plastic, masonry and ceramics are particularly energy-intensive (Gutowski et al., 2013). Apart from the large amount of energy embodied in these products, significant environmental pollution occurs during processing, where toxic gases and effluents are discharged with harmful effects on aquatic and marine life, as well as contributing to atmospheric pollution (Cole, 1998). The greenhouse gas emissions during the manufacture of these products include carbon dioxide (CO_2), sulphur dioxide (SO_2), nitrogen dioxide (NO_2) and methane (CH_4), and they induce global warming with enormous climate change impacts. This is particularly the case with CO_2, which accounts for more than 50 per cent of greenhouse gases (González and Navarro, 2006). The environmental implications of built assets transcend embodied energy and attendant greenhouse gas emissions; they extend to energy consumption in the use of buildings (Praseeda et al., 2017). Indeed, energy in use predominates the total energy consumption in the life cycle of a building, and is estimated at between 80 and 90 per cent of the total (GEA, 2012; Unalan et al., 2014). It is also estimated that 83 per cent the total CO_2 emissions in a building occur during the operational phase (Urge-Vorsatz, 2007; Hui et al., 2010; UNEP, 2020; Röck et al., 2020). Designing and constructing built assets to use alternative or renewable energy, thereby reducing the use of conventional fossil fuel energy, will adduce significant environmental impacts (Akadiri et al., 2012). Thus, in general, the construction industry has the capacity to influence the amount and intensity of natural resource consumption, particularly concerning embodied and in-use energy.

This is where the concepts of renewable and non-renewable natural resources are introduced into the sustainability debate (Güney, 2019), as this is also central to the weak and strong sustainability debates. As argued, sustainability is not about curtailing the growth and development necessary to reducing global poverty (Curwell, 2002); rather, the emphasis is on pursuing growth and development sustainably.

Under weak sustainability, technological innovations can ameliorate the impacts of the built environment on the natural environment (Pacala and Socolow, 2004); this is particularly the case where renewable resources, which are inexhaustible, can substitute for natural resource use in buildings (Khan, 1995; Schulze et al., 2020). For example, consuming woods from sustainably managed forests and using recycled woods in construction can reduce the rate of deforestation significantly (Schulze et al., 2020). Renewable energy is an area where unfolding technological innovations have driven down costs and enhanced quality, allowing substitution of conventional energy sources by alternative energy sources such as solar and wind energy (Lotfabadi, 2015). In the United Kingdom (UK), renewables generated more electricity than fossil fuel for the first time in the third quarter of 2019 (CarbonBrief, 2019). On a critical note, technological innovation can be natural resource-intensive, as is the case with solar energy, where solar cells and semiconductors depend on silicon and the batteries component relies on lithium (Cohen et al., 2019).

While potential consumption of renewable resources is growing, consumption of non-renewable natural resources is still rising (UNCHS Habitat, 1993), and undermining the regenerative capacity of the natural environment (Barbier, 1993; Khanna et al., 1999), which has major implications for the construction industry. As forest products become more scarce, their prices rise, and so do the distances travelled to explore and to distribute the materials to areas where they are needed (UNCHS Habitat, 1993; Mancini et al., 2013). Thus, while the construction industry is a major contributor to global environmental degradation (Krausmann et al., 2009), it also presents an opportunity for mitigating global environmental degradation and the attendant effects (IRP, 2020).

MAINSTREAMING SUSTAINABILITY INTO THE CONSTRUCTION PROCESS

Under weak sustainability, mainstreaming the three pillars of sustainable development into construction projects will allow infrastructural project developments with less pressure on natural resources, thereby lessening the natural resource intensity of the industry while, at the same time, adhering to the equity and justice prerequisites of sustainability (Ochieng et al., 2014). Economic efficiency is promoted as a critical component of decoupling natural resource consumption intensity and economic growth (Horvath, 2004; Yu et al., 2013). In construction project delivery, the efficiency drive in resource use has witnessed several innovations in construction project management, such as lean construction and just-in-time construction strategies (Nahmens and Ikuma, 2012). These measures aim to achieve productive efficiency that ensures elimination of wastes (Hurt and Thomas, 2009; IRP, 2020). In other words, if all resources, be they human capital, financial capital or natural capital, are deployed efficiently, this leads to sustainable outcomes (IRP, 2020). Additional to reductions in materials consumption, economic and process efficiency also contribute to productivity and profitability, which fuels further innovation.

Sustainable project management applied in the design and construction of built assets can mitigate cost overruns and project delays (Hurt and Thomas, 2009). Sustainable project management is the 'planning, monitoring and controlling project delivery and supporting processes, with consideration for the environmental, economic, and social aspects of the lifecycle of project resources, processes, deliverable and effects, aimed at realising benefits to for stakeholders, and performed in a transparent, fair, and ethical way that includes proactive stakeholder participation' (Armenia et al., 2019, p. 2). The emphasis is on forward planning, consideration of long-term impacts, and delivering for clients and end-users; hence, cost control is of critical importance (Robichaud and Anantatmula, 2011).

The emphasis on life cycle impacts of projects is also a process efficiency drive to ensure that construction strategies take a long-term view of costs, explaining why sustainable buildings typically have lower running and maintenance costs. These are achieved through sustainable design strategies and innovative use of sustainable materials and equipment (Pullen et al., 2012). Even where upfront costs of sustainable projects are high, it usually takes a relatively short time to recover such additional costs. In addition, there are indirect benefits to clients, end-users and society (Qualk and McCown, 2009). In particular, projects designed or refurbished to the attributes of economic sustainability can significantly extend or prolong both the physical and the economic life of a built asset (Akadiri et al., 2012). The economic life of a building is exhausted the moment the cost of maintenance exceeds revenue flow from the asset (Salvado et al., 2019). The benefits of designing to sustainable attributes also extend to low maintenance and running costs, enhancing returns on investment, affordability and clients' or end-users' retention (Salvado et al., 2019).

Various design and construction management strategies exist to drive economic sustainability in the delivery of projects, but for effective outcomes, sustainability has to be designed into the project right at the very beginning (Ahmad and Thaheem, 2018; Bragança et al., 2014). Conscious of post-construction running costs in terms, efforts are deployed to see the potential for alternative energy sources such as solar energy, requiring site optimisation and reorientation in the direction where solar energy can easily be captured (Cansino et al., 2011). Using stack effects and a passive ventilation system will avoid mechanical processes and hence significantly reduce the use of energy (Hughes and Abdul Ghani, 2010). In addition, increasing ceiling heights and using live plants can enhance cooling, and allow in daylight (Guimarães et al., 2013). Similarly, plants and innovative designs are used to cool buildings to reduce the amount of energy in use to reduce running costs (Wright et al., 2002). Furthermore, implementing sustainability-focused site analysis will provide information not only on what materials can be reused and recycled, but also on the ecosystem which would be displaced, resulting in savings on building costs and less disruption to the ecosystem (LaGros, 2008). This is particularly the case in the refurbishment of built assets. The economic sustainability pillar and the environmental pillar can be seen to be inextricably linked and complementary, as economic sustainability measures also enhance environmental sustainability and vice versa; thus, synergies can be found between actions under the two pillars. Other strategies, which include the use of life cycle cost analysis (LCCA) and building information modelling (BIM) (Azhar and Brown, 2009; Wong and Fan, 2013), are used to achieve economic sustainability; these are further elaborated upon in the discussion of the tools for achieving economic sustainability.

Another dimension to economic sustainability is that involving the opportunities projects provide for employment, boosting local employment and income generation in the process (Nijaki and Worrel, 2012). However, for full effects, a sustainable site analysis is necessary (LaGros, 2008); this will indicate materials that could be sourced locally, and a local skills audit will reveal the types of construction skills and competences available locally. If project designers consider local materials and skills availability during design, the local projects could be designed for maximum effects in terms of employment and local income-generating opportunities (Akadiri et al., 2012). The Crossrail project in London is a case in point where a decision was taken at the conception stage of the project to boost the local economy (de Silva and Paris, 2015), and the following were the contributions to the local economy:

- At least £42 billion estimated to be generated.
- 55 000 jobs supported.
- 96 per cent of work awarded to businesses in the UK.
- 62 per cent of suppliers based outside London.
- 62 per cent of Tier 1 suppliers are small and medium-sized enterprises.
- 72 per cent of Tier 2 suppliers are small and medium-sized enterprises.
- 1.5 million additional people to access central London within 45 minutes when the railway fully opens.
- 3 million square feet of high-quality office, retail and residential space at 12 sites.
- More than 1 million square feet of improved public space across 40 sites.
- Delivery of 57 000 homes was supported by the project.

What this demonstrates is that the construction industry is well positioned to drive economic sustainability, given the scope to drive down costs through design and management efficiency and innovations (Osmani et al., 2008; Kivimaa and Martiskainen, 2018) and, above all, to leverage employment and income earning opportunities for the local economy (Mossman et al., 2011). However, to do this effectively, sustainability policies and practices should be mainstreamed into projects at the conceptualisation stage (Valdes-Vasquez and Klotz, 2013).

The 'green building' concept has been a strategy for driving environmental sustainability in the construction industry where built assets impose less burden on the natural environment (Giesekam et al., 2016). The emphasis is on reducing natural resource consumption not only at the building stage but also at the post-construction stage, to optimise functionality when in use (Jiao et al., 2012). Thus, energy and water conservation in buildings is a key aspect of the green building initiative (Meireles and Sousa, 2020). This explains the emphasis on airtight buildings and the insulation initiatives embarked on by UK government targeting sources of heat escapes in buildings such as walls and roofs (Pullen, 2010). The adaptive reuse of old industrial buildings and intensive development of brownfield sites to create more living spaces avoids building on green fields (De Sousa, 2002; Clifford et al., 2019). Green building can facilitate reductions of between 24 per cent and 50 per cent in total energy used in buildings, including embodied and energy in use; reductions of between 33 per cent and 39 per cent in CO_2 emissions; about 40 per cent reduction in water consumption; and as much as 70 per cent in solid wastes (UNEP, 2019). This can be achieved through various innovations and construction management processes that will allow growth in construction output, yet decoupled from increased consumption of environmental resources (UNEP, 2011). Indeed, process innovations can result in significant reductions in natural resource use together with associated wastes (Daoud et al., 2020).

A series of innovations are in operation to ensure environmental sustainability of building construction and built assets (Foxon and Pearson, 2008). Starting with sustainable site analysis, the environmental and social impacts assessment of sites is to provide vital information to feed into design (Valdes-Vasquez and Klotz, 2013). Sustainable site analysis reveals information on the microclimate to inform decisions on building orientations in order to take advantage of designs that allow different energy scenarios and use of renewable resources (Thormark, 2006). This has resulted in numerous innovations such as harnessing alternative energy sources: solar energy, wind energy and geothermal energy. There has also been much rainwater harvesting (Angrill et al., 2017). Other initiatives implemented to reduce the natural resource consumption intensity of construction include the mainstreaming of the 'Four Rs' strategy: reuse, recycle, reduce and recovery of materials into construction management policies to reduce waste generation (Yang et al., 2017). This has seen a considerable reduction in construction wastes going into landfills (Oyedele et al., 2013).

Legislation has also played a part in the efforts to achieve environmental sustainability (Lee and Yik, 2004; Sebopela and Odeku, 2014). The UK's use of legislation and economic incentives to drive environmental sustainability is instructive. The Climate Change Act 2008 is a case in point; it places the government under a legal duty to reduce greenhouse gas emissions by 80 per cent below 1990 levels by 2050 (Averchenkova et al., 2021). This has obliged the government to adopt and promote measures that encourage sustainable innovations throughout project life cycles, including materials, the supply chain, demolition, and education and training.

SOCIAL SUSTAINABILITY

Construction activities are a social process with substantial social impacts at various stages of the life cycle of a project (Akadiri et al., 2012; Dong and Ng, 2015). The social impacts associated with construction activities manifest themselves in various ways, with construction workers being the first to be impacted upon (Tadesse and Israel, 2016), and extending to local communities and end-users of construction products (Valdes-Vasquez and Klotz, 2013). Construction activities can place significant burdens on workers and communities, particularly where these infringe on the social rights and well-being of workers and communities. Therefore, social sustainability requires a safe working environment and conditions for construction workers, equitable access to the opportunities available throughout the construction process, while striving to eliminate or at least minimise the disruptive impacts of construction activities on local communities and consumers of construction products (Zuo et al., 2012).

Notwithstanding the benefits the construction industry confers on society in providing the necessary growth-inducing infrastructure and services, there is also a 'dark side', and this has earned construction the unenviable reputation as the most dangerous industry in the world (CIOB, 2015). For example, in the UK, 1.4 million construction workers suffer from work-related illness, such as the 2446 mesothelioma deaths due to past asbestos exposure in 2018. In addition, 111 workers died on construction sites in 2019/20, while 581 000 workers sustained injuries at work, according to the Labour Force Survey; and 69 208 injuries to employees were reported under the Reporting of Injuries, Diseases and Dangerous Occurrences Regulations (RIDDOR) (HSE, 2020). As a result, 28.2 million working days were lost and the total injuries and ill health-related cost to the sector in 2017/18 were estimated at £15 billion (HSE, 2020). The personal costs to construction workers and their dependants are in lost income, and where death occurs, it is dependants and the construction industry that suffer. Accidents and deaths in the construction industry can be minimised through design, if the interests of construction workers are considered along with those of clients and end-users (Behm, 2005). Here lies the importance of social sustainability to economic sustainability, demonstrating that neglect of equity and well-being in construction activities does incur significant economic effects in terms of low productivity and cost escalation.

The construction industry is also compromised concerning equality and diversity (Dainty and Bagilhole, 2005). Racial and gender discrimination is common, and in the particular case of gender discrimination, the prevailing culture in the industry is one where it is seen as 'a man's world' (Fielden et al., 2000). Powell and Sang (2013) evaluated the UK construction industry from an ethical perspective, and found that the industry is still failing to embrace 'the ideal of social justice, equality and inclusivity for all'. This observation followed from an earlier finding areas of the UK construction sector where minorities feature in the construction process, they are poorly paid, poorly treated and confronted by 'numerous cultural and structural barriers'. Provision of minimum welfare facilities for workers has the potential to increase productivity and retention of workers. This is particularly the case where such a policy is adequately resourced and underpinned by effective health and safety measures to create a conducive and safe working environment. Apart from the well-being benefits to workers, reductions in absenteeism and worker retention, are enhanced productivity and growth of the industry (Hafner et al., 2015; Isham et al., 2020); this is particularly the case given the importance of tacit knowledge in construction practice (Pathirage et al., 2008; Addis, 2016).

Buildings built to high environmental standards, and delivered on time and within budget, which fail to adhere to the 'ideals of social justice, equality and inclusivity for all' cannot be deemed sustainable. Sustainable buildings are those built efficiently, eliminating all wastes, delivered on time and within budget, and adhering to ideals of social justice, equality and inclusivity for all stakeholders, including end-users (Bal et al., 2013). One of the emerging social issues where social sustainability continues to provide solutions is security (Armitage, 2016). Where security is articulated and mainstreamed into the design strategy, building security can be enhanced. This is particularly the case where the need to create opportunities for social capital formation is designed into buildings to afford opportunities for social interaction. In buildings and communities where this applies, issues of burglary, vandalism and other anti-social activities have been successfully designed out of buildings (Cozens and Love, 2015). In other words, social sustainability encourages a community's social cohesion. Aesthetically, socially sustainable buildings also confer benefits on the community, attracting visitors and contributing to the local economy, enhancing the quality of life and well-being of the community.

Generally, it is evident that a building cannot be truly sustainable without implementing social sustainability together with other pillars of sustainability (Hendiani and Bagherpour, 2019). Buildings built to social sustainability attributes should eliminate building-related sickness and their attendant effects: fatigue, absenteeism, and other conditions normally associated with poor indoor air quality, thermal discomfort and poor lighting (Ghaffarianhoseini et al., 2018; McGill et al., 2015). These socio-health issues can be designed out of buildings using sustainable principles (Marzbali et al., 2011; Cozens and Love, 2015).

EXAMPLES OF SUSTAINABLE PROJECTS

Recognising sustainable development as a dynamic and increasingly unfolding concept, particularly as global environmental degradation and climate change effects continue to manifest themselves differently, requiring dynamic policy-making in response (Adger et al., 2003), the construction industry has been responding with various adaptation and mitigation measures, as well as measures to encourage the uptake of equity and justice in the use of natural capital by the industry (Ortiz et al., 2009). Selected projects are presented here to highlight the response of the construction industry to natural resource depletion and climate change.

Crossrail, UK

This is a megaproject and one of the largest indivisible projects embarked upon by the UK government since 2009 (Buck, 2016). Crossrail's 2016 sustainability report demonstrates that sustainability was mainstreamed into the project at the conception stage. The decision was taken to have 84 per cent of all machines fitted with pollutant-reducing emission controls, accounting for 11 per cent of the baseline emissions during construction. Similarly, 98 per cent of the 7.9 million tonnes of excavated materials was reused (Dempsey, 2016). This avoided their dumping in landfills, acquisition of new filling materials, and the associated energy and carbon emission implications given the number of journeys to landfills and to deliver new filling materials to the site.

Economic sustainability was interpreted and incorporated into construction strategies, policies and management. Aside from the stringent efficiency measures, Crossrail demonstrates how a project can be used to effect local and national economic development, an aspect of economic sustainability not often considered in project development. To ensure maximum economic benefits to local communities, a decision was adopted that saw 96 per cent of contracts awarded to UK contractors; and for a wider regional impact, 62 per cent of suppliers were based outside of London (Dempsey, 2016). Many of these firms were established and mentored on the back of this project, and their capacities were enhanced.

Regarding social sustainability, Crossrail delivered 573 apprenticeships, and of these, 27 per cent were female, which contributes to bridging the gender gap in the construction industry (Dempsey, 2016). This was possible because of a deliberate policy to have a certain percentage of jobs set aside for women, and this policy was worked on assiduously to realise. Additional to the 4544 local jobs created, over 15 000 people were trained either on the job, or at specially created off-site training facilities (Dempsey, 2016).

It is evident that a holistic sustainability approach to project delivery was adopted on the Crossrail project. The key to this is having sustainability mainstreamed into construction strategies, policies and management at the conception stage of the project development; it must not be an afterthought.

Other Examples

Evidence abounds showing the significant strides the construction industry has been making to implement sustainability strategies across a range of projects around the world. Many buildings around the world are built to the stringent building sustainability ratings, which include the Leadership in Energy and Environmental Design (LEED) developed in the USA (Clevenger, 2008); Building Research Establishment Environmental Assessment Method (BREEAM) developed in the UK (BRE, 2006); German Sustainable Building Certificate (DGNB – GeSBC); Green Globes, formulated in Canada; and Gold Star by Australia. The emerging and developing countries have also developed their systems. These include: the Pearl Rating System for Estidama by the United Arab Emirates; the Green Rating for Integrated Habitat Assessment by India; and the Saudi Arabia Green Building Rating System. A new consciousness in these countries is driving new infrastructure procurements and the refurbishment of existing infrastructure stocks to these sustainability rating standards (Chen and Lee, 2013). The examples of major buildings in Canada, the UK and the United States (US) which are now presented show that sustainability concerns are being taken seriously in construction.

One Angel Square, Manchester, UK

This building received an 'Outstanding' BREEAM rating, scoring 95.5 per cent (BRE, 2013); it is the headquarters of the Co-operative Group. It is 'powered by a pure plant oil fed Combined Heat and Power system and utilises rapeseed oil which is grown on The Co-operative's own farm land'. The sustainable features led to an 80 per cent reduction in carbon emissions and a 40–60 per cent reduction in energy consumption of a building of its kind housing 3000 workers. Other features of the building include the ultra-modern information technology (IT) heat recovery systems, low-energy lighting, and greywater and rainwater harvesting and recycling. The excess energy generated is put back onto the power grid. Yet other features include

LED lighting and the 50 per cent reduction in energy consumption as a result of embedded sustainability measures to cut the operating cost of the building by 30 per cent.

The Crystal, London, UK

This building has both LEED Platinum and BREEAM Outstanding ratings. The Crystal offers a fossil fuel-free vision of the future. Run entirely on electricity – the majority of which is generated by photovoltaic solar panels – the building is lit by a combination of LED and fluorescent lights, which are switched on or off depending on the amount of daylight present. The building's roof collects rainwater, while sewage is treated, recycled and reused onsite.

One Bryant Park, New York City, US

The first high-rise building to get LEED Platinum certification, the Bank of America Tower in Manhattan is one of the world's greenest skyscrapers (Cook, 2013). DePaola and Mueller-Lust (2008) detail the sustainability attributes of this building. It has arrays of water tanks located throughout the floors that can store over 329 000 gallons of water for irrigating plants and flushing toilets, thereby reducing stormwater contribution by 95 per cent (Guduri et al., 2009). In addition to the installation of CO_2 monitors, waterless urinals and LED lighting, the building also has its own power generation plant that produces 4.6 megawatts of electricity with zero emissions (Cook, 2013). Greywater treatment occurs on site that feeds the cooling towers and is vaporised into the atmosphere, mimicking what trees do naturally in extracting water from the aquifer, and vaporising it. At night, when the power generated is not needed, it is used for ice-making in the 44 storage tanks located in the basement of the building (Cook, 2013); this avoids using energy during peak demand in the working hours to cool the building, lessening pressures on the national grid. In addition, indoor air quality in the building is better than outdoor air quality, as 95 per cent of particulates in the air coming into the building are filtered out. Sustainability was mainstreamed into the building at the beginning, and this made it possible to recycle 83 per cent of related construction wastes. In total, 50 per cent of the materials used in constructing the building (including 60 per cent of the steel) were recycled products (Guduri et al., 2009).

Manitoba Hydro Place, Winnipeg, Canada

This building incorporates the most sustainable attributes and features of any building (Al-Kodmany, 2018). It makes use of passive design and natural ventilation to make it one of North America's most energy-efficient office buildings; it runs on 60 per cent less energy than a conventional building. About 80 per cent of the air in the building is recycled to provide fresh air. The building has a geothermal system to heat and cool it, roof gardens and triple-glazed windows. Cold air flowing into the building is preheated and channelled through a circular heat recovery loop from the exhaust air, and the combination of preheated exhaust air and passive solar gain in the atria contributes to the fresh air in the building (Kuwabara et al., 2013). Fresh air enters the building passively through occupant-controlled operable windows in the double façade and is exhausted through the solar chimney.

These examples demonstrate that by mainstreaming sustainability principles into built environment policies, processes and practices, considerable scope exists to decouple growth in its activities from intensive consumption of natural resources (UNEP, 2011).

CONCLUSION

This chapter analyses the main thrust of sustainability economics, which is the theoretical propositions and the cautions offered on the rate at which natural resources are liquidated or depleted to satisfy current lifestyles. While accepting that the relationship between natural capital and man-made capital is complementary, the discussion rejects the notion of a perfect substitution possibility between natural and human-made capital. The main reason is to do with the view that very little is known about the mechanisms and processes underpinning ecosystems' services to humankind and other flora and fauna. The value of the ecosystem provides services critical to human existence and well-being that cannot be substituted by human-made capital. Linking ecosystem services to well-being raises equity and justice concerns implicit in the consumption of natural resources, particularly intergenerational as well as intragenerational equity considerations which are externalities to the market mechanisms and denote the strong sustainability approach to the sustainability question. Under strong sustainability, a strict limit is placed on natural resource extraction for economic growth and development. Applying the strong sustainability approach to the construction industry calls for a limit to construction activities, given the natural resource consumption intensity of the industry and the pollution attendant to construction activities. This creates a dilemma of how to meet the needs of the rapidly growing global population, particularly the challenge of bridging the growing poverty gaps within and between countries.

However, it has also been shown in the discussion that while the construction industry is a major natural resource consumer, and accounts for significant proportions of the factors responsible for environmental degradation, it also affords the greatest opportunities, relative to other sectors of the economy, to begin reversing global environmental degradation trends. Alternative views to strong sustainability rely on technological and process innovations to argue that this will enable significant efficiency in natural resource consumption, such that these will lead to reductions in natural resource consumption by the construction industry. There is evidence that the construction industry has the capacity to mainstream sustainability into policies and practices that will allow it to deliver projects with minimal impact on the environment, while also achieving equity and justice throughout its operations. As illustrated by the Crossrail project in the UK, construction can be used effectively to effect social sustainability, maximising employment and equal opportunities as well as health and safety measures, while delivering projects that meet client and user requirements. This emphasises the criticality of all built environment stakeholders, including companies, professions and practitioners as well as developers, the government and institutions, to the drive for sustainable development.

As also revealed, construction is known as a polluting and dangerous industry. This calls for investments in training to build the necessary skills and competences to implement sustainability in construction projects. Similarly, investment in research and development for innovative technologies holds the future for effective decoupling of construction activities and natural resource consumption. The construction industry has a duty to adopt sustainability because the very survival of the industry depends on availability of natural resources, and their depletion could impose constraints on project viability. Therefore, unless sustainability becomes the logo of global construction, the rapid depletion of natural resources occasioned by the industry will quickly become its undoing.

REFERENCES

Ackerman, F., Heinzerling, L. (2004). *Priceless: On knowing the price of everything and the value of nothing*. New York Press, New York.
Addis, M. (2016). Tacit and explicit knowledge in construction management. *Construction Management and Economics*, 34(7–8), 439–445, DOI: 10.1080/01446193.2016.1180416
Adger, W.N., Huq, S., Brown, K., Conway, D., Hulme, M. (2003). Adaptation to climate change in the developing world. *Progress in Development Studies*, 3(3), 179–195.
Ahmad, T., Thaheem, M.J. (2018). Economic sustainability assessment of residential buildings: A dedicated assessment framework and implications for BIM. *Sustainable Cities and Society*, 38, 476–491.
Akadiri, P.O., Chinyio, E.A., Olomolaiye, P.O. (2012). Design of a Sustainable Building: A Conceptual Framework for Implementing Sustainability in the Building Sector. *Buildings*, 2, 126–152.
Al-Kodmany, K. (2018). *The Vertical City: A Sustainable Development Model.* WIT Press, Ashurst, Southampton.
Angrill, S., Segura-Castillo, L., Petit-Boix, A., Rieradevall, J., Gabarrell, X., Josa, A. (2017). Environmental performance of rainwater harvesting strategies in Mediterranean buildings. *Int Journal of Life Cycle Assessment*, 22, 398–409.
Arif, M., Bendi, D., Toma-Sabbagh, T., Sutrisna, M. (2012), Construction waste management in India: an exploratory study. *Construction Innovation*, 12(2), 133–155. https://doi.org/10.1108/14714171211215912
Armenia, S., Dangelico, R.M., Nonino, F., Pompei, A. (2019). Sustainable project management: a conceptualization-oriented review and a framework proposal for future studies. *Sustainability*, 11, 2664.
Armitage, Rachel (2016). Crime prevention through environmental design. In: *Environmental Criminology and Crime Analysis*. Crime Science Series. Routledge, Abingdon, pp. 259–285.
Averchenkova, A., Fankhauser, S., Finnegan, J.F. (2021). The impact of strategic climate legislation: evidence from expert interviews on the UK Climate Change Act. *Climate Policy*, 21(2), 251–263.
Ayres, R.U. (2008). Sustainability economics: where do we stand? *Ecological Economics*, 67, 281–310.
Ayres, R.U., Jeroen, C.J.M., van den Bergh, Gowdy, J.M. (2001). Strong versus weak sustainability: economics, natural sciences, and 'consilience'. *Environmental Ethics*, 23(2), 155–168. https://doi.org/10.5840/enviroethics200123225.
Azhar, S., Brown, J. (2009). BIM for sustainability analyses. *International Journal of Construction Education and Research*, 5, 276–292. http://dx.doi.org/10.1080/15578770903355657.
Bal, M., Bryde, D., Fearon, D., Ochieng, E. (2013). Stakeholder engagement: achieving sustainability in the construction sector. *Sustainability*, 6, 695–710.
Barbier, E.W. (1993). Economic aspects of tropical deforestation in Southeast Asia. *Global Ecology and Biogeography Letters*, 3(4–6), 215–234.
Bar-On, Y.M., Phillips, R., Milo, R. (2018). The biomass distribution on Earth. PNAS, 115(25), 6506–11. https://www.pnas.org/content/115/25/6506
Barry, B. (1997). Sustainability and intergenerational justice. *Theoria*, 44(89), 43–64.
Baumgärtner, S., Quaas, M. (2009). What is sustainability economics? Working Paper Series in Economics, No. 138, Leuphana Universität Lüneburg, Institut für Volkswirtschaftslehre, Lüneburg. https://www.econstor.eu/bitstream/10419/30222/1/609497626.pdf.
Baumgartner, S., Quaas, M. (2010). What is sustainable economics? *Ecological Economics*, 69, 445–450.
Behm, M. (2005). Linking construction fatalities to the design for construction safety concept. *Safety Science*, 43(8), 589–611.
Bertinellia, L., Stroblb, E., Zoua, B. (2008). Economic development and environmental quality: a reassessment in light of nature's self-regeneration capacity. *Ecological Economics*, 66(2–3), 371–378.
Boyd, R., Caporale, T. (1996). Scarcity, resource price uncertainty, and economic growth. *Land Economics*, 72(3), 326–335.
Bragança, L., Vieira, S.M., Andrade, J.B. (2014). Early stage design decisions: the way to achieve sustainable buildings at lower costs. *Scientific World Journal*, Article ID 365364. https://doi.org/10.1155/2014/365364.
Brand, F. (2009). Critical natural capital revisited: ecological resilience and sustainable development. *Ecological Economics*, 68(3), 605–612. doi:10.1016/j.ecolecon.2008.09.013.

BRE (2006). Building research establishment environmental assessment method. Building Research Establishment, Garston, UK. http://www.breeam.org.
BRE (2013). The Co-Op's new HQ achieves highest ever BREEAM score. News from BRE. http://www.breeam.com/newsdetails.jsp?id=923.
Brekke, K.A. (1997). *Economic Growth and the Environment: On the Measurement of Income and Welfare*. Edward Elgar Publishing, Cheltenhasm, UK and Northampon, MA, USA.
Brundtland, G.H. (1987). *Our Common Future*. The Brundtland Report. Oxford University Press, Oxford.
Buck, M. (2016). Crossrail project: finance, funding and value capture for London's Elizabeth line. *Civil Engineering*, 170(CE6), 15–22. http://dx.doi.org/10.1680/jcien.17.00005.
Cansino, J.M., Pablo-Romero, M.P., Román, R., Yñiguez, R. (2011). Promoting renewable energy rources for heating and cooling in EU-27 countries: assessment of the sustainable building steering mechanisms in selected EU member states. *Energy Policy*, 39, 3803–3812.
CarbonBrief (2019). Analysis: UK renewables generate more electricity than fossil fuels for first time. https://www.carbonbrief.org/analysis-uk-renewables-generate-more-electricity-than-fossil-fuels-for-first-time.
Chen, H., Lee, W.L. (2013). Energy assessment of office buildings in China using LEED 2.2 and BEAM Plus 1.1. *Energy and Buildings*, 63, 129–137.
CIOB (2011). Carbon Action 2050: Your Action Plan. Aston.
CIOB (2015). *Modern Slavery: The Dark Side of Construction*. Chartered Institute of Building (CIOB), UK.
Clevenger, C. (2008). Leadership in Energy and Environmental design (LEED). US Green Building Council. http://seedconsortium.pbworks.com/f/LEED.pdf.
Clifford, B., Ferm, J., Livingstone, N., Canelas, P. (2019). *Understanding the Impacts of Deregulation in Planning*. Palgrave Macmillan, London.
Coase, R.H. (1960). The problem of social cost. *Journal of Law and Economics*, 3, 1–44.
Cohen, F., Hepburn, C.J., Teytelboym, A. (2019). Is natural capital substitutable? *Annual Review of Environment and Resources*, 44, 425–448.
Cole, R.J. (1998). Energy and greenhouse gas emissions associated with the construction of alternative structural systems. *Building and Environment*, 34(3), 335–348.
Colquhoun, I. (2004). Design out crime: creating safe and sustainable communities. *Crime Prevention and Community Safety*, 6, 57–70.
Cook, R. (2013). Sustainable skyscrapers and the well-being of the city. In: Madhavan G., Oakley B., Green D., Koon D., Low P. (eds), *Practicing Sustainability*. Springer, New York, pp. 25–29.
Costanza, R., Cumberland, J., Daly, H., Goodland, R., Norgaard, R. (1997). *An Introduction to Ecological Economics*. International Society for Ecological Economics and St Lucie Press, Boca Raton, FL.
Costanza, R., d'Arge, R., de Groot, R., Farberk, S., Grasso, M., et al. (1996). The value of the world's ecosystem services and natural capital. *Nature*, 21, 387.
Costanza, R., de Groot, R., Braat, L., Kubiszewski, I., Fioramonti, L., et al. (2017). Twenty years of ecosystem services: how far have we come and how far do we still need to go? *Ecosystem Services*, 28(Part A), 1–16.
Cozens, P., Love, T. (2015). A review and current status of crime prevention through environmental design (CPTED). *Journal of Planning Literature*, 30(4), 393–412.
Curwell, S. (2002). Hazardous Building Materials: A Guide to the Selection of Environmentally Responsible. London: E&FN Spon Press.
Dainty, A., Bagilhole, B. (2005). Guest editorial. *Construction Management and Economics*, 23(10), 995–1000.
Daly, H., Farley, J. (2010). *Ecological Economics: Principles and Applications*: 2nd edition. Island Press, Washington, DC.
Daoud, A.O., Othman, A.A.E., Ebohon, O.J., Bayyati, A. (2020). Overcoming the limitations of the green pyramid rating system in the Egyptian construction industry: a critical analysis. *Architectural Engineering and Design Management*. DOI: 10.1080/17452007.2020.1802218.
Dempsey, A. (2016). Crossrail 2016 Sustainable Report. https://2577f60fe192df40d16a-ab656259048fb93837ecc0ecbcf0c557.ssl.cf3.rackcdn.com/assets/library/document/e/original/ea753_sustrpt2016onlinefinal1.pdf.

DePaola, E.M., Mueller-Lust, A.D. (2008). One Bryant Park, New York. *Structural Engineering International*, 18(1), 35–39. DOI: 10.2749/101686608783726768.

de Silva, M., Paris, R. (2015). Delivering Crossrail, UK: a holistic approach to sustainability. Engineering Sustainability. *Proceedings of the Institution of Civil Engineers*, 168(ES4), 151–158. http://dx.doi.org/10.1680/ensu.1400037.

De Sousa, C.A. (2002). Measuring the public costs and benefits of brownfield versus greenfield development in the Greater Toronto area. *Environmental Planning B: Planning and Design*, 29(2), 251–280.

Dietz, S., Neumayer, E. (2007). Weak and strong sustainability in the SEEA: concepts and measurement. *Ecological Economics*, 61, 617.

Dionisio, F., Gordo, I. (2006). The tragedy of the commons, the public goods dilemma, and the meaning of rivalry and excludability in evolutionary biology. *Evolutionary Ecology Research*, 8, 321–332.

Dong, Y.H., Ng, S.T. (2015). A social life cycle assessment model for building construction in Hong Kong. *International Journal of Life Cycle Assessment*, 20, 1166–1180.

Ekins, P., Simon, S., Deutsch, L., Folke, C., De Groot, R. (2003). A framework for the practical application of the concepts of critical natural capital and strong sustainability. *Ecological Economics*, 44, 165–185.

FAO, UNEP (2020). *The State of the World's Forests 2020: Forests, Biodiversity and People*. FAO and UNEP, Rome.

Fedrigo-Fazio, D., Schweitzer, J-P., ten Brink, P., Mazza, L., Ratcliff, A., Watkins, E. (2016). Evidence of absolute decoupling from real world policy mixes in Europe. *Sustainability*, 8(517), 2–22.

Fielden, S.L., Davidson, M.J., Gale, A.W., Davey, C. (2000). Women in construction: the untapped resource. *Construction Management and Economics*, 18(1), 113–121.

Foxon, T., Pearson, P. (2008). Overcoming barriers to innovation and diffusion of cleaner technologies: some features of a sustainable innovation policy regime. *Journal of Cleaner Production*, 16(1), S148–S161.

Fullerton, D., Leicester, A., Smith, S. (2010). Environmental taxes. In: Institute for Fiscal Studies (ed.), *Dimensions of Tax Design*. Oxford: Oxford University Press, pp. 423–547.

Gaines, S.E. (1991). The polluter pays principle: from economic equity to environmental ethos. *Texas International Law Journal*, 26, 463–496.

GEA (2012). *Global Energy Assessment – Toward a Sustainable Future. International Institute for Applied Systems Analysis*, Vienna, Austria; and Cambridge University Press, Cambridge, UK and New York, USA.

Ghaffarianhoseini, A., AlWaer, H., Omrany, H., Ghaffarianhoseini, A., Alalouch, C., et al. (2018). Sick building syndrome: are we doing enough? *Architectural Science Review*, 61(3), 99–121. DOI: 10.1080/00038628.2018.1461060.

Giesekam, J., Barrett, J.R., Taylor, P. (2016). Construction sector views on low carbon building materials. *Building Research and Information*, 44(4), 423–444. http://dx.doi.org/10.1080/09613218.2016.1086872.

Glotzbach, S., Baumgartner, S. (2012). The relationship between intragenerational and intergenerational ecological justice. *Environmental Values*, 21(3), 331–355.7.

Golub, A., Mahoney, M., Harlow, J. (2013). Sustainability and intergenerational equity: do past injustices matter? *Sustainable Science*, 8, 269–277.

González, M.J., Navarro, J.G. (2006). Assessment of the decrease of CO_2 emissions in the construction field through the selection of materials: Practical case study of three houses of low environmental impact. *Building and Environment*, 41, 902–909.

Gowdy, J.M., McDaniel, C. (1999). The physical destruction of Nauru: an example of weak sustainability. *Land Economics*, 75, 333–338.

Guduri, A.K., Devineni, B., Manchikatla, A., Koehn, E. (2009). Sustainable green features of Bank of America Tower, NY. Proceedings of 2009 ASEE Gulf-Southwest Annual Conference, Baylor University, Waco, TX.

Guimarães, R.P., Carvalho, M.C.R., Santos, F.A. (2013). The influence of ceiling height in thermal comfort of buildings: a case study in Belo Horizonte, Brazil. *International Journal for Housing Science*, 37(2), 75–86.

Güney, T. (2019). Renewable energy, non-renewable energy and sustainable development. *International Journal of Sustainable Development and World Ecology*, 26, 389–397.

Guo, S., Zheng, S., Hui, Y., Hong, J., Wu, X., Tang, M. (2019). Embodied energy use in the global construction industry. *Applied Energy*, 256. https://doi.org/10.1016/j.apenergy.2019.113838.

Gutowski. T.G., Sahni. S., Allwood, J.M., Ashby, M.F., Worrell, E. (2013). The energy required to produce materials: constraints on energy-intensity improvements, parameters of demand. *Philosophical Transactions of the Royal Society A*, 371, 20120003. http://dx.doi.org/10.1098/rsta.2012.0003.

Hafner, M., van Stolk, C., Saunders, C., Krapels, J., Baruch, B. (2015). Health, wellbeing and productivity in the workplace: a Britain's Healthiest Company summary report. RAND. https://www.rand.org/pubs/research_reports/RR1084.html.

Hartwick, J. (1978). Substitution among exhaustible resources and intergenerational equity. *Review of Economic Studies*, 45, 347–354.

Helm, D., Pearce, D. (1990). Assessment: economic policy towards the environment. *Oxford Review of Economic Policy*, 6(1), 1–16.

Hendiani, S., Bagherpour, M. (2019). Developing an integrated index to assess social sustainability in construction industry using fuzzy logic. *Journal of Cleaner Production*, 230, 647–662.

Hong, J., Shen, G.Q., Guo, S., Xue, F., Zheng, W. (2016). Energy use embodied in China's construction industry: a multi-regional input–output analysis. *Renewable and Sustainable Energy Reviews*, 53, 1303–1312.

Horvath, A. (2004). Construction materials and the environment. *Annual Review of Environmental Resources*, 29, 181–204.

HSE (2020). *Summary Statistics for Great Britain 2020*. https://www.hse.gov.uk/statistics/overall/hssh1920.pdf.

Huang, B., Zhao, F., Fishman, T., Chen, W., Heeren, N., et al. (2018). Building material use and associated environmental impacts in China 2000–2015. *Environmental Science and Technology*, 52(23), 14006–14014.

Hugh, C., Finn A.C., Nahiid, S., Stephens, B. (2017). The invisible harm: land clearing is an issue of animal welfare. *Wildlife Research*, 44(5), 377–391.

Hughes, B.R., Abdul Ghani, S.A.A. (2010). A numerical investigation into the effect of Windvent louvre external angle on passive stack ventilation performance. *Building and Environment*, 45(4), 1025–1036.

Hui, Y, Shen, Q., Fan, L.C.H., Wang, Y., Zhang, L. (2010). Greenhouse gas emissions in building construction: a case study of One Peking in Hong Kong. *Building and Environment*, 45(4), 949–955. https://doi.org/10.1016/j.buildenv.2009.09.014Get.

Hurt, M., Thomas, J.L. (2009). Building value through sustainable project management offices. *Project Management Journal*, 40(1), 55–72. doi:10.1002/pmj.20095.

Imbabi, M.S., Carrigan, C., McKenna, S., (2012). Trends and developments in green cement and concrete technology. *International Journal of Sustainable Built Environment*, 1, 194–216.

IPBES (2019). Global assessment report on biodiversity and ecosystem services of the Intergovernmental Science-Policy Platform on Biodiversity and Ecosystem Services. E.S. Brondizio, J. Settele, S. Díaz, and H.T. Ngo (eds). IPBES Secretariat, Bonn. https://doi.org/10.5281/zenodo.3831673.

IPCC (Intergovernmental Panel on Climate Change) (2018). Summary for policymakers. In: *Global Warming of 1.5°C: An IPCC Special Report*. Geneva: World Meteorological Organization. https://www.ipcc.ch/sr15/.

IRP (2020). Resource efficiency and climate change: material efficiency strategies for a low-carbon future. Hertwich, E., Lifset, R., Pauliuk, S., Heeren, N. (eds). A report of the International Resource Panel. United Nations Environment Programme, Nairobi.

Isham, A., Mair, S., Jackson, T. (2020). Wellbeing and productivity: a review of the literature. Centre for the Understanding of Sustainable Prosperity. CUSP Working Paper Series No. 22. University of Surrey, Guildford.

Jiao, Y., Lloyd, C., Wakes, S. (2012). The relationship between total embodied energy and cost of commercial buildings. *Energy and Buildings*, 52, 20–27.

Khan, M. (1995). Sustainable Development: the key concepts, issues and implications. *Sustainable Development*, 3(2), 63–69.

Khanna, P., Babu, P.R., George, S.M. (1999). Carrying capacity as a basis for sustainable development a case study of National Capital Region in India. *Progress in Planning*, 52(2), 101–166.

Kivimaa, P., Martiskainen, M. (2018). Innovation, low energy buildings and intermediaries in Europe: systematic case study review. *Energy Efficiency*, 11, 31–51.
Klenow, J., Rodríguez-Clare, A. (2005). Chapter 11 Externalities and growth. *Handbook of Economic Growth*, 1(Part A), 817–861.
Koontz, T.M., Thomas, C.W. (2006). What do we know and need to know about the environmental outcomes of collaborative management. *Public Administration Review*, 66(s1), 111–121. https://doi.org/10.1111/j.1540-6210.2006.00671.x.
Krausmann, F., Gingrich, S., Eisenmenger, N., Erb, K.-H., Haberl, H., Fischer-Kowalski, M. (2009). Growth in global materials use, GDP and population during the 20th century. *Ecological Economics*, 68(10), 2696–2705.
Krebs, A. (1997). Discourse ethics and nature. *Environmental Values*, 6(3), 269–279. Accessed 19 January 2021 at http://www.jstor.org/stable/30301579.
Kuwabara, B., Auer, T., Akerstream, T., Pauls, M. (2013). Manitoba Hydro Place: design, construction, operation – lessons learned. PLEA2013 – 29th Conference, Sustainable Architecture for a Renewable Future, Munich, 10–12 September.
LaGros Jr, J.A. (2008). *Site Analysis: A Contextual Approach to Sustainable Site Land Planning and Site Design*. John Wiley & Sons, Hoboken, NJ.
Lawson, B. (2006). *How Designers Think: The Design Process Demystified*, 4th edn. Elsevier / Architectural Press, Oxford, UK and Burlington, MA, USA.
Layke, J., Mackres, E., Liu, S., Aden, N., Becqué, R., et al. (2016). *Accelerating Building Efficiency: Eight Actions for Urban Leaders*. World Resource Institute, Washington, DC.
Lee, W.L, Yik, F.W.H. (2004). Regulatory and voluntary approaches for enhancing building energy efficiency. *Progress in Energy and Combustion Science*, 30(5), 477–499.
Lenzen, M., Treloar, G.I. (2002). Embodied energy in buildings: wood versus concrete – reply to Börjesson and Gustavsson. *Energy Policy*, 30, 249–244.
Liu, B., Wang, D., Xu, Y., Liu, C., Luther, M. (2018). Embodied energy consumption of the construction industry and its international trade using multi-regional input–output analysis. *Energy and Buildings*, 173, 489–501.
Lotfabadi, P. (2015). Solar considerations in high-rise buildings. *Energy and Buildings*, 89, 183–195.
Mancebo, F. (2013). *Development Durable*, 2nd edn. Armand Colin, Paris.
Mancini, L., De Camillis, C., Pennington, D. (2013). *Security of supply and scarcity of raw materials: Towards a methodological framework for sustainability assessment*. European Commission Joint Research Centre Institute for Environment and Sustainability.
Marzbali, M.H., Abdullah, A., Razak, N.A. (2011). A review of the effectiveness of crime prevention by design approaches towards sustainable development. *Journal of Sustainable Development*, 4(1), 160–172.
McCormack, M.S., Treloar, G.J., Palmowski, L., Crawford, R.H. (2007). Modelling direct and indirect water consumption associated with construction. *Building Research and Information*, 35(2), 156–162.
McGilivray, M., Clarke, M. (eds) (2006). *Understanding Human Well-being*. United Nations University Press, Geneva.
McGill, G., Oyedele, L.O., McAllister, K. (2015). An investigation of indoor air quality, thermal comfort and sick building syndrome symptoms in UK energy efficient homes. *Smart and Sustainable Built Environment*, 4(3), 329–348
Medema, S.G. (2020). ‘Exceptional and unimportant’? Externalities, competitive equilibrium, and the myth of a Pigovian tradition. *History of Political Economy*, 52(1), 135–170.
Meireles, I., Sousa, V. (2020). Assessing water, energy and emissions reduction from water conservation measures in buildings: a methodological approach. *Environmental Science and Pollution Research*, 27, 4612–4629.
Millenium Ecosystem Assessment (MEA) (2003). *Ecosystems and Human Well-being: A Framework for Assessment*. Washington, DC: Island Press.
Mora, C., Tittensor, D.P., Adl, S., Simpson, A.G.B., Worm, B. (2011). How many species are there on Earth and in the ocean? *PLoS Biology*, 9(8), e1001127. DOI: 10.1371/journal.pbio.1001127.
Mossman, A., Ballard, G., Pasquire, C. (2011). The growing case for lean construction. *Construction Research and Innovation*, 2(4), 30–34.

Munda, G. (1997). Environmental economics, ecological economics, and the concept of sustainable development. *Environmental Values*, 6(2), 2013–2033.

Nahmens, I., Ikuma, L.H. (2012). Effects of lean construction on sustainability of modular homebuilding. *Journal of Architectural Engineering*, 18(2), 155–163.

Neumayer, E. (2003). *Weak versus Strong Sustainability: Exploring the Limits of Two Opposing Paradigms*. Edward Edgar Publishing, Cheltenham, UK and Northampton, MA, USA.

Neumayer, E. (2007). A missed opportunity: The Stern Review on climate change fails to tackle the issue of non-substitutable loss of natural capital. *Global Environmental Change*, 17, 297–301.

Nijaki, L.K., Worrel, G. (2012). Procurement for sustainable local economic development. *International Journal of Public Sector Management*, 25(2), 133–153.

Ochieng, E.G., Wynn T.S., Zuofa, T., Ruan, X., Price, A.D.F., Okafor, C. (2014). Integration of sustainability principles into construction project delivery. *Architectural Engineering Technology*, 3(116), 1–5. doi: 10.4172/2168-9717.1000116.

Ofori, G., Chan, P. (1998). Procurement methods and contractual provisions for sustainability in construction. In: Proceedings of Construction and the Environment: CIB World Building Congress, Gävle, 7–12 June.

Ortiz, O., Castells, F., Sonnemann, G. (2009). Sustainability in the construction industry: a review of recent developments based on LCA. *Construction and Building Materials*, 23(1), 28–39.

Osmani, M., Glass, J., Price, A.D.F. (2008). Architects' perspectives on construction waste reduction by design. *Waste Management*, 28, 1147–1158.

Oyedele, L., Regan, M., Ahmed, A., Ebohon, O.J., Elnocklay, A. (2013). Reducing waste to landfill in the UK: identifying impediments and critical solutions. *World Journal of Science, Technology and Sustainable Development*, 10(2), 131–142.

Pacala, S., Socolow, R. (2004). Stabilisation wedges: solving the climate problem for the next 50 years with current technologies. *Science*, 305, 968–972.

Pathirage, C., Amaratunga, D., Haigh, R. (2008). The role of tacit knowledge in the construction industry: towards a definition. University of Salford Manchester. http://usir.salford.ac.uk/9814/.

Pearce, D.W., Barbier, E., Markandya, A. (1990). *Sustainable Development: Economics and Environment in the Third World*. Earthscan Publications, London.

Powell, A., Sang, K.J.C. (2013). Equality, diversity and inclusion in the construction industry. *Construction Management and Economics*, 31(8), 795–801.

Praseeda, K.I., Reddy, B.V.V., Mani, M. (2017). *Life-Cycle Energy Assessment in Buildings: Framework, Approaches, and Case Studies*. Elsevier Science Direct. https://www.sciencedirect.com/science/article/pii/B9780124095489101885?via%3Dihub.

Pullen, S. (2010). An analysis of energy consumption in an Adelaide suburb with different retrofitting and redevelopment scenarios. *Urban Policy and Research*, 28(2), 161–180.

Pullen, S., Chiveralls, K., Zillante, G., Palmer, J., Wilson, W., et al. (2012). Minimising the impact of resource consumption in the design and construction of buildings. Proceedings of the Annual Conference of the Architectural Science Association (ASA '12), Queensland.

Qualk, J.D., McCown, P. (2009). The cost-effectiveness of building green. Looking beyond initial costs to the true cost of green-building ownership. Accessed 7 February 2021 at http://ezinearticles.com/?The-Five-Elements-ofGreen-Design&id=2417950.

Rezai, A., Foley, D.K., Taylor, L. (2012). Global warming and economic externalities. *Economic Theory*, 49, 329–351.

Robichaud, L.B., Anantatmula, V.S. (2011). Greening project management practices for sustainable construction. *Journal of Management in Engineering*, ASCE, January.

Röck, M., Saade, M.R.M., Balouktsi, M., Rasmussen, F.N., Birgisdottir, H., et al. (2020). Embodied GHG emissions of buildings – the hidden challenge for effective climate change mitigation. *Applied Energy*, 258, 114107.

Salvado, F., de Almeida, N.M., Vale e Azevedo, Á. (2019). Historical analysis of the economic life-cycle performance of public school buildings. *Building Research and Information*, 47(7), 813–832.

Schulze, E.D., Sierra, C.A., Egenolf, V., Woerdehoff, R., Irslinger, R., et al. (2020). Forest management contributes to climate mitigation by reducing fossil fuel consumption: a response to the letter by Welle et al. *GCB Bioenergy*, 10.1111/gcbb.12754, 13(2), 288–290.

Sebopela, M.C., Odeku, K.O. (2014). A comparative analysis of regulatory frameworks on climate change. *Mediterranean Journal of Social Sciences*, 5(20), 3078.

Sen A. (1999). *Development as Freedom*. New York: Alfred Knopf.

Solow, Robert M. (1993). Sustainability: An economist's perspective. In: Robert Dorfman and Nancy (eds), *Economics of the Environment*. Norton, New York.

Suzuki, M., Oka, T. (1998). Estimation of life cycle energy consumption and CO_2 emission of office buildings in Japan. *Energy and Buildings*, 28(1), 33–41.

Tadesse, S., Israel, D. (2016). Occupational injuries among building construction workers in Addis Ababa, Ethiopia. *Journal of Occupational Medicine and Toxicology*, 11, 16. https://doi.org/10.1186/s12995-016-0107-8.

Thormark, C. (2006). The effect of material choice on the total energy need and recycling potential of a building. *Building and Environment*, 41, 1019–1026.

Tilton, J.E., Lagos, G. (2007). Assessing the long-run availability of copper. *Resource Policy*, 32(1–2), 19–23.

Turner, R.K. (ed.) (1993). *Sustainable Environmental Economics and Management*. Belhaven Press, London.

Unalan, B., Tanrivermis, H., Bülbül, M., Ciaramella, A. (2014). Impact of embodied carbon in the life cycle of buildings on climate change for a sustainable future. Conference: 40th IAHS World Congress on Housing – Sustainable Housing Construction. Funchal, Madeira.

UNCHS Habitat (1993). *Development of National Technological Capacity for Environmentally Sound Construction*. UN-Habitat, Nairobi. https://digitallibrary.un.org/record/624386/export/xe.

UNEP (2011). Decoupling Natural Resource Use and Environmental Impacts from Economic Growth. UNEP, Nairobi.

UNEP (2019). *Towards a Zero-Emission, Efficient, and Resilient Buildings and Construction Sector: Global Status Report*. https://www.worldgbc.org/sites/default/files/UNEP%20188_GABC_en%20%28web%29.pdf.

UNEP (2020). *2020 Global Status Report for Buildings and Construction: Towards a Zero-Emission, Efficient and Resilient Buildings and Construction Sector*. Nairobi. https://globalabc.org/news/launched-2020-global-status-report-buildings-and-construction.

Urge-Vorsatz, D., Danny Harvey, L.D.D., Mirasgedis, S., Levine, M.D. (2007). Mitigating CO_2 emissions from energy use in the world's buildings. *Building Research and Information*, 35(4), 379–398.

Valdes-Vasquez, R., Klotz, L.E. (2013). Social sustainability considerations during planning and design: framework of processes for construction projects. *Journal of Construction Engineering Management*, 139(1), 80–89.

Wright, J.A., Loosemore, H.A., Farmani, R. (2002). Optimization of building thermal design and control by multi-criterion genetic algorithm. *Energy and Buildings*, 34(9), 959–972.

Wong, K., Fan, Q. (2013), Building information modelling (BIM) for sustainable building design. *Facilities*, 31(3–4), 138–157.

WWF (World Wildlife Fund) (2019). 'How many species are we losing?' Our work, how much is being lost? https://wwf.panda.org/our_work/biodiversity/biodiversity/.

Yang, H., Xia, J., Thompson, J.R., Flower, R.J. (2017). Urban construction and demolition waste and landfill failure in Shenzhen, China. *Waste Management*, 63, 393–396.

Yu, Y., Chen, D., Zhu, B., Hu, S. (2013). Eco-efficiency trends in China, 1978–2010: DECOUPLING environmental pressure from economic growth. *Ecological Indicators*, 24, 177–184.

Zuo, J., Jin, X., Flynn, F. (2012). Social sustainability in construction – an explorative study. *International Journal of Construction Management*, 12(2), 51–63.

12. International construction data: a critical review

Jim Meikle and Asheem Shrestha

INTRODUCTION

Construction data is used – and abused – by politicians, policy-makers, industry, journalists, analysts and researchers. All these users need data, but to use it effectively they need to understand the nature of construction data and its limitations. This chapter is specifically about international construction data.

Construction is an important part of all economies. It consists of the production, extension, modification, repair and maintenance of structures. It can be undertaken by individuals, construction firms or other organisations; its products are essential for housing, work, education, and so on, and for the infrastructure that connects and services people, businesses and communities; and, directly and indirectly, it provides employment for a significant proportion of the population. Most construction is location-specific, is made to order, and takes a relatively long time to produce; generally, it is not made in a factory, has few standard final products and is not available off the shelf (that is not true of all construction; off-site prefabrication provides some of these latter attributes). Construction is also long-lasting; for example, much of United Kingdom (UK) housing is over 50 years old and some of it is much older, although most will have had significant elements repaired or replaced, often many times.

Construction is difficult to measure reliably, particularly in developing countries. It consists of very many projects of all types and sizes, some small and simple, some large and complex, and everything in between. As noted, it is not only undertaken by construction companies, but also by individuals and households and by public and private enterprises on their own account; much is undertaken formally and officially recorded but some is not. All – or nearly all – of it should be included in national accounts; the exception is repair and maintenance undertaken by households on their own dwellings.

If national construction data, for the reasons outlined above, is difficult to measure, international data is even more problematic. This chapter is primarily concerned with sources that present data on a number of countries. However, it starts with a general discussion of the concepts used for construction data, how this is processed and presented, and how accurate and reliable it is. Subsequent parts of the chapter describe spatial and temporal deflators, how raw data is brought to common bases, and present and comment on international construction data sets. A final section provides concluding remarks including some recommendations.

The main types of international data considered in this chapter are measures of construction output, resources, company performance and price data. These are selected because they are important for policy-makers, analysts and researchers, but also because there is extensive published data available on them. They are the focus of most international construction sector data sets, and the data that is most often sought by users. Other data sets are mentioned in the chapter, including data on capital and building stock, and physical measures of construction activity. The chapter aims to cover the main sources of international construction data but does not claim to be comprehensive. It has been written by data users rather than data producers; it is, therefore, mainly concerned with what is available rather than being a treatise on statistical methods.

SOURCES OF CONSTRUCTION DATA

The principal source of guidance on the structure of international construction data is the International Recommendations for Construction Statistics published by the United Nations (UN, 1998). This sets out how construction data should be collected and presented. The introduction to that document provides a useful overview of the statistical issues peculiar to construction and sets out definitions for the producers and products of construction. The recommendations are underpinned by the UN System of National Accounts (SNA) published by the United Nations Statistical Commission; the latest edition, the eighth, was published in 2009 (UN, 2009). The SNA defines the framework used to represent economic activity in countries and, within it, the International Standard Industrial Classification of All Economic Activities (ISIC) provides a structure for industrial production (UN, 2008).

Most, but not all, countries in the world have construction data on the websites of their national statistical institutions, and these should be the starting point for individual country researchers. One notable exception is the United States (US), which does not have a national statistical office; its statistical responsibilities are dispersed across government departments. Although almost all national statistics formally comply with international statistical conventions, it is unrealistic to assume that there is universal consistency in the methods used for their collection, treatment and publication.

There are a number of international organisations that publish economic and other statistics, including data on construction, for example, the United Nations and its agencies, the World Bank, the Statistical Office of the European Union (Eurostat), the Organisation for Economic Co-operation and Development (OECD), and regional development banks, including the African, Asian and Inter-American Development Banks. Most of these, however, are secondary collectors; they generally take data from national organisations and process it for publication; their merit is that they bring the national data together, they do not radically alter the basic data.

There are also independent sources of international data, universities and research organisations, and private firms. A number of these are mentioned in this chapter. Much data from public organisations is freely available; data from private organisations often has to be paid for and some is very expensive (there is rarely a strong correlation between cost and quality).

DATA QUALITY

The quality of construction data is varied and variable. For example, national construction statistics in the UK have shortcomings discussed by Cannon (1994), Briscoe (2006) and Green (2019). Commenting on UK data, David Pearce, in his *Social and Economic Value of Construction* (Pearce, 2003, p. 59), concluded that: ‘It is clear … that different sources produce different results and can generate, at best, uncertainty and, at worst, confusion.’

A number of writers on construction economics comment on the quality and reliability of the data they use, although many do not. Ofori (1990) notes the lack of consistency generally in the definition and presentation of statistics on construction. Reviews of European construction data (Meikle and Grilli, 1999) and African construction data (K'akumu, 2007; Meikle, 2011) highlight the inconsistencies among measures of construction activity. *Poor Numbers*, by Morten Jerven, catalogues the shortcomings of African national statistical institutions and African statistical data generally (Jerven, 2013). Data quality has a number of dimensions, including:

- Content. Are all the components that should be there included in the data?
- Reliability. Are the rules and conventions of data collection, analysis and presentation complied with?
- Comparability. Can data sets over time or across countries be compared?
- Confidence. Are potential or real errors and inconsistencies in data acknowledged by their authors?

These dimensions are discussed briefly below. It should also be noted that most recently published data is subject to revision. The pressures of publication deadlines and the demands of data users require national statistical institutions and other providers to publish data faster than they might ideally wish to do. This tends to mean that the most up-to-date data is not necessarily the most reliable. There are also revisions from time to time when there are significant changes to statistical rules or methods. These latter revisions typically only involve changes to a relatively small number of historical values and are not necessarily reflected in long-term time series.

Content

The international recommendations on the preparation and presentation of construction statistics (UN, 1998) provide the framework for most construction statistics that are intended for inclusion in national accounts. International construction data produced as market intelligence is not subject to the same rules and in many cases will comprise data that is collected nationally with less attention to whether definitions and measurements are consistent.

National data on construction activity is usually available at some degree of disaggregation, by producer, type of work, Standard Industrial Classification (SIC) code or some other breakdown. However, aggregate data – as it typically appears in international data sets – is most often presented as a single value for each country. This conceals as much as it reveals: it includes all types and sizes of construction, so that the data presented is an average of a wide range of work, some highly productive, some not, some undertaken by large international contractors and some by individual households.

Reliability

Almost 40 years ago, Andrew Kamarck, a former Director of Economics at the World Bank, noted that there is a spurious precision in much national and international economic statistics (Kamarck, 1983), particularly in developing countries. For example, the reliability of data on GDP per capita in US dollars in many countries depends on knowledge of national values for economic activity, population and currency conversions, and often some of these are fairly imprecise estimates. Therefore, the resulting data has significant error bands around it, although these are often unacknowledged.

Blades and Roberts (2002), in the context of OECD countries, report that many economic sectors can be observed and recorded fairly accurately: power generation, heavy industry, rail and air transport, government services, banking and telecommunications are given as examples. They note that the problem areas are activities such as home repairs, retail trade, taxis, trucking, cafes and restaurants.

Official construction data is collected directly from industry surveys or indirectly via estimates or data collected for other purposes. Surveys are rarely comprehensive and usually involve estimates, particularly for the activities of smaller projects and smaller firms. The reliability of surveys depends on the knowledge of all the firms to be included so that sampling frames can be constructed and samples can be scaled up to represent the entire sector. Moreover, since many construction projects are of long duration and it is expenditure on work in progress that should be measured as construction output, data collection can be problematic. Despite guidance from statistical offices, survey forms are often completed by relatively junior members of responding organisations and there is scope for inconsistency in how, and how well, they are completed. It is sometimes not clear whether the value of output recorded represents the cost of work to the supplier (typically, a contractor) or the price that the customer should pay for that work, or the amount of money that has been paid by the customer to the supplier. There is also scope for wrong categorisation of activity, for example, by type of work or SIC code.

Construction activity on work by households, including self-build housing, can be estimated from household expenditure surveys, tax returns or other administrative sources. Construction activity on small projects and by small firms or individual tradesmen is usually based on such sources. There is also grey, or cash-in-hand, economic activity in construction, often for small repair and maintenance work, that will be estimated where it cannot be derived from secondary sources. This last is discussed by Chancellor et al. (2019), Schneider et al. (2010) and Blades and Roberts (2002).

Most inaccuracies in data are not intentional but there are reasons for deliberately overstating or understating economic data. It can be overstated, usually for political reasons, in order for an economy or sector to appear larger than it really is, to boast that it is higher up a league table than it should be, or that it is the largest in the world or the largest in the region. It can also be understated in order to appear smaller than it is so as to fall below thresholds and be eligible for particular grants or loans, or other forms of assistance or privileges.

Comparability

Value data on construction activity is collected in current price terms, the prices of the day, but it is presented in constant, as well as current, price terms and this necessarily involves a process of statistical adjustment – deflation – bringing prices at different points in time to a common base date. This allows volume or quantity comparisons over time to be made. The reliability of constant price data depends on the quality of the deflators used and how they are applied. Temporal deflators are discussed in more detail in the next section. Cross-country comparisons present problems because of different currencies and price level differences between countries, and that is also discussed below.

While irregularities in national data can be acceptable because the same irregularities are maintained over time, international data exercises involve comparisons between national data sets where concepts, conventions and conditions can differ, sometimes significantly. For example, rules for floor area measurement differ; the structure of housing in North America, predominantly timber, is markedly different from that in Asia or Latin America, which is predominantly concrete (Abergel et al., 2017); climatic, seismic and ground conditions vary and can have major impacts on regulations, construction techniques and building costs. In international comparisons there is always a tension between 'comparability' and 'representativity'; that is, the more a building is typical of its country's building industry, the less it may be comparable to one of the same notional type in another country (Groak, 1992). A telling phrase from international economic statistics definitions is that 'construction is comparison resistant' (OECD, 2007).

A particular issue of comparability, over time and across countries, is that there can be significant quality differences in products and in construction works. This issue is recognised and addressed in a number of situations, and there are statistical techniques that take account of quality changes over time, for example in cars or computers where changes are relatively rapid and production volumes are high. However, it is particularly difficult to accurately measure quality differences in construction.

Confidence

In the physical sciences, it is expected that reports on measurements will note the confidence there is in the data and the error bands that should be put around it. Economic data – for example, gross domestic product (GDP) growth or unemployment figures – is frequently reported to a supposed accuracy of one or two decimal places with no indication of the confidence that can be placed in these figures. It is also, as noted previously, frequently revised.

Therefore, users should be cautious about the reliance they place on published construction data, and general economic data. This is particularly true when basic data has been manipulated, for example, converted from current to constant prices, presented in a common currency, or expressed as per capita or per worker values. These adjustments involve temporal and spatial deflation or the use of other factors that are not themselves always reliable, particularly in developing countries.

DEFLATORS, NOMINAL AND REAL DATA

In order to compare construction value data over time and between different countries, it is necessary to bring it to a common basis, to permit assessment of volume or quantity changes. To measure changes over time, it is necessary to adjust for price changes and to measure differences between countries; adjustment must be made for currency and price level differences. How well this is done influences the quality of intertemporal and international comparisons. The methods of bringing current values to constant prices and making international price level comparisons are similar in general but different in detail.

Temporal Price Indices

Adjusting for price changes over time involves the application of price indices, the purpose of which is to convert values in prices of the day (current prices) into prices at a common base date (constant prices). The process of adjustment is called deflation, and current prices are known as nominal prices while constant prices are referred to as real prices. Discrete price indices for different products are familiar; for example, they measure trends in consumer or retail prices or house prices. Indices – or deflators – are also required to adjust economic activities from current to constant prices, for example, construction output or all economic output (GDP). Temporal price indices are applicable within a country.

There are methodological and data issues with the production of construction price indices and deflators, and no entirely satisfactory approach has been developed for them in any country. Methodological aspects of construction price indices are discussed in documents published by Eurostat and the OECD (Eurostat, 1996; OECD and Eurostat, n.d.). These cover the situation in more than 20 countries and demonstrate the problems experienced, the range of approaches used and their limitations.

The essential components of price indices and deflators are lists of items and weights and prices for each item. The diversity of construction makes it difficult to come up with simple lists of items and weights and the nature of construction makes it difficult to come up with reliable prices. The issue of weights is partly addressed by having a range of indices covering different types of work – for example, residential, non-residential and civil engineering work – and within each of these, subindices covering individual project types, or new work and repair and rehabilitation work. Some work types, for example, new housing and roads, can be represented by 'standard' projects but others cannot.

The greater problem is almost certainly determining prices for construction work. The prices required are purchaser prices, the prices paid by the final customer for the work. Unlike most consumer products, such as cars or refrigerators, there are no – or very few – standard products and publicly listed prices; the prices of construction products – projects – cannot be easily observed. As indicated earlier, construction projects are large, often unique, take a long time to deliver, and often the real price is unknown until long after it is complete; and construction projects and their prices are particular to a specific place and time.

Often, the approach to determining prices for construction projects is to use construction experts, for example, quantity surveyors, to provide estimated prices for complete projects or types of work, and to do that on a regular basis so as to estimate price trends for these types of work. Ideally, this should be done using a number of experts, but the cost of doing that often makes it prohibitive. The advantage of using construction experts is that, if well chosen, they will have a good understanding and current knowledge of construction prices nationally; the downside is that the prices are based on opinion, not direct observation.

Spatial Price Indices

The issues with volume adjustments across countries are less familiar than those over time, but they are important for international comparisons. Volume comparisons of GDP and other economic aggregates, including construction investment, involve comparisons free of price level differences and in the same currency units. The conventional method of making international price comparisons is to use published exchange rates. There are different versions of these – for example, tourist rates and commercial rates – and while many currencies are easily convertible, others are not, and officially quoted exchange rates do not represent what these currencies exchange for on the black market. And exchange rates are single rates applicable to all parts of an economy. Exchange rates represent the price of money in the money markets; they can be influenced by political events or events in financial markets. Commercial exchange rates can vary significantly in relatively short periods of time.

While measuring prices over time is a national or subnational issue, measuring price and real quantity differences between countries is an international issue. Temporal indices are national and their make-up and methodology are designed to suit each country's needs and resources, but spatial indices – currency convertors – need to conform to internationally agreed conventions. Spatial price adjustment indices have all the methodological problems of temporal price indices, plus they have to address the problems of international comparability. The features that make each country's items special – their representativity – need to be reconciled with the need for comparability of the items between countries. The spatial price indices that have been devised are purchasing power parities (PPPs) and these are calculated for different components of economies as well as for whole economies.

PPPs are designed to adjust for price level differences between countries that are annual averages and national averages and are usually produced some time after the period to which they apply (not least because, to be credible, they need international agreement from all the countries involved). They are useful in the presentation of international economic data and the examination of particular issues between countries, for example, eligibility for regional assistance in Europe or international aid programmes, or for comparisons of poverty levels. In the specific case of construction PPPs, they can be used to compare components of aid programmes or for other broad sectoral comparisons. However, they do have limited utility for industrial market analysis where data that relates to specific times, places and types of work is often needed.

The main sources of official PPPs are international agencies. Eurostat[1] was the first agency to establish a regular official international price comparison programme, including the compilation of construction PPPs, in the early 1970s. The European Union (EU) uses PPPs to determine eligibility for, and to allocate, regional support funds. The Eurostat construction methodology uses bills of quantities priced by national experts in each country; PPPs are calculated by European Commission officials and agreed by the countries. The Eurostat programme currently covers 37 member, associate and candidate countries. The OECD[2] followed Eurostat in the mid-1970s, using a similar approach. Eurostat works on an annual pricing programme, and OECD on a three-year programme.

In 2000, the World Bank took over the management of the International Comparison Program (ICP) that had been managed since 1970 by the United Nations. The first UN ICP construction survey included ten countries and in six subsequent surveys the number rose to more than 100. The first World Bank survey, in 2005, included 146 countries; the second, in 2011, 198 countries; and the latest, in 2017, 176 countries. The development of construction PPPs on the ICP and more generally is discussed in Best and Meikle (2015a, 2019). The World Bank now implements a rolling programme of data collection.

There are other ways of comparing construction price levels between countries that remove some of the disadvantages of using exchange rates or PPPs. In his doctoral thesis, Rick Best uses a CityBLOC basket of goods approach (Best, 2008); Chan (2019) uses the local price for

Table 12.1 *Statistical concepts for construction output*

Concept	Source	Content
Contractors' construction output	Construction statistics	Construction output by construction contractors classified to construction SICs
Construction value-added	National accounts (production version)	Value-added (crudely: labour, profit and depreciation) of all activities by all producers of construction classified to construction SICs, informal construction activity and self-build construction by households
Construction gross output	National accounts (input–output tables)	The above plus intermediate consumption (goods and services purchased from other parts of the economy), including double counting of subcontracting by firms classified to construction SICs
Gross fixed capital formation in construction	National accounts (expenditure version)	Construction expenditure on capital works (new work and major refurbishment) by all public and private entities and households, including related professional services

key construction resources such as hours of skilled labour or cubic metres of concrete to bring international construction prices to a common base.

CONSTRUCTION OUTPUT DATA

There are a number of sets of construction output data produced by national statistical institutions and reproduced internationally. Table 12.1 presents the main concepts of such data and their sources, and gives brief descriptions of their content.

Value data is published in current and constant price terms and in national currency units or in a standard currency, for example, USD or euros; when standard currencies are used, these may be calculated using PPPs or commercial exchange rates. None of the concepts in Table 12.1 includes all construction activity. Gross construction output and gross fixed capital formation (GFCF) in construction, construction investment, are the nearest but the former excludes work by organisations not registered to construction and the latter, all construction repair and maintenance work.

Different concepts are used for different purposes. Contractors' construction output is mainly used by industry to indicate market size and market trends in contracting, and as the basis for market forecasts. Construction value-added is used in most official economic reports because it is the construction component of GDP. With some notable exceptions (Bon and Pietroforte, 1990) construction gross output is rarely used by researchers, although input–output tables can provide interesting insights into the structure of construction output. GFCF in construction is used when national investment is being studied; it is also the measure of construction expenditure in the World Bank's ICP.

In international data sets, contractors' construction output is usually given as a single number for each country, although a few sources provide breakdowns for some countries into residential, non-residential and civil engineering; and new work and repair and maintenance; but national definitions often differ in detail and comparisons should be treated with caution. There is a useful description of the current content and methods for British contractors' output in Green (2019).

EuroConstruct[3] is a network of research institutions and consulting organisations, established in 1974. Twice a year, it organises an international conference on short-term forecasts of contracting activity in 19 European countries. Forecasts are provided for general economic and demographic indicators and for construction broken down into different types of residential buildings, non-residential and civil engineering work (EuroConstruct, n.d.). National definitions of what is included and what is excluded in the different types of work and total construction vary. The members of EuroConstruct are also sources of data-based construction consultancy.

AsiaConstruct[4] is a similar network, established in 1995, and involving 15 Asia-Pacific countries in its most recent (2019) conference but with a more limited remit. Its events are annual and it produces separate country reports on current construction activity and issues, but it does not produce collated data or forecasts. The measures of construction activity used by AsiaConstruct vary from country to country; value of contracts awarded and value of output are both used. The detailed breakdown of construction activity in each country also varies (AsiaConstruct, n.d.).

Construction value-added is the output metric used in most official reports and in many academic papers. In international data sets it is usually given as a single figure. Most papers on the relationship between construction and economic development use this concept (Ofori, 1990; Strassmann, 1970; Turin, 1973; Wells, 2007). In the past, the source for much of this data was UN Housing and Construction Statistics, produced in hard copy, but the construction-related data from the UN nowadays is more limited. The main source is the UN Analysis of Main Aggregates (AMA) database (UN, 2019) that comprises sectoral breakdowns of GDP, including construction, and values for GFCF and other economic variables for most countries. Data is in current and constant prices in US dollars and in most cases starts from 1960. Similar data for particular groups of countries is produced by organisations such as the OECD (n.d.) and the African Development Bank (AfDB, 2020). In many international data sets, including the World Bank's World Development Indicators and the Asian Development Bank's Data Library, construction output is combined with manufacturing and is not available as a separate sector.

Long-term data on construction value-added in current and constant price terms is published online in Groningen University's Groningen Growth and Development Centre GGDC 10-Sector Database.[5] This has data on ten economic sectors, including construction, on 38 countries, in national currencies from the 1950s or 1960s to the early 2010s (GGDC, 2017). The GGDC database does not just reproduce data from national statistical institutions, it also reviews and revises data where necessary; there are caveats on the data for some countries, including China, India and African countries in the database (Timmer et al., 2015). Figure 12.1 illustrates construction activity for the US and China produced using GGDC data; it shows value-added in current and 2005 constant prices for broad components of the economy, including construction.

The US demonstrates relative stability in terms of economic structure over time, but with the volume of financial services increasing and that of construction decreasing over the period. On the other hand, China shows dramatic changes in structure since the 1960s, particularly the decline in agriculture, and the growth in manufacturing and, to a much lesser extent, construction.

The EU KLEMS database, funded by the European Union, is a more extensive version of the GGDC database. The initials refer to capital (K), labour (L), energy (E), materials (M) and services (S). The original version of KLEMS is hosted by GGDC and the most recent version by the Vienna Institute for International Economic Studies. The latter covers 28 EU countries and Australia, Canada, Japan, South Korea and the US, and the data runs from the mid-1990s

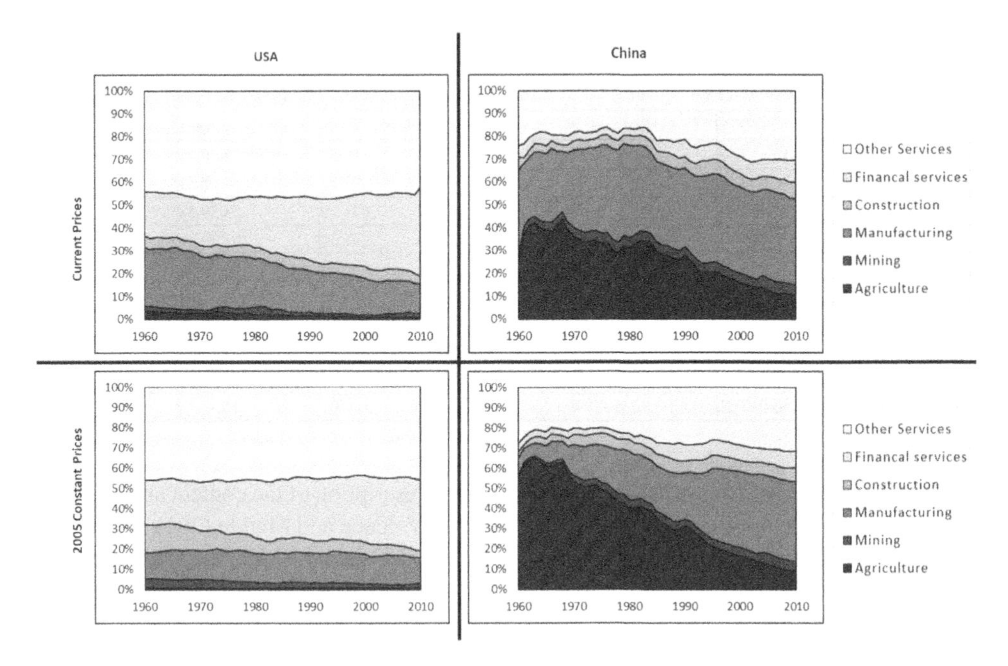

Source: Compiled from the GGDC 10-Sector Database (GGDC, 2017).

Figure 12.1 *Construction output in the USA and China, 1960–2010*

to the mid- or late 2010s. EU KLEMS extends the number of countries, brings the data nearer to the present day and provides much more detail (EU KLEMS, 2019). The GGDC 10-Sector Database and EU KLEMS together cover 62 countries.

Input–output and supply-and-use tables lay out in tabular form construction purchases from other industries and groups of industries (categorised by ISIC) plus the value added to these purchases by construction. National tables can have 100 or more columns and rows, but international data sets have reduced numbers of columns and rows so that most differences between national table structures can be resolved. Some national tables – such as those for Australia, Singapore and South Africa – have more than one set of columns and rows for construction, although international data sets have a single column and a single row for all construction for each country. The result is that industry breakdowns of the inputs to construction are provided in a standard format so that the economic structures of industries can be compared.

The OECD publishes tables with 36 columns and rows for 64 countries (OECD, 2018); Eurostat tables have 64 columns and rows (Eurostat, n.d.). There is also a World Input–Output Database (WIOD, 2016). Input-output tables are also compiled and published by the Asian Development Bank (ADB, 2018). Ranko Bon and a few others have written on construction and economic development using national input–output tables (Bon and Pietroforte, 1990), and they were considered but rejected as a basis of national weights in the development of the ICP 2011 construction methodology development (Meikle, 2019).

GFCF in construction, construction investment, is set out in the expenditure versions of national GDP. It has been used by a number of construction researchers (Andrews et al., 1972;

Table 12.2 Key data on selected countries (values in USD billions)

Country	GDP		Construction expenditure		Construction per	Construction as a
	In PPPs	Using xrates	In PPPs	Using xrates	capita	% of GDP
China	19 617.38	12 134.57	10 618.59	3 632.82	7 659.11	29.92
India	8 050.53	2 552.54	2 356.73	373.23	1 760.51	14.62
USA	19 519.42	19 519.42	1 641.29	1 641.29	5 050.28	8.41
Indonesia	2 893.57	1 015.42	1 616.82	245.38	6 109.28	24.17
Turkey	2 266.51	852.67	729.37	136.84	8 993.46	16.05
Korea, Rep.	2 105.89	1 623.90	564.04	259.54	10 982.09	15.98
Japan	5 172.97	4 859.95	520.58	490.59	4 105.84	10.09
France	2 994.45	2 592.73	504.87	311.45	7 550.02	12.01
Germany	4 381.79	3 665.87	504.64	362.33	6 105.01	9.88
Brazil	3 017.72	2.062.84	474.15	149.01	2 281.43	7.22

Source: Compiled from IPC (2017).

Wells, 1985; Girardi and Mura, 2014). There is a useful description of the content and methods for GFCF in construction for a group of countries in de Valence and Meikle (2019).

Table 12.2 sets out GDP and construction expenditure data for the countries with the largest domestic construction expenditures (in USD PPP) in 2017. The table shows GDP and construction expenditure in PPPs and using exchange rates ('xrates'). Construction per capita is calculated using expenditure based on PPPs, while construction as a percentage of GDP is calculated using figures based on exchange rates because PPP expenditures are not additive (Meikle, 2019).

The figures illustrate the effect of using PPPs rather than exchange rates. GDP and construction expenditure in developing countries are significantly understated using exchange rates (in India by a factor of more than three in the case of GDP, and six in the case of construction), while they tend to be overstated in high-income countries. The per capita figures for construction in China, Indonesia and Turkey are remarkably high; the figures for construction as a percentage of GDP are also remarkable for China and Indonesia.

National construction output includes work by registered construction contractors, by unregistered firms or individuals, and by households (and, in the case of construction GFCF, capital works by organisations not registered to construction SICs). In most developed countries, work by registered contractors is the dominant mode but there is also significant work undertaken by the other modes, mainly relatively small-scale work. In many developing countries, a fairly high proportion of work, particularly residential construction, is undertaken by unregistered firms and individuals and households.

The term 'informal construction' – or black or grey or cash-in-hand construction – is used to describe a range of situations where some aspects fall outside the formal regulatory environment (Wells, 2007). This can refer to compliance with planning or building regulations or standards, conditions of engagement, or payment of taxes or other charges. It is more prevalent in developing countries although it is by no means absent in developed economies.

Construction output in the national accounts should include all economic activity on construction. However, there are always questions with informal construction data in the national accounts: is it included, is all of it included, and is it accurately measured? Work that is excluded, but should be included, will reduce both construction values and total economic values in national statistics. There are studies of the extent of informal construction activity in countries (Blades and Roberts, 2002; Chancellor et al., 2019; Schneider et al., 2010) but these do not necessarily indicate how much of that activity is included in national accounts. Blades (1975) was one of the first to study that issue and, more recently, Meikle (2011).

CONSTRUCTION PRICE DATA

The most reliable data on topics such as construction output and expenditure is usually from official sources. However, construction cost and price data is more commonly obtained from private sector sources, although temporal and spatial price adjustment factors – essential for official data – are inherent in constant price and standard currency data. There are useful discussions on the issues and methods around international construction cost comparisons in Best (2008) and Best and Meikle (2015b).

Since 1979, The European Council of Construction Economists (CEEC) has priced a standard office building, originally in six and most recently (2019) in 11 European countries. The exercise produces total costs and costs per m^2 for 11 'elements' and for the overall project. The pricing is based on a set of agreed quantities for a model building; therefore, the costs are comparable but not necessarily representative of what would be built in each country (CEEC, 2019). The CEEC results are also calculated as work type-specific PPPs.

Since at least the 1980s, some construction professional services firms have published international construction price information in hard copy, and latterly online, largely as a marketing exercise. A few years ago, there were half a dozen or more firms doing that, but nowadays the numbers have reduced; a web search revealed two: Turner & Townsend (T&T) (2019) and Arcadis (2020). There seem to be a number of reasons for this, including perhaps a concern about giving company knowledge away for nothing, and a wish to project the firms as more than just construction cost consultants. The Turner & Townsend and Arcadis publications provide information on construction and development in around 100 major locations around the world, including:

- prices for common construction resources (labour, materials and equipment);
- prices for construction elements or components (for example, partition, flooring or cladding systems); and
- prices for complete projects (for example, costs per m^2 for different building or work types).

Prices are given for a particular date and particular locations, and adjustment factors for other locations may also be given. In all cases, prices will reflect construction in each location: they will be representative but not necessarily comparable. Recent T&T publications include efforts to produce construction PPPs. The data provided in these can be variable, and sometimes apparently similar items indicate markedly different prices in different sources (Best, 2008). Rider Levett & Bailey (RLB),[6] another firm that would have provided free price information in the past, has established a European network and developed a web-based construction price information network that provides fee-paying members with access to a range of price data.

International construction price information is also published in construction journals and price books. *Engineering News-Record* (*ENR*)[7] in the US and *Building* magazine[8] in the UK publish occasional features on resource and other price data for numbers of countries. National hard copy construction price books have been around since the nineteenth century; the first Spon's UK construction price book appeared in 1870. International price books are more recent and may already have had their day. E&FN Spon published an *International Construction Cost Handbook* in 1988 (Davis Langdon & Everest, 1988), followed by similar books on Europe, with the latest, third edition in 2000 (Davis Langdon & Everest, 2000); Asia Pacific, the fifth edition in 2015 (Langdon & Seah, 2015); Africa, the second edition in 2005 (Franklin & Andrews, 2005b), the Middle East, the second edition in 2005 (Franklin & Andrews, 2005a); and Latin America, the first edition in 2000 (Franklin & Andrews, 2000). The format of the books is similar, with a description of each country's economy and construction industry followed by lists of prices for resources, work items and types of work. The books also include summary tables of selected price information.

The problem with hard copy construction price information is that by the time it is compiled, printed and published, it is out of date; and much the same applies to web-based documents. The usefulness of these kinds of data is to provide snapshots at a particular point in time and to allow examination of price trends over time (they tend to have data on similar resources, work items and project types in different editions). The data they provide is, typically, representative of each country rather than comparable across countries.

PHYSICAL QUANTITATIVE DATA

Construction Labour

There are two main concepts commonly used for construction labour: employment and the workforce. Employment, as the term implies, refers to individuals in full-time employment; the workforce refers to all those working in construction, including individuals engaged on a limited time or task basis and those not in formal employment. The units used for employment are typically numbers of workers – usually total numbers per year – and days or hours worked per year. Because of the casual basis of much construction employment, and practices such as labour-only contracting, labour employment data in many countries is fairly unreliable and often not comparable across countries. The GGDC 10-Sector Database is a source for employment numbers for the same sectors and countries as the value-added data. Figure 12.2 presents employment data for the USA and China.

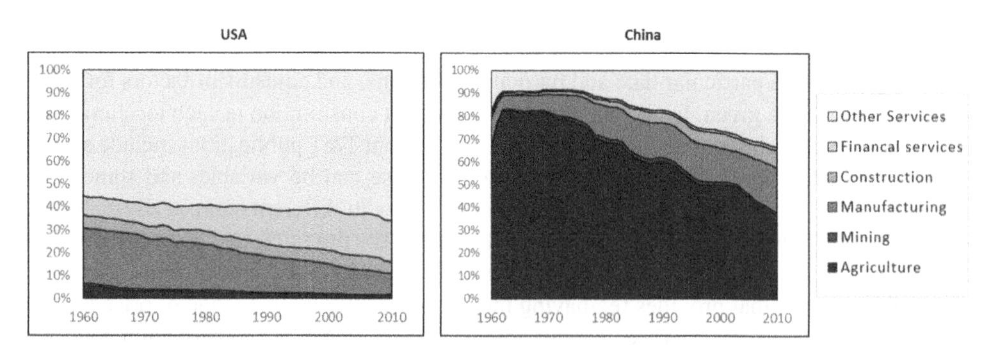

Source: Compiled from the GGDC 10 sector database (GGDC, 2017).

Figure 12.2 Construction employment in the USA and China

The trends in output (from Figure 12.1) and employment in some sectors are different, indicating differential movements in quantities or prices or both (the employment data indicates changes in quantity only). In the US, the price of manufacturing and financial services workers (their wages) has increased over the period; the prices of workers in other services have declined. In China, the price of agricultural workers has gone down while that of financial services workers has gone up; the price of construction workers has probably declined slightly.

Construction employment data, along with construction output data, are the basic requirements for analyses of construction labour productivity, and productivity comparisons are a staple of reports and commentaries by politicians, journalists and others. However, they are fraught with difficulties, not least that useful breakdowns of both employment and output are usually not available. Therefore, most international construction productivity comparisons are comparisons of 'all construction' labour productivity, a metric that is an average of all types of work by all types of producers and conceals those types that are productive and those that are not. Ive et al. (2004, p. 121), in a report on international construction competitiveness for the UK government, noted that:

> For construction, the reported 'productivity gaps' for levels between the UK and the US, France and Germany appear quite large but may be within the margins of error of the data used; further work removing sources of error might either widen or narrow these reported gaps; and definitive findings, even as to national rank orders in terms of productivity levels, are simply not possible at this stage, in the present state of the data.

Cement Consumption

Cement is one of the few construction materials for which fairly reliable production and trade data is available, that is used predominantly in construction, and has a relatively short shelf life. Therefore, it has the potential to provide a useful indicator of construction volume, particularly where the bulk of construction is of modern materials. The best-known current source of data on cement production, trade and consumption is the Global Cement Report (see Global Cement Report, 2019), an annual publication. This provides a two- to six-page report on around 170 countries plus an Excel database, giving annual figures for production, imports, exports and consumption since 1990. Comprehensive historic data on cement internationally is reviewed in Low and Tan (1993); this presents and analyses data on cement production, trade and consumption and its relationship to GDP for 154 countries and territories from 1967 to 1987.

Table 12.3 sets out cement consumption (domestic production plus imports less exports) per capita and per million USD construction expenditure, based on PPPs for the ten largest countries (by population) in the UN high- and low-income groups in 2017. It also lists the percentage of the urban population in each country. (The urbanisation figures are included because, particularly in low income countries, modern construction is likely to be concentrated in urban areas.)

The data in the table for cement per capita broadly aligns with the income categories, with higher per capita consumption in the more developed and more urbanised countries, and lower consumption and urbanisation in the low income countries. The broad trend in cement consumption per million US dollars construction expenditure based on PPPs seems to be from lower figures in high-income countries to higher figures in lower-income countries. This possibly reflects greater complexity of work in high-income countries and relative simplicity

Table 12.3 Cement consumption and urbanisation, 2017

High income				*Low income*			
Country	Cement per capita, kg	Cement per USDm (PPPs) construction, T	Urban pop. (%)	Country	Cement per capita, kg	Cement per USDm (PPPs) construction, T	Urban pop. (%)
USA	302.04	59.81	82.06	Ethiopia	82.8	127.27	20.31
Japan	344.12	83.81	91.54	Congo, DR	37.47	66.44	43.88
Germany	346.49	56.75	77.26	Tanzania	91.65	249.18	33.05
France	260.67	34.52	80.18	Uganda	64.14	114.51	23.2
UK	219.50	34.87	83.14	Mozambique	99.83	341.55	35.46
Italy	324.76	50.82	70.14	Nepal	193.29	213.15	19.34
Korea, Rep.	1057.01	96.25	81.50	Madagascar	27.38	120.58	36.52
Spain	243.38	67.45	80.08	Niger	26.85	142.54	16.35
Argentina	247.02	76.58	91.75	Burkina Faso	83.88	289.18	28.74
Poland	429.76	119.76	60.11	Mali	144.23	439.85	41.57

Source: Cement data from the Global Cement Report (2019); construction and population data from (ICP, 2017); urbanisation data from World Development Indicators (World Bank, 2020).

and higher cement intensity in lower-income countries. The table suggests that using a range of measures to check data credibility or raise queries on data has value.

DATA ON CONSTRUCTION FIRMS

There are two main types of data on construction firms: league tables, for example, of the 'top' so-many firms, that rank companies by, say, turnover; and databases, that collate financial data on firms. National versions of both are available in many countries but there are a limited number of versions that are international or, at least, multinational. League tables tend to be annual, compiled by publishers or based on surveys. Financial databases are usually live and contain a number of years' data. They are based on formal financial returns to national authorities.

The best-known and longest-running league tables are published regularly in the US publication *Engineering News Record* (*ENR*) and have been since 1978. The early days of modern international contracting are described in the first chapter of Strassman and Wells's (1988) book, *The Global Construction Industry*, using *ENR* data, and subsequent chapters describe the situation in a range of countries that have firms operating internationally. Perhaps the most striking fact looking at the *ENR* data in recent years is the rise of Chinese firms. Figure 12.3 shows the share of contracting activity of different geographic groups of large firms since 2005. The figure illustrates the changing importance of the world's major contractors. In 2005, US and European firms undertook more than half of total revenues of the top 225 firms; by 2010, Chinese firms had overtaken US firms; and by 2018, they had by far the largest single share and almost half the total.

International Construction magazine published by KHL Group[9] produces an annual league table of worldwide sales figures for the world's largest construction contractors. The data is compiled from a variety of sources including the magazine's own estimates and converted to US dollars at exchange rates. The data represents the total sales of firms, domestic and international.

Note: * 'Europe' refers to firms from France, Germany, Italy, Spain and the UK.
Source: Compiled from *ENR* (2019).

Figure 12.3 *Revenues of construction contracting firms, 2005–2018*

Company financial data is collated for firms that are required to submit financial information to central authorities. The FAME database (Forecasting Analysis and Modelling Environment) published by Bureau van Dijk[10] provides financial data for UK companies using Companies House returns. The FAME data has been used to examine the use of trade credit by construction companies (Ive and Murray, 2013), and Murray and Kulakov (2019) have used it to investigate the make-or-buy decisions of contractors. Bureau van Dijk is also the publisher of international company financial data, including Amadeus (Europe), Oriana (Asia Pacific), and Osiris, a financial performance database for globally listed companies. However, it should be noted that as these platforms span regulatory borders, to make accounts comparable, concepts are aggregated to make them match, and as a result some detail, and to some extent accuracy, is lost.

CONSTRUCTION STOCK

This chapter is substantially about the production and maintenance of construction works but it is also important to recognise the stock of construction works. Stock data in value and volume terms exists, and both are now briefly described.

Capital stock data represents the value of capital assets in a country each year; it is a net value: existing stock plus additions to stock less losses from stock. Although broadly half of capital investment (GFCF) is construction works, construction nowadays typically represents 60–80 per cent of total capital stock. That is because construction is generally longer-lasting

Table 12.4 Investment: stock ratios for selected countries, 2011 (US$ billion based on PPPs)

Country	Construction investment	Total capital stock	Investment/ stock ratio	Country	Construction investment	Total capital stock	Investment/ stock ratio
High-income				*Upper-middle-income*			
USA	1295	51 847.91	2.50	Brazil	458.22	12 414.24	3.69
Japan	432.53	22 118.35	1.96	Russia	282.83	13 132.20	2.15
Germany	342.56	14 784.84	2.32	Mexico	329.68	6 840.12	4.82
France	338.44	11 692.67	2.89	Iran	234.16	4 508.05	5.19
UK	231.45	9 601.11	2.41	Turkey	171.27	4488.26	3.82
Lower-middle-income				*Low-income*			
China	6 230.30	52 581.84	11.85	Bangladesh	150.24	1 219.32	12.32
India	1 627.20	20 351.85	8.00	Ethiopia	21.22	233.28	9.10
Indonesia	1 001.74	11 143.64	8.99	Congo, DR	7.92	116.33	6.81
Pakistan	73.50	1 479.87	4.97	Myanmar	38.04	238.65	15.94
Nigeria	30.00	1 640.05	1.83	Tanzania	33.20	392.75	8.45

Sources: Construction data from ICP (2011); capital stock data from Penn World Tables version 9.0, Feenstra et al. (2015).

than other types of capital investment (for example, machinery and equipment and information technology software). Pearce (2003) presents a long-term profile of capital stock in the UK; construction represents 89 per cent of the total in 1760 and 74 per cent in 1980.

Capital stock data for construction broken down into residential and non-residential and by type of purchaser is available in the EU KLEMS database, but only for EU countries and a few other high- income countries. A single figure for all capital stock is available for virtually all countries in the UN AMA database. A number of recent publications analyse construction capital stock data (Ruddock and Ruddock, 2019; Lopes et al., 2019). Figures for construction investment, total capital stock ratios for some higher- and lower-income countries in 2011, are set out in Table 12.4. The value data is in USD billions based on PPPs.

The table indicates that the investment/stock ratio increases steadily from high-income countries to low-income countries. This is because existing stock in the lowest-income countries is relatively small and current construction investment is relatively high as a proportion of that stock. It should be noted that these figures represent new capital construction to total stock of all types, so the construction ratios will be lower as construction will represent, say, 60 to 80 per cent of the total stock.

The quantity and nature of buildings and works in a country is an indication of its stage of development and a predictor of its future patterns of demand. National data on building stock varies from country to country and is rarely directly comparable. The European Union Building Stock Observatory (2020) is a rare example of a publicly available repository of quantitative and qualitative data on the building stock of a group of countries. The primary rationale for the Observatory is energy use in buildings, but its potential uses are wider than that. Commercial sources of building stock information include Guidehouse Insight's Global Building Stock Database[11] which provides access to data on building floor space for six regions and over 200 countries; it also provides forecasts of future floor space production.

In many building stock databases it is not clear whether floor areas cited are built area, lettable area, useable area, treated area or something else, and terms have different meanings in different places; and whether measurements are within or outside external walls, include balconies, voids, stairwells, common areas or car parks. There are international conventions on measurement for different building types such as those published by the International Property Measurement Standards Coalition (IPMSC, 2014, 2016, 2018, 2019) but these are by no means universally observed and it is necessary for users of floor area data to check carefully what rules have been followed.

Housing data is an important subset of construction-related data and a topic of interest to a wider group of researchers than construction researchers; it deserves a chapter of its own. A major attraction of housing, in terms of data, is that housing units can be counted; a major difficulty, however, is that these units are so diverse, within countries but particularly between countries. Comparing physical measurements, m^2 floor area, for example, raises two main questions: are the rules of measurement the same? and are the characteristics of housing (floor area, features, quality, and so on) the same? There are international efforts to standardise definitions and descriptions such as floor area measurement (IPMSC, 2016), but the reality is that local conventions often prevail and are likely to do so for some time.

Moreover, the nature of housing varies across different countries. The average floor area of a new housing unit in the US in 2017 was $202m^2$, in Australia $186m^2$ and in the UK, $90m^2$ (CommSec, 2018); it is unlikely that these areas are all on the same basis, but these are large differences regardless. There are also national differences in built form, construction, number of rooms per unit, thermal insulation standards, and so on. The number of houses produced each year also varies from country to country, depending on population and household growth, housing longevity and replacement rates, cultural norms and the prevailing national policy agenda. Japan and the UK have similar levels of GDP per capita and Japan's population is declining while the UK's is increasing. However, Japan builds in the order of 1 million housing units each year for a population of 128 million (781 houses per 100 000 people) and has done so for many years; UK housing newbuild has averaged around 200 000 annually in recent years for a population of 67 million (299 per 100 000). The difference in these house-building rates is marked.

CONCLUDING REMARKS

This chapter largely describes the sources and content of international data on construction. This final section of the chapter provides a brief commentary on the state of international construction data, some implications of that for the construction economics research community, and recommendations for future work.

Construction is one of the largest components of national economies but it is dangerous to treat it as monolithic or homogenous. There is a wide range of construction firms and activities, and analyses and prescriptions that are appropriate for global firms and megaprojects are often not appropriate for small firms and local projects. However, the variety of construction is not reflected in most international databases where output, employment, resource use and other variables are often represented as a single aggregate value.

Furthermore, not only is construction itself an aggregate in many international databases, but also in some it is aggregated with other sectors of the economy. For example, in the World Bank World Development Indicators, construction is aggregated with manufacturing, and in the United Nations Analysis of Main Aggregates, gross fixed capital formation includes both machinery and equipment and construction. Users must bear in mind this issue of aggregation and the many features of international construction data discussed in this chapter when working with international databases.

Much official data on construction activity has shortcomings, although that criticism is not unique to construction. General economic data is also unreliable, if only because of shortcomings in the construction data. Economic – and construction – data in many developing countries is particularly unreliable. In construction economics studies and reports there is often insufficient comment on the error bands around data; it is unrealistic to cite values for GDP or components of GDP to the precision that it is typically expressed without acknowledging the uncertainty that exists around these numbers.

The starting point of any national analysis of construction should be country-specific data from national statistical agencies, but cross-country analyses, whether based on national or international data, raise issues of consistency and comparability. Models and theories of construction economics should be subject to rigorous scrutiny and, where they are based on data, any shortcomings in that data should be acknowledged.

To address some of the problems of data quality, wherever possible, it is prudent to investigate construction data using different sources and different concepts. Growth rates of output or employment can be checked against general population or economic growth rates. Trends in output per capita can be checked against employment or cement consumption. And international comparison data as indicated in the tables and diagrams in this chapter can help to test the plausibility of data.

The fact that construction is an unusual economic sector needs to be recognised. The development of construction data calls for statisticians who understand construction, supported by construction researchers who understand statistics. While it is difficult, in the short term, to standardise the coverage of construction in official statistics, it should be possible to improve the transparency of what is included in and excluded from construction data, and how value data is converted from nominal to real prices. Wherever possible, the construction economics research community should foster relationships with national and international statistical agencies to develop and improve construction statistics.

The focus of many international data-based construction exercises is construction volumes rather than construction values. Thus, the quality of spatial and temporal deflators is critical. Development of deflators that use similar data sources and are broadly consistent internationally is a key area for statistical agency and research community collaboration. A key element of improved deflators is consideration of quality differences. Construction deflators should be matched to measures of construction output. For example, PPPs are spatial deflators of construction expenditure and are not necessarily appropriate as deflators of construction value-added.

An institutional focus for construction data is desirable, and that should involve public, industrial, academic and other organisations and individuals interested in, and knowledgeable about, construction statistics. This can be a largely virtual institution but it needs leadership, a minimal physical base, some administrative support and funding. It is not practical to set out a fixed plan here, but whatever is done, it should take in existing organisations and probably be based in one of these. A construction data resource centre should be considered, linked to the foregoing. This could maintain records of knowledgeable and interested individuals and institutions, lists of available databases and publications with notes on their contents. It could also have a public face via a website, blogs, newsletters, and the like.

There is a need to explain construction statistics to different groups, including international agencies, governments, industry and the press, so that the sector, its activities and its performance are well presented and not misrepresented. Education for construction professionals should include explanations of construction data and sources and statistical concepts and methods which are relevant for the analysis and application of the data. Research on related topics should be encouraged.

This chapter has identified a range of sources and types of data, some of which have not been exploited as much as they could have been by the construction economics research community. Relatively little use has been made by construction researchers of some of the databases presented here, but a number provide interesting long-time series. Examples include the ICP Results, the GGDC and EU KLEMS databases, and capital stock data from different sources.

Although construction data is often used by construction researchers, there is relatively little direct work on its scope and quality, or the definitions and terminology used to describe it. That should be encouraged. New types and sources of data on construction are emerging and should be investigated. Important public policy-related topics include investigations of the role of construction in the economy, the relationship between construction activity and national development, and construction productivity measurement. These are not new, but it is important that the research that has been undertaken is kept under review in the light of data developments and improvements.

NOTES

1. https://ec.europa.eu/eurostat/web/purchasing-power-parities/data/database.
2. https://data.oecd.org/conversion/purchasing-power-parities-ppp.htm.
3. https://www.euroconstruct.org/.
4. http://www.asiaconst.com/.
5. https://www.rug.nl/ggdc/productivity/10-sector/?lang=en.
6. http://constructioncosts.eu/partner/rider-levett-bucknall/.
7. https://www.enr.com/.
8. https://www.building.co.uk/.
9. https://www.khl.com/.
10. https://www.bvdinfo.com/.
11. https://guidehouseinsights.com/subscription-services/building-stock-database.

REFERENCES

Abergel, T., Dean, B., and Dulac, J. (2017). *Towards a Zero-Emission, Efficient, and Resilient Buildings and Construction Sector: Global Status Report 2017* (Vol. 22). Paris: UN Environment and International Energy Agency.

ADB (2018). *Economic Indicators for Eastern Asia: Input–Output Tables*. http://dx.doi.org/10.22617/TCS189778-2.

AfDB (2020). Socio Economic Database, 1960–2021. African Development Bank Group. https://dataportal.opendataforafrica.org/nbyenxf/afdb-socio-economic-database-1960-2021.

Andrews, J., Hatchett, M., Hillebrandt, P., Jaffar, J., Kaplinski, S., and Oorthuys, H. (1972). Construction and development: a framework for research and action. Discussion paper for the International Bank for Reconstruction and Development (World Bank). London: Building Economics Research Unit (BERU), University College London.

Arcadis (2020). *Rethinking Resilience, International Construction Costs 2020*. https://www.arcadis.com/en/global/our-perspectives/international-construction-costs-2020/.
AsiaConstruct (n.d.). http://www.asiaconst.com/.
Best, R. (2008). The development and testing of a purchasing power parity method for comparing construction costs internationally. PhD thesis, University of Techology Sydney, Sydney. http://hdl.handle.net/10453/20285.
Best, R., and Meikle, J. (2015a). The international comparison program and purchasing power parities for construction. In R. Best and J. Meikle (eds), *Measuring Construction: Prices, Output and Productivity* (pp. 45–58). London: Routledge.
Best, R., and Meikle, J. (2015b). *Measuring Construction: Prices, Output and Productivity*. London: Routledge.
Best, R., and Meikle, J. (2019). *Accounting for Construction: Frameworks, Productivity, Cost and Performance*. New York: Routledge.
Blades, D.W. (1975). *Non-Monetary (Subsistence) Activities in the National Accounts of Developing Countries*. Paris: OECD.
Blades, D., and Roberts, D. (2002). *Measuring the Non-Observed Economy* (Vol. 5). Paris: OECD.
Bon, R., and Pietroforte, R. (1990). Historical comparison of construction sectors in the United States, Japan, Italy and Finland using input–output tables. *Construction Management and Economics*, 8(3), 233–247.
Briscoe, G. (2006). How useful and reliable are construction statistics? *Building Research and Information*, 34(3), 220–229.
Cannon, J. (1994). Lies and construction statistics. *Construction Management and Economics*, 12(4), 307–313.
CEEC (2019). Office Cost Model. European Council of Construction Economists. https://www.ceecorg.eu/downloads/.
Chan, T.K. (2019). Cost Ratios and Technology Choice. In R. Best and J. Meikle (eds), *Accounting for Construction: Frameworks, Productivity, Cost and Performance* (pp. 129–153). New York: Routledge.
Chancellor, W., Abbott, M., and Carson, C. (2019). Measuring construction industry activity and productivity: The impact of the shadow economy. In R. Best and J. Meikle (eds), *Accounting for Construction: Frameworks, Productivity, Cost and Performance* (pp. 74–86). New York: Routledge.
CommSec (2018). Australian home size hits 22 year low. CommSec Economic Insights. https://www.commsec.com.au/content/dam/EN/ResearchNews/2018Reports/November/ECO_Insights_191118_CommSec-Home-Size.pdf.
Davis Langdon & Everest (2000). *Spon's European Construction Costs Handbook*. London: E. & F.N. Spon.
de Valence, G., and Meikle, J. (2019). Construction output as gross fixed capital formation. In S. Gruneberg (ed.), *Global Construction Data* (pp. 18–43). New York: Routledge.
EC, IMF, OECD, UN, and WB (2009). *System of National Accounts 2008*. New York: United Nations.
ENR (2019). Top global contractors. Engineering News Record (ENR) https://www.enr.com/toplists#Top%20Global%20Contractors.
EU Building Stock Observatory (2020). EU Buildings Database. https://ec.europa.eu/energy/en/eu-buildings-database.
EU KLEMS (2019). Growth and productivity accounts. Vienna Institute for International Economic Studies (wiiw). https://euklems.eu/download/.
EuroConstruct (n.d.). https://www.euroconstruct.org/.
Eurostat (1996). *Methodological Aspects of Construction Price Indices*. https://op.europa.eu/en/publication-detail/-/publication/6e9c8c74-95eb-42b0-bd3a-b0367b15e348/language-en.
Eurostat (n.d.). https://ec.europa.eu/eurostat/web/purchasing-power-parities/data/database.
Davis Langdon & Everest (1988). *Spon's International Construction Cost Handbook*. London: E. & F.N. Spon.
Feenstra, R.C., Inklaar, R., and Timmer, M.P. (2015). The next generation of the Penn World Table. *American Economic Review*, 105(10), 3150–3182. www.ggdc.net/pwt.
Franklin & Andrews (2000). *Spon's Latin America Construction Cost Handbook*. Abingdon: E. & F.N. Spon.

Franklin & Andrews (2005a). *Spon's Middle East Construction Costs Handbook*. Abingdon: Taylor & Francis.
Franklin & Andrews (2005b). *Spons African Construction Costs Handbook*. Abingdon: Taylor & Francis.
GGDC (2017). GGDC 10 Sector Database. Groningen University. https://www.rug.nl/ggdc/productivity/10-sector/.
Girardi, D., and Mura, A. (2014). The construction–development curve: evidence from a new international dataset. *IUP Journal of Applied Economics*, 13(3), 7–26.
Global Cement Report (2019). https://www.cemnet.com/Publications/Item/182291/the-global-cement-report-13th-edition.html.
Green, B. (2019). The challenges of measuring British construction output. In R. Best and J. Meikle (eds), *Accounting for Construction: Frameworks, Productivity, Cost and Performance* (pp. 46–73). New York: Routledge.
Groak, S. (1992). *The Idea of Building: Thought and Action in the Design and Production of Buildings*. London: E. & F.N. Spon.
ICP (2011). World Bank International Comparison Program 2017. https://databank.worldbank.org/source/icp-2017.
ICP (2017). World Bank International Comparison Program 2017. https://databank.worldbank.org/source/icp-2017.
IPMSC (2014). *IPMS for Office Buildings*. https://ipmsc.org/standards/office/.
IPMSC (2016). *IPMS for Residential Buildings*. https://ipmsc.org/standards/residential/.
IPMSC (2018). *IPMS for Industrial Buildings*. https://ipmsc.org/standards/industrial/.
IPMSC (2019). *IPMS for Retail Buildings*. https://ipmsc.org/standards/retail/.
Ive, G., Gruneberg, S., Meikle, J., and Crosthwaite, D. (2004). *Measuring the Competitiveness of the UK Construction Industry* (Vol. 1). London: Department of Trade and Industry (DTI).
Ive, G., and Murray, A. (2013). *Trade Credit in the UK Construction Industry: An Empirical Analysis of Construction Contractor Financial Positioning and Performance* (Vol. 118). London: Department for Business, Innovation, and Skills.
Jerven, M. (2013). *Poor Numbers: How We are Misled by African Development Statistics and What to Do About It*. London: Cornell University Press.
K'akumu, O.A. (2007). Construction statistics review for Kenya. *Construction Management and Economics*, 25(3), 315–326.
Kamarck, A.M. (1983). *Economics and the Real World*. Philadelphia, PA: University of Pennsylvania Press.
Langdon & Seah (2015). *Spon's Asia-Pacific Construction Costs Handbook* (5th edn). Boca Raton, FL: CRC Press.
Lopes, J.P., Oliveira, R.A., and Abreu, M.I. (2019). Estimating the built environment stock in Cape Verde. *Engineering, Construction and Architectural Management*, 26(5), 814–826.
Low, S.P., and Tan, O.B. (1993). *The Global Cement Industry*. Singapore: Singapore University Press, National University of Singapore.
Meikle, J. (2011). Measuring construction activity in the national accounts of African countries. ICP working paper, unpublished.
Meikle, J. (2019). A review of the 2011 construction survey and results from the World Bank International Comparison Program. In R. Best and J. Meikle (eds), *Accounting for Construction: Frameworks, Productivity, Cost and Performance* (pp. 155–180). New York: Routledge.
Meikle, J.L., and Grilli, M. (1999, 27–29 October). Measuring European construction output: problems and possible solutions. Paper presented at the Second International Conference on Construction Industry Development and First Conference of TG29 on Construction in Developing Countries, Singapore.
Murray, A., and Kulakov, A. (2019). Make/buy decisions in international construction industry firms. In *Global Construction Data* (pp. 149–166). New York: Routledge.
OECD (2007). Glossary of Statistical Terms. https://stats.oecd.org/glossary/glossary.pdf.
OECD (2018). Input–Output Tables (IOTs). OECD Stat. https://stats.oecd.org/Index.aspx?DataSetCode=IOTSI4_2018.

OECD (n.d.). Purchasing power parities (PPP). https://data.oecd.org/conversion/purchasing-power-parities-ppp.htm.
OECD and Eurostat (n.d.). Sources and methods: Construction Price Indices. OECD and Eurostat.
Ofori, G. (1990). *The Construction Industry: Aspects of its Economics and Management*. Singapore: Singapore University Press.
Pearce, D. (2003). *The Social and Economic Value of Construction: The Construction Industry's Contribution to Sustainable Development*. London: nCRISP.
Ruddock, L., and Ruddock, S. (2019). Wealth measurement and the role of built asset investment: an empirical comparison. *Engineering, Construction and Architectural Management*, 26(5), 766–778.
Schneider, F., Buehn, A., and Montenegro, C.E. (2010). Shadow economies all over the world: New estimates for 162 countries from 1999 to 2007. https://openknowledge.worldbank.org/handle/10986/3928.
Strassmann, W.P. (1970). The construction sector in economic development. *Scottish Journal of Political Economy*, 17(3), 391–409.
Strassman, W.P., and Wells, J. (1988). *The Global Construction Industry: Strategies for Entry, Growth and Survival*. London: Unwin Hyman.
Timmer, M., de Vries, G.J., and de Vries, K. (2015). Patterns of structural change in developing countries. In J. Weiss and M. Tribe (eds), *Routledge Handbook of Industry and Development* (pp. 79–97). London: Routledge.
Turner & Townsend (2019). *International Construction Market Survey 2019*. https://www.turnerandtownsend.com/en/perspectives/international-construction-market-survey-2019/.
Turin, D. (1973). *The Construction Industry: Its Economic Significance and its Role in Development*. London: Building Economics Research Unit (BERU), University College London.
UN (1998). *International Recommendations for Construction Statistics*. New York: United Nations Publications.
UN (2008). *International Standard Industrial Classification of all Economic Activities (ISIC)*. New York: United Nations Publications.
UN. (2009). *System of National Accounts 2008*. New York: United Nations.
UN (2019). National Accounts – Analysis of Main Aggregates (AMA). United Nations Statistical Division. https://unstats.un.org/unsd/snaama/index.
Wells, J. (1985). The role of construction in economic growth and development. *Habitat International*, 9(1), 55–70.
Wells, J. (2007). Informality in the construction sector in developing countries. *Construction Management and Economics*, 25(1), 87–93.
WIOD (2016). World Input–Output Database. http://wiod.org/database/wiots16.
World Bank (2020). World Development Indicators. https://data.worldbank.org/indicator/SP.URB.TOTL?view=chart.

13. Measuring and comparing construction costs in different locations: methods and data

Rick Best

INTRODUCTION

In 2012 the Business Council of Australia (BCA) published a report (BCA, 2012) that suggested that the cost of building infrastructure in Australia was significantly higher than that in the United States (US), and therefore productivity in the Australian construction industry was lower than that in the US. These claims were reported in at least two Australian national newspapers (Forrestal and Dodson, 2012; Hepworth, 2012) and mentioned in the Australian parliament. The general tenor of the commentary on the BCA report was to attack the industry, and particularly the relevant trade unions, with claims of inefficiency and poor performance.

Best (2012) showed that the report had several major flaws which influenced the results presented in the BCA report. The problems he identified related to the data used and the method of comparison, specifically the way that costs in US dollars (USD) were converted to Australian dollars (AUD) so that costs could be directly compared. In addition, he noted that higher costs do not automatically indicate lower productivity. Best showed that using data from other equally reliable sources and/or a different method for currency conversion produced very different results that showed that Australian costs might be on par with (or even considerably lower than) comparable US costs. This example illustrates several significant points:

- Governments, policy-makers and industry organisations are interested in such comparisons.
- The results of such exercises can vary markedly depending on the methodology employed.
- The complexities and potential pitfalls associated with such studies are often not well understood.

International comparisons of construction costs are notoriously difficult. The World Bank (2013, 2020), through its International Comparison Program (ICP), describes construction as 'comparison-resistant' due to the heterogeneous nature of construction activities and projects, and the difficulty of gathering suitable cost data from a large number of countries.

This chapter summarises the major issues that arise when comparative studies of construction cost[1] are carried out, with particular emphasis on the cost data used and how costs recorded in different currencies are converted to a common base that enables direct comparisons.

THE COMPARISON QUESTION

If a client wants to build a facility in another country, using money that is available in their home currency, then the question is fairly simple: if the home currency is Australian dollars (AUD) and the facility is to be built in Malaysia (for example) then the money market exchange rate can be used to convert the expected cost in Malaysian ringgit (MYR) to AUD so the client knows how many AUD they need to spend. The amount will fluctuate with movements in the exchange rate; while these rates typically do not move much in the short term, clients do need to be aware of, and allow for, potential shifts during the design and build period. If, however, the question is whether it is more, or less, expensive to build in Malaysia or Australia, then the question becomes much more complex.

Economics is largely concerned with the allocation of scarce resources and how those resources can be employed so that maximum benefit is achieved; governments want to see their countries' resources directed into activities that will produce maximum benefit for their people. For that reason, construction industries in many countries are routinely examined to see whether they are operating efficiently and maximising output. As there are no absolute measures of production efficiency, the usual approach involves benchmarking performance against competitors or similar industries. If a national government wants to gauge its construction industry's performance and/or efficiency the only option is to compare it against those of other countries. The heterogeneity of construction output (buildings of many types, sizes and levels of complexity, plus infrastructure projects of endless diversity) means that there are few options for measuring the volume of output other than to measure it in terms of money. The value of construction is inevitably measured as amounts of a country's own currency, and comparisons of value can only be made if the amounts are expressed in the same base currency; it makes no more sense to compare an amount of money in Australian dollars directly with, say, an amount in Malaysian ringgit than it does to compare an elephant with a chainsaw.

Money market exchange rates do not provide an appropriate mechanism for converting different currencies to a single base, for two reasons: one is that fluctuations in exchange rates will suggest different cost relationships at different points in time (and in some cases these can be quite dramatic[2]); the other is that exchange rates do not reflect differences in price levels in different places. If, for example, someone in the United Kingdom (UK) is considering taking a job in Canada they will want to know how the salary offered (say, CN$80 000) compares to their current salary (say, £45 000). If the Canadian salary is converted to GBP at CN$1 = £0.58, then the promised salary appears to be on a par with the person's UK salary. However, such a comparison does not account for differences in price levels between the two countries, and thus, if general living costs in the UK account for 90 per cent of the current salary (leaving £4500 of disposable income available) and similar living costs in Canada would account for only 80 per cent of the salary (leaving disposable income of CN$16 000, or £9280 at the current exchange rate) then the Canadian salary looks more attractive, as the person can live and have nearly twice as much surplus to save or to spend on discretionary purchases.

Cassel (1921) introduced the idea of purchasing power parity (PPP) to address these problems. PPP provides a conversion method based on what can be bought in different places using local currencies. A simple, and perhaps simplistic, example is the Big Mac Index. In 1986 *The Economist* first published, somewhat light-heartedly, a comparative index based on a single, manufactured commodity, the Big Mac hamburger, which has essentially the same specification in around 120 countries (Segal, 2019). As the product is basically identical in each place, its value should be the same in all places and thus its price reflects the purchasing power of the local currency in each place. If the price of a Big Mac in the US were US$3.00 and in Australia AU$4.00, then that implies an exchange rate of US$1 = AU$1.33, and that rate (sometimes called the 'real' exchange rate) has virtually no connection to current market rates (referred to as 'nominal' exchange rates).

The underlying theory is the Law of One Price (Runeson, 2015), which suggests that if an item can be bought more cheaply in country A than in country B then traders will buy in country A and sell in country B and make a profit. Over time, international trade will see the price of the item in the two countries level out as prices rise in A (due to increased demand) and fall in B (due to increased supply). In practice this rarely happens, even over long periods of time, even for goods that are easily traded (Rogoff et al., 2001). Services, such as haircuts and accountancy, are not tradeable and the same is true of most construction output (that is, built facilities). While many components of buildings are traded internationally, a majority of the cost of most buildings lies in labour and materials such as aggregate and bricks which are locally produced. Other factors such as market exchange rates, tariffs, and transport and transaction costs affect prices of traded goods, while the prices of local products are not affected by exchange rates or tariffs and are generally less affected by transport costs.

THE NATURE OF CONSTRUCTION PROJECTS AND COSTS

The other complicating factor is that many products and services, notably the non-tradeable items, are not even close to being identical in different locations. This is equally true of buildings and many services such as hairdressing. Items in two locations may be functionally similar yet be quite different in nature; a house in Norway, for example, is not the same as a house in southern Italy, yet both serve similar functions and can be classified as 'European houses'; and while having a haircut in India is generally a different experience to having one in Sydney or Dublin, it should produce a very similar result.

Construction projects, whether buildings or infrastructure such as highways and bridges, may be functionally similar but vary greatly in terms of design, materials and construction technology employed. Many characteristics of projects are location-specific, with differences in design arising from differences in regulations, climate, ground conditions (including the possibility of seismic activity), technical capacity and material availabilities. Resource mixes will also vary between countries; those with plenty of cheap labour will tend to employ more people and invest less capital, while countries with high labour costs are likely to show higher levels of investment in machines and technology and thus use less labour.

In any attempt to compare costs between countries it is necessary to identify and collect the cost of items (that is, goods and services, as well as completed projects) that are generally representative of consumption in the various countries and are comparable in nature. With construction the problem is in finding products (buildings or components of buildings) that are both similar enough for comparison yet typical enough of products in use in the locations of interest. The more locations (cities, regions or countries) that are included, the more difficult it becomes to find such products. In the case of the ICP around 200 countries are surveyed and the problems of finding comparable yet representative items, whether complete buildings, components or sub-assemblies, or even basic inputs such as cement and timber, are multiplied.

Even if the cost data that needs to be collected is known, the problem of getting reliable and comprehensive data on construction costs is not easy to solve. National statistical offices, and their capacity, vary greatly between countries as poorer countries often have more pressing concerns (such as provision of housing and clean water) than collecting and processing statistical information. Even such offices in richer countries tend not to collect fine-grained data on construction costs, but may focus more on macro-level data such as total value of residential or non-residential construction. That data varies in composition from country to country, so that even where the data is readily available it may not be comparable due to differences in what is collected and how it is aggregated and recorded (de Valence, 2019a, 2019b).

While it would always be preferable to collect prices actually paid for construction work, whether these be component prices, materials prices or whole project prices, such data is very hard to acquire. The price that a client pays to a contractor is generally made up of the cost of labour, materials, plant and equipment, and site overheads (referred to as 'preliminaries' in some countries), plus the contractor's margin which includes company overheads and profit. However, total project cost can include fees for design, legal work, project management, land purchase, demolition, development applications, environmental assessment and more, so decisions must be made about what is included and what is not. Even if an effort is made to look for the final contract value on completed projects, accurate data is seldom available and often takes a long time to appear. The other option is to use estimated project costs (if available) or use some sort of input costs. Estimates are, of course, estimates; that is, they are predictions of the cost to build (plus margins) over a period of time in the future, which could be anything from a few months to a few years. Given that construction projects routinely run over budget, sometimes in spectacular fashion, the use of estimates is far from ideal. In most cases there must be compromises, and cost data will often consist of averages, and even averages of averages.

There is a further complication: no matter what methodology is adopted there are no 'correct' answers against which results can be tested. Yet, for all the complexities, comparisons of cost and productivity will be made, and it is necessary to look for ways to achieve the most robust results possible.

SUMMARY OF EXISTING METHODS

Given the extent of variation in construction output it is not feasible to try to gather costs for all sorts of construction. Instead, suitable proxies which represent all construction, or sectors of the industry, have to be found in the way that a basket of items (goods and services) is used as a proxy for all consumption in the production of a consumer price index (CPI). In the Big Mac Index (BMI) mentioned previously the hamburger is a proxy for consumption, although the purpose of the BMI was only to assess whether different currencies were under- or overvalued against some base currency (typically USD). Research suggests that the BMI worked reasonably well for that purpose (Ong, 2002; Pakko and Pollard, 2003), but Langston and Best (2005) tested a variety of conversion factors for construction and concluded that while the BMI seemed to work well in some locations it did not in others, and therefore should be used with considerable caution in that context.

Comparisons of construction cost go back at least to 1949 when the Building Industry Productivity Team from the UK visited the US to look at how the UK industry might learn from its US counterpart and thus improve performance and lower costs (AACP, 1950). The team compared construction costs by comparing the cost of buildings of similar size used for similar purposes, and by comparing buildings built to the same design and specification constructed under similar circumstances. Costs were in local currencies and were converted to a common base using the exchange rate current at the time of their visit (US$4.00 = £1.00). When the results were published in 1950 the exchange rate was only US$2.80 = £1.00, and

Table 13.1 *Cost/m² for selected building types, Sydney, Australia (in AUD)*

Project type	Australian Institute of Quantity Surveyors (AIQS)	Turner & Townsend (T&T)	Rider Levett Bucknall (RLB)
3 star hotel	4324	3000	3350–4200
Shopping mall	1923	2950	2100–4400
Basic warehouse/factory unit	774–968	880	770–1240
Regional hospital	5594	4100	3750–5000
Prestige CBD office	5800	5850	3750–5500

Sources: AIQS (2019), Turner & Townsend (2019) and RLB (2019).

had that exchange rate been used it would have made a significant difference to the team's findings. This demonstrates the need for other conversion factors in these circumstances, as fluctuations in exchange rates do not reflect changes in relative costs to build.

Cost comparisons may be based on output, intermediate or input prices (costs). Input costs are typically prices paid to suppliers for materials, labour and equipment; output costs are prices paid by clients to contractors for completed works; while intermediate prices are prices for components or assemblies that are incorporated into structures.

Using Output Prices

Output prices may be prices paid for specific projects (the standard projects approach) or average prices for different building types (for example, three-star hotels, simple warehouses) based on cost per unit area (for example $/m²). As it is virtually impossible to find identical buildings in each location that can be compared, in the standard projects approach the costs of one or more hypothetical building project are priced by experts using detailed bills of quantities (BQs), in which each project is broken down into a number of detailed work items that are individually priced. Project costs are aggregated, generally with the various projects weighted according to the volume of each type of building constructed in each location. This method is currently used by the Organisation for Economic Co-operation and Development (OECD)–Eurostat programme[3] to produce purchasing power parities (PPPs) for construction.

Prices are best estimates of actual prices paid and are therefore forecasts of the cost to build and not actual out-turn prices. Projects should be both reasonably representative and comparable across all locations and can be used to collect data for different construction types and 'all construction' by using a range of project/building types. This approach does not work well with construction other than newbuild, and can be costly to implement as experts have to be paid to price and this mostly means that projects are priced by only one respondent in each location.

Data on cost per unit area for various building types is available for many countries, but data from different sources often displays significant variances, as illustrated in Table 13.1.

There are several possible explanations for the variances in the costs shown above. It is likely that a major factor is that the building description seldom gives much detail of what is included and what is not. It is also difficult to determine whether the costs given are for comparable projects even where descriptions are similar. With regard to the data for 'shopping malls', for example, the Turner & Townsend typical project is described as 'large shopping centre including mall', while the AIQS description is 'shopping: regional complex up to three storeys including air conditioning, malls, public toilets and rest rooms, standard finishes, escalators and lifts, shopfronts excluding shop fitouts and carparking', while RLB simply says 'mall'. The three are superficially much the same but there is no way of knowing how similar they are in detail. International comparisons based on this sort of data must be treated with caution as not only may buildings of similar description in each country be quite different at a detailed level, but there is also ample scope for large variations in how projects are measured and what has been included in published construction costs.

The BCA example discussed early in this chapter provides a very good example of the potential for misleading results to be announced and treated with more respect than they deserve; one finding in that study was reported as follows in *The Australian* newspaper (Hepworth, 2012): '[in Australia] compared with the US, airports are 90 per cent more expensive to deliver'.

This is quite misleading, as the cost data used was for airport terminals, not airports in general, and as any international traveller can attest, airport terminals around the world vary considerably in size and design, and they are not built in great numbers year on year so there are relatively few examples to compare. Terminal buildings actually provide an excellent example of how much buildings of the same general type can vary, even within one country, as regional air terminals are very different to the massive buildings, such as Terminal 3 at Beijing Capital International airport, that now serve large centres with high volumes of international traffic. In addition, large terminals in particular tend to enclose very large volumes of space with relatively few upper floors, making cost measurement in $/unit area rather unhelpful; it may even be more appropriate to revert to the now seldom-used measure of $/cubic unit for such buildings.

The data used in the BCA study came from an annual international construction survey carried out by a firm of international construction cost consultants (Turner & Townsend, 2012). A number of large firms apart from Turner & Townsend (such as Faithful+Gould) with offices in various countries that used to publish construction cost data from around the world are no longer doing so, perhaps due to some larger quantity surveying (QS) firms being subsumed into even larger multidisciplinary firms with the result that traditional QS services have been replaced by broader project management services. Collecting and collating cost data from many offices diverts human resources from other tasks and may simply not be considered by the firms to be worth the time and effort required.

Variances in costs for various building types may occur due to differences in specification for functionally similar buildings in different places, as well as basic differences in the way floor areas (see Table 13.2) are measured, and what is included or excluded from the project costs that are reported. While it would be beneficial to have international cost databases that are consistent in their content, the reality is that there is little consistency in approach between countries. Even within single countries there have often been changes in the data that national statistical offices collect and how results are reported (for example, Green, 2019). This is well illustrated in a study conducted by the Council of European Construction Economists reported by Wright and Stoy (2008). Table 13.2 shows the differences in gross floor area (GFA) for an identical building as measured using local measurement conventions in a number of European countries.

The variances in Table 13.2 are likely attributable to differences in measurement of areas such as lift wells and other void spaces, thickness of walls and similar. The large difference in areas for Spain and Denmark occur because the convention in those locations was not to include basement areas in GFA calculations. More recent developments in international

Table 13.2 *Gross floor area for an identical building measured in various European countries using local measurement conventions/rules (UK base)*

Country	Gross floor area (m^2)	Variance (%)
UK	2585	-
Switzerland	2875	+11
Holland	3007	+16
France	3412	+32
Finland	2758	+7
Denmark/Spain	1800	-30

Source: Wright and Stoy (2008).

measurements standards (ICMS, 2019; IPMS, 2020) are an attempt to address these sorts of differences, but adoption of uniform measurement practices may still be some way off, even in Europe, while worldwide alignment of measurement conventions is not likely in the foreseeable future.

If the aim is to assess the relative cost of building between countries, it can be argued that the use of average costs for certain building types (such as schools or hospitals) does not provide credible results even if consistent and comparable costs are available, as the buildings used, while functionally similar, are often far from identical (for an early example, see Sebestyen, 1978). Standard projects, while generally not exactly what is built in each location, are arguably more directly comparable, and thus the cost to build such projects in different locations should provide better comparisons. Some past studies have included comparisons not only of identical buildings but also of various designs adjusted to reflect typical design and specifications in the different locations (for example, AAPC, 1950; Lynton, 1993; DLC, 1999). This sort of triangulation is a form of sensitivity analysis that provides a better understanding of the relative effects of variations in design and specification on comparative costs.

While not strictly a cost comparison, the study done by Langston and Best (2001) comparing the contractor performance on high-rise office projects in various countries uses actual (out-turn) cost to complete as one of the parameters in the performance index that they devised. Costs were brought to a common base (USD) using nominal exchange rates, ICP PPPs and the Big Mac Index, but each produced rather different results; this further illustrates the need for dependable currency conversion factors.

Using Intermediate Costs

Intermediate prices are prices for items of work, or assemblies or composite elements such as concrete piers, areas of finishes, concrete footings or an electrical outlet. These rates are available from price books and tenders, for example, but specific rates depend on the pricing approach of particular contractors and the circumstances of particular projects. It is preferable, when prices from successful tenders are used, that all prices come from the same tender; even then there is no guarantee that overheads and profit have been evenly distributed across the items in the tender, as tenderers may weight some items to enhance their cashflow. Price book prices do not reflect the specific context of projects or the methods used by different contractors.

Prices combine the cost of all inputs (materials, labour and so on) and may be aggregated into a cost for a basket of construction components (BOCC); such a basket was used in the 2005 round of the ICP (Walsh and Sawhney, 2004); however, countries were asked to price tightly specified identical components rather than to extract prices from published prices or tenders, and thus it became a combination of intermediate and input costs. In any case it is necessary to find items of output that are sufficiently similar in all locations that are both representative and comparable. The 2005 ICP approach increased comparability of items, while careful selection of items, based on field work that identified common construction components in a number of countries, was intended to ensure the representativeness of the items selected for the basket.

Furthermore, the intention was to weight the different components so that the relative impact of each component on the total basket cost reflected its contribution to project costs. In the 2005 ICP round this proved to be problematic, and that concern, along with other perceived shortcomings, led to a review of the process and its eventual replacement with a method based on input costs. Subsequent work by Walsh (2012) suggested that variations in the weights had negligible impact on final results.

Using Input Costs

In the input costs approach, a basket of resources is priced. Inputs include construction materials (possibly including some components, such as windows, that are usually manufactured offsite), various types or classes of labour, and some items of construction plant and equipment. Basket costs are adjusted to include estimates of site-specific and general overheads and builders' profit so that basket costs are as close as possible to actual prices paid by clients.

The items in a basket may be unweighted, have fixed weights, or be weighted differently in different locations according to the relative level of use of each item in each place. If an unweighted basket is used, then price relatives have to be manipulated statistically to produce PPPs; this method is currently in use in the ICP. Fixed- and variable-weight baskets can be used to produce costs for whole baskets, and these can be compared directly. Best (2008) used a fixed-weight basket based on a suburban hotel, a building type that was assumed to be both reasonably representative and comparable in the six target cities in that study (the cities were Sydney, Brisbane and Melbourne in Australia, Auckland in New Zealand, Singapore, and Phoenix in the US). Deriving variable weights for many locations is a complex exercise and, as was shown in the 2005 ICP round (based on the BOCC), proved to be largely unworkable in practice. Further research is required in order to know whether fixed, variable or unweighted baskets produce significantly different and/or better results.

Comparable input prices should be relatively straightforward to collect, although there are always local variants that must be accounted for and, in practice, there are often differences (sometimes large differences, in the order of 200 to 300 per cent) between prices supplied by different respondents. Prices are obtained for standard measured units (for example, tonnes, m^2, m^3, days) of readily available inputs (for example, concrete, timber, steel, skilled labour). As input prices are usually prices paid by contractors to their suppliers, they are not end-user (out-turn) prices and margins should be added. All such prices are context-dependent, that is, they depend, at least to some extent, on the quantity used, the scale of the project, how and where it is used, who is doing the work, and who is the customer for the work. Thus, prices collected have to be some sort of average that accounts for these variations, and list prices published by suppliers will very often be considerably higher than actual prices paid when contractors' discounts are applied.

The approach has some important features: it is similar to other price index methodologies, prices are at current levels, and different baskets can be used to gather prices for different types of construction (the important difference being in the inputs selected and their relative weights[4]). Basket prices must be adjusted to account for margins (overheads and profit) to bring them closer to prices paid by clients to contractors (out-turn prices). Items in the basket have to be as comparable and representative as possible of construction in different locations, and as straightforward to price as possible. While it may appear to be relatively easy to implement, in practice collecting reliable, comparable cost data is difficult, and collecting more than one set of costs from any one country can be expensive.

The World Bank adopted this approach for construction in the 2011 and 2017 rounds of the ICP; Best and Meikle (2015) describe the development of the ICP basket in detail, based on their work for the World Bank in the period 2009 to 2012. However, there are variants within the ICP: the OECD and Eurostat still use BQs for standard projects, while the Commonwealth of Independent States (CIS) collects input prices for around 100 items (materials, labour, energy) and feeds relevant prices from this price pool into a set of cost models for different project types.

The equation for the RTM (ICP, 2005, p. 203) is:

$$P_{mk} = [\left(1+\frac{a}{100}\right)\sum_{j=1}^{m} PM_j QM_j + \sum_{k=1}^{l} PE_k QE_k + W \times F_k + \left(1+\frac{S+A+B+P}{100}\right)] \times I \times VAT$$

where:

P_{mk} is the predicted price that a purchaser would pay for a project of type k that includes m types of materials from the list of materials that have been specifically priced and a is the estimated proportion of other materials that are not on the list of materials that are specifically priced.

PM and QM are the prices and quantities of the m types of materials from the list of materials that have been specifically priced.

PE and QE are the prices and quantities of the l types of energy that have been specifically priced.

F_k the estimated number of work days required to build a project of type k.

W is the daily wage rate.

S is the percentage of wage costs paid by employers for their employees' social security.

A is the rate of consumption of fixed capital (basically depreciation) expressed as a percentage of wage costs.

B is other costs expressed as a percentage of wage costs.

P is operating surplus expressed as a percentage of wage costs.

I is the cost of 'engineering services'; this includes design fees, supervision, technical advice and similar costs expressed as a percentage of all other costs.

VAT is the rate of value added tax.

The ICP (2005) notes some concerns with the method, including the handling of the operating surplus and consumption of fixed capital, which is included in the model because it is usual in CIS countries for contractors to own their own equipment rather than hiring it, as is common practice in many other countries. Best and Meikle (2015) identified several other issues, including the potential errors in estimates of construction time; and the question of how, or whether, general and site overheads are accounted for, which could represent 20 per cent or more of the total cost of a project.

Most of the remaining 160 countries are now asked to price the new basket of materials, labour and equipment (see World Bank, 2015, for details of the process and the survey forms). Data is collected via a survey of participating countries, with in-country construction cost experts (for example, quantity surveyors) generally providing the input prices. Suitable people and/or agencies must be identified, and individuals trained so they properly understand the intent and processes of the survey. They also must be able to respond to queries during data validation so that the data set produced is as robust and reliable as possible. Meikle (2019) describes and analyses the process and the results of the 2011 round. The results of the 2017 round were published in May 2020 (World Bank, 2020).

Other Examples of Cost Studies Using Input Costs

Some consultancies with offices in various countries (such as Turner & Townsend) still collect and publish cost data, including supply costs for some basic inputs, such as structural steel and concrete, as well as some labour and plant costs, and unit rates for work items such as excavation, painting and tiling, from their international offices, but such data is generally limited in scope and is produced largely for marketing purposes. Gathering and collating the data is a cost to the firm, and even when the data is being supplied by a firm's own offices, getting data often requires repeated requests and a good deal of cajoling by whoever is managing the data collection exercise. Emmett and Langston (2019) discuss the evolution and implementation of Turner & Townsend's annual *International Construction Market Survey* and the various difficulties that are regularly encountered in the process.

Many countries have commercially published construction price books; for example, Spon (UK), RSMeans (US) and Rawlinsons (Australia). Such books often include some international cost data but once again it is generally limited, as the focus is on domestic costs/prices for use by estimators and quantity surveyors in their local markets.

There have been various attempts to implement a basket of goods approach to cost comparisons, some of which predate the current ICP basket method. In 2003, Davis Langdon Consultancy in London carried out a limited test of such a method (DLC, 2003) using a basket comprising 35 materials and four classes of labour. Items in the basket were weighted using national input/output (I/O) tables for a handful of countries. At the time, I/O data was not readily available in many countries and that approach was not pursued. However, this weighting method is worthy of further exploration now that more I/O data is routinely available.

In 2007, Best tested a small basket as a pilot for a larger study (Best, 2008) that used a fixed-weight basket of inputs to produce construction-specific PPPs for six cities: three in Australia plus Phoenix (US), Auckland (New Zealand) and Singapore. These baskets were weighted based on analysis of a full priced BQ from a successful tender for a three-star hotel. Cost significant items were selected based on their value in the BQ. The basket was dubbed a BLOC, or 'basket of locally obtained commodities'. Langston (2015a, 2015b) expanded the idea and refined it into the citiBLOC. He used this to compare construction costs for four countries, and for five major cities within each country. He tested the citiBLOC against several other methods, including the Big Mac Index, and concluded that the citiBLOC produced the most consistent outcomes.

Best and Langston both used their baskets to express project costs as a number of BLOCs, thus creating a dummy 'currency' specific to construction. If a typical project in one location (Seattle, for example) cost 1 million BLOCs and the same project in another location (Sydney, for example) cost 1.1 million BLOCs, then this relationship can be used to convert construction costs to a single real currency. For example, if a simple warehouse in Seattle costs US$2 million and a similar building in Sydney costs AU$2.1milllion, then using the 1.0:1.1 ratio, the Australian cost can be expressed in USD: 2.1 x 1.1 = US$2.31 (million). On that basis it can be said that it is around 12 per cent more expensive to build a warehouse in Sydney than it is in Seattle, based on the cost of resources used to build that might otherwise be used to purchase other things. This is an example of cost conversion based on purchasing power rather than market (nominal) exchange rates. Turner & Townsend use data from their annual survey (Turner & Townsend, 2019) to compute construction PPPs for 63 cities across 39 countries using a similar method.

More recently Langston (AIIB, 2019) followed this approach in producing a specific basket for road construction. A basket of items relevant to that type of engineering construction (the roadBLOC) was created, and then priced in nine cities stretching from Sydney to Moscow. Given the nature of engineering construction, the roadBLOC was weighted to reflect a resource mix of 30 per cent labour, 40 per cent materials and 30 per cent plant and equipment. The cost of a section of four-lane arterial road was expressed as a number of roadBLOCs in each location; again this produced costs that could be directly compared without involving nominal exchange rates. Whether the roadway could be used as a proxy for all engineering construction is unknown, and further testing is required, perhaps involving BLOCs for other types of infrastructure projects such as power stations, dams or bridges, in order to test its potential as a proxy for all engineering construction.

Other Building Activity

There are two areas of activity that may add additional layers of complexity to cost comparisons: repair and maintenance of existing structures, and unrecorded or informal construction.

In many countries a significant proportion of total activity is devoted to work on existing buildings and other structures. For example, in Italy the expenditure on work to the country's building stock includes a great number of structures, including a very large number of houses, that are hundreds, and in some notable cases thousands of years old. Even in countries that do not have a similarly large stock of ageing permanent buildings, the value of repairs and maintenance should not be ignored in measures of construction expenditure, but it is often not recorded, for several reasons. One reason is that much of it, at a domestic level, is done as 'do-it-yourself' work by householders; other work is done by very small firms, often with one or two employees, whose income may or may not be fully and accurately reported. The Australian Bureau of Statistics estimate that in 2011–12 repairs and maintenance to buildings represented a little over 10 per cent of total income from 'trade services, building and construction', under the headings 'Houses, Other residential building and Non-residential building' (ABS, 2013). The heterogeneity of new buildings has been noted, but the diversity in the nature of repair and maintenance work is even greater, so much so that it is probably best left out of performance and/or cost comparisons. However, it is a complicating factor in national accounts as the work has value that may be unaccounted for in total construction expenditure.

As a result, best estimates of the value of work in this sector are added to national accounts, but how accurate these estimates are is hard to assess.

Informal construction includes a lot of repair and maintenance work, but it also includes new buildings that are constructed without any sort of formal contract, in poorer countries often by owner-builders, and is thus missed in national accounts. While there is a construction component in the shadow economy in most countries (Chancellor et al., 2019) it is not of great importance in industry comparisons. The efficiency of such work does not have much bearing on how efficient the industry is, nor on how efficiently resources are being used, but again, if the value of such work is not included in national accounts then the total value of construction activity is understated. In countries where there is a lot of informal construction, such as in most developing countries, the understatement of total activity can be quite significant.

SUMMARY

The foregoing discussion is centred on the issues that must be considered when seeking an answer to a very specific question: is it more expensive to build in one location or another (or perhaps a series of other locations)? While it may be of little more than passing interest to most, it is a key question for governments and policy-makers in several ways:

- Higher project costs in one location may be an indicator of lower efficiency/productivity and thus lead to efforts to improve performance through benchmarking against industries in other places that appear to be more efficient.
- Improved performance means more efficient utilisation of scarce resources with greater volumes of construction being completed with the use of the same or fewer resources.
- Governments are often the major clients for construction in their respective countries; a more efficient industry means greater returns on government spending and thus more benefit from the use of public money.
- Productivity is not the only factor affecting construction costs: higher real costs may result from, *inter alia,* compliance with local regulations, or transaction costs arising from outdated or overly complex business models. Comparative cost studies that negate price-level differences between countries provide a way of identifying where construction costs are higher or lower in real terms, and once higher costs are identified, reforms that can ameliorate the problems can be considered.

Construction, along with machinery and equipment, is part of gross fixed capital formation (GFCF) in the ICP. While only one part of the ICP, construction's contribution to gross domestic product (GDP) in most countries is substantial and thus it is important in the overall ICP. The ICP is primarily concerned with producing GDP (that is, whole-economy) PPPs which are used to establish which countries are those with lowest GDP per capita in real terms, that is, based on purchasing power. Once again the focus is on resource allocation: the World Bank can provide only a finite amount of aid to countries in need of assistance, and the ICP PPPs are the tool that enables the bank to allocate funding to places where it will (theoretically) do the most good by targeting countries with the lowest real GDP per capita (that is, the countries that are poorest in real terms).

Over the past 70 years a range of methods have been used in attempts to make valid comparisons of the cost to build and of construction industry productivity at project and sectoral levels. The methods fall into these main categories:

- Standard projects.
- Building area models.
- Baskets of basic inputs.
- Baskets of components or assemblies.
- Matched building models.

There are variations in the detail of individual comparison exercises that lie within these categories, such as the CIS Resource Technological Model approach which combines basic input costs (for example, steel, fuel, labour) supplied by member countries with hypothetical standard project models, while the Eurostat–OECD standard projects approach utilises BQs that contain discrete items of work that include materials, labour and so on (for example, areas of brickwork, quantities of concrete placed into footings) for a set of projects that are priced in each country. Some studies have been based on identical projects, others on functionally similar projects, and some have combined the two by testing variations of standard projects in an attempt to account for differences in local design and building practices, client expectations and regulatory environments. In all project-based studies it is the identification of comparable yet representative projects that is the major challenge, although the challenges of collecting reliable price data should not be underestimated.

The exercises based on baskets of items (inputs/resources or construction components) avoid some, but not all, of the difficulties associated with project-based methods. Collecting multiple observations of costs in many places remains a problem, not least because it can be an expensive exercise if consultants have to be paid to provide data. Comparability and representativeness are still of concern; in the 2005 ICP round (which used the BOCC approach) relatively few components were identified that were sufficiently similar in their construction across the large number of countries were reasonably representative of construction in each location. In the end, the BOCC exercise relied as much on basic inputs such as sand and cement as on components/assemblies.

Whatever the approach, conversion of local currencies to a single base is the key concern. It has been amply demonstrated that nominal exchange rates do not enable robust comparisons, because they do not reflect differences in price levels in different locations, and they are influenced by a range of factors such as interest rate movements, changes in fiscal and monetary policy and economic upswings and downturns. Purchasing power parities provide better conversion factors; these are often available as industry- or sector-specific indices that should better reflect real price differences, based on the volume of construction that can be bought with units of different currencies.

CONCLUSION

Construction industry comparisons are not easy and there is no consensus as to how they may best be made. The World Bank has put a lot of effort into developing a solid methodology for producing PPPs that include construction, but cost conversion is only one part, albeit a crucial part, of the puzzle. The great variety in the structures that are created around the world makes the notion of a standard project almost meaningless, and the idea of comparing construction industries across many countries is equally problematic. Factors such as differences in labour (skill and availability), materials (imported and locally produced), and availability and use of heavy equipment, mean that construction in many countries is carried out in quite different ways. Where studies have been limited to relatively similar situations (for example, the UK and the US) useful insights into different techniques and processes have been gained and have prompted industry improvements.

While there are great strides being made in digital technologies such as big data, artificial intelligence (AI), cloud computing and building information modelling (BIM), the nature of the major problems in construction cost comparisons means that such technologies are not likely to be of much assistance at least in the short to medium term. The lack of uniformity in construction output and the difficulties associated with obtaining reliable cost data from multiple sources in any location are key concerns, and it appears that for the time being human input and judgement will continue to be the primary tools for gathering, interpreting and adjusting cost data. Modern communication tools have streamlined many processes, and perhaps, in time, there will be AI applications that can process and interpret data. However, firms tend to protect data for commercial reasons and to maintain client privacy, so there is likely to be limited scope for automated data collection at the level of detail required for such purposes as the production of PPPs. Once again the potential cost of developing such systems is likely to outweigh the benefits that would accrue to the relatively few people and agencies who are concerned with cost comparison and similar exercises.

While much has been done in developing comparison methods, much remains to be done. Data collection and validation, including the selection and training of those providing data, needs further refinement, as does the weighting of components, which may be items in a basket of resources, or components, or project types, or even the basic headings of residential, non-residential and engineering construction in total output. Equally important is the need for better education and thus better understanding of comparisons by policy-makers and industry generally, so that undue weight is not given to the conclusions drawn from studies that are not well grounded in theory.

NOTES

1. While it can be argued that 'cost' is what we pay for something, and 'price' is the amount paid to us for something, the terms are used here in a general sense and are more or less interchangeable.
2. The fluctuation of the value of the Australian dollar against the US dollar over the past 20 years illustrates this well: in September 2001 the AUD was worth US$0.49, in July 2008 it was US$0.97, but by November 2010 it had dropped to US$0.65. Just eight months later, in July 2011, AU$1 was buying US$1.10 and by February 2020 the AUD had dropped back to just below US$0.60. At different points in time, cost and productivity comparisons (based on cost) that were done using these rates would have produced very different results, yet it is unlikely that the relative cost to build in the two countries would have moved in a similar way to that of the exchange rate. Similarly, the euro (EUR) to AUD exchange rate increased by 10 per cent in just nine weeks in early 2010 (ExchangeRates, 2020a), and by over 14 per cent in nine months during 2017 (ExchangeRates, 2020b). Once again, comparisons based on these fluctuating rates would show markedly different results if conducted just weeks or months apart.
3. While the OECD and Eurostat use a similar approach there are still differences; for example, Eurostat uses a North European and a South European house, and the OECD uses North American,

Japanese and Australia/New Zealand houses. The emphasis in this example is on representativeness of the projects that are priced, with some trade-off in respect of comparability.

4. Engineering construction typically uses a smaller range of materials, has larger proportions of key materials such as concrete and steel, and employs more equipment (excavators, graders, rollers and similar) than the construction of buildings. Similarly, residential buildings use a different mix of resources compared to non-residential buildings (including high-rise residential).

REFERENCES

AACP (1950) *Productivity Team Report: Building. Report of a visit to the USA in 1949 of a productivity team representing the building industry*. London: Anglo-American Council on Productivity.

ABS (2013) *Private Sector Construction Industry 8772.0. (Canberra: Australian Bureau of Statistics)*. https://www.ausstats.abs.gov.au/ausstats/subscriber.nsf/0/9798157B9BCEDB58CA257B960014FEBA/$File/87720_2011-12.pdf.

AIIB (2019) *Asian Infrastructure Finance 2019: Bridging Borders: Infrastructure to Connect Asia and Beyond*. Beijing: Asian Infrastructure Investment Bank.

AIQS (2019) Building Cost Index. *Building Economist*, June, pp. C1–C3.

BCA (2012) *Pipeline or Pipe Dream? Securing Australia's Investment Future*. Melbourne: Business Council of Australia. https://www.bca.com.au/pipeline-or-pipe-dream-securing-australias-investment-future.

Best, R. (2008) Development and testing of a purchasing power parity method for comparing construction costs internationally. Unpublished PhD thesis. http://works.bepress.com/rick_best/.

Best, R. (2012) International comparisons of cost and productivity in construction: a bad example. *Australasian Journal of Construction Economics and Building*, 12 (3), pp. 82–88.

Best, R. and Meikle, J. (2015) The International Comparison Program and purchasing power parities for construction: Chapter 4 Appendix. In: R. Best and J. Meikle (eds), *Measuring Construction* (pp. 59–76). Abingdon: Routledge.

Cassel, G. (1921) *The World's Money Problems*. New York: E.P. Dutton & Co.

Chancellor, W., Abbott, M. and Carson, C. (2019) Measuring construction industry activity and productivity: The impact of the shadow economy. In: R. Best and J. Meikle (eds), *Accounting for Construction* (pp. 74–85). Abingdon: Routledge.

de Valence, G. (2019a) Accounting for the built environment. In: R. Best and J. Meikle (eds), *Accounting for Construction* (pp. 14–30). Abingdon: Routledge.

de Valence, G. (2019b) Comparing construction in national industrial classification systems. In: R. Best and J. Meikle (eds), *Accounting for Construction* (pp. 31–43). Abingdon: Routledge.

DLC (1999) *A Framework for International Construction Cost Comparisons*. Davis Langdon Consultancy. London: Department of the Environment, Transport and the Regions.

DLC (2003) An initial test exercise on a basket of goods approach to construction price comparisons. Davis Langdon Consultancy, London, unpublished.

Emmett, G. and Langston, C. (2019) Comparative construction cost data for industry: a case study of Turner & Townsend's experience. In: R. Best and J. Meikle (eds), *Accounting for Construction* (pp. 181–191). Abingdon: Routledge.

ExchangeRates (2020a) Euro to Australian dollar spot exchange rates for 2011. www.exchangerates.org.uk/EUR-AUD-spot-exchange-rates-history-2011.html.

ExchangeRates (2020b) Euro to Australian dollar spot exchange rates for 2017. www.exchangerates.org.uk/EUR-AUD-spot-exchange-rates-history-2017.html.

Forrestal, L. and Dodson, L. (2102) High costs could kill big projects. *Australian Financial Review*, 7 June.

Green, D. (2019) The challenges of measuring British construction output. In: R. Best and J. Meikle (eds), *Accounting for Construction* (pp. 46–72). Abingdon: Routledge.

Hepworth, A. (2012) Local project costs 40pc above the US, says Business Council of Australia. *Australian*, 7 June.

ICMS (2019) *ICMS: Global Consistency in Presenting Construction and Other Life Cycle Costs*, 2nd edition. International Construction Measurement Standards Coalition (ICMSC). https://icms-coalition.org/the-standard/.

ICP (2005) *ICP Operation Manual*. International Comparison Program. https://pubdocs.worldbank.org/en/858081487090309311/ICPOperationalManual2005.pdf.

IPMS (2020) *International Property Measurement Standard*. https://ipmsc.org/standards/ (accessed 3 July 2020).

Langston, C. (2015a) Performance measures for construction. In: R. Best and J. Meikle (eds), *Measuring Construction* (pp. 157–181). Abingdon: Routledge.

Langston, C. (2015b) Refining the citiBLOC index. In: R. Best and J. Meikle (eds), *Measuring Construction* (pp. 184–202). Abingdon: Routledge.

Langston, C. and Best, R. (2001) An investigation into the construction performance of high-rise commercial office buildings worldwide based on productivity and resource consumption. *International Journal of Construction Management*, 1 (1), 57–76.

Langston, C. and Best, R. (2005) Using the Big Mac Index for comparing construction costs internationally. In: A. Sidwell (ed.), *Proceedings of QUT Research Week 2005*. Queensland University of Technology, July.

Lynton (1993) *The UK Construction Challenge*. London: Lynton.

Meikle, J. (2019) A review of the 2011 construction survey and results from the World Bank International Comparison Program. In: R. Best and J. Meikle (eds), *Accounting for Construction* (pp. 155–179). Abingdon: Routledge.

Ong, L.L. (2002) *The Big Mac Index: Applications of Purchasing Power Parity*. Basingstoke: Palgrave Macmillan.

Pakko, M. and Pollard, P. (2003) Burgernomics: a Big Mac guide to purchasing power parity. *Federal Reserve Bank of St Louis Review*, 85 (6), 9–28.

RLB (2019) *Riders Digest 2019*. Brisbane: Rider Levett Bucknall.

Rogoff, K., Froot, K. and Kim, M. (2001) The Law of One Price over 700 years. IMF Working Paper WP/01/174.

Runeson, G. (2015) Background to purchasing power parity indices. In: R. Best and J. Meikle (eds), *Measuring Construction* (pp. 12–24). Abingdon: Routledge.

Sebestyen, G. (1978) Comparison of construction costs in the United Kingdom and Hungary. *Habitat International*, 3 (1–2), 65–69.

Segal, T. (2019) What is the Big Mac Index? *Investopedia*. https://www.investopedia.com/ask/answers/09/big-mac-index.asp (accessed 29 June 2020).

Turner & Townsend (2012) *International Construction Cost Survey 2012*. www.turnerandtownsend.com/en/perspectives/international-construction-cost-survey-2012/.

Turner & Townsend (2019) *International Construction Market Survey 2019*. www.turnerandtownsend.com/en/perspectives/international-construction-market-survey-2019/.

Walsh. K. (2012) Presentation to workshop at Bond University. 4 May. Unpublished.

Walsh, K. and Sawhney, A. (2004) International comparison of cost for the construction sector: an implementation framework for the basket of construction components approach. Report submitted to the African Development Bank and the World Bank Group, June.

World Bank (2013) *Measuring the Real Size of the World Economy*. doi: 10.1596/978-0-8213-9728-2.

World Bank (2015) *Operational Guidelines and Procedures for Measuring the Real Size of the World Economy*. http://pubdocs.worldbank.org/en/777881487094209758/OG-eBook.pdf.

World Bank (2020) *Purchasing Power Parities and the Size of World Economies: Results from the 2017 International Comparison Program*. Washington, DC: World Bank. Doi: 10.1596/978-1-4648-1530-0.

Wright, M. and Stoy, C. (2008) The CEEC code of measurement for cost planning: introduction and practical examples. FIG Working Week, Stockholm, 14–19 June. https://www.fig.net/pub/fig2008/papers/ts07j/ts07j_04_wright_stoy_3056.pdf.

14. New trends in international construction

Hongbin Jiang

INTRODUCTION

International construction is in a state of significant transformation now. The future looks different, and it requires innovation, integration, collaboration and an understanding of the trends that will shape the industry to succeed (Flanagan, 2019). The construction industry and infrastructure are crucially important for most economies, and international construction has truly transformative impacts on people's lives, firms' business and nations' policies across different countries. Globally, the need for infrastructure investment is forecast to reach US$94 trillion by 2040, and to meet this investment need, the world will have to increase the proportion of gross domestic product (GDP) it dedicates to infrastructure to 3.5 percent (Global Infrastructure Hub, 2017). The market is large and the industry is complicated. The objectives of this chapter are to review the major literature on international construction which comprises part of the body of knowledge of construction economics, and to discuss its underpinnings and future trends with a proposed analytical framework.

A comprehensive review of the literature produced from studies on international economics and the construction industry covered works including Myers (2003, 2017), Ofori (2003, 2015, 2019), Hillebrandt (2000) and Peh and Low (2013). Based on the review of the mainstream literature and the existing analytical frameworks for international construction, this chapter analyses and presents the profiles of four major attributes of international construction, that is, growth capacity driven by ownership, market coverage driven by locational factors, business modality driven by internalization, and sector presence driven by specialty. This is followed by the proposal of a new framework which encompasses the driving factors of the transformation. The framework has six dimensions: the endogenous catalysing factors (Catalyzer), the hindering factors (Hinderer), the ownership essentials (Owner), the regulatory factors (Regulator), the disruptive exogenous factors (Disruptor) and the synergizing factors (Synergizer). This CHORDS framework facilitates the analysis of the future trends of international construction.

International construction is facing many game-changers in the disruptive environment. The disruptions of technology, environmental issues, energy transition, and so on, have big impacts on the long-term global trends which include demographic and social changes, shift of global economic power, rapid urbanization, climate change and resource scarcity (World Economic Forum, 2016; GIH, 2017). Due to the infrastructure financing deficits in many countries, the difficulties in fundraising and the investors' appetite for the sector's growth have made the task of deploying capital more complex. This will eventually force all stakeholders to transform themselves in a disruptive way. In addition, the fast growth of the required construction industry in developing countries also requires a richer, more complex knowledge base (Ofori, 2012, 2019).

Currently, some major global infrastructure schemes are under way, and this chapter discusses the Belt and Road Initiative advocated by China from 2013, the Global Infrastructure Programme launched by the United Kingdom (UK) Prosperity Fund in 2018, and the Blue Dot

Network established by the United States (US) government in 2019. The chapter discusses the potential impacts of the schemes in international construction. Finally, six future trends are identified, and suggestions for further research and implications for the development of the field of knowledge are made.

REVIEW OF THE LITERATURE

The literature on international construction has long been complementary to that on international economics. A body of literature has been built up by the application of international economics and theories relating to multinational corporations (MNCs) to the construction industry. To analyze international construction as a specialized area in international economics, a framework should be built upon international business and MNCs theories, and the literature on construction economics.

International Business and MNCs Studies

International business is a distinct strand of general economics as it studies transnational business and internationalization. It has been extensively studied since the past century, and became more influential in the trend of globalization in recent decades and the even more recent potential of deglobalization. The major streams of studies in international economics included the industrial organization and monopolistic advantages theory (Hymer, 1976; Kindleberger, 1973), Vernon's product cycle model (Vernon, 1979), Kojima's comparative advantage theory (Kojima, 1985) and the location theory (Dunning, 1973; Caves, 1974; Buckley and Casson, 1985). The studies on the strategic and structural perspective of internationalization include the network approach (Johanson and Mattsson, 1988) and the organization theory (Ghoshal and Westney, 1993).

Another strand of MNCs and internationalization theories was built up from Coase's transaction cost theory; it includes the works of transaction cost analysis and market versus hierarchies theory (Williamson, 1975), and the internalization theory (Buckley and Casson, 1996; Rugman, 1981). For sophisticated business motivation and activities, Porter's (1980, 1990) competitive development stages theory and the diamond framework, and Dunning's (1988) investment development paths theory, made great efforts to explain the phenomenon of international business. Dunning's (2000) eclectic paradigm, also known as the "ownership, location and internalization" (OLI) model, provided a powerful explanatory framework for MNCs' international business development, and was perceived as an envelope for most of the theories (Dunning, 2000).

Construction Economics

Construction economics as a distinct discipline and a branch of economics (Ofori, 1994) has been contributed to by many scholars over several decades (Hillebrandt, 2000; Wells, 1985; Ofori, 1988, 1990; Myers, 2017). These studies apply the principles, theories and techniques of economics to the key players in construction, the construction process and the construction industry, and ultimately to how they support economy growth (Hillebrandt, 1984). The construction industry plays a fundamental role in the economy and in socio-economic development (Wells, 1985; Ofori, 1988), and it is a growth engine for the whole economy. Hence, construction economics involves research on macroeconomic, industry and company perspectives. The intervention of government in construction has become more important in developing countries, especially in the wave of globalization and greater involvement of private sector in recent decades. Ofori (1994) described the domain, the conceptual structure, the analytical approaches and the future development of construction economics, which serves as a good guide for research subsequently. Ofori (2015) provided an overview of the research on construction economics over the past four decades, and he also contributed thoughts on the construction industry in developing countries.

De Valence (2006) summarized the debate on construction economics, and categorized the topics under macroeconomics, microeconomics and industry economics. He cited Ofori's distinction of construction industry economics and construction project economics. Hillebrandt (1985) described the unique features of the construction industry and its products and processes. He also elaborated on the structure of the market, the profile of demand, the price determination by contractors, as well as the firms' operational approaches and their development over time. It could be argued that Hillebrandt's works (1985, 2000) emphasized the project-based nature of construction, and are devoted to microeconomic analysis for project economics (De Valence, 2006). Construction project economics is concerned with project planning, cost management, procurement and implementation, and life cycle costing, and financing and investment analysis. Construction industry economics concerns macroeconomic growth linking with the construction industry, and the application of various economics theories to construction. Hence, by applying macroeconomics and microeconomics to construction, and delivering the research at industry level and project level, construction economics can be considered as an area of research with a matrix structure, where the two dimensions influence each other.

In addition to the above, Carassus (2004) presented a meso-economic approach to analyze the operation and function of the construction industry within the economy. The study analyzed nine major countries and provides valuable insights on their construction industry profiles. Research on the construction company is embedded in all these studies. The firm's behavior, driven by its endogenous growth capacity and influenced by exogenous factors, has an impact on both the industry and the projects. The firm perspective is more important when analyzing international construction, and strategy and decision-making are crucial at the firm level in international business.

From the macroeconomic perspective, Bon (1992) is among authors who have made the proposition that the share of construction activities and the stages of economic development show an inverted U-shaped relationship. Later researchers tested the proposition, and more found that the hypothesized relationship is to be interpreted as explaining variation within the developed economies over time (Choy, 2011). Ruddock and Lopes (2006) suggested that this proposition deals with long-term, secular development patterns, rather than short- to medium-term associations. In this century, national economic development is challenged by many disruptive factors, such as the new technology revolution, digitalization and artificial intelligence (AI) adaptive innovation. In addition, the financial crisis, trade wars and the trend of deglobalization will have significant impacts on the economies. This will eventually drive the construction industry to a new age. During this process, the construction industry has a special role to play in influencing the fluctuations in the national economy in a desired manner. Hence, a generic relationship between construction activities and a country's stages of economic development is not fully predicted, even without considering the impact of destructive events, such as Covid-19. Studies in this regard should take these disruptive factors into account.

Construction economics needs continuous research and further theoretical development. Myers (2017) pointed out that in the past decade the construction industry has experienced major systemic shocks, or disruptors, that are complete game-changers affecting many aspects of the economic system. He mentioned three such events that have had profound implications for construction economics. The first is the global collapse of the banking system in 2007 to 2008, which led to a worldwide series of deep and protracted economic recessions. Currently, a similar or even worse situation is happening, from the outbreak of Covid-19 and the trend of deglobalization, and these may lead to another deep recession worldwide. The second major event Myers (2017) mentioned is the direct impact of sustainable development on the construction industry. The third is the development of new technology, especially the digitalization in information and communication technology (ICT) sector. He provided insights into the general landscape, and international construction is not separately considered in his work.

International Construction

International construction has attracted increasing interest from researchers. Based on the economic theories on MNCs, foreign direct investment (FDI) and transaction cost theory, international construction is becoming a unique field in mainstream economics in light of its own peculiar features. Researchers have analyzed the internationalization of construction enterprises in particular countries, and the construction activities across countries. Some recent works include Butković et al. (2014), Saloranta (2015) and Brooks et al. (2020). Research on international construction includes studies on the relationship between economic growth and investment in construction between countries, the various business modalities adopted by international firms, investment models of cross-border construction, and location and sector presence of transnational construction. The core of the business modalities and investment models adopted by multinational enterprises is their internalization strategies. Ofori (2003) examined the applicability of various analytical frameworks of international business to international construction, and made comments on the implications, especially on Porter's diamond model and Dunning's eclectic paradigm.

One important earlier work on international construction is by Seymour (1987), who applied Dunning's eclectic paradigm to international construction. Öz (2001) analyzed the international operations of Turkish contractors using Porter's diamond framework. By further extending Porter's diamond and the generalized double diamond framework (Chang et al., 1998), Ericsson et al. (2005) introduced a new framework, the 'construction industry competitiveness hexagon', to analyze construction industry competitiveness. Peh and Low (2013) presented a study on the organization design of international construction business, and discussed the globalization of construction with various organizational and operational issues. Jin et al. (2013) developed a framework based on the balanced scorecard to measure the performance of international construction companies. More recent studies include Amiri and Dausman (2018), who examined the progressive evolution of international construction and identified the key drivers for the internationalization of the construction industry; and Ye et al. (2018) who studied the diversification strategies and modalities in international construction business. Given the disruptive influence of the financial sector in recent decades, Henckel and

McKibbin (2017) discussed the massive fiscal stimulus during the global financial crisis, and the nature and the role of infrastructure spending. International construction may generate large economic returns by increasing market access and reducing regional segregation.

Derived from Dunning's eclectic paradigm and Seymour (1987), Low and Jiang (2003) analyzed the internationalization of Chinese construction companies, and proposed a structural index of factors identifying "truly global" companies. In this work, they introduced the fourth attribute for internationalization of construction, specialty advantage, in addition to Dunning's OLI model, and presented the OLI+S analytical framework for international construction. The OLI+S model was used to analyze international construction at both country level and firm level, and it was followed by Low and Jiang (2004, 2006), Low et al. (2004) and Mat Isa et al. (2017). Mat Isa et al. (2017) further developed the model, and provided empirical and theoretical insights on how the OLI+S model addresses firms' entry decisions to penetrate international markets.

For the studies of international construction in developing countries, Ofori (2007, 2012, 2015, 2019) has made significant contributions. In his most recent work, Ofori (2019) shared insights regarding the project support systems and social responsibility. He also noted that construction in developing countries requires a richer, more complex knowledge base. This is especially relevant in the current global construction market where the construction firms from the emerging countries are playing more proactive roles. Other researchers have also contributed to this school of thought. Lopes et al. (2002) analyzed and reviewed the impact of growth theory on the relationship between economic development and construction economics in developing countries. Looking into the changing global environment, Ofori (2007) presciently identified the new issues and research required in developing countries. These issues include: private sector involvement, internationalization of construction in the advent of globalization, the formation of regional economic blocs and common markets, the global consensus on the need to fight poverty, and the concerns with sustainable development, threats of global pandemics and the influence of cultural issues.

Another stream of studies on international construction is the contributions from the international institutions, including multilateral organizations, international financial institutes, national bilateral aid agencies and large consulting firms (GIH, 2017; WEF, 2016; ADB, 2017; AIIB, 2019; PwC, 2020). The institutional research focuses on infrastructure investment and construction activities in developing countries, and mainly contributes to two streams: one on the macroeconomic studies and their implications on the construction industry, and the other on the analysis of the construction process and policy implications at national and regional levels. The latter focuses on knowledge sharing of international best practices, particularly for the developing countries. The development of theory is not the priority in such institutional research.

In summary, the research on international construction economics is increasingly an important and specialized field, and provides contributions and lessons learned to the mainstream of economics. A multidisciplinary analytical framework is needed, across the studies of general economics, MNCs, FDI, the construction industry, international business and financing. This is even more pertinent given the increasing impacts on international construction from the rapidly evolving and disrupting global environment. As an important segment of this area of research, governments and international organizations are promoting significant initiatives, such as the Belt and Road Initiative, the Global Infrastructure Hub and the Blue Dot Network, and these will generate a momentum for international construction that researchers and industry should further comprehend.

AN ANALYTICAL FRAMEWORK FOR INTERNATIONAL CONSTRUCTION

This section describes how to use the OLI+S model as an attribute framework to profile the nations and top firms in international construction, and then proposes a new analytical framework which encompasses the driving factors in the transformation of international construction.

Profiling International Construction Using the OLI+S Model

Growth capacity driven by ownership essentials

Business growth is essentially driven by the firm's ownership advantages. As suggested by many studies (Buckley et al., 1978; UNCTAD, 2019), the ratio of a firm's international revenue over total revenue is the most meaningful indicator of its capacity in international business. Another important indicator of a firm's ownership strength is the value of its market capitalization. A firm's financing capacity reflects how the firm uses its ownership to leverage financial resources for its growth. The ratio of net debt over the aggregate of net debt and equity indicates the capital structure of comparison between total debt and the total capital which can be used by the firm as a source of financing. Profitability can reveal the firm's ability to produce earnings for the owners and also is an indicator of its success in its operations.

Market coverage driven by locational factors

The locational specificity of construction projects distinguishes international construction as a services business from the others. The international firm needs strong abilities to efficiently manage logistics and communications, and to effectively acquire local management knowledge and business information. The market coverage should be considered in the context of the global construction market, where the different national market volumes and investment demand determine the potential of market share. The market coverage of the firm reflects the market share it obtains, which is driven by its locational advantages. Two indicators of market coverage are considered in this chapter: the market coverage by region, which shows how a regional market is shared by firms from different nations; and the market coverage by nation, which indicates how the firms from the particular country occupy the different regional markets. Market coverage by region uses the ratio of the aggregate international revenue of one nation's firms in a particular regional market over the total revenue in this particular regional market; while market coverage by nation is estimated by the ratio of the aggregate international revenue of one nation's firms in a particular regional market over the total revenue generated by this nation's firms.

Business modality driven by internalization

How business modality of international construction is created is largely driven by the firm's internalization strategy. Internalization is considered with two transaction chains for cross-border construction (Jiang, 2006): the investment transaction chain and the procurement transaction chain (Figure 14.1). The first explains how the investment evolves from non-equity to equity involvement along with the associated risks being internalized; while the latter demonstrates how the service is provided with different contractual or procurement arrangements along with the associated risks being internalized. However, due to the lack of data, it is difficult to quantitatively measure the business modality.

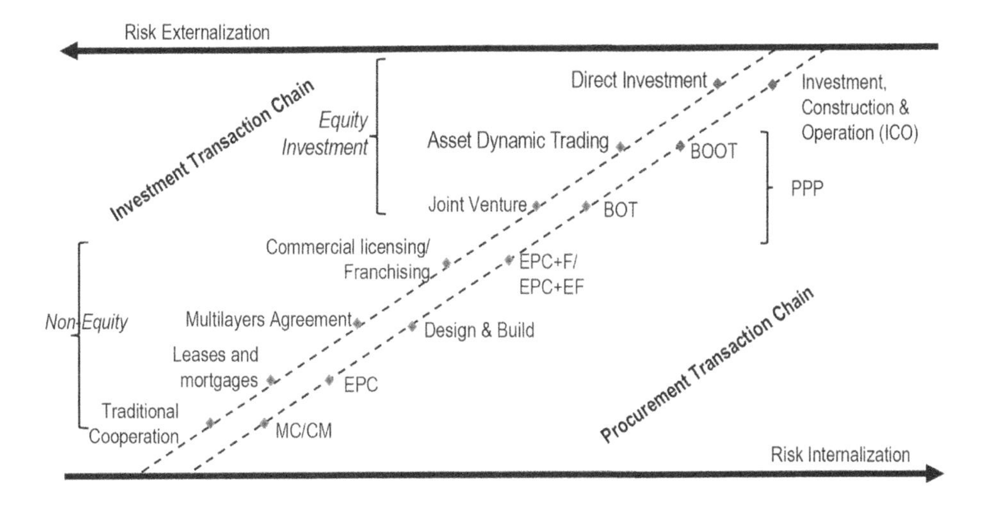

Figure 14.1 *Two transaction chains in international construction*

In recent decades, the public–private partnership (PPP) and concessional arrangement has been promoted in the global market, and this requires the construction firms to be able to make considerable capital investment together with the diversified management and technical capabilities. Furthermore, extending construction business into operation and facility management has also become common among the international firms. The operation and management of built assets require different sets of expertise and investment capability. The integration of investment, construction and operation (ICO) is increasingly seen in some countries. Lastly, it is not uncommon nowadays that the traditional engineering, procurement and construction (EPC) contract is mandated with financing (known as EPC+F), and sometimes the EPC+F has to be leveraged by a certain amount of equity injection into the project company (EPC+EF). Therefore, capital investment has become a key driver for internationalization of construction. Business modality is driven by the firm's investment capability. In this regard, the ratio of capital expenditure over revenue of a firm is adopted to illustrate the firm's capital investment capacity. This is to attempt to provide an indication of the firm's capability in transforming its business modalities driven by its internalization advantages.

Sector presence driven by specialty

The OLI+S model introduces another important attribute of international construction, which is the specialty of a firm. The more diversified the specialties a firm possesses, such as across energy, transportation and water sectors, the more business expansion it is capable of pursuing, but this does not necessarily mean that it definitely achieves high performance in terms of revenue or profits. A firm with high specialty strength in a particular sector may also achieve the same or higher performance. Hence, a firm's specialty capacity is an important attribute of its internationalization, rather than it being only driven by performance. One may argue that the specialty is part of the ownership capability of a firm with the required employees' expertise, management capability and technical capacity. However, as the recent trend shows, construction firms can quickly build up their specialty strength through other means, such as through mergers and acquisitions (M&A), or licensing in entering a new market. The different specialties a firm possesses may allow the firm to diversify its business across the sectors with large differences, for example, from general building to the information and communication technology (ICT) sector. Therefore, adding specialty to the OLI model is rather to reveal and to emphasize this particular characteristic of international construction. Specialty advantages may help a firm to diversify its sector presence to further increase its market coverage and to gain growth. In this chapter, the ratio of the international revenue generated from a particular sector over the total international revenue of the firm is used to represent a firm's sector presence.

Specialty advantage for a firm's internationalization is more important in the advancement of new technologies, such as the digitalization and artificial intelligence (AI) technology in recent decade. *Engineering News-Record* (*ENR*) defines the telecommunication sector to include transmission lines, cabling, towers and antennae, data centers, and so on (*ENR*, 2019). In fact, more construction companies are penetrating into this technology-intensive sector. In the new technology era, a firm requires more specialty capacity to meet the challenges and to explore the opportunities in the forthcoming large demand of construction services for the ICT sector. Figure 14.2 illustrates the transformation of construction services with the disruption of new technology. Specialty capacity driving the sector presence of firm is particularly important in international construction. The international firms often capture more business opportunities with their monopolistic advantages when they work in other countries if they have accumulated enough specialty capacity domestically.

The CHORDS Framework for International Construction in Transformation

New infrastructure, such as smart city, smart grid, smart factory and smart medical care facility, demonstrate the many changes which have been occurring over the past decade. The construction industry is being transformed to include the new infrastructure as part of its mandates. Like Porter's diamond which explains the factors that drive competitive advantages for one economy or one business over the other, a factor-based framework is needed to analyze the influential factors of construction transformation. In a disruptive business environment, the most influential factors are the exogenous disruptive factors (defined as the Disruptor) and the endogenous catalyzing factors (the Catalyzer), which influencing the transformation of construction industry at most. The regulatory factors also play key roles to administer and supervise the industry (the Regulator). The global business environment becomes more fractured, and international construction always has a multidimensional business nature with various stakeholders interacting. Hence the synergizing factors among the stakeholders will not be any less important (the Synergizer). Nevertheless, the obstacles or hindering factors in the transformation should also be taken into account (the Hinderer). Finally, the transforma-

Figure 14.2 Transformation of infrastructure to new infrastructure

tion is motivated by ownership capability, and all the factors in turn have an impact on the ownership advantages (the Owner).

Therefore, an analytical framework of CHORDS (the Catalyzer, the Hinderer, the Owner, the Regulator, the Disruptor and the Synergizer) is proposed as a tool for analyzing the transformation of international construction. The studies of construction economics contribute insights into the key driving factors. With contributions by industry experts, academia, and international organizations, the World Economic Forum (WEF, 2016) presents a framework highlighting major measures to support the industry transformation with best practices and innovative approaches. This chapter incorporates some of these measures which relate to international construction into the CHORDS framework.

In summary, an analytical framework for international construction is established by firstly profiling the nations and top firms with four sets of quantitative indicators with the OLI+S model, in terms of growth capacity (ownership), market coverage (locational), business modality (internalization) and sector presence (specialty), and then analyzing the key factors influencing the four dimensions in the transformation of international construction using the qualitative CHORDS framework (Figure 14.3). Then the dominant factors identified in the CHORDS framework are realigned with the four dimensions in the OLI+S model (see Table 14.2 later in this chapter).

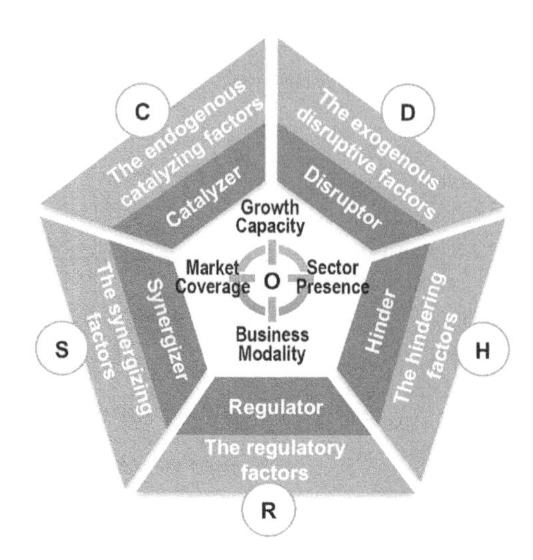

Figure 14.3 *CHORDS framework for international construction in transformation*

PROFILES OF NATIONS AND TOP FIRMS IN INTERNATIONAL CONSTRUCTION

Sources of Data

This section presents an analysis of the current profile of international construction with the four major attributes: growth capacity, market coverage, business modality and sector presence at both both the national level and for the top firms. The data are from various sources, including *ENR* (2019), GIH (2017), WEF (2016), UNCTAD (2019), and companies' annual financial reports.

Growth Capacity: International Revenue, Market Capitalization and Leverage Ratios

The *ENR* (2019) reports that the top 250 international contractors obtained $487.29 billion in their international revenue from projects outside their home countries and $1.148 trillion from domestic projects in 2018. Their international revenue had fallen by 10.4 percent from five years ago (*ENR*, 2019). The large contractors are competing for works aggressively and facing increasingly tough conditions in contracts. Many large international contractors took actions to limit exposure to risks in international markets. Figure 14.4 shows the top 20 nations with the largest international revenues of their firms among the top 250 international contractors. It shows that firms from the UK, Greece, United Arab Emirates (UAE) and Luxembourg almost fully relied on their business outside their home countries, and those from Spain, Germany, Austria, Italy, Sweden and Portugal had over 75 percent of their turnover from other countries. China had the largest revenue from the international market, at US$ 119 billion in 2018; however, it had the lowest ratio of international to total revenue, at 14 percent.

These large firms have been continuously seeking growth abroad. For the top 250 international contractors in *ENR* (2019), the total international revenue was 30 percent of the total

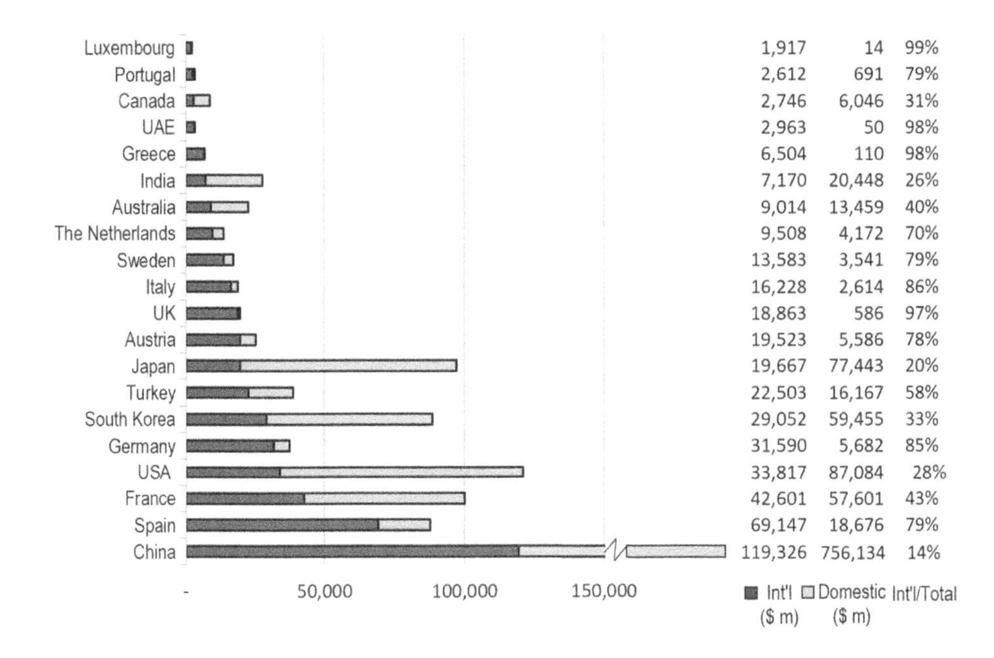

Source: Data from *ENR* (2019).

Figure 14.4 *Top 20 nations with the largest international revenue of firms*

revenue in 2018. In another study on the top 100 listed international construction firms, the total international revenue represents 36 percent of the total revenue in 2018 (excluding the Chinese firms which are mainly domestic markets-focused) (Deloitte, 2019). The market capitalization (market cap) is an indicator of a firm's size, and it is also the value determined by the market. It is difficult to obtain the valuation data for non-listed companies. Figure 14.5 shows the top 20 nations with the largest aggregate market cap of their listed construction firms among the top 100 in 2018. The firms from China, Japan, the US and France have the highest aggregate market cap. To evaluate a firm's growth capacity, the ratio of total market cap to total revenue shows the firms' attractiveness to investment. The lower the ratio, the higher the potential that they were undervalued by the market, and therefore the higher the attractiveness to investment. As shown in Figure 14.5, the firms from the Netherlands (13 percent), China (21 percent), the UAE (27 percent) and Finland (28 percent) showed higher attractiveness to investment, while the firms from India, Mexico and Turkey have further room to improve on their profiles in this regard.

A firm's leverage ratios, such as the ratio of net debt over the aggregate of net debt and equity, can be used to assess the sustainability of the firm's growth capacity. To maintain its sustainable growth capacity abroad, an international construction firm normally allocates light asset investment abroad, and maintains a healthy leverage ratio. This healthy leverage ratio helps the firm to ensure the requirement of working capital and cash flow for projects. This ratio indicates the capital structure of total debt and total capital which can be used as the source of financing. As shown in Table 14.1, among the top 20 largest listed construction com-

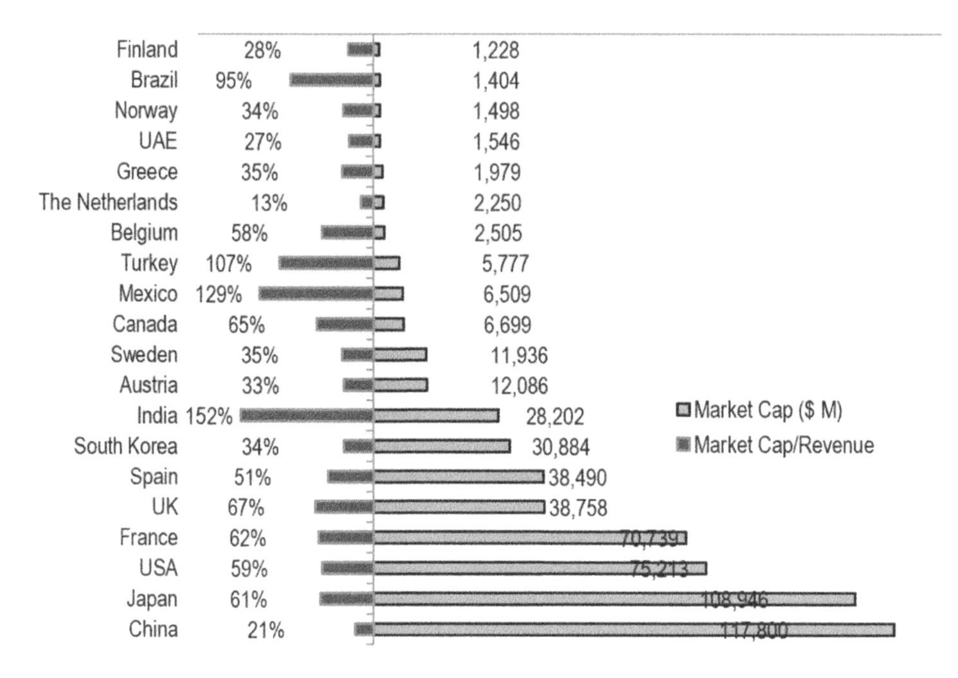

Source: Data from Deloitte (2019).

Figure 14.5 Top 20 nations with the largest market cap of their listed firms

panies in the survey, Samsung from South Korea, and Strabag from Austria, had better cash position comparatively, while CRCC from China, and Eiffage from France, had the signs of high leverage. In general, European firms and Japanese firms have better position to leverage more financial resources, and therefore have better growth capacity. Profitability reveals the ability of a firm to produce earnings for the firm's ownership and also indicates its operational success. ACS Group from Spain, Vinci from France, Skanska AB from Sweden, China State Construction Engineering Corporation from China, and Daiwa from Japan, have a good return on equity; while nine of these 20 firms were above the average of the top 100 firms in the survey, which was 12.4 percent in 2018.

Market Coverage: Regional Market Shares by Nations and Concentration Ratios

GIH (2017) estimates that the global infrastructure investment needs to be an average of $3.7 trillion per year between 2016 and 2040. To meet this required investment, the world will need to increase the proportion of gross domestic product (GDP) it dedicates to infrastructure to 3.5 percent, compared to the 3.0 percent expected under the existing trends. Asia will dominate the global infrastructure market and accounts for about 54 percent of global infrastructure investment needs to 2040, compared to 22 percent for the Americas, the next largest region. GIH (2017) also reveals that four countries account for more than half of global infrastructure investment requirements to 2040: China, the US, India and Japan. China alone is estimated to account for 30 percent of global needs. These data indicate the growing opportunities for inter-

Table 14.1 *Top 20 international firms with selected financial data*

Rank	Company	Net debt/ (Net debt + Equity) (%)	Return on Equity (%)	Capital exp. ($M)	Capital exp./ Sales (%)
1	China State Construction Engineering Corp. (CSCEC)	10.0	15.8	2 903	1.6
2	China Railway Group Ltd. (CREC)	26.0	9.3	3 249	2.9
3	China Railway Construction Corp. Ltd. (CRCC)	69.0	11.0	4 861	4.4
4	China Communications Construction Company (CCCC)	42.0	8.9	5 471	7.4
5	Vinci, France	45.0	16.0	2 466	4.8
6	Metallurgical Corporation of China Ltd, China	38.0	7.9	832	1.9
7	Actividades De Construccion Y Servicios S.A. (ACS), Spain	0.0	21.5	1 125	2.6
8	Bouygues, France	25.0	13.9	2 560	6.1
9	Daiwa House Industry Co., Japan	5.0	15.6	2 501	7.3
10	Samsung C&T Corp., South Korea	-86.0	8.7	538	1.9
11	Shanghai Construction Group (SCG), China	7.0	6.7	542	2.1
12	Lennar Corp., USA	32.0	11.6	123	0.6
13	Aecom, USA	39.0	3.3	81	0.4
14	Eiffage, S.A., France	66.0	13.6	818	4.1
15	Skanska Ab, Sweden	2.0	15.9	316	1.6
16	Sekisui House, Japan	-2.0	10.9	560	2.9
17	Fluor Corp., USA	19.0	7.6	211	1.1
18	Larsen & Toubro Ltd. (L&T), India	10.0	13.7	37	0.2
19	Strabag, Austria	-36.0	10.1	755	4.2
20	Obayashi Corp., Japan	11.0	13.4	652	3.8

Source: Data from Deloitte (2019).

national construction firms to consolidate their presence in the global marketplace. Similarly, *ENR* (2019) shows that the two largest regional markets are in Asia and Europe.

With respect to market coverage, Chinese and Spanish firms were the most active players in the Asian market, while Spanish and French companies had the majority share of the European market in 2018 (Figure 14.6). Spanish firms also have dominant positions in Latin American, Caribbean and American markets. In addition to the traditional market in Asia and Africa, Chinese firms had become active in Middle East and Latin America.

In terms of locational strategies, Chinese, Japanese and Korean firms have larger market shares in Asian countries in their neighborhood, with over 44 percent of their international revenues from this regional market. As a whole, the firms from these three countries took a total of 58 percent of the Asian market share. Spanish firms are more diversified in terms of locational coverage, and they are active in most of the regional markets; that is, Asia, Europe, the US and Latin America. Turkish companies took the Middle East and Europe as their primary markets, accounting for 38 percent of their total international revenues in 2018. This analysis also shows that the international construction market is still oligopolistic, with a 62 percent concentration ratio, which means the top five countries with the largest aggregate international revenues take up 62 percent of the total international revenue under the survey. For each of the regional markets, all of the concentration ratios are well above the general threshold of oligopoly at 60 percent: for example, Asia at 80 percent, Africa at 81 percent, the Middle East at 64 percent, Europe at 74 percent, the US at 79 percent, Canada at 94 percent, and Latin America at 80 percent.

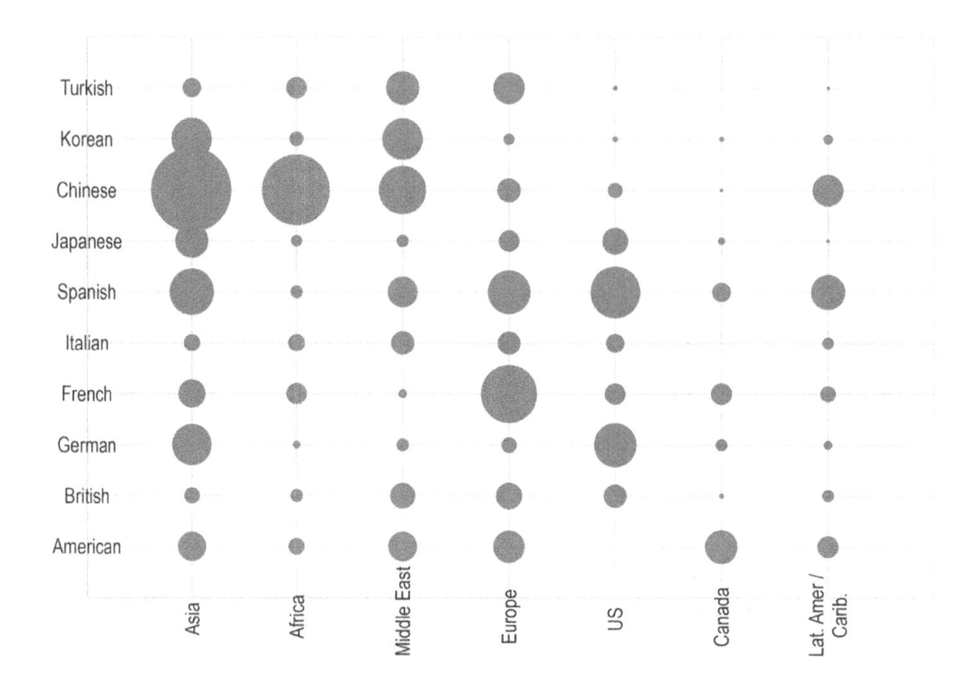

Note: The size of the bubble shows the amount of international revenues in 2018.
Source: Data from *ENR* (2019).

Figure 14.6 Top ten nations with market coverage in different regions

Business Modality: Cross-Border Investment and Capital Expenditure Ratios

As implied in the two transaction chains in international construction (Figure 14.1), firms may adopt different business and investment strategies by internalizing or externalizing their risk exposures. The United Nations Conference on Trade and Development (UNCTAD, 2019) analyzed the value of cross-border M&A sales and purchases, and the greenfield foreign direct investmant (FDI) projects worldwide. This chapter selects the data for the three sectors which are closely related to international construction: (1) electricity, gas and water; (2) construction; and (3) transport, storage and communications. Figure 14.7 shows the investment trends in international construction from 2009 to 2018. The total value of cross-border M&A sales for the three construction sectors have steadily increased in recent years, with an amount of US$92 billion in 2018. The assets for transport, storage and communications took over 50 percent of the total transaction value. The total value of cross-border M&A purchases in the three sectors reported was at US$39 billion in 2018. This implies that international construction is attracting more investment from other sectors. Nevertheless, greenfield projects are still dominating the investment activities in international construction, and the aggregate FDI value of the three sectors was at US$217 billion in 2018. For the FDI in greenfield projects, the electricity, gas and water sector, and the general construction sector, took up the majority, which was over 80 percent of the total amount in 2018. For a sector-wise observation, the transactions in transport, storage and communications sectors are in a dominant position for both cross-border

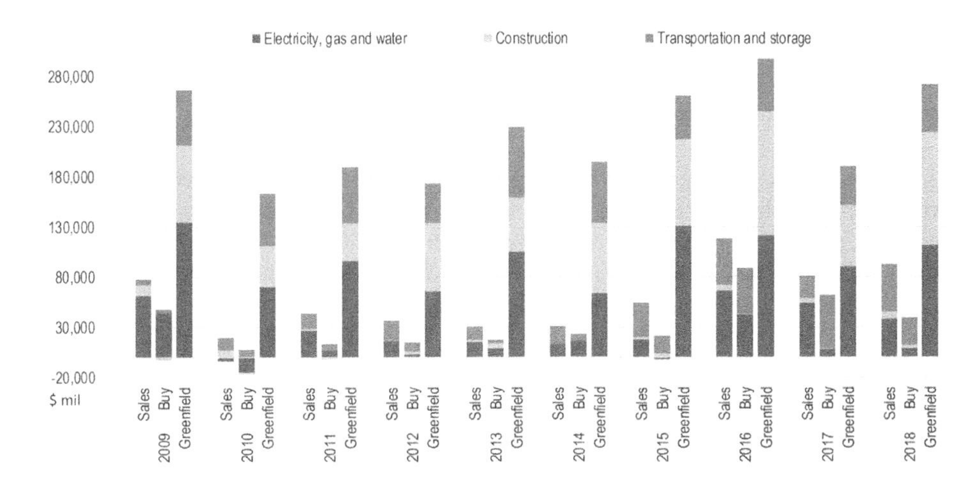

Notes: The cross-border M&A sales are calculated on a net basis as follows: Sales of companies in the host economy to foreign MNCs minus Sales of foreign affiliates in the host economy. The data cover only those deals that involved an acquisition of an equity stake of more than 10 percent. Data refer to the net sales by the region/economy of the immediate acquired company. Cross-border M&A purchases are also calculated on a net basis as follows: Purchases of companies abroad by a country-based MNCs minus sales of foreign affiliates of the country-based MNCs.
Source: Data from UNCTAD (2019).

Figure 14.7 Cross-border investment of international construction

M&A sales and purchases, and the sales of assets for electricity, gas and water are significantly larger than those of other assets.

How to best structure the business modality to cope with the global market opportunities is the key issue for a firm's internalization strategy. As mentioned earlier, the ratio of capital expenditure over international revenue is a proximity of the firm's capability for transforming its business modality with its internalization advantages. Table 14.1 shows that the top 20 listed construction firms had considerable capital expenditure in 2018. The four Chinese listed construction companies, the two French firms and one Japanese firm presented the higher amount of capital expenditure comparatively. CCCC from China had the highest capital expenditure at US$5.47 billion. Both CCCC from China and Daiwa House Industry Co. from Japan had capital expenditure of over 7 percent of their total sales in 2018, and Bouygues from France at 6.1 percent. This is consistent with the observation that the large construction conglomerates tend to utilize capital investment to lead their business expansion, rather than focus on construction works only.

As reported in GIH (2017), the construction sector needs strong capital investment to meet the requirement of infrastructure spending in both domestic and international markets, especially in the emerging markets. Government spending has been limited for the past decade, and PPP modality has been widely used to finance and develop major infrastructure in many countries. In addition, some international firms are transforming themselves to take over the full life cycle ownership of projects with the ICO approach. Moreover, concessional arrangements are expected to cover not only toll road, water, wastewater and urban rail, but also other social facilities. Some large construction companies earned over 55 percent of their revenue from non-construction activities in 2018 (Deloitte, 2019).

Sector Presence: International Revenue from Major Sectors by Nation

In terms of sector presence, general building, hazardous waste and power sectors show the footprints of most of the large international construction firms (*ENR*, 2019). As the most traditional service, general building works consists of 24 percent of the total of all companies under the survey. The power sector presents the largest portion taken by all international companies, accounting for 31 percent of the total international revenues from all sectors (Figure 14.8). The firms in the power sector generally provide a diverse range of services, ranging from design and development to construction, maintenance and operation. This sector is dominated by highly specialized companies from China, Spain and France. The water and waste management sector focuses on the treatment, collection, recycling and management of different types of waste, and these services are usually outsourced by municipal entities with concessional arrangements. However, this sector shows relatively low international revenues.

For country-wise observation, firms from Germany, the US, France and Spain have all the expertise, with presence in most sectors including the telecommunication sector, and their total international revenue accounted for 97 percent of the sector's total, which was US$6.9 billion in 2018. Firms from China, Japan, South Korea and Turkey had a relatively low presence in this technology-intensive sector. Most of the international firms have a relatively low presence in telecommunication and industrial and petroleum sectors. This implies that construction companies have not reaped the benefits of industrial process and product innovations; they

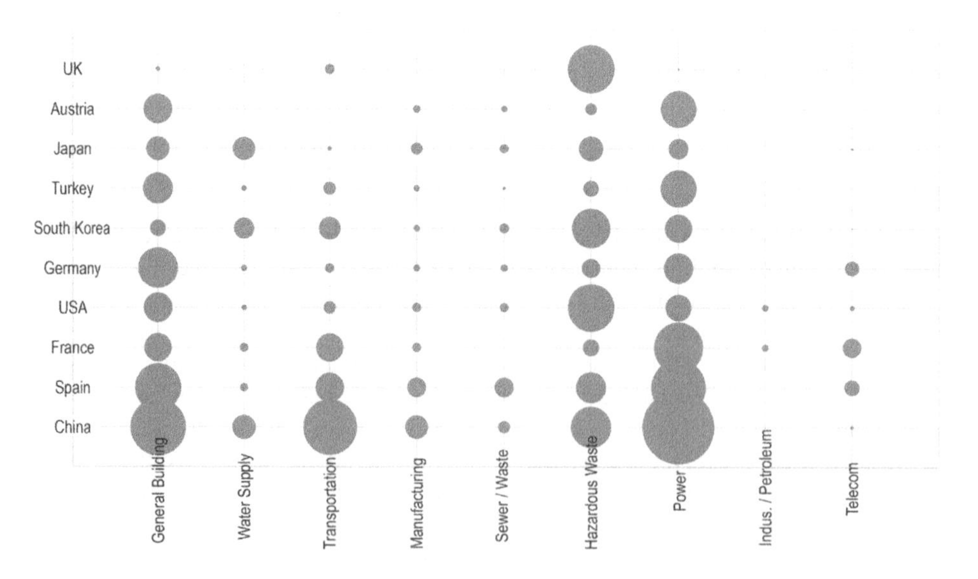

Note: The size of the bubble shows the amount of international revenues in 2018.
Source: Data from *ENR* (2019).

Figure 14.8 Top ten nations with different sector presence

still need further enhancement of their specialty capacity to take up the latest technological opportunities. New technologies in ICT are developing rapidly, with increasing business opportunities globally. The various forms of service products, such as smart city, smart grid, smart community, smart factory, smart medical care facility, have triggered a new wave of innovation in international construction. To capture the new opportunities, international construction firms tend to have more presence in the ICT sector. Currently, international firms from Germany, France and Spain seem to dominate this market segment, while Asian firms show little presence. It is worth mentioning the New Infrastructure Initiatives (NII) promoted in China in 2020, which have a strong focus on telecommunication and digitalization construction. The NII are one of the most important national strategies in China to stimulate and restructure the economy during the post-pandemic age (Jiang and Ma, 2020; Ren et al., 2020), and the Chinese construction companies will probably take this opportunity to upgrade their specialty capacity equipped with new technologies, to expand into more markets in international construction.

ANALYSIS OF INTERNATIONAL CONSTRUCTION IN TRANSFORMATION

Following the analysis of the four attributes of international construction, the CHORDS framework, built up with the major factors in WEF (2016) and other important considerations, is applied to analyze international construction in transformation.

The Exogenous Disruptive Factors: Disruptor

Many opportunities are available through the application of new technologies, new materials and tools into construction. This is perceived as one of the key disruptive factors. New technologies in the digital space, such as the three-dimensional (3D) printing, artificial intelligence (AI) technology, building information modelling (BIM), will not only improve productivity and quality of work, and reduce project delays, but can also enhance the management and increase the added value along the value chain in international construction. This is the key enabler and facilitator for many other technologies used in construction, and also the driver for business expansion of construction firms into telecommunication and other digital-related infrastructure construction.

Other disruptive factors may come from the financial market. The disruption or fluctuation in financial market requires innovative business models and targeted consolidation, for example, the different partnership modalities and ICO approach. To gain the support of society at large, especially in developing countries, the international firms need to work collectively, by engaging local communities during operations to address potential disruptions due to political uncertainties in the locations. Lastly, the emphasis in the recent decade on sustainability and resilience of infrastructure increasingly drives international construction works to address the environmental issues and climate change, especially in some low-lying countries. This factor will continue to impact the industry and transform the services provided by international firms.

The Endogenous Catalyzing Factors: Catalyzer

The transformation of the construction industry has many catalyzing factors which are endogenous, no matter whether the firms act actively as the game-changers or not. An appropriate framework and updated process for project management is necessary to unlock the potential lying in new technologies. For example, the process can be facilitated by combining robotics and 3D technology via a parametrically designed 3D model, and by adopting the lean approach in construction and operation. Rigorous project monitoring on performance indicators (such as cost, time, and quality, health and safety and environmental considerations), strategic workforce planning, smart hiring, enhanced retention of people are also emerging in international firms during the transformation. Companies need to focus on attracting, retaining and developing talent, and establishing company culture conducive to innovation and improved skills. This is more necessary for the international construction firms which often work in an environment of lack of local skilled personnel.

Industrialization is another catalyzing factor for international construction. When international contractors work in multiple locations for similar construction projects, standardized, modularized and prefabricated construction components may significantly increase efficiency. The new business modalities as mentioned earlier, such as PPP and ICO, also require firms to adopt front-loaded and cost-conscious design and project planning. Innovative contracting models with balanced risk-sharing become essential. Finally, the accumulated retained earnings from construction activities motivate the contractors to invest and venture into new sectors.

The Regulatory Factors: Regulator

The construction industry is very much affected by policy and regulation, and thus needs to optimize its interaction with local governments. To monitor political developments, international firms need to maintain constructive communication with public agencies, and to implement effective advocacy strategies. Local regulatory agencies can enhance competition and productivity by simplifying and harmonizing building codes and standards. By setting and enforcing time limits for regulatory and environmental approvals, governments can reduce project delay. In addition, they need to minimize entry barriers to international firms, and provide appropriate support to academia and companies to promote research and development (R&D) and new technology application. As one of the most important regulatory issues, corruption in construction needs strict implementation of transparency and anti-corruption standards. Innovative and whole life cycle-oriented procurement procedures with fairness and transparency to international firms need to be formulated and applied by the governments.

The Hindering Factors: Hinderer

The transformation also faces some new obstructive factors in addition to the traditional issues in construction. Essentially, international construction firms often operate on a light-asset basis. This may hinder the firms in their effort to seek large amounts of financing. The mismatch of heavy initial investment and long-term return is also an issue when financing a project. There is relatively less readiness of the capital market for the infrastructure project with its low return and longer payback period. Meanwhile, the tighter global banking regulations have the effect of reducing interest in investing in longer-term cross-border infrastructure. The uncertainty of local tax regimes in developing countries often increases the risk of infrastructure, and has an impact on the return from it. Insufficiency of economies of scale in some sectors, such as solar power and clean-energy projects, also hinders the transformation of the business model. Some items of infrastructure often have unattractive and unreliable prospects as business operating models, such as the leakage and inappropriate metering in wastewater treatment systems, and this reduces the attractiveness of investment with using PPP or ICO modalities. Due to the poor transparency in some developing countries, it is difficult to evaluate the bankability of the projects, and the financial viability of projects becomes crucial at earlier stage.

Another factor which needs to be monitored from time to time is the imposition of international economic sanctions. Comprehensive sanctions affect a wide range of commercial activities with regard to an entire country or sector. International construction works in countries or sectors subject to comprehensive sanctions might be prohibited. Moreover, international sanctions are complex, and they can also evolve very rapidly.

The Synergizing Factors: Synergizer

International construction is one of the most fragmented business sector and it relies on a seamless interplay between all stakeholders along the value chain of a project life cycle. To reduce complexity and improve cooperation, international firms need to enhance coordination across the value chain and jointly define standards and agree on them. Given the disruptive effect of new technology, the application of high-technology and digitalized products needs an integrated approach across sectors. Furthermore, the new technology provides a new way of collaborating and sharing information between stakeholders. This will transform the existing business model to a more sustainable way with optimized life cycle value across different industry players. Synergy among the different stakeholders can improve productivity and help to optimize existing processes. The firm's internationalization strategies for entering more locations and more sectors with synergized allocation of resources can help it to achieve efficiency and increase economies of scale. From a global perspective, international and regional cooperation initiatives and schemes provide more synergies for international construction across countries.

Based on the analysis of CHORDS framework, key considerations for international construction are highlighted in Table 14.2.

KEY GLOBAL SCHEMES IN INTERNATIONAL CONSTRUCTION

International construction supports other industries and has direct impacts on the global economy. It is significantly affected by politics and regulation, and synergizing with major global schemes are crucial for international firms. In recent years, major global schemes are changing the international political and economic structure, and these include: the Belt and Road Initiative (BRI) led by China since 2013, the Global Infrastructure Programme (GIP) launched by the UK Prosperity Fund in 2018, and the Blue Dot Network (BDN) established by the US government in 2019. There are other initiatives, including the Infrastructure Transaction and Assistance Network, America Crece, and Asia Enhancing Development and Growth through Energy initiated by the US government, Germany's National Initiative

Table 14.2 CHORDS framework for international construction in transformation

CHORDS	Growth capacity	Market coverage	Business modality	Sector presence
Owner (the key OLI+S indicators)	Int'l revenue vs total revenue Market cap Financial leverage ratio Profitability	Country market volume Market coverage by regions Market coverage by nations Concentration ratio	Cross-border M&A sales Cross-border M&A purchases Greenfield FDI projects Capital expenditure	International revenue by sectors
Disruptor (the exogenous disruptive factors)	Advanced building and finishing materials, autonomous construction equipment High-tech in construction e.g. 3D printing, AI, etc. Sustainability and resilience requirement for infrastructure	Coordinated communication with civil society Political uncertainties in the project countries	Digital technologies and big data transform the service model Enhanced management of subcontractor and supplier Disruption in financial market requires innovative business model and targeted consolidation (e.g. ICO) and partnerships	Digital technologies and big data application bring more market in ICT and digital related infrastructure New infrastructure requires construction in connection with many other sectors
Catalyzer (the endogenous catalyzing factors)	An appropriate framework for project management Strategic workforce planning, smart hiring, enhanced retention Rigorous project monitoring (cost, time, quality)	Industry-wide employer marketing Continuous training and knowledge management on local employees	Standardized, modularized and prefabricated components Front-loaded and cost-conscious design and project planning Innovative contracting models with balanced risk-sharing	Motivation of investing the retained earnings accumulated from construction activities into new sectors
Regulator (the regulatory factors)	Harmonized building codes / standards and simplified permit processes Promotion and funding of R&D, technical adoption and education	Effective interaction with the public sector Indigenous market openness to international firms Strict implementation of transparency and anti-corruption standards	Actively managed and staged project pipeline with reliable public funding Innovative and whole-life-cycle oriented procurement with fairness and transparency to international firms	Connection of standards for construction across sectors

CHORDS	Growth capacity	Market coverage	Business modality	Sector presence
Hinderer (the hindering factors)	Financing capacity of light asset basis Tighter global banking regulations	Impact of de-globalization on global economy Uncertainty of local tax regimes International sanction on particular countries and the related lines of business Poor transparency and corruption issue	Financing difficulty with the light asset basis and mismatch of heavy initial investment and long-term return Readiness of capital market for infrastructure with low return and longer period of payback	Insufficiency of economies of scale in some sectors (e.g. solar power and clean-energy projects) International sanction in particular sectors and the related lines of business
Synergizer (the synergizing factors)	Enhanced collaboration way with others More data exchange, bench-marking and best practice sharing More global and regional schemes in international construction	Internationalization strategy to more locations with synergized allocation of resources to efficiently increase scale	Sustainable service models with optimal life cycle value across different industry players	Integration of new technology across sectors Internationalization strategy to more sectors by synergizing local resources Cross-industry collaboration along the value chain

on Energy Transition, and the most recent one, China's New Infrastructure Initiatives (Development Finance Corporation, 2019a; WEF, 2016; Jiang and Ma, 2020).

Belt and Road Initiative (BRI), 2013

China's BRI, launched in 2013, provides support on infrastructure needs and strengthens economic and financial connectivity in Belt and Road countries. This proposition is the most comprehensive strategy proposed by China for international cooperation and integration of infrastructure with logistics, trade, communications and culture. BRI is to construct a wide range of projects, to connect China with other countries through the Silk Road Economic Belt and the Maritime Silk Road. The Chinese government later on announced the Digital Silk Road to be included in the BRI (Wang and Liu, 2020; State Information Center, 2020).

The BRI is part of China's economic reform domestically. Through the cross-border projects, the BRI may help China to internationalize its currency in order to diversify the associated financial risks in investment, to address the surplus domestic industrial capacity, to increase China's exports and facilitate trade, and further enhance geo-economic relationships. It also helps to establish China's global infrastructure construction capability. China also intends to use the BRI to showcase its growing technical excellence and demonstrate its compliance with globally recognized standards for technology, safety and sustainability. However, the BRI has come under criticism and opposition in other countries, such as being considered to be accelerating energy consumption and deteriorating environmental sustainability; bringing in sovereign debt trap for some developing countries; and also lacking in transparency and institutionalization (Yin, 2019).

The BRI's geographical scope is constantly expanding. As of July 2019, it covered over 130 countries with 195 government-to-government agreements signed. The total trading volume between China and BRI countries is over US$6 trillion, and the total FDI to BRI countries is over US$90 billion, which is largely for infrastructure development (State Information Center, 2020). The significance of BRI is not only to build up new logistic, communication and supply chains, but also to establish a global network of low-cost production and financing facilities, which will gradually form an appropriate protocol for investment, construction and production through improved infrastructure. This network will ultimately be integrated into the global supply chain, so as to benefit end-users in the rest of the world.

Global Infrastructure Programme (GIP), 2018

Launched by the UK Prosperity Fund in 2018, the Global Infrastructure Programme (GIP) aims to improve the provision of sustainable and resilient infrastructure as a critical enabler for economic development in middle-income countries (IPA International, 2018). This is part of the effort of the UK government in putting the construction industry on the national agenda to develop a long-term view, and to put the UK at the forefront of global construction. It will help to address the global infrastructure investment gap through the facilitation of the Prosperity Fund, which aims to reduce poverty across the world by removing barriers to economic growth. An important feature of GIP is that it has full operational guidance based on industry best practices. It helps partner countries to adopt and adapt the UK's leading infrastructure methodologies for business case development (Five Cases), project planning (Project Initiation Route-map) and digital construction (BIM). These methodologies take into account the full life cycle of infrastructure, to ensure the desired outcomes embedded during the project design, so as to deliver high social value and return on investment and ensure whole-life performance and cost efficiencies.

The GIP aims to adapt and disseminate these world-class infrastructure methodologies, to help the development of major infrastructure in Brazil, Colombia, Indonesia, Mexico, South Africa and Vietnam. The programme is expected to deliver an estimated £1.18 billion in primary benefits over the first ten years. In addition, through increased opportunities for international business, the programme is also expected to deliver secondary benefits of an estimated £452 million in UK exports over ten years (HM Government, 2020). Through the initial impact in these pathfinder countries, the overall long-term impact of the GIP is expected to be much higher, and to result in permanent transformation globally

Blue Dot Network (BDN), 2019

Led by the US International Development Finance Corporation (DFC), the Blue Dot Network was jointly launched by the US, Japan (Japanese Bank for International Cooperation) and Australia (Department of Foreign Affairs and Trade) in November 2019 (DFC, 2019b). The BDN is an initiative to bring multiple stakeholders together, to promote high-quality, trusted standards for global infrastructure development in an open and inclusive framework (DFC, 2019b). The BDN will evaluate and certify infrastructure projects based upon adherence to internationally accepted principles and standards to promote market-driven, transparent and financially sustainable infrastructure development in the Indo-Pacific region and around the world. The certification of the BDN will primarily focus on sustainability and transparency of infrastructure projects. The approved projects will be marked as a "Blue Dot" by setting universal standards of excellence, which will attract private capital to the projects in developing countries. The BDN also provides private and public financing institutions with additional confidence and information when they evaluate the projects for funding, and this will help to mitigate financing risks. The high-standard projects with BDN certification will have high value for money, and ensure the projects are delivered under safeguards and standards, so as to avoid the lowest-cost bidding which is prevalent in developing countries. In the long-term perspective, the endorsement of the BDN will create a solid foundation for infrastructure global trust standards, and also reinforce the need for the establishment of global trust standards in other sectors.

As the functioning core of the BDN, the DFC with the help of the US Congress has raised about $60 billion from the private sector. Three months after its announcement, the BDN was already in action, with US–Japan investing $10 billion in a liquid natural gas (LNG) project (*EuroAsian Times*, 2020). The BDN is a new initiative, and the detailed protocol has yet to be announced.

Table 14.3 presents a comparison of the three key global schemes based on the information publicly available so far.

FUTURE TRENDS OF INTERNATIONAL CONSTRUCTION

The global disruptive forces are reshaping the future of the construction industry, and international construction has a vital role to play in the transformation of the industry. Based on the

Table 14.3 *Comparison of the three key global schemes*

Schemes	Est.	Initiator	Participants	Objectives	Investment Platforms	Ancillary Institutions	Collaborating Funding Agencies	Market Position
B&R	2013	China	136 countries and 30 international organizations with 195 signed cooperation agreements (As of July 2019)	To promote policy coordination, facility connectivity, unimpeded trade, financial integration, and people to people bond	Silk Road Fund China-LAC Cooperation Fund China-Africa Development Fund China-ASEAN Investment Cooperation Fund	China Export & Credit Insurance Corporation Multilateral Investment Guarantee Agency Multilateral Development Financing Cooperation Centre	China Development Bank Export-Import Bank of China Asian Infrastructure Investment Bank New Development Bank Commercial Banks	The total FDI of China to BRI countries is over US$ 90 billion (as of July 2019)
Global Infrastructure Programme	2018	UK	Brazil, Colombia, Indonesia, Mexico, South Africa, Vietnam	To assist the countries to develop the economic and social infrastructure needed for their sustainable and inclusive economic growth	UK Prosperity Fund		Asian Infrastructure Investment Bank	Program budgeted at £25 million, and expected to deliver an estimated £1.18 billion in primary benefits over the first 10 years
Blue Dot Network	2019	US	US, Japan, Australia, India, and other partners will be announced; a multi-stakeholder initiative with governments, private sector, and civil society	To serve as a globally recognized certification system for infrastructure with the focus on the Indo-Pacific region	US International Development Finance Corporation (DFC)	The Millennium Challenge Corporation (MCC) Australian Department of Foreigner Affairs and Trade	The Export-Import Bank, US-Japan Bank for International Cooperation Commercial Banks Private investors	The certification standards to be refined, the inclusiveness of the plan to be determined, and the DFC has raised $60 billion from private sectors

Sources: SIC (2020), DFC (2019b) and IPA International (2018).

Figure 14.9 *The future trends of international construction*

analysis in this chapter, six major trends (Figure 14.9) are identified and expected to emerge in connection with international construction.

Consolidated Connectivity in Physical Domain, Information and Financial Flows

In 2016, the Group of 20 (G20) launched the Global Infrastructure Connectivity Alliance (GICA) in order to promote a coherent approach to connectivity. GICA collaborates with the Global Infrastructure Hub, Organisation for Economic Co-operation and Development (OECD), World Bank, Asian Development Bank, Asian Infrastructure Investment Bank, United Nations, and other international organizations and countries. Consolidated connectivity is an important trend of international construction, not only involving the physical connection of most infrastructures completed globally, but also involving the connection of people, organizations and technology. Connectivity is a high policy agenda of most countries and global organizations. The BRI, the BDN, the GIP, America Crece and others were all initiated for better connectivity between nations, societies and people. Connectivity has three attributes: physical domain, information and financial flows (GICA, 2018). Physical connectivity needs a large volume of construction in transportation, power and energy, telecommunication, information technology facilities, and so on, all of which are either the traditional or the emerging sectors in international construction. These also serve as the backbone of the global or regional supply chain. Information connectivity supports worldwide collaboration and knowledge sharing, which is the most important synergizing factor for international construction. Financial flow for connectivity is critical in international construction. The efforts to increase capital availability are being made by international financial institutions, the top nations and

MNCs, although there are still some factors hindering the financial connections, as discussed earlier.

Continued High-Tech Disruption in Process Transformation, Smart Facilities, and ICT Expansion

The most disruptive factor in recent decades is the new technology revolution across all industries. The construction industry, as one of the most traditional ones, sits at a major confluence buffeted by major influencing factors in adapting to the transformation. There are three major trends due to the high-tech disruption in international construction: process transformation, various smart facilities (Figure 14.2) and the expansion of the ICT sector. New technology has altered the construction process, from digitalization in planning and design to robot-based monitoring and construction management. The business process is also transforming with a more integrated approach to gain efficiency. The advancement of high technology also shifts the infrastructure development from the "green" to the "smart" concept. The smart initiatives have brought many new elements into energy, transport, grids, buildings, municipal utilities, such as smart city, smart grid and smart transport. This will redefine the infrastructure sector to include the so-called "new infrastructure," and ultimately transform personal lives, mobility and business. As discussed earlier, international firms are enhancing their specialty capability to meet the requirement for expansion into the ICT sector. This will result in changes to the comparative advantages between firms from the industrialized countries and those from the emerging countries. The large international market in developing countries is the arena for them to compete in future.

Demanding Resilience in Environmental, Operational and Financial Sustainability

Sustainable development has been an important agenda for governments and industries in the past decade, in order to mitigate carbon emissions and to address various ecological and environmental issues. The trend is to promote both sustainability and resilience. In the context of the industry, sustainable development is to protect the environment from damage due to the construction and operation of items, while resilience seeks to protect infrastructure operation from the deteriorating environment. Application of new technology will facilitate the process, and the utilization of renewable energy will continue to be important in international construction. The works in some environmentally vulnerable countries will move towards using disaster-resistant materials, adaptive design and resilient infrastructure. The attribute of resilience in international construction is also reflected in financial sustainability, including project financial viability and firm investment efficiency. Governments in developing countries have an agenda to keep public debt sustainable, and this results in the change of construction business models, such as from EPC to EPC+F, PPP and ICO.

Synergized Diversification in Business Models, Sector Presence and Financing Modality

International construction tends to pursue more diversification in terms of business models and sector presence. Large construction firms have diversified their business models by providing non-construction services along the life cycle of infrastructure, and this includes land development, concession arrangement, operation and maintenance services, and facility management. This increases the synergies within and between construction companies, and the efficiency of utilizing their competitive advantages and knowledge on life cycle management. Diversification is also seen in their expansion into other sectors. Mergers and acquisitions are increasing when international firms enter new sectors or new markets abroad. The new business modalities may require higher investment than construction activity, and this requires the firms to have greater financing capacity. On the other hand, a more diversified financing system has been built up and a multi-tiered capital market is emerging for international construction. Equity financing, various guarantee and insurance facilities, sovereign or sub-sovereign backed financing, project bonds and funds, and many other financing vehicles, are available for international construction projects.

Improved Standardization in Technical Standards, Procurement and Legal Procedures

Cross-border construction activities are often impeded by the lack of standardization. Harmonized building codes, engineering design and construction standards have been promoted in some countries. China is making efforts to establish commonly accepted industrial standards, to boost trade and infrastructure development in the BRI countries, and has concluded many agreements with some other BRI countries to ramp up standardization. It is necessary to accelerate the connection of construction standards across different sectors, given the rapid transformation in construction. The US Blue Dot Network initiative aims to promote high-quality, trusted standards for global infrastructure development in an open and inclusive framework, and once implemented it will facilitate cross-border partnerships in international construction. In addition to technical standards, streamlining the procurement and legal procedures is also emerging as an important agenda in international construction. This will reduce time delays and cost escalation, and mitigate many of the risks associated with investment in and implementation of projects. Moreover, innovative and whole life cycle-oriented procurement procedures with fairness and transparency to international firms is now attractive in the countries where the number of PPP and other investment-led projects has increased.

Sophisticated Oligopoly of Firms and Nations in the Midst of Globalization and Deglobalization

The construction industry has a varied structure of concentration in different countries. The competition is normally among the indigenous firms in their domestic markets, where the industry is made up of a large number of small or medium-sized firms. However, competition for large international projects happens among large international construction firms, while a large number of local firms tend to be their subcontractors or suppliers. This is particularly true in developing countries, unless the government imposes restrictions on market entry. Oligopoly in international construction has been seen for a long time, as it has been led by firms from the industrialized world working in the developing countries. A recent trend is for firms from the emerging economies to increasingly play leading roles in other developing countries. At the national level, the international construction market is also an oligopoly, with a high concentration ratio, not only worldwide but also in most regional markets. This has significant implications for government policy-making and firm business strategy, such as market entry policy, incentive and tax issues, and so on.

After decades of the globalization with large stimulation in international construction, there is an increasing debate on the trends of globalization and deglobalization. Some evidence has been seen on the decrease in trade, capital flows and movement of people over the past few years (Alicia, 2018), and the strategic competition between the US and China has fostered the deglobalization trend further. This may bring concerns and downward pressure to international construction, but how the key global schemes will go and how they will impact international construction in the long term, and whether globalization and deglobalization will lead to a compromise situation of regionalization, are yet to be seen.

CONCLUDING REMARKS

It has been over a decade since Ofori (2007) presciently identified the new issues in construction, and some of the issues have become more critical. The global consensus on sustainable development needs further enhancement; globalization now coexists with the challenges of deglobalization and the ongoing formation of regional economic blocs; and internationalization of construction is faced by the transformation of the whole industry, and threats of global pandemics such as the outbreak of Covid-19. The construction industry is at the core of the economy in most nations, and international construction will become a more influential vehicle in the current disruptive political and economic environment globally.

This chapter reviews the current profile of international construction, in terms of growth capacity, market coverage, business modalities and sector presence. The key driving factors underlying the major attributes are analyzed using the CHORDS framework, which encompasses the factors under the categories of the Catalyzers, Disrupters, Hinderers, Regulators and Synergizers. Six major trends are expected to emerge: (1) consolidated connectivity in the physical domain, information and financial flows; (2) continued high-tech disruption in process transformation, smart facilities and ICT expansion; (3) demands for resilience in environmental, operational and financial sustainability; (4) synergized diversification in business models, sector presence and financing modality; (5) improved standardization in technical standards, procurement and legal procedures; and (6) sophisticated oligopoly among firms and nations in the midst of globalization and deglobalization. International construction is in a state of transformation. There may be no absolute winners or losers in market competition, but it is important to continue the innovations driven by the disruptive factors, to enhance the collaboration between international firms, local firms, governments and international organizations, so as to realize the optimization of benefits for all stakeholders. The factors in the CHORDS framework are driving the movement of innovation, collaboration and optimization in the transformation of international construction.

The study of international construction economics will be an important and specialized field, which will make significant contributions and provide lessons to mainstream general economics and construction economics. There could be further research including the historical study of international construction and its relationship with capital flow; how deglobalization, if it continues, will influence the international market for construction; the potential of the CHORDS framework for analyzing the disruptive business situation for construction firms; and how to integrate the strategies for international construction at the international, national and firm levels during the post-pandemic age. The different voices in the current global arena may not necessarily lead to a fractured world, and can be expected to build up into a harmonized international construction market and business, just like the different sounds in chords.

REFERENCES

ADB (2017). *Meeting Asia's Infrastructure Needs*. Retrieved from https://www.adb.org/ publications/ asia-infrastructure-needs.

AIIB (2019). *Asian Infrastructure Finance 2019 Bridging Borders: Infrastructure to Connect Asia and Beyond*. Retrieved from https://www.aiib.org/en/news-events/asian-infrastructure-finance/bridging .html.

Alicia, G.H. (2018). From globalization to deglobalization: zooming into trade. *Las Claves De La Globalización 4.0*, 33–42.

Amiri, Ahmad Farid, and Dausman, D.C. (2018). The internationalization of construction industry – a global perspective. *International Journal of Engineering Science Invention (IJESI)*, *7*(8 Ver V), 59–68.

Bon, R. (1992). The future of international construction. Secular patterns of growth and decline. *Habitat International*, *16*(3), 119–128.

Brooks, T., Scott, L., Spillane, J.P., and Hayward, K. (2020). Irish construction cross border trade and Brexit: practitioner perceptions on the periphery of Europe. *Construction Management and Economics*, *38*(1), 71–90.

Buckley, P.J., and Casson, M. (1985). *The Economic Theory of the Multinational Enterprise*. London: Palgrave Macmillan.

Buckley, P.J., and Casson, M. (1996). An economic model of international joint venture strategy. *Journal of International Business Studies*, *27*(5), 849–876.

Buckley, P.J., Dunning, J.H., and Pearce, R.D. (1978). The influence of firm size, industry, nationality, and degree of multinationality on the growth and profitability of the World's largest firms, 1962–1972. *Weltwirtschaftliches Archiv*, *114*(2), 243–257.

Butković, L.L., Bošković, D., and Katavić, M. (2014). International marketing strategies for Croatian construction companies. *Procedia – Social and Behavioral Sciences*, *119*, 503–509.

Carassus, J. (2004). The construction sector system approach: an international framework. In CIB W055-W065 "Construction Industry Comparative Analysis" Project Group. April.

Caves, R.E. (1974). Causes of direct investment: foreign firms' shares in Canadian and United Kingdom manufacturing industries. *Review of Economics and Statistics*, *56*(3), 279.

Chang, M.H., Rugman, A.M., and Verbeke, A. (1998). A generalized double diamond approach to the global competitiveness of Korea and Singapore. *International Business Review*, *7*(2), 135–150.

Choy, C.F. (2011). Revisiting the "Bon curve." *Construction Management and Economics*, *29*(7), 695–712.

De Valence, G. (2006). Guest Editorial – The future of construction economics. *Construction Management and Economics*, *24*(7), 661–668.

Deloitte (2019). *GPoC 2018 Global Powers of Construction*. Deloitte. Retrieved from https://www2 .deloitte.com/content/dam/Deloitte/at/Documents/real-estate/2017-global-powers-of-construction .pdf.

Development Finance Corporation (DFC) (2019a). *Coordination Report*. Retrieved from https://www .dfc.gov/sites/default/files/2019-08/CoordinationReport_Shelby_7_31_19.pdf.

Development Finance Corporation (DFC) (2019b). The launch of multi-stakeholder Blue Dot Network. Retrieved from https://www.dfc.gov/media/opic-press-releases/launch-multi-stakeholder-blue-dot -network.

Dunning, J.H. (1973). The determinants of international production. *Oxford Economic Papers*, *25*(3), 289–336.

Dunning, J.H. (1988). The eclectic paradigm of international production: a restatement and some possible extensions. *Journal of International Business Studies*, *19*(1), 1–31.

Dunning, J.H. (2000). The eclectic paradigm as an envelope for economic and business theories of MNE activity. *International Business Review*, *9*(2), 163–190.

Engineering News-Record (*ENR*) (2019). Engineering News Record – The Top 250. enr.com.
Ericsson, S., Henricsson, P., and Jewell, C. (2005). Understanding construction industry competitiveness: the introduction of the Hexagon framework. *Proceedings of the 11th Joint CIB International Symposium Combining Forces – Advancing Facilities Management and Construction through Innovation*, 13–16 June , Helsinki, pp. 188–202.
EuroAsian Times (2020). What is the US–India "Blue Dot Network" that aims to counter Chinese BRI initiative? Retrieved from https://eurasiantimes.com/what-is-the-us-india-blue-dot-network-that-aims-to-counter-chinese-bri-initiative/.
Flanagan, R. (2019). Trends shaping the global construction industry: the race to the future. In Laryea, S., and Essah, E. (eds), *WABER 2019 Conference Proceedings*, August. WABER Conference.
Ghoshal, S., and Westney, D.E. (eds) (1993). *Organization Theory and the Multinational Corporation*. London: Palgrave Macmillan.
GICA (2018). Why connectivity matters. Retrieved from https://www.gica.global/sites/gica /files/Discussion-Paper-Why-Connectivity-Matters-May-10-2018.pdf.
Global Infrastructure Hub (GIH) (2017). *Global Infrastructure Outlook – A G20 Initiative*. Retrieved from https://outlook.gihub.org/.
Henckel, T., and McKibbin, W.J. (2017). The economics of infrastructure in a globalized world: issues, lessons and future challenges. *Journal of Infrastructure, Policy and Development*, *1*(2), 254.
Hillebrandt, P.M. (1984). *Analysis of the British Construction Industry*. London: Palgrave Macmillan.
Hillebrandt, P.M. (1985). *Economic Theory and the Construction Industry*. London: Palgrave Macmillan.
Hillebrandt, P.M. (2000). *Economic Theory and the Construction Industry* (3rd edn). London: Palgrave Macmillan.
HM Government (2020). *Prosperity Fund Business Case: Global Infrastructure Programme*. Retrieved from https://assets.publishing.service.gov.uk/government/uploads/system/ uploads/attachment_data/file/884463/Business_Case_for_the_Global_Infrastructure_Programme.odt.
Hymer, S.H. (1976). *The International Operations of National Firms: A Study of Direct Foreign Investments*. Cambridge, MA: MIT Press.
IPA International (2018). Launch of the Global Infrastructure Programme. Retrieved from https://ipa.blog.gov.uk/2018/08/01/launch-of-the-global-infrastructure-programme/.
Jiang, H. (2006). Chinese multinational construction firms in international and domestic markets: a re-examination of the eclectic paradigm. National University of Singapore.
Jiang, H., and Ma L. (2020). Investing and financing tools and risk factor analysis. In X. Xu (ed.), *China's New Infrastructure: The New Structural Force of the Digital Age* (pp. 468–491). Beijing: People's Publishing House.
Jin, Z., Deng, F., Li, H., and Skitmore, M. (2013). A practical framework for measuring performance of international construction firms. *Journal of Construction Engineering and Management*, *139*(9), 1154–1167.
Johanson, J., and Mattsson, L.G. (1988). Internationalisation in Industrial Systems – a network approach. In N. Hood and J-E. Vahlne (eds), *Strategies in Global Competition* (pp. 303–321). London: Routledge.
Kindleberger, C.P. (1973). *International Economics* (5th edn). Homewood, IL: Richard D. Irwin.
Kojima, K. (1985). Japanese and American direct investment in Asia: a comparative analysis. *Hitotsubashi Journal of Economics*, *26*(1), 1–35.
Lopes, J., Ruddock, L., and Ribeiro, F.L. (2002). Investment in construction and economic growth in developing countries. *Building Research and Information*, *30*(3), 152–159.
Low, S.P., and Jiang, H. (2003). Internationalization of Chinese construction enterprises. *Journal of Construction Engineering and Management*, *129*(6), 589–598.
Low, S.P., and Jiang, H. (2004). Estimation of international construction performance: analysis at the country level. *Construction Management and Economics*, *22*(3), 277–289.
Low, S.P., and Jiang, H. (2006). Analysing ownership, locational and internalization advantages of Chinese construction MNCs using rough sets analysis. *Construction Management and Economics*, *24*(11), 1149–1165.
Low, S.P., Jiang, H., and Leong, C.H.Y. (2004). A comparative study of top British and Chinese international contractors in the global market. *Construction Management and Economics*, *22*(7), 717–731.

Mat Isa, C.M.B., Saman, H.M., and Preece, C.N. (2017). Development of OLI+S entry decision model for construction firms in international markets. *Construction Economics and Building*, *17*(4), 66–91.
Myers, D. (2003). The future of construction economics as an academic discipline. *Construction Management and Economics*, *21*(2), 103–106.
Myers, D. (2017). *Construction Economics: A New Approach*. New York: Routledge.
Ofori, G. (1988). Construction industry and economic growth in Singapore. *Construction Management and Economics*, *6*(1), 57–70.
Ofori, G. (1990). *The Construction Industry: Aspects of Its Economics and Management*. Singapore: National University Press.
Ofori, G. (1994). Establishing construction economics as an academic discipline. *Construction Management and Economics*, *12*(4), 295–306.
Ofori, G. (2003). Frameworks for analysing international construction. *Construction Management and Economics*, *21*(4), 379–391.
Ofori, G. (2007). Guest editorial: Construction in developing countries. *Construction Management and Economics*, *25*(1), 1–6.
Ofori, G. (ed.) (2012). *New Perspectives on Construction in Developing Countries*. New York: Routledge.
Ofori, G. (2015). Nature of the construction industry, its needs and its development: a review of four decades of research. *Journal of Construction in Developing Countries*, *20*(2), 115–135.
Ofori, G. (2019). Construction in developing countries: need for new concepts. *Journal of Construction in Developing Countries*, *23*(2), 1–6.
Öz, Ö. (2001). Sources of competitive advantage of Turkish construction companies in international markets. *Construction Management and Economics*, *19*(2), 135–144.
Peh, L.C., and Low, S.P. (2013). *Organization Design for International Construction Business* (Vol. 9783642351). Berlin, Heidelberg: Springer Berlin Heidelberg.
Porter, M.E. (1980). *Competitive Strategy*. New York: Free Press.
Porter, M.E. (1990). *The Competitive Advantage of Nations*. New York: Free Press.
PwC (2020). *Increasing private sector investment into sustainable city infrastructure*. Retrieved from https://www.pwc.com/gx/en/industries/assets/pwc-increasing-private-sector-investment-into-sustainable-city-infrastructure.pdf.
Ren, Z., Ma J. and Xi Y. (2020). *New Infrastructure: China's New Economic Engine under Global Changes*. Beijing: CITIC Press Group.
Ruddock, L., and Lopes, J. (2006). The construction sector and economic development: the "Bon curve." *Construction Management and Economics*, *24*(7), 717–723.
Rugman, A.M. (1981). *Inside the Multinationals: The Economics of Internal Markets*. London: Croom Helm.
Saloranta, J. (2015). Internationalization of Finnish construction companies: an interview-based study in the Swedish construction industry. Thesis, Department of Management Studies, School of Business, Aalto University.
Seymour, H. (1987). *The Multinational Construction Industry*. London: Routledge.
State Information Center (SIC) (2020). *Six Years of Belt and Road Initiative*. Retrieved from https://www.yidaiyilu.gov.cn/xwzx/gnxw/102792.htm.
UNCTAD (2019). *World Investment Report 2019*. New York: United Nations Publications.
Vernon, R. (1979). The product cycle hypothesis in a new international environment. *Oxford Bulletin of Economics and Statistics*, *41*(4), 255–267.
Wang, W. and Liu Y. (2020). *Digital Belt and Road: Progress, Challenges and Practice Plan*. Retrieved from http://www.china.com.cn/opinion/think/2020-05/25/content_76086649.htm.
Wells, J. (1985). The role of construction in economic growth and development. *Habitat International*, *9*(1), 55–70.
Williamson, O.E. (1975). Markets and hierarchies: analysis and antitrust implications. A study in the economics of internal organization. Retrieved from https://ssrn.com/abstract=1496220.
World Economic Forum (WEF) (2016). *Shaping the Future of Construction: A Breakthrough in Mindset and Technology*. Retrieved from https://www.bcgperspectives.com/Images/ Shaping_the_Future_of_Construction_may_2016.pdf.

Ye, M., Lu, W., Flanagan, R., and Ye, K. (2018). Diversification in the international construction business. *Construction Management and Economics*, *36*(6), 348–361.
Yin, W. (2019). Integrating Sustainable Development Goals into the Belt and Road Initiative: would it be a new model for green and sustainable investment? *Sustainability*, *11*(24), 6991.

15. Economics of trust in construction

Anita Cerić

INTRODUCTION

Trust is essential to civil society, a factor granting legitimacy to political institutions, and an indicator of social cohesion (Putnam, 1993; Fukuyama, 1995; Murphy, 2006). Trust is often viewed as a foundation for social order (Lewicki et al., 1998, p. 438). Granovetter (1985) offered two main reasons why social relations may discourage malfeasance: individuals with whom one has a continuing relation have an economic motivation to be trustworthy, so as not to discourage future transactions; and departing from pure economic motives, continuing economic relations often become overlaid with social content that carries strong expectations of trust and abstention from opportunism.

Steve Knack, a senior economist at the World Bank who has been studying the economics of trust, suggests that it is worth $12.4 trillion a year to the United States economy (Forbes, 2010). The reason for this is that trust operates in all sorts of ways, from saving money that would have to be spent on security, to improving the functioning of the political system. Above all, trust enables people to do business with each other. Doing business is what creates wealth (Forbes, 2010). Economists mostly use the theory of repeated games to analyse institutions that support social order. The literature often revisits Plato's question, 'Who will guard the guardians?' The answer is that, in a world of self-interest, there is no disinterested third party to enforce society's rules, so some of those rules must be self-enforcing (Gibbons, 2000).

The aim of this chapter is to demonstrate the importance of trust in construction economics through application of the well-known theory in economics: principal–agent theory. According to the principal–agent theory, information asymmetry between principal and agent occurs due to opportunistic behaviour and self-interest. Information asymmetry causes communication risk, which is one of the most common risks in any project. There are different strategies for minimizing information asymmetry between project participants (see, for example, Schieg, 2008). This chapter focuses on the role of trust as one of the key strategies for minimization of information asymmetry between project participants in the construction phase of the project. This result comes from an empirical research study conducted from 2010 to 2014 involving project managers with considerable experience and expertise in the field (Cerić, 2016).

The principal–agent theory is part of the new institutional economics and has been receiving attention in various fields of application, including construction management. In the next section, the interconnections between the main concepts of the new institutional economics are investigated. This is important, as the main concepts of the subject are the principal–agent theory, transaction cost theory, asymmetric information and economic governance. In all these theories and concepts, trust plays an important role.

Trust inherently involves risk, but risk can be reduced by means of communication, and communication leads to better cooperation. Therefore, trust enhances cooperation (Cook et al., 2005). According to Williams (2001, p. 377), interpersonal trust is an important social resource that facilitates cooperation and enables coordinated social interaction. Trust is not a control mechanism, but a substitute for control (Rousseau et al., 1998, p. 399). According to Murphy (2006), the concept of trust is organized into three broad categories: trust as a rationally calculated and transaction cost-reducing input for exchange relationships; trust as a structurally embedded characteristic of inter- and intraorganizational relations; and trust as an emergent phenomenon constructed through social performances. Bromiley and Cummings (1995) claim that trust reduces transaction cost, and they suggest that the inclusion of trust would expand and extend the research framework of transaction cost economics. They also developed a trust concept addressing calculative and non-calculative components.

Based on a review of the literature, Li et al. (2012) argue that trust tends to matter the most in the following contexts: when the uncertainty is high; when the vulnerability of control (such as failure of formal contract) is high; when the stakes (such as financial loss) of unmet expectation or control failure are high; and when long-term interdependence (such as a reciprocal relationship) is high. Investigating trust and its different dimensions in construction projects is of great importance. The duality of trust and control depends on the context in which they occur. Trust and control, as well as institutional and personal trust, can simultaneously complement and substitute for each other (Welter, 2012).

The origin and meaning of 'trust' in English is associated with veracity, integrity, and other virtues of someone or something. Coming from *traust* in Old Norse, it harks back to help, confidence, protection and support that were essential in olden times. In today's English, trust is often associated with fidelity, faithfulness and belief. It is also pertinent to look into trust from the vantage point of other concepts that are of central importance in this chapter, which are communication and risk. Communication comes from *communicatio* in Latin, which stands for sharing and dividing out, as well as joining, uniting and participating in something. This originates from Latin *communis*, or making common. On the other hand, risk comes from *risque* in French, which stands for running into harm or danger. In Latin, *fides* stands for trust. Its other meanings are faith, belief, confidence, loyalty and promise of protection. Having already touched on communication in Latin, risk or *periculum* has a number of other meanings, including danger, peril, trial, insecurity, jeopardy, attempt and hazard. It is pertinent to note that Latin words for trust and risk are often opposites of each other.

The importance of research on trust is well recognized in most management disciplines. For example, in the last two decades, the Academy of Management Review has published two special issues on trust; two research journals dedicated to trust research have been established: *Journal of Trust Research* (Taylor & Francis) and *Journal of Trust Management* (Springer); Russell Sage Foundation (RSF) has published the RSF Series on Trust with focus on the conceptual and empirical research into the role of trust in the social, economic and political context. The First International Network on Trust (FINT) was created in 2001; it is a group of academics and practitioners from around the globe who are interested in the study of trust. *International Project Management Journal* published a special issue on 'Trust and Governance in Megaprojects' in May 2021. And so on, and so forth. Trust has become one of the important subjects in the field of construction management in the last two decades, for it is increasingly clear that external authority is rarely sufficient to guarantee successful completion of construction projects.

The overlap of the principal–agent theory and transaction cost theory within the new institutional economics shows that links between economics, sociology and psychology are essential for building an understanding of construction projects. Construction projects are essential to the development of the field of construction economics. Through links between these three social sciences, it is possible to understand better how construction projects are managed. This is why these economic theories have such an important place in the management literature. And this is why trust is essential in understanding relations between project participants.

The rest of the chapter is organized as follows. First, an analysis of the literature on the new institutional economics is presented. Next, the application of the principal–agent theory to construction projects is described. Special focus is given to trust in minimization information asymmetry between project participants. The chapter ends by offering a framework for exploring different forms of trust in construction projects and guidance for future research focusing on reputation, trust and reciprocity.

NEW INSTITUTIONAL ECONOMICS

The foundation of the new institutional economics was provided by Ronald Coase (1937), where he investigated why economic activities take place through different institutions, such as firms, and not only through markets. The new institutional economics (Williamson, 1985), comprises the principal–agent theory and the transaction cost theory, as well as the property rights theory and the contract theory (Jäger, 2008). The transaction cost theory and the principal–agent theory are characterized by many similarities, but also by some differences (Williamson, 1988). According to Williamson (1993), there are three issues that create a need for trust: the bounded rationality or limited information that constrains the decision-making capabilities of individuals; the value of trust as a highly specific asset that improves the efficiency of particular transactions; and the potential for opportunistic behaviour by the other agents in a transaction.

The central part in the principal–agent theory, sometimes referred to as the agency theory (Eisenhardt, 1989), is the relationship between the orderer as the principal and the contractor as the agent. Both parties are individuals who act out of their self-interest. Neither party is fully informed about the motivations of the other party, and both parties can therefore act opportunistically. According to the transaction cost theory, as well, every transaction involves a social relationship between individuals engaged in the exchange of goods or services. Each transaction takes place in an environment in which collective rules apply. Economic governance concerns these rules in firms and other institutions.

George Akerlof, Michael Spence and Joseph Stiglitz received the Nobel Prize in Economic Sciences in 2001 for their work on information asymmetry. Akerlof's (1970) 'market for lemons' is one of the best-known examples of adverse selection effects, in which used cars of different levels of quality are traded between buyers and sellers. It provides one of the best-known applications of information asymmetry in economics based on the principal–agent theory. Supervising the exchange of goods and services in order to control opportunistic behaviour involves a variety of costs such as screening and monitoring by the principal, to signalling and reputation by the agent. These are often referred to as agency costs. Due to asymmetric information the problems that arise are those of adverse selection, moral hazard and hold-up. Contracts that arise under threat of opportunism are referred to as relational contracts. They revolve around trust between the parties involved, because asymmetric information cannot be removed by contracts alone.

In the following text, the literature review on the new institutional economics is presented to show how much attention the subject has received in the construction management field. This

Table 15.1 Top construction management journals

Journal	Acronym	Publisher
Automation in Construction	*AIC*	Elsevier
Building Research and Information	*BRI*	Taylor & Francis
Construction Innovation	*CI*	Emerald
Construction Management and Economics	*CME*	Taylor & Francis
Engineering, Construction and Architectural Management	*ECAM*	Emerald
International Journal of Project Management	*IJPM*	Elsevier
Journal of Construction Engineering and Management	*JCEM*	ASCE

Source: Bröchner and Björk (2008).

is important, because the main concepts of the new institutional economics being investigated are the principal–agent theory, transaction cost theory, asymmetric information and economic governance. In all these concepts, trust plays an important role. Principal–agent theory describes the situation where the two parties have different interests and there is asymmetric information between them. Trust plays an important role in minimization of information asymmetry in the construction project phase, due to the many participants who are not bound by any contract. Trust also may decrease transaction cost or agency cost in projects. Some studies have shown that trust, when used for governance, reduces transaction costs in organizations and improves performance (for example, Das and Teng, 1998; Gulati and Nickerson, 2008).

For the literature review on the new institutional economics, the seven leading journals in the field identified by Bröchner and Björk (2008, p. 742) were searched. Their research focused on the preferences of the authors contributing to the construction management field. They followed the most-cited authors, whose preferences they investigated by an opinion survey. In the process, they identified 45 journals in the field, from which they identified the seven leading ones. The seven leading journals in construction management identified by Bröchner and Björk (2008, p. 742), together with their publishers, are shown in Table 15.1. It should be noted that these journals, 12 years after the research by Bröchner and Björk (2008), are still the best journals in the field according to Scientific Journal Rankings.

For purposes of this literature analysis, the online archives of the journals in Table 15.1 were searched for the following leading keywords: new institutional economics, transaction cost theory, principal–agent theory, economic governance, and asymmetric information. Again, these are the central concepts of the theoretical framework investigated. The analysis of construction management literature proceeded in three distinct steps. First, the papers in the leading journals containing the above keywords were identified by literature search (these keywords appeared anywhere in the papers). Second, the identified papers were analysed to identify the keywords listed by the authors. Third, the identified keywords were analysed for their interconnections, which suggest connections between the underlying concepts, as well. Keywords are important as they define the field, topic and research issue covered by the paper. Also, the keywords are important for search engines and databases and journal websites. Analysis of keywords listed by the authors and their interconnections give us representation on how the research community perceive investigated concepts.

It should be noted that the archives of the journals go back to different years. In this case, the *JCEM* archive goes back to 1930, *BRI* to 1973, *CME* and *IJPM* to 1983, *AIC* to 1992, *ECAM* to 1994, and *CI* to 2001. Therefore, the present review is slightly biased towards the journals with further-reaching archives. However, as will be shown below, the bulk of the literature cited

Table 15.2 Incidence of leading concepts in papers in selected journals

Journal Concept	*AIC*	*BRI*	*CI*	*CME*	*ECAM*	*IJPM*	*JCEM*
New institutional economics	0	4	3	6	3	10	5
Principal–agent theory	1	5	7	23	17	71	25
Transaction cost theory	27	31	16	184	68	200	78
Asymmetric information	10	15	6	38	30	77	94
Economic governance	0	4	0	5	0	1	1

Table 15.3 Incidence of leading concepts in keywords in selected journals

Journal Keyword	*AIC*	*BRI*	*CI*	*CME*	*ECAM*	*IJPM*	*JCEM*
New institutional economics	0	0	0	0	0	0	0
Principal–agent theory	0	1	0	2	2	6	3
Transaction cost theory	0	1	2	20	3	13	7
Asymmetric information	1	1	0	0	0	1	2
Economic governance	0	2	0	4	0	0	1

falls within the last decade. Therefore, the historical reach of the archives does not appear to be of great relevance in this case.

Table 15.2 presents a summary of the search conducted in the seven leading journals. Altogether, keywords central to the theoretical framework of the new institutional economics appeared 1065 times. As noted above, there were five such keywords, of which the new institutional economics occurred 31 times, principal–agent theory 149 times, transaction cost theory 604 times, asymmetric information 270 times, and economic governance 11 times. As explained above, these keywords correspond to the main concepts of the theoretical framework in question. There are some overlaps in this count, as several of these concepts can sometimes be found in the keywords listed in the same paper. However, these overlaps are not large.

Table 15.3 shows all the leading keywords found among the keywords listed in papers that appeared in the seven selected journals. Altogether, there are 72 such keywords that appear in 68 papers as follows: one in *AIC*, five in *BRI*, two in *CI*, 26 in *CME*, five in *ECAM*, 20 in *IJPM*, and 13 in *JCEM*. As was noted above, this shows that there have been few overlaps in the results of the search.

The papers whose keywords contain one or more of the five leading keywords are presented in Table 15.4. Again, there are 68 of them. As in the previous tables, they are organized by leading journals in the field of construction management and economics, identified by Bröchner and Björk (2008).

As can be seen from Table 15.4, most of the papers identified by literature search come from the third millennium. The only exceptions here are Reve and Levitt (1984), De Witt (1986) and Winch (1989). These are the forerunners in the construction management field with respect to the new institutional economics. In particular, all three early papers refer to the transaction cost theory. The peak was in 2015, when seven papers bearing the keywords central to the theoretical framework of the new institutional economics appeared in print. The next peak can be found in 2016, when five papers were published. This shows that the research in question is by and large recent.

Table 15.4 Papers cited

Journal	Papers cited
AIC	Xu et al. (2019)
BRI	Lützkendorf and Speer (2005), Sha (2013), Qian et al. (2016), Sha and Wu (2016)
CI	Finne (2006), Ross (2011)
CME	Winch (1989), Lai (2000), Costantino et al. (2001), Chang and Ive (2002), Rahman and Kumaraswami (2002), Bridge and Tisdell (2004), Turner (2004), Chang (2007), McCann (2006), Boukendour (2007), Chang and Ive (2007b), Ive and Chang (2007), Jin and Doloi (2008), Lai et al. (2008), Petrovic-Lazarevic (2008), Yung and Lai (2008), Ho et al. (2009), Roehrich and Lewis (2010), Sha (2011), Wu et al. (2011a), Chang (2013a, 2013b), Rajeh et al. (2015a), Sha (2019)
ECAM	Li et al. (2014), Rajeh et al. (2015b), Carbonara et al. (2016), Wang and Shi (2019), Zhang et al. (2020)
IJPM	Reve and Levitt (1984), De Witt (1986), Turner and Müller (2003), Zaglouhl and Hartman (2003), Müller and Turner (2005), Chang and Ive (2007a), Jin and Zhang (2011), Li et al. (2012), Bond-Bernard et al. (2013), Chang (2013c), Chang (2013d), Chang (2015), De Schepper et al. (2015), Lu et al. (2015), Xiang et al. (2015), Liu et al. (2016), Teo and Bridge (2017), Zheng et al. (2017), Tang et al. (2020)
JCEM	Lee et al. (2009), Jin (2010, 2011), Wu et al. (2011b), Xiang et al. (2012), Li et al. (2013), Zerjav et al. (2013), Hosseinian and Carmichael (2013), Chang (2014), Chang and Chou (2014), Chang and Qian (2015), Szentes and Eriksson (2016), Yao et al. (2020)

Table 15.5 shows all interconnections between the leading keywords from the new institutional economics, as well as interconnections between them and a number of associated keywords. Altogether, there are 18 of them. As all connections are bidirectional, the table is triangular in structure. The values on the diagonal show all cases in which a keyword appears by itself, where the keywords in question are connected only to themselves.

Figure 15.1 offers a network representation of interconnections between keywords in Table 15.5. In other words, it is the graph equivalent to the matrix of keyword interconnections. The cases where keywords are connected only to themselves are represented by loops.

In summary, Figure 15.1 represents a mapping of the collective appreciation of the new institutional economics by the research community in construction management over some two decades. As can be seen from this mapping, the principal–agent theory and transaction cost theory are weakly connected in the network of keywords. The connection is provided by only two of the 68 papers identified. The same holds for the two theoretical frameworks and the Coase model, which is central to new institutional economics; as well as the connections between the Coase model, the property rights theory and the transaction cost theory. The keywords transaction cost theory and property rights theory are connected only in one paper. The same holds with the principal–agent and Coase model.

The connections between the principal–agent theory, asymmetric information and adverse selection are also weak. In fact, the triangle connecting the three keywords in the network appears in only one paper identified by the literature search conducted in this investigation. The situation is markedly different in the case of the transaction cost theory and its connections to economic governance, hold-up and opportunism. This part of the network is interconnected more richly than the other. For example, opportunism and the transaction cost theory are connected five times. In addition, a large number of cases can be found where the transaction cost theory is connected only to itself. Again, the interconnections between the two theoretical frameworks are largely neglected.

It is pertinent to briefly review the network shown in Figure 15.1 in terms of network connectivity as conceived in network analysis (Rodrigue et al., 2009, p. 31). The Gamma index

Table 15.5 Interconnections between leading and associate keywords from the new institutional economics

Keywords	1	2	3	4	5	6	7	8	9	10	11	12	13	14	15	16	17	18	Sum
1. Adverse selection																			0
2. Agency costs																			0
3. Asymmetric information	1		3																4
4. Coase model																			0
5. Contract theory			1																1
6. Economic governance						5													5
7. Hold-up																			0
8. Monitoring																			0
9. Moral hazard																			0
10. New institutional economics																			0
11. Opportunism																			0
12. Principal–agent theory	1		1	1	3				1		1	5							13
13. Property rights theory				1															1
14. Relational contracts																			0
15. Reputation																			0
16. Screening																			0
17. Signalling																			0
18. Transaction cost theory				2	2	4	4				5	2	1					28	48
Sum	2	0	5	4	5	9	4	0	1	0	6	7	1	0	0	0	0	28	72

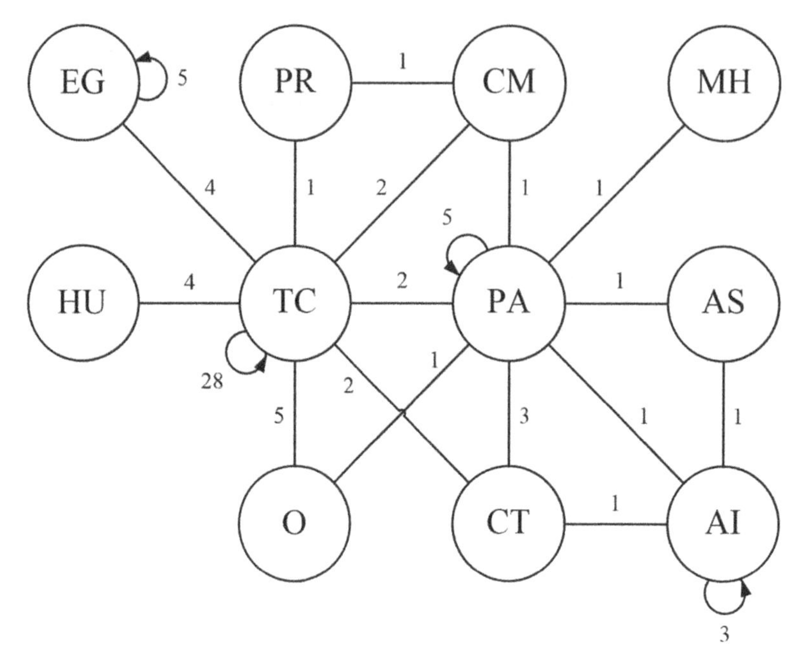

Note: AI: asymmetric information; AS: adverse selection; CM: Coase model; CT: contract theory; EG: economic governance; HU: hold-up; MH: moral hazard; O: opportunism; PA: principal–agent theory; PR: property rights theory; TC: transaction cost economics.

Figure 15.1 Network of interconnections between keywords in the new institutional economics

of network connectivity is a ratio between the actual number of links (e) and the potential number of links given by the number of nodes (v) in a network. It varies between zero and one. In terms of New Institutional Economics, all the concepts represented by the keywords in the network ought to be connected. The potential number of links is equal to $1/2\ v(v-1)$. In the case here, $e = 16$. The potential number of links is therefore $1/2\ 11(11-1) = 55$. In other words, the Gamma index is equal to $16/55 = 0.29$ or 29 per cent. Thus, the mapping in Figure 15.1 shows relatively low connectivity in comparison with the potential connectivity of the network.

Analysis of the interconnections between the new institutional economics concepts in the literature shows that the most distinguished among all papers is the paper by Lai et al. (2008), which examines the Coase model and its role in the construction management field. The literature reviewed in the paper is exemplary, and the paper offers a Coasian research agenda for the field. However, it is evident from Figure 15.1 that a great number of papers present empirical work without a proper theoretical foundation. In other words, the connectivity among all the concepts represented by the keywords in the network ought to be much higher.

APPLICATION OF PRINCIPAL–AGENT THEORY TO CONSTRUCTION PROJECTS AND THE IMPORTANCE OF TRUST IN MINIMIZATION OF INFORMATION ASYMMETRY

In this section the application of the principal–agent theory to construction projects is presented. Information asymmetry causes communication risk, which is one of the most important risks that appears in every phase of the construction project. Information asymmetries and opportunistic behaviour are more prominent on construction projects than in some other types of projects. There are several reasons for this: construction projects involve a great number of participants over a long period of project duration; among the large number of participants in the construction phase there are no contractual relationships; government often makes decisions on major construction investments; and the tendency towards involvement in corruption is more prominent than in other industries.

Information asymmetries apply whenever the principal and the agent are not in possession of the same information at the same time. According to Turner and Müller (2004) the project owner and contractor form the key relationship in construction projects. The delegation of tasks forms a principal–agent relationship between the project owner and contractor, wherein the principal (project owner) depends on the agent (contractor) to undertake a task on the principal's behalf (Müller and Turner, 2005). One can act on the assumption that agents try to maximize their own benefit even when it may involve significant damage to the client (Schieg, 2008). This problem is characterized by three issues of risk concerning the principal–agent relationship: adverse selection, moral hazard and hold-up. Adverse selection occurs when the principal does not have the exact qualifications of the agent before the contract is signed. In the case of moral hazard, the principal cannot be certain that the agent will fully act on the principal's behalf after the contract is signed. Hold-up occurs when the principal has invested some resources in the belief that the agent will behave appropriately, but the agent acts opportunistically after the contract is signed (Schieg, 2008).

The relationship between the project owner and contractor is extended, and includes their respective project managers, as illustrated in Figure 15.2. The project owner is the overall principal and all the others are agents directly or indirectly employed by the project owner. These four participants are crucial in every construction project. The contractor is the principal with respect to their project manager. It is important to note that the project owner's and contractor's project managers play important roles in any construction project even though they are not in a contractual relationship with each other. They can be praised or blamed for success or failure on the project, and thus they have a great moral responsibility (Corvellec and Macheridas, 2010). It is commonly assumed that all participants in the project will work smoothly together in order to achieve the same goal. However, there is a potential conflict of interests between the participants because they all have their self-interests, too.

The principal–agent framework presented in Figure 15.2 is applied to construction projects using different research techniques from the exploratory survey, Delphi method and different multi-attribute decision-making methods (Cerić, 2016). Due to space limitations, only the main findings enhancing the role of trust in minimization of information asymmetry are presented in this chapter. A survey of project managers (Cerić, 2014a) with considerable experience in managing large and complex projects around the globe showed that the main strategy for the minimization of communication risk caused by information asymmetry in the construction phase is to build trust among project participants. As the principal–agent theory

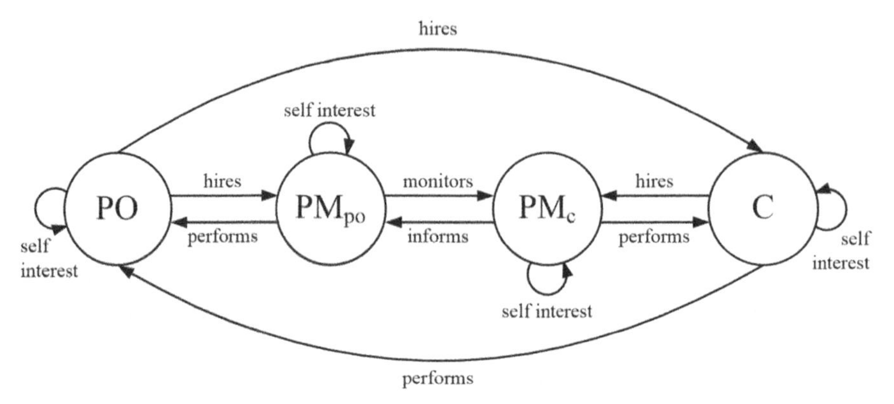

Note: PO: project owner, C: contractor, PMpo: project owner's project manager, PMc: contractor's project manager.
Source: Cerić (2016).

Figure 15.2 Principal–agent theory framework for construction projects

shows, even contractual relationships can be guided by opportunistic behaviour between the project parties. However, such behaviour is even more difficult to assess, let alone to control, in non-contractual relationships. For instance, project managers forming the project team can either collude to the detriment of the project owner or the agents, or they can be involved in internal conflict detrimental to the project owner. Thus, it is not surprising that the main strategy in the management of large projects is found by empirical research to be trust (Cerić, 2014a). Although contracts play an important role in such projects, trust appears to be the only viable way forward towards successful performance on projects.

Schieg (2008) offered the following information asymmetry minimization strategies which are most promising in the management of construction projects: bureaucratic control (contracts), information systems, incentives (bonuses), corporate culture, reputation and trust. The project managers involved in the ranking of these strategies had considerable experience on sizable construction projects across a large number of countries. The ranking of strategies established by this research shows that trust is considered to be the winning strategy, followed by bureaucratic control (contracts), information systems, reputation, corporate culture and incentives, in that order (Cerić, 2014a). In short, contracts and the remaining strategies do have a role to play, but they can be successful only up to a point.

As Figure 15.3 shows, in construction projects the non-contractual relationships start to dominate the contractual ones (Cerić, 2014b). In the case of large and complex projects that require considerable time for completion, the project team becomes increasingly autonomous from the project owner, as well as the contractor, designer, and consultants as agents. Even though contracts will always play an important role in construction projects, the need for trust between project parties grows exponentially with project size.

In general, the relationship between project participants is controlled by means of contract (Bower and Skountzos, 2003). The contract expresses the intentions of the two parties, and so the roles and responsibilities of both sides are evident in case any dispute arises (Simister and Turner, 2003). However, the ultimate success of any construction project is questionable if there is no trust, even when compelling control systems, including contractual documents, are being utilized (Zaghloul and Hartman, 2003). Therefore, building trust in construction projects

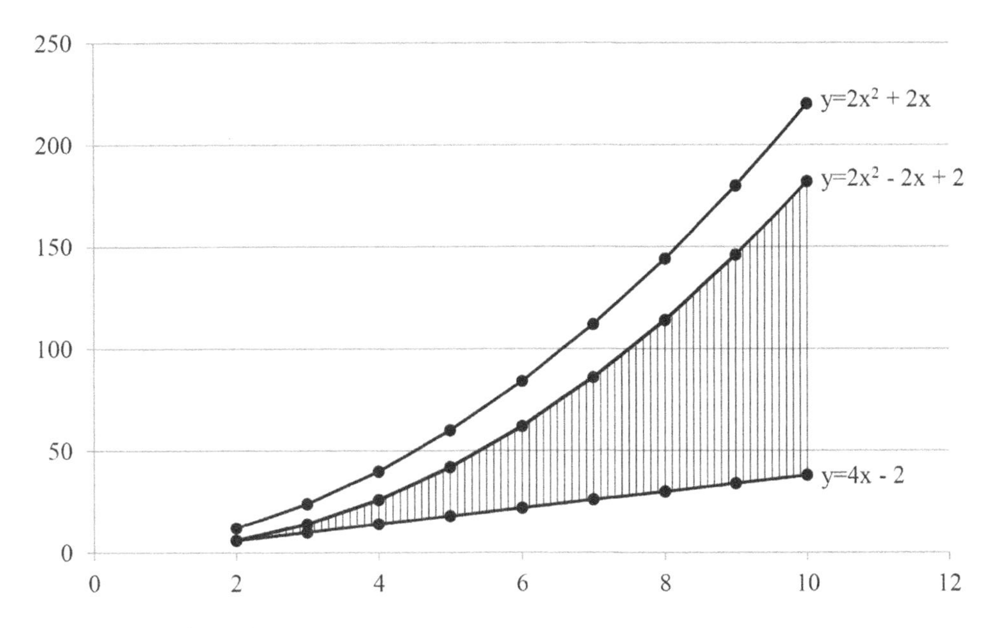

Source: Cerić (2014b, 2016).

Figure 15.3 *Project parties and relationships*

is very important in order to achieve project goals. Murphy (2006) also emphasized that, in contrast to the transaction cost perspectives, where trust is a rationally calculated input, organizational theorists and economic sociologists generally view trust as a structurally embedded asset of property relationships, organizations, and networks that helps to mobilize resources, enable cooperation, and shape interaction patterns within economies, industries and firms.

In construction projects, which involve many different participants, and where non-contractual relationships in the construction phase dominate, building trust is of great importance. Again, this gap itself cannot be addressed by additional contractual arrangements, but project parties can endeavour to increase the trust between project parties. As Zaghloul and Hartman (2003, p. 421) point out, there is a cost to mistrust:

> With the absence of trust in business relationships, there is a significant need for good and powerful control system to manage and administrate the contracting process. However, even with the existence of this powerful control system (the contract documents), with the absence of trust, the success of any project or business relationship is always questionable.

In addition, there are agency costs allocated to the monitoring of contract.

To find out how much the concept of trust has gained attention in the field of construction management and economics, again, seven leading journals were searched using the same methodology presented in the previous section, The results show that, up to 2015, the research on trust in the field of construction management and economics put strong emphasis on partnering. This form of collaboration between firms sometimes goes under the name of alliancing. Although partnering is of great importance in contemporary construction, and although it stands to reason that interfirm trust dominates the research in construction management and economics, other forms of trust need to be given greater attention in the future.

Since 2015, the research on partnering has significantly decreased, but other issues connected with trust have emerged. They include: relational behaviours, communication, collaboration, incentives, building information modelling, and blockchain technology. It is pertinent to observe that the research community in the field of construction management and economics goes along with the findings from the survey of project managers presented earlier, where they ranked information system strategy very high in connection to minimization of information asymmetry. Additionally, the project managers argue that establishing communication protocols would play an important role in minimizing information asymmetry, and increase the efficiency of communication on construction projects. Following lesson learned from practitioners, Cerić (2019) proposed the implementation of 'blockchain' as the record-keeping technology for establishing communication protocols in projects, in order to improve communication among project participants and increase transparency. Thus, this will increase trust, too.

FRAMEWORK FOR FURTHER RESEARCH ON TRUST IN CONSTRUCTION PROJECTS: TRUST, RECIPROCITY AND REPUTATION

Further research on trust should focus on the study of different dimensions of trust related to intrafirm and interfirm relationships. Research on trust in construction projects needs to focus on both empirical research and its theoretical foundations. Field research is crucial to enable better understanding of the different dimensions of trust over the duration of the project. This is where researchers and practitioners must work together and improve current theory and practice in the field of construction management and economics. On trust, researchers and practitioners in the field need to establish a close relationship conducive to the advancement of both theory and practice. The empirical research would help in better understanding of reciprocal relationships and trustworthy behaviour between project participants. At the core of a behavioural explanation are the links between the trust an individual has in others, the investments others make into trustworthy reputations, and the probability of using reciprocity norms (Ostrom, 2003). Trust is crucial in economic and social exchange; thus, researchers from disciplines such as economics, sociology and psychology should be involved in studying trust.

The principal–agent theory framework for construction projects includes interpersonal, interfirm and intrafirm relationships (Figure 15.4). Trust involves project participants interacting at interpersonal and interfirm levels. Trust involved in interfirm relationships falls mainly into the domain of economics, but both sociology and psychology can contribute to the study; trust involved in intrafirm relationships relies mainly on sociology, but economics and psychology are also relevant to its study; and trust involved in interpersonal relationships falls mainly in the domain of psychology, but both sociology and economics can also contribute to studying it.

Projects, as temporary organizational frameworks, deserve special attention in this regard. Various types of trust are deeply embedded in construction projects. In particular, interpersonal, intrafirm and interfirm trust become intertwined in intraproject trust. Although interpersonal trust underlies all these types of trust, the interaction among them needs to be addressed

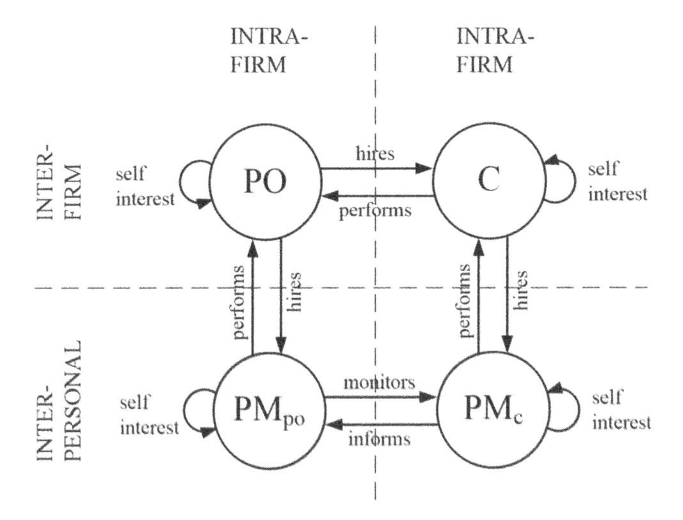

Note: PO: project owner, C: contractor, PMpo: project owner's project manager, PMc: contractor's project manager.
Source: Cerić (2014c, 2016).

Figure 15.4 *Remapping of the principal–agent theory framework for construction projects showing key relationships between project parties*

by future research. This applies especially to the project dynamics, where the relationships between project parties continually change, as does the expertise each party brings to the project. The cultural dimension of trust also needs to be taken into consideration when different dimensions of trust are concerned. Intercultural and intracultural trust need to be carefully differentiated in this context.

As Williams (2001, p. 377) points out, interpersonal trust is an important social resource that facilitates cooperation and enables coordinated social interaction. Thus, the study of interpersonal trust has helped social scientists to better understand the dynamics of cooperation (Lewicki et al., 2006, p. 991). Interpersonal trust underlies all other forms of trust. Also, trust within groups leads to trust between groups, which include both private and public organizations. Within economics, economic sociology and organizational theory, trust reflects the alliance strategies used in interfirm and organizational relations and agents (Murphy, 2006).

The interplay of trust and reciprocity in building interpersonal, intrafirm and interfirm organizational commitment would be worth exploring in the construction projects context. Trust and reciprocity are expectations or beliefs about the intentions and trustworthiness of others, and are the common elements in most definitions of collective trust (Lewicki and Brinsfield, 2011). The scholars share broad definitions of trust which they generally consider as the willingness to take some risk in relation to other individuals in the expectation that the others will reciprocate (Ostrom, 2003). According to Welter (2012), reciprocity signals to both trustor and trustee that the trust they extend to each other will be returned. In this regard, trust is based on a perception of the probability that other agents will behave in a way that is benevolent.

Reputation is a key information asymmetry minimization strategy for minimization of the adverse selection problem according to principal–agent theory. Reputation plays an important role in interfirm relationships and should be investigated in future research. The longer one spends building a reputation, the less likely one is to risk it by taking one quick gain (Ensminger, 2001, p. 199). Also, reputation is connected to professionalism. Ofori and Cerić (2018, p. 430) observe that trust is an important aspect of professionalism, and that a new concept of professionalism in the built environment is required. They note that if the construction industry is to be able to meet the high aspirations set for it in terms of the nature and quality of its output and its performance, the professionals should be technologically sophisticated, autonomous, self-regulating, and sensitive to local needs and cultures. Trust should be their watchword; they should exhibit it towards, and merit it from, all partners and stakeholders.

CONCLUSIONS

The proposition that trust is of great importance to construction projects comes both from empirical research and from theoretical foundations of associated fields of social science, including economics, sociology and psychology. Different types of trust are part of construction projects. Most importantly, interpersonal, intrafirm and interfirm trust become intertwined in intraproject trust. As a survey of project managers with considerable experience in managing large and complex projects around the world shows, the central strategy for the minimization of communication risk caused by information asymmetry during the construction phase is to develop and maintain trust between all project participants. Trust is the key to project success.

As far as construction projects are concerned, it is the project managers as agents who play most important roles in the construction phase of a project that often takes a considerable amount of time to bring to completion. Even though they are principals to their agents, the project owner and the contractor play only subsidiary roles during this particular phase of project execution. In fact, neither the project managers nor their teams are bound by contracts. Their cooperative behaviour stems from their contractual obligations to their principals, professional norms and mutual trust. A more profound understanding of that interaction will be of value to other fields in which project managers play central roles in the execution of complex projects. Therefore, monitoring the monitors is a concern of wider importance in the management and economics fields.

To date, the main interest of researchers in construction management and economics associated with the principal–agent theory has to do with contracts and risk management. At the same time, a good number of papers in the field explore the new institutional economics concepts empirically without proper theoretical foundations. The same observation holds for research concerning trust: theoretical foundations are typically missing.

In future, the literature review presented in this chapter should be extended in several interrelated ways. In addition, the overlap between the principal–agent theory and transaction cost theory within the framework of the new institutional economics should be explored with greater care in the years to come. A literature analysis connecting these two theoretical frameworks would be helpful in pinpointing the substantial potential commonalities and differences between the two in the context of construction management. More to the point, such a review would need to concentrate on the difference between markets and firms, as well as a wide variety of other institutions involved in construction project management and economics.

Eventually, both economic theory and management practice will benefit from research on trust in construction projects. Given that it is the leading strategy in information asymmetry minimization in the construction phase, it should receive greater attention from researchers in the field. The principal–agent theory demonstrates that even contractual relationships can be affected by opportunistic behaviour between project parties. However, behaviour of this kind is even more difficult to establish and control in non-contractual relationships that prevail in large and complex construction projects. Better collaboration among researchers concerned with various aspects of construction projects will be invaluable to construction management and economics, and project management, but it will also be of great value to the field of construction economics, and of management as a whole. Better understanding of construction projects requires joint research projects integrating several interrelated social sciences. In the final analysis, this will be reflected in practice, as well.

REFERENCES

Akerlof, G. (1970). The market for lemons: quality, uncertainty, and the market mechanism. *Quarterly Journal of Economics*, *84*(3), 488–500.

Bond-Bernard, T.J., Steyn, H. and Fabris-Rotelli, I. (2013). The impact of a call centre on communication in a programme its projects. *International Journal of Project Management*, *31*(7), 1006–1016.

Boukendour, S. (2007). Preventing post-contractual opportunism by an option to switch from one contract to another. *Construction Management and Economics*, *25*(27), 723–727.

Bower, D. and Skountzos, F. (2003). Partnering, benchmarking and incentive contracts. In R. Turner (ed.), *Contracting for Project Management* (pp. 81–104). Aldershot: Gower Publishing.

Bridge, A.J. and Tisdell, C. (2004). The determinants of the vertical boundaries of the construction firm. *Construction Management and Economics*, *4*(3), 233–236.

Bromiley, P. and Cummings, L.L. (1995). Transactions costs in organizations with trust. In R. Bies, B. Sheppardand and R. Lewicki (eds), *Research on Negotiations in Organizations*, (Vol. 5, pp. 219–247). Greenwich, CT: JAI Press.

Bröchner, J. and Björk, B-C. (2008). Where to submit? Journal choice by construction management authors. *Construction Management and Economics*, *26*(7), 739–749.

Carbonara, N., Costantino, N., and Pellegrino, R. (2016). A transaction costs-based model to choose PPP procurement procedures. *Engineering, Construction and Architectural Management*, *23*(4), 491–510.

Cerić, A. (2014a). Minimising communication risk in construction: a Delphi study of the key role of project managers. *Journal of Civil Engineering and Management*, *20*(6), 829–838.

Cerić, A. (2014b). The nemesis of project management: the gaping non-contractual gap. *Procedia – Social and Behavioral Sciences*, *119*, 931–938.

Cerić, A. (2014c). Communication risk and trust in construction projects: a framework for interdisciplinary research. In A. Raidén and E. Aboagye-Nimo (eds), Proceedings of the 30thAssociation of Researchers in Construction Management (ARCOM) Conference (pp. 835–844). Reading, United Kingdom: Association of Researchers in Construction Management.

Cerić, A. (2016). *Trust in Construction Projects*. London: Routledge.

Cerić, A. (2019). Blockchain strategy for minimizing information asymmetry in construction projects. In I. Završki, A. Cerić, M. Vukomanović, M. Huemann and D. Ronggui (eds), *14th Organization, Technology and Management in Construction Conference* (pp. 494–506). Zagreb: Croatian Association for Construction Management, University of Zagreb, Faculty of Civil Engineering.

Chang, C.Y. (2007). The determinants of the vertical boundaries of the construction firm: comment. *Construction Management and Economics*, *24*(3), 229–232.

Chang, C.Y. (2013a). A critical review of the application of TCE in the interpretation of risk allocation in PPP Contracts. *Construction Management and Economics*, *31*(2), 99–103.

Chang, C.Y. (2013b). When might a project company break up? The perspective of risk-bearing capacity. *Construction Management and Economics*, *31*(12), 1186–1198.
Chang, C.Y. (2013c). A critical analysis of recent advances in the techniques for the evaluation of renewable energy projects. *International Journal of Project Management*, *31*(7), 1057–1067.
Chang, C.Y. (2013d). Understanding the hold-up problem in management of megaprojects: the case of the Channel Tunnel rail link project. *International Journal of Project Management*, *31*(4), 628–637.
Chang, C.Y. (2014). Principal–agent model of risk allocation in construction contracts and its critique. *Journal of Construction Engineering and Management*, *140*(1), 04013032-1–9.
Chang, C.Y. (2015). Risk-bearing capacity as a new dimension to the analysis of project governance. *International Journal of Project Management*, *33*(6), 1195–1205.
Chang, C.Y. and Chou, H.-Y. (2014). Transaction-cost approach to the comparative analysis of user-pay and government-pay public–private partnership systems. *Journal of Construction Engineering and Management*, *140*(9), 04014039-1–13.
Chang, C.Y. and Ive, G. (2002). Rethinking the multi-attribute utility approach based procurement route selection method. *Construction Management and Economics*, *20*(3), 275–284.
Chang, C.Y. and Ive, G. (2007a). The hold-up problem in the management of construction projects: a case study of Channel Tunnel. *International Journal of Project Management*, *25*(4), 394–404.
Chang, C.Y. and Ive, G. (2007b). Reversal of bargaining power in construction projects: meaning, existence and implications. *Construction Management and Economics*, *25*(8), 845–855.
Chang, C.Y. and Qian, Y. (2015). An econometric analysis of holdup problems in construction projects. *Journal of Construction Engineering and Management*, *141*(6), 04015004-1–8.
Coase, R.H. (1937). The nature of the firm. *Economica (new series)*, *4*(16), 386–405.
Cook, K.R., Hardin, R. and Levi, M. (2005). *Cooperation without Trust?* Russell Sage Foundation Series on Trust, Vol. 8. New York: Russell Sage Foundation.
Corvellec, H. and Macheridis, N. (2010). The moral responsibilities of project selectors. *International Journal of Project Management*, *28*(3), pp. 212–219.
Costantino, N., Pietroforte, R. and Hamill, P. (2001). Subcontracting in residential and commercial construction: an empirical study. *Construction Management and Economics*, *19*(4), 439–447.
Das, T.K. and Teng, B.-S. (1998). Between trust and control: developing confidence in partner cooperation in alliances. *Academy of Management Review*, *23*(3), 491–512.
De Schepper, S., Haezendonck, E. and Dooms, M. (2015). Understanding pre-contractual transaction costs for public–private partnership infrastructure projects. *International Journal of Project Management*, *33*(4), 932–946.
De Witt, A. (1986). Cost-effective owner project management. *International Journal of Project Management*, *4*(2), 77–81.
Eisenhardt, K.M. (1989). Agency theory: an assessment and review. *Academy of Management Review*, *14*(1), 57–74.
Ensminger, J. (2001). Reputations, trust, and the principal agent problem. In K. Cook (ed.), *Trust in Society* (pp. 185–201). New York: Russell Sage Foundation.
Finne, C. (2006). Publishing building product information: a value net perspective. *Construction Innovation: Information, Process and Management*, *6*(2), 79–96.
Forbes (2010, July 21). *The Economics of Trust*. Retrieved from https://www.forbes.com/2006/09/22/trust-economy-markets- tech_cx_th_06trust_0925harford.html#32eb5e3f2e13 (accessed 5 July, 2020).
Fukuyama, F. (1995). *Trust: The Social Virtues and the Creation of Prosperity*. New York: Free Press.
Gibbons, R. (2000). Trust in social structures: Hobbes and Coase meet repeated games. In K. Cook (ed.), *Trust in Society* (pp. 332–353), Russell Sage Foundation Series on trust, Vol. 2. New York: Russell Sage Foundation.
Granovetter, M. (1985). Economic action and social structure: the problem of embeddedness. *American Journal of Sociology*, *91*, 481–510.
Gulati, R. and Nickerson, J.A. (2008). Interorganizational trust, governance choice, and exchange performance. *Organization Science*, *19*(5), 688–708.
Ho, S.P., Lin Y-H., Wu, H.L. and Chu, W. (2009). Empirical test of a model for organizational governance structure choices in construction joint ventures. *Construction Management and Economics*, *27*(3), 315–324.

Hosseinian, S.M. and Carmichael, D.G. (2013). Optimal incentive contract with risk-neutral contractor. *Journal of Construction Engineering and Management*, *139*(8), 899–909.
Ive, G. and Chang, C.Y. (2007). The principle of inconsistent trinity in the selection of procurement systems. *Construction Management and Economics*, *25*(7), 677–690.
Jäger, C. (2008). *The Principal–Agent Theory within the Context of Economic Sciences*. Norderstadt: Books on Demand GmbH.
Jin, X-H. (2010). Determinants of efficient risk allocation in privately financed public infrastructure projects in Australia. *Journal of Construction Engineering Management*, *136*(2), 138–150.
Jin, X-H. (2011). Model for efficient risk allocation in privately financed public infrastructure projects using neuro-fuzzy techniques. *Journal of Construction Engineering Management*, *137*(11), 1003–1014.
Jin, X-H. and Doloi, H. (2008). Interpreting risk allocation mechanism in public–private partnership: an empirical study in a transaction cost perspective. *Construction Management and Economics*, *26*(7), 707–721.
Jin, X-H. and Zhang, G. (2011). Modelling optimal risk allocation in public private partnership using artificial neural networks. *International Journal of Project Management*, *29*(5), 591–603.
Lai, L.W.C. (2000). The Coesian market–firm dichotomy and subcontracting in the construction industry. *Construction Management and Economics*, *18*(3), 355–362.
Lai, L.W.C., Ng, F.W.N., and Yung, P. (2008). The Coase theorem and a Coasian construction economics and management research agenda. *Construction Management and Economics*, *26*(1), 29–46.
Lee, H., Seo, J., Park, M., Ryu, H., and Kwon, S. (2009). Transaction-cost based selection of appropriate general contractor–subcontractor relationship type. *Journal of Construction Engineering Management*, *135*(11), 1232–1240.
Lewicki, R. and Brinsfield, C. (2011). Measuring trust beliefs and behaviours. In F. Lyon, G. Möllering and M. Saunders (eds), *Handbook of Research Methods on Trust* (pp. 29–39). Cheltenham, UK and Northampton, MA, USA: Edward Elgar Publishing.
Lewicki, R.J., McAllister, D.J. and Bies, R.I. (1998). Trust and distrust: new relationships and realities. *Academy of Management Review*, *23*(3), 438–458.
Lewicki, R.J., Tomlinson, E.C. and Gillespie, N. (2006). Models of interpersonal trust development: theoretical approaches, empirical evidence, and future directions. *Journal of Management*, *32*(6), 991–1022.
Li, H., Arditi, D. and Wang, Z. (2012). Transaction-related issues and construction project performance. *Construction Management and Economics*, *30*(2), 151–164.
Li, H., Arditi, D. and Wang, Z. (2013). Factors that affect transaction cost in construction projects. *Journal of Construction Engineering Management*, *139*(1), 60–68.
Li, H., Arditi, D. and Wang, Z. (2014). Transaction costs incurred by construction owners. *Engineering, Construction and Architectural Management*, *21*(4), 444–458.
Li, P.P. (2012). When trust matters the most: the imperatives for contextualising trust research. *Journal of Trust Research*, *2*(2), 101–106.
Liu, J., Gao, R., Cheah, C.Y.J. and Luo, J. (2016). Incentive mechanism for inhibiting investors' opportunistic behavior in PPP projects. *International Journal of Project Management*, *34*(7), 1102–1111.
Lu, W., Zhang, L. and Pan, J. (2015). Identification and analyses of hidden transaction costs in project dispute resolutions. *International Journal of Project Management*, *33*(3), 711–718.
Lützkendorf, T. and Speer, T.M. (2005). Alleviating asymmetric information in property markets: building performance and product quality as signals for consumers. *Building Research and Information*, *33*(2), 182–195.
McCann, P. (2006). On the supply-side determinants on the regional growth. *Construction Management and Economics*, *24*(7), 681–693.
Murphy, T. (2006). Building trust in economic space. *Progress in Human Geography*, *30*, 427–450.
Müller, R. and Turner, J.R. (2005). The impact of principal–agent relationship and contract type on communication between project owner and manager. *International Journal of Project Management*, *23*(5), 398–403.
Ofori, G. and Cerić, A. (2018). A new professionalism in construction: importance of trust. In C. Egbu and G. Ofori (eds), *CIB: TG 95 Conference: International Conference on Professionalism and Ethics in Construction* (pp. 425–435). London: London South Bank University.

Ostrom, E. (2003). Toward a behavioural theory linking trust, reciprocity, and reputation. In E. Ostrom and J. Walker (eds), *Trust and Reciprocity: Lessons from Experimental Research* (pp. 19–79). New York: Russell Sage Foundation.
Petrovic-Lazarevic, S. (2008). The development of corporate social responsibility in the Australian construction industry. *Construction Management and Economics*, *26*(2), 93–101.
Putnam, R.D. (1993). *Making Democracy Work: Civic Traditions in Modern Italy*. Princeton, NJ: Princeton University Press.
Qian, Q.K., Fan, K. and Chan, E.H.W. (2016). Regulatory incentives for green buildings: gross floor area concessions. *Building Research and Information*, *44*(5–6), 675–693.
Rahman, M.M. and Kumarswamy, M.M. (2002). Joint risk management through transactionally efficient relational contracting. *Construction Management and Economics*, *20*(1), 45–54.
Rajeh, M.A., Tookey, J.E. and Rotimi, J.O.B. (2015a). Developing a procurement path determination chart SEM-based approach. *Construction Management and Economics*, *33*(11–12), 921–941.
Rajeh, M.A., Tookey, J.E., and Rotimi, J.O.B. (2015b). Estimating transaction costs in the New Zealand construction procurement: a structural equation modelling methodology. *Engineering, Construction and Architectural Management*, *22*(2), 242–267.
Reve, T. and Levitt, R.E. (1984). Organization and governance in construction. *International Journal of Project Management*, *2*(1), 17–25.
Rodrigue, J-P., Comtois, C. and Slack, B. (2009). *The Geography of Transport Systems* (2nd edition). New York: Routledge.
Roechrich, J.K. and Lewis, M.A. (2010). Towards a model of governance in complex (product-service) inter-organizational systems. *Construction Management and Economics*, *28*(11), 1155–1164.
Ross, A. (2011). Supply chain management in an uncertain economic climate: a UK perspective. *Construction Innovation: Information, Process and Management*, *11*(1), 5–13.
Rousseau, D.M., Sitkin, S.B., Burt, R.S. and Camerer, C. (1998). Not so different after all: a cross-discipline view on trust. *Academy of Management Review*, *23*(3), 393–404.
Schieg, M. (2008). Strategies for avoiding asymmetric information in construction project management. *Journal of Business Economic Management*, *9*(1), 47–51.
Sha, K. (2011). Vertical governance of construction projects: an information cost perspective. *Construction Management and Economics*, *29*(11), 1137–1147.
Sha, K. (2013). Professionalism in China's building sector: an economic governance perspective. *Building Research and Information*, *41*(6), 742–751.
Sha, K. (2019). Incentive strategies for construction project manager: a common agency perspective. *Construction Management and Economics*, *37*(8), 461–471.
Sha, K. and Wu, S. (2016). Multilevel governance for building energy conservation in rural China. *Building Research and Information*, *44*(5–6), 619–629.
Simister, S. and Turner, R. (2003) Standard form of contract. In R. Turner (ed.), *Contracting for Project Management* (pp. 59–63). Aldershot: Gower Publishing.
Szentes, H. and Eriksson, P.E. (2016). Paradoxical organizational tensions between control and flexibility when managing large infrastructure projects. *Journal of Construction Engineering and Management*, *142*(4), 05015017-1–10.
Tang, Y., Chen, Y., Hua, Y. and Fu, Y. (2020). Impacts of risk allocation on conflict negotiation costs in construction projects: does managerial control matter? *International Journal of Project Management*, *38*(3), 188–199.
Teo, P. and Bridge, A.J. (2017). Crafting an efficient bundle of property rights to determine the suitability of a public–private partnership: a new theoretical framework. *International Journal of Project Management*, *35*(3), 269–279.
Turner, J.R. (2004). Farsighted project contract management: incomplete in its entirety. *Construction Management and Economics*, *22*(1), 75–83.
Turner, J.R. and Müller, R. (2003). On the nature of the project as a temporary organization. *International Journal of Project Management*, *21*(1), 1–8.
Turner, J.R. and Müller, R. (2004). Communication and co-operation on projects between the project owner as principal and the project manager as agent. *European Management Journal*, *22*(3), 327–336.

Wang, Q. and Shi, Q. (2019). The incentive mechanism of knowledge sharing in the industrial construction supply chain based on a supervisory mechanism. *Engineering, Construction and Architectural Management*, *26*(6), 989–1003.
Welter, F. (2012). All you need is trust? A critical review of the trust and entrepreneurship literature. *International Small Business Journal*, *30*(3), 193–212.
Williams, M. (2001). In whom we trust: group membership as an affective context for trust development. *Academy of Management Review*, *26*(3), 377–396.
Williamson, O.E. (1985). *The Economic Institutions of Capitalism: Firms, Markets, Relational Contracting*. New York: Free Press.
Williamson, O.E. (1988). Corporate finance and corporate governance. *Journal of Finance*, *43*(3), 567–591.
Williamson, O.E. (1993). Calculativeness, trust and economic organisation. *Journal of Law and Economics*, *36*, 453–486.
Winch, G. (1989). The construction firm and the construction project: a transaction cost approach, *Construction Management and Economics*, *7*(4), 331–345.
Wu, J., Kumaraswamy, M.M. and Soo, G.K.L. (2011a). Dubious benefits from future exchange: an explanation of payment arrears from 'continuing clients' in Mainland China. *Construction Management and Economics*, *29*(1), 15–23.
Wu, J., Kumaraswamy, M.M. and Soo, G.K.L. (2011b). Regulative measures addressing payment problems in construction industry: a calculative understanding of their potential outcomes based on gametric models. *Journal of Construction Engineering Management*, *137*(8), 566–573.
Xiang, P., Huo, X. and Shen, L. (2015). Research on the phenomenon of asymmetric information in construction projects – The case of China. *International Journal of Project Management*, *33*(3), 589–598.
Xiang, P.C., Zhou, J., Zhou, X.Y. and Ye, K.H. (2012). Construction project risk management based on the view of asymmetric information. *Journal of Construction Engineering and Management*, *138*(11), 1303–1311.
Xu, Q., Chong, H.Y., and Liao, P.C. (2019). Collaborative information integration for construction safety monitoring. *Automation in Construction*, *102*, 120–134.
Yao, M., Wang, F., Chen, Z. and Ye, H. (2020). Optimal incentive contract with asymmetric cost information. *Journal of Construction Engineering and Management*, *146*(6), 04020054-1–13.
Yung, P. and Lai, L.W.C. (2008). Supervising for quality: an empirical examination of institutional arrangements in China's construction industry. *Construction Management and Economics*, *26*(8), 723–737.
Zaghloul, R. and Hartman, F. (2003). Construction contract: the cost of mistrust. *International Journal of Project Management*, *21*(6), 419–424.
Zerjav, V., Hartmann, T., and Javernick-Will, A. (2013). Internal governance of design and engineering: the case of multinational firm. *Journal of Construction Engineering Management*, *138*(1), 135–143.
Zhang, H., Yu, L. and Zhang, W. (2020). Dynamic performance incentive model with supervision mechanism for PPP projects. *Engineering, Construction and Architectural Management*. Retrieved from https://doi.org/10.1108/ECAM-09-2019-0472.
Zheng, L., Lu, W., Chen, K., Chau, K.W. and Niu, Y. (2017). Benefit sharing for BIM implementation: Tackling the moral hazard dilemma in inter-firm cooperation. *International Journal of Project Management*, *35*(3), 393–405.

16. The builders of cities: prospects for synergy between labour and the built environment

Edmundo Werna and Jeroen Klink

INTRODUCTION

> Odera is a young Kenyan from Nairobi who completed secondary schooling in 1985. After 'tarmacking' for eight years (local word for walking up and down the tarmac road looking for a job) his family prevailed on him to join his uncle who was a plumber. Odera learned on the job and became an accomplished plumber, but he feels that his job is unrewarding in terms of pay and recognition, as exemplified by the fact that he lives in a one-room shack in the Nairobi slums. Odera does not miss a chance to remind his family and friends to work hard in their studies so that they do not end up like him. (ILO, 2001)

The Report for the 2001 International Labour Organization (ILO) Tripartite Meeting on the Construction Industry opened with the above quotation. Some 20 years have passed since then. Still, the urban construction workforce faces challenges, such as those described in the above quotation. It is important to understand the connection between urbanization, construction and its workers. The chapter reveals gaps in theory and policy, with recommendations to address them.

Cities are literally built by construction workers. Construction is a major provider of employment in urban areas, especially for the poor. Further to employment on sites, the industry provides a large number of other jobs, for example in the production of building materials, equipment and post-construction maintenance of built items. Also, a large number of construction workers render various services to low-income people. For all these reasons, construction labour has a significant impact on urban development and on poverty alleviation. As the world continues to rapidly urbanize, the number of urban construction workers, already significant, is bound to increase. Concomitantly, the trends in urban development have an impact on construction and its workers. However, the research and related policy frameworks tend to separate construction workers and urban settlements.

The construction industry has changed significantly in the past few decades. More is to come, with the spread of the fourth industrial revolution. In parallel, considering the problems which already exist in cities, and given the forecast of continuous increase in urbanization (UN-Habitat, 2016), urban paradigms continue to be the object of intense debates, with conceptual blueprints coming from intergovernmental and private sector organizations. (In this chapter, 'intergovernmental organizations' refers to organizations whose members are national governments, such as the United Nations and its agencies.)

This chapter contributes in many different ways to the field of construction economics. It:

- brings together the fields of construction economics and urban economics, establishing innovative connections between them with implications for theory and practice;
- analyses how major sets of international urban policies (fail to) bring on board the construction industry and its impacts;

- analyses labour as a key issue of construction economics, while highlighting that many current problems faced by workers are due to the structure of the industry itself; and
- contributes to further research on construction economics and its linkages with urban economics and the agency of workers.

The chapter first brings together elements of the literature on construction and urban economics with special attention to labour. It argues that cities are territorialized networks of the construction economy. This has not been well explored at the policy level. The next section states that the urban policies of the intergovernmental organizations do not address whether and how construction is supposed to implement the development agenda that is proposed. The potentials and challenges faced by construction workers are neglected, although there is a generic emphasis on more and better jobs. Although private sector policies are better aligned with the role of construction, labour is narrowly regarded as an internal affair of the industry. However, the following section explains that the problems of the workers exist to a large extent because of the structure of construction. They can lead to a dysfunctional industry.

Moreover, despite globalization and international finance, a significant share of construction is still anchored in cities through its labour pools, dense subcontracting patterns, and the circulation of information and innovation in networks. Acknowledging this contributes to the design of urban policies that address the role of construction labour in building cities as better places to work and live. Although specific countries and conditions of some types of workers (such as informal, female, migrant) are mentioned as illustrations, the focus here is on construction labour in general and throughout the world.

The chapter notes three modes of urbanization: smart-classy (in high-end neighbourhoods), alternative (in low-income neighbourhoods), and ordinary (between the two extremes of types of neighbourhoods). It inserts the analysis of labour in the context of construction in these modes. It highlights the risks of not addressing the challenges faced by workers.

The concluding section provides recommendations for a research and policy agenda with emphasis on the agency of workers and its beneficial effects for the construction industry, and the cities they produce.

CONSTRUCTION ECONOMICS AND URBAN ECONOMICS: SILENCES AND CONVERSATIONS

Construction Economics: Fleshing Out the Production of Space

Construction economics has gradually opened up its perspectives, in the light of a continuous search for balance between analytical rigour and political and social relevance. Originally grounded in microeconomic-oriented work on the firm and industry, it has broadened its perspective to include the coverage of concepts such as value chains, actor–network theory and the socio-economic and cultural dimensions. Considering its entanglements with microeconomics (and management studies), construction economics has a long record in investigating the firm in a context of economic restructuring (Chancellor et al., 2015). This has generated a vast literature on project management, the incorporation of information technologies, as well as transaction cost approaches regarding the efficiency of in-house versus outsourcing of specific activities. Influenced by new institutional economics (Williamson, 1985; Moulaert,

2005), this strand of work has gradually incorporated a relational stance toward the firm, considering its network interactions involving workers, subcontractors, suppliers, customers, financiers and governments (Ruan et al., 2013).

Construction economics has also generated contributions to the understanding of industrial market structures, including housing provision, entry barriers and the competitiveness of construction in specific places (Ball, 1981; De Valence, 2012). This literature has also produced quantitative studies on input–output matrices and income/employment multipliers (Kofoworola and Gheewhala, 2008), considering how specific actors – as part of professional communities – share and circulate norms, expectations and information, and consequently learn and innovate in a traditionally low-value added industry (Ruan et al., 2013). Somewhat stretching the perspective of market structures, research on value chains has influenced construction economics (McDermott and Khalfan, 2012). Value chain management impacts upon both individual firms and traditionally sheltered national construction groups that operate in an increasingly global economy. This has generated work on the limits and potentials of national sectoral strategies, particularly with re-emerging global trade and finance. The value chain perspective also enables promising linkages between construction economics and traditional work on innovation and the knowledge economy (Schumpeter, 1934; Lim et al., 2010). Such ideas have provided fresh, albeit relatively unexplored, perspectives on workers' agency in construction.

While this broadened perspective of construction economics has stimulated the field's innovation, it has also increased expectations regarding its potential to contribute to an ambitious urban development agenda. As is discussed in the next section, the policy literature on smart, sustainable and inclusive cities has left blind spots regarding exactly how construction is supposed to contribute to these goals (Lawrence and Werna, 2009). There is another, related challenge: the remarkable silence of the policy literature on the city economy and urban labour. Cities – as built environment and the production of space – are omnipresent in construction economics. Nevertheless, there is little work on the explicit entanglements of construction economics with the role of cities as privileged arenas for the generation and appropriation of value associated with the location of production in space (including the construction cluster and its pooled labour force). While cities could be interpreted as the territorialized networks of construction and its products, construction economics has not explored the potential of these entanglements between urban areas and its own field. It is argued in this chapter that a more fine-grained reading of the entanglements between construction economics, as a stylized representation of how space is produced, and urban economics, as the field concerned with production in space, provides insights regarding the role of labour, governments and private enterprise in reshaping cities to be more inclusive and sustainable.

Urban Economics: Production in Space

The traditional concern of urban economics is location theory, structured around the optimization of distance from the city centre and its interaction with land prices. Neoclassical economics claims that location choices represent a trade-off between transportation costs to the central business district (CBD) and land prices. Krugman (1991) was one of the first to problematize the neoclassical synthesis of the city, suggesting a way out. In his view, the theory took the CBD for granted, offering no explanation for its existence. Krugman mobilized the Marshallian (1920) concept of economics of agglomeration in order to come to grips with the CBD.

Economics of agglomeration is based on three dimensions that underpin contemporary debates in orthodox urban economics: labour pooling (an abundant and qualified labour force that reduces costs and increases productivity); networks of specialized suppliers that generate horizontal and vertical interactions in clusters; and positive technological externalities. The last of these relates to the unpriced circulation of norms, expectations and information in urban-metropolitan economies, which are transformed in dynamic localized learning and innovation systems that generate value added through new products and processes. Positive technological externalities are related with the tacit nature of information, gradually transformed by cities in new processes and products. This happens, for instance, through information sharing and skills development in labour markets, among suppliers and clients in value chains, as well as the interaction between academic institutions and the private sector regarding applied science and technology (Storper and Venables, 2005).

Much of the debates in orthodox urban economics are structured around these dimensions of economics of agglomeration and link to the field's normative-political agenda (Camagni, 2005; Glaeser, 2011). Nevertheless, while cities are engines of economic development, they simultaneously concentrate the bulk of negative externalities: unemployment, lack of decent work, unaffordable and badly located housing in risky areas, congestion, and pollution (Brueckner, 2011). Orthodox urban economics has received criticism, both from inside as well as from authors influenced by political economy. The first argument noted that urban economics was silent on the varying forms of agency in construction, considering that landowners, builders, developers and contractors, as well as finance capital, are not price-takers that passively react to price signals transmitted in competitive markets (Pirounakis, 2013; Malpezzi and Wachter, 2001). Contemporary real estate economics has responded to these demands and articulates a broad range of fields such as construction economics itself, law and urban planning, corporate finance, land economics and civil engineering (Dipasquale and Wheaton, 1996).

Urban political economy provided a structural critique on the neoclassical city (Harvey, 2012), based on the assumption of unequal social relations between capital and labour. This led to a radically different perspective on economies of agglomeration and labour pooling, as synonymous with social segregation and differentiation in urban labour markets in terms of wages and working conditions. Dual and divided labour markets in localized construction systems were the backbone of city economies. Moreover, this critical perspective had implications for the investigation of issues such as labour productivity and industry competitiveness. For instance, 'business as usual' in fact meant downloading social-economic and environmental responsibilities and aggressive cost reduction of suppliers in order to boost productivity and competitiveness. An alternative approach would recognize that labour productivity and competiveness could emerge as a result of different corporate strategies, for instance by improving labour and environmental standards while increasing the value added of products. Each of the two above-mentioned idealized corporate strategies, and their institutional embedding in a set of accepted norms and conventions of 'doing business', or in the words of Storper (1997), 'worlds of production', affect labour differently. The former indicates cut-throat competition and a race to the bottom. The latter triggers superior trajectories, increasing value added, production and better working conditions and wages. Nevertheless, most construction (here termed 'ordinary') is hybrid, complex, and does not follow the neat distinctions between these two worlds of production.

A Potential for Dialogue? Cities as Territorialized Networks of Construction

Analysing the North American building industry, Buzzelli and Harris (2006, p. 894) use Marshallian agglomeration economics to describe cities as the industrial districts of construction. This metaphor is useful, considering that homebuilders as 'numerous small, specialized firms interact frequently within a rich, embedded market network; subcontracting is the norm; networks and firm boundaries are fluid'. The dense, interpersonal networks and circulation of non-codified information, norms and expectations between workers, landowners, builders, contractors, developers and financiers also explains the resilience of local builders in a context of globalization driven by transnational groups (Halbert and Rouanet, 2014).

While the metaphor of cities as the industrial districts of house-building is debatable, the essential point is to understand cities as territorialized networks of construction. This echoes Markusen's (1999) earlier work on the relations between firms, industry and value chains, on the one hand, and urban territories on the other. This perspective connects with the debate in the previous section on the relations between construction firms' strategies towards competitiveness, labour productivity and the strength of urban economies. In Markusen's (1999) perspective, the industrial district is also only one among a series of possible territorialized firm interactions. Figure 16.1 illustrates other possible 'regional worlds of production' and firm strategies to build competitiveness and boost productivity (Storper, 1997). Figure 16.1(a) is a stylized representation of the territorial interactions in clusters of specialized suppliers, subcontractors and firms that are concentrated in the economic agglomeration of industrial districts. It is where small and medium-sized establishments and their labour pools, embedded in urban space, share and circulate information, and build up experience and a common base of technical-professional knowledge that set the stage for dynamic learning and innovation through the successive implementation of projects.

Figure 16.1(b) illustrates a 'hub-and-spoke' territorial interaction in value chains. It represents a hybrid setting: a network composed of small local business units as well as large headquarters firms, located both in and outside specific cities. As a consequence of the hierarchical relations that also involve larger, externally located firms, the potential for information sharing and dynamic accumulation of technical skills, learning and incremental innovation in cities will depend on active mobilization of local actors – local governments, labour unions and business associations – around an agenda of high-end competitiveness and productivity.

Figure 16.1(c) indicates a case of territorial interaction resembling a 'satellite platform'. Local branch offices buy from large-scale external suppliers, perform standardized assemblage operations, and deliver to external headquarters firms that are situated in 'command and control' positions of hierarchical value chains. Considering that leading external headquarters corporations and subcontractors are keen on downloading costs and responsibilities for upholding environmental and labour standards to local establishments, this pattern triggers predatory, cost-cutting strategies by local branches. This is distant from the gradual build-up of local skills, learning and innovation systems that characterize the labour pool and territorial interaction in clusters and industrial districts. In the specific examples of Figure 16.1, the potential for high-end strategies that maintain labour and environmental standards and increase competitiveness, value added and productivity in cities as territorialized networks of construction tends to decrease when moving from (a) to (c). Consonantly, challenges increase for a policy framework providing construction with appropriate incentives to build affordable, smart and sustainable cities.

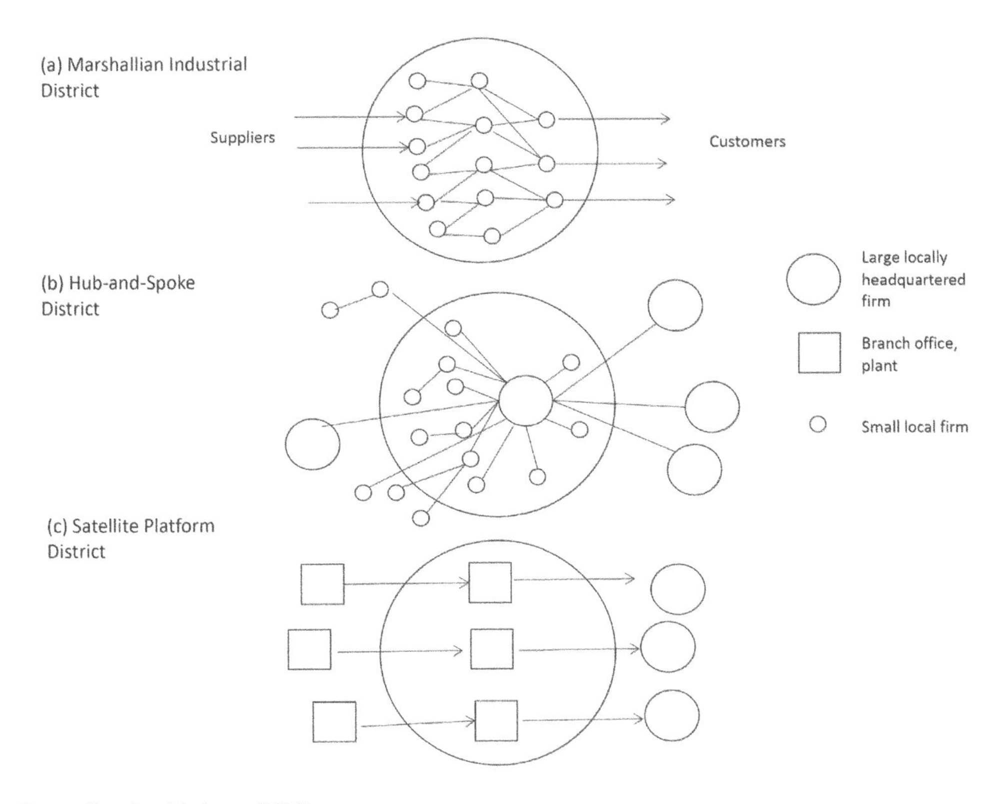

Source: Based on Markusen (1999).

Figure 16.1 Cities as territorialized networks of building and construction

The next sections flesh out two implications of the argument of cities as territorialized networks of construction, with an emphasis on labour. First, there is considerable variation in the structure and strategies adopted by construction: 'high-end' strategies (that is, delivering to customized segments that generate high value added and potential for dynamic learning and innovation) versus cut-throat cost reduction (structured around cost reduction through internal lowering and/or downloading of labor and environmental standards to suppliers); income segmentation and target groups; degree of standardization or customization of products (articulating either scale or scope); as well as technologies, materials and types of labour mobilized. Most of current understanding is based on partial readings of high-end, customized construction for smart cities, and of alternative, community-driven construction. However, between the two, 'ordinary cities' (Robinson, 2006) have been built based on a myriad of pragmatic, hybrid arrangements that need better understanding (see Table 16.1 for a summary of the argument).

Second, understanding the connections between urban and construction economics, as well as the complex interface between labour, business and governments as potential innovation supporters and actors of change in construction, are necessary to achieve the goals of the global urban development agendas as established by international agencies (Lawrence and Werna, 2009). However, more frequently, these urban policy documents are not grounded in

Table 16.1 Cities as territorialized networks of construction

	Marshallian industrial districts (clusters)	Hub-and-spoke districts	Satellite platform districts
Potential for development of new skills, competences and dynamic learning in urban labour markets	High	Medium	Low
Relations among firms in production chains	Networked	Hierarchy and networks	Hierarchy
Potential for territorial local innovation systems	High	Medium	Low
Predominant firm strategies toward labour productivity and competitiveness	High-end strategies: successive stages of innovation, higher value added and labour and environmental standards	Hybrid strategies	Cut-throat, predatory subcontracting and downloading
Worlds of construction	Smart, ordinary and alternative construction		

Note: The last row does not allocate a specific industrial structure because, as argued in the text, there is not a one-way relation.

an understanding of the role of construction in reshaping cities as better places to work and live.

TRENDS IN URBANIZATION AND CONSTRUCTION

This section provides background on urban policies and construction. The section starts with the two overarching sets of recent public urban policies: Sustainable Development Goal 11 (SDG 11) and the New Urban Agenda (NUA), both endorsed by the members of the United Nations (UN). Afterwards, the private sector-driven urban thinking of the World Economic Forum (WEF) and International Federation of Real Estate Developers (FIABCI) is reviewed, as a counterpart to the government-led SDG 11 and NUA. Most of the WEF companies with a stake in urbanization are not from the real estate sector, while members of FIABCI are exclusively from this sector. Therefore, the two organizations and their agendas are complementary. Finally, the section presents a summary of trends in construction considering interfaces with urbanization.

Urban Policies

Sustainable Development Goal 11 (SDG 11)

The SDGs are a set of 17 global goals agreed upon in 2015 by the UN General Assembly as a plan for a sustainable future for all, to be achieved by 2030. Although other SDGs are in some ways related to urbanization (for example, Goal 6: Clean Water and Sanitation; Goal 9: Industry, Innovation and Infrastructure; and Goal 13: Climate Action), the core urban SDG is number 11: 'Make cities and human settlements inclusive, safe, resilient and sustainable' (UN, 2015).

The main contributions of the SDGs are their clear targets and indicators. Yet, achievement depends on what member-states will effectively do. SDG 11 encompasses the view of a city with good environmental conditions, appropriate and equitable distribution of urban facilities, adequate access to urban services, good governance, resilience and safety. There is no suggestion on how these targets will be achieved. SDG 11 and the other SDGs resemble a set of intentions rather than policies.

There are a few points on construction in the SDG 11 statement related to the environment: sustainable, resilient and resource-efficient buildings; retrofitting; and utilization of local materials. This is an important trend in the urbanization–construction nexus. In addition, SDG 11 has statements about appropriate and equalitarian access to construction products, such as housing, and water and sanitation systems. However, it does not propose how construction will achieve this. SDG 11 does not appear to be grounded in an understanding of urban and construction economies.

New Urban Agenda

Every 20 years, the United Nations holds a conference on human settlements, the most important urban event worldwide. Habitat I took place in Vancouver, Canada in May 1976; and Habitat II in Istanbul in June 1996. Habitat III was held in Quito, Ecuador in October 2016. It produced the New Urban Agenda (NUA). The agenda is similar to SDG 11 in its conception of a city as having several attributes (such as being related to the environment and urban facilities, and requiring good governance). However, the list of attributes and level of detail in the statements are greater. Also, NUA provides specific recommendations, such as compact, polycentric growth, mixed-use streetscapes, transit-oriented development and prevention of sprawl.

None of the above documents explains what kind of construction is needed. The NUA gives some hints on construction in resemblance to SDG 11. There are clear references regarding green construction and statements about appropriate and equalitarian access to construction products, but no indication of how to shape the sector to achieve them (Habitat III, 2016). Similarly to SDG 11, the NUA is delinked from urban and construction economies.

A noteworthy element introduced by the NUA vis-à-vis SDG 11 is the notion of the 'smart city', derived from efforts to expand the reach of the fourth industrial revolution. It does not give details on how this feature will combine with those of an environment-friendly and socially equitable city. Neither does it elaborate on the type of construction necessary to build a 'smart city'. Nevertheless, this provides an important link to a paradigm on how cities and construction should develop.

World Economic Forum

While the WEF has a not-for-profit status, it preponderantly promotes private businesses. The WEF organizes the yearly high-profile, global event in Davos, and it also organizes fora at regional and national levels, produces and disseminates reports and implements projects. The WEF has an 'urban arm': the Global Future Council on Cities and Urbanization, and a number of associated initiatives. There is a strong emphasis on promoting the fourth industrial revolution as a driver of urbanization. This revolution is the advent of 'cyber-physical systems'. Examples include, *inter alia*, genome editing, new forms of machine intelligence, breakthrough materials, and approaches to governance that rely on cryptographic methods such as blockchain. The White Papers published by the Global Future Council on Cities and Urbanization are emblematic: examples are 'Agile Cities: Preparing for the Fourth Industrial Revolution', 'Data Driven Cities: 20 Stories of Innovation' and 'Industrial Internet of Things:

Safety and Security Protocol' (in WEF, n.d.). Thus, while the NUA tentatively broached the smart city paradigm, the WEF clearly embraced it.

The implication for construction is that the sector should equip itself with fourth-generation techniques to deliver products for such envisioned cities. The WEF actually goes beyond mere implications. It has a platform called 'Shaping the Future of Cities, Infrastructure and Services' which gives detailed recommendations, such as the Open Systems Interconnection model; smart buildings with sensors to help reduce energy use and ensure timely repair; big data; road charging infrastructure for electric vehicles; and Nonlinear Channel Optimisation Simulator online, among others (WEF, 2019). In short: smart construction for smart cities.

The WEF also pays attention to the urban environment, arguing that smart construction includes environmental protection. The WEF also expresses concern with social issues, such as urban migration and affordable housing. The White Paper on migration suggests solutions which include construction – such as housing, and even jobs in construction for migrants (WEF, 2017). However, there are no details on how the construction industry is to achieve this. In turn, the White Paper on affordable housing gives detailed suggestions on construction, almost all aligned with the 'smart construction' paradigm (WEF, 2018).

International Federation of Real Estate Developers

FIABCI (the French acronym of the International Federation of Real Estate Developers) is a worldwide networking organization for the real estate industry. Its mission is to provide access and opportunities for real estate professionals interested in conducting international business. Its members comprise the top international developers' firms. They offer construction products which will fit into the urban fabric. They aim for profit, therefore need to follow closely (and often intervene on) how cities develop. At the same time, as customers, they can influence what construction companies should offer. Thus, FIABCI's interests and activities connect cities and construction. FIABCI members seem to be more concerned with the sophistication of the built space and less with 'smart' technology, although the latter is welcome. FIABCI also promotes green construction, blended into the 'classy' approach. Environmental concerns seem to have become a (fortunate) dogma, due either to genuine concern for the environment, or to greenwashing (driven by marketing or to avoid taxes).

It is pertinent to note that in addition to promoting its own classy products, FIABCI has engaged with UN-Habitat on a campaign promoting a broader range of construction products and methods. This entails a set of reports called *The City We Want is Affordable* (FIABCI, 2016), and the City Prosperity Initiative (FIABICI and UN-Habitat, n.d.).The latter includes a range of proposals and examples of construction methods and modes of provision. The City Prosperity Initiative – Perception Index (CPI-PI), in turn, aims to assess sustainable urban development in terms of citizens' happiness. It could provide valuable ideas to city planners and real estate developers, with implications for construction.

The reports mentioned above and the CPI-IP are useful for educating society and are part of a corporate social responsibility strategy of FIABCI. However, they do not include a clear direction to urban development and/or construction.

Urban Construction

SDG 11 and the NUA clearly emphasize a 'green' approach to construction. They also include a plethora of social issues, which need construction as part of the solution, but do not give details. The WEF, in turn, is adamant about 'smart cities', which is also present in the NUA and FIABCI, although more discreetly. The WEF and FIABCI mention care for the environment, with some specific ideas. Moreover, they acknowledge the social *problematique* in cities, in various ways. The WEF suggests solutions to affordable housing; almost all the proposals are in line with the 'smart construction' paradigm. In contrast, it does not discuss in any detail how construction will help social stakeholders. Through its reports, FIABCI provides a large range of construction topics and examples, mostly as a menu of cases, although they do not propose a clear direction.

What does this entail for construction? The normative agendas (SDG 11 and SDGs) are not linked to an understanding of urban and construction economies. The WEF is fine-tuned with urban economics (particularly through the link between the knowledge economy and positive technological externalities). FIABCI establishes dialogues with construction economics (especially its emphasis on the firm and industrial structures). However, it is difficult to find out the extent of the true concern of these institutions with building just and inclusive cities.

There is much more in the literature and in terms of players concerning cities and urban areas than SDG 11, the NUA, the WEF and FIABCI. However, as the largest sets of public policies and private bodies with a stake in urbanism, they have a significant impact on the dissemination of ideas. SDG 11 and the NUA have broader penetration because the United Nations (UN) member states are signatories. International associations of local authorities (such as United Cities and Local Governments, and Local Governments for Sustainability), also disseminate aligned policy documents.

The private organizations have a more limited outreach to middle- and low-income actors. Nevertheless, large cities in the Global South have adopted strategies of urban entrepreneurialism, aspiring to global or regional status structured around high-end, high-tech construction to become gateways for international investors. While 'first movers' aspiring to global status, such as Abu-Dhabi, Beijing, Dubai and Shanghai, have paved the way, this has now spread to include many other cities. Another feature of this competition, in both North and South, is the increasing attention-grabbing buildings and public spaces prepared for global events (sports and fairs). The bottom line is a trend towards a combination of smart-classy construction with a green touch. The challenge is that such a model is still far from providing construction which will substantially improve the living standards of hundreds of millions of urban residents in poverty (Watson, 2009). There is also evidence that even the middle classes have limited access to this model.

A counterpoint to the smart-classy WEF and FIABCI is a movement called by different terms including 'alternative development' and 'appropriate technology'. It originated in the 1960s from the understanding around the growing hiatus between developed and developing countries. The type of industrialization of the former when applied to the latter failed to fulfil the hopes of fostering development and solving basic problems (Werna, 1996). While initially challenging the modernist and post-modernist approaches of the 20th century, now it equally challenges the smart-classy model by proposing urbanism based on local materials, traditional construction techniques, small-scale production and community-based participation. The ethos is 'small is beautiful' (Schumaker, 1973). This has evolved since the pioneering work of Turner and colleagues (Turner, 1976; Turner and Fischter, 1972). It has included several types of self-help schemes, sites-and-services projects and slum upgrading (as opposed to eviction of the residents). This approach has been widely applied in low-income communities. Virtually every government in the Global South and every donor have come on board, to different degrees (Lizarralde, 2014; Cohen, 2015). This movement also championed environmental protection and green construction, much before the WEF, FIABCI and others. The NUA also advocates many of its principles.

There is evidence that substandard settlements throughout the world have benefited from such ideas, although many problems remain (Curley, 2012; Ward, 2019). It is not clear how the alternative development approach (and particularly the finance and institutional arrangements that underpin it) relate to the urban and construction economies of the vast majority of the built environment; nor how it could represent a solution to the constructed amenities needed by the poor.

As mentioned above, construction that generates the majority of our cities is between the smart-classy and the alternative development models. In these 'ordinary' cities (Robinson, 2006), construction is driven by what is possible to build given the local circumstances: the materials available, techniques and matching resources, finance and building codes.

The public sector often uses such pragmatic solutions. Due to the diversity of problems and limited resources, answers require a variety of modes of provision: public–private partnerships, co-operatives, direct administration, subcontracting and so on (Keivani and Werna, 2001a, 2001b; Werna et al., 2000).

Despite the difficulty in projecting the future, it can be expected that construction is bound to become greener; otherwise the planet will not be able to bear the burden construction imposes on it. The expansion of the smart-classy model will depend on economic growth. High-income settlements are likely to continue to pursue this model. Smart construction in low-income communities is still in its infancy (see Abanda et al., 2021 for an exception). In turn, alternative development continues to benefit low-income settlements, but still with little evidence that it will become 'the' paradigm for urbanism and construction. Except for advocating theorists (such as Atkinson, 2008) who claim that this is the only type of development that will prevent a global environmental catastrophe.

Ordinary construction and urbanization are likely to remain strong, entangled between the smart-classy and the alternative approaches. Meanwhile, it is pertinent to consider what is happening with the construction workers.

WHO ARE THE BUILDERS?

This section presents a profile of the urban construction workers, preparing the ground for the following section which puts the labour issues together with the analyses of construction and urbanization. It reveals challenges for workers to find and maintain employment, have adequate working conditions, and be part of labour governance. The ILO (2001) argues that the structure of the construction industry plays a role in this situation, which leads to contradictions with the ideals of a well-trained and strong workforce, and building smart and inclusive cities. As noted in the introduction, the focus here is on construction labour in general and throughout the world.

Employment

Urban construction is a major employer; not only in sites, but also in services, the supply of materials and components, and post-construction maintenance. Considering that the ILO figure for the global urban workforce is 1 792 799 000, and that construction's share is generally between 7 and 10 per cent, it follows that there are some 125 495 000 to 179 280 000 urban formal construction workers (ILO, n.d.-a). Considering the informal workers, the figure increases considerably. According to Mella and Savage (2017, p. 3) in their report on construction in developing countries: 'Challenges in the [construction] sector include high informality (76.5 per cent)'.

Urban construction includes many different professions, such as bricklayers, plumbers, electricians and carpenters. This list increases with the level of technological sophistication: building information modelling (BIM) technicians, drone operators, and so on. The structure of the workforce is pyramidal, as in many other sectors, with a smaller number of highly skilled and better-paid workers at the top, a large number of low-skilled and poorer workers at the bottom, and intermediate degrees. Work tends to be more complex in higher-income products and services. This does not mean that those who work in the provision of lower-income products and services are necessarily low-skilled. Given fluctuations in employment, many highly skilled workers find themselves periodically having to work on low-paid, poorer projects (or are unemployed during some periods). At the same time, classy projects also include a significant amount of simple physical activities such as cleaning which are carried out by relatively unskilled workers.

According to the latest publication of the ILO on labour in construction (ILO, 2015a), despite some increases, the participation of women in construction remains lower than that in many sectors. Women are sometimes employed as part of a family work unit, often without receiving direct payment. The percentage of all women paid workers in construction – which denotes the share of paid women working in the construction sector, out of female workers in all sectors – in selected regions is: Africa, 5.5 per cent; Asia, 14.6 per cent; Latin America, 5.5 per cent; North America, 11.7 per cent; and Western Europe, 7.5 per cent. The share of women workers out of all workers in the construction industry: Africa, data not available; Asia, 7.5 per cent; Latin America, 0.5 per cent; North America: 2 per cent; and Western Europe, 1 per cent. Still according to the ILO (2015a), the rate of women's participation in the construction sector in Asia is higher than in other regions. High shares in some countries – such as Kazakhstan, 23.5 per cent; Singapore, 22.9 per cent; Mongolia, 21.1 per cent; and Ethiopia, 17.8 per cent – can be partly explained by women working in administrative, human resources, clerical and other office-related and technical areas of work. Women represent up to 50 per cent of the workforce on some sites in India. Most of them (92 per cent) are engaged in load-carrying, the rest being in semi-skilled work, including plastering or mixing concrete; 93.6 per cent are engaged as casual labourers.

While products of construction are fixed, production takes place on a project-by-project basis with sites constantly changing. Therefore, the labour force has to be mobile. Construction has a long tradition of employing migrant labour. During urban growth, construction has provided migrants from the countryside with an entry point into the labour market, frequently being the only significant alternative to farm labour for those without any particular skill or education. When the pool of surplus labour from the countryside dries up, or due to shortage of local labour for other reasons, recruitment from overseas may occur (ILO, 2001; Lawrence and Werna, 2009; Mella and Savage, 2017; Werna, 2007). Migrant construction workers are generally from less-developed and lower-wage economies with labour surpluses (ILO, 2001). This supports the earlier argument regarding the dual character of labour markets in urban construction.

As work in a given project is limited, workers tend to look for new projects, ideally nearby. Despite globalization and internationalization of part of construction, more than 95 per cent of its activity is still undertaken by firms from within the country, the region or the neighbourhood (ILO, 2001). Therefore, an important share of the interfirm construction networks composed of headquarter companies, subcontractors and their labour pools are still territorially embedded in cities. This provides perspectives for policies aimed at developing higher-end trajectories towards competitiveness and productivity through information-sharing in value chains, training schemes, as well as increasing labour's role in learning-by-doing and innovation through the implementation of projects (see Table 16.1).

Despite large numbers of workers, job continuity is low given the project-based nature of construction. Moreover, the industry is prone to periods of boom and bust, with impacts on (un)employment. During downturns, workers may resort to unemployment insurance in countries which provide it. Otherwise, they need to seek employment in poorer construction sites or in other sectors. Flexibility in hiring is a broad trend, with casual contracts. Employers have also pushed for 'zero hours' contracts, with no obligation to provide minimum working hours. Workers have to be constantly on the look-out for assignments, in different enterprises. Flexibility has also increased the hiring of workers as 'self-employed' or 'one-person enterprise' as opposed to contracted workers. In extreme cases, a site may have only 'enterprises'. This scenario allows firms to avoid employment obligations, such as payment into a social security fund (ILO, 2001; Lawrence and Werna, 2009; Mella and Savage, 2017; Werna, 2007, 2016).

While a large share of urban construction is labour-intensive and provides opportunities to enter the labour market, there are challenges in terms of employment stability and decent income. This echoes the earlier discussion on neoclassical labour pooling, which stresses cost reduction associated with hiring and firing. However, there is more to firm competitiveness and labour productivity than cost reduction. Aggressive strategies of downloading costs and responsibilities have mixed results, while corporate approaches structured around the maintenance of labour and environmental standards that generate value-added have not been mainstreamed (Amin, 1994; Cox, 1997; Shimbo, 2020).

Workers' insecurity has generated a toll on the competitiveness of the industry and its contribution to urban development. It is debatable whether insecurity has increased productivity, considering that competition is offset by stress and its effects on output (ILO, 2016; Lelo et al., 2019; WHO, 2003). Moreover, as exemplified in the epigraph to this chapter, many workers lack motivation to remain or move up in the industry, and entrants do not consider the industry as attractive. For women, opportunities are even worse (ILO, 2001, 2015a; Mella and Savage, 2017; Norberg and Johansson, 2020; Zitzman, 2021).

The above contrasts with the fact that employers often complain about the lack of skilled labour (ILO, 2015a; Mella and Savage, 2017). If this is a problem now, it will become more serious in light of the 'smart construction, smart cities' utopia, which requires more sophisticated labour. Unless employers move to a radically automated industry; which may turn into a dystopia, as explained subsequently.

Part of the problem may be due to inappropriate technical training institutions (ILO, 2001, 2015a; Mella and Savage, 2017). Working conditions are also a cause: demotivation due to the image of the industry, no time for reskilling in light of the constant look-out for new work, long working hours and frequent job-switching to other sectors. Additionally, women face cultural barriers. The practice of 'triangular employment relationship' (contractors – subcontractors and labour agents – workers) imposes a considerable barrier to training, with a narrowing of skill development. The heightened division of labour into ever more specialized trades, implicit in subcontracting, limits the range of skills that can be acquired in any one enterprise (ILO, 2001, 2015a; Lawrence and Werna, 2009; Mella and Savage, 2017; Werna, 2007). This means that versatile craftsmen and supervisory workers are difficult to train (ILO, 2001).

The observation regarding the deficit of skilled labour is a reflection of the industry's structure and how it organizes the interactions among suppliers, subcontractors, leading firms and their labour pools. In locally clustered interactions that mobilize a large number of small-scale specialized suppliers, (sub)contractors and developers, the potential for the development of a diversified base of skills and competences that set the stage for further learning-by-doing and incremental innovation is higher (Figure 16.1(a) and Table 16.1). Conversely, endogenous skills development is more challenging in settings whereby local builders are hierarchically inserted into a satellite type of territorial interaction, which is coordinated by external contractors that download costs and responsibilities in the procurement of relatively standardized products. The consequence is that the predatory subcontracting, cost-cutting and low value added that characterizes satellite interaction between local and external suppliers, subcontractors and lead firms (Figure 16.1(c) and Table 16.1) tends to increase construction's deficiencies, in terms of a low and undiversified base of skills and competences of its labour pool, and relatively higher vulnerability to the boom-and-bust of the business cycle.

Working Conditions

Conditions vary depending on the level of enforcement of labour rights at national and city level. The labour market also exerts pressure in one direction or another (that is, workers willing, or not, to accept poor working conditions depending on the ease of finding jobs). In countries with weak law enforcement, the more specialized workers are less vulnerable, while those at the bottom of the pyramid bear the brunt of poor conditions of work. In the 'triangular employment relationship' and related casualization, workers' rights are often unclear and they enjoy less protection from the law than those who are directly employed; informal workers, by definition, are outside the boundaries of the law (ILO, 2001). Examples of deficits in working conditions of urban construction workers include underpayment, delays, long working periods, occupational accidents and diseases, bonded labour, child labour, and discrimination against migrants and female workers.

Payments below what the law stipulates and/or longer working hours without proper compensation are illegal. Given the numbers of informal or casual workers, and/or the lack of control, such practices happen. Delays in payment are also an extended practice (Wells, 2016/18; Wells and Prado, 2019). Overwork is one of the causes, *inter alia*, of occupational safety and health deficits. Construction is one of the most dangerous occupations (ILO, 2001, 2015a; Mella and Savage, 2017). Again, the structure of the industry plays a role in this. Subcontracting, on a piecework basis, intensifies pressures while increasing the difficulties to coordinate the work and ensure site safety. It is estimated that 95 percent of serious accidents involve workers employed by subcontractors (ILO, 2001, 2015a; Lawrence and Werna, 2009; Mella and Savage, 2017; Werna, 2007). Most of them are on temporary contracts, which in a context of fluctuating demand encourages long working hours to make the most of it. They are also less likely than workers in permanent contracts to gain the training required to work safely, and are in a weaker position to refuse unsafe work. A construction worker with a fixed-term contract is three times more likely to suffer an occupational accident than one with a permanent contract. Informal workers are particularly vulnerable (Lawrence and Werna, 2009; Mella and Savage, 2017). As to social protection, evidence indicates that many employers do not contribute to social security funds for workers on temporary contracts. Hence, workers who are most in need do not receive the necessary care (ILO, 2001, 2015a; Lawrence and Werna, 2009; Mella and Savage, 2017; Werna, 2007).

Despite the global campaigning against bonded and child labour, there are still cases in construction. For example, many cities in the Middle East have large numbers of migrants in situations defined as bonded labour, such as restrictions which make it impossible for workers to change jobs, and to move around during their free time; sometimes, their passports are confiscated (Buckley, 2014; Kumar and Fernandes, 2017). Child labour is found in peri-urban activities related to the production of building materials and low-income housing construction (Lawrence and Werna, 2009). Migrant workers are particularly at risk, regarding not only bonded labour but also other working deficits. They often have to accept precarious conditions in order to survive. There is also evidence regarding deficits in the rights of female workers, which points to a lack of equal treatment compared to their male counterparts and existence of various forms of harassment (Lawrence and Werna, 2009; Kumar and Fernandes, 2017).

Governance

According to the ILO, social dialogue with employers and governments has traditionally been a powerful means for workers to collectively bargain for better wages and working conditions (ILO, n.d.-b; Van Empel and Werna, 2010). However, the current high proportion of temporary, casual, informal and unemployed workers makes it difficult to organize and engage in dialogue (ILO, 2001, 2015a; Mella and Savage, 2017). Zero-hours contracts and the transformation of workers into one-person enterprises generate further challenges. Workers' rights should be respected and enforced irrespective of the level of dialogue. Reality varies according to the capacity of the government. Cities usually depend on central governments to enforce workers' rights. Building and Wood Workers International is the global confederation of construction workers. Its constituents have often successfully engaged in dialogue with its counterparts at both international and national levels. The challenge is that unionized workers are the minority.

While social dialogue usually takes place at the international, national or provincial level, Van Empel and Werna (2010) argue for its promotion at the city level. This offers an opportunity for local actors, notably associations of informal workers and/or businesses, co-operatives, grassroots associations and social movements. Furthermore, the national agenda does not apply to all municipalities, while specific labour market issues in a given city cannot be envisaged in national regulatory frameworks. Local dialogue is also more flexible and enables rapid responses. Cities concentrate tacit information networks that are strategic in building up working experience, technical knowledge and dynamic social learning and experimentation with different construction methods (not only cost-driven but also higher-end productivity and involvement, for example through solidarity economics). Social dialogue adds value to urban governance. There is evidence of successful urban social dialogue in construction, for example in the expansion of the metro in Hong Kong, and the engagement of informal construction workers in Dar-es-Salaam (Van Empel and Werna, 2010). Nevertheless, these are still exceptions.

In summary, while urban construction provides employment for many and is also often an important entrance into the labour market, several challenges remain. The condition of the urban construction workforce supports the neoclassical labour pooling argument with regard to hiring and firing. However, productivity should be seen from a broader perspective. The structure of the construction industry has contributed to challenges faced by workers to obtain and maintain employment and decent working conditions, and to be part of labour governance. This contradicts the fact that employers complain about a lack of skilled labour. The industry structure has created a mismatch between the aforementioned challenges and the idea of a smart, inclusive city.

THE LABOUR–CONSTRUCTION–URBANIZATION NEXUS

The section is divided according to three types of urbanization discussed before: smart-classy, alternative and ordinary. There are no sharp divisions among them. Still, there are sufficient differences to allow a comparative analysis of urban labour pools as territorial networks of construction.

Smart-Classy Construction/Urbanization

As noted above, smart-classy construction and its top-quality products do not necessarily imply highly skilled, well-paid and decent work, as reflected in bonded labour in smart buildings in some prime cities. Since 2014, there have been recurring complaints and discussions at the ILO Governing Body about the working conditions in construction of the venues for the Football World Cup 2022 in Qatar (ILO, 2015b, 2017).

Joan Clos, the Secretary-General of Habitat III, when referring to affluent communities, repeatedly emphasized that gated communities were a sign of bad urbanization.[1] However, there is a more disturbing phenomenon: compulsory gated communities of poor construction workers, bonded to their contracts; migrant workers who are not allowed to leave their dormitories except for going to the worksites. The extremes meet in many cases: gated rich communities built by workers who live in gated/bonded poor communities (Kumar and Fernandes, 2017). The lack of attention on such bonded communities supports our point that the international agenda is oblivious to the construction workers.

The green component of the smart-classy approach will need continuous strengthening as it is estimated that the built environment is responsible for the bulk of environmental problems (UN Environmental Programme, 2019).

This brings us to 'green jobs' in construction. There is burgeoning research on the subject. and an entire programme at the International Labour Office (Afolabi et al., 2018: Araújo et al., 2018: Werna, 2012). The workforce will need to be trained or retrained on green technology. An important caveat is that green jobs are not necessarily decent jobs. Green building sites frequently have workers still facing inappropriate conditions (O'Neill, 2009). Companies may use the 'green label' as a smokescreen to pretend that all is good inside. Despite its high visibility, smart-classy construction frequently entails command-and-control, hierarchical networks and low spin-off for local labour. It tends to download responsibilities through subcontracting while reaching competitiveness through cost-cutting. This links with what has been labelled above as satellite platforms (Table 16.1).

In sum, smart-classy cities do not necessarily generate decent work. Simultaneously, the current challenges faced by the workforce may eventually mean that there are not enough well-trained, well-paid and motivated workers to continue to develop in accordance with this model. The dystopian alternative would be smart construction and urbanism intensifying its trajectory of automation. This is a direction pointed to by the fourth industrial revolution, and could lead to mass unemployment of construction workers. Unless the wealth generated by automated production is evenly distributed (which is unlikely), there will be significant problems. Furthermore, 'smartness' and automation are not only targeting construction; they are for all sectors. Their advent raises the economic issues of growing inequality, and insufficient effective demand for the products made by machines in scenarios of depressed incomes and unemployment. Thus, the smart-classy concept is surrounded by doubts regarding its capacity to include all socio-economic sectors as viable consumers.

Alternative Construction/Urbanization

Based on the concept of 'small is beautiful' (Schumaker, 1973), large-scale enterprises or several layers of subcontracting are virtually non-existent in this approach. Emphasis is on co-operatives, small and micro enterprises, and self-employed workers under only one layer of contracting. There is an emphasis on traditional skills, sometimes combined with innovation, and simple employment and working relations. The whole idea is to go back to vernacular ways of building. International agencies and national and local governments embraced this approach, especially to upgrade low-income settlements. Working conditions and respect for labour rights are usually in place. Yet, there is little opportunity for workers to 'grow'. No or slow growth tends to be the essence of the approach. An important advantage (including in the private market) of the alternative approach is that construction techniques are usually labour-intensive, providing much-needed jobs, including to local workers. This often comes together with training (formal or informal), providing an entrance into the labour market.

Alternative development also encompasses non-paid participation of families in construction. There are two problems from the labour perspective. First, the residents are workers in different sectors (drivers, cooks, sellers, cleaners, and so on). When they come home from their occupations, they still have to work as builders rather than rest. Second, they do not have the technical knowledge to work as builders. This can generate unemployment in construction and, at the same time, end up consolidating poorly built low-income settlements. Moreover, the claim that self-help brings communities (politically) together can be counteracted by the fact that paid work through community contracting also does that (Tournee and van Esch, 2001). The usual justification is that self-help is the only way for the urban poor to have access to housing. Yet, its mainstreaming would ultimately consolidate a divide whereby part of the population can afford to pay for their housing, and the rest are left on their own to have substandard shelter.

Even recognizing that the alternative approach had good intentions, it did not take off beyond small-scale solutions in low-income settlements, and its dissemination through international development agencies. The question is: how can it be used effectively in a world populated by billions? Alternative construction could be considered a small-scale, underdeveloped industrial district or fragile cluster, located outside the formal market. As shown in Figure 16.1(a) and Table 16.1, clusters provide potential for higher-end trajectories in terms of labour specialization, skills development and gradual territorialized learning and innovation in cities.

Although it has not delivered in terms of scale, alternative construction provides promising perspectives for what is called progressive labour pooling in cities. Nevertheless, it requires a policy framework aimed at high-end solutions through a variety of strategies that strengthen clusters, community businesses and co-operatives, instead of the institutionalization of precarious and informal construction. The experimentations with funding mechanisms aimed at the upscaling and professionalization of slum upgrading are more relevant alternatives.

Ordinary Construction/Urbanization

Between the smart-classy and alternative constructions lies the majority of the built territory of cities, with construction techniques which are somehow in between. Ordinary urbanization is far from being monolithic. There are many modes of construction. This was called 'pluralistic' in the 1990s (Keivani and Werna, 2001a, 2001b; Werna et al., 2000). It has spread ever since. Ordinary construction also incorporates hybrids of small-scale, underdeveloped clusters and what we called 'modern', hierarchical satellite platforms (Figure 16.1(c)), as well as variegated arrangements of hub-and-spoke worlds of construction (Figure 16.1(b)). Without appropriate policies, it has generated the fragmented, disconnected labour markets that characterize ordinary cities as territorialized networks of construction. Most of the labour challenges explained above take place in ordinary construction and urbanization. This explains, at least partially, why most cities in the Global South are deteriorated, and filled with low-quality buildings. It does not mean that workers are unskilled or do not perform their job well. It actually reflects the employment conditions, and what workers are told to do, and under low levels of motivation. To a large extent, the future of ordinary construction and urbanization depends on whether the labour force will improve their skills, will gain ground to use them to the maximum, and be motivated to do so with decent work. More than anywhere else, the ordinary settlements may be the battleground where urban construction workers will win or lose their rights, well-being and morale.

CONCLUSION

The challenges faced by workers can affect the current ideals of sustainability and equality in construction and urbanization. Such challenges, to a significant extent, are related to the structure of the industry but they are not addressed by urban policies. There are important missing links related to construction labour in the understanding and operationalization of policies that address challenges of sustainable and inclusive urbanization. The implications are highlighted in this chapter. Such missing links reveal a potential area for future research related to construction economics. If the construction industry continues to undermine labour, what are the scenarios? How can the industry ultimately work with ever-increasing decent work deficits? What are the implications for the future of cities, as products of the construction industry, considering that construction workers are both the producers as well as, together with their families, the users? Is it realistic to predict a construction industry that is highly automated with a minimum number of (well-trained) workers, and what are the social implications?

In terms of governance, an important part of the solution is for unions to reconsider their frequently 'one size fits all', 'Fordist' strategy, in order to come to grips with, as well as more aggressively capture potential advantages from, an industry that has changed towards a plural, hybrid and territorialized network anchored in cities. The huge numbers of informal and casual workers make it virtually impossible for unionized workers alone to change the status quo; they need to be together. In addition, strong governments with the political will to enforce workers' rights and a better income distribution are key to trigger change. A third option is for the entrepreneurs to create strategic awareness that a race to the top is in everyone's best interest, as opposed to a race to the bottom. All these alternatives can happen in parallel. The city is an ideal locus, for the reasons mentioned above. A research question could focus on different scenarios in terms of workers' organizations and social dialogue, and their impact on the development of the construction industry.

There is much potential to flesh out the entanglements between labour and construction, on the one hand, and agglomeration economies, on the other, through labour markets, and horizontal and vertical relations in production chains and clusters. What was labelled as alternative construction represents undeveloped clusters that could benefit from an urban policy framework structured around market creation initiatives such as upscaling community contracting, organizational-managerial and institutional strengthening of alternative forms of local economic development (solidarity co-operatives), and selective incentives in public procurement. For instance, the Brazilian My House My Life programme experimented with the professionalization and upscaling of community building through a specific line of subsidized funding for construction managed by co-operatives.

The dynamic learning and innovation effects that spread out from smart construction to related sectors has remained small. For one, this could be stimulated through the emergence of new services, consultancy or research related to social and environmental challenges. Less frequently explored is the potential of other market extension initiatives: for example, by developing programmes that recognize the self-defeating character of contractor strategies that download costs and responsibilities for maintaining social-environmental standards in hierarchical value chains, and stimulate the alternative involvement of both labour and small-scale suppliers in building better cities as superior places to work and live. In that sense, the experiences of industrial cluster programmes aimed at stimulating managerial-productive restructuring of small and medium-sized construction firms, and strengthening their labour force, provides useful insights. These are also important avenues for further research related to construction economics. Furthermore, the articulation of market creation and extension schemes, within a broader policy framework that recognizes ordinary cities as territorial networks of construction and their labour pools, would increase the potential to trigger higher-end trajectories of competitiveness and increased value-added through successive cycles of learning, development of additional skills, innovation and improved employment conditions.

Public–private partnerships (PPPs) have attracted increasing attention as a driving mechanism for construction and urbanization. If steered in the right direction, PPPs could incorporate on-the-job training. However, it is also important to recognize that many PPPs have generated mixed results; therefore, they are not a panacea. (It is beyond the scope of this chapter to elaborate on PPPs.)

The first section of this chapter noted the actor–network theory, which opens up space for the agency of construction workers. It has led to burgeoning practical applications focusing on a systemic approach to value chains, which is used extensively by the ILO. The advantage is that it unveils the challenges that limit the development of construction (or other) value-chains. This means that workers' issues also surface, counteracting the fact that their agency is often disregarded. The evaluation of the impact of the systemic approach to construction value chains is another important line of investigation. Finally, a combined effort of additional research and policy innovation based on the above recommendations, that explicitly incorporates labour's agency in consolidating better construction and urban spaces, provides important contributions to filling in some of the missing links analysed in this chapter.

ACKNOWLEDGEMENTS

The authors wish to express their gratitude to the following colleagues: Dr Jill Wells, for her work on the effects of the structure of the construction industry on the workers (ILO, 2001), which was used in this chapter; Professor George Ofori, Professor Monica Haddad and an anonymous peer reviewer for their comments and suggestions.

NOTE

1. Testimony from Edmundo Werna, who represented the International Labour Office throughout the process leading to Habitat III.

REFERENCES

Abanda, F.H., Weda, C., Manjia, M.B. and Pettang, C. (2021). BIM and M&E systems for the performance of slum upgrading projects in sub-Saharan Africa. *International Journal of Digital Innovation in the Built Environment* 10(1), 1–17.

Afolabi, A.O., Ojelabi, R.A., Tunji-Olayeni,P.F., Fagbenle, O.I. and Mosaku, T.O. (2018). Survey datasets on women participation in green jobs in the construction industry. *Data in Brief* 17, April, 856–862.

Amin, A. (ed.) (1994). *Post-Fordism: A Reader*. Oxford, UK and Malden, MA, USA: Blackwell.

Araújo, N., Cardoso, L., Brea, J.A.F. and Araújo, A.F. (2018). Green jobs: the present and future of the building industry: evolution analysis. *Social Sciences* 7(12), 266. https://doi.org/10.3390/socsci7120266.

Atkinson, A. (2008). Cities after oil – 3: Collapse and fate of cities. *Cities* 12(1), 79–106.

Ball, M. (1981). The Development of capitalism in housing provision. *International Journal of Urban and Regional Research* 5, 145–177.

Brueckner, J. (2011). *Lectures on Urban Economics*. Cambridge, MA, USA and London, UK: MIT Press.

Buckley, M. (2014). On the work of urbanization: migration, construction labor, and the commodity moment. *Annals of the American Association of Geographers* 104(2), 338–347.

Buzzelli, M. and Harris, R. (2006). Cities as the industrial districts of housebuilding. *International Journal of Urban and Regional Research* 30(4), 894–917.

Camagni, R. (2005). *Economia Urbana*. Barcelona: Antoni Bosch.

Chancellor, W., Abbott, M. and Carson, C. (2015). Factors promoting innovation and efficiency in the construction industry: a comparative study of New Zealand and Australia. *Construction Economics and Building* 15(2), 63–80.

Cohen, M. (2015). John F.C. Turner and housing as a verb. *Built Environment* 41(3), 412–418.

Cox, K.R. (ed.) (1997). *Spaces of Globalization: Reasserting the Power of the Local*. New York and London: Guilford Press.

Curley, A. (2012). Self-help housing organisations. Elsevier (ed.), *International Encyclopedia of Housing and Home* (pp. 292–296). London: Elsevier.

De Valence, G. (2012). The significance of barriers to entry in the construction industry. *Australian Journal of Construction Economics and Building* 7(1), 29–75.
Dipasquale, D. and Wheaton, W.C. (1996). *Urban Economics and Real Estate Markets*. Upper Saddle River, NJ: Prentice Hall.
FIABCI (International Association of Real Estate Developers) (2016). *The City We Want is Affordable* (4 vols). https://fiabci.org/. Retrieved 18 June 2020.
FIABCI and UN-Habitat (n.d.). Methodology for the City Prosperity Initiative – Perception Index. http://www.perceptionindex.org/Public/Methodology. Retrieved 4 February 2021.
Glaeser, E.L. (2011). *Triumph of the City: How Urban Space Makes Us Human*. London: Pan Books.
Habitat III (United Nations Conference on Housing and Sustainable Urban Development) (2016). *New Urban Agenda*. Quito: Habitat III.
Halbert, L. and Rouanet, H. (2014). Filtering risk away: global finance capital, trans-scalar territorial networks and the (un)making of city-regions: an analysis of business property development in Bangalore, India. *Regional Studies* 48(3), 471–484.
Harvey, D. (2012). *Rebel Cities: From the Right to the City to the Urban Revolution*. London and New York: Verso.
International Labour Organization (ILO) (2001). The construction industry in the twenty-first century: its image, employment prospects and skill requirements. Report for Tripartite Meeting. Sectoral Activities Department. Geneva: ILO.
International Labour Organization (ILO) (2015a). Good practices and challenges in promoting decent work in construction and infrastructure projects. Issues paper for discussion at the Global Dialogue Forum on Good Practices and Challenges in Promoting Decent Work in Construction and Infrastructure Projects, 19–20 November. Geneva: ILO.
International Labour Organization (ILO) (2015b). Complaint concerning non-observance by Qatar of the Forced Labour Convention, 1930 (No. 29), and the Labour Inspection Convention, 1947 (No. 81), made by delegates to the 103rd Session (2014) of the International Labour Conference under article 26 of the ILO Constitution. 325th Session of the Governing Body, Geneva, 29 October – 12 November. Geneva: ILO.
International Labour Organization (ILO) (2016). Workplace stress: a collective challenge. Report. Geneva: ILO.
International Labour Organization (ILO) (2017). Complaint concerning non-observance by Qatar of the Forced Labour Convention, 1930 (No. 29), and the Labour Inspection Convention, 1947 (No. 81), made by delegates to the 103rd Session (2014) of the International Labour Conference under article 26 of the ILO Constitution. 331st Session of the Governing Body. Geneva, 26 October–9 November. Geneva: ILO.
International Labour Organization (ILO) (n.d.-a). ILOSTAT. Retrieved from https://ilostat.ilo.org/data/.
International Labour Organization (ILO) (n.d.-b). Social dialogue. https://www.ilo.org/ifpdial/areas-of-work/social-dialogue/lang--en/index.htm. Retrieved 8 February 2021.
Keivani, R. and Werna, E. (2001a). Refocusing the housing debate in developing countries from a pluralist perspective. *Habitat International* 25(2), 191–208.
Keivani, R. and Werna, E. (2001b). Modes of housing provision in developing countries. *Progress in Planning* 55 (2), 66–118.
Kofoworola, O.F. and Gheewhala, S. (2008). An input–output analysis of Thailand's construction sector. *Construction Management and Economics* 26(11), 1227–1240.
Krugman, P. (1991). *Geography and Trade*. Leuven, Belgium and Cambridge, MA, USA: Leuven University Press and MIT Press.
Kumar, S. and Fernandez, M. (2017). *The Urbanisation–Construction–Migration Nexus in Five Cities in South Asia: Kabul, Dhaka, Chennai, Kathmandu and Lahore*. Final Report. Research of the London School of Economics and Political Sciences, funded by DfID. London: LSE.
Lawrence, R. and Werna, E. (eds) (2009). *Labour Conditions for Construction: Building Cities, Decent Work and the Role of Local Authorities*. Oxford: Wiley-Blackwell.
Lelo, D., Yusof, S., PurbaJan, J.H.V. (2019). Influence of work safety and work stress on productivity. *Proceedings, International Conference on Industrial Engineering and Operations Management*. August, Bangkok. pp. 3602–3609.

Lim, J., Schultman, F. and Ofori, G. (2010). Tailoring competitive advantages derived from innovation to the needs of construction firms. *Journal of Construction Engineering and Management* 136(5), 568–580.

Lizarralde, G. (2014). *The Invisible Houses: Rethinking and Designing Low-Cost Housing in Developing Countries*. Abingdon: Taylor & Francis.

Malpezzi, S. and Wachter, S.M. (2001). The role of speculation in real estate cycles. *Journal of Real Estate Literature* 13(2), 141–164.

Markusen, A.R. (1999). Four structures for second tier cities. In A.R. Markusen, L. Yong-Sook and S. Digiovanna (eds), *Second Tier Cities: Rapid Growth Beyond the Metropolis*. (1st edn, pp. 21–41). London, UK and Minneapolis, MN, USA: University of Minnesota Press.

Marshall, A. (1920). *Principles of Economics*. London: Macmillan.

McDermotti, P. and Khalfan, M. (2012). Achieving supply chain integration within the construction industry. *Australasian Journal of Construction Economics and Building* 6(2), 44–54.

Mella, A. and Savage, M. (2017). *Construction Sector Employment in Low Income Countries*. ICED (Infrastructure and Cities for Economic Development) Report, DfID (Department of International Development, UK). London: DfID.

Moulaert, F. (2005). Institutional economics and planning theory: a partnership between ostriches? *Planning Theory* 4(1), 21–32.

Norberg, C. and Johansson, M. (2020). 'Women and "ideal" women': the representation of women in the construction industry. *Gender Issues* 38, 1–24.

O'Neill, R. (2009). Greenwash. *Hazards Magazine* 108(October–December). http://www.hazards.org/greenjobs/greenwash.htm.

Pirounakis, N.G. (2013). *Real Estate Economics: A Point to Point Handbook*. London and New York: Routledge.

Robinson, J. (2006). *Ordinary Cities: Between Modernity and Development*. London: Routledge.

Ruan, X., Ochieng, E.G., Price, A.D.F and Egbu, C. (2013). Time for a real shift to relations: appraisal of social network analysis applications in the UK construction industry. *Australasian Journal of Construction Economics* 13(1), 92–105.

Schumaker, E.F. (1973). *Small is Beautiful – A Study of Economics as If People Mattered*. London: Abacus.

Schumpeter, J.A. (1934). *The Theory of Economic Development: An Inquiry into Profits, Capital, Credit, Interest, and the Business Cycle*. New Brunswick, NJ: Transaction Books.

Shimbo, L.Z. (2020). O concreto do capital. Os promotores do valor imobiliário nas cidades brasileiras. Unpublished thesis. São Carlos: Instituto de Arquitetura e Urbanismo de São Carlos, Universidade de São Paulo. Tese de Livre Docência.

Storper, M. (1997). *The Regional World: Territorial Development in a Global Economy*. New York: Guilford Press.

Storper, M. and Venables, A.J. (2005). O burburinho: a força econômica da cidade. In C.C. Diniz and M.B. Lemos (eds), *Economia e Território* (1st edn, pp. 21–56). Belo Horizonte: Editora UFMG.

Tournee, J. and Van Esch, V. (2001). *Community Contracts in Urban Infrastructure Works: Practical Lessons from Experience*. Geneva: ILO.

Turner, J. (1976). *Housing by People: Towards Autonomy in Building Environments, Ideas in Progress*. London: Marion.

Turner, J. and Fischter, R. (1972). *Freedom to Build: Dweller Control of the Housing Process*. New York: Macmillan.

United Nations (UN) (2015). *Sustainable Development Goals*. https://www.un.org/sustainabledevelopment/sustainable-development-goals/.

United Nations Environment Programme (2019). *Global Environment Outlook 6: Healthy Planet, Healthy People*. Report. Nairobi: UN Environment.

United Nations Human Settlements Programme (UN-Habitat) (2016). *World Cities Report*. Nairobi: UN-Habitat.

Van Empel, C. and Werna, E. (2010). Labour oriented participation in municipalities: How decentralized social dialogue can benefit the urban economy and its sectors. Sectorial Policies Department Working Paper 280. Geneva: ILO.

Ward, P. (2019). Self-help housing. In Orum, A.M. (ed.), *The Wiley-Blackwell Encyclopedia of Urban and Regional Studies* (pp. 502–511). Oxford: Wiley-Blackwell.
Watson, V. (2009). Seeing from the South: refocusing urban planning on the globe's central urban issues. *Urban Studies* 46(11), 2259–2275.
WEF (World Economic Forum) (2017). Migration and its impact on cities. https://www3.weforum.org/docs/Migration_Impact_Cities_report_2017_HR.pdf.
WEF (World Economic Forum) (2018). *White Paper on Migration and Housing.* http://www3.weforum.org/docs/WEF_Making_Affordable_Housing_A_Reality_In_Cities_report.pdf.
WEF (World Economic Forum) (2019). *Transforming Infrastructure: Frameworks for Bringing the Fourth Industrial Revolution to Infrastructure.* Community Paper. Shaping the Future of Cities. Infrastructure and Services Initiative. Geneva: WEF. http://www3.weforum.org/docs/WEF_Technology_in_Infrastructure.pdf.
WEF (n.d). *Global Future Council on Cities and Urbanization.* https://www.weforum.org/communities/the-future-of-cities-and-urbanization/articles?page=2.
Wells, J. (2016/18). *Protecting the Wages of Migrant Construction Workers, Parts One and Two.* Engineers Against Poverty Reports. http://engineersagainstpoverty.org/.
Wells, J. and Prado, M.G. (2019) *Protecting the Wages of Migrant Construction Workers, Part Three: What can be Learned from Systems of Wage Protection in China, EU, US and Latin America.* Engineers Against Poverty Report. http://engineersagainstpoverty.org/.
Werna, E. (1996). *Business as Usual: Small-Scale Contractors and the Production of Low-Cost Housing in Developing Countries.* Aldershot: Avebury.
Werna, E. (2007). Tendencias de la Mano de Obra. Special issue about the construction industry. *Economia Exterior* 40, 133–142.
Werna, E. (2012). Green jobs in construction. In G. Ofori (ed.), *Contemporary Issues in Construction in Developing Countries* (1st edn, pp. 158–200). Abingdon: Spon.
Werna, E. (2016), From Blade Runner to Habitat IV – how livelihoods can make or break the city of tomorrow – Parts I and II. URBANET (News and Debates on Municipal and Local Governance, Sustainable Urban Development and Decentralization). http://www.urbanet.info/blade-runner-to-habitat-iv-city-of-tomorrow. Retrieved 6 February 2021.
Werna, E., Abiko, A.K., Coelho, L.O., Simas, R., Keivani, R., et al. (2000). *Pluralism in Housing Provision: Lessons from Brazil.* New York: Nova Science.
Williamson, O. (1985). *The Economic Institutions of Capitalism.* New York: Free Press.
World Health Organization (WHO) (2003). Work organization and stress: systematic problem approaches for employers, managers and trade union representatives. Paper written for the WHO by S. Leka, A. Griffiths and T. Cox. Protection Workers' Health Series No. 3. Geneva: WHO.
Zitzman, L. (2021). Women in construction: the state of the industry in 2021. Bigrentz website/blog. https://www.bigrentz.com/blog/women-construction. Retrieved 7 February 2021.

17. Economic principles of bidding for construction projects

Samuel Laryea

INTRODUCTION

In construction procurement, clients award most contracts through a bidding process which uses the principle of market competition to obtain an economic price from competing contractors. As this chapter will demonstrate, price formation in construction contracts usually occurs in a bidding process largely dictated by economic considerations. The textbook approach to describing the way that the price of a construction project is established reveals a complex process that is not well articulated in the literature. This is perhaps to be expected, given that much of the information relating to this aspect of construction procurement is commercially sensitive.

While clients use competition to try and achieve the best possible price, competitive tendering procedures alone do not always lead to best value, particularly in instances where selection is solely based on price (Rintala et al., 2008). However, better value may be procured if tenderers compete on price and quality as provided for in the International Organization for Standardization standard ISO 10845-1. The raises a distinction between competitive tendering and lowest-price tendering. The study by Smith and Bohn (1999) showed that for clients, periods of high competition may yield bid prices that would appear on the face of it to be exceptional value. However, ultimately, the lowest bids may not prove to be such bargains, especially where this leads to claims and insolvencies. This is one reason why bidding processes for construction projects need to be carefully configured to arrive at sustainable outcomes for each party.

The majority of clients in the construction industry seek to procure the services of contractors who offer the best price and are capable of delivering the intended project outcomes and value for money. Some clients interpret 'best price' as the lowest price. However, recent literature warns against this practice (see, for example, Watermeyer, 2018). Contractors aim to make a profit from their contracts once a job is won and executed. To achieve their aims, the bidding practices of clients and contractors are usually informed by certain principles and considerations that relate to their economic needs. The economic needs of the parties are outlined below.

ECONOMIC CONTEXT OF THE BIDDING PROCESS

The bidding process for construction projects is a complex buyer–supplier relationship in which clients and contractors aim to satisfy their respective economic needs. In the construction economy, the bidding process is an economic activity at the core of the contracting market through which construction contracts and subcontracts are placed (Chang, 2015). The application of various economic theories in construction contracting and the construction bidding process has been discussed by researchers in construction economics who apply theories from the mainstream field of economics to explain economic behaviour in construction bidding and other areas of the construction industry. Some of the theories that have been applied to explain economic behaviour in construction bidding include price theory, game theory, decision theory, portfolio theory, economics of information theory, asymmetric information theory, auction theory, transaction cost theory, contract theory, utility theory, prospect theory, choice theory, bounded rationality and bidding theory. A discussion of the application of mainstream economic theories in developing our understanding of the economics of the construction bidding process and construction economics in general can be found in works such as Kangari and Riggs (1989), Winch (1989), Myers (2003), Ofori (1994), Bon (2001), Chang (2015) and Ahmed et al. (2016).

These microeconomic theories frequently come into play in the bidding processes used by clients and contractors to arrive at a bidding price for a construction project. The way that the bidding process plays out in the construction industry, including how the forces of demand and supply dictate outcomes and value-for-money considerations, broadly aligns with the economic behaviour of buyers and sellers in other industrial sectors outside construction. This is particularly so when factors such as price, reputation and current circumstances are considered. Therefore, it may be argued that key principles of bidding economics in construction are largely applicable to other industrial sectors, and vice versa, although the nature of competition, production information, risk and on-site production conditions in the construction industry can be identified as primary economic issues and distinguishing factors of bidding considerations in construction contracts.

On a macro level, bidding is dictated by the economic forces of demand and supply in competitive market environments where the client is buyer and the contractor is seller (see Mankiw, 2021). The transaction is complex and acted upon by certain economic factors that ultimately influence a bidding price. This includes the quality of the project information or degree of project definition at the time of bidding. Sichtmann (2007) explains the way that industrial buyer–seller relationships are influenced by information economics. In the construction industry, asymmetric information has long been established as one of the key factors characterising commercial relationships in construction projects (Xiang et al., 2012). Construction is an information-intensive industry where the information provided by clients is a major determinant of a contractor's bidding price.

The microeconomic theory of the behaviour of individual competitive markets suggests that a bidding price may be dependent on the market or competitive environment in which it takes place (as explained in *Microeconomics* by McConnell et al., 2021). However, it would also depend on a firm's particular circumstances (explained in Laryea and Hughes, 2011). Usually, the client/employer must be convinced that:

1. It will receive its project at a reasonable market price and that it will not pay more than it is contractually and ethically obliged to do.
2. It will receive the project within a reasonable period of time.
3. It will receive the quality and scope of work represented in the contract.

On the other hand, contractors tend to focus on at least the following:

1. The contractor must get work at the right price, or it will succumb in the medium/long run.

2. The contractor must perform its current and future work at the lowest cost, using the most effective and efficient means and methods.
3. There will be flawless legal management to maintain excellent client/employer relations while getting rewarded with what it is legally and ethically entitled to.

These microeconomic needs ultimately shape the bidding behaviour and practices of clients and contractors. It should be noted that bidding is also impacted upon by the macroeconomic environment of construction projects. In growing markets and booming economic times, there is likely to be higher demand for construction projects and contractors are likely to bid higher margins. However, in recessions and difficult economic times, there is likely to be higher competition arising from lower demand for construction projects and margins are likely to be lower (Hillebrandt, 2000). Another factor influencing the bidding process of construction projects is risk. The research by Laryea and Hughes (2008) provides comprehensive insights into how contractors take risk into account in the bidding process. In the next section, the subject of risk will be more broadly considered in a general sense; and then specific consideration will be given later on in the chapter to how risk influences pricing levels in the bidding process.

RISK AND CONSTRUCTION PROJECTS

Risk is one of the key factors influencing decisions about the procurement and contract strategy for construction projects. All construction projects have inherent risks. These have to be effectively evaluated, followed by sound decisions based on the evaluation, and appropriate action as a result of these decisions. If the monetary loss resulting from risk events is not considered or is underestimated due to associated uncertainties, a construction enterprise may suffer a loss and eventually fail (Paek et al., 1993, p. 743).

The unique nature of the construction industry, construction projects, and how work is organised in the industry, makes contracting different from other industries. This is mainly the result of peculiar factors relating to the economic, contractual, political and physical environments within which products are manufactured. They include: necessity to price product before production, competitive tendering, low fixed-capital requirements, preliminary expenses, delays to cash inflows, tendency to operate with too low a working capital, seasonal effects, fluctuations and their effects, government intervention, activity related to development, uncertain ground conditions, unpredictable weather, and no performance liability or long-term guarantees (Calvert et al., 1995, p. 124; Hughes and Hillebrandt, 2003, 1, pp. 508–510).

Laryea and Hughes (2006) explain that contractors have used various ways to survive risks in construction industry since the early 19th century. Most contactors adopted the strategy of speculative house building in the 19th and 20th centuries to sustain their labour force and business costs through the different economic cycles of contracted work. In modern times, there is a growing tendency for contractors to use their positive cash flows to invest in projects, rather than house building. More recently, successful contractors have been diversifying into businesses whose cycles counteract those of construction (Hughes and Hillebrandt, 2003). Contractors are minimising risk by declining work they perceive as too risky, subcontracting large portions of their work to others, and apportioning risk in wage structures. In essence, they are passing on risk to others.

Traditionally, contractors use intuition and experience to judge allowances to cover the risks. Conceptual research by Tah et al. (1993, p. 284), introducing a fuzzy set model for a contractor's project risks contingency allocation, suggests that such decisions are influenced mainly by the estimator's perception of risks and management's view of the future and their desire to avoid an overrun situation. In contemporary times, classical and conceptual risk models that contractors can use when bidding have proliferated, but they are not used in practice. The reasons found by Akintoye and MacLeod (1997, p. 36) in a questionnaire-based survey of 30 general contractors in the United Kingdom (UK) include the sophistication of the techniques, doubts over whether they are applicable in practice, and the consideration that the majority of risks are contractual or construction-related, which makes them fairly subjective and better dealt with based on experience.

Non-empirical research proposing a construction risk management system to substitute for the traditional intuitive approach of contractors' risk assessment, by Al-Bahar and Crandall (1990, p. 534), defines risk as: 'exposure to the chances of occurrences of events adversely or favourably affecting project objectives as a consequence of uncertainty'. Hackett and Statham (2016, p. 14) conceptualise risk as: 'the possible loss resulting from the difference between what was anticipated and what finally happened'. Essentially, risk results from uncertainty (Loosemore et al., 2006, p. 8); construction projects are surrounded by uncertainties. Therefore, it is important for contractors and contractual parties to properly recognise risk and account for it, as risk forms the basis of almost everything in construction: pricing policies, contract strategies, procurement routes, and health and safety.

Contractors normally plan their tenders to satisfy two main objectives which may sometimes seem incompatible. They aim to offer the most attractive price, whilst minimising their risks and liabilities under a potential contract (Marsh, 2000, p. 53). However, getting work is the overriding factor, and they employ several means to respond to risks not covered by the contract, bonds or insurance. Contractors systematically add a premium to their bids to account for both the level of risk in a project and their lack of enthusiasm to do a job when they do not need the work. This was found by Neufville and King (1991, p. 668) when they performed a two-way empirical investigation on 30 contractors in the United States of America (USA) to identify how risk and need for work influence the behaviour of bidders. The study also found that, for contractors, the risk premium can be apportioned as high as 3 percent of the total project cost. The risk premium covers both the transactional costs of organising the market and reasonable profit margins. Owing to the competition factor, there is only a certain extent to which risk allowances can be apportioned in markup.

Markup, the sum of contingencies and profit, is calculated as a percentage of the sum of overhead and direct costs for materials, labour and plant. In a study, 29 contractors in the USA indicated the important factors influencing markup as: project characteristics, project documents, company characteristics, bidding situation, economic situation and client characteristics (Liu and Ling, 2005, p. 391). Appropriate contingency allowances are added to the markup where necessary. This contractor contingency is the estimated value of the extraordinary risks that will be encountered in a project. Extraordinary risks are those project risks that are not covered by the contract, bonds or insurance. A study of 12 small to medium-sized contractors in the USA found that in times when competition is high, contractors do not typically include contingency in tenders (Smith and Bohn, 1999, p. 107). A high number of bidders and low workload will guarantee almost no contingency in the bid. Where contingencies were used, eight factors influencing the contingency levels were: workload, contract size, project complexity, number of bidders, owner's reputation, bidder mentality, clarity of contract documents and time frame for bidding (Smith and Bohn, 1999, p. 106). The contractors mostly used a percentage of the total cost approach and their intuition and previous contract knowledge to determine the contingency allowance. A final point to note is that payment mechanisms included in contracts also attract risk pricing.

PRICE FORMATION IN CONSTRUCTION CONTRACTS

The way that contractors and clients negotiate and agree on a price is complex and not well articulated in most of the literature. Most standard textbooks provide descriptive and experiential accounts of the way that price formation occurs in construction. However, few empirical accounts exist apart from the ethnographic study of tender processes by Laryea (2008). This section provides a detailed account of pricing strategies in the construction industry, and the principles used by contractors to calculate the price of a construction project.

Pricing Approaches in Construction

The pricing approaches in construction can be categorised into client pricing and contractor pricing strategies. Client pricing strategies relate to payment mechanisms adopted to remunerate contractors for work done. Contractor pricing strategies refer to the methods by which a business calculates the amount to charge for a product or service (ISO 10845-1).

Watermeyer (2012) explains two types of pricing strategies in the International Federation of Consulting Engineers (FIDIC) and the Institution of Civil Engineers (New Engineering Contract, NEC3) families of contracts, which are price-based and cost-based. Price-based pricing strategies include lump sum, bill of quantities, price list/schedule and activity schedule. Cost-based pricing strategies include cost reimbursable and target cost. In the target cost pricing strategy, a target price is agreed and on completion of the works the difference between the target price and the actual cost is apportioned between the employer and contractor on an agreed basis (see Laryea, 2016 for examples of target cost contract cases including examples of share ratios/formulae used in practice).

In the pricing of projects procured under relational contracting, a framework contract pricing strategy may be adopted where the contractor's fee for profit and overheads would be fixed for the duration of the framework agreement (Watermeyer, 2013). Nowadays, many clients are increasingly adopting the use of framework contracts for engaging contractors in longer-term relationships to create a commercial relationship of mutual benefit. One of the key objectives of the UK Government's Construction Sector Deal (HM Government, 2018) was to 'Promote long-term, collaborative relationships with industry to reduce transaction costs in procurement and maximise innovation'. However, contractors should be mindful of the economic risks in collaborative contracting approaches, and ensure their pricing takes the asymmetrical nature of the client–contractor relationship into account to minimise their risk exposure.

The pricing strategies in construction may also be categorised into cost-based and market-based approaches. In cost-based pricing, the price of a product is established by adding a profit margin to the total cost of the product. In market-based pricing, the price of the product is based on the required size of market needed to achieve a certain level of profit. Mochtar and Arditi (2001) used a similar terminology to describe the pricing philosophies of 400 US contractors. However, Skitmore et al. (2006) referred to the two pricing strategies or approaches as full-cost (cost-based) pricing and neoclassical microeconomic (market-based) pricing. It may be appropriate to consider these approaches as pricing models or methods of arriving at a price rather than as pricing strategies.

The survey of the top 400 US contractors by Mochtar and Arditi (2001, p. 410) showed 11 factors that influence a contractor's pricing strategy: project size/complexity, financial goals of company, company's strengths and weaknesses, expected future project from the owner, need for work, owner's characteristics, project location, demand/economic conditions, competition, owner's consultant characteristics, and subcontractors' characteristics. The main objectives of contractors when adopting a pricing strategy are to win the work, ensure a healthy cash flow and achieve a reasonable profit (Tah et al., 1994; Skitmore and Wilcock, 1994; Shash, 1998; Rooke et al., 2004; Hughes et al., 2015). Such strategies include back-end loading and front-end loading where, in the former, contractors can artificially reduce the rates for work at the beginning of the project and add a corresponding proportion to the rates for work later on in the project. Alternatively, in the latter, they can increase the rates at the beginning of a project with a corresponding decrease in the rates at the end.

The survey of nine small builders in Australia by Skitmore and Wilcock (1994) found that some contractors tried to limit their tender preparation costs by reducing pre-tender planning and risk appraisal to an absolute minimum. This would reduce the bid price and increase competitiveness. They also found that contractors would rate preliminary items based only on what they felt the architect would eventually want on site, rather than what was actually specified in the bills of quantities. The study also showed that some of the contractors actually limited their tender preparation costs by using experience to price certain items rather than a detailed rate analysis. The study involving seven construction projects and interviews with contractors by Rooke et al. (2004) showed that contractors bid at prices that reflect the expectation that the ultimate price of the job will be inflated by claims. Such claims are planned carefully, and the value of the expected return is calculated with a degree of accuracy that allows the contractor to bid at prices that, if quantities were to remain unchanged, would render a negative profit.

The survey of 30 specialty contractors in the US by Shash (1998) found that some general contractors negotiate prices with subcontractors after they are awarded a contract. They would include a given subcontractor's quotation in their bids, but in an effort to increase their own profit they tend to persuade the subcontractor to reduce their prices. In response, subcontractors either avoid working for such a contractor or inflate their quotations by 5–10 per cent. When they take the job because they need it, they: (1) reduce, if at all possible, the amount of equipment needed at any given time; (2) rework the specifications to allow for an alternative; (3) reduce their workforce; (4) manage the project more efficiently; and (5) assign more skilled workers to the project to improve productivity.

These studies demonstrate that contractors employ a range of pricing strategies when formulating their bids. The pricing level in a bid submission is influenced by factors including pricing strategy, the extent of competition and how keen a firm is to win the work.

The Bidding Process for Construction Projects

Clients of the construction industry widely adopt competitive bidding as a process to obtain a market-related price from contractors (Hackett and Statham, 2016). However, a major problem with current competitive procurement and tendering practices is the wasted effort and resources when several companies are bidding for the same work and inevitably some will be wasting resources (Laryea et al., 2013). Multiplying this up throughout the construction industry would reveal the significant amount of wasted resources. This appears to go against the grain of sustainable procurement theory, as some firms will inevitably be wasting resources that could be used elsewhere in the wider economy. The benefits of competitive procurement are well established, but this procurement process has been shown not to be the most successful, often ending in dispute, wasting even more resources (Hughes et al., 2006). Research is needed to propose alternative procurement approaches that reduce bidding cost and minimise the current wasted effort. It can be argued that if construction companies were able to reduce their bidding costs, their overhead costs might reduce and cost of construction could reduce.

The issue of wasted resources and achieving better value in construction procurement was a subject of consideration in the UK Government Construction Strategy (HM Government, 2011, p. 16), which set out some principles for an alternative approach towards procurement. This was designed to eliminate the wastefulness of teams completing and costing a series of alternative designs for a single project, only one of which will be built. This has been the subject of continuing dialogue, seeking a more collaborative, integrated model that nonetheless maintains competitive tension and the ability to demonstrate value for money. Two propositions were put forward by industry teams. One effectively pitches framework contractors against a challenging cost benchmark, with the understanding that if no member of the framework can beat the benchmark, then the project will go to tender without the initial framework contractors being permitted to bid. This provides opportunities for new entrants. The other proposition offers a guaranteed maximum price underwritten by insurance, which also extends to protection against defects. Both propositions assume the full engagement of an integrated team, with designers, other professional consultants and constructors offering an integrated proposition, and with key trade contractors and manufacturers involved in developing the design.

A subsequent Cabinet Office (2014) publication on *New Models of Construction Procurement* proposed three 'new' procurement models, namely cost-led procurement, integrated project insurance, and two stage open-book. More recently, the UK government launched the *Construction Playbook* (HM Government, 2020) to define expectations, aims and procedures for the procurement of projects and programmes from the construction sector. The intention is to be more strategic and less transactional in the procurement of projects. Given the evolving global economic conditions, clients' demand will increase for more strategic construction procurement approaches that deliver better value and outcomes. Contractors will have to respond innovatively with greater competitiveness and more efficient supply models.

The remainder of this section looks at the nature of the real-life bidding process in practice. The bidding process is complex and involves multiple activities and participants from both client and contractor sides (see Figure 17.1).

Standard textbooks reveal two main objectives for a bidding process which are (1) to obtain a suitable price and programme; and (2) to select a suitable contractor for a project. This means that the pricing and selection processes occur in the same bidding process.

Figure 17.1 shows three basic stages of the bidding process and various activities performed by contractors to determine a bidding price. The bid preparation process involves a series of complex activities requiring detailed information to perform. One key challenge of the real-life bidding process is that it is often characterised by incomplete information, which often leads to various tender queries and clarifications that the bidding contractors use to fill in gaps in

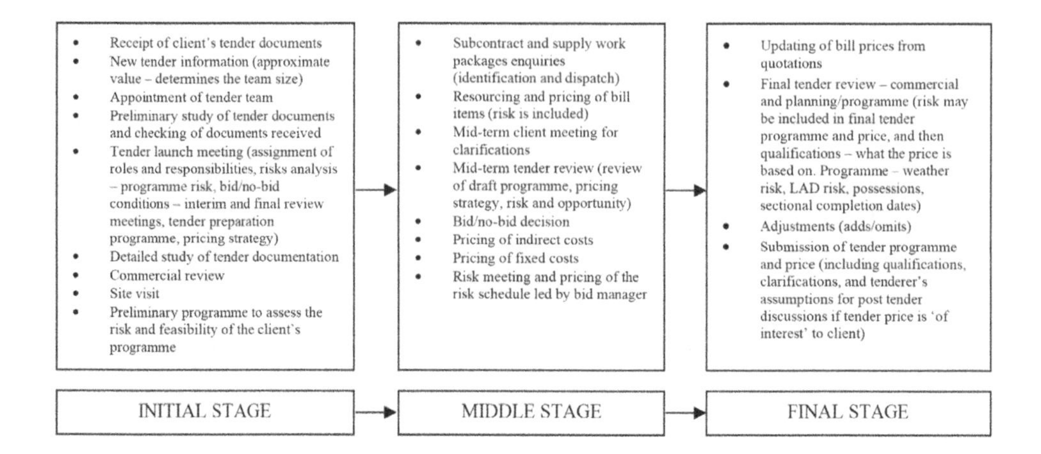

Figure 17.1 Stages of a typical tender process

the information required to price their bids (this is explained in a study on tender documents by Laryea, 2011).

A significant feature of the bidding process is that there is significant reliance on subcontractors' quotations, which constitute a significant proportion of a contractor's bidding price. The study by Laryea (2009) discusses in detail the processes involved in the soliciting and use of subcontract and supply quotations in the formulation of a bidding price. The way contractors and subcontractors collaborate in the tender pricing process is discussed by Laryea and Lubbock (2014) based on data from 94 UK subcontractors. Laryea's (2008) study showed that the pattern of activities involved in a contractor's bidding process does not follow the typical S-curve (flatter at the beginning and end, steeper in the middle). The classic S-curve is described as having three parts: a gentle rise, a steep slope, and a gradual path to the asymptote (Cioffi, 2005). In a contractor's bidding process, the pattern of the estimator's activity is loaded at the beginning (with the tender documents being studied to gain an understanding of the scope of works in order to prepare supply and subcontract quotes and price the job well), it slows down in the middle (as the estimator waits for invited quotations) and is loaded towards the end (with quotations having come in, addenda, tender query responses, meetings, and submission time nearing). Thus, the pattern of basic activity is steeper at the beginning and end, and flatter in the middle. Hence, bidding activities may not model the typical S-curve behaviour of activities in a project.

Bidding is a complex process and interactions between the actors are as indicated in Figures 17.1 and 17.2. The outcomes are influenced by the factors discussed above, and the time available for the bidding activities, quality of information, supply chain relationships and quotations, as well as the project-specific information. Having to take a wide range of factors into account to process a single bidding price within a relatively short period of time, the poor quality of project information in some cases, and having to do this in the context of a competitive process, makes bidding a complex, challenging and risky process. Contractors, particularly, face greater risk exposure due to the risky nature of the construction business which often manifests itself in the high business failure of firms in the industry. For construction clients, there tends to be a relatively lower level of risk exposure, as a range of financial protection instruments such as performance and retention bonds are built into contracts to

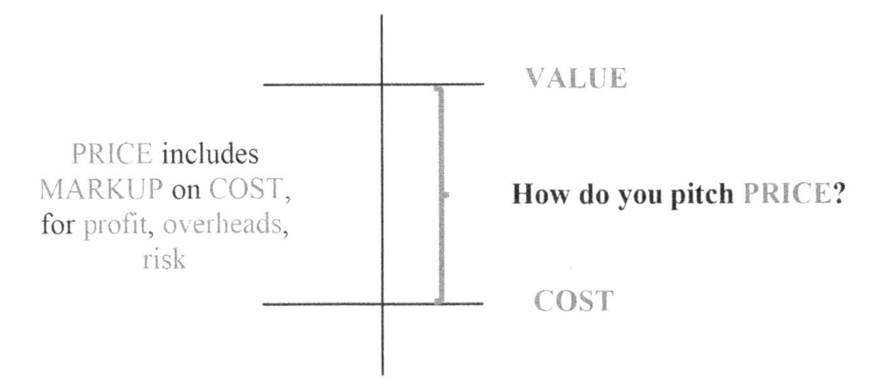

Source: Laryea and Hughes (2011).

Figure 17.2 Relationship between cost, price and value in tendering

provide security in the event of occurrence of risks pertaining to performance and defects in the construction phase.

The construction industry combines the price formation and contractor selection processes into one single tendering process. That is one reason why competition often has a significant impact on the pricing levels of contractors who are keen to win. Clients use competition as a strategy to award work and obtain the best price. In response, contractors employ various strategies to bid their lowest price to enhance their competitiveness. Thus, competition from other suppliers in the market is one of the key economic factors influencing pricing levels of contractors in the bidding process. Construction projects involve risk; the research by Laryea and Hughes (2008) explains how contractors take risk into account when calculating the price for a construction project. Thus, risk is another major factor influencing the pricing levels of contractors. The relationship between risk and price is an important variable in the bidding process. When clients invite tenders, contractors process their bidding price through three distinct stages, explained below.

Bid Pricing Processes and Components

A contractor's bidding process comprises three stages: the commercial review, estimating and adjudication stages. Details of each process are explained below.

Commercial review process

The study by Laryea and Hughes (2009) provides detailed insights into the commercial review process used by contractors to identify commercial risks in a client's conditions of contract in order to make the bid/no-bid decision, or to enable them to provide a response in the bid submission. Jaselskis and Talukhaba (1998) discuss some of the factors influencing bidding considerations.

When inviting bids from contractors, clients often present their terms and conditions to the market. These are often referred to as the conditions of contract, and examples of such contracts include FIDIC or NEC. An empirical study of 83 specialist contractors in the UK construction industry showed that in practice such contracts are rarely used without amendments (Hughes et al., 1998). When clients present their amended standard-form contract terms and conditions to contractors during the tender process, contractors analyse the proposed terms and conditions using a process called 'commercial review'. This process is often carried out by the commercial department and it helps contractors to respond appropriately to the contractual risks.

The research by Hughes et al. (1998) showed that nearly all of the 83 specialist contractors qualified their bids because of non-standard conditions. The remainder added qualifications during the negotiations leading to the signing of the contract. The contractors reported that they respond in two main stages to standard contract amendments. First, the work content is priced and negotiations firm this up. Second, the risk elements are negotiated. The contractors either sought to amend the clauses or negotiated a suitable payment for taking the risk. Mechanisms used for responding to amendments were: qualifying bids (80 per cent of contractors), seeking legal advice (55 per cent), consulting trade associations (30 per cent) and negotiating (58 per cent). Some firms (15 per cent) simply ignored non-standard conditions in their bid proposals. Non-standard clauses were found to range fairly evenly across the following themes: payment, others, programme, set-off, attendances, coordination, design liability, insurances, arbitration, liquidated damages, and retention. The attempt to quantify the degree to which standard forms are amended was based on a three-point scale: 'rarely', 'sometimes' and 'usually' (Hughes et al., 1998, pp. 87–89). JCT forms were 'usually' amended, whereas the FCEC Blue form was only 'sometimes' amended.

The research by Laryea (2008) investigated the extent to which clients amend standard-form contracts in practice, and the locus of the amendments. The study found an average of 52 amendments that clients had introduced into the standard conditions of contract for the four cases examined. In all four case studies, the locus of amendment by clients related mostly to financial and payment issues, and contractual and legal arrangements. These mainly centred around issues related to liquidated damages, time for completion, payment terms, minimum amount of interim certificates, percentage of value of goods and materials to be included in interim certificates, retention, defects liability period, PCG, performance bond, service from utilities companies, weather conditions, taxes, contract document definitions and warranties.

None of the contractors in the study chose to inflate their prices as a response to the amendments, as this was likely to reduce their chances of winning the work. Nevertheless, they were mindful of the contractual risks and liabilities in the proposed terms and conditions. Therefore, the contractors preferred to deal with the amendments using contractual rather than price mechanisms. They preferred to qualify and clarify the risks as part of their tender submission for post-tender negotiation.

Estimating and tender preparation

Estimating is the stage in which the actual project costs are considered. According to Hall (1972, pp. 7–8), each price build-up comprises the cost of materials (based on quotations from suppliers and merchants), unloading (including allowances for unloading), waste (2.5–15 per cent), labour, plant and sundry items. Buchan et al. (2003) explain that on receipt of the instructions to tenderers, specifications, drawings, bill of quantities and standard form of contract from the client, the contractor's estimating task is to price the project and produce an estimate.

Harrison (1981) explains some of the basic features of the estimating and tendering process. The author brought together a group of estimators with a collective experience of over 100 years. He defined an estimate as: 'a reasonably accurate calculation and assessment of the probable cost of carrying out defined work under known conditions'. The estimate involves both calculation and assessment, and often both scientific fact and human judgement of circumstances and probabilities must be brought together in its production (Harrison, 1981). The main difference between an estimate and a tender is highlighted in a textbook definition by Smith (1986, p. 34):

> an estimate is the preliminary assessment of the net cost of carrying out a specified amount of work whereas a tender is the final price offered to the client by the contractor and is the sum of money for which the contractor is prepared to carry out the work and will include not only the estimate but also a margin for overheads and profit.

Thus, the estimate comprises the net cost of the contractor's own measured work, the net cost of the items in the preliminaries section, the value of prime cost and provisional sums together with attendances, and the value of the domestic subcontractors' quotations. This amount does not include management costs, overheads, profit, allowances for risk and other required tender adjustments. It is when the latter charges are added that the estimate converts into a tender (Brook, 2017).

Harrison (1981) described a tender as 'an offer to carry out defined work under stated conditions for a stated reimbursement'. Converting an estimate into a tender is a delicate task that is often undertaken by management at an adjudication meeting, because a decision often has to be made on the markup that produces a bid that is both competitive and profitable to the contractor.

The elements of a bidding price are now explained.

Costs

Contractor bids often include direct and indirect costs, as described from the practical experiences of Carr (1988). Direct costs are physically traceable to the activity in an economic manner, and not counted if the activity is not performed; for example, labour, plant and materials. Indirect costs are mainly business costs that are not physically traceable, and counted even if the activity is not performed; for example, job site superintendent and tower crane. Direct cost estimates are theoretically assumed to be 'risk-free' (Tah et al., 1993, p. 282), where a risk-free estimate is described as the most likely estimate of the known scope of works (Mak and Picken, 2000, p. 131). Project costs may also be classified into variable and fixed costs (Carr, 1988). If a cost changes in proportion to a change in volume or quantity, it is variable; for example, project insurance, building permit, project supervision, security, crane and job site offices. If a cost remains unchanged in total despite wide fluctuations in volume or quantity, it is fixed; for example, installation of electric and telephone service to the job site and installation of shops, hoist, parking and warehouse facilities. Most construction activities have a mixture of variable and fixed costs. Lump-sum bids give an owner fixed prices, whereas unit rate bids give the owner a set of variable costs. Using a typical building estimate, Brook (2017, p. 109) broke down the elements of a project estimates as: labour (23 per cent), plant (5 per cent), materials (28 per cent) and subcontractors (44 per cent). However, this would differ between projects, depending on variables including project characteristics, geographical region, distance from manufacturers, productivity of labour force and wage rate. Subcontracting can also vary depending upon the structure of the company

Preliminaries

ISO 6707-2 Buildings and civil engineering works – Vocabulary – Part 2: Contract and communication terms, defined preliminaries as preliminary section of a contract (3.1.1), setting out the basic elements on which the parties have agreed. Examples of basic elements are the names of parties, location, scope and sums payable. ISO 6707-2:1993 (3.4.32) defined preliminaries as part of a bill of quantities or specification that describes not the work itself but associated matters, for example, site use and facilities, security, supervision, health and safety requirements, accommodation, and attendance on employer's staff.

Overheads

According to Harrison (1981, p. 3), overheads are: 'those costs incurred in the operation of a business which are not directly related to the individual items of production'. There are two main types of overheads. First, site overheads, which include site supervisory staff, site buildings, temporary roads and services. Second, head office (general) overheads, which include all costs incurred by the business as a whole which cannot be related directly to an individual contract. They include head office staff and buildings, and a wide range of other costs including those relating to personnel. Sometimes, company overheads may be included as a percentage in the markup. The contradictions between preliminaries overheads and markup can be resolved if site overheads are included in preliminaries (or general items) and company overheads are included in the markup.

Markup

Markup is an addition to costs in order to calculate price. It is made up of two parts: contingency for risk, and market premium. Markup is commonly understood in the construction management literature as the sum of overheads, profit and contingencies. However, the views vary slightly among authors. The main point of departure hinges on whether markup includes overheads. Most authors, such as Shash and Abdul-Hadi (1992), Shash (1998), Tah et al. (1994) and Hughes et al. (2006) define markup to include overheads. However, a few others, such as Liu and Ling (2005) and Paek et al. (1993), view markup as profit plus contingencies. Tah et al. (1994, p. 31) attempts to clarify this by explaining that: 'direct costs of a project comprise labour, plant, material and subcontractor costs. Indirect costs consist of site overheads, general overheads, profit, and allowances for risks. When indirect costs exclude site overheads they are often termed the markup.'

Profit margin

Harrison (1981, p. 3) defines profit as 'the amount by which the excess of the assets of a company over its liabilities grows over time'. In this case, growth in real terms was considered as distinct from that resulting from inflation. Profit may be in the form of assets, such as buildings. A profitable company may go into liquidation if sufficient assets are not in the form of cash, or easily convertible into cash.

Contingency allowances

ISO 6707-2 defines the contingency sum as a sum of money budgeted for, or included in a contract to cover, construction work that may be required but cannot be foreseen or predicted with certainty. This should be explained alongside a provisional sum, which is a sum of money that is included in a contract for work that is foreseen but cannot be accurately specified at the time the tender documents are issued (ISO 6702-2). This concept has a long history of definition. Willis (1929, p. 13) explains that, when pricing:

> besides the usual sum of money inserted in the bill of quantities for contingencies, many surveyors or estimators are in the habit of inserting provisional measured quantities additional to the measurements made from the drawings. Such a half-hidden contingency sum may in some cases be justified by a rough and incomplete nature of the drawings from which the quantities have been prepared, but it is more often a confession by the surveyor of lack of skill or confidence in his work.

Thus, there may be a client contingency and contract contingency. According to Hackett and Statam (2016), there is a hidden premium in every contractor's tender intended to charge the client for risk borne. A study of 12 US contractors by Smith and Bohn (1999, p. 106) showed that contractors may indeed 'buffer' their bids whenever they feel uncertain about the cost of an individual work item. They did not view this as contingency, but as an adjustment to working conditions. The contractors also suggested that, in addition to adjusting unit rates to account for anticipated work difficulties and risks, contingency was a percentage added to the total project cost. The only contractor who indicated the use of contingency in bids tried to pitch it to a sufficiently low value in order to be competitive; 11 out of 12 contractors claimed that they do not price visible contingencies in bids because of competition.

Adjudication

Estimating is often followed by an adjudication process, which is the stage where the directors of a firm take a commercial view of the estimated cost in the context of the firm's particular circumstances, market conditions and risk. Management will ultimately try to pitch the bidding price between cost and value in order to win the work. The adjudication or consideration of the tender by senior management involves adding to the estimate a further element for overheads and profit. The process may often appear to be a technical exercise. However, management often considers all factors affecting the company's well-being. The factors considered in adjudicating a bid include the accuracy and confidence regarding estimator's work; workload – to avoid overstretching the company's resources; market competition – consciousness of other bids from competitors; project reports – conditions of contract, terms of quotations from suppliers and subcontractors, assessment of risk; and resources required for the project (scheduling the firm's resources efficiently). The estimator may suggest to management to reduce the allowances added for risk based on the confidence obtained from knowing that the estimate has been made to a high degree of accuracy (Harrison, 1981, p. 2).

Based on his 30 years of experience as an estimator, Harrison (1981, p. 2) noted that there were companies which spent much time compiling statistics of their own and their competitors' relative competitive positions, and took heed of this information in their tendering. However, as the potential number of competitors rose, the difficulty and cost of obtaining and maintaining such information tended to become greater than any potential benefits. He argued that based on his many years of involvement with the finalization of tenders, for a wide range of types and sizes of contracts, it seemed that decisions on the final tender price were rarely as logical and reasoned as those who based tender decisions on past records would have others believe. This dispels the common belief underlying the maintenance of such records: namely, that tendering is a logical and rational process.

Pitching the tender price

A final tender price has to be decided once the estimate has been completed and risks have been assessed (Harrison, 1981). Three major factors are considered:

1. The probable cost of carrying out the works and the extent to which this could vary, depending on circumstances which can be foreseen but the likelihood of which cannot be assessed with any accuracy, that is, the extent of risk.
2. The minimum price at which this contract is likely to benefit the company.
3. The price level at which the firm: (a) will certainly get the work; or (b) is likely to get the work.

The critical and logical evaluation of (2) above should often inform the minimum price at which the tender should be submitted even if the price under (3) is lower. However, if the price assessed in (3) is greater than that in (2), a decision will need to be made as to whether it is more advantageous to be almost certain to obtain a contract at a lower price, or to have less likelihood of winning it at a higher price. A fine subjective judgement of the situation will often be required at this stage, and management experience can play a vital role (Harrison, 1981).

RELATIONSHIP BETWEEN COST, PRICE AND VALUE

The relationship between cost, price and value plays an important role in determining the bidding behaviour and pricing of contractors (see Laryea and Hughes, 2011); this is shown in Figure 17.2. Hughes et al. (2015) observed that contractors know about costs (how much they pay for their resources), price (how much they sell their products/outputs for) and value (how much it is worth to the buyer). However, quite often, most of the existing literature in tendering and estimating textbooks has focused on costs without giving much attention to value, which may in fact form the basis of a contractor's tender pricing strategy. This means that clever contractors would indeed pitch their tender price above cost, but below value, in order to improve their chances of winning a job. This bidding exercise often involves the 'estimating' and 'adjudication' processes that are explained in the section on price formation in construction contracts. Hence, tendering may depend on the level of expertise in the contractors' estimating department.

However, the elemental microeconomic theory of the behaviour of individual competitive markets (Lipsey, 1979, p. 93) suggests that the price clients will be willing to pay for construction work will depend not only on their available resources, but also on what other sellers (contractors) in the market are willing to offer the same product for. Thus, tendering may be very dependent on the market or the competitive environment in which it takes place. Neufville and King (1991, p. 659) identified 'need-for-work' as one form of market inefficiency in their experimental study of 30 US contractors on risk, and this illustrated the impact of 'need-for-work' premiums in bidding. In other words, this means that contractors who are hungry for work may undercut the market price, and in the process influence the subsequent pricing behaviour of their competitors who had lost work because of this. Thus, in recessionary periods, competent contractors who need work may be forced to pitch their prices very low and exploit other techniques, such as contract claims, to recover losses that result from accepting risks, as explained in a tendering costs survey involving 16 UK contractors by Hughes et al. (2006).

INFLUENCE OF RISK ON PRICING LEVELS

Risk plays an important role in the economic consideration of clients and contractors in the bidding process of construction projects. The general principle of risk allocation in construction is to allocate a risk to the party in the best position to control the risk (Smith et al., 2021; Hughes et al., 2015). However, this is not always the case in practice, as some clients will seek to pass on significant risk to contractors. Contractors acknowledge the risk that they should price. However, in practice, they may be unable or unwilling to make appropriate allowances for the risk. Risk apportionment influences a contractor's pricing strategy, but other complex microeconomic factors also affect price. Therefore, there is a complex relationship between risk and price in bids, which is empirically illuminated in Laryea and Hughes (2008) and Laryea and Hughes (2011). Instead of pricing contingencies that may increase price and affect competitiveness, contractors tend to adopt the more pragmatic approach of pricing some of their risk using contractual rather than price mechanisms to reflect commercial imperatives.

Based on the live observational study of tender processes in different contractor organisations, it was found that four key considerations inform the economic and strategic approach of contractors when it comes to risk pricing in the bidding process of construction projects (see Laryea, 2008). These reveal a pragmatic approach dictated by the desire to win work.

First, not all of the risk in construction projects is normally priced by contractors, due to the severity of competition for work within the industry. This situation leads some contractors to cope by exploiting all other necessary means. Research by Smith and Bohn (1999) and Rooke et al. (2004) showed that some contractors would plan to exploit opportunities in the construction phase of a contract, such as claims, to compensate for unusual risks imposed by the optimistic bids that they offer to clients in order to win the jobs. They would also plan to shift down some of the risk to suppliers and subcontractors in their supply chains, and apportion some in wage structures. Hence, the incorporation of opportunity in risk assessment and accountability provides a means to balance risk and reward to enhance a contractor's bidding competitiveness. The risk and price interface in the tendering processes of contractors involves a complex interaction of several, sometimes conflicting, factors that contractors must often weigh skilfully and try to price strategically to the mutual satisfaction of themselves and their clients.

Second, in practice, risk apportionment appears to occur at three stages of the bidding process of contracts (Laryea and Hughes, 2011). The first stage is the initial stage, when bid teams perform qualitative and quantitative assessment of risks during risk workshops. The second stage is during the actual estimating and pricing, when estimators price some of the risk so intuitively or tacitly that even they sometimes do not realize that it is being included. The third stage is when the adjudication team often adjust the estimate to make it reflect the prevailing market conditions and the firm's particular circumstances. The different teams or individuals involved in the various stages of the bidding process influence or decide the pricing at different stages. Most of the past research in this area has focused on the initial stage. However, if a contractor is keen to win a contract, priced risks may be excluded from the final settlement, depending on a set of complex microeconomic factors. The adjustments for risk or competitiveness could take considerable time to decide, but the actual arithmetic involved in reducing or increasing the bid price tends to be simpler than the sophisticated prescription in analytical pricing models proposed by researchers (Laryea, 2008).

Third, a careful examination of the way that tender prices are put together would reveal that contractors fundamentally include risk allowances in their bidding price for the 'normal' risk that they expect to occur. This is sometimes done intuitively or tacitly, using a single fixed percentage or lump sum based on the experience of the estimator and planner. Contractors expect their competitors to also price for normal risk in their bids. Therefore, they would not feel disadvantaged at the point of settling the tender. The key area where contractors differ is how they deal with 'residual' risk, which may be presented separately in a formal risk schedule considered at the 'settlement' meeting. Where possible, the contractors prefer to 'qualify' this 'risk pot' for post-tender negotiation because of concerns with the competition. Therefore, once risks have been fundamentally included in the building up of prices, contractors generally tend not to further account for risk in their price because of competition.

Fourth, due to the increasing nature of the competition for work, some contractors prefer to adopt contractual ways of approaching risk in their tender submissions. They carry out a commercial review and try to offer the lowest price, in order to gain an advantage over their competitors. However, they also compile a list of commercial issues they may wish to negotiate with the client if the tender price is of interest. Thus, some of the current mechanisms used by contractors for responding to risk such as qualifications, clarifications and assumptions have not been captured sufficiently in the literature. Hence, apart from using contingencies, other mechanisms used by contractors to price risks are qualifications, clarifications and assumptions in the tender programme and price.

CONCLUSION

The examination of economic principles of bidding for construction projects demonstrates that two key considerations driving the commercial behaviour of the parties in a bidding process are their respective economic needs and the risk exposure arising from the construction project. Generally, the client needs to be convinced that it will get the project at the right price, time and quality. Likewise, the contractor needs to get the work at the right price in order not to succumb to financial stress in the medium to long run. Price is an overriding factor for both parties, which must also deal with the risk associated with the construction project in the bidding and construction phases. Risk is an important consideration for both clients and contractors in the bidding process. For clients, risk significantly informs the procurement and contract strategies adopted for a project. For contractors, risk significantly influences the bidding approach and pricing levels.

The principal economic principle or strategy adopted by clients to achieve an economic price from contractors is market competition, which serves as the framework used to regulate the tendency for contractors to inflate their prices. Ultimately, clients want to achieve the intended outcomes for their projects with the best value for money. This is one reason for the increasing use of relationship-based contracting approaches, such as framework contracts which clients use as a strategy to enter into longer-term contractual relationships with contractors.

The principal approach used by contractors to determine their bidding price is the systematic calculation of costs using an estimating process. Thereafter, the management of a firm conducts a strategic adjudication of the estimate based on the market situation and firm's particular circumstances. The account in this chapter reveals the practice where contractors use a combination of price and contractual mechanisms for risk pricing. This is done for reasons of competitive advantage. Ultimately, the relationship between cost, price and value informs the final bidding price that is negotiated and agreed by the client and contractor. The microeconomic theory of the behaviour of individual competitive markets suggests that the price clients will be willing to pay for construction work will depend not only on their available resources, but also on what other sellers (contractors) in the market are willing to offer the same product for. Thus, a bidding price may be dependent on the market or competitive environment in which it takes place.

The construction industry is likely to experience significant change in the future of the bidding process in the medium to long term. The current bidding process in construction has been criticised for the extent of wasted resources associated with the way that bidding process for construction projects is organised. In current practice, several construction companies are bidding for the same work, and inevitably some will be wasting resources that could be used elsewhere. Research on viable alternative models is required to address the shortcomings of current competitive procurement models. If construction companies were able to reduce their bidding costs, their overhead costs would reduce and the cost of construction could reduce.

Further studies should develop a better understanding of aspects of the bidding process relating to the client's leadership role in achieving project economy; improving the quality of project information used for construction bidding; developing reliable cost norms to achieve quicker pricing of projects; redesigning the bidding process to make it more efficient and less wasteful; and exploring new supply models to enhance contractor competitiveness and efficiency in construction pricing and production.

REFERENCES

Ahmed, M.O., El-adaway, I.H., Coatney, K.T. and Eid, M.S. (2016) Construction bidding and the winner's curse: game theory approach. *Journal of Construction Engineering and Management*, 142, 2. https://doi.org/10.1061/(ASCE)CO.1943-7862.0001058.

Akintoye, A.S. and MacLeod, J.M. (1997) Risk analysis and management in construction. *International Journal of Project Management*, 15(1), 31–38.

Al-Bahar, J.F. and Crandall, K.C. (1990) Systematic risk management approach for construction projects. *Journal of Construction Engineering and Management*, 116(3), 533–546.

Bon, R. (2001) The future of building economics: a note. *Construction Management and Economics*, 19, 255–258.

Brook, M. (2017) *Estimating and Tendering for Construction Work*, 5th edn, Taylor & Francis, Abingdon.

Buchan, R.D., Fleming, F.W.E. and Grant, F.E.K. (2003) *Estimating for Builders and Surveyors*, 2nd edn, Oxford: Butterworth-Heinemann.

Cabinet Office (2014) *New Models of Construction Procurement*, London.
https://assets.publishing.service.gov.uk/government/uploads/system/uploads/attachment_data/file/325011/New_Models_of_Construction_Procurement_-_Introduction_to_the_Guidance_-_2_July_2014.pdf.

Calvert, R.E., Bailey, G. and Coles, D. (1995) *Introduction to Building Management*, 6th edn, Oxford: Heinemann.
Carr, R.I. (1988) Cost estimating principles. *Journal of Construction Engineering and Management*, 545–551.
Chang, Chen-Yu (2015) A festschrift for Graham Ive. *Construction Management and Economics*, 33(2), 91–105. DOI: 10.1080/01446193.2015.1039044
Cioffi, D.F. (2005) A tool for managing projects: an analytic parameterization of the S-curve. *International Journal of Project Management*, 23(3), 215–222.
Hackett, M. and Statham, G. (2016) *The Aqua Group Guide to Procurement, Tendering and Contract Administration*, 2nd edn, Chichester: Wiley-Blackwell.
Hall, D.S.M. (1972) *Elements of Estimating*, London: B.T. Batsford.
Harrison, S. (1981) Estimating and tendering – some aspects of theory and practice. Estimating Information Service, Chartered Institute of Building, No. 41.
Hillebrandt, P.M. (2000) *Economic Theory and the Construction Industry*, London: Palgrave Macmillan.
HM Government (2011) *Government Construction Strategy*, May, London.
https://www.gov.uk/government/publications/government-construction-strategy.
HM Government (2018) *Construction Sector Deal*, London. https://assets.publishing.service.gov.uk/government/uploads/system/uploads/attachment_data/file/731871/construction-sector-deal-print-single.pdf.
HM Government (2020) *The Construction Playbook: Government Guidance on Sourcing and Contracting Public Works Projects and Programmes*, London: Cabinet Office. https://bit.ly/3rz0kqR
Hughes, W., Champion, R. and Murdoch, J. (2015) *Construction Contracts: Law and Management*, 5th edn, Routledge, London.
Hughes, W.P. and Hillebrandt, P.M. (2003) Construction industry: historical overview and technological change. In: Mokyr, Joel (ed.), *The Oxford Encyclopaedia of Economic History*, Oxford: Oxford University Press, 504–512.
Hughes, W., Hillebrandt, P.M., Greenwood, D. and Kwawu, W. (2006) *Procurement in the Construction Industry: The Impact and Cost of Alternative Market and Supply Processes*, London: Routledge.
Hughes, W., Hillebrandt, P. and Murdoch, J.R. (1998) *Financial Protection in the UK Building Industry: A Guide for Clients and the Construction Supply Chain*, London: Thomas Telford Publications.
Jaselskis, J.E. and Talukhaba, A. (1998) Bidding considerations in developing countries. *Journal of Construction Engineering and Management*, 125(3), 185–193.
Kangari, R. and Riggs, L.S. (1989) Construction risk assessment by linguistics. *IEEE Transactions on Engineering Management*, 36(2), 126–131.
Laryea, S. (2008) How contractors take account of risk in the tender process: theory and practice. Unpublished PhD thesis, School of Construction Management and Engineering, University of Reading, UK. https://isni.org/isni/0000000426756017.
Laryea, S. (2009) Subcontract and supply enquiries in the tender process of contractors. *Construction Management and Economics*, 27(12), 1219–1230.
Laryea, S. (2011) Quality of tender documents: case studies from the UK. *Construction Management and Economics*, 29(3), 275–286.
Laryea, S. (2016) Risk apportionment in target cost contracts. *Proceedings of the Institution of Civil Engineers – Management, Procurement and Law*, 169(6), 248–257. https://doi.org/10.1680/jmapl.15.00046
Laryea, S., Alkizim, A. and Ndlovu, T. (2013) The increasing development of publication on sustainable procurement and issues in practice. In: Smith, S.D. and Ahiaga-Dagbui, D.D. (eds), *Procs 29th Annual ARCOM Conference, 2–4 September 2013*, Reading, UK, Association of Researchers in Construction Management, 1285–1294.
Laryea, S. and Hughes, W. (2006) The price of risk in construction projects. In: Boyd, D. (ed.), *Procs 22nd Annual ARCOM Conference, 4–6 September 2006*, Birmingham, UK, Association of Researchers in Construction Management, 553–561.
Laryea, S. and Hughes, W. (2008) How contractors price risk in bids: theory and practice. *Construction Management and Economics*, 26(9), 911–924.
Laryea, S. and Hughes, W. (2009) Commercial reviews in the tender process of contractors. *Engineering, Construction and Architectural Management*, 16(6), 558–572.

Laryea, S. and Hughes, W. (2011) Risk and price in the bidding process of contractors. *Journal of Construction Engineering and Management*, 137(4), 248–258. https://doi.org/10.1061/(ASCE)CO.1943-7862.0000293.
Laryea, S. and Lubbock, A. (2014), Tender pricing environment of subcontractors in the United Kingdom. *Journal of Construction Engineering and Management*, 140(1). https://doi.org/10.1061/(ASCE)CO.1943-7862.0000749.
Lipsey, R.G. (1979) *An Introduction to Positive Economics*, 5th edn, London: Weidenfeld & Nicolson.
Liu, M. and Ling, Y.Y. (2005) Modelling a contractor's markup estimation. *Journal of Construction Engineering and Management*, 131(4), 391–399.
Loosemore, M., Raftery, J., Reilly, C. and Higgon, D. (2006) *Risk Management in Projects*, 2nd edn, London: Taylor & Francis.
Mak, S. and Picken, D. (2000) Using risk analysis to determine construction project contingencies. *Journal of Construction Engineering and Management*, 126(2), 130–136.
Mankiw, N.G. (2021) *Principles of Economics*, 9th edn, Boston, MA: Cengage.
Marsh, P.D.V. (2000) *Contracting for Engineering and Construction Projects*, 2nd edn, Aldershot: Gower.
McConnell, C., Brue, S. and Flynn, S. (2021) *Microeconomics*, 22nd edn, New York: McGraw-Hill Education.
Mochtar, K. and Arditi, D. (2001) Pricing strategy in the US Construction industry. *Construction Management and Economics*, 19, 405–415.
Myers, D. (2003) The future of construction economics as an academic discipline. *Construction Management and Economics*, 21, 103–106.
Neufville, R. and King, D. (1991) Risk and need-for-work premiums in contractor bidding. *Journal of Construction Engineering and Management*, 117(4), 659–673.
Ofori, G. (1994) Establishing construction economics as an academic discipline. *Construction Management and Economics*, 12(4), 295–306.
Paek, J.H., Lee, Y.W. and Ock, J.H. (1993) Pricing construction risk: fuzzy set application. *Journal of Construction Engineering and Management*, 109(4), 743–756.
Rintala, K., Root, D., Ive, G. and Bowen, P. (2008) Organizing a bidding competition for a toll road concession in South Africa: the case of Chapman's peak drive. *Journal of Management in Engineering*, 24, 146–155. 10.1061/(ASCE)0742-597X(2008)24:3(146).
Rooke, J., Seymour, D. and Fellows, R. (2004) Planning for claims: an ethnography of industry culture. *Construction Management and Economics*, 22(6) 655–662.
Shash, A.A. (1998) Bidding practices of sub-contractors in Colorado. *Journal of Construction Engineering and Management*, 124(3), 219–225.
Shash, A.A. and N.H. Abdul-Hadi (1992) Factors affecting a constructor's margin-size decision in Saudi Arabia. *Construction Management and Economics*, 10, 415–429.
Sichtmann, C. (2007) Buyer–seller relationships and the economics of information. *Journal Of Business Market Management*, 1, 59–78.
Skitmore, M., Runeson, G. and Chang, X. (2006) Construction price formation: full-cost pricing or neoclassical microeconomic theory? *Construction Management and Economics*, 24, 773–783.
Skitmore, M. and Wilcock, J. (1994) Estimating processes of smaller builders. *Construction Management and Economics*, 12, 139–154.
Smith, G.R. and Bohn, M.C. (1999) Small to medium contractor contingency and assumption of risk. *Journal of Construction Engineering and Management*, 125(2), 101–108.
Smith, N.J., Merna, T. and Jobling, P. (2021) *Managing Risk in Construction Projects*, 3rd edn, Chichester: Wiley Blackwell.
Smith, R.C. (1986) *Estimating and Tendering for Building Work*, London: Longman.
Tah, J.H.M., Thorpe, A. and McCaffer, R. (1993) Contractor project risks contingency allocation using linguistic approximation. *Journal of Computing Systems in Engineering*, 4(2–3), 281–293.
Tah, J.H.M., Thorpe, A. and McCaffer, R. (1994) A survey of indirect cost estimating in practice. *Construction Management and Economics*, 12(31), 31–36.
Watermeyer, R.B. (2012) A framework for developing construction procurement strategy. *Proceedings of the Institution of Civil Engineers – Management, Procurement and Law*, 165(4), 223–237. http://dx.doi.org/10.1680/mpal.11.00014.

Watermeyer, R. (2013) Unpacking framework agreements for the delivery and maintenance of infrastructure. *Civil Engineering*, January/February, 21–26.
Watermeyer, R. (2018) Client guide for improving infrastructure project outcomes. School of Construction Economics and Management, University of the Witwatersrand and Engineers Against Poverty.
Willis, A.J. (1929) *Some Notes on Taking Off Quantities*, London: Architectural Press.
Winch, G. (1989) The construction firm and the construction project: a transaction cost approach. *Construction Management and Economics*, 7, 331–345.10.1080/01446198900000032.
Xiang, P, Zhou, J., Zhou, X. and Ye, K. (2012) Construction project risk management based on the view of asymmetric information. *Journal of Construction Engineering and Management*, 138, 11.

18. Procurement and delivery management

Ron Watermeyer

INTRODUCTION

The Nature of Infrastructure Projects

Although many infrastructure projects are similar in nature, each project is unique. This is due to several important project variables employed in delivering the required infrastructure, including:

- what is delivered, differences between locations where the infrastructure is delivered, the client's value proposition for projects, stakeholder influences, resources employed, constraints, and processes and procurement practices; and
- different combinations of funders, clients and built environment professionals, site conditions, materials and technologies and general contractors, specialist contractors, skills and workforces.

Infrastructure projects need to be planned, specified, procured and delivered. Once decisions are made on what the project needs to deliver, who will deliver it, and how will it be funded and governed, the remaining decisions centre on how it will be managed through to completion. Such management takes place within a project-specific environment which continuously involves the management of risk events, which may be foreseen or unforeseen, and that have the potential to negatively impact on project outcomes during the protracted delivery process (Watermeyer, 2019).

Risk-taking is necessary in the delivery of infrastructure projects. Sources of risk in infrastructure projects include commercial and legal relationships, economic circumstances, human behaviour, natural events, weather, inherent and unforeseeable site conditions, political circumstances, community unrest, technology and technical issues. Risks can also manifest in weak clients who are not capable of making timeous decisions and who have difficulty in providing information timeously or paying promptly. Risks also arise due to differences between the actual prices paid in terms of the contract and those estimated at the time of tender and changes to requirements during the execution of projects to enhance quality, performance in use, or the usefulness of outputs or to address shortcomings in design. The parties to a contract face choices on how to deal with inherent project risks. Risks can be transferred or accepted. Accordingly, a central issue that needs to be dealt with in infrastructure projects is the financial liability related to the uncertainty of information when decisions are made, particularly in the early stages of a project, and future events (Watermeyer and Philips, 2020).

Value for Money in an Infrastructure Context

'Value for money' in common usage refers to something that is well worth the money spent on it. British Standard (BS) 8534:2011 defines value for money as the 'optimum combination of whole-life cost and quality to meet the user's requirement'. The World Bank (2016) suggests that value for money is the 'effective, efficient, and economic use of resources'. The United Kingdom (UK) National Audit Office (2010) and the South African National Treasury (2015) define value for money as 'the optimal use of resources to achieve intended outcomes'. Accordingly, value for money in a construction context can be regarded as the most desirable possible outcome from the use of resources (finances, people, equipment, plant, materials, and so on) that can be drawn upon, given expressed or implied restrictions or constraints such as risks and costs. Value for money is realised when the value proposition that was set for the project at the time that a decision was taken to invest in a project is as far as possible realised. It is about maximising actual outcomes and impacts and spending money well and wisely (Watermeyer, 2018).

Underlying value for money is an explicit commitment to ensure that the best results possible are obtained from the money spent, or maximum benefit is derived from the resources available. It is about striking the balance between the three 'E's' – namely, economy, efficiency and effectiveness – whilst being mindful of a fourth 'E', equity, as indicated in the results chain as presented in Figure 18.1. Economy, efficiency and effectiveness relate to the primary objectives of a project, whereas equity relates to the secondary objectives of the project, that is, what can be promoted or leveraged through the delivery of the project (Watermeyer, 2013b).

The critical starting point in delivering value for money through infrastructure projects is, in the first instance, to align such projects with strategic objectives, priorities, budgets and plans. Thereafter, during the planning phase, objectives and expected outcomes need to be clearly articulated, as well as parameters such as the timelines, cost and levels of uncertainty, that is, the value proposition or promise of measurable benefits resulting from the project. This frames the value-for-money proposition that needs to be implemented at the point in time that a decision is taken to proceed with a project; that is, it establishes economy and identifies equity. The end point is to compare the projected outcomes against the actual outcomes to confirm the effectiveness of the project in delivering value for money.

Implementation sits between economy and effectiveness in the results chain framework. It needs to be executed efficiently in order to minimise time delays, scope creep and unproductive costs, and to mitigate the effects of uncertainty on objectives, so as to maintain the value proposition formulated at the outset of the project. This necessitates that the client exercises due care and reasonableness during implementation. Failure to do so may result in substandard or unacceptable performance, which results in a gap between intended and achieved outcomes. This gap puts value for money for a project at risk (Watermeyer, 2018).

REASONS FOR POOR PROJECT OUTCOMES

All too often, worse-than-expected project outcomes are experienced. The gap between what was planned and what was achieved can be significant. For example, Merrow (2012) considers a project to have failed if the schedule slips or the project overspends by more than 25 per cent, the execution time is 50 per cent longer, or there are severe and continuing operational prob-

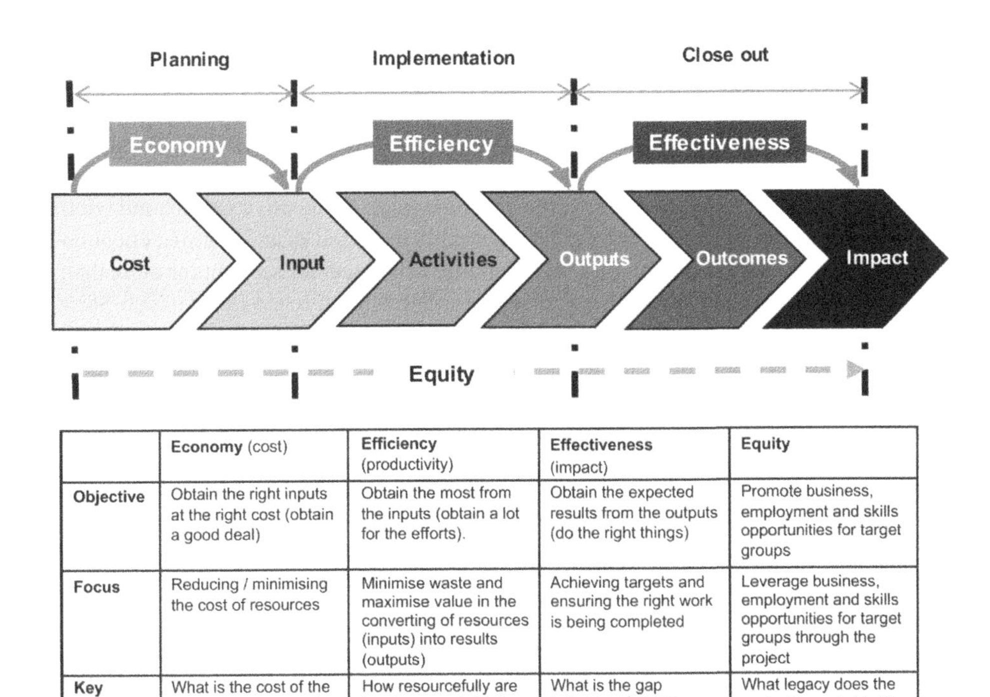

	Economy (cost)	Efficiency (productivity)	Effectiveness (impact)	Equity
Objective	Obtain the right inputs at the right cost (obtain a good deal)	Obtain the most from the inputs (obtain a lot for the efforts).	Obtain the expected results from the outputs (do the right things)	Promote business, employment and skills opportunities for target groups
Focus	Reducing / minimising the cost of resources	Minimise waste and maximise value in the converting of resources (inputs) into results (outputs)	Achieving targets and ensuring the right work is being completed	Leverage business, employment and skills opportunities for target groups through the project
Key question	What is the cost of the resources consumed and the value of the output delivered?	How resourcefully are inputs converted into outputs and subsequent outcomes?	What is the gap between what has been achieved and what was intended?	What legacy does the project leave behind?

Source: Based on Watermeyer (2013b).

Figure 18.1 The relationship between the four 'Es' in the value-for-money concept

lems into the second year of the project. His findings, following a review of a large database of international projects, shows that megaprojects (projects having a project value of US$1 billion or more) around the world fail by a staggering rate of 65 per cent, compared to a failure rate of 35 per cent for projects under the US$500 million mark. McKinsey & Company (2017) assert that, on average, projects with budgets above US$1 billion are delivered one year behind schedule, and run 30 per cent over budget.

Denico et al. (2020) conducted an extensive systematic literature review of 6007 titles and abstracts and 86 full papers, which explored the megaproject management literature to reveal the main causes and cures of poor megaproject performance as found in the literature. This review identified 18 predominant concepts clustered together under six themes: (1) decision-making behaviour; (2) strategy, governance and procurement; (3) risk and uncertainty; (4) leadership and capable teams; (5) stakeholder engagement and management; and (6) supply chain integration and coordination. Failure of governance, or poor procurement and delivery management within one or more of the predominant concepts, all of which are controlled by the client, can result in project failure.

The UK Office of Government Commence in 2005 identified eight common causes of project failure, the root causes of which related to lack of governance and poor procurement and delivery management practices (Watermeyer 2018). The Report of the Independent

Inquiry into the Construction of Edinburgh Schools (2017) criticised the public body for not ensuring that due diligence was undertaken at an appropriate level to confirm that the requirements of the public–private partnership contract were actually delivered in accordance with the terms of that contract. The Report of the Independent Inquiry into the Construction of the DG One Complex in Dumfries (Dumfries and Galloway Council, 2018) found that Dumfries and Galloway Council failed to provide effective strategic and executive project and contract management support to the level that would be normally expected of an informed client body.

Foster (2008) examined infrastructure in 24 countries that together account for more than 85 per cent of the gross domestic product, population and infrastructure aid flows of sub-Saharan Africa. Foster found that countries typically only manage to spend about two-thirds of the budget allocated to investment in infrastructure. Watermeyer and Phillips (2020) found that in South Africa, on average over the 2015/16 to 2017/18 financial years, state-owned enterprises and public entities spent not more than 75 and 65 per cent of their respective budgeted amounts, while the state as a whole spent not more than 85 per cent of the available budget. In the 2017/18 financial year, none of the metropolitan councils spent more than 80 per cent of their capital budgets, while 76 per cent of municipalities spent less than 80 per cent of such budgets. Watermeyer and Phillips attribute these poor infrastructure outcomes to a lack of governance and poor client procurement and delivery management practices.

APMG International (a multinational project management accreditation body) points out (APMG, 2016) that 'many project failures have their ultimate origin in defects in the identification, assessment and preparation (appraisal) of the project, poor structuring and poor management of the tender process, or poor contract management'. This body identifies several common project management and governance factors that compromise project outcomes. These include the lack of management capacity and proper skills (lack of skilled resources and lack of funds to hire advisors), the lack of continuity/frequent changes in the project team, lack of clear project ownership and leadership and a lack/absence of a champion. They also include failures in taking and managing decisions (insufficient delegation of powers, external interference), stakeholder identification, communication (inside, outside and to the public), and matching the government's strategic objectives or changes in the government objectives, as well as political rush and unrealistic time scales. Again, most, if not all, of these factors are within the control of the client.

McKinsey & Company (2017) conducted interviews with 27 people with a collective project delivery experience exceeding 500 years to gain a better understanding of ultra-large projects. They identified four mindsets (lead as a business; take full ownership of outcomes; make your contractor successful; and trust your processes but know that leadership is required), and eight practices that define the 'art' of project leadership (define purpose, identity and culture; assemble the right team; carefully allocate risk and align incentives; work hard on relationships with stakeholders; invest in your team; ensure timely decision-making; adopt forward-looking performance management; and drive desired behaviours consistently). McKinsey & Company concluded that to successfully deliver projects client teams need to have not only the 'hard' technical abilities but also the 'soft' organisational and leadership elements of project delivery. The absence of these mindsets and practices within client teams contributes to poor outcomes.

The Infrastructure Client Group (2017) identified five key organisation features in their proposed approach to delivering high-performing infrastructure, namely governance, organisation, integration, capable owner and digital transformation. These features are within the control of the client organisation.

PRINCIPAL FUNCTION OF THE CLIENT

Clients need to identify what is valuable to the organisation (core values), as well as what values should be reflected in their infrastructure projects. This is important, as values serve as broad guidelines in all situations where choices need to be made regarding which path to follow. They enable those responsible for delivering projects to understand the difference between what is desirable and what is not. They also serve to shape the outcome of the investment in an infrastructure project (Watermeyer, 2018).

A supply chain can be regarded as the sequence of tasks that provides products or services to the organisation. The supply chain for infrastructure can be viewed as the flow of information from one set of tasks to the next, with decision points or gates at the boundaries between tasks, which provide the opportunity for ensuring that the proposed project remains within agreed mandates, aligns with the purpose for which it was conceived, and can progress successfully to the next task. Procurement binds the participants in the supply chain (professionals, general contractors, specialist and subcontractors) and defines the obligations, liabilities and risks that link the parties together in a process that needs to deliver value for money.

The client's business case, vision, values and project priorities collectively make up the client's value proposition for a project. Activities associated with the planning, designing, manufacturing, fabrication, construction, installation and commissioning need to translate the client's value proposition into project outcomes which impact on the three aspects of sustainability (economic, environmental and social) and result in a product. In order to achieve project value (Watermeyer and Phillips, 2020), clients need to (see Figure 18.2):

- plan – decide on what needs to be done, how it is to be resourced and achieved and in what time frames, and set a budget;
- specify – define the client's functional and other requirements for the project clearly and precisely;
- procure – obtain project resources (internal and external) to execute project activities with care and effort; and
- oversee delivery – observe and define the execution of the project to realise the client's value proposition associated with a business case (promise of measurable benefits resulting from the project).

Clients can influence project outcomes through client leadership at both a programme level and a project level, governance, and when contracting external resources through the adoption of suitable procurement strategy and tactics. Those tasked with project activities need to manage project activities through project management controls (see ISO 21502:2020 and ISO 10845-1:2020) within budget, schedule, quality, and safety, health and environmental parameters, as indicated in Figure 18.2.

Client leadership is vital to the success of a project as it establishes the culture within which the project is delivered. Vision and values are two important tools in the exercising of client leadership. Vision provides direction and a sense of purpose, while values provide trust and appropriate behaviours for successful outcomes. Client leadership creates an enabling environment for resources responsible for delivering projects to perform at their best (Watermeyer, 2018).

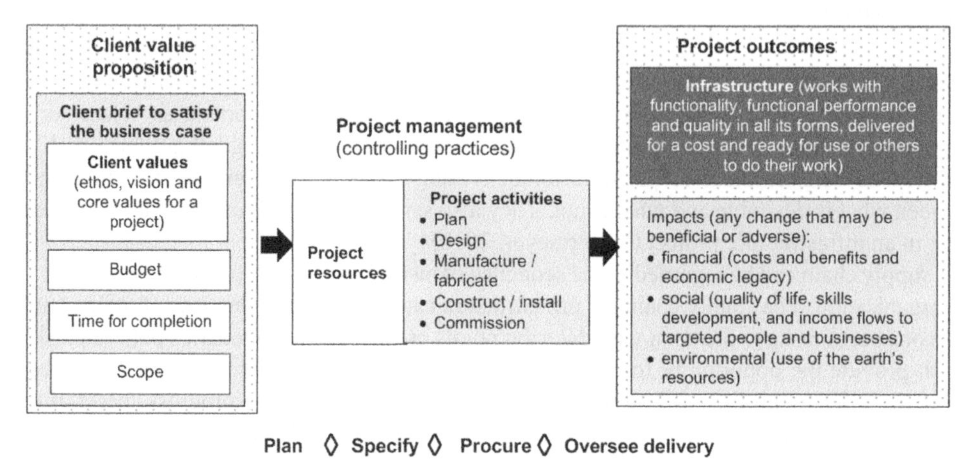

Source: Based on Watermeyer (2019).

Figure 18.2 Translating the client value proposition into project outcomes

PRINCIPAL ROLE PLAYERS IN THE DELIVERY OF INFRASTRUCTURE

The principal role players in the delivery of infrastructure are the client delivery management team led by a client delivery manager, the delivery team and stakeholders (see Figure 18.3). The client delivery manager needs to provide effective leadership and direction to the delivery team and meaningfully engage with internal and external stakeholders. The delivery team, on the other hand, needs to deliver the required work and manage the interfaces between the client delivery management team and stakeholders in doing so. The client delivery management team accordingly performs a buying function. The delivery team, on the other hand is responsible for supplying the goods and services which are necessary to deliver construction work projects, and as such performs a supply function. Both teams are driven by different objectives in the buying and supply exchange (Watermeyer, 2018).

Stakeholders may include parties external to the project such as financiers, regulators, asset management, end-users (beneficiaries of the business case) and affected communities. Stakeholder inputs are critical, particularly in the early stages of a project where the cost of effecting changes to accommodate requirements is low.

Governance at the highest level within an organ of state needs to focus on the attainment of the organisation's core purpose over the long term, and on ensuring that the right project purpose is achieved in the right way. Governance at a programme and project level provides a way for senior management and notable stakeholders to exercise oversight and ensure the intended strategic outcomes of projects are realised in a coordinated and integrated manner.

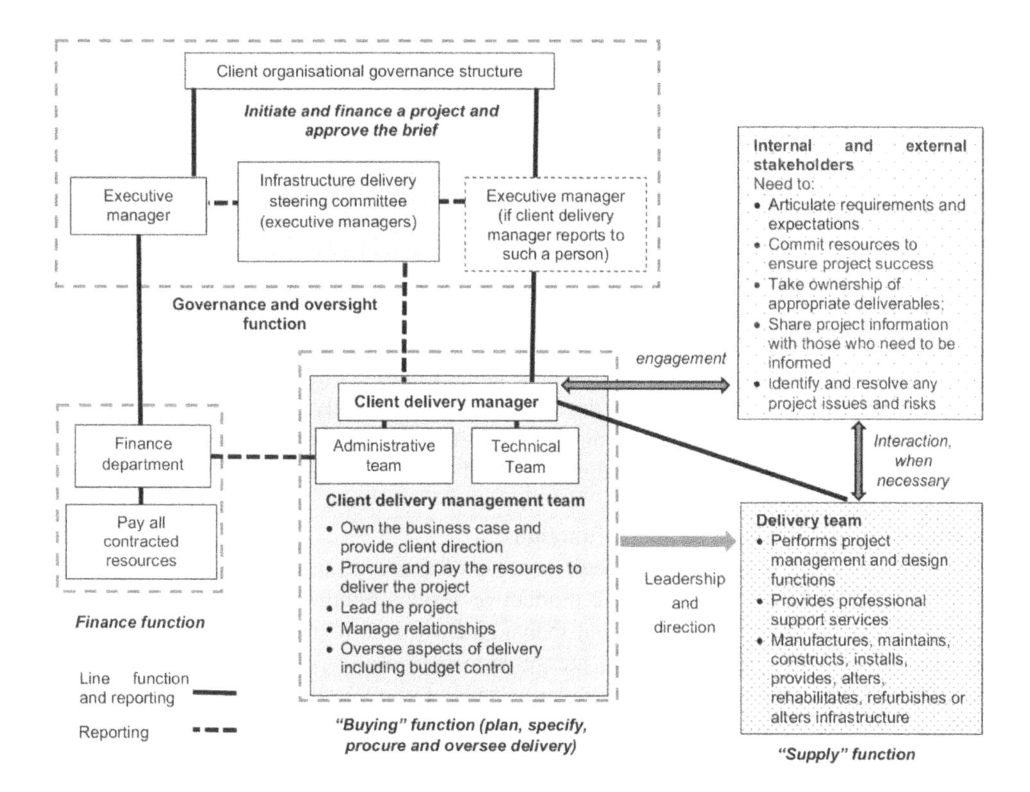

Source: Based on Watermeyer (2018).

Figure 18.3 Functions of the principal role players in infrastructure delivery

PROCUREMENT

Introduction

Procurement can be narrowly or broadly defined. For example, ISO 20400:2017 defines procurement as 'the activity of acquiring goods or services from suppliers' whereas ISO 6707-2:2017 defines it as 'the process which creates, manages and fulfils contracts relating to the provision of goods, services and construction works or disposals, or any combination thereof'. The latter definition recognises that there are commonly three phases to the procurement process (Watermeyer, 2018):

- a planning phase during which decisions are made as to what, where and when goods and services are required, how the market is to be approached, and what is the number, type, nature and timing of the required contracts;
- an acquisition phase during which contracts are entered into following the execution of a selection procedure; and

- a contract management phase during which compliance with contractual requirements, changes in requirements and risk events which manifest during the execution of contracts are managed.

Arrowsmith (2010) has found that regulatory rules generally focus on the acquisition phase, as legal rules and other regulatory measures become important tools of policy. This results in the narrow view that public procurement is 'government's activity of purchasing goods and services it needs to carry out its functions' (Arrowsmith, 2010) or 'the acquisition of goods, construction or services by a procuring entity' (UNCITRAL, 2014).

Procurement, covering the aforementioned three phases, is one of the ten knowledge areas or subject groups in the Project Management Institute's PMBOK®Guide (plan, conduct and control procurement), and the ISO 21500:2012 guide to project management (plan procurements, select suppliers and administer procurements), respectively. Procurement comprising the aforementioned stages is not only an integral part of project management practice, but is also an integral part of architectural, engineering and quantity surveying practice.

Locating procurement as a strategic function

Infrastructure procurement is the strategic process of sourcing a product or service, which includes the identification of a specific product or service requirement, the establishing of payment terms, the putting in place and management (or administration) of contracts, the minimisation of risks, identification of measures to secure cost savings, while focusing on value and return on investment. Procurement of general goods and services for consumption (purchasing) is commonly a back office or administrative function often linked to a finance department, whereas procurement relating to the delivery of infrastructure needs to be a front office or strategic function linked to the department or director responsible for delivering projects and services (Watermeyer and Phillips, 2020).

The APM (2017), the chartered body for the project profession in the UK, warns against the handling of procurement by a specific purchasing resource or department, rather than being a central competency within portfolio, programme and project management. They point out that to do so in complex projects frequently leads to 'unforeseen issues developing, leading to time cost and quality over runs'. Accordingly, infrastructure procurement needs to be recognised as a central competency of those responsible for delivering infrastructure programmes and projects. It needs to be delinked from any centralised purchasing or buying department established within an organisation, and located with the delivery management function.

Standardising procurement systems

There is a need to standardise procurement processes, methods and procedures for the delivery of infrastructure in a generic and flexible manner which supports and does not frustrate infrastructure delivery. This enables those engaged in a range of infrastructure delivery activities to perform their duties, within the confines of the procurement policies of their organisations, in a uniform and generic manner, and enables procurement documents to be readily compiled in a uniform and generic manner. It also enables curricula to be developed to equip those engaged in a range of infrastructure delivery activities and the public sector to readily develop an internal procurement skills base, which is not lost when members of staff move between different departments or levels of government or public entities.

The universally applicable principal tasks associated with a procurement process are: to establish what is to be procured, to decide on a procurement strategy, to solicit tender offers, to evaluate tender offers, to award a contract and to administer a contract. A procurement system needs to be built around a number of procurement objectives relating to good governance and the use of procurement to promote other objectives. It also needs to establish requirements for methods and procedures, procurement documents, procurement policy and procurement governance arrangements which reflect these system objectives (Watermeyer, 2011a; and ISO 22058).

The ISO 10845 family of standards for construction procurement

The ISO 10845 family of standards for construction procurement establish standard procurement processes, procedures and methods. These standards are framed around the following system objectives:

- the procurement system is fair, equitable, transparent, competitive and cost-effective; and
- the procurement system may, subject to applicable legislation, promote objectives additional to those associated with the immediate objective of the procurement itself.

The objective of the ISO 10845 series of construction procurement standards (Parts 1 to 8) is to create a framework for the development and implementation of procurement systems that facilitates fair competition, reduces the possibility of abuse, improves predictability of outcomes and allows the demonstration of best-value or cost-effective outcomes (Watermeyer, 2011b).

ISO 10845-1 (processes, methods and procedures) describes generic procurement processes around which an organisation can develop a procurement system, establishes basic requirements for the conduct of those involved in an organisation's procurement activities, establishes the framework for the development of an organisation's procurement policy, and establishes generic methods and procedures that are used in soliciting tender offers and awarding contracts.

Most multilateral organisations, international agencies and legislators often deal with the procurement of goods, services and construction works separately. Typically, different sets of rules, guidelines and regulations are established for these different categories of procurement. This compartmentalised approach is confusing and makes procurement unnecessarily complex, particularly on projects which require a facility to be delivered in a fully functional state, which invariably involves all three categories of procurement. The ISO 10845 series of standards introduces an alternative approach of first developing generic procurement methods which may be used in combination with each other, and thereafter enabling the identification of the methods best suited to particular categories or types of procurement. Combinations of the generic procedures and methods, with or without eligibility criteria, can be used to simulate the bespoke procurement methods adopted by multilateral organisations, international agencies and legislators for a particular category of procurement. At the same time, this approach provides employers with a wide range of options in the pursuit of best-value procurement outcomes.

ISO 10845-2 (formatting and compilation of procurement documentation) establishes a uniform format for the compilation of calls for expressions of interest, tender and contract documents, and the general principles for compiling procurement documents for supply, services and construction contracts, at both main and subcontract levels. This standard is based on the principle that each subject within a procurement document can only be addressed once, and only in one component document.

ISO 10845-4 establishes what is required for a respondent to submit a compliant submission, makes the evaluation criteria known to respondents, and establishes the manner in which the procuring entity conducts the process of calling for expressions of interest. ISO 10845-3 establishes what a tenderer is required to do to submit a compliant tender, makes the evaluation criteria known to tenderers, establishes the manner in which the client conducts the process of offer and acceptance, and provides the necessary feedback to tenderers on the outcomes of the process.

ISO 10845-2 enables ISO 10845-3 (standard conditions of contract), ISO 10845-4 (standard conditions for the calling for an expression of interest) and standard forms of contract to be readily referenced in procurement documents. Parts 1 to 4 of ISO 10845 can be readily incorporated into procurement systems by reference in organisational policies or in legislation (Watermeyer, 2011a).

ISO 10845-1 describes a number of techniques and mechanisms associated with targeted procurement procedures, that is, the process used to create a demand for services or goods (or both) from, or to secure the participation of, targeted enterprises and targeted labour in contracts in response to the objectives of a secondary procurement policy. These techniques and mechanisms are designed to promote the participation of targeted enterprises and targeted labour in contracts. Key performance indicators (KPIs) relating to the engagement of enterprises, joint venture partners, local resources and local labour in contracts are needed to implement many of these procedures. Parts 5 to 8 of ISO 10845 establish KPIs to measure the outcomes of a contract in relation to the engagement of target groups, and to establish a target level of performance for a contractor to achieve or exceed in the performance of a contract.

Some of the parts of the ISO 10845 family of standards were revised in 2020. The major enhancements relate to the expansion of the provisions for organisational procurement policies and framework agreements, and improved guidance on managing and controlling procurement processes.

A procurement system that is capable of being audited contains rigorous and documented processes, produces comprehensive and complete procurement documents and evaluation reports, discloses information and contains infrastructure-specific codes of ethics, is most likely to curb corruption (Watermeyer and Phillips, 2020).The ISO 10845 standards provide a solid platform to do so.

Standard forms of contract

ISO 6707-2:2017 defines 'conditions of contract' as 'the document that contains the detailed provisions incorporated in a contract, laying down the rights and duties of the parties, the functions of the people connected with the contract and the procedures for administering the contract'. Standard forms of contract, which are usually published by an authoritative industry organisation, provide fixed terms and conditions which are deemed to be agreed and are not subject to further negotiation or amendment when applied to a particular tender.

Standard forms of contract enable tenderers to take into account the allocation of risks embedded in such contracts when preparing tenders for infrastructure projects, and enables tenders to be evaluated on a comparative basis. There is also no need for tenderers who are familiar with a particular form of contract to price risks arising from uncertainties as to how particular issues will be viewed or handled in terms of the contract, unless such contracts are substantially amended.

Standard forms of contract can be drafted around significantly different objectives and principles; for example, master–servant relationship or collaboration between two experts, risk-sharing or risk transfer, independent or integrated design, short-term relationships based on one-sided gain or long-term relationships focused on maximising efficiency and shared value. Forms of contract may also support open-book approaches to the costing of changes due to the occurrence of risk events, foster collaborative working relationships, provide pricing structures that align payments to results and reflect a balanced sharing of performance risk, and deal with delays and disruptions efficiently and effectively. Standard forms of contract tend to be country-specific as they are developed by bodies within a particular country. There are, however, two international families of standard forms of contracts, namely those published by the International Federation of Consulting Engineers (FIDIC) and the Institution of Civil Engineers (New Engineering Contract, NEC). These standard forms of contract cover a range of contract types, contracting strategies and approaches to managing risks (Watermeyer, 2018).

Procurement Strategy and Tactics

Project outcomes in infrastructure projects are sensitive not only to the decisions made during the planning, design and execution of such projects, but also to the manner in which resources are structured and procured to deliver such projects. There are a number of different approaches to procuring goods and services and any combination thereof, each of which can result in different outcomes. Procurement strategy is all about the choices made in determining what is to be delivered through a particular contract, the contracting arrangements, how secondary procurement objectives are to be promoted, and which selection method will be employed to solicit tender offers. A procurement strategy commonly comprises (Watermeyer, 2018; and ISO 22058):

- a packaging strategy which organises work packages into contracts or orders issued in terms of a framework agreement;
- a contracting strategy which governs the nature of the relationship which the client wishes to foster with the contractor, which in turn determines the risks and responsibilities between the parties to the contract and the methodology by which the contractor is to be paid;
- a targeting strategy which leverages business, employment and skills opportunities for target groups through the contracts that are entered into; and
- a selection method used to solicit tender offers from the market.

Procurement strategies are also formulated around procurement objectives which may relate to either the delivery of the product (primary objectives) or what can be promoted through the delivery of the product (secondary objectives), that is, broader societal objectives. Procurement strategy is also informed by spend, organisational, market and stakeholder analyses.

A client, where new or refurbished construction works is required, also needs to answer basic questions relating to (see Figure 18.4):

- the financing of the project on a buy or make basis; and

- if the decision is to make, whether or not design responsibilities and/or responsibilities for the management of interfaces between direct contracts are to be retained or transferred.

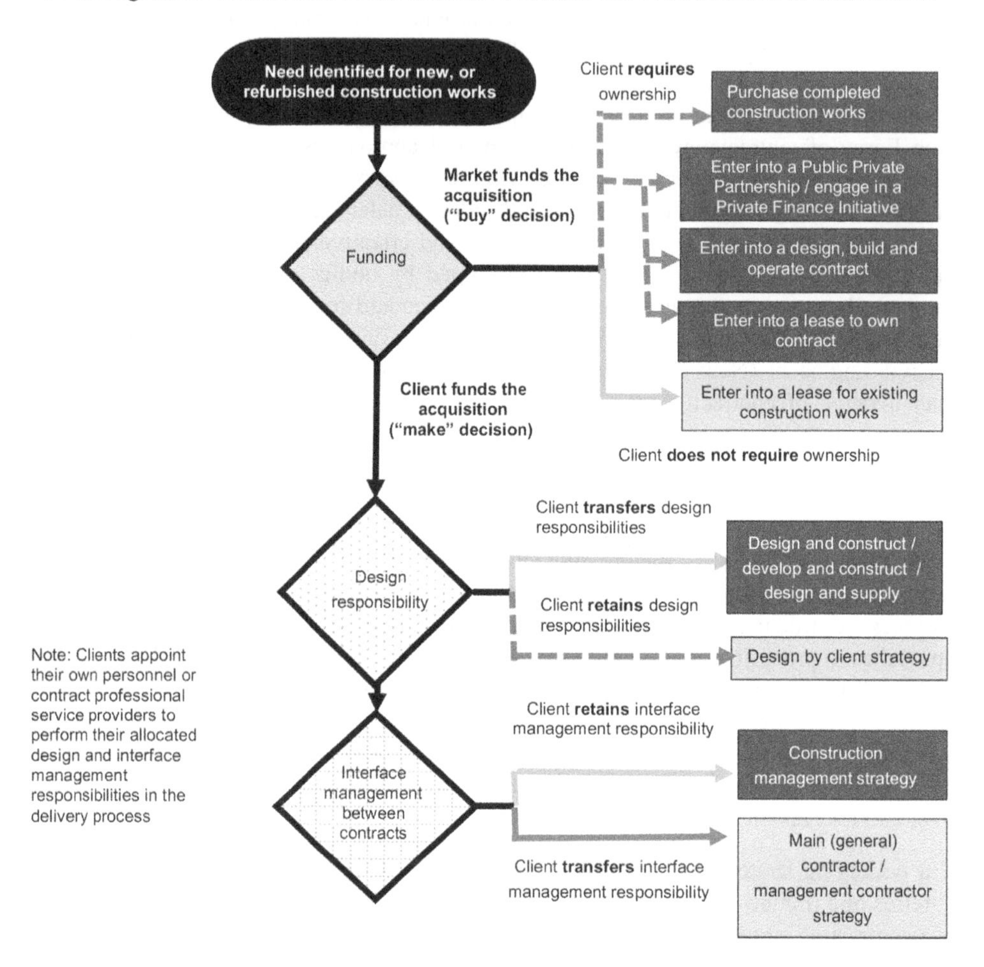

Source: Watermeyer (2018).

Figure 18.4 Common project delivery options for new or refurbished construction works

Procurement strategy determines the number of contracts that need to be procured and overseen, as well as the capacity and capabilities of the client delivery management team which needs to be put in place to oversee the delivery of the required construction works (Watermeyer, 2018).

Procurement tactics are required to successfully implement procurement strategies and in so doing improve project outcomes. They also impact upon the cost-effectiveness of specific procurement transactions. Procurement tactics commonly relate to (ISO 22058):

- the publicity to attract the right level of interest from the market;
- the sequencing and timing of the issuing of tenders and orders; and

- the setting up of procurement documents to solicit tender offers and to enter into contracts, focusing on the selection of a contractor (the other party to a contract) who is most likely to deliver best value or a cost-effective solution through the performance of the contract, and the setting up of the terms and conditions of contracts not only to allocate specific risks between the parties to a contract but also to incentivise performance to achieve best results.

DELIVERY MANAGEMENT

Overview

Delivery management (the critical leadership role played by a knowledgeable client to plan, specify, procure and deliver infrastructure projects efficiently and effectively, resulting in value for money) is required to translate the value proposition associated with a business case into project outcomes (see Figure 18.2). Project value is the outcome of client decision-making to achieve an optimal balance of the project benefits, risks and costs.

The activities directly associated with infrastructure delivery take place within product controls which focus primarily on budget, time, function, quality, conflict (positive and negative) and health, safety and the environment. However, product controls are informed by delivery constraints which in turn are shaped within an underlying context. The underlying context describes aspects which are beyond the system boundaries for delivery constraints and product controls which are prevalent within the region where the infrastructure is required. Part of the task of those entrusted with the infrastructure delivery is the management necessary to regulate and manage boundary-spanning actions and flows (Hughes, 1989; Watermeyer, 2018; see Figure 18.5). The delivery team are fully capable of managing the product controls at the interface with activities. However, the client delivery management team is best placed to manage the interface between product controls and delivery constraints and the boundary between delivery constraints and the underlying context.

Workflow may be regarded as the sequence of activities with explicit start and end points to describe a task. Activities are punctuated by decisions. For an infrastructure project to progress meaningfully, its objectives and their achievement need to be closely allied to the decision structure. Decision points (controls or decision gates) form the major boundaries to activities and provide an opportunity to authorise proceeding with an activity, within a process, or the commencing of the next process, to confirm conformity with requirements before completing processes, or to provide information which creates an opportunity for corrective action to be taken.

Control systems are necessary to regulate work in relation to its context which may from time to time change to match performance against objectives. Such systems deal with the boundary between project context and project activity. Control systems accordingly involve the comparing of progress against requirements, objectives or targets and, where necessary, taking some corrective action. They enable the taking of steps to change the performance of the activity to bring it closer to what was planned, or changing the plan so that it more closely reflects the changed situation brought about by the departure from the plan (Watermeyer, 2018).

A stage is a collection of logically related activities in the delivery or project life cycle of the infrastructure project that culminates in the completion of a major deliverable. The

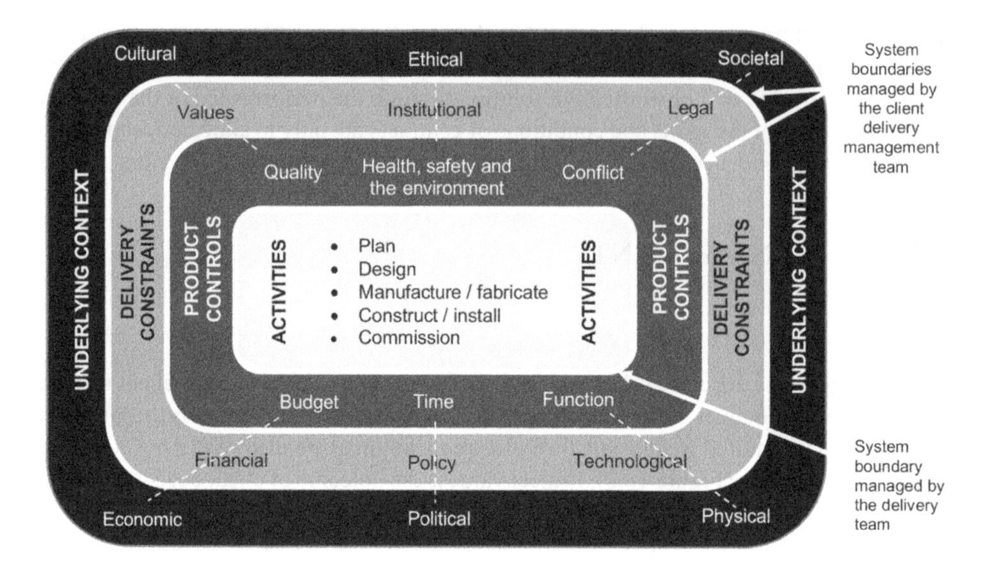

Source: Watermeyer (2018).

Figure 18.5 *The underlying context, delivery constraints and product controls within which infrastructure projects are delivered*

workflow for the delivery of infrastructure projects comprises several applicable stages, which commence with the development of a business case and end with the close-out of contracts and the project, and deal with the work flow associated with the planning, design and execution of infrastructure projects. The key deliverables associated with each task inform the decisions which are made at each decision gate. Control frameworks need to be developed around the stages within the project life cycle to enable the client to test the viability of a project during its conceptualisation, and to monitor and control projects against the client's value proposition as it progresses. Client organisations need to delegate persons to accept or reject the end-of-stage deliverables (Watermeyer, 2018).

Plan

High-level strategic plans and associated business cases need to be translated into a detailed client brief and scope for a project, and ultimately into implementation plans. Such planning is an integral part of delivery management and the work of the client delivery management team.

The demand for infrastructure delivery should be managed through service life and infrastructure management plans. Service life plans need to be aligned with the organisation's spatial development framework and strategic plans, and should be based on an assessment of current performance against desired levels of service or functionality. Infrastructure management plans need to identify and prioritise infrastructure projects within a pipeline of identified projects to meet the organisation's strategic objectives and operational commitments within available resources over a three- to five-year horizon (DHET, 2019a).

Whole-life costs should inform decisions that are taken. All significant and relevant initial and future costs and benefits associated with infrastructure throughout its lifetime from inception to construction, operation and disposal should be considered. It is imperative that realistic life cycle cost estimates are made and robust control budgets (estimates of the outturn capital cost of a project) are established for individual projects ahead of implementation to facilitate the delivery of projects within a budget.

Implementation plans for individual projects or a programme of projects need to be developed. These plans should provide information on objectives, scope, budget, schedule, key success factors, key performance indicators, the adopted procurement strategy, projected budget, cash flow and allocation of resources; as well as identify major risks, including those relating to health and safety and the environment. Such plans should also include time management, procurement, quality and communication plans. An implementation plan for a programme of projects, if correctly formulated, provides the client organisation with a baseline against which the annual delivery management performance can be measured against a number of metrics (DHET, 2019a).

Specify

The client delivery management team needs to specify the client's functional and other requirements for the project clearly and precisely, and to communicate these requirements to the delivery team. Requirements can be specified in a number of ways, ranging from prescriptive specifications (input specifications) where requirements are prescribed in detail to performance specifications as described in ISO 19208:2016 (outcome-based specifications), or any combination thereof.

Quality has a pragmatic interpretation captured in the term 'fitness for purpose', which embraces a balance of features such as the architectural aesthetics and functionality, material and functional robustness, maintainability, user comfort, environmental sustainability and life cycle costs, all of which are generally benchmarked against the cost of the delivered product. Design quality is much more than style or appearance. Good design is critical to the success of any infrastructure project. Just as important is the need to communicate these design ideas to stakeholders. Client delivery management team leadership is needed in the quest for design quality within budgetary constraints.

Procure

Clients need to obtain internal and external project resources to advance and implement projects. Careful thought needs to be given to the assembly and composition of the client delivery management team which supports the client delivery manager, as it is important to have not only synergy between members of the team, but also the necessary technical knowledge, skill and competence as well as the necessary organisational and leadership skills. Individuals external to the client organisation may be required. Such persons can be contracted on a part-time or full-time basis to provide technical support and expertise as members of the client delivery management team, to fill any gaps. Such individuals need to effectively function as staff members of the client organisation.

Clients and their procurement practices are the drivers of industry behaviour, performance and the delivery of project value. The delivery of infrastructure requires specialist procurement know-how and practices that address the project-specific nature of construction products, which are delivered over an extended period by a wide array of role players, frequently with little certainty about the ultimate product cost. For each project, or cluster of projects, it is the client's infrastructure procurement processes that need to bring together different design professionals, contractors and subcontractors to deliver specific client requirements in the most effective and efficient way possible. Procurement arrangements bind this supply chain and define the obligations, liabilities and risks that link the parties together in a process that is required to deliver value for money. Fundamental to this is the procurement of, and payment for, the required external resources, and the management of those relationships to maximise value (DHET, 2019b).

It is imperative that an organisation has in place procurement policies, procedures and systems which support infrastructure delivery and enable the client delivery manager to adopt a strategic and tactical approach to procurement.

Oversee Delivery

The client delivery management team, led by a client delivery manager, needs to oversee delivery. This, apart from providing effective leadership and direction to the delivery team, and meaningful engagement with internal and external stakeholders throughout the project cycle, needs to ensure that (Watermeyer, 2018):

- specified requirements will achieve the required benefits of the business case and provide value for money;
- momentum is maintained for the investment appropriate to the needs of the stakeholders and the delivery team for the efficient delivery of outcomes;
- requirements are translated into project or, if relevant, programme purpose, delivery principles and roles before the detail;
- value is added through the establishment of relationships and the incorporation of best practice;
- a clear governance structure, founded on the principles of honesty, accountability and integrity, is established and maintained;
- interface management occurs which aligns all stakeholder organisations so as to maximise the potential of the project or programme to deliver on the required outcomes;
- high-level progress is monitored, focusing on prognosis rather than detailed progress; and
- expenditure can be accounted for, and assets can be readily capitalised, in financial accounting systems.

A client delivery team needs to inculcate an interventionist risk management culture within the client organisation in all phases of the project, and decisively deal with any deviation from the way work was planned, designed, budgeted or scheduled, irrespective of the cause, and manage the impact of such change. Control budgets which incorporate contingencies and allowance for inflation need to be set, monitored and actively managed to ensure that project costs remain within authorised amounts, and that timeous approval for any required increase in authorised amounts is obtained. Up to data needs to be gathered and analysed to enable any course correction can be made sooner rather than later (Watermeyer, 2018).

CASE STUDY: DELIVERY OF TWO NEW UNIVERSITIES IN SOUTH AFRICA

Background

South Africa's National Development Plan 2030, 'Our future – make it work' (National Planning Commission, 2012) recognises higher education as one of the major drivers of economic development, and as being critically important for educating and training people with high-level skills so that positive social mobility changes can be realised. In accordance with this, the Plan set out to augment enrolments from 1.1 million in 2014 to 1.6 million by 2030 so as to improve access and success, particularly for those groups previously disadvantaged on the grounds of race, gender and disability.

Until 2011, South Africa had just 24 universities located in seven of its nine provinces. In September of that year, government decided to establish two new universities in the two provinces which lacked them: Northern Cape and Mpumalanga. Given the complex scope and challenging time frames, the government's Department of Higher Education and Training (DHET) appointed the experienced Campus Development and Planning Unit at the University of the Witwatersrand (Wits) to lead the delivery of the project. This was based on the proven capability of this unit to successfully deliver a R1·5 billion (about US$140 million) Wits capital projects programme (Laryea and Watermeyer, 2014). Work immediately began to carry out feasibility studies, institutional and spatial planning and to formally establish the two universities through legislation (see Table 18.1).

What is unusual about this project is that a 295-page close-out report, complete with 97 reference documents, was produced in 2018 (NUPMT, 2018). This report provides a succinct anatomy of the project from the adoption of the business case to the handover and close-out of the first delivery phase. This close-out report has been placed in the public domain on a website (www.wits.ac.za/ipdm/) as a resource for all those involved in research, teaching or delivering infrastructure projects.

Project Outcomes

Buildings were refurbished, repurposed and ready to receive the first start-up intake of students at the start of the 2014 academic year: 127 students at the Sol Plaatje University (SPU) in Kimberley and 169 at the University of Mpumalanga (UMP) in Nelspruit. The second intake in February 2015 increased the total number of student enrolments to 337 at SPU and 828 at UMP. The third intake of 2016 planned to significantly increase the student population to 700 students at SPU and 1255 students at UMP. This increase in student population required new teaching and residence facilities to accommodate the increased enrolments at a cost of approximately US$95 million (see Table 18.1).

A construction plan was created which allocated delivery management oversight to staff at the universities to oversee the construction of new buildings for the third intake. It became evident during the latter half of 2014 that the universities lacked the human resources to oversee this. The Wits team were accordingly required to oversee the delivery of the construction of the new facilities. The new facilities for the 2016 intake were built within budget, slightly below well-established cost norms, with less than 2 per cent difference between the cost at the start of construction and the final cost. This was despite 70 per cent of the works

Table 18.1 *An overview of the first phase development of the two new universities*

	Period			
	2012–13 getting started	2014 1st academic year	2015 2nd academic year	2016–17 3rd and 4th academic year
Subphase	1: Feasibility and establishment	2: Mobilising for construction	3: Delivering construction	4: Handover and close out
Key milestones	President of the Republic of South Africa announces: the seats of the new universities (July 2012) Interim Council and names of universities (July 2013) Universities formally established by notice in Government Gazette (August 2013)	Appointment of full councils of both universities (August 2014) First vice chancellor appointed to UMP (November 2014)	First vice chancellor appointed to SPU (April 2015)	SPU and UMP take over full responsibility for further infrastructure delivery from 1 April 2016
Principal activities	Site selection and land assembly Record of intention securing the land Academic and institutional planning Spatial Planning Establishment of universities Architectural competitions Procurement for renovation work for 1st academic year Repurpose / refurbish existing buildings Procurement of Furniture, Fittings and Equipment (FFE) for 1st academic year	Procurement of professional team Procurement of project managers Procurement of main contractors Design development Construction start in October to complete for 3rd academic year Repurpose refurbish for 2nd academic year Procurement of FFE for 2nd academic year	Continue first phase build for start of 3rd academic year Procurement of FFE for new buildings for 3rd academic year Establish delivery management capacity at each university Complete and furnish new buildings for start of 3rd academic year	Finalise furnishing and handover of new buildings Start of 3rd academic year in February 2016 Start 2nd phase build for 4th academic year (2017) managed by each university Handover capacity and infrastructure responsibility Finalise close out process, July 2017
No of Students	-	UMP 169 SPU 127	UMP 828 SPU 337	UMP 1255 (3rd year) SPU 700 (3rd year)
Expenditure	R 57 171 599 (≈US$6m)	R 271 621 431 (≈US$27m)	R 925 341 707 (≈US$4m)	R370 365 758 (≈US$25m)

not capable of being priced when construction commenced (Laryea and Watermeyer, 2020). In the physical construction of the universities, local content was promoted, particularly targeting those previously excluded from working on projects due to the apartheid system; 545 construction staff and workers were given approximately 40 000 hours of structured workplace learning. One of the buildings received a commendation at the World Architectural Festival. All in all, it was an exceptional project outcome.

So, what was exceptional about this project? Firstly, the time taken between the political decision to develop a new university and the receiving of the first intake of student was extremely short: just 28 months (see Table 18.1). Secondly, the necessary academic facilities and residences were delivered in a cost-efficient and effective manner, and they were delivered within the constraints of public sector procurement legislation whilst supporting the development of the surrounding community. This was due to strong governance arrangements being in place which ensured value-led decision-making, an innovative procurement strategy, and a competent client team being assembled, including a client delivery manager who had single-point accountability for delivering the client's value proposition and who provided strong leadership. What is especially important to note is that all of these enablers were within the control of the client (Laryea and Watermeyer, 2020).

Delivery Management Arrangements

A client delivery manager provided chief executive officer-level leadership in performing the buying function and was supported by a technical and administrative team which delivered strategic advice, and procurement, technical and administrative support services. A project steering committee provided constant direction, while a technical integration committee focused on the integration of the planning activities, progress and thinking, and budget and procurement approvals. A delivery team comprising a project management, design, support services and supply teams performed the supply function at each of the two universities. Wits entered into contracts with the delivery team through its procurement system, which differentiated between the procurement of infrastructure and goods and services for general consumption. The DHET periodically transferred monies to Wits to meet the university's contractual obligations in terms of a memorandum of agreement.

Project governance took place through a Project Steering Committee (PSC) and a Technical Integration Committee (TIC). The PSC originally included representatives from the DHET, Wits, the University of Johannesburg, University of Pretoria, National Institute for Higher Education (NIHE) and the Premier's Office in each province. It was subsequently expanded to include representatives of the new universities. The PSC met 15 times between March 2012 and January 2016, and provided oversight and guidance to the development of both universities until its last meeting in January 2016. The TIC, which met 50 times between February 2012 and March 2016, integrated the planning work and thinking of the DHET, the client delivery management team and, following their establishment, the new universities. The monthly TIC Contracts Committee (DHET and Wits) meetings dealt with budget and procurement approvals. This committee continued to meet beyond March 2016 in order to finalise outstanding contractual commitments. It met a total of 71 times and enabled the development of budgets and the unfolding contractual commitments to be met that resulted in peak expenditure levels of approximately R134 million (about US$10 million) per month.

Procurement Arrangements

Wits adopted a procurement policy which enabled the procurement of all the goods and services necessary to provide facilities as delivered to be occupied and used as functional entities. This policy was based on the provisions of ISO 10845 parts 1 and 2 and the Construction Industry Development Board (CIDB) Standard for Uniformity in Construction Procurement which was fully aligned with the provisions of ISO 10845 parts 3 and 4. This policy included provisions for the implementation of framework agreements (Watermeyer, 2013a) and a control framework for procurement which linked milestones in the procurement process to governance activities (Watermeyer, 2018).

A decision was taken early on in the project for the client to retain design and interface management responsibilities rather than transfer them to contractors. This was driven by a number of considerations mainly surrounding the client's brief. Firstly, the content of the academic programme: the DHET made some initial assumptions, which were modified by interim university councils and finalised by the appointed university councils (see Table 18.1). Accordingly, the detailed requirements for academic spaces were a moving target. Secondly, it was the client delivery management team's view that such an approach would provide more flexibility given the uncertainties in requirements, and enable the use of expertise within universities to ensure that the design of the teaching spaces would be aligned with current and future best practice. The team was also of the view that such a strategy would better serve the architectural design competition approach that was adopted, to create a superior design that would not only be responsive to spatial requirements but also result in architectural landmarks symbolic of intellectual aspiration.

Some 143 procurements were undertaken resulting in 219 contracts, the majority of which were framework agreements. Framework agreements were based on the NEC3 Engineering and Construction Contract (options C (target contract) and F (management contract)) and the NEC3 Engineering and Construction Short Contract, the NEC3 Professional Service Contract (option G (term contract)) and NEC3 Supply Contract and the NEC3 Supply Short Contract. Approximately 700 orders were issued in terms of these framework agreements prior to the handing over of delivery management responsibilities to SPU and UMP. At SPU approximately 19 000 furniture items were procured from 200 unique items. At UMP approximately 7000 furniture items were procured from 250 unique items. The total certified expenditure amounted to R1.62 billion (about US$280 million).

SPU and UMP took over the framework agreements in 2016 that were put in place during 2014 by Wits, and maintained the programme momentum (about US$132 million/annum) through the issuing of orders in terms of these framework agreements.

Reflections on the Reasons for Successful Project Outcomes

Table 18.2 identifies the key drivers leading to the successful management of the first phase of the delivery of the two new universities presented in the case study, mapped against the cures and causes of poor megaproject performance identified by Denico et al. (2020). It can be immediately seen that the design of the delivery management system intuitively took account of most of the themes and dominant concepts developed for megaproject performance, despite the new university projects having a financial value well below that commonly regarded as a megaproject.

Some of the core team members responsible for delivering the first phase of the new universities, drawing upon their experiences, developed a client maturity model for delivering infrastructure projects for the DHET (DHET, 2019b). This model is designed to equip a client organisation with the means to improve its infrastructure delivery management capability, so that it can consistently produce successful delivery results for client organisations. It was also designed to enable organisations to identify a process improvement pathway along which they may choose to travel, and to obtain a view of where they currently are. Four key processes were identified, namely: client governance and organisational practices; client leadership practices at a programme level; client leadership practices at a project level; and infrastructure procurement practices. The most appropriate responses to five levels of descriptors within

Table 18.2 Key drivers for the successful infrastructure delivery in the first phase of the new universities project

Cures for and causes of poor megaproject performance (Denico et al., 2020)		The key drivers of the successful management of the delivery of the first phase of SPU and UMP
Theme	Dominant concepts	
Decision-making criteria	optimism bias (delusion): executives are overly optimistic and thus overestimate benefits and underestimate costs strategic misrepresentation (deception): executives strategically misrepresent the truth and seek to satisfy their own interests escalating commitment: executives continue to follow the pattern of behavior leading to unsuccessful outcomes rather than follow an alternative course of action	strong governance arrangements which enabled sound decision-making at an appropriate level an overarching vision articulated at an early stage of the project which guided and informed decision making availability of building space and cost norms to forecast annual budgetary requirements and to manage the delivery of buildings within a budget
Strategy, governance and procurement	sponsor, client, owner, operator: associated with the roles and responsibilities of these entities throughout the project life cycle, with particular emphasis on the front-end stage governance: linked to the delegation of authority formally and informally, at the organisational and individual levels delivery model strategy: related to the strategy adopted by firms to organise themselves in combination with partners and suppliers, and combining in-house and external capabilities to best organise and deliver the project	clearly defined roles and responsibilities between governance structures and those performing a 'buying' and a 'supply' function management of project delivery as an enterprise rather than an ad hoc collection of contracts with clear delegations of authority procurement led by the client delivery manager as a strategic function which put in place resources to deliver infrastructure over a term with early contractor involvement and the forging of long term relationships
Risk and uncertainty	technological novelty: first- of- a- kind technologies have frequently being introduced in large innovative projects and are associated with risks flexibility: the ability to be adaptive and responsive to changing and uncertain circumstances complexity: the underlying factor of megaprojects that can be defined by the large number of parts and its relationships among each other and with the external environment.	adoption of a collaborative culture based on the belief that collaboration and teamwork across the whole supply chain optimises project outcomes appointment of the delivery team to execute the works as the project unfolded on a design by client basis using framework agreements which provided continuity of work and flexibility in adapting to changing requirements packaging of the work in such a way that the development of the two universities could be managed incrementally in a controlled manner
Leadership and capable teams	project leadership: the need for project champions, dedicated leaders who are committed to the success of the project competencies: competencies and skills that individuals forming project teams need to possess capabilities: the abiliy that firms have to produce specific products or services relying upon collective organisational knowledge	client delivery manager had single-point accountability for delivering the client's value proposition and provided strong (CEO level) leadership competent client delivery-management team being assembled with expert skills adoption and implementation of innovative procurement strategy and tactics aligned to the client delivery objectives which enabled suitably qualified entities to be procured and procurement related risks to be mitigated

Cures for and causes of poor megaproject performance (Denico et al., 2020)		The key drivers of the successful management of the delivery of the first phase of SPU and UMP
Theme	Dominant concepts	
Stakeholder engagement and management	institutional context: the set of formal organisational structures, rules, and informal norms stakeholder fragmentation: the number of parties, which often results in an intense level of interaction among involved stakeholders community engagement: the processes and engagement activities by which the project involves the local population in the project	project steering committee provided effective stakeholder oversight and guidance while the technical integration committee integrated evolving project requirements client delivery management team proactively engaged stakeholders buying and supply teams embraced stakeholder expectations by securing project related business, employment and skills opportunities for local communities
Supply chain integration and coordination	program management: associated with systems, procedures, and tools to monitor, control, consolidate, optimise, and achieve benefits from a number of individual inter-related projects commercial relationships: linked to the establishment of formal relationships with the organisations delivering projects and subprojects, as well as the management of those interfaces throughout several phases of the project systems integration: related to the technical and managerial capabilities required to integrate several components produced by different parties in order to deliver an operational asset to the client	setting of control budgets for projects within the programme and the accessing of contingencies to fund risk events on a stepped-access basis framework agreements that incentivised performance of the delivery team in order to secure future orders (call-offs) and enabled long-term relationships which focused on maximising efficiency and shared value provision of effective project management services overseen by the client delivery management team which enabled interfaces between suppliers, service providers, contractors and end users to be effectively managed

25 practices associated with the identified key processes enable organisations to determine the maturity of an organisation on a self-assessment basis within one of five levels: ad hoc (an improvised process capability); basic (a basic, disciplined process capability); structured (a fully established and institutionalised process capability); integrated (full integration with other organisational processes resulting in synergistic benefits); and optimised (processes focus on continuous improvement and adoption of lessons learned and best practices).The pre-requisite for achieving a structured level of maturity are that client organisations have in place:

1. a client delivery manager who is an appropriately qualified and experienced built environment practitioner and able to lead the infrastructure projects and programmes;
2. an effective organisational governance structure which enables the appointed client delivery manager to exercise chief executive officer-level leadership;
3. appropriate delegations of authority which enable decisions to be made swiftly to ensure delivery in an accountable manner; and
4. enabling polices which facilitate or do not preclude the application of sound infrastructure procurement practices to achieve desired outcomes and value.

The DHET oversight monitoring team during 2019 obtained the self-assessment maturity scores of all 26 South Africa universities where the annual infrastructure expenditure did not exceed US$60 million and was commonly below US$10 million. The self-assessed levels of maturity were compared against delivery capability, as determined by DHET staff and the oversight monitoring team, based on consideration of a number of indicators as well as interactions with the universities, and in terms of their expenditure record against their allocated annual budgets. A good correlation between delivery capability, expenditure record and maturity level were obtained, except where underperforming universities overestimated their maturity. (An underestimation of maturity is indicative that an organisation needs to undergo a process of becoming more aware of its weaknesses before it can start to improve.)

CONCLUSION

Clients need to plan, specify, procure and oversee the delivery of construction projects through a supply chain process which involves buying and supply responsibilities. The delivery of infrastructure projects within this fragmented environment, which is fraught with uncertainty, is held together by the client's vision, leadership and procurement practices. Kershaw and Hutchison (2009) expressed the view that: 'the role of the client is the single most important factor in determining the success of construction projects and capital works programmes, regardless of their size, complexity and location'. The case study of the two new universities and recent research outputs reinforce this view.

Clients can improve infrastructure project outcomes, irrespective of the size, complexity and location of a project, should they put in place effective governance processes which ensure that the institution takes ownership of infrastructure delivery, and such delivery is managed as an enterprise and creates an enabling environment for the effective performance of the buying function.

Clients are likely to consistently produce successful project outcomes should they put in place infrastructure procurement and delivery management systems which:

1. provide effective governance processes which ensure that:
   a. there is proper management and control of infrastructure delivery including demand management;
   b. all the various parts of the organisation which play a role in infrastructure delivery work together in a coordinated, efficient and effective manner; and
   c. infrastructure delivery is, wherever possible, managed as a long-term and strategic system of individual yet interlinked projects aimed at meeting the organisation's objectives;
2. establish clear delegation of authority to enable timeous decision-making and individual and organisational accountability for infrastructure delivery;
3. provide for the assignment of single-point accountability to a suitably qualified and experienced built environment practitioner (client delivery manager) to provide executive-level leadership in the planning, specifying, procuring and overseeing of infrastructure delivery;
4. provide control frameworks which include decision gates to enable risks to be appropriately and proactively managed, and render the system capable of being audited;
5. provide for a range of approaches to packaging, contracting, pricing and procurement, enabling:
   a. the formulation and implementation of suitable infrastructure procurement strategies and tactics to be adopted by the client delivery manager for different categories of infrastructure expenditure to achieve value for money; and
   b. focus to be on likely outturn (final costs) rather than lowest price;
6. make appropriate use of the skill and expertise of built environment professionals in the management and mitigation of risks associated with infrastructure delivery; and
7. recognise that infrastructure procurement is a central competency of those responsible for delivering infrastructure.

It is also advantageous for clients to adopt standard generic procurement processes, methods and procedures such as those provided in ISO 10845. Such standards establish a way of doing things that provides order and a platform for the methodical planning of a means of proceeding. They also generate confidence in the procurement system and provide a platform to achieve fair competition, reduce the possibilities for abuse and the scope for corruption, and improve the predictability in procurement outcomes.

REFERENCES

Arrowsmith, S. (ed.) (2010). Public Procurement Regulation: An Introduction. www.nottingham.ac.uk/pprg/documentsarchive/asialinkmaterials/publicprocurementregulationintroduction.pdf.

APMG International (2016). APMG PPP Certification Program Guide. https://ppp-certification.com/ppp-certification-guide/about-ppp-guide.

Association for Project Management (APM) (2017). *APM Guide to Contracts and Procurement.* Apmknowledge.

Denico, J., Davies, A. and Krystallis, I. (2020). What are the Causes and Cures of Poor Megaproject Performance? A Systematic Literature Review and Research Agenda. *Project Management Journal*, 51(February).

Department of Higher Education and Training (DHET) (2019a). University Macro-Infrastructure Framework: Infrastructure Management Guidelines for Universities. www.dhet.gov.za/Manuals/DHET%20MIF%20guidelines%20(second%20edition)%20with%20Annexures.pdf.
Department of Higher Education and Training (DHET) (2019b). University Macro-Infrastructure Framework: Infrastructure Management Guidelines for Universities – Annexure 16: A Client Maturity Model for Delivering Infrastructure Projects. www.dhet.gov.za/Manuals/Annexure%2016%20Maturity%20mode%202020.pdf.
Dumfries and Galloway Council (2018). Report of the Independent Inquiry into the Construction of the DG One Complex in Dumfries. DG_One_Inquiry_Report_Bookmarks.pdf (dumgal.gov.uk).
Foster, V. (2008). Overhauling the Engine of Growth: Infrastructure in Africa. World Bank, September.
Hughes, W. (1989). Organisational Analysis of Building Projects. Thesis (PhD) Liverpool Polytechnic.
Infrastructure Client Group (2017). From Transaction to Enterprise: A New Approach to Delivering High Performance Infrastructure. Institution of Civil Engineers, May.
Kershaw, S. and Hutchison, D. (eds) (2009). *Client Best Practice Guide. Institution of Civil Engineers.* London: Thomas Telford.
Laryea, S. and Watermeyer, R.B. (2014). Innovative Construction Procurement at Wits University. *Proceedings of the ICE-Management, Procurement and Law*, 167(5), 220–231.
Laryea, S. and Watermeyer, R. (2020). Managing Uncertainty in Fast-Track Construction Projects: Case Study from South Africa. *Proceedings of the Institution of Civil Engineers – Management, Procurement and Law*, 173(2), 49–63.
McKinsey & Company (2017). The Art of Project Leadership: Delivering the World's Largest Projects. McKinsey Capital Projects and Infrastructure Practice. September.
Merrow, Edward (2012). Why Megaprojects Fail. *Project Manager Magazine*, 18–21.
National Audit Office (2010). *Analytical Framework for Assessing Value for Money.* National Audit Office (UK).
National Planning Commission (2012). National Development Plan 2030, 'Our Future – Make It Work'. Presidency of the Republic of South Africa.
National Treasury (2015). *Standard for Infrastructure Procurement and Delivery Management.* National Treasury, South Africa.
New Universities Project Management Team (NUPMT) (2018). Close Out Report of the New Universities Project in Mpumalanga and the Northern Cape, 1 November 2011 to 31 July 2017. University of the Witwatersrand. https://www.wits.ac.za/ipdm/guides/close-out-report/.
United Nations Commission on International Trade Law (UNCITRAL) (2014). *UNCITRAL Model Law on Public Procurement.* New York: United Nations.
Watermeyer, R.B. (2011a). Standardising Construction Procurement Systems. Report. *Structural Engineer*, 89(20), October.
Watermeyer, R.B. (2011b). Building Trust – A Platform for Best Practice Construction Procurement. Special Report. ISO Focus +, 24 to 26 September.
Watermeyer, R.B. (2013a). Unpacking Framework Agreements for the Delivery and Maintenance of Infrastructure. Civil Engineering, January/February.
Watermeyer, R.B. (2013b). Value for Money in the Delivery of Public Infrastructure. West Africa Built Environment Research Conference, Accra, Ghana, August.
Watermeyer, R. (2018). Client Guide for Improving Infrastructure Project Outcomes. School of Construction Economics and Management, University of the Witwatersrand and Engineers Against Poverty.
Watermeyer, R.B. (2019). The Critical Role Played by the Client in Delivering Infrastructure Project Outcomes. Civil Engineering, January/February.
Watermeyer, R. and Phillips, P. (2020). Public Infrastructure Delivery and Construction Sector Dynamism in the South African Economy. NPC Economy Series – Background Paper. National Planning Commission.
World Bank (2016). *Procurement Regulations for IFP Borrowers.* Washington, DC.

19. The economics of housing policy and construction: developing a responsive supply sector

Suraya Ismail

INTRODUCTION

Most housing policies attempt to overcome the problems of severe housing shortages and unaffordability created by the prevailing market conditions. The underlying assumptions of utility-maximizing households and profit-maximizing firms create market inefficiencies such as speculative purchases, substandard housing, high vacancy rates and rapid house price escalations, with the resultant effect of decreased affordability for a significant proportion of society. Housing affordability is an outcome of comparing both house prices and rentals with household incomes within a specific market area. This infers an analysis into the demand and supply of local housing conditions which is shaped by a country's national institutions and social norms. The policy responses usually cover: (1) the demand side, for example, tenure policies, taxation, interest and mortgages; and (2) the supply side, for example, land, property rights, spatial planning and the firms in the construction and maintenance of the built form sector.

Policy interventions to improve housing affordability based on demand-side initiatives tend to concentrate on negotiating house prices once the house is received by the consumer at the end of the production process (Gibb, 2011; Marom and Carmon, 2015). This is based on the notion that government assistance should be given directly to homeowners; a case of targeted subsidies to those who need it (Ruprah, 2010; World Bank, 2006, 2009, 2010), rather than assistance given to housing projects (or firms). However, these measures are unsustainable as they may contribute towards increase in prices when the supply of houses is relatively inelastic (DCLG, 2017). Demand-side interventions appear to accept the relatively inelastic supply curve and are predominantly aimed at enabling consumers to 'afford' houses as they become increasingly more expensive. These include policies allowing consumers to borrow more (for example, increasing the ratio for loan to house value, LTV), or those that subsidize the costs of housing (in the form of housing grants). As some evidence suggests, these demand-side interventions tend to support and buffer inefficiencies within the supply sector, which increases costs for both consumers and governments (Apgar, 1990; Jones and Watkins, 2009; DCLG, 2017; Glaeser and Gyourko, 2018).

This chapter takes a different approach, because the evidence suggests that it may be more efficient to employ strategies which enhance capacity on the supply side (Bramley, 2007; Marom and Carmon, 2015; Hansson, 2017). It follows that if housing displays a relatively inelastic supply curve, any increase in the factors of demand will effect a steeper price increase. The more elastic the supply curve for the housing industry is, the more affordable it becomes to consumers. Therefore, it is imperative to develop sustainable and responsive housing supply sectors in order to enhance the affordability of houses to a wider proportion of society. This suggests that only then will demand-side assistance prove efficacious in the provision of competitive house prices to consumers, or the 'target group' (Turner and Whitehead, 2002; UN-Habitat, 2009; Ismail et al., 2019).

The problem of inadequate supply of housing is located within the perspectives of both the institutional arrangement and the governance of firms in the construction project coalition. It involves analysing the national construction business systems as manifested by the temporal clustering of firms within the procurement routes. This is premised on the view that improvements in the institutional and governance structures are needed at the project level (Ive and Gruneberg, 2000; Winch, 2001) to increase the responsiveness of the construction industry to both the supply of housing need and demand (Khazanah Research Institute, 2015; Ismail et al., 2019).

This chapter is divided into three main sections. The first section examines the affordability crisis as it emerges in both developed and emergent economies; and the ensuing discussion on why the measurement of affordability should exclude macroprudential measures. The second section illustrates the role of government intervention in building an enabling policy framework to address the affordability problem; the importance of distinguishing policies for the social and markets sectors, and the utilization of housing ratios as indicators or measurements of affordability. The third section focuses on the development of a responsive housing sector and the role of the construction industry in facilitating successful supply-side strategies to overcome the challenges of housing affordability. It is suggested that the improvements at the level of construction projects will reduce the inefficiencies of existing organizational constructs for production, and inevitably increase the general affordability of houses. The chapter ends with some concluding remarks and suggestions for further research.

THE HOUSING AFFORDABILITY CRISIS: THE DIFFERENT REALITIES AROUND THE WORLD

The factors that contribute to the issue of affordability are manifold, and contingent on whether countries fall within the emergent or developed economies status (Oxley, 2004). Official statistics from most developed economies suggest that houses are increasingly seen as one of the most valuable assets in society (Demographia, 2018, 2015). For example, houses in the United Kingdom were valued at £5.5 trillion, accounting for 62 per cent of the UK's net worth at the end of 2015, rising 48.7 per cent from 20 years prior (Demographia, 2018). Incoming stock or new-built houses into these markets will tend to generate higher prices due to what is seen as its 'profit-making', potential and consequently this reduces general affordability. On the other hand, houses in emergent economies can sometimes experience limited capital appreciation. This can be due to many factors, such as inadequate local infrastructure, substandard quality of houses (leading to negative equity for owners), and even the absence of property rights, especially in the informal housing sector (De Soto, 2000; Buckley and Kalarickal, 2005; ACHR, 2018). The provision of good-quality housing with decent prices and rental levels remains a major challenge for governments of both emergent and developed economies, albeit due to different reasons based on a country's distinctive characteristics and development trajectories.

In emergent economies the extent of housing problems reflects to a large degree the manner in which governments, both national and local, have attempted to address these issues (World

Bank, 2009). Caught in a rush to lift their countries out of developing-nation status, political leaders, administrators and planners alike have given housing programmes low priority in the total development strategy (Johnstone, 1976; Angel, 2000). In many countries, the provision of housing is usually left to private developers, who invariably cater to the growing demand from middle- to upper-income households, often with minimal consideration for the less affluent (Johnstone, 1976). The result is an increasing gap between most urban dwellers' need for housing and the ability of the housing system to provide it at an affordable price. This problem is further exacerbated under globalization and the increasing influences of neoliberal practices, where speculative buyers with access to stronger currencies tend to price out the majority of the local populace (Georgiera, 2017).

The shift from rural to urban living is largely the result of positive global and national economic growth. Additionally, the gradual demise of agriculture in rural areas has intensified high unemployment rates and created inward migration to urban areas (World Bank, 2009). The high rate of urbanization has now reached a point where half of the global population lives in towns and cities (UN-Habitat, 2020). Furthermore, it is estimated that by the year 2050, this will increase to approximately 6 billion people (two-thirds of the world population) (UN-Habitat, 2007).

As these new urban centres grow, the locus of global poverty will also move into towns and cities, especially within the burgeoning informal settlements and slums (Fay, 2004; Gulyani and Talukdar, 2008; Asian Coalition of Housing Rights, 2018). This is due, in part, to demand for housing becoming increasingly income-elastic and price-inelastic for those at the bottom of the income distribution. In other words, scarcity of resources pushes up prices, which results in making housing increasingly financially unattainable. This has led to a variety of responses. Some cities support informal self-help policies and strategies (bottom-up approaches) designed to obviate further slum creations. These strategies do not necessarily perceive the informal supply sector as an inferior solution. In fact, they are often viewed as a practicable solution where local councils, in partnership with representatives from the informal sector, propose special apparatus and governance structures to assist communities in creating improved living conditions in informal shelters. On the other hand, certain governments (UN-Habitat, 2009) are concerned with transmitting property rights that inevitably come with ownership, and are therefore insistent that the housing system should be formalized. This is with the intention of creating equities for households, a key feature of asset-based welfare policies. However, this policy has created its own set of complications, ranging from 'slum' social housing to homelessness (Chen and Shin, 2019).

The problem of housing affordability afflicts developed economies as well, but in different ways (Angel, 2000; Sorek et al., 2015; Demografia, 2018). The post-war period saw a range of housing policies that changed the provision of housing from being seen as a basic right to viewing it as a key instrument for wealth creation, as part of asset-based welfare policies. This resulted from the liberalization of the financial markets, which enhanced the ability for households not only to access the home ownership ladder but also to accumulate houses as assets (Maclennan and Miao, 2017). Studies on the possibilities of releasing equity from housing in Europe concluded that this solution is difficult, especially among vulnerable households in less optimal areas, and non-existent for the non-homeowners, even in mature economies (Toussaint, 2011; Doling and Elsinga, 2013).

It could be observed that, in general, central governments are withdrawing from developing housing policies (Johnstone, 1976; Katz and Turner, 2003). As a result, housing markets have become increasingly vulnerable to global economic forces; on the other hand, housing policies are left largely to regional authorities and resources (Katz and Turner, 2003; Salim, 2011; Hansson, 2017). This implies that the policy instruments of devising taxes and subsidies at the national level becomes less evident and more difficult to institutionalise. The deregulation of financial markets and the attendant commodification of houses creates affordability challenges for people in the lower income groups, who live in increasingly remote suburban locations under poor living conditions, and are often susceptible to local labour market shocks (Maclennan and Miao, 2017).

The Price Mechanism: Between New Stock and Existing Stock

Most developed countries have shown a steady increase in the house price index (HPI), and some with double-digit compound annual growth rates (CAGRs) (Global Housing Watch, 2020). If housing is viewed as a generator of asset-based income, then the CAGR of household wages will never be commensurate with the rapid price escalation of housing as an investment. This is the paradox of housing affordability when housing is both an asset and a shelter. There is an asymmetry between the interests of existing house owners and those of prospective home buyers. Figure 19.1 shows the theory of housing prices, where the prices of the existing stock signal the prices of incoming stock; this is known as the 'comparative sales approach' (Ismail et al., 2019). This is especially true for developed and mature cities where the number of additional incoming units is disproportionately small when compared to the existing stock. The price that developers are willing to pay for new land will increase as most feasibility studies use the existing stock's last-transacted sale price, less the housing project's costs and profits, as the residual to bid for the price of land. The hedonic price models (Rosen, 1979; Roback, 1982) and those integrated with the spatial equilibrium models (Gyourko, 1991; Gabriel et al., 2003) are outcome price models and are different from the input-based approach as illustrated below. However, both these methods are complementary to the approach adopted by Harvey and Jowsey (2004).

Rapid price escalations supported by easy credit programmes from financial institutions pose systemic risks and affect the wider economy including homeowners (Ball, 1994). This is best illustrated by the 2007–2009 United States subprime crisis, which demonstrated the highly detrimental effects of loans extended at high financing costs (Jaffee, 2008; Duca and Muellbauer, 2013). A sharp correction in property prices can affect asset values, hence the emergence of the expression 'underwater' mortgages (loans exceeding the market value of homes).

Although housing finance plays an important role in ensuring sustainable demand for housing, it is also equally critical that financing is not seen as the central instrument in creating affordability. Typically, financial instruments ostensibly make house financing cheaper, by lowering monthly instalment payments via reducing the cost of credit or lengthening the loan period, as well as by providing subsidies in the form of downpayments. While these measures are inclined to make housing appear affordable, or ensure sustainable demand for housing, they tend to increase the household debt burden and extend it over a longer duration. In other words, increasing the financialization of housing is not necessarily the same as enhancing its affordability.

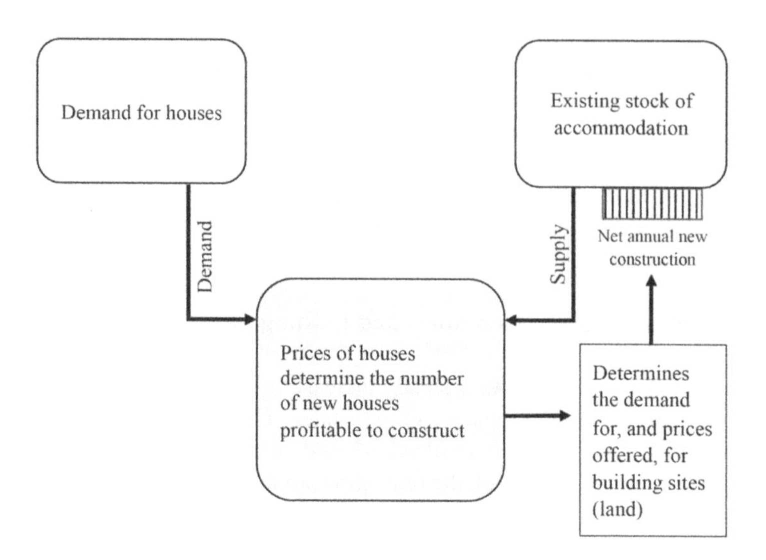

Source: Adapted from Harvey and Jowsey (2004).

Figure 19.1 House prices and the dominance of the existing stock

Measuring Housing Affordability through Housing Indicators

The efficacy of the housing sector can be measured by several types of indicators. For example, the house price index (HPI) and the repeated transaction index (RTI) identify trends in house price movements while the changes in loan-to-value (LTV) ratios and the number of mortgage possessions measure the ability of home buyers to finance their purchases over a stipulated period of time. Amongst these, housing affordability indicators attempt to track changes in the price or cost of attaining a housing unit relative to a person's ability to pay, that is, housing cost-to-income ratios.

Policy-makers and housing sector analysts use several house price-to-income ratios to observe the affordability of housing for both home buyers and renters. Depending on the market segment of interest, different prices and income indicators are used. For example, to estimate the affordability of social housing rental rates, local governments in the UK compare the weekly cost of renting a social housing unit with the average weekly salary from paid work for the bottom 10 per cent of income earners in their local population (ONS, 2016). In terms of market rental affordability, other methods such as a housing wage approach may be used. The housing wage approach involves calculating the ratio of the full-time minimum wage against the rental rate of a modest one- or two-bedroom apartment in the locality (Belsky et al., 2005). A selected list of indicators that are used to measure housing affordability is presented in Tables 19.1 and 19.2.

The lists of measures in Tables 19.1 and 19.2 are not exhaustive. However, there are three broad approaches which are commonly used to determine housing affordability for homeowners: the median multiple, the housing cost burden and the residual income. These are presented in Table 19.3.

Table 19.1 *House renter affordability measures*

Measure	Definition	Data used
Ability to afford private renting	Affordable if rent payable is up to 25% of their gross household income	Rent payable Gross household income
Affordability of social housing	Compares average cost of social housing rent for local authorities and 10th percentile gross salary	Cost of social housing rent 10th percentile gross salary
Residual incomes	Measures income a household has left over after they have paid for housing costs Residual income = income - rent - income support applicable amount + housing benefit	Household income Rental payment Income support Housing benefit
Rent ratio (used to compare both renting and buying)	Purchase price of a house divided by annual rent of a similar home	House price Annual rental payment
Housing wage approach	The rent of a standard, modest-quality rental unit with either one or two bedrooms in an area is compared to the multiples of full-time minimum wage work it would take to afford (at 30% of income) that apartment	Rental payment Full time minimum wage work

Source: Ismail et al. (2019).

The housing cost burden and the residual income approach both measure the ability of households to purchase houses based on mortgage requirements. The housing cost burden approach caps an affordable amount at 30 per cent of the household income, while the residual income approach estimates the maximum amount of housing loan a household can attain based on the household income after accounting for expenditure on other, non-housing related items. In this way, both the housing cost burden and residual income approach are influenced by factors or policies related to housing finance. This includes the amount of deposit required for housing loans, the effective rate of interest and the maximum tenure for mortgages.

These housing costs also include depreciation and maintenance costs, and sometimes transportation costs. In this indicator, the monthly repayments or rents are not the only factor in determining affordability because there are other hidden costs to owning a house. Some studies (Haas et al., 2006; Angel et al., 2011; Mattingly and Morissey, 2014) also recommend giving opportunity costs a numerical value (examples of such opportunity costs are the time taken in congestion, the level of pollution and crime rates).

The inability to find affordable units may force many households to make sacrifices regarding decent housing quality and/or location. Even after making these trade-offs, housing costs are often still substantial, and result in households making compromises with regard to spending on other necessities or the repair or maintenance of their homes (Glaeser and Gyourko, 2018). The former cutback might seriously undermine the well-being of low-income households and the latter will, in turn, create run-down properties. Questions have arisen about the validity of income-based measures of housing affordability for individual households; some households may either have other sources of income not captured in the available statistics (such as savings or in-kind incomes), or be willing to allocate a higher proportion of income to housing, due to factors such as neighbourhood preference or family size (Stone, 2006).

In addition, affordability standards pertaining to households can differ across time, period and location. For example, Glaeser and Gyourko (2018) highlight that in some high-cost cities, high-income households can spend up to 40–50 per cent on housing costs and still have generous resources for other expenditures. On the other hand, in a low-income city, 30 per

Table 19.2 Home buyer affordability measures

Measure	Definition	Data used
House price index	A ratio that shows how much the cost of housing has changed between two periods, by pricing a basket of house characteristics of the 'average' house transacted in the current period and comparing this price with the price of the same basket of house characteristics in the base year	House price based on transactions
National Association of Home Builders Housing Market Index	Tracks sentiment among participants in the housing industry. Based on a monthly survey of home builders. They are asked to rate current sales of single-family homes and sales expectations for the next six months and to rate traffic of prospective buyers. Scores for responses to each component are used to calculate a seasonally adjusted overall index. A reading above 50 indicates that more builders view conditions as good than poor.	Rating of current sales of home and expectations for next six months Rating of traffic of prospective buyers
Conventional mortgage home price index/repeat sales method	Measure price appreciation while holding constant property type and location, by comparing the price of the same property over two or more transactions. Exclude outlier properties, i.e., transactions that suggest substantial physical change through renovation or deterioration, and properties that have appreciated at exceptionally large or small rates (perhaps due to data entry errors), relative to a normal range of appreciation rates	Repeat mortgage transactions for single-family properties Actual sales prices and appraised values for the same homes over time
House prices to income ratios	A ratio of average or median house prices to average or median gross or disposable income in a given geographical area	Mean and median house price Mean and median gross and disposable income
Ability to afford home ownership	Affordable if a house costs 3.5x the annual gross household income (single earner) or 2.9x the gross household income (dual income)	House price (mean and lower quartile) Gross household income Other relevant information: Deposit required Lender multiple Interest rate Period of loan Mortgage payment protection insurance Building insurance Saving or capital available
Annual income necessary to purchase a house	Commonly used by housing associations in setting the affordable price and rent levels for shared ownership products and assessing eligibility Can be used to calculate for purchasing of shared ownership flat	House price Rental payment Monthly mortgage payment
Access to finance	Indicators of access-to-finance related affordability constraints, including changes in the loan to value (LTV) ratio, the size of deposit paid by first time buyers and the deposit as a proportion of annual income	LTV ratio Deposit paid by first time buyer Deposit as % of annual income Loan approval rate
Duration to save for minimum downpayment	The time (in months) required for people earning the median income to afford a median-priced home in the county	% of down payment House price % of saving Household income

Measure	Definition	Data used
Repossessions	The fall in the number of mortgage possession actions. May be due to: (1) low interest rates, which decreased the average cost of mortgages for those who can afford them; or (2) the situation making it more difficult for first-time buyers to purchase a house.	Number of mortgage and landlord possession actions
Supply–demand mismatch approach	Number of households with incomes at, or below, a particular level is compared with the number of rentals with rents that are affordable at 30% of the threshold income	Household income Rental payment
House building	Housing construction statistics consider the impact of new housing developments on prices	Number of completed dwellings
Housing market activity	Refers to number of residential property sales (as % of the total privately owned housing stock)	Number of residential property units sold as a percentage of private sector dwellings
Social housing stock	Compares % of social housing stock and 10th percentile salary	Social housing prevalence 10th percentile gross weekly salary
Vacant social housing	Measures vacant social housing out of the total social housing stock	Vacant social housing as a percentage of all social housing and housing affordability ratio
Social housing shortfall	Measures number of households on a local authority's social housing waiting list minus vacant social housing. It is expressed as a percentage of the total social housing stock	Indices of social housing shortfall, waiting lists and vacant dwellings Number of property sales under the Right to Buy scheme and number of additional social rented properties provided
% of ownership with/without mortgage	Measures the proportion of owner-occupied homes that did not have a mortgage	Rate of home ownership Number of owner-occupied homes that did not have a mortgage

Source: Ismail et al. (2019).

cent of income spent on housing may be excessive. Therefore, the amount of money allocated for housing costs can be subject to the circumstances and priorities of households. However, despite some reservations, studies undertaken across different cities and countries (at various levels of development and growth) concluded that 30 per cent of income spent on housing remains an insightful rule of thumb in assessing housing costs (Glaeser and Gyourko, 2018).

Table 19.3 Comparison of housing affordability indicators

Approach	Definition of housing affordability	Advantages/disadvantages
Median multiple	Median house price of three times or less than the median gross annual household income	Easy to calculate Cross-country comparison over time is possible Excludes the role of finance
Housing cost burden	Housing expenditure (for example, rent or mortgage) that is less than 30 percent of household income	Accounts for the role of finance and non-housing expenditures of households Cross-country comparison is possible but may be affected by differences in cost of living and financial systems
Residual income	Residual income (i.e., total income minus non-housing costs) that is sufficient to service monthly mortgage obligations	Accounts for the role of finance and the household's spending patterns Requires detailed data on household income and expenditures as well as housing costs Limited cross-country comparability

Sources: BNM (2017) and Ismail et al. (2019).

The median multiple approach excludes the role of finance when measuring housing affordability. It considers only the price of housing relative to income, whereby a house is considered affordable if it is less than three times a household's median annual income. In this chapter, it is suggested that the median multiple approach, which is discussed next, may prove to be the most effective indicator for housing affordability, given the tendency for both the residual income and housing cost burden approaches to skew house prices higher (Angel, 2000; Khazanah Research Institute, 2015).

The Median Multiple

The price-to-income ratio, or commonly known as the 'median multiple' was developed in 1988 by the then United Nations Centre for Human Settlement (UNCHS) and the World Bank under the Housing Indicators Programme, and was later used in the UN-HABITAT Housing Indicators Programme, focusing on monitoring the performance of dwellings.

The median multiple assumes that as house prices increase relative to income, only a smaller proportion of households can afford to buy housing, holding all other factors constant. More importantly, deviations of this indicator from global norms is seen to signal distortions in the housing market. When the value is excessively high, these distortions may indicate that the housing sector is restricted in its ability to supply sufficient housing to meet effective demand. Under these circumstances, it has been found that housing quality is depressed below levels typically found in countries with well-functioning and responsive markets.

On the other hand, abnormally low values signal insecurity of tenure, and can lead to a reduced willingness of the population to invest in housing, and thereafter create a lower-than-necessary quality of housing (Angel et al., 1993; Angel, 2000). Subsequent empirical research by the UNHCS and the World Bank utilizing international data (and adapted by Demographia) found that the 'global norm' for affordability was 'three times', meaning that if the median price for the whole of a housing market was three times the median gross annual household income, this signals a well-functioning market (Angel et al., 1993; Angel, 2000). The main advantage of the median multiple is its non-reliance on macroprudential measures, and therefore affordability thresholds cannot be distorted by innovative financing that renders monthly repayments as 'affordable' but in reality increases the total debt repayments.

ROLE OF GOVERNMENT: BUILDING AN ENABLING POLICY FRAMEWORK FOR THE HOUSING SECTOR

It is important to recognize that housing is not an initiative that 'belongs' to one discrete sector or entity (Angel, 2000). Many examples can be given to illustrate this point. Housing production is part of the construction sector; residential development is contingent on property rights (Oxley, 2004), spatial structure plans and development approvals (Turner and Whitehead, 2002; Ismail et al., 2019); housing investment is a part of overall capital formation; residential property is part of the real estate sector (Harvey and Jowsey, 2004); household income is dependent on the labour structure; housing subsidies are part of social welfare expenditures; and housing finance is part of the financial sector (O'Sullivan, 2003). It is pertinent to note that this list is not exhaustive. It is possible to consider many other factors; for example, even the interplay of housing and transportation for the effective allocation of units to residents of a market area can also be integral to the larger policy environment of housing.

Each of the items indicated above sits within its own web of policies. Negotiating an autonomous housing policy from such a heterogeneous range of other policy concerns is complex. One method suggested is to form a performance objective for the country and include the other policies within the 'housing policy environment' (Angel, 2000). 'Housing policy environment' is defined as a set of policies or government interventions that have a critical and measurable effect on the performance of the housing sector. These policies motivate, enable and constrain housing initiatives and actions. The housing policy environment consists of a set of interventions in the housing sector by different government agencies and, ideally, should be pursued as a 'comprehensive housing agenda' (Angel 2000).

A comprehensive housing agenda should be seen as a move towards guiding and managing the housing sector, recognizing housing as both a basic need and effective demand (Oxley, 2004; Ismail et al., 2019). Often, polices tend to be formulated in isolation from the larger housing environment. For example, policies that are specifically developed for social housing may fail to recognise how this is disconnected from the larger market sector. Indeed as Angel (2000) argues that there is little merit in a housing policy that solely focuses on the poor, in the hope that 'the market' will take care of the rest. This was also previously raised by Ramirez (1978), who noted that 'universal housing queues' frequently make it difficult for the poor to have decent housing; while in turn the 'not-so-poor' remain ill-housed. This suggests that a narrow focus on the welfare aspects of housing is often insufficient, resulting in imbalances in, and mismanagement of, the housing sector. Therefore, it is crucial to move away from the notion that the provision of shelter is necessarily a drain on government's fiscal considerations, and instead to view housing as an important and productive sector, where policies carry serious repercussions for overall national economic performance and societal well-being (Angel et al., 1993).

By adopting this approach, the enabling policy environment would usually consist of: (1) demand-side policies, for example, tenure policies (ownership or rent), fiscal and finance policies; (2) supply-side policies, such as productivity of the construction sector, maintenance of public buildings, price determination policies; and (3) cross-cutting policies which include spatial planning guidelines, and data policies. Policies should not be concerned just with the end product (that is, providing good-quality housing with decent prices). It is perhaps more important to focus on aligning different policy prescriptions which enhance the role of specific institutions and processes that create the end product.

Demarcating Housing Need and Demand, at the Right Spatial Scale

Notwithstanding its uniqueness, as in all markets, a well-functioning housing sector is one in which prices are determined by the interplay of demand and supply, and the equilibrium price clears the market. However, the high rate of urbanization, incidence of urban poverty, wage stagnation for the lower- and middle-income groups, rapid house price escalation, the commodification and financialization of housing, have rendered the market price for housing with minimum/decent housing standards unattainable for a significant proportion of society. As a result, many households are priced out from the market and find themselves demonstrating a housing need. Aggregate housing need has been defined as 'the quantity of housing that is required to provide accommodation of an agreed minimum standard and above for a popu-

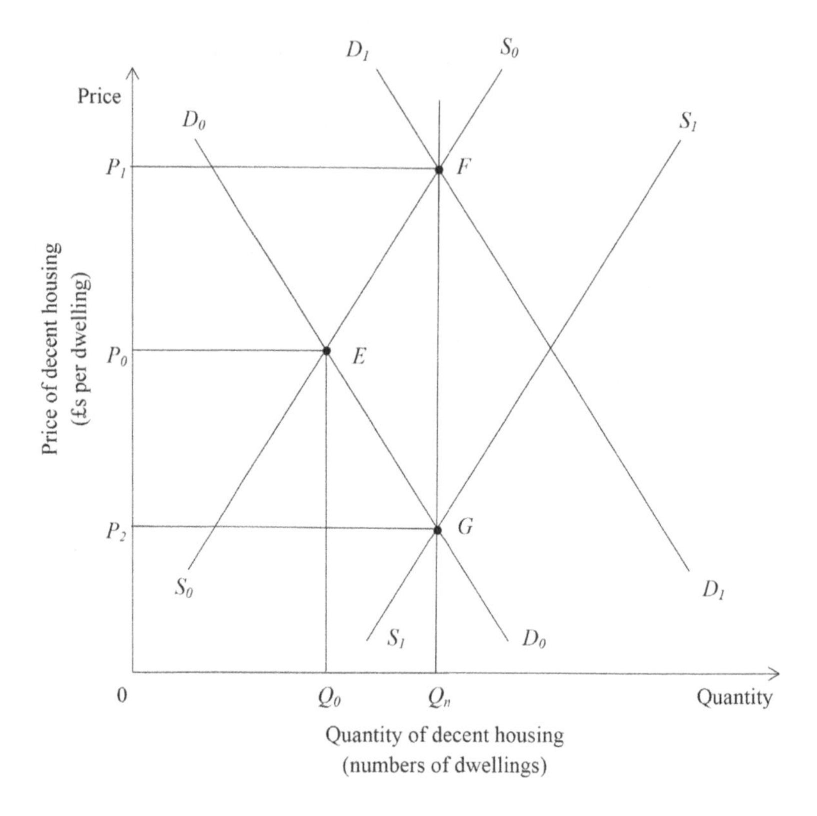

Source: Adapted from Oxley (2004).

Figure 19.2 *Housing need*

lation given its size, household composition, age distribution and so on, without taking into account the individual household's ability to pay for the housing assigned to it' (Robinson, 1979, pp. 56–57). This is illustrated in Figure 19.2, which shows points at which market failure occurs, and where the appropriate rectification ought to take place.

In a given area, say a metropolis, a city or a town, the demand and supply curves for decent housing (D_0D_0 and S_0S_0) intersect at E to give an equilibrium quantity of Q_0. Q_n is the total quantity of decent housing needed. Normally, this quantity is determined by performing a housing need and demand assessment (HNDA) for the given area. However, at the market equilibrium of E, there is an unmet housing need equal to the difference of quantity Q_n to Q_0. How would a government respond to this unmet housing need? In most cases, a combination of steps below is taken:

1. Increase the demand for decent housing (through housing allowances and housing vouchers) so that a new demand curve of D_1D_1, with a higher price of P_1 and quantity Q_n.
2. Increase the supply of decent housing (for example, making land available for house building, or providing subsidies or tax concessions to suppliers of decent housing) that will

move the supply curve to S_1S_1 with the new equilibrium fetching the price lower price of P_2 and quantity Q_n.
3. Government direct intervention to supply quantity Q_oQ_n, with pricing based on other household indicators (for example, at an affordable rate of three times median multiple, or housing cost of 30 per cent).

Formulating the right policies and strategies depends significantly on the ability to differentiate between welfare needs and market failures (Oxley, 2004). The former requires more direct government involvement in addressing housing needs of the right target groups (redistributive measures), while the latter requires the government to provide an environment where market failures (for example, a situation of unsold stock, mismatch of supply and effective demand) can be corrected (Ismail et al., 2019). The in-depth analysis on the supply of, and demand for, both sectors must be contextualized within the spatial dynamics of households. The spatial dynamics of housing refers to the analysis of households within an ecosystem of the home (first place), workplace (second place) and social places (third place) (Oldenberg, 1999). The connectivity of these three places forms the functional boundary of the households. Therefore, as opposed to the conventional practice of analysing and planning housing within a local council's administrative boundaries, a more meaningful analysis would be to map the functional boundary of households (which may surpass the town or city's administrative boundaries).

Housing Indicators as Measurements of Government Policies

The outcome of government policies can be measured through housing indicators. Housing indicators are important, since they are an instrument that can both monitor and model the housing sector. As shown in Figure 19.3, the rent-to-income ratio, house price-to-income ratio, and house price index, are important signals to stakeholders in different sectors of the market. For example, to a household in the social renting sector, the rent-to-income ratio in the private renting sector is a clear signal (and motivation) to migrate from social renting to private renting. Moreover, for the government, this rent-to-income ratio is a signal for market efficiency (where private renting is affordable), so that it provides a tangible basis for making exit policies on social housing (distributive welfare). Therefore, such indicators are both critical and instructive in the continuous decision-making of policy trajectories.

An institutional conceptualization of housing market efficiency is defined in terms of the ability of the market institution to adapt its structure and to provide the outcomes that the economy requires (Arvanitidis, 2015). This is different from neoclassical perspectives, where efficiency is measured in terms of marginal utilities or when the price system creates negative externalities, and government should only interfere when an 'economic problem' becomes a 'social problem' (Heilbroner, 1995). The contrast between the United States (US) housing policy and those of Western European countries remains instructive. There is a high level of social housing stock in the Netherlands (33 per cent), France (17 per cent), Denmark (20 per cent), and the United Kingdom (18 per cent) (Housing Europe, 2015). In the US, it ranges between 1 and 4 per cent, depending on the city (Metcalf, 2018).

A resilient and effective housing market should be flexible and effective for society to make informed choices on whether to own or rent a home. Renting is a viable option, and generally should be promoted alongside, not in direct competition to, home ownership. However, in

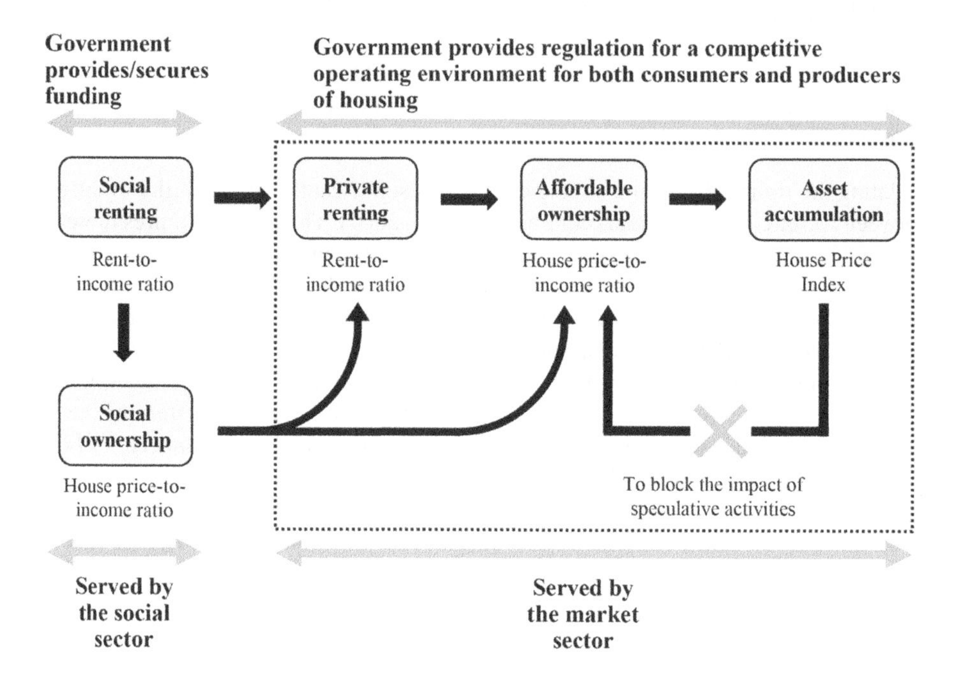

Source: Adapted from Ismail et al. (2019).

Figure 19.3 Housing indicators as instruments of policies

either case, the prerequisite is to ensure there is an adequate supply of houses that reflects genuine affordability.

DEVELOPING AN EFFICACIOUS SUPPLY SECTOR: A PREREQUISITE FOR COMPETITIVE HOUSE PRICES

How does one influence the supply of houses to become more elastic, so that an increase in demand will not lead to drastic increments in house prices? Increasing the elasticity of supply depends on the responsiveness of the housing supply sector to provide houses in a timely manner. This, turn, is dependent upon the supply of land, government and planning restrictions, and the time taken for the housing supply sector to build new houses and to refurbish existing stock. Land use regulations and buildable land (Bramley, 2007; Glaeser and Ward, 2009; Gurran and Whitehead, 2011) are important supply-side prerequisites, but equally critical is the capacity of firms to deliver housing projects that exhibit the appropriate levels of affordability (Gibb, 2011; Hansson, 2017).

Cities and towns experience different forms of urbanization, growth, physical planning strategies and economic priorities, as well as having a different composition of firms in the local housing supply sector. One of the main concerns in supply-side initiatives is the ability of the local housing supply sector to deliver good-quality housing, at a price accessible even to people in the lower income groups. In other words, the more efficient the construction industry is in providing affordable homes in a timely manner, the more responsive supply is to the overall socio-economic profile of society. This is against the backdrop of the availability of land and an enabling spatial planning environment.

Any demand-based policy intervention that disregards the efficacy of the construction industry to produce houses at competitive prices first will only lend support to the development of non-competitive housing products. It is also important to reiterate that competitive prices for houses induce lower bid prices for land, since it is a derived demand. Therefore, a reduction in the general costs of houses due to the efficiency of the construction industry will, *ceteris paribus*, bring about a reduction in the bid price for land and, in turn, play a significant role in decreasing the prices of homes.

A Responsive Supply Sector: Between Finland and Pakistan?

The housing supply sector's responsiveness can be determined by considering the range of housing units transacted in different price brackets, compared to the distribution of the nations' household incomes. This is best depicted in the market median multiple and the down-market penetration ratios.

Down-market penetration focuses on the affordability of the lowest-price new house provided by the private housing development sector, without subsidies from the government (Angel et al., 1993). This indicator is rooted in the filtering model of housing consumption, in which low-income households are generally restricted to informal housing or 'filtered-down' older formal sector housing. The down-market penetration ratio captures the phenomenon that, in certain formal markets, the supply of housing is generally targeted at high-income groups but ignores low-income consumers. However, in other markets, private sector providers are able to produce housing within the affordability thresholds of lower-income households with no, or limited, public incentives, subsidies or other assistance. In general, the down-market penetration ratio is compared with the median multiple to indicate the extent to which the market supplies to households which fall below the median income.

Figures 19.4a and 19.4b depict the contrast between Finland and Pakistan. In Finland, housing distribution caters for the population's need and there is effective demand at almost all income levels. This is due to the successful public–private partnership policy in Finland, in which the government prioritizes housing for all residents. However, in Pakistan, low-cost housing makes up 1 per cent of total housing provision, but caters to 68 per cent of total households with less than PKR30 000 monthly income. On the other hand, 56 per cent of the property available caters to the upper 12 per cent of households with a monthly income which exceeds PKR100 000. This suggest that there is a significant mismatch between supply and demand in the case of Pakistan. The supply of houses is skewed significantly to the more affluent segments of society, rather than ensuring sufficient supply to society as a whole, as in the case of Finland.

Figure 19.5 illustrates a typical income distribution of a population in a country, where the Y-axis is the percentage of the households plotted against the X-axis of monthly household incomes. Household distribution is demarcated by the B40, M40 and T20 (bottom 40 per cent, middle 40 per cent and top 20 per cent of income distribution) classifications (World Bank, 2016). Further demarcation is placed at the median household at the 5th income decile (the midpoint). This midpoint is important, since it has been utilized as a measure for well-being or

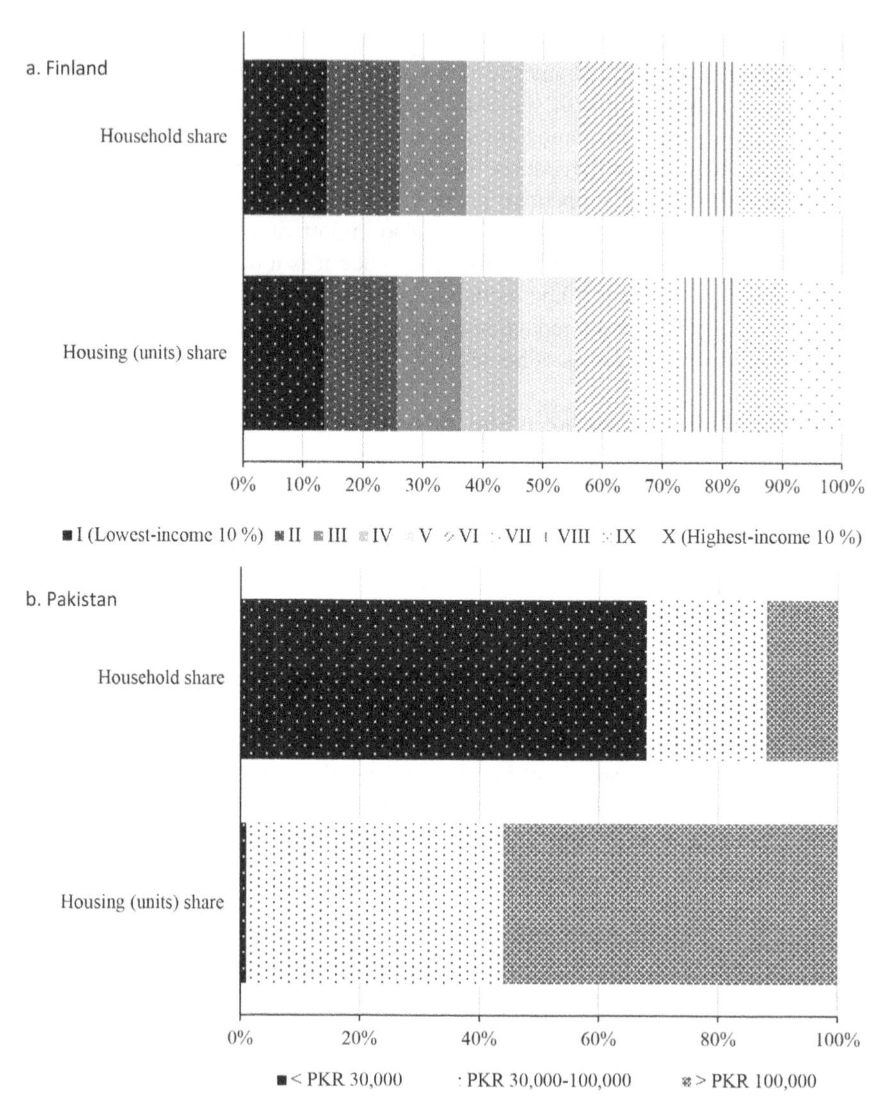

Source: CEIC (n.d.) and author calculations.

Figure 19.4 Household shares and housing units

the general standard of living for a household, and what is taken to determine relative poverty is pegged to it (Ravallion and Sen, 1996).

The bottom part of Figure 19.5 shows the demarcation of the supply sector in the provision of homes. It depicts low-income earners, which encompass up to 40 per cent of the household distribution, as being in the social sector and require government assistance for the provision of shelter (the area on the left of the vertical line). This can be a high number of households in many countries. Therefore, governments have the option of increasing the down-market pen-

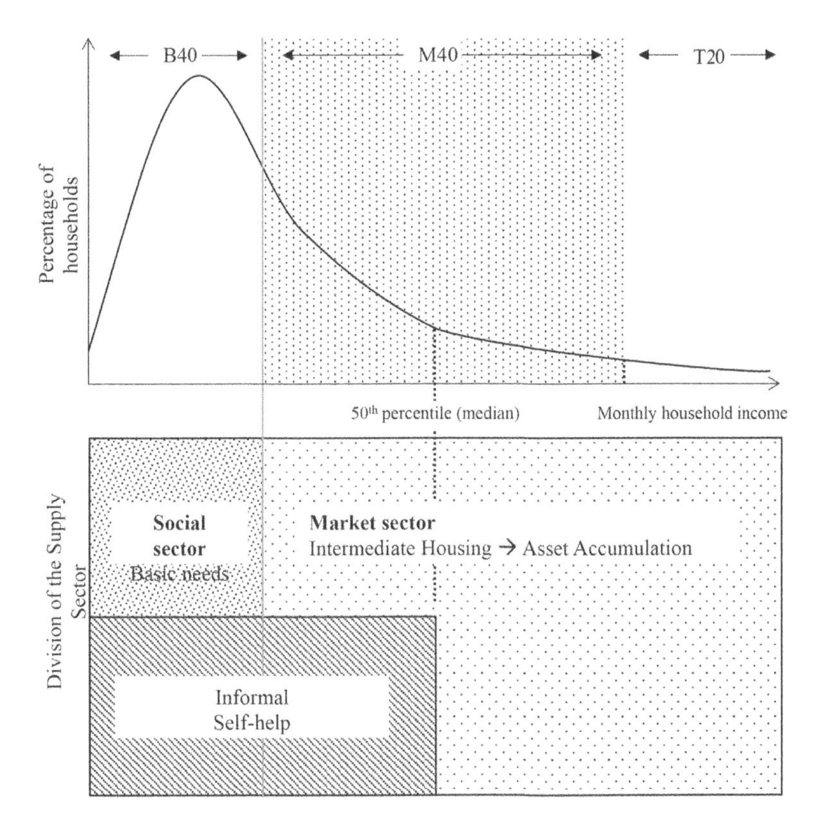

Figure 19.5 *Household distribution and the division of the supply sector*

etration ratio (or shifting the vertical line more to the left) where the private sector provision reaches the lower income distribution (B10 or B20).

The move to enhance the efficiency of the construction industry (Ofori, 1984, 1991) is one of the key factors for making supply more elastic as well as improving the down-market penetration ratio. However, this is not a prerequisite specifically for housing, but also for the general well-being of the built environment: 'If a serious attempt is to be made to improve the urban situation the capacity and efficiency of the construction industry will have to improve significantly' (Government of Ghana, One Year Development Plan, 1971, p. 141, cited in Ofori, 1989).

The Construction Industry and the National Construction Business Environment

The production of goods and services in the construction industry is predominantly based on the contracting system, whereby contracts form the basis of alliances for companies that come together during the construction of the built items. In many countries, the method of production predominantly follows the traditional general contracting procurement route, whereby there is a clear separation of the design and construction phases. This fragmentation of design and construction reduces efficiency and raises the costs of production. A result of fragmentation of the construction process has been the low level of investment in research and development within firms in the construction industry of various countries (Dulaimi et al., 2002).

A critical factor is to recognize that construction projects exist within the national construction business environment. The national construction business system is the distinctive configuration of actors and institutions within which the construction industry in each country has evolved historically (Winch and Campagnac, 1995). The actors are different types of supply-side firms (designers, contractors, specialist trades contractors, and so on) and clients (developer firms, government agencies and utility companies), while the institutions are the bodies that regulate or otherwise control the practices of the industry and its interactions with its clients. These may be state bodies, bodies created by the industry for self-regulation purposes, or other forms of regulatory and control bodies. These features (for example, the role played by specific institutions such as government, or by a class of firms such as contractors or housing developers) serve to distinguish one national construction business sector from another.

National construction business systems provide the institutional context in which construction procurement takes place. Clients operating within a specific construction business system then have a suite of procurement routes from which the most appropriate is selected for the project they are embarking upon. The operation of a procurement route on a project is then influenced by the values and behaviours of those working on the project. The actors, because of tradition and their education, and through their experience, approach the project with certain perceptions and practices.

There are different approaches to attempts to explain why problems occur when construction products and services are delivered to clients. Some researchers have focused upon systems and processes, including the use of flow-charts, critical-path analysis and process mappings (Davis and Newstrom, 1989; Curtis et al., 1991); while others have represented authority and responsibility relationships such as linear responsibility analysis and organizational structures (Masterman, 2001; Franks, 1998; Loosemore, 1999; Bennet and Grice, 1990; Hughes and Murdoch, 2001).

Research in project management has been concerned mainly with process and technique (Bennet, 1991; Morris, 1994; Walker, 1996). The latter aims at increasing efficiency, the former is concerned with developing an understanding of why construction projects are organized with certain configurations and how best to design them. The major works adopt a systems approach with the focus on how the parts are interrelated and interact in seeking clients' objectives (Walker and Chau, 1999). The focus is to determine the most appropriate project organization structure for a specific project, and management functions which should be carried out to realize the objectives of the client are understood (Morris, 1994; Walker, 1996), but these ideas do not incorporate the economics of structures and how the functions should be provided. Morris (1994) also observed that procedural approaches cannot overcome the problems of externalities, institutions and strategic issues.

Organizational arrangements among project participants have come to be recognized as one of the most important factors affecting the performance of the construction industry (Miller and Lessard, 2000; Winch, 2001). However, these developments have not been accompanied by more sophisticated theoretical approaches. These concerns are not new, and had been raised by earlier scholars such as Turin (1975) who opined that: 'construction practices lend themselves to easy generalization and mythical half-truths which, under the cloak of ... common sense, disguise serious and sometimes fundamental misconceptions' (Turin, 1975, p. xi; see also Ofori, 1984).

This makes it imperative to explore other forms of analysis outside project management models which have been the predominant means of understanding the functioning of the construction industry. Most economic analysis of the industry has been concerned primarily by the outcomes of processes, that is, measuring its efficiency in terms of output production. Not much has been said about the industry's production process and how this is crucial within any economic analysis of the industry. This is where the theory of transaction costs economics (TCE) may provide a more comprehensive analytical lens towards understanding the organization of the production processes at the project level. The project-level analysis can subsequently be expanded and utilized for the governance of the construction industry at the national level.

Various studies (Winch, 1989, 2001; Winch and Campagnac, 1995; Chang and Ive, 2000; Gruneberg and Ive, 2000) have attempted to explain the governance of construction projects by adopting a TCE approach, postulating that, firstly, the parties to the contract have limited knowledge of each other and behave opportunistically, that is, the behavioural assumptions within the TCE framework suggest that the identities or 'knowledge' of parties in medium- or long-term construction contracts are important (as opposed to spot contracts where knowledge of contracting parties does not matter) to prevent opportunistic behaviour (Gruneberg and Ive, 2000)

Second is the context in which procurement takes place, that is, within the TCE's environmental contingencies of high uncertainty, low frequency and low asset specificity. The construction industry is characterized by low transaction frequencies (uneven work orders) and a high level of project uncertainty (Winch, 1989). In the competitive construction industries of developed countries, asset specificity is low due to the high levels of technical diffusion whereby the specific investment of one firm does not create an added advantage over another (Winch, 1995). This is quite different from most situations in emergent economies, where in an uncompetitive and uneven sector (low technical diffusion), the capacity of individual or small coalition of firms to invest in specific technologies (for example, the industrial building system in Malaysia) can create a significant advantage.

The move to explain the workings of the construction project from the TCE lens stems from the nature of the contracting system adopted in the construction industry. It makes the selection of partners in the 'production cluster' or 'temporary coalitions' matter. Therefore, the interfirm relations are as important, if not crucial, to analyse. Eccles (1981) regards the construction project as a network of firms working together for the purpose of the project. Cherns and Bryant (1984) also refer to the project organization as a temporary multi-organization whose articles of association are the contracts. Pryke (2004) suggested that the construction industry appears to be evolving in procurement and management systems that appear to oscillate between the market and hierarchy models. Work packages are contracted out through 'market' contracts, but subsequently managed in a hierarchical context within the environment of the temporary project coalition through a nexus of contracts (Stinchcombe and Heimer, 1985; Pryke, 2004). These contracts act as governance structures (Stinchcombe and Heimer, 1985) which appear to instil workable order between contracting parties.

Structuring the Alliance of Firms: Procurement Routes as Governance Structures

In the current procurement approaches, there is a diverse array of methods and mechanisms for acquiring the building. Figure 19.6 highlights five functional forms to distinguish how procurement routes differ, based on: (1) functional forms (Franks, 1998; Skitmore and Marsden, 1988; Masterman 2001; Winch, 2001; Murdoch and Hughes, 2008); and (2) governance structures ('market' – competitive tendering, or 'hierarchy' – in-house operations) available for structuring the temporary clustering of the firms within the project duration, or post-project period (especially for the maintenance of social housing).

As Williamson (1985, 1996) suggests, the organizational construct of the entities is not assumed as given, but rather is derived, based on the properties of the transaction, especially when viewed within the behavioural assumptions of bounded rationality and opportunism:

1. Asset specificity: are parties to the contract expected to make specific investments (say on innovation) in the housing project; and if so, how are residual claimants to the profits apportioned?
2. Uncertainty: can the *ex post* rewards and risks of the project be quantified *ex ante*, or before parties enter the contract?
3. Frequency: if this transaction is frequently transacted between parties, can one party (say, the client) internalize this function?

Consequently, the organizational imperative is to ensure that the organization of transactions economizes on bounded rationality while simultaneously safeguarding against the hazards of opportunism. A transaction is made the micro-analytic unit of analysis; a transaction occurs when a good or service is transferred across a technological separable interface (Williamson, 1985). In construction, it refers to different and separable tasks or services attached to a contractor or a designer (Winch, 1989, 2001; Gruneberg and Ive, 2000). Therefore, this is a dimensionalizing exercise of transactions (which differ in their attributes) to the appropriate governance structures (which differ in costs and competencies) in a transaction-economizing manner. Viewed in this way, procurement routes can be 'a complex suite of relational contracts which charter the project coalition for a particular project' (Winch, 1996, p. 15).

The way a transaction is organized depends on its characteristics. For example, if one kind of transaction occurs frequently in similar ways, people develop routines to manage it effectively. If a transaction is unusual, then parties may need to bargain about its terms, which raises the costs of carrying out the transaction. This is the basic notion of transaction cost economics; it is the properties of transactions that determine the efficient governance structures. In short, the transaction is the basic unit of analysis; but governance is an effort to craft 'order', and thereby mitigate conflict and realize mutual gains (Commons', 1968, triple variables of 'conflict, mutuality and order').

Some conceptual underpinnings to the selection of the appropriate governance structures are now considered; they are explained through the five functional forms.

Methods of funding

Who will finance the facility? The funding for social housing in developing countries is generally derived from government expenditure. In more developed economies, particularly those in Western Europe, there are instances where public–private initiatives finance the provision of social housing (UN-Habitat, 2009). In lieu of this, the project is conceived as both a construc-

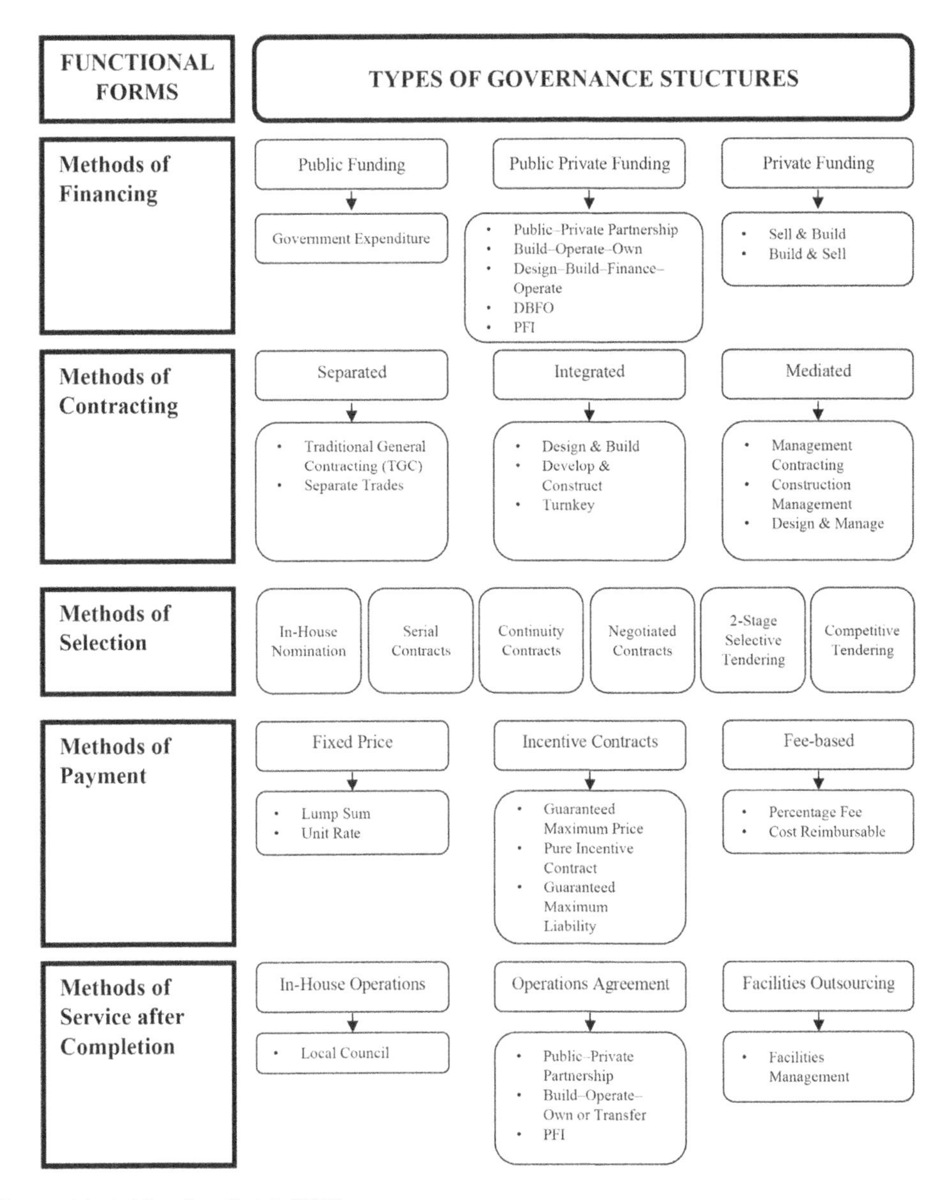

Source: Adapted from Ismail et al. (2019).

Figure 19.6 *The functional forms and the types of governance structures for housing projects*

tion and maintenance contract due to the need to keep maintenance costs low. For example, in the Philippines and Indonesia, the financial mortgage systems are generally inaccessible for lower-income groups. Under these circumstances, private developers have developed innovative financing schemes that establish links between the non-bankable communities and the financial sector (Khazanah Research Institute, 2015). In these contexts, the mode of governance structure includes the financing as well as the service after project completion. Within these segments of the national construction business system, the clustering of financial firms with the conventional construction sector firms is evident (Salim, 2011; NEDO, 2017).

Methods of contracting

What forms of contracts bind the client with the design and construction stages of the constructed facility? The contract form typically follows the type of procurement routes currently available in a given country. These methods of contracting can be classified into three main categories: separated, integrated and mediated.

In most emerging economies, the level of technological absorption in the construction industry is low (Ofori, 1985, 1991). If a project in this national construction business system requires innovative solutions, then these transactions are best situated in an integrated mode where team interdependences will be high and the separate measurement of performance of each party will be difficult to measure (Alchian and Demsetz, 1972). Uncertainties regarding costs, prices and quantities tend to lead to transactions being parked under one entity, rather than the separation of contracts between design and construction.

On the other hand, in most developed economies where technology resides in specialist firms, then the mediated or even separated route will provide a good workable order for all parties in the temporary project coalition. This is because the associated rewards and risks can be ascertained *ex ante* and negotiated to reflect the appropriate level of residual or profit claimants.

Methods of selection

How shall the client choose the contractors for the job? The methods available range from competitive tendering, negotiated contracts, continuity contracts and serial contracts, to in-house nomination. The selection of parties to the coalition is of utmost importance, considering that parties to the contracts have a multitude of information and technical asymmetries. This is within the context of an industry which is plagued by adversarial relationships and social norms of claims and counter-claims in project disputes (Murdoch and Hughes, 2008). *Ex ante* considerations in contractual relationships dictate the importance of partner selection as a starting point to minimize contractual hazards. Sometimes it is more prudent to engage with the same group of firms for different projects, due to the familiarity already established. This is referred to as relational contracting, where the relationship survives longer than the contract duration (Eccles, 1981). This infers that competitive tendering (or price signals) might not be appropriate when the identity of partners matters to minimize opportunistic behaviour. It is important to note that firms frequently form alliances with similar partners not due to monopolistic considerations, but in order to attain economic efficiency (Coase, 1993 [1937]; Williamson, 1985, 2002).

Methods of payment

How shall the client pay the contractor: a price estimated at the outset or the actual costs incurred by the contractor? There are, largely, three types of method of payment: fixed-price contracts, incentive contracts and fee-based contracts. There is a wide variety of such contracts, but what unites them is the attempt to provide positive incentives within the contract to motivate performance through gain-sharing between parties.

Methods of service after completion

Does the housing complex need maintenance services after the completion of the facility? This is critical in stratified buildings where other contractors may also be involved in the maintenance of the buildings after the duration of the contract.

Trade-offs

There are trade-offs between the types of governance selected. Market contracting is formidable in presenting high-powered incentives – for example, in rewarding early completion of projects – but not necessarily conducive in providing adaptable solutions when there are frictions. Hierarchy reverses this relationship: it is unable to present high-powered incentives, but it is highly stable in uncertain conditions due to its discretionary propensities. Hybrid structures preserve the autonomy of firms within the coalition but provide transaction-specific safeguards to diffuse opportunistic behaviour by contracting parties (Stinchcombe and Heimer, 1985; Menard, 2004).

CONCLUDING REMARKS AND SUGGESTIONS FOR FURTHER RESEARCH

This chapter has explored how affordability could be increased through supply-side measures. This is not to devalue the importance of demand-side measures, but rather to illustrate how a supply-side approach can foster the creation of a relatively elastic supply curve. Therefore, the sequencing of policy prescriptions (between demand and supply measures) is required for the intervention to achieve its intended aims. Housing indicators are utilized as tools to monitor the performance of the housing market. The importance of the construction industry in contributing to the housing agenda is also discussed. The premise is that improvements in the institutional and governance structures are needed at the project level to increase the responsiveness of the construction industry to both housing need and demand. Further research can be directed towards providing a better understanding of how distinct temporary coalitions adopt different governance structures to attain the objectives of their projects. For example, the rise of co-operatives of homeowners that partner directly with building contractors in some countries will likely see the emergence of new forms of governance structures.

Despite the inherent differences between the levels of economic and political development, as well as the institutional context of countries, housing affordability is usually an outcome of policy interventions and the attendant market responses. The underlying assumptions of rational choice theory – utility-maximizing households and profit-maximizing firms – underpins the realities of market responses. Although economic justifications are important, they still tend to be focused on marginal utility and not the collective or common good. The conceptions of possible policy interventions will be quite limited if affordability considerations and concerns are not discussed within the wider agenda of social development and furthering democratic practices.

The conceptual foundations of the different approaches governments adopt to address the problems of housing affordability are of equal importance due to the different institutional arrangements they entail. For example, a liberal interventionist approach in housing maintains that the state can usefully take action to compensate for the inefficiencies of the market, and directly involve itself in the building of houses. On the other hand, a public choice perspective proposes that states may fail to improve on market outcomes if they have insufficient regard for the strengths of markets and the weaknesses of government; hence, some governments will attempt to create pseudo-market entities in order to mimic the advantages of the market. In both instances, the coordination of the government machinery will be based on administrative competence, and not high-powered price signals. Further research may investigate the types of tools and techniques (for example, the use of data science, big data, and the like) that governments can utilize to supplant price signals as the main instrument of coordinating resources.

Finally, it should be borne in mind that one of the major limitations in developing and implementing effective policies is their dependence on decision-making within the political process. All policies are generally subordinate to these considerations.

REFERENCES

Alchian, A.A. and Demsetz, H. (1972) Production, information costs, and economic organization. *American Economic Review*, 62(5), 777–795.
Angel, S. (2000) *Housing Policy Matters: A Global Analysis*. New York: Oxford University Press.
Angel, S., Mayo, S.K. and Stephens, W.L. (1993). The Housing Indicators Program: a report on progress and plans for the future. *Netherlands Journal of Housing and the Built Environment*, 8(1), 13–48.
Angel, S., Parent, J., Civco, D.L., Blei, A. and Potere, D. (2011) The dimensions of global urban expansion: estimates and projections for all countries, 2000–2050. *Progress in Planning*, 75(2), 53–107.
Apgar, William C., Jr (1990) Which housing policy is best? *Housing Policy Debate*, 1(1), 1–32.
Arvanitidis, P.A. (2015) *The Economics of Urban Property Markets: An Institutional Economics Analysis*. London: Routledge.
Asian Coalition of Housing Rights (ACHR) (2018) A Report on Housing Policies in the Asia Region for UCLG GOLD Global Report on Housing Policy. www.achr.net.
Ball, L. (1994) Credible disinflation with staggered price-setting. *American Economic* Review, 84(1), 282–289.
Bank Negara Malaysia (BNM) (2017) *Annual Report 2016*. Kuala Lumpur: Bank Negara Malaysia.
Belsky, E.S., Goodman, J. and Drew, R.B. (2005) Measuring the nation's rental housing affordability problems. Joint Center for Housing Studies, Graduate School of Design and John F. Kennedy School of Government, Harvard University.
Bennett, J. (1991) *International Construction Project Management, General Theory and Practice*. Oxford: Butterworth-Heinemann.
Bennet, J. and Grice, A. (1990) Procurement systems for buildings. In Brandon, P.S. (ed.), *Quantity Surveying Techniques: New Directions*. Oxford: BSP Professional Books, pp. 243–261.
Bramley, G. (2007). The sudden rediscovery of housing supply as a key policy challenge. *Housing Studies*, 22(2), 221–241.
Buckley, R. and Kalarickal, J. (2005) Housing policy in developing countries: conjectures and refutations. *World Bank Research Observer*, 20(2), 233–257.
CEIC (n.d.) CEIC Database.
Chang, C.Y. and Ive, G. J. (2000) A comparison of two ways of applying a transaction cost approach: the case of construction procurement routes. Bartlett Research Paper No. 13, University College London.
Chen, Y-L. and Shin, H.B. (eds) (2019) *Neoliberal Urbanism, Contested Cities and Housing in Asia*. The Contemporary City Series. New York: Palgrave Macmillan.
Cherns, A.B. and Bryant, D.T. (1984) Studying the client's role in construction management. *Construction Management and Economics*, 2(2), 177–184.
Coase, R.H. (1993) [1937]. *The Nature of the Firm: Origin, Evolution and Development*. New York: Oxford University Press.
Commons, J. (1968) *Legal Foundations of Capitalism*. Madison, WI: University of Wisconsin Press.
Curtis, B., Ward, S. and Chapman, C. (1991) *Roles, Responsibilities and Risks in Management Contracting*. London: Construction Industry Research and Information Association (CIRIA).
Davis, K. and Newstrom, J.W. (1989) *Human Behavior at Work*. Singapore: McGraw Hill.

Demographia (2015) 11th Annual Demographia International Housing Affordability Survey: 2015. HTTP://www.demographia.com/dhi.pdf.
Demographia (2018) 14th Annual Demographia International Housing Affordability Survey: 2018. https://www.demographia.com/dhi.pdf.
Department of Communities and Local Government (DCLG) (2017) *Fixing Our Broken Housing Market*. London: HMSO.
De Soto, H. (2000) *The Mystery of Capital: Why Capitalism Triumphs in the West and Fails Everywhere Else*. New York: Basic Books.
Doling, J. and Elsinga M. (2013) *Demographic Change and Housing Wealth: Home-owners, Pensions and Asset-based Welfare in Europe*. Dordrecht: Springer Netherlands.
Duca, J. and Muellbauer, J. (2013) Tobin LIVES: integrating evolving credit market architecture into flow of funds based macro-models. In Winkler, B., van Riet, A. and Bull, P. (eds), *A Flow-of-Funds Perspective on the Financial Crisis*, Vol. 2. Frankfurt am Main: Palgrave Macmillan, pp. 11–39.
Dulaimi, M.F., Ling, F.Y.Y, Ofori, G. and De Silva, N. (2002) Enhancing integration and innovation in construction. *Building Research and Information*, 30(4), 237–247.
Eccles, R. (1981) The quasifirm in the construction industry. *Journal of Economic Behavior and Organizations*, 2, 335–357.
Fay, M. (2004) *The Urban Poor in Latin America*. Washington, DC: World Bank.
Franks, J. (1998) *Building Procurement Systems*, 3rd edition. Chartered Institute of Building. Harlow: Longman.
Gabriel, S.A., Manik, J. and Vikas, S. (2003) Computational experience with a large-scale, multi-period, spatial equilibrium model of the North American natural gas system. *Networks and Spatial Economics*, 3(2), 97–122.
Georgieva, V. (2017). The effect of foreign investment on housing prices in major Canadian cities. Masters thesis, Department of Economics, University of Ottawa.
Gibb, K. (2011). Delivering new affordable housing in the age of austerity: housing policy in Scotland. *International Journal of Housing Markets and Analysis*, 4(4), 357–368.
Glaeser, E.L. and Ward, B.A. (2009). The causes and consequences of land use regulation: evidence from Greater Boston. *Journal of Urban Economics*, 65(3), 265–278.
Glaser, E. and Gyourko, J. (2018) The economic implications of housing supply. *Journal of Economic Perspectives*, 32(1), 3–30.
Global Housing Watch (2020). *Global Real House Price Index Data*. International Monetary Fund.
Gruneberg, S. and Ive, G.J. (2000). *The Economics of the Modern Construction Firm*. Basingstoke and London: Springer.
Gulyani, S. and Talukdar, D. (2008) Slum real estate: the low-quality high-price puzzle in Nairobi's slum rental market and its implications for theory and practice. *World Development*, 36(10), 1916–1937.
Gurran, N. and Whitehead, C. (2011) Planning and affordable housing in Australia and the UK: a comparative perspective. *Housing Studies*, 26(7–8), 1193–1214.
Gyourko, J. (1991) Impact fees, exclusionary zoning, and the density of new development. *Journal of Urban Economics*, 30(2), 242–256.
Haas, P.M., Makarewicz, C., Benedict, A., Sanchez, T. and Dawkins, C. (2006) *Housing and Transportation Cost Trade-offs and Burdens of Working Households in 28 Metros*. Chicago, IL: Center for Neighbourhood Technology.
Hansson, A.G. (2017) City strategies for affordable housing: the approaches of Berlin, Hamburg, Stockholm, and Gothenburg. *International Journal of Housing Policy*, 19(1), 95–119.
Harvey, J. and Jowsey, E. (2004) *Urban Land Economics* (6th edition). London: Palgrave Macmillan.
Heilbroner, R. (1995) *The Crisis of Vision in Modern Economic Thought* (with William S. Milberg). New York: Cambridge University Press.
Housing Europe (2015) *The State of Housing in the EU 2015*. Brussels.
Hughes, W.P. and Murdoch, J.R. (2001) *Roles in Construction Projects: Analysis and Terminology*. Birmingham: Construction Industry Publications.
Ismail, S., Wai, C.C.W., Hamid, H.A., Mustapha, N.F., Ho, G., et al. (2019) *Rethinking Housing: between State, Market and Society: A Special Report for the Formulation of the National Housing Policy (2018–2025), Malaysia*. Kuala Lumpur: Khazanah Research Institute.

Ive, G.J. and Gruneberg, S.L. (2000) *The Economics of the Modern Construction Sector*. Basingstoke: Macmillan Press.
Jaffee, D.M. (2008) The US subprime mortgage crisis: issues raised and lessons learned. Commission on Growth and Development Working Paper No. 28. Washington, DC: World Bank.
Johnstone, M.A. (1976) Housing policies in third world cities: a review. *Australian Geographical Studies*, 14, 133–166.
Jones, C. and Watkins, C. (2009) *Housing Markets and Planning Policy*. Oxford: Wiley-Blackwell.
Katz, B. and Turner, M.A. (2003) *Rethinking local affordable housing strategies: lessons from 70 years of policy and practice*. The Brookings Institute, Centre on Urban and Metropolitan Policy.
Khazanah Research Institute (2015) *Making Housing Affordable*. Kuala Lumpur: Khazanah Research Institute.
Loosemore, M. (1999) Bargaining tactics in construction disputes. *Construction Management Economics*, 17(2), 177–188.
Maclennan, D. and Miao, J. (2017) Housing and capital in the 21st Century. *Housing, Theory and Society*, 34(2), 127–145.
Marom, N. and Carmon, N. (2015) Affordable housing plans in London and New York: between marketplace and social mix. *Housing Studies*, 30(7), 993–1015.
Masterman, J.W.E. (2001) *Introduction to Building Procurement System* (2nd edition). London: E. & F.N. Spon.
Mattingly, K. and Morrissey, J. (2014) Housing and transport expenditure: socio-spatial indicators of affordability in Auckland. *Cities*, 38, 69–83.
Menard, C. (2004) The economics of hybrid organizations. *Journal of Institutional and Theoretical Economics*, 160(2), 345–376.
Metcalf, G. (2018) Sandcastles before the tide? Affordable housing in expensive cities. *Journal of Economic Perspectives*, 32(1), 59–80.
Miller, R. and Lessard, D. (2000) *The Strategic Management of Large Engineering Projects: Shaping Institutions, Risks and Governance*. Cambridge, MA: MIT Press.
Morris, P.W.G. (1994) *The Management of Projects*. London: Thomas Telford.
Murdoch, J. and Hughes, W. (2008) *Construction Contracts: Law and Management*. London: Taylor & Francis.
National Economic and Development Authority (NEDO) (2017) Philippines Development Plan (PDP) 2017–2022.
ONS (2016) *Housing Summary Measures Analysis 2015*. Newport: UK Office for National Statistics.
Ofori, G. (1984) Improving the construction industry in declining developing countries. *Construction Management and Economics*, 2(2), 127–132.
Ofori, G. (1985) Managing construction industry development. *Construction Management and Economics*, 3(1), 33–42.
Ofori, G. (1989) Housing in Ghana: the case for a central executive agency. *Habitat International*, 13(1), 5–17.
Ofori, G. (1991) Programmes for improving the performances of contracting firms in developing countries: a review of approaches and appropriate options. *Construction Management and Economics*, 9(1), 19–38.
Oldenburg, R. (1999) *The Great Good Place: Cafes, Coffee Shops, Bookstores, Bars, Hair Salons, and other Hangouts at the Heart of a Community*. Boston, MA: Da Capo Press.
O'Sullivan, E. (2003) Bringing a perspective of transformative learning to globalized consumption. *International Journal of Consumer Studies*, 27(4), 225–350.
Oxley, M. (2004) *Economics, Planning and Housing*. Basingstoke, UK and New York, USA: Red Globe Press.
Pryke, S. (2004) Analysing construction project coalitions: exploring the application of social network analysis. *Construction Management and Economics*, 22(8), 787–797.
Ramirez, R. (1978) The housing queue. Paper presented at a special seminar on People's Participation and Government Inputs in Low Cost Housing, Bouwcentrum International, Rotterdam.
Ravallion, M. and Sen, B. (1996) When method matters: monitoring poverty in Bangladesh. *Economic Development and Cultural Change*, 44(4), 761–792.
Roback, J. (1982) Wages, rents, and the quality of life. *Journal of Political Economy*, 90(6), 1257–1278.

Robinson, R. (1979) *Housing Economics and Public Policy*. Basingstoke and London: Springer.
Rosen, S. (1979) Wage-based indexes of urban quality of life. In Mieszkowski, P. and Straszhaim M. (eds), *Current Issues in Urban Economics*, Vol. 1. Baltimore, MD: Johns Hopkins Press, pp. 77–104.
Ruprah, I. (2010) The transparency, incidence, and targeting efficiency of housing programs in Latin America. Working Paper OVE/WP-04/10. Washington, DC: Inter-American Development Bank.
Salim, W. (2011) *Indonesia: State of Low-Income Housing Policy and Finance. Legislative, Institutional, Planning, and Regulatory Frameworks*. For the World Bank Office of Jakarta.
Skitmore, R.M. and Marsden, D.E. (1988) Which procurement system? Towards a universal procurement selection technique. *Construction Management and Economics*, 6(1), 71–89.
Sorek, C. Begachina, K. and Charlwood, L. (2015) *Housing Review 2015; Affordability, Sustainability and Livability*. Bratislava: Habitat for Humanity
Stinchcombe, A. and Heimer, C. (1985) *Organization Theory and Project Management: Administering Uncertainty in Norwegian Offshore Oil*. Oslo: Norwegian University Press.
Stone, M.E. (2006) What is housing affordability? The case for the residual income approach. *Housing Policy Debate*, 17(1), 151–184.
Toussaint, J. (2011) Housing assets as a potential solution for financial hardship: households' mental accounts of housing wealth in three European countries. *Housing Theory and Society*, 28(4), 320–341.
Turin, D.A. (1975) Introduction. In Turin, D.A. (ed.), *Aspects of the Economics of Construction*. London: Godwin, pp. viii–xvi.
Turner, B. and Whitehead, C. (2002). Reducing housing subsidy: Swedish housing policy in an international context. *Urban Studies*, 39(2), 201–217.
UN-Habitat (2007) *Annual Progress Report 2006*. Nairobi: United Nations Human Settlements Programme.
UN-Habitat (2009) *Financing affordable social housing in Europe*. Nairobi: United Nations Human Settlements Programme.
UN-Habitat (2020) *Annual Progress Report 2019*. Nairobi: United Nations Human Settlements Programme.
Walker, A. (1996) *Project Management in Construction*. Oxford: Blackwell Science.
Walker, A. and Chau, K.W. (1999) The relationship between construction project management theory and transaction cost economics. *Engineering, Construction and Architectural Management*, 6(2), 166–176.
Williamson, O. (1985) *The Economic Institutions of Capitalism*. New York: Free Press.
Williamson, O. (1996) *The Mechanisms of Governance*. New York: Oxford University Press.
Williamson, O.E. (2002). The theory of the firm as governance structure: from choice to contract. *Journal of Economic Perspectives*, 16(3), 171–195.
Winch, G. (1989) The construction firm and the construction project: a transaction cost approach. *Construction Management and Economics*, 7, 331–345.
Winch, G.M. (1995) Project management in construction: towards a transaction cost approach. Le Groupe Bagnolet Working Paper No. 1, University College London.
Winch, G. (1996) The Channel Tunnel: le Projet du Siecle. Working paper, University College London, Bartlett School of Architecture.
Winch, G. (2001) Governing the project process: a conceptual framework. *Construction Management and Economics*, 19, 193–207.
Winch, G.M. and Campagnac, E. (1995) The organization of building projects: an Anglo/French comparison. *Construction Management and Economics*, 13, 3–14.
World Bank (2006) *Thirty years of shelter lending. What have we learned?* Washington, DC: World Bank.
World Bank (2009) *Reshaping Economic Geography. The 2009 World Development Report*. Washington DC: World Bank.
World Bank (2010) *Systems of Cities. Harnessing Urbanisation for Growth and Poverty Alleviation. The World Bank Urban and Local Government Strategy*. Washington, DC: World Bank.
World Bank (2016) *Poverty and Shared Prosperity 2016: Taking on Inequality*. Washington DC: World Bank.

20. A review of stakeholder management in construction

Ezekiel Chinyio, Sukhtaj Singh and Subashini Suresh

INTRODUCTION

A stake can be an interest, a right or ownership, where an 'interest' is a situation in which individuals will be affected by a decision or certain action; a 'right' could be legal or moral in nature; and 'ownership' is the legal authority on an asset (Bourne and Walker, 2006). Clarkson (1994) posited that without an element of risk, there is no stake. Similarly, Olander (2007) described stakes as the actual or perceived benefits or risks or harms due to an organisation because of their activities. Organisations and individuals can have stakes. Organisations are represented by departments or units that are in turn represented by individuals who operate on the basis of their beliefs, interests and motivations (Eskerod and Jepsen, 2013). Where there are several people or organisations in an undertaking, their many stakes can either align or be in conflict with each other (Orts and Strudler, 2009). The need to resolve conflicting stakes is the thrust of stakeholder management.

A stakeholder connotes someone with a stake. According to Clayton (2014) and Freeman and Reed (1983), the Stanford Research Institute started exploring the terms 'shareholder' and 'stockholder' in 1963; and according to Mitchell et al. (1997), the concept originated from the idea of corporate social responsibility. Clayton (2014, p. 6) defined a stakeholder as 'the holder of a wager' and explained how the *Oxford English Dictionary* tracked the roots of the word back to the year 1708. Ramirez (1999) concurs that the term 'stakeholder' originated in 1708, and defined it as 'a person who holds the stake or stakes in a bet' (p. 101). Freeman (1984) defined a stakeholder as: 'any group or individual who can affect or is affected by the achievement of the firm's objectives'. This popular and often-quoted definition makes writers refer to Freeman as the (modern) protagonist of stakeholder management.

There are diverse definitions and conceptions of stakeholders and stakeholder theory, and these come from many disciplines that include strategic management, business management, marketing, human resource management, finance, and many others (Miles, 2017). The many definitions of a stakeholder in literature are in two main forms:

- The broad definitions (for example, Mitchell et al., 1997; Phillips, 2003; Orts and Strudler, 2002), opine that virtually anyone or anything can be a stakeholder, including interested organisations and businesses, tourist organisations, research organisations, knowledge organisations and other development interests (Storvang and Clarke, 2014); animals; plants; non-living entities, such as ecology or the natural environment; and even future generations (Amaeshi, 2010). Many stakeholders may be interrelated and tend to have joint stakes (Freeman, 2017).

- The narrow definitions restrict the consideration of stakeholders to mainly the persons or organisations that concern an organisation's core economic interests (Phillips, 2003; Freeman and Reed, 1983; Mitchell et al., 1997; Olander, 2007).

Freeman's (1984) treatise popularised stakeholder theory and management, especially in the (project) management sector. Thus, Freeman is often described as the father of stakeholder theory, and the concept of stakeholder management was formally recognised rather more recently by professional bodies such as the Project Management Institute and the International Project Management Association (Worsley, 2017).

Stakeholders in the Construction Industry

Walker et al. (2008a) provided a definition of stakeholders which sets the tone for the discussion of the topic in this chapter: 'Stakeholders are individuals or groups who have an interest or some aspect of rights or ownership in the project, and can contribute to, or be impacted by, either the work or the outcomes of the project' (p. 73). Construction projects tend to have many stakeholders. Indicatively, the stakeholders in construction will include: clients, owners, users, organisations, legal authorities, project managers, designers, contractors, subcontractors, insurance companies, suppliers, process and service providers, facilities managers, competitors, banks and funding bodies, regulatory agencies, neighbourhoods (local communities), government bodies, customers, visitors, regional development agencies, the media, consultants, environmentalists and the natural environment (Mitchell et al., 1997; Chinyio and Akintoye, 2008; Aaltonen and Sivonen, 2009; Heravi et al., 2015).

Those with interests in and concerns related to a construction project could be many (see Box 20.1) and may include other stakeholders not listed above. For example, noise from the construction processes of a particular project can be a major concern to neighbours (Eskerod and Jepsen, 2013), which may necessitate a response from the project team. It is also conceivable that many stakes in construction would often conflict, hence the need for stakeholder management in this sector.

The chapter discusses stakeholder theory, the types and characteristics of stakeholders, the management of stakeholders, its enablers and critical success factors; and potential areas for future research. As the body of literature on the subject matter is quite large, an indicative rather than exhaustive coverage of the subject matter is intended.

BOX 20.1 CONSTRUCTION STAKEHOLDERS ARE OFTEN MANY

Many stakeholders would be affected by, for example, a town centre development that is introducing tram lines into their transport network. Such a development would disrupt the usual bus routes and timetables and change some commuting patterns while the construction is going on. The local authority, regulatory bodies, hauliers, visitors, businesses along the new tram line, and so on, would also have their interests in this project, which will affect them in different ways.

In England and Wales, most major developments warrant obtaining planning permission, which involves a go-ahead decision. When a client or developer applies to a local authority for planning permission, the authority will often carry out a consultation where people and businesses around the site of the proposed development, as well as members of the public that have an interest, are given the chance to comment on or even oppose the proposed scheme. The local authority would factor the views of stakeholders from this consultation exercise into its considerations in making a (not) go-ahead decision. This illustrates the presence of many stakeholders in a construction project and how their interests matter.

STAKEHOLDER THEORY

Given that an organisation has a responsibility to its shareowners and to other people and groups (Philips, 1997), stakeholder theory aims to solve the two major issues of: how to manage people fairly and efficiently; and determining the extent of an organisation's moral responsibilities beyond making profits and adding economic value (Orts and Strudler, 2009). That is, the main thrust of stakeholder theory is an organisation's relationships with its stakeholders. The literature refers to three dimensions of stakeholder theory: the descriptive, instrumental and normative.

The Descriptive/Empirical Theory

The descriptive aspect of stakeholder theory describes and explains the specific characteristics of an organisation, such as its nature, how its managers think about operational issues, how its board members think about corporate affairs, profits and how the organisation is actually managed. The descriptive aspect reflects and explains the past, present and future conditions of relations between an organisation and its stakeholders; and corresponds to observed reality (Donaldson and Preston, 1995).

The Instrumental Theory

The instrumental theory guides an organisation to adopt (or not) certain principles and practices in order to achieve (or avoid) certain results. The instrumental aspect argues that considering the interests of other stakeholders (or at least pretending to do so) may increase value to shareholders (Orts and Strudler, 2002). It justifies itself by showing evidence between stakeholder management and organisational performance. This theory links stakeholder management with the accomplishment of traditional organisational objectives such as growth, making profit and being stable (Donaldson and Preston, 1995). However, the instrumental aspect does not explain in detail the link between stakeholder management and organisational performance (Donaldson and Preston, 1995). According to Orts and Strudler (2002), researchers who favour the instrumental aspect assume, in the absence of persuasive evidence, that good ethical behaviour towards stakeholders will yield good economic results. Orts and Strudler (2002) further argued that this is not always true because ethics and economic self-interest do sometimes conflict with each other.

The Normative Theory

Donaldson and Preston (1995) argued that the normative aspect is the core of stakeholder theory. This strand of the theory states that if an organisation wants to achieve or avoid certain results, then it should do certain things (or not) because they are the right (or wrong) things to do. Phillips (1997) proposed a concept of 'fair play', or 'fairness', in the normative aspect; that is, anyone affected by a firm's operations is considered to be a stakeholder and the organisation is morally obliged to address their interests.

The Three Strands of Stakeholder Theory

According to Donaldson and Preston (1995), the three aspects of stakeholder theory (descriptive, instrumental and normative) have different values but are interrelated. Three nested circles are used to depict their relatedness: the descriptive is represented by the outer circle, the instrumental is in the middle, while the normative is the innermost circle. The descriptive aspect is supported by the instrumental perspective because of its instrumental and predictive value. The normative aspect is then the central core due to the descriptive accuracy of the theory. The normative aspect posits that managers and their agents act as if all stakeholders' interests have intrinsic value. Adopting moral values and obligations in this manner makes the normative aspect the fundamental base of stakeholder theory.

Heravi et al. (2015) opined that stakeholder theory should focus on the stakeholders that contribute towards the decision-making processes of an organisation and on the subsequent (secondary) stakeholders also. Irrespective of the view an individual may hold on stakeholder theory, the key point is to identify legitimate and valid stakeholders and address their concerns (Bourne and Walker, 2006).

Shortcomings of Stakeholder Theory

Stakeholder theory provides a guide on how to perceive stakeholders and act in relationships with them. However, some shortcomings of the theory have been highlighted. They include:

1. The theory is somewhat vague and overly broad (Orts and Strudler, 2009).
2. The theory does not provide a normative justificatory framework for its foundation (Phillips, 1997; Donaldson and Preston, 1995).
3. The theory does not assist in the identification of stakeholders (Phillips, 1997).
4. The theory does not often suggest a concrete method in decision-making while balancing the diverse interests of stakeholders (Orts and Strudler, 2002, 2009).
5. The theory focuses on the interest of only human participants and pays less attention to the natural environment (Orts and Strudler, 2002).

TYPES OF STAKEHOLDERS

Stakeholders can be: internal/external; inside/outside; direct/indirect (Lester, 2017); primary/secondary; vested/non-vested (Chinyio and Olomolaiye, 2010; Jergeas et al., 2000); voluntary/involuntary (Clarkson, 1995); strategic/moral (McManus, 2002); critical/distal (Miles, 2017);

visible/invisible; and upstream/downstream (Walker et al., 2008a). Internal stakeholders are actively involved in project execution and may form project coalitions or support projects financially (Cleland and Ireland, 2006; Olander, 2007), whereas external stakeholders are those who are affected by the actions taken in respect of the project (Chinyio and Olomolaiye, 2010). Examples of invisible stakeholders are family members and social networks who can offer support and advice without being involved directly in a project (Walker et al., 2008a).

The continuous support of 'primary stakeholders' may be crucial for the survival of an organisation, hence they should be given high attention (Jepsen and Eskerod, 2009). Cleland and Ireland (2006) categorised owners, investors, suppliers and functional groups into this category. On the other hand, 'secondary stakeholders' are affected by, or can affect, an organisation's actions; they include the media, environmental organisations and local communities (Jepsen and Eskerod, 2009). Box 20.2 illustrates how primary or even secondary stakeholders can influence a project significantly.

BOX 20.2 COUL LINKS GOLF COURSE PROJECT REFUSED PLANNING PERMISSION

A controversial plan to build an 18-hole championship golf course in the Highlands has been refused planning permission by the Scottish government.

About 32 acres of the planned course was proposed for dunes at Coul Links at Embo, near Dornoch. Highland councillors gave the project the go-ahead last June, before Scottish ministers called in the planning application for further scrutiny.

Following a public inquiry, the government has refused permission. In their decision, Scottish ministers said the plan would have supported economic growth and rural development. But they agreed with government-appointed planning officials' findings that the golf course would have 'significant' effects on rare plantlife, wintering and breeding birds and the dunes themselves.

The government said the 'likely detriment to natural heritage is not outweighed by the socio-economic benefits of the proposal'.

Planning Minister Kevin Stewart said: 'This proposal does not comply with the relevant provisions of the Highland Wide Local Development Plan, and runs contrary to Scottish Planning Policy's emphasis on protecting natural heritage sites and world class environmental assets.

'The Scottish government has considered the reporter's findings carefully and agree with the recommendation that planning permission should be refused.'

A group of conservation charities opposed the project, including RSPB Scotland, Butterfly Conservation Scotland and the Scottish Wildlife Trust.

Source: https://www.bbc.co.uk/news/uk-scotland-highlands-islands-49733098.

'Vested stakeholders' have an interest in the success of a project and can influence it by, for example, withholding or giving resources. On the contrary, 'non-vested stakeholders' have no control over the allocation of a project's resources (Jergeas et al., 2000). According to McManus (2002), strategic stakeholders are those who can affect the organisation's objectives, while moral stakeholders are those affected due to the activities of a project.

Some stakeholders can belong to more than one of the foregoing groups. In general, stakeholders range from those who are actively involved in an organisation's projects or activities, to those who merely have an interest (Jergeas et al., 2000). Analytical and intuitive skills are needed by project managers for identifying both the key and many stakeholders, so that their influence on projects' activities can be anticipated (Walker et al., 2008b). Although stakeholders can be many (broad view), project managers tend to list and concentrate on only the crucial ones (Jepsen and Eskerod, 2009; Young, 2013). Karlsen (2002) suggested that clients who define and finance a project, and end-users who decide its usefulness, are the most prominent stakeholders. Worsley (2017) considered three project tags on the basis of their stakeholders: stakeholder-neutral (stakeholders that have very little interest or influence); stakeholder-sensitive (projects that can have an impact on stakeholders and should be monitored closely); and stakeholder-led (where stakeholders have very high power and influence on the directions of projects).

The prioritisation of stakeholders is discussed subsequently under tracking. Meanwhile, there will often be several stakeholders in a project, and when their interests are not aligned, there will be a potential for overt or covert conflicts which can manifest themselves as opposition, hostility, sabotage, antagonism, and so on. Usually, the antagonistic stakeholders are fewer in number, but they can sometimes be extremely vocal or even disruptive.

Construction Economics and Stakeholder Management

Myers (2017) argues that the science of economics concerns choice in a circumstance of scarce resources. For instance, land and money, which are types of resources, may be in short supply, so a choice must be made when these factors are competing against each other. In stakeholder management, the availability of land, money and materials, or other resources necessary for the project, can influence the direction in which a project will proceed. A further perspective of economics is that several wants or demands by stakeholders can outweigh the supply available, which again reinforces the need for choice (Myers, 2017). This demand–supply perspective is very common in construction settings, where a project's many stakeholders would have many and diverse demands (wants), and it is always difficult to supply all these, which triggers the need for choice. The objectives of economics and stakeholder management do overlap, and the two concepts supplement each other.

Construction and other built environment activities contribute highly to the gross domestic product (GDP) of any country. Therefore, the built environment sector stands out as a viable source of economic enhancement. Since stakeholders' actions can add to project cost, time and other disruptions, the effective management of stakeholders can forestall or minimise their potential negative impacts on projects and, by extension, the adverse effect on the national economy. The introduction to, and discourse of, construction economics are covered in other chapters of this book. Thus, it suffices to point out here that stakeholder management is a dimension of, and contributor to, construction economics.

CHARACTERISTICS OF STAKEHOLDERS

Stakeholders have certain characteristics through which they affect organisations and projects. These characteristics include power, legitimacy and urgency (Mitchell et al., 1997); these are now discussed.

Power

Mitchell et al. (1997) perceived power as the ability to influence an organisation. Power is the ability to persuade, induce or coerce the activities of others (Chinyio and Olomolaiye, 2010). It is used to obtain favoured pay-offs where interests are opposed (Willer et al., 1997). Power can be used to exert influence; however, influence can be exercised without power (Willer et al., 1997).

There are different types of power. For instance, Mitchell et al. (1997) identified coercive, utilitarian and normative power. Bourne and Walker (2005) compiled from the literature seven types of power: coercive, connection, reward, legitimate, referent, information and expert; and also three dimensions of power: position, personal and political. Stakeholders can team up to increase their combined power (Jepsen and Eskerod, 2009).

Legitimacy

Legitimacy is 'a generalised perception or assumption that the actions of an entity are desirable, proper, or appropriate within some socially constructed system of norms, values, beliefs, and definitions' (Suchman, 1995, p. 574). Suchman (1995) further explained that legitimacy is an assumption or perception of an observer pertaining to an organisation, and it is created subjectively but possessed objectively. Legitimacy as a held view may involve some risk, which may be relatively beneficial or harmful to an organisation (Olander, 2007).

Suchman (1995) described three conceptualisations of legitimacy: pragmatic (which is similar to power), cognitive (which is habitual) and moral (which is normative). Mitchell et al. (1997) and Yang et al. (2014) adopted Suchman's definition, arguing that the social system within which legitimacy is achieved is based on various analyses, the most common of which are the individual, organisational and societal. However, they argued that legitimacy could be defined and justified differently at the various levels of a social organisation.

Urgency

Urgency is the extent to which stakeholders' claims require attention or instant action. Mitchell et al. (1997) argued that urgency is composed of two attributes: time sensitivity and criticality. Time sensitivity is the point after which stakeholders would not accept a delay to attend to their relationships or claims. Criticality is how important a relationship or claim is to a stakeholder. Urgency provides an informed basis of responding to stakeholders' expectations.

Overview of the Attributes

The stakeholders' attributes are considered (or should be considered) while engaging with them. Olander (2007) acknowledged that all three attributes were important, but that power had an edge over the other two and affects the decision-making process of a project; hence, managers should especially pay attention to stakeholders that have high power. According to Bourne and Walker (2005), the nature of power, its source, influence and how it is exercised to manipulate or to contribute to cooperative relationships, shape all procurement strategies and the relationships that develop from them. The stakeholders who possess the attribute of legitimacy are risk-bearers who may try to get power, either by themselves or by forming coalitions with other stakeholders. Urgency could be considered as a factor while prioritising stakeholders.

STAKEHOLDER AND PROJECT MANAGEMENT

Stakeholder management is 'the systematic identification, analysis and planning of actions to communicate with, negotiate with and influence stakeholders' (APM, 2006, p. 20). According to Eskerod and Jepsen (2013), stakeholder management involves the continuous development of relationships with stakeholders to achieve a successful project outcome. By inclination, stakeholder management is proactive and helps organisations to deal with issues in a timely manner.

Stakeholder management is a crucial part of project management. The importance of stakeholder management is emphasised by some prominent bodies. For example, the Project Management Institute (PMI), in its PMBOK guide of 2017 (PMI, 2017), discussed the process of stakeholder management and considered it to have four stages: stakeholder identification, stakeholder engagement planning, managing of stakeholder engagement, and monitoring stakeholder engagement. The guide recommends that stakeholder identification and engagement should be commenced as soon as possible after the project charter is approved, and the project manager is appointed. Another promoter of stakeholder management is the International Organization for Standardization (ISO). The ISO 21500 series discusses stakeholder management (PMI, 2021), and the ISO 14000 series (such as ISO 14001) specifically offers guidance to organisations that face pressure from stakeholders to provide evidence of their improved environmental practices (Simpson and Sroufe, 2014).

Projects can be perceived as temporary coalitions of stakeholders that have been formed to create something (Jepsen and Eskerod, 2009). A project and its stakeholders can be treated as a network in which individuals interact and exchange information and resources in order to yield outcomes (Karlsen, 2002). Cleland and Ireland (2006) argued that the management of project stakeholders warrants that they (projects) are explicitly described in terms of individuals and groups that have interests in them.

Supportive and influential stakeholders can contribute their knowledge and insights to projects (Walker et al., 2008a) and enhance successful project delivery (Jepsen and Eskerod, 2009). Effective stakeholder management boosts trust and stimulates collaborative efforts that increase relational wealth (McManus, 2002). Thus, stakeholder management should be considered a crucial part of an organisation's strategic activities (Cleland and Ireland, 2006). McManus (2002) identified inadequate stakeholder management as a critical factor leading to project failure; in his study, this factor ranked higher than lack of user involvement, lack of resources, unrealistic expectations, lack of executive support, lack of information technology (IT) management and unclear project objectives.

As stakeholders have different priorities and objectives, the unequal distribution of resources among them can intensify conflicts between them (McManus, 2002). Hence, stakeholder management and economic theory (which was discussed before) requires a balancing of the competing claims on resources between the project and the organisation, the project and other projects, and different parts of the project (Bourne, 2005), but the elements of project complexity and uncertainty often make it difficult to achieve this balance (Turner and Muller, 2003). Thus, unanticipated and conflicting interests of stakeholders, if not appropriately managed, can lead to a project's failure (Bourne and Walker, 2005). This argument reinforces the need for a conscious, proactive and structured approach to stakeholder management in organisations. Many researchers have proved that project stakeholder management is critical to making a project successful (Offenbeek and Vos, 2016). Cleland and Ireland (2006) argued that there should be a formal and organised process for identifying and managing stakeholders.

The Stakeholder Management Process

A formal process of stakeholder management is required, especially in projects with long duration which often undergo significant changes. Moreover, a formalised approach would ensure that the reactions of stakeholders to decisions and impacts of these on a project are determined precisely; as well as how the stakeholders will interact with each other to influence a project (Cleland and Ireland, 2006). Furthermore, Karlsen (2002) explained that stakeholder management processes assist in managing stakeholders in several ways, such as ensuring a good balance between rewards to or contributions from stakeholders.

A stakeholder management process involves determining project goals and what constitutes project success, and proceeding to manage stakeholders towards achieving these two aspects. In addition to identifying and analysing stakeholders, a management process will involve stakeholder engagement. Processes for stakeholder management are reported or proposed in the literature. Some examples are:

- Develop a stakeholder map of a project; prepare a chart of specific stakeholders; identify the stakes of stakeholders; prepare a 2x2 power versus stake grid; conduct a process level stakeholder analysis; conduct a transactional level stakeholder analysis; determine the stakeholder management capability of your organisation; and analyse the dynamics of stakeholders (Elias et al., 2002).
- Plan, identify and analyse stakeholders; communicate with them; act accordingly and follow up (Karlsen, 2002).
- Identify and prioritise stakeholders; develop a stakeholder management strategy on this basis (Bourne and Walker, 2006).
- Identify and gather information on stakeholders; identify their missions, strengths and weaknesses; identify your stakeholder strategy; predict stakeholder behaviour (that is, underlying forces, key stakes, potential influence); and modify and implement your strategy accordingly (Cleland and Ireland, 2006).
- Identify important stakeholders; identify their characteristics (their contributions and associated rewards, and their power in relation to the project); and determine which strategy to use to influence each stakeholder (Jepsen and Eskerod, 2009).
- Identify stakeholders; gather information about them; analyse their influence; and decide a strategy for communicating with them (Young, 2013).

The idea of a stakeholder management process is to ensure that stakeholders are identified and engaged with, in order to achieve project success.

Stakeholder Engagement

According to Bourne and Walker (2006), stakeholder engagement is a formal process of relationship management by which project teams or organisations interact with their stakeholders in order to align stakeholders' goals with those of their projects or organisations. According to Worsley (2017) it is not stakeholders that are managed, but the engagement with them. Indeed, stakeholder management is about relating with stakeholders.

A stakeholder engagement plan should be treated as a crucial part of a risk management plan, but the two plans do not mean the same thing (Bourne and Walker, 2006). A project, and by extension an organisation, should have stakeholder engagement principles and tactics. The principles (the instrumental aspect of stakeholder theory) are formulated and rolled out in an organisation, while the tactics of implementation (the normative aspect) can be subject to the personality and style of a project manager. Stakeholders could initially be identified at an aggregated level, but may be engaged at an individual level (Eskerod and Jepsen, 2013).

Ineffective stakeholder engagement can lead to project delays and an increase in costs due to problems with planning, logistics and production (Walker et al., 2008b). In the construction industry, for instance, powerful stakeholders, such as a client or project sponsor, are critically considered in decision-making; while the opinions of stakeholders with less power or influence, such as small subcontractors and suppliers, may sometimes be discounted.

The strategies for engaging stakeholders could be:

- Proactive versus reactive: a proactive inclination should be preferable as an organisation would want to minimise any negative consequences arising from stakeholders (Chinyio and Akintoye, 2008; Olander and Landin, 2008).
- Collaborative versus distancing: a collaborative approach warrants frequent engagement with stakeholders and involving them in some decision-making activities (Yang and Shen, 2014; Steyn, 2014).

A collaborative approach creates a community spirit (Lehtinen and Aaltonen, 2020) and has a higher potential of win–win outcomes, or at least yielding higher satisfaction to all stakeholders. In contrast, a distancing approach does not often involve stakeholders in decision-making, and liaises with them only when it is absolutely necessary. To illustrate this point, a house builder using the collaborative approach would hold meetings with members of the community about the development to find out how the project would have an impact on them, and vice versa. Through this interaction, some important issues could be picked up and relevant adjustments made. In contrast, a house builder with a distancing approach would only contact the community about the project when the planning authorities indicate that it is mandatory for them to do so. A distancing approach may find any input from external stakeholders less welcoming.

Some tactics are warranted in stakeholder engagement. Some of these tactics are:

- Communication: Communication with, and between, stakeholders is vital for understanding and meeting relevant stakes. Without communication, it may be difficult to know the expectations of stakeholders; this underlines its importance, and the stress usually put on it. There are different means of communication: face to face, over the phone, in writing, via email, text, and so on. Some stakeholders prefer certain channels of communication, so it is worthwhile to establish and use the preferred mode for each stakeholder. In general, a combination of communication channels would often be needed.
- Face-to-face. Face-to-face interaction facilitates communication. Some stakeholders value this approach highly because it makes them feel they are highly appreciated, which in turn

makes them open up quickly and engage with the project team and other stakeholders without inhibition.

- Negotiation. The use of negotiations can resolve conflicting stakes between stakeholders (Eskerod and Jepsen, 2013). Negotiations involve bargaining, making trade-offs and using other tactics, which are outside the scope of this chapter.
- Timing. Engaging the right stakeholder at the right time is crucial. Some stakeholders may require constant updates, while others may not. Getting the timing right with stakeholders can make the engagement efficient.
- Information management. Information overload is unnecessary and information underload is not good. It is ideal to provide the right (amount of) information to stakeholders. A system that enables an organisation to provide sufficient information to stakeholders when needed is necessary. This system should work well with the previous tactic of timing.
- Behavioural attributes. The use of tact in dealing with stakeholders is vital (Lester, 2017). Stakeholders differ, and so should the ways for engaging them.

The enablers of stakeholder management, which are discussed further below, illustrate how these strategies are deployed for efficacy.

Difficulties of Stakeholder Management

The project environment is usually high in uncertainty, complexity and ambiguity; this makes stakeholder management more difficult because the project and its uncontrolled factors are interrelated (Yang et al., 2011b). The limited duration and unique nature of some projects demand extra efforts from project managers to build effective project teams, and stimulate and engender trust among project stakeholders (Grabher, 2002).

On some occasions, some stakeholders coalesce due to their common interests. However, these coalitions of stakeholders may not be stable throughout a project, and are often difficult to track as project managers do not often have the time, resources or capabilities to do so effectively (Jepsen and Eskerod, 2009). Even when stakeholders do not form mini coalitions, their individual interests or levels of power can change over time. Thus, tools for stakeholder management are worthwhile.

Consequences of not Managing Stakeholders

The inadequate management of stakeholders can lead to suspicion and conflict among stakeholders which may eventually lead to increased costs and project delays (McManus, 2002). According to Bourne and Walker (2006), antagonistic stakeholders can create immense trouble for a project manager in many ways; for instance, by changing the scope of the project, by changing the technical direction or by reducing funding. The actions or interventions of stakeholders can even lead to project collapse. Thus, project managers have to understand the influence of stakeholders and their potential threats. The disadvantages of not managing stakeholders properly include:

- Inadequate resources (in both quality and quantity) may be assigned to stakeholder engagement (Jergeas et al., 2000; Karlsen, 2002).
- Project costs may increase (McManus, 2002); for example, owing to suppliers not providing discounts.

- There may be a lack of a clear and comprehensive definition of project success (Jergeas et al., 2000), as different stakeholders might have different opinions about it.
- A project can experience negative public reaction (Jergeas et al., 2000; Karlsen, 2002), such as resistance to a new airport, dam or road scheme. Some projects may attract unfavourable news in the media (Karlsen, 2002), while some may even fail to start if the resistance to them is extreme (Jergeas et al., 2000).
- There may be poor communication (Karlsen, 2002) owing to failure to identify the preferred channels and frequencies of information exchange with the stakeholders.
- Some unsatisfied stakeholders may be unwilling to work with the organisation on future projects (Jergeas et al., 2000).
- Possible time delays may occur (McManus, 2002) due to late responses or unanticipated interruptions from some stakeholders (Jergeas et al., 2000).
- The end results of a project may be unsatisfactory to some stakeholders (Jergeas et al., 2000).

The large number of potential disadvantages portend negative consequences for a project. For example, in the United States, the California Department of Motor Vehicles had to stop a project for modernising driving licences and its application for registration process midway after spending $45 million, due to reasons that included the lack of support and involvement of some key stakeholders, and poorly managed relationships with some external stakeholders (McManus, 2002). In another example, a 27 km highway project in New Zealand worth NZ$245 million was delayed due to inability to resolve stakeholder issues (Elias et al., 2002). Therefore, it is crucial for project managers and organisations to manage stakeholders adequately to avoid potential negative consequences.

ENABLERS OF STAKEHOLDER MANAGEMENT

Project managers come across various stakeholders that have different interests and perceptions (Davis, 2014). It is unlikely that all expectations of all the stakeholders on a project will be achievable (McManus, 2002). Moreover, the construction industry has a feeble record of stakeholder management (Yang et al., 2009). The adoption and deployment of the 'Clarkson principles' of stakeholder management is worthwhile for the construction sector. These principles are: (1) acknowledge and monitor stakeholders; (2) listen and openly communicate with all stakeholders; (3) be sensitive to the stakeholders' concerns; (4) recognise the interdependencies among stakeholders; (5) work with stakeholders in a cooperative manner; (6) observe their human and other relevant rights; and (7) be wary of potential conflicts (https://www.valuebasedmanagement.net/methods_clarkson_principles.html).

The behaviour of stakeholders towards a project may change over time (Elias et al., 2002), and according to Jawahar and Mclaughlin (2001) the importance of stakeholders depends on an organisation's current needs. Thus, some stakeholders may get more priority than others at a given time. This dynamism is a challenge for project managers, who will need to constantly prioritise stakes among the complex issues they are facing in relation to achieving the objectives of a project (Offenbeek and Vos, 2016; Olander, 2007). This makes stakeholder analysis an important part of the stakeholder management process because it enables project managers to formulate or deploy strategies to address relevant stakes in a timely manner (Bourne and

Walker, 2006). For example, Cleland and Ireland (2006) explained that stakeholder analysis during the planning stage assists in formulating strategies for managing the stakeholders throughout the life cycle of a project. Stakeholder analysis, in this regard, informs the decisions, objectives and plans of a project (Jepsen and Eskerod, 2009).

Stakeholder Analysis and Tracking

Stakeholder analysis involves listing, classifying and assessing the influence of stakeholders (Lester, 2017). Mapping tools are used to depict, and also track, the characteristics of stakeholders; that is, their power, urgency, (level of) interest, and so on. Several mapping tools are highlighted in the literature, and some of these are reviewed in the following sections.

A Stakeholder Matrix

A commonly used tool to map stakeholders is a 2×2 matrix, or table which involves two key variables. Variants of this matrix or table differ in the variables considered. In each case, the variables are depicted, and in most of them their magnitudes are assessed. Some examples of the matrices or tables are:

- the power–predictability matrix (Newcombe, 2003);
- the 'help potential'–'harm potential' matrix (Eskerod and Jepsen, 2013);
- the vested interest–impact index, which measures the potential influence of stakeholders through a simple arithmetical computation (Bourne and Walker, 2005);
- the stakeholder impact index (Olander, 2007);
- the stakeholder influence matrix (Young, 2013).

The Stakeholder Circle™ is a visualisation tool which depicts the power possessed by a particular stakeholder, or a group of stakeholders, as well as the influence they can exert on a project (Walker et al., 2008b). The Stakeholder Circle™ operates through five stakeholder management and engagement steps: identify, prioritise, visualise, engage and monitor stakeholders. In the second step (prioritise), project managers rate stakeholders on the basis of three attributes: power, proximity and urgency. Power and proximity are rated on a scale of 1 (low) to 4 (high), while urgency is rated on a scale of 1 (low) to 5 (high). The tool is then able to generate a visual representation that plots stakeholders within concentric circles to reflect their respective attributes.

The tools discussed here are effective but tend to provide a cross-sectional position of stakeholders. There is a need to keep updating the position of stakeholders and issues, which can be tedious. However, relevant IT technology can facilitate a more dynamic and longitudinal approach to stakeholder analysis and management. Through computerised simulations or application of models, stakeholders can be continuously tracked live, or semi-live.

Greater Use of IT Software for Stakeholder Analysis and Management

New tools for stakeholder management are worth considering (Storvang and Clarke, 2014) and IT software is increasingly being developed for stakeholder management specifically. Examples of these include:

- AGS stakeholder management (https://www.airiodion.com/stakeholder-tool/).
- Borealis (https://www.boreal-is.com/documentation/).
- Darzin (Yang et al., 2011a) (https://www.darzin.com/stakeholder-management-ultimate-guide).
- Jambo (www.jambo.cloud).
- Tractivity (https://www.tractivity.co.uk).
- Vuelio (https://www.vuelio.com/).

These IT platforms are also compendiums of knowledge on the application of the various tools, that preserve information for future use. An example of the platforms, Jambo, is presented in Box 20.3.

BOX 20.3 JAMBO

Managing stakeholders and your communications with them can become an overwhelming task. As projects grow and the number of stakeholders you need to communicate with increases, you can quickly lose track of what has been said and to whom. Specialized stakeholder relationship management (SRM) software helps make the stakeholder management process more collaborative, efficient and scalable, while also allowing users to search through their data fast.

Jambo is an SRM software tool used by teams worldwide to record, manage and report on communications with stakeholders and any issues or commitments associated with them. All information in Jambo is stored securely in the cloud in real time, so teams can feel confident in knowing that they are viewing the most up-to-date stakeholder information.

Jambo has features and workflows made specifically for managing stakeholders. The dashboard analytics allow users to visually understand their projects; the issues management tool helps to identify the impact and priority of stakeholder or project issues; and the commitments management tool helps to monitor the status of stakeholder commitments (also known as promises) to ensure they are fulfilled and never forgotten. Jambo offers smart reporting tools so that users can quickly run concise or detailed project insight reports. These reports can be easily shared with internal teams, decision-makers or external stakeholders to help build project understanding and gain buy-in. Jambo also has a task management feature that helps stakeholder managers to stay on track by allowing them to assign tasks to themselves and other team members. Task progress can be monitored in the dashboard and users will receive email prompts when tasks are due.

Thanks to secure cloud collaboration, all information in Jambo is centralized, searchable and accessible from any location with internet access. This means that teams can instantly access the history of engagement with their stakeholders, so they are always prepared before communicating with them; an important component in successful stakeholder relationship management.

For more information on Jambo, please visit www.jambo.cloud.

Overview of the Stakeholder Trackers

The tracking tools outlined above involve mapping stakeholders, but do not necessarily assist in identifying stakeholders. Also, the assignment of values to the stakeholders' attributes (vested interest, impact, power, legitimacy, urgency, position, proximity) is done by, and depends on the experience of, a project manager. Furthermore, any matrix or software used to track stakeholders should be updated regularly, as their attributes tend to be dynamic. While a dynamic analysis will not solve all the problems, it will facilitate the use of up-to-date strategies and tactics for communicating with and managing stakeholders, organising complex situations, understanding coalitions of stakeholders, and optimising the use of resources and time for obtaining maximum benefit.

Use of Building Information Modelling (BIM)

Building information modelling (BIM) is a model for delivering construction projects. It operates by holding project information in one place which is then accessible by all relevant stakeholders; usually the ones that are internal to a project. As information is centralised, its accuracy is not disturbed due to, say, transmission between stakeholders. Whenever information is updated in the model, all other stakeholders have the privilege of seeing it when they like. Information can be accessed quickly via BIM. Therefore, BIM aids communication and minimises the dearth of relevant information on the project; all of which augurs well for stakeholder management. Some of the challenges associated with stakeholder management concern the lack of information, or the provision of wrong information. BIM avoids this challenge through the relatively quick and accurate provision of information, which deters stakeholder dissatisfaction.

BIM was developed fundamentally for information management. However, the way it functions facilitates the exchange of information (engagement) between some project stakeholders. Thus, BIM is an enabler of stakeholder management.

Collaborative Procurement

The general outlook of procurement in the construction sector was previously adversarial, which demanded shifting to collaborative relationships (Hasanzadeh et al., 2014). Many calls for change were made towards more cooperation between those involved in delivering projects. In the United Kingdom (UK), these calls were buoyed by the Latham (1994) and Egan (1998) reviews of the 1990s (Hosseini et al., 2016). Entering the 21st century, construction has witnessed an increase in the use of collaborative forms of procurement (Oyegoke et al., 2009).

A form of procurement which the UK embraced was the private finance initiative (PFI), which is a type of public–private partnership (PPP) (Jayasuriya et al., 2020). In PFI/PPP, a public sector client would contract with a private sector concessionaire, consisting of designers, financiers, constructors and facilities managers; these key stakeholders come together to develop a composite solution for the client. By using PFI/PPP, many relationships that would otherwise have been managed separately are bundled together, and the client deals with one concessionaire; and in turn, members of the concessionaire have their supply chains (secondary stakeholders).

If managed well, PFI/PPP and other forms of collaborative procurement should minimise or avoid many stakeholder conflicts (Hasanzadeh et al., 2014). For example, a potential conflict between the designer (architect or engineer) and the contractor is not likely because these major project participants belong to one organisation (the concessionaire). Similarly, a potential conflict between an architect and a project quantity surveyor is avoided because all of them are working for the same concessionaire organisation (Jayasuriya et al., 2020).

Partnering, alliancing and other forms of collaborative engagement bring stakeholders closer together in a project environment (Oyegoke et al., 2009), where any conflicting stakes can be negotiated more quickly, amicably and without fear (Chan et al., 2006; Conley and Gregory, 1999). Some people argue that partnering and alliancing should not be considered as types of procurement. However, it is clear that they are used for project delivery and involve collaborative engagements (McDermott and Khalfan, 2006). In summary, collaborative procurements enhance engagements among stakeholders (Oyegoke et al., 2009) and the effective management of their stakes (Eriksson and Nilsson, 2008). It will not be surprising if stakeholder management is explicitly referred to, or recommended, in modern collaborative forms of procurement.

Standard Forms of Contract

Standard forms of contract for construction projects are increasingly proffering tactics for stakeholder engagement, albeit implicitly. For example:

- The New Engineering Contract (NEC4) provides for an 'early warning system' through which the client and contractor are obliged to immediately notify the other party as soon as they become aware of any thing or risk that will have a negative impact on the project. Upon this notification, a meeting to which other relevant stakeholders are invited is arranged to forestall, curtail or manage the matter. This approach is intrinsically proactive and inherently collaborative; the relevant knowledge of each of the stakeholders is utilised to solve problems.
- The International Federation of Consulting Engineers (FIDIC) Standard Form has an advance warning clause where each party is requested to advise the other party as well as the engineer in advance of known or probable future events that may affect the project in terms of time, price and resources. An example of these future events would be an anticipated strike action by workers and possible disruption by other stakeholders. After notification, the engineer could ask the contractor to submit a proposal to avoid or minimise the event or circumstance.

These contract provisions enable a proactive approach as well as wider consultation with stakeholders in the course of problem resolution. In this way, no stakeholder is sidelined, and project decisions reached will not attract resistance as the stakeholders would have been involved in the resolution. Thus, these two examples demonstrate how standard forms of contract are enablers of stakeholder management during and around the project execution stage.

In the UK, the construction sector has a court that is partly devoted to it, which is referred to as the Technology and Construction Court (TCC). The TCC takes up issues of a 'highly technical' nature that are too complicated for jurors. Many disputes in construction are heard in the TCC. Meanwhile, fewer cases on the NEC are going to court (Murphy et al., 2014), suggesting that conflicts and, by extension, stakeholder issues are being managed in a better way where this standard form of contract is used. Conflict avoidance and minimisation, quick problem resolution and managing stakeholders will increasingly feature in standard forms of contract. The phrase 'stakeholder management' may not be used in these contracts, but the implications of future provisions would encourage greater engagement between stakeholders.

Soft Landings

There is often a gap between: (1) design intention and actual performance of a building; and (2) what is built versus what users want. The Soft Landings initiative, which is managed by the Building Services Research and Information Association, was designed to close this gap. Through Soft Landings the project success criteria, especially on energy and environmental performance, are more fully established at the outset and used to guide project delivery and outcomes, as well as to ensure a smooth transition from the construction phase to the operational phase of a building. Users of buildings and their needs are considered in the design and construction phases.

The concept of Soft Landings facilitates the identification of many stakeholders including users of buildings, making sure that the diverse needs of stakeholders are met, and that stakeholders are satisfied. Thus, Soft Landings is a salient contributor to stakeholder management.

Critical Success Factors (CSFs) in Stakeholder Management

Managing stakeholders demands attention, time and costs. For example, a project manager may need to take time to develop an information flyer that will keep the neighbouring community updated about the progress of a project. The preparation and distribution of a paper flyer will cost money and take time. Thus, the costs and time consequences of stakeholder management impinge on the economics of an organisation and industry at large.

The literature identifies directly or indirectly the critical success factors of stakeholder management. Table 20.1 presents some of these factors, which together indicate the breadth of stakeholder management. It is possible to miss out on one or more of these factors in the course of a project; doing so would undermine stakeholder management. Hence the use of enablers, especially the IT-based tools and their inbuilt prompts, is worthwhile. However, not many organisations are fully using these tools.

SUBJECTS FOR FURTHER RESEARCH

Some areas of future research on the stakeholder management process are now considered:

- Benefits: the tangible and intangible benefits of stakeholder management can be studied further. Vuori and Vesalainen (2018) studied and quantified the impact of multiple stakeholders on the research and development (R&D) processes of small and medium-sized enterprises (SMEs). Similar studies on the general impacts of stakeholder management on an organisation could be undertaken in all sectors such as construction. These impacts could be studied over time and project phases as well as in respect of procurement types and project performance.

Table 20.1 *Critical success factors for stakeholder management*

Critical success factor	References
Managing stakeholders with social responsibilities	1
Exploring stakeholders' needs and constraints to the project	1, 3, 4, 8, 20
Communicating with and engaging stakeholders properly and frequently	1, 2, 3, 5, 6, 8, 9, 11, 12
Understanding stakeholders' interests	1, 6
Using formalised procedure for identifying stakeholders	1, 2, 8, 18, 19, 20, 21, 22, 23
Keeping and promoting a good relationship with stakeholders	1, 2, 3, 6, 7, 10, 11
Analysing conflicts and coalitions among stakeholders	1
Accurately predicting the influence of stakeholders	1, 3, 22
Formulating appropriate strategies for the management of stakeholders	1, 20
Assessing the attributes of stakeholders	1, 17
Effectively resolving conflicts among stakeholders	1, 6
Formulating a clear statement of project missions (common goals, objectives and project priorities)	1, 2, 3,
Predicting stakeholders' reactions to the implementation of strategies	1
Analysing changes in stakeholder influences and relationships	1
Assessing stakeholder behaviour	1
Analysing the effects of decisions made about stakeholders	3, 17
Build and maintain a base of trust	3, 6, 9, 13
Early involvement of key project participants in stakeholder management	11, 14, 15, 16
Proactive interaction with affected stakeholders to mitigate conflicts that could potentially arise	3
Identifying the necessary contributions required from stakeholders	20
Identifying the benefits desired by stakeholders	20
Assessing the power of stakeholders in this project	20, 22

Note: References: 1 = Yang et al. (2009); 2 = Jergeas et al. (2000); 3 = Olander and Landin (2008); 4 = Bourne (2005); 5 = Bakens et al. (2005); 6 = Aaltonen et al. (2008); 7 = Rowlinson and Cheung (2008); 8 = Bourne and Walker (2005); 9 = Hartman (2000); 10 = Olander (2006); 11 = Yong and Mustaffa (2013); 12 = Al-Khafaji et al. (2010); 13 = De Schepper et al. (2014); 14 = Erkul et al. (2016); 15 = El-Gohary et al. (2006); 16 = Missonier and Loufrani-Fedida (2014); 17 = Donaldson and Preston (1995); 18 = McManus (2002); 19 = Olander (2007); 20 = Jepsen and Eskerod (2009); 21 = Walker et al. (2008a); 22 = Bourne and Walker (2006); 23 = Cleland and Ireland (2006).
Source: Adapted from Singh et al. (2017).

- Transaction cost (economics) is a potential area of research in construction generally and stakeholder management specifically. The time and cost of managing stakeholders can be investigated. The impact of using software applications in this regard can also be studied.
- Time: the exact time spent on managing stakeholders can be estimated in appropriately designed research.
- Cost: the cost of managing stakeholders is seldom reckoned and set out as an entry in the overheads of an organisation. This can also be researched, either on its own or as part of the transaction cost analyses.
- Use of technology: the impact of modern technology on the process of managing stakeholders can also be researched; this will include the use of social media in stakeholder management.
- The use of stakeholder management to harvest ideas that will benefit an organisation could be explored.

- ‘Stakeholder coalitions’ is another area of possible research. Some relevant questions include: Are coalitions formed deliberately or accidentally? Who leads these coalitions, and how? Are coalitions one-off, transient or continuous?

Recommendations and Looking Forward

There is scope for the greater development and uptake of software for construction stakeholder management. This advancement will enhance knowledge management and the efficacy of stakeholder management which will benefit the construction sector. The (near) absence of post-project reviews in construction is often voiced by practitioners. Use of stakeholder management software would provide an option where some project activities can be recorded and later retrieved for the purpose of any subsequent studies or reviews.

There is also scope for the development of the soft/art aspect of stakeholder management. An area worth exploring is personality. Different stakeholders, both persons and organisations, have different predominant personalities. It is also possible for two stakeholders to have a common demand, but require different tactics to be used to engage them on the basis of their different personalities. An understanding of these personalities will facilitate the engagements between stakeholders and minimise frictions and disputes between them. (Useful contacts for exploring psychological aspects in stakeholder management and other built environment aspects can be found at www.psycon.info.)

CONCLUSION

That stakeholders can ‘affect or be affected’ indicates that they have influence. An organisation will have a circle of stakeholders, and these can influence each other. Some stakeholders will be closer to the epicentre of the organisation or project than others. Some stakeholders’ influence will be more potent than that of others. Some stakeholders’ influences are short-term, while others are long-term.

The influence and management of stakeholders is an integral part of human life. For example, children and their parents influence each other from the time the children are born. Thus, stakeholder engagement and management is as old as human existence. The concept then extended to organisations and their operations. However, the formal recognition of the concept in business settings came in the 18th century, and was made popular in the management sector of the modern era by Ronald Freeman. Since then, organisations, including those in the construction sector, have been growing increasingly more conscious of their stakeholders, and applying tools and tactics to engage with them.

Stakeholders are managed instinctively, but a formalised approach is much more effective and efficient. Thus, organisations are being advised to develop and/or follow a formal stakeholder management approach and deploy their relevant strategies and tactics.

Stakeholders have varying interests, urgency and levels of power. These attributes enable them to influence an organisation. The attributes are dynamic, and some stakeholders can team up on the basis of these features. When stakeholders coalesce, their level of influence can be increased. So, an organisation needs to track its stakeholders, interpret their potential current and future influences, and engage with them on that basis. IT tools are available for tracking and managing stakeholder engagements. There are still some unknowns in stakeholder engagement which provide room for research. These include the time and cost spent on stakeholder engagement, and the delineation of the tangible and intangible benefits of stakeholder engagement, including the contributions of technology in this regard.

Economics is interested in how value is generated from production activities. In construction, the consideration would be: what products are generated, how sustainable they are, and what wider value they create to diverse stakeholders. Thus, construction economics examines construction production, activities and stakeholders, and their contributions in value generation. So stakeholder management effectively flags up the role of economics in construction production. For example, a major current concern with construction activities is the relatively high carbon footprint. Efforts are being made to lower the carbon footprint in view of its negative consequences. For example, the Carbon Trust is offering technical and financial advice to construction and other companies on approaches they could adopt to expedite the move to a low-carbon economy (https://www.carbontrust.com/). The contributions of individual stakeholders to the minimisation of the carbon footprint, considering a balancing of their life cycle costs and benefits, is a relevant subject of discussion.

REFERENCES

Aaltonen, K., Jaakko, K. and Tuomas, O. (2008) Stakeholder salience in global projects. *International Journal of Project Management.* 26(5), pp. 509–516.

Aaltonen, K. and Sivonen, R. (2009) Response strategies to stakeholder pressures in global projects. *International Journal of Project Management.* 27(2), pp. 131–141.

Al-Khafaji, A.W., Oberhelman, D.R., Baum, W. and Koch, B. (2010) Communication in stakeholder management. In: Chinyio, E. and Olomolaiye, P. (eds), *Construction Stakeholder Management.* Chichester: Wiley-Blackwell, pp. 159–173.

Amaeshi, K. (2010) Stakeholder management: theoretical perspectives and implications. In: Chinyio, E. and Olomolaiye, P. (eds), *Construction Stakeholder Management.* Chichester: Wiley-Blackwell, pp. 13–40.

Association for Project Management (APM) (2006) *APM Body of Knowledge* (5th edn). Princes Risborough: Association for Project Management.

Bakens, W., Foliente, G. and Jasuja, M. (2005) Engaging stakeholders in performance based building: lessons from the Performance-Based Building (PeBBu) Network. *Building Research and Information.* 33(2), pp. 149–158.

Bourne, L. (2005) Project relationship management and the Stakeholder Circle™. PhD Thesis, RMIT University, Melbourne.

Bourne, L. and Walker, D.H.T. (2005) Visualising and mapping stakeholder influence. *Management Decisions.* 43(5), pp. 649–660.

Bourne, L. and Walker, D.H.T. (2006) Visualising stakeholder influence – two Australian examples. *Project Management Journal.* 37(1), pp. 5–21.

Chan, A.P.C., Chan, D.W.M., Fan, L.C.N., Lam, P.T.I. and Yeung, F.Y. (2006) Partnering for construction excellence – a reality or myth? *Building and Environment.* 41, pp. 1924–1933.

Chinyio, E.A. and Akintoye, A. (2008) Practical approaches for engaging stakeholders: findings from the UK. *Construction Management and Economics.* 26(6), pp. 591–599.

Chinyio, E. and Olomolaiye, P.O. (2010) Introducing stakeholder management. In: Chinyio, E. and Olomolaiye, P.O. (eds), *Construction Stakeholder Management.* Chichester: Wiley-Blackwell, pp. 13–40.

Clarkson, M.B.E. (1994) A risk based model of stakeholder theory. *Proceedings of the Second Toronto Conference on Stakeholder Theory.* Centre for Corporate Social Performance and Ethics, University of Toronto, Canada.

Clarkson, M.E. (1995) A stakeholder framework for analyzing and evaluating corporate social performance. *Academy of Management Journal.* 20(1), pp. 92–118.

Clayton, M. (2014) *The Influence Agenda: A Systematic Approach to Aligning Stakeholders in Times of Change*. Basingstoke: Palgrave Macmillan.

Cleland, D.I. and Ireland, L.R. (2006) *Project Management: Strategic Design and Implementation* (5th edn). New York: McGraw-Hill.

Conley, M.A. and Gregory, R.A. (1999) Partnering on small construction projects. *Journal of Construction Engineering and Management*. 125(5), pp. 320–324.

Davis, K. (2014) Different stakeholder groups and their perceptions of project success. *International Journal of Project Management*. 32(2), pp. 189–201.

De Schepper, S., Dooms, M. and Haezendonck, E. (2014) Stakeholder dynamics and responsibilities in public–private partnerships: a mixed experience. *International Journal of Project Management*. 32(7), pp. 1210–1222.

Donaldson, T. and Preston, L.E. (1995) The stakeholder theory of the corporation: concepts, evidence, and implications. *Academy of Management Review*. 20(1), pp. 65–91.

Egan, J. (1998) Rethinking Construction: The report of the construction task force. 54315_RETHINKING CONST3 (constructingexcellence.org.uk) (accessed 8 February 2020).

Elias, A.A., Cavana, R.Y. and Jackson, L.S. (2002) Stakeholder analysis for R&D project management. *R&D Management*. 32(4), pp. 301–310.

El-Gohary, N.M., Osman, H. and El-Diraby, T.E. (2006) Stakeholder management for public private partnerships. *International Journal of Project Management*. 24(7), pp. 595–604.

Eriksson, P.E. and Nilsson, T. (2008) Partnering the construction of a Swedish pharmaceutical plant: case study. *Journal of Management in Engineering*. 24(4), pp. 227–233.

Erkul, M., Yitmen, I. and Çelik, T. (2016) Stakeholder engagement in mega transport infrastructure projects. *Procedia Engineering*. 161, pp. 704–710.

Eskerod, P. and Jepsen, A.L. (2013) *Project Stakeholder Management*. Farnham: Routledge.

Freeman, R.E. (2017) Five challenges to stakeholder theory: a report on research in progress. In: Wasieleski, D.M. and Weber, J. (eds), *Stakeholder Management*. Bingley: Emerald Publishing, pp. 1–20.

Freeman, R.E. (1984) *Strategic Management: A Stakeholder Approach*. New York: Cambridge University Press.

Freeman, R.E. and Reed, D.L. (1983) Stockholders and stakeholders: a new perspective on corporate governance. *California Management Review*. 25(3), pp. 88–106.

Grabher, G. (2002) Cool projects, boring institutions: temporary collaboration in social context. *Regional Studies*. 36(3), pp. 205–214.

Hartman, F.T. (2000) The role of trust in project management. *PMI Research Conference 2000: Project Management Research at the Turn of the Millennium*. Paris, France and Newtown Square, PA, USA: Project Management Institute.

Hasanzadeh, S.M., Hosseinalipour, M. and Hafexi, M.R. (2014) Collaborative procurement in construction projects performance measures, Case Study: Partnering in Iranian construction industry. *Procedia – Social and Behavioural Sciences*. 119, pp. 811–818.

Heravi, A., Coffey, V. and Trigunarsyah, B. (2015) Evaluating the level of stakeholder involvement during the project planning processes of building projects. *International Journal of Project Management*. 33(5), pp. 985–997.

Hosseini, A., Wondium, P.A., Bellini, A., Tune, H., Haugseth, N., et al. (2016) Project partnering in the Norwegian construction industry. *Energy Procedia*. 96, pp. 241–252.

Jawahar, I.M. and Mclaughlin, G.L. (2001) Toward a descriptive stakeholder theory: an organisational life cycle approach. *Academy of Management Review*. 26(3), pp. 397–414.

Jayasuriya, S., Zhang, G. and Yang, R.J. (2020) Exploring the impact of stakeholder management strategies on managing issues in PPP projects. *International Journal of Construction Management*. 20(6), pp. 666–678. DOI: https://doi.org/10.1080/15623599.2020.1753143 (accessed 20 January 2021).

Jepsen, A.L. and Eskerod, P. (2009) Stakeholder analysis in projects: challenges in using current guidelines in the real world. *International Journal of Project Management*. 27(4), pp. 335–343.

Jergeas, G.F., Williamson, E., Skulmoski, G.J. and Thomas, J.L. (2000) Stakeholder management on construction projects. *AACE International Transactions*. PM.12.1–PM.12.6.

Karlsen, J.T. (2002) Project stakeholder management. *Engineering Management Journal*. 14(4), pp. 19–24.

Latham, M. (1994) *Constructing the Team*. London: HMSO.
Lehtinen, J. and Aaltonen, K. (2020) Organising external stakeholder engagement in inter-organizational projects: opening the black box. *International Journal of Project Management*. 38(2), pp. 85–98.
Lester, A. (2017) *Project Management, Planning and Control – Managing Engineering, Construction and Manufacturing Projects to PMI, APM and BSI Standards* (7th edn). Oxford: Butterworth-Heinemann.
McDermott, P. and Khalfan, M.M.A. (2006) Achieving supply chain integration within construction industry. *Australian Journal of Construction Economics and Building*. 6(2), pp. 44–54.
McManus, J. (2002) The influence of stakeholder values on project management. *Management Services*. 46(6), pp. 8–14.
Miles, S. (2017) Stakeholder theory classification, definitions and essential contestability. In: Wasieleski, D.M. and Weber, J. (eds), *Stakeholder Management*. Bingley: Emerald Publishing, pp. 21–47.
Missonier, S. and Loufrani-Fedida, S. (2014) Stakeholder analysis and engagement in projects: from stakeholder relational perspective to stakeholder relational ontology. *International Journal of Project Management*. 32(7), pp. 1108–1122.
Mitchell, R.K., Agle, B.R. and Wood, D.J. (1997) Towards a theory of stakeholder identification and salience: defining the principle of who and what really counts. *Academy of Management Review*. 22(4), pp. 853–886.
Murphy, S.E., Spillane, J.P., Hendron, C. and Bruen, J. (2014) NEC contracting: evaluation of the inclusion of dispute review boards in lieu of adjudication in the construction industry in the United Kingdom. *Journal of Legal Affairs and Dispute Resolution in Engineering and Construction*. 6(4), 04214002.
Myers, D. (2017) *Construction Economics: A New Approach* (4th edn). London: Routledge.
Newcombe, R. (2003) From client to project stakeholders: a stakeholder mapping approach. *Construction Management and Economics*. 21(8), pp. 841–848.
Offenbeek, M.A.G. and Vos, J.F.J. (2016) An integrative framework for managing project issues across stakeholder groups. *International Journal of Project Management*. 34(1), pp. 44–57.
Olander, S. (2006) External stakeholder analysis in construction project management. PhD thesis, Lund University, Sweden.
Olander, S. (2007) Stakeholder impact analysis in construction project management. *Construction Management and Economics*. 25(3), pp. 277–287.
Olander, S. and Landin, A. (2008) A comparative study of factors affecting the external stakeholder management process. *Construction Management and Economics*. 26(6), pp. 553–561.
Orts, E.W. and Strudler, A. (2002) The ethical and environmental limits of stakeholder theory. *Business Ethics Quarterly*. 12(2), pp. 215–233.
Orts, E.W. and Strudler, A. (2009) Putting a stake in stakeholder theory. *Journal of Business Ethics*. 88(4), pp. 605–615.
Oyegoke, A.S., Dickinson, M., Khalfan, M.M.A., McDermott, P. and Rowlinson, S. (2009) Construction project procurement routes: an in-depth critique. *International Journal of Managing Projects in Business*. 2(3), pp. 338–354.
Phillips, R.A. (1997) Stakeholder theory and a principle of fairness. *Business Ethics Quarterly*. 7(1), pp. 51–66.
Phillips, R.A. (2003) Stakeholder legitimacy. *Business Ethics Quarterly*. 13(1), pp. 24–41.
Project Management Institute (PMI) (2017) *A Guide to the Project Management Body of Knowledge* (6th edn). Newtown Square, PA.
Project Management Institute (PMI) (2021) Organisational project management is not a privilege of lage companies. https://www.pmi.org/learning/library/project-management-try-iso-21500-5972 (accessed 8 February 2021).
Ramirez, R. (1999) Stakeholder analysis and conflict management. In: Buckles, D. (ed.), *Cultivating Peace: Conflict and Collaboration in Natural Resource Management*. Ottawa: International Development Research Centre, pp. 101–126.
Rowlinson, S. and Cheung, Y.K.F. (2008) Stakeholder management through empowerment: modelling project success. *Construction Management and Economics*. 26(6), pp. 611–623.
Simpson, D. and Sroufe, R. (2014) Stakeholders, reward expectations and firms' use of the ISO14001 management standard. *International Journal of Operations and Production Management*. 34(7), pp. 830–852.

Singh, S., Chinyio, E. and Suresh, S. (2017) The potential of BIM for stakeholder management in infrastructure projects. *Proceedings of 13th International Postgraduate Research Conference*. Salford, UK, 14–15 September, pp. 552–563. https://usir.salford.ac.uk/id/eprint/43913/1/13th%20IPGRC%202017%20Full%20Conference%20Proceedings.pdf (accessed 10 June 2021).
Steyn, M. (2014) Organisational benefits and implementation challenges of mandatory integrated reporting: perspectives of senior executives at South African listed companies. *Sustainability Accounting, Management and Policy Journal*. 5(4), pp. 476–503.
Storvang, P. and Clarke, A.H. (2014) How to create a space for stakeholders' involvement in construction. *Construction Management and Economics*. 32(12), pp. 1166–1182.
Suchman, M.C. (1995) Managing legitimacy: strategic and institutional approaches. *Academy of Management Review*. 20(3), pp. 571–610.
Turner, J.R. and Muller, R. (2003) On the nature of the project as a temporary organization. *International Journal of Project Management*. 21(1), pp. 1–8.
Vuori, V. and Vesalainen, J. (2018) SMEs as extended enterprises: a 360-degree model for profiling SMEs stakeholder involvement in R&D. *Journal of Enterprising Culture*. 26(3), pp. 225–250.
Walker, D.H.T., Bourne, L. and Rowlinson, S. (2008a) Stakeholders and the supply chain. In: Walker, D.H.T. and Rowlinson, S. (eds), *Procurement Systems: A Cross-Industry Project Management Perspective* (1st edn). Abingdon: Taylor & Francis, pp. 70–100.
Walker, D.H.T., Bourne, L.M. and Shelley, A. (2008b) Influence, stakeholder mapping and visualisation. *Construction Management and Economics*. 26(6), pp. 645–658.
Willer, D., Lovaglia, M.J. and Markovsky, B. (1997) Power and influence: a theoretical bridge. *Social Forces*. 76(2), pp. 571– 603.
Worsley, L.M. (2017) *Stakeholder-Led Project Management: Changing the Way we Manage Projects*. New York: Business Expert Press.
Yang, J., Shen, P.Q., Bourne, L., Ho, C.M. and Xue, X. (2011a) A typology of operational approaches for stakeholder analysis and engagement. *Construction Management and Economics*. 29(2), pp. 145–162.
Yang, J., Shen, G.Q., Ho, M., Drew, D.S. and Chan, A.P.C. (2009) Exploring critical success factors for stakeholder management in construction projects. *Journal of Civil Engineering and Management*. 15(4), pp. 337–348.
Yang, J., Shen, G.Q., Ho, M., Drew, D.S. and Xue, X. (2011b) Stakeholder management in construction: an empirical study to address research gaps in previous studies. *International Journal of Project Management*, 29(7), pp. 900–910.
Yang, R.J., Wang, Y. and Jin, X.H. (2014) Stakeholders' attributes, behaviors, and decision-making strategies in construction projects: importance and correlations in practice. *Project Management Journal*. 45(3), pp. 74–90.
Yang, R.J. and Shen, G.Q.P. (2014) Framework for stakeholder management in construction projects. *Journal of Management in Engineering*. 31(4) pp. 1–14.
Yong, Y.C. and Mustaffa, N.E. (2013) Critical success factors for Malaysian construction projects: an empirical assessment. *Construction Management and Economics*. 31(9), pp. 959–978.
Young, T.L. (2013) *Successful Project Management* (4th edn). Philadelphia, PA: Kogan Page.

Websites

www.airiodion.com/stakeholder-tool/ (Accessed on 15 June 2020)
www.bbc.co.uk/news/uk-scotland-highlands-islands-49733098 (Accessed on 21 June 2020)
www.boreal-is.com (Accessed on 23 June 2020)
www.darzin.com (Accessed on 16 June 2020)
www.jambo.cloud (Accessed on 18 June 2020)
www.tractivity.co.uk (Accessed on 22 June 2020)
www.valuebasedmanagement.net/methods_clarkson_principles.html (Accessed on 17 October 2020)
https://www.vuelio.com/ (Accessed on 20 June 2020)

21. The global construction market

Weisheng (Wilson) Lu and Meng Ye

INTRODUCTION

Construction is a hugely important industry in scope and scale (Jewell et al., 2014). Construction materialises the built environment and influences human health, social behaviour and economic activities; some of the works (for example, the Pyramids, the Great Wall and the Colosseum) have become the cultural identity and civic pride of humankind (Pearce, 2003). Construction generates a vast number of employment opportunities. For instance, over 55 million people were employed in construction in China in 2018 (National Bureau of Statistics of China, 2019). Around 7.2 million were employed in the industry in the United States (US Bureau of Labor Statistics, 2020) and 13.5 million in the European Union in 2019 (Eurostat, 2020). Construction is used to boost economies, especially during the recession times. For example, in the aftermath of World War II, the US-sponsored Marshall Plan was launched to aid reconstruction of cities, industries and infrastructure in Western Europe (Hogan, 1987). In stimulating the recovery from the global financial crisis of 2008, most of the Chinese government's 4 trillion *yuan* stimulus package was spent on railways, highways, and other infrastructure projects (Tang, 2020). Construction contributes significantly to national economies. Across Organisation for Economic Co-operation and Development (OECD) countries, construction consistently contributes 5 to 10 per cent of gross domestic product (GDP) even without taking into account intermediate inputs such as materials, plant and equipment. Construction consultancy company Davis Langdon (2012) in their report estimated that construction yields US$4.6 trillion and contributes 6.6 per cent of the global GDP annually. By 2030, according to a study from Global Construction Perspectives and Oxford Economics (2015), the global construction market will have grown by US$8 trillion. Clearly, construction is an enormously important industry that cannot be ignored by any economy (Hillebrandt, 1984).

After briefly outlining the global construction industry, this chapter shift the focus to the global construction market. As Hillebrandt (2000, p. 9) argues, 'While considering many of the matters of interest to economists … it is not the industry but the market which is relevant'. A market is 'any organisation whereby the buyers and sellers of a particular commodity keep in touch with each other and determine the price of the commodity' (Hillebrandt, 2000, p. 9). Analysis of a market involves consideration of demand, supply, pricing, equilibrium, competition, and many other aspects. The heterogeneities of the construction market make this analysis a complex undertaking. In construction, a commodity typically goes through several consecutive stages, from two to ten years of inception, design and construction, to 30 years or longer of operation and maintenance, and then demolition. A construction commodity cannot be purchased 'off the shelf'. Rather, clients procure construction from the market via different methods (design–bid–build, design–build, and build–operate–transfer). Demand is cyclical, being heavily contingent on 'the state of the economy and government policies' (Hillebrandt, 1984, p. 2), while supply is normally delivered through a project; an endeavour with a clear start and end (Sydow et al., 2004; Turner and Müller, 2003). Suppliers in the construction market – architects, surveyors, engineers, contractors, specialty contractors, manufacturers and suppliers, and operators/facility managers – are contracted to complete parcels of the project along its life cycle. These players belong to distinct professional bodies, and at times operate in silos. The majority of the companies in the construction market have fewer than ten employees, while the rest are large regional or multinational companies; somewhat following the Pareto principle.

While the list of features of heterogeneity could be longer, and it is still making sense in many places, the global construction market has witnessed a sea change over the past five decades or so. Once a local undertaking using indigenous materials, technologies and labour, construction today is conducted in a global marketplace. According to *Engineering News-Record* (*ENR*), in 2019 its top 250 global contractors had a total of US$1.9 trillion contracting revenues (Tulacz and Reina, 2020). Nearly 25 per cent of this revenue was obtained from construction projects undertaken by contractors outside their home countries (Tulacz and Reina, 2020). Facilitated by economic globalisation, companies are expanding across continents or regions to pursue growth, sometimes because they have become too large for their domestic home market, or to mitigate the risks of overreliance on individual and cyclical markets, or for other purposes. Advanced technology, fast transportation, convenient communications, effective knowledge transfer, integrated markets and trade liberalisation have all helped to shape a competitive international construction marketplace (Lu et al., 2013). Companies rise and fall in this market. While the old-line international contractors from the US, Japan, the United Kingdom (UK), France, Germany, Spain, Sweden and Italy are still active, contractors from China, South Korea and Turkey have become strong contenders in the field (Low and Jiang, 2003), with contractors from India, Russia, Brazil and Australia now active in the market as well. In a high-profile example, the Chinese government's Belt and Road Initiative promotes the internationalisation of construction by developing the supportive infrastructure. A reasonably comprehensive account of its evolution of the global construction market is yet to be seen in the literature.

Therefore, the chapter aims to sketch the historical development and status quo of the global construction market through empirical analyses and, based on these, to extrapolate its future trends. It is particularly challenging to achieve any precision in the extrapolation and forecasting of market trends amid drastic geopolitical change brought about by the Belt and Road Initiative, Brexit, the Trump administration's policy shift on trade relationships, and the rise of populism, among other factors. Some hold the view that globalisation will encounter a backlash in favour of more localisation in response to this change, which will only be exacerbated by the COVID-19 pandemic (Antràs, 2020; Elliott, 2020; Goodman, 2020). National borders are closed and international airports have been shut down, global sourcing of materials has virtually been halted, and the free movement of labour has ceased. The global construction market will see a 'new normal' that is shaped by the impact of the pandemic and the rethinking of globalisation. However, what will this 'new normal' look like?

The remainder of this chapter is structured as follows: subsequent to this introductory section, scope definitions of several concepts are provided in the next section with a view to forming a framework for the subsequent analyses. The following section presents an analysis of the historical development and status quo of the global construction market, based predominantly on empirical data from the United Nations (UN) and the *ENR*. A trend analysis of the global construction market is undertaken under three areas: emerging construction force, kaleidoscopic diversification and uncertain economic globalisation. Conclusions are drawn and topics for future research are recommended in the final section.

SCOPE DEFINITIONS

Construction is a complex concept. A range of seemingly synonymous terms are used to discuss it: 'construction industry', or 'construction sector'; 'construction', or 'architecture, engineering and construction (AEC)' or 'architecture, engineering, construction and operation (AECO)'; and, in the context of this chapter, 'international construction' or 'global construction'. As Dainty et al. (2007, p. 4) has pointed out, 'precisely what constitutes the "construction industry" is in itself subject to a range of different boundary definitions'. Therefore, the scope, similarities and differences of these terms are outlined here in order to proceed with the discussion meaningfully.

Construction Sector, Construction Industry or AEC Industry

Often used interchangeably in the literature, 'industry' and 'sector' have slightly different meanings. The term 'sector' refers to a large segment of the economy, whereas 'industry' refers to a much more specific group of companies or businesses. According to Jewell (2011), the economy is divided into several large sectors, each including several industries.

Narrow and broad definitions of the construction sector identified by Pearce (2003) have been used in many subsequent studies (for example, Dainty et al., 2007; Lu et al., 2013). The narrow definition focuses on on-site assembly and repair of buildings and infrastructure. It involves specialised construction activities, often collectively known as 'contracting'. This can be perceived as equivalent to 'the construction industry', in which the players are contractors and subcontractors who primarily undertake construction works onsite. The broad definition extends to off-site activities such as quarrying of raw materials, manufacturing of building materials, sales of construction products, architecture and engineering and other types of professional consultancy, and plant and equipment production or supply. It also extends to construction-related financing, accounting, information and communication technology (ICT), and legal services. Therefore, according to the broad definition of the construction sector the players include architects, engineers, other consultants, investors, plant manufacturers, accountants, lawyers and other professional service providers.

Standard industrial classification systems reveal the intricate relationships between construction-related concepts. Figure 21.1 is developed from the UN's (2008) International Standard Industrial Classification of All Economic Activities (ISIC). The narrow definition of the construction industry is mainly included in Section F comprising construction of buildings (ISIC 4100), construction of roads and railways (ISIC 4210), construction of utility projects (ISIC 4220), construction of other civil engineering projects (ISIC 4290), demolition (ISIC 4311), site preparation (ISIC 4312), electrical, plumbing and other construction installation activities (ISIC 4321, 4322 and 4329), building completion and finishing (ISIC 4330), and other specialised construction activities (ISIC 4390). As mentioned above, these activities may also be referred to as 'contracting'. According to Pearce (2003), the narrowly defined construction industry contributes around 5 per cent of the total GDP of a nation.

Another term in wide use is the largely self-explanatory 'architecture, engineering, and construction (AEC) industry' and, more recently, the 'architecture, engineering, construction and operation (AECO) industry'. As shown in Figure 21.1, the AEC industry combines traditional contracting (ISIC Section F) with construction professional services (ISIC Section M) to provide architectural, engineering, drafting, building inspection and surveying and mapping services. It mainly consists of three kinds of players: architects, engineers and contractors, including consultants to their clients in addition to contractors. By integrating these players into one industry, they are expected to be more efficient in achieving the common goal of bringing a construction project to fruition.

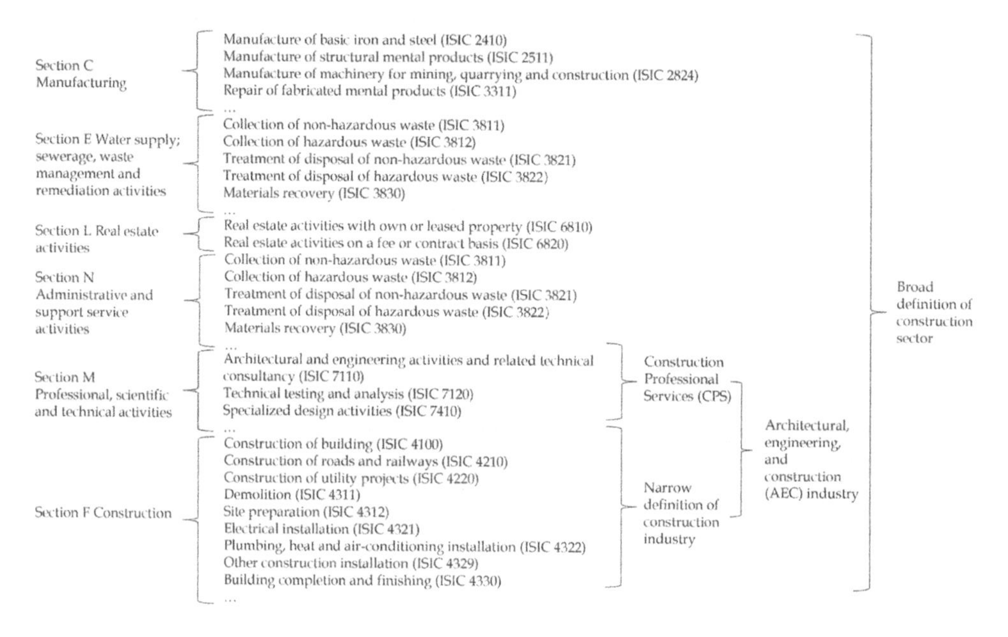

Source: United Nations (2008).

Figure 21.1 *Industrial classification of construction businesses*

The broad definition of the construction sector as shown in Figure 21.1 includes on-site and off-site activities. It encompasses manufacturing activities (ISIC Section C), water supply and waste management (ISIC Section E), real estate activities (ISIC Section L), professional and technical activities (ISIC Section M), administrative and support services activities (ISIC Section N), and construction works (ISIC Section F). If this broad definition of the construction industry is considered, it contributes around 10 per cent of a nation's GDP (Pearce, 2003).

Production or Service

'Production' and 'service' are different types of business following different business paradigms. The production process, in its simplest form, transforms inputs (raw materials, semi-finished goods or subassemblies) into finished products or goods, tests and packages the finished products, and ships them for sale 'off the shelf' in the market. Unlike most manufactured goods, demand for services is extrinsic; they are bespoke and dependent on each client's requirements (Jewell et al., 2014). A widely accepted paradigm holds that they have four defining characteristics: (1) intangibility, meaning a service cannot be seen, touched, tasted or smelled; (2) heterogeneity, meaning it is composed of parts of different kinds having widely dissimilar elements or constituents; (3) inseparability of the production of a service from its consumption; and (4) perishability, meaning unused service capacity cannot be stored for future use (Zeithaml et al., 1985). A strategic trend over the past decades of economic globalisation has been a shift of production from the developed to the developing and emerging world, first to the 'Four Asian Tigers' – Hong Kong, Singapore, South Korea, and Taiwan – and later to China and then Thailand and Vietnam, to exploit the cheaper labour and material inputs there. However, the developed world remains dominant in knowledge-intensive service sectors such as finance, design, ICT, and other professional services.

Construction has features of both production and service. The production cycle for construction outputs is long; from conception, it can take many years to reach project completion. The production process itself is quite standard, involving transformation of materials, semi-finished goods or subassemblies (for example, prefabricated components) into a final product. Construction production also relies on professional services, including architecture, urban planning and landscape architecture; civil, structural, mechanical and electrical engineering; surveying; as well as construction-related accountancy, legal and ICT services. Two points are worthy of mention here. First, unlike those in the manufacturing industry, construction outputs are attached to the land and are not readily transferable goods (Hillebrandt, 2000). This means that, in construction, whether production or services are outsourced or offshored, they ultimately have to be brought back to a locality. Second, many contracting firms responsible for materialising construction products have relisted themselves on the stock market as service firms (Goodier et al., 2006), arguing that they are providing construction services.

Indigenous, International or Global

Traditionally, construction is a local business, using mainly local labour and materials, and complying with local standards and regulations. This is largely due to the site-specific, unique and purpose-built, fixed nature of construction outputs. As a result, among industries, construction has been described as conservative and slow to respond to globalisation. However, intensive urbanisation and global economic growth have given rise to demand for international construction. Ngowi et al. (2005) define this as the part of the construction business which is resident in one country, and performing in another country either through ownership of an overseas business or through acquisition. The *ENR* has provided a ranking of the top 225 (250 in recent years) international construction companies since the 1970s. In its methodology, Tulacz and Reina (2019) state that the ranking is based on general construction contracting revenues generated from projects outside each firm's respective home country. The international construction firms' revenues measure their presence in international commerce (Reina and Tulacz, 2005).

Another concept in use is that of the 'global construction market', which is used to represent the overall construction market on a global scale. The *ENR* ranks the top 250 global contractors by total revenues in both international and domestic markets; most sizable global contractors have both international and domestic construction businesses. This chapter focuses on the global construction market, and both international and domestic construction businesses are considered.

Market Segment and Geographic Dispersal

The global construction market can also be perceived from two dimensions: market segments and geographic spread. As delineated above, even the narrow definition of the construction industry (that is, contracting business) can be divided into several segments. The *ENR* also adopts such a taxonomy, classifying the contracting business into ten market segments: general building, manufacturing, power, water supply, sewerage/solid waste, industrial process, petroleum, transportation, hazardous waste and telecommunication (see Figure 21.2). Classification is based on the type of final construction outputs. For example, 'general building' includes commercial buildings, offices, stores, educational facilities, government buildings, hospitals, hotels, apartments and housing; 'transportation' includes airports, bridges, roads, marine facilities, piers, railroads and tunnels; and 'telecommunication' comprises transmission lines, cabling, towers and antennae, and data centres. Although they are all contracting business, different market segments draw on unique technologies and operational processes (Reina and Tulacz, 2005).

Rarely explored are the globalisation strategies of construction firms that guide their expansion into different geographical areas. In analysing the global construction market, there is no standard agreed taxonomy for geographical areas. Researchers tend to adopt the *ENR* convention, classifying regional markets into six categories: North America, Latin America and the Caribbean, Asia and Australia, the Middle East, Africa, and Europe (see Figure 21.3). Unlike other types of business that can create the demand for their new products or services,

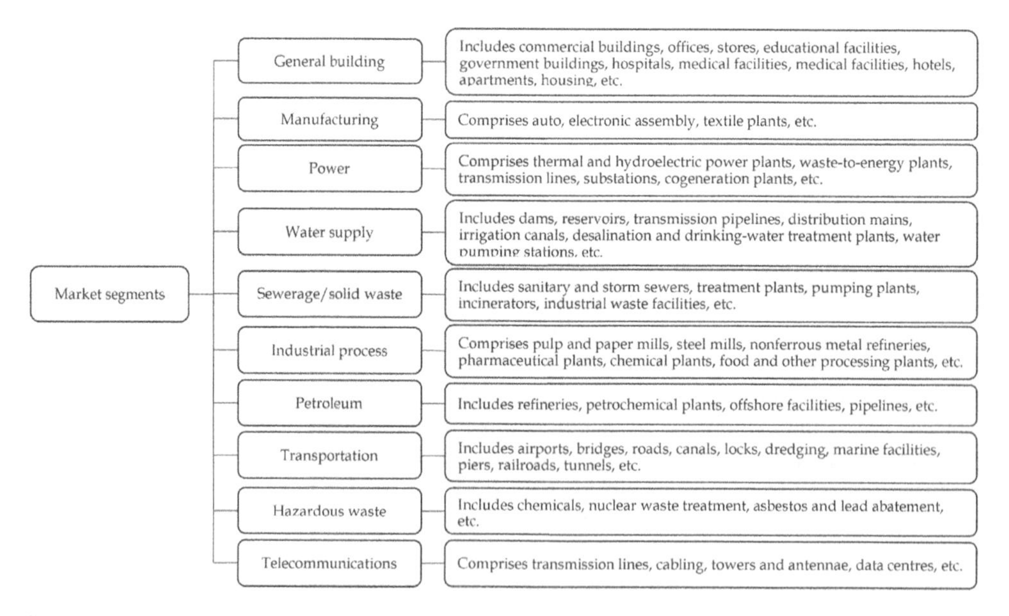

Source: ENR report, Reina and Tulacz (2005).

Figure 21.2 Market segments of contracting business defined by ENR

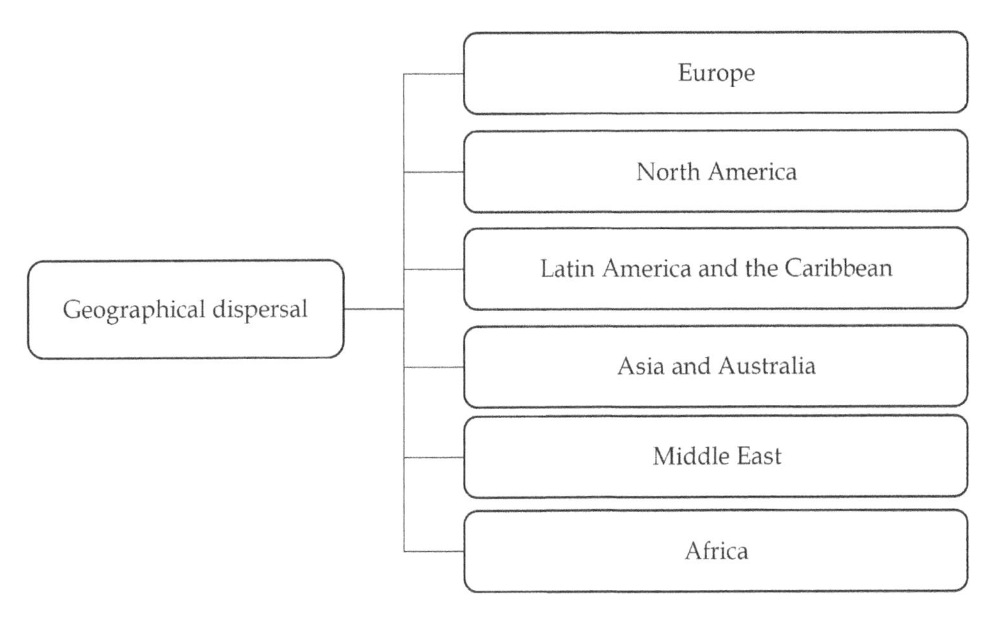

Source: ENR report, Reina and Tulacz (2005).

Figure 21.3 Geographic dispersals of the global construction market defined by ENR

international construction companies tend to react to construction demand passively (Kim and Reinschmidt, 2011). Many large construction firms thus flow with the rise and fall of market demands in different continents. In this chapter, the *ENR*'s taxonomy of geographic dispersal is adopted in discussing the global construction market.

Given that the aim of the chapter is to examine the past, present and future of the global construction market, we adopt an underlying longitudinal perspective. The examination follows different geographic dispersals, and delves into different market segments. While the focus is on contracting, other emerging trends such as AEC integration that are happening in the market are considered to provide one-stop construction services.

HISTORICAL DEVELOPMENT, STATUS QUO AND FUTURE OF THE GLOBAL CONSTRUCTION MARKET

A Fast-Growing but Fluctuating Global Market

Construction and overall economic activities data were searched and downloaded from the Economic Statistics Branch of the UN Statistics Division (UNSD) (http://data.un.org/). The branch maintains a database of detailed official national accounts statistics from 1970 onwards for more than 200 countries and areas of the world. The data includes the gross value added by construction and total economic activities from 1970 to 2018. According to the US Bureau of Economic Analysis (2020), the contribution of private industry or the government sector to overall GDP is the value added of an industry, also referred to as 'GDP by industry'. Value added by construction activities can thus be called 'construction GDP', as it is here. The UNSD uses the definition of 'construction' applied in ISIC Section F (see Figure 21.1). The GDPs of individual countries were aggregated into the *ENR*'s geographical areas (see Figure 21.3) but 'Asia and Australia' were further divided into two areas: 'Asia' and 'Australia and New Zealand'.

Figure 21.4(a) presents construction GDP in the seven geographic locations over 48 years from 1970 to 2018. Overall, there is an upward trajectory with variation among locations. North America and Europe consistently provided the largest construction markets. Europe remained the largest construction market from the 1970s to the early 1990s when it was overtaken by Asia. After stable growth in the 1990s, it grew rapidly in the 2000s until the 2008 global financial crisis. Since then, European construction GDP has fluctuated radically in line with economic variations. In North America, construction GDP enjoyed a steady increase from 1970 until the financial crisis. The market was seriously affected, but recovered from 2010.

Among the geographic areas, Asia in particular has experienced growing demand for construction. Construction GDP in Asia increased from 1970 to 1995, dropped significantly with the 1997 Asian financial crisis, and started to rebound from 2002. The Asian market has become the biggest market in recent years. Top markets within Asia are China, Japan, Korea and India, which have seen strong year-on-year growth. The Middle Eastern market is also significant. Although its overall construction GDP is below that of Latin America and the Caribbean, it is an attractive market to construction companies, which is boosted by a rising oil price, and hit by oil shocks such as the ones in the 1970s, 2009, and the 2020 break-up of the oil production cut agreement. Africa, Latin America and the Caribbean, and Australia and New Zealand, have witnessed growth in construction GDP since 2007, but the rates of change are less impressive than those in Europe, North America and Asia.

Figure 21.4(b) plots the total GDP by all economic activities in the seven geographic areas over the same period of time, with a view to examining the correlation between construction, crude oil price and the overall economy. Table 21.1 examines the correlations between the construction GDP, crude oil price and total GDP for each country or region all over the world. It can be seen that construction has a close association with economic development and the crude oil price. Construction markets have fluctuated more or less at the same pace in line with economic cycles in various countries. Construction may foster economic growth through two ways: (1) materialising the built environment in support of economic activities; and (2) boosting economies as an investment vehicle via government expenditure.

From a Bipolar to a Multipolar Market

Figure 21.5 plots the construction GDP of each country on a world map, longitudinally comparing the development of construction markets across continents. Owing to the long period considered, snapshots of 1970, 1980, 1990, 2000, 2010 and 2018 are provided as representatives. The circle size represents the construction GDP by country. Figure 21.5 can be read in conjunction with Figure 21.4. It appears that global construction has developed from a bipolar market (that is, Europe and North America only) to a multipolar market that includes emerging economies such as India, and China, the Middle East, South Korea and Australia.

Figure 21.5 shows that in 1970 the biggest construction markets were the US and Western Europe. In 1980, from a low base, construction markets developed in Canada, Mexico, Brazil, Nigeria, the Middle East, China, India and Australia. Among continents, Europe had the

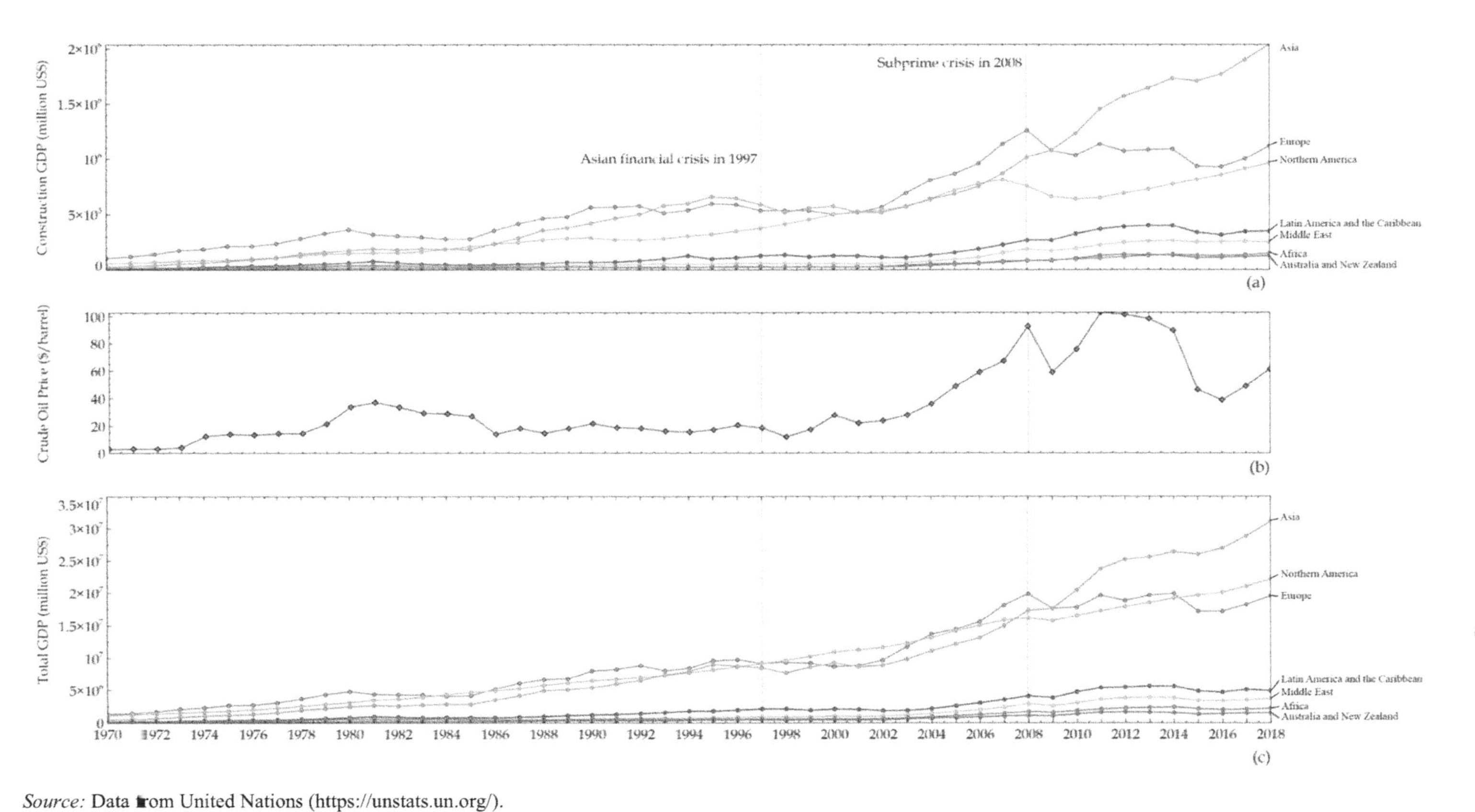

Source: Data from United Nations (https://unstats.un.org/).

Figure 21.4 Construction GDP in seven geographic areas, 1970–2018

Table 21.1 Correlations between construction GDP, crude oil price and total GDP for each country all over the world

		1	2	3
1	Construction GDP	1		
2	GDP	0.9928***	1	
3	Crude oil price	0.1002***	0.1013***	1

*** p-value < 0.001

Source: Based on data from United Nations (https://unstats.un.org/).

largest construction GDP in 1980, with Western European countries such as the UK, Germany, France, Spain and Italy contributing the largest shares. Japan's construction market started to take off in line with its economic development in the 1980s and 1990s.

In 1990, the biggest construction markets were in the US, Japan and the European continent. The construction market of South Korea saw a rapid increase in 1990. Among individual countries, the US and Japan were the biggest construction markets in 2000. When the Japanese asset price bubble burst in 1992, the construction market just stagnated along with the overall economy, as evident in the size of the circles over Japan being unchanged from 1990 to 2000. The year 2000 saw unprecedented growth of China's construction GDP, a trend which has

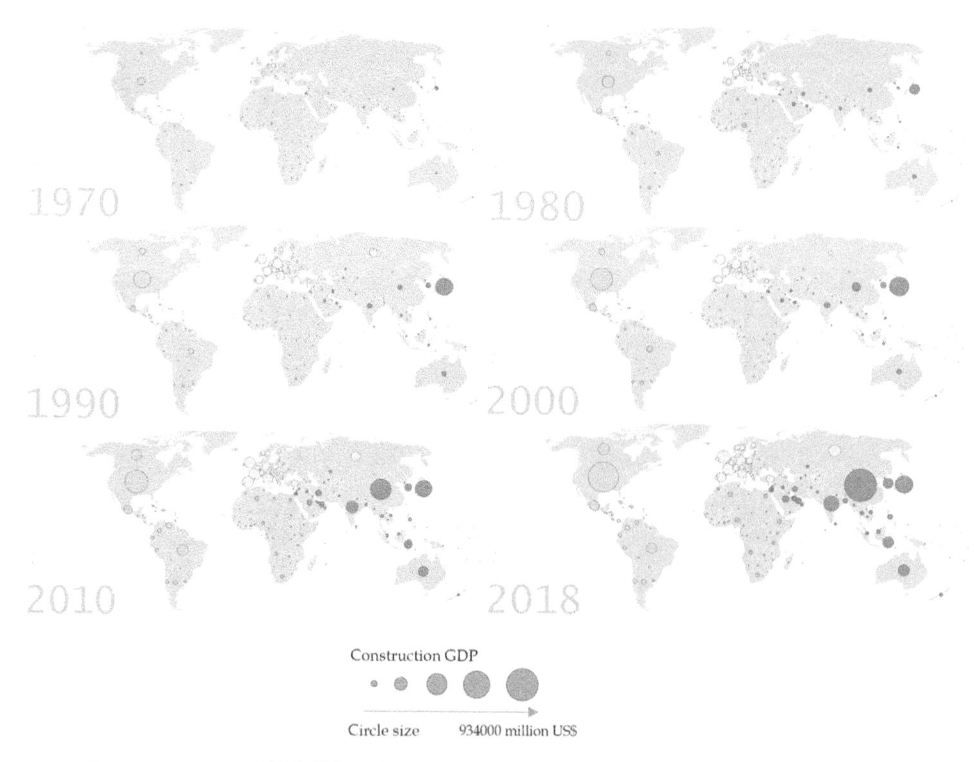

Source: Based on data from United Nations (https://unstats.un.org/).

Figure 21.5 Snapshots of construction GDP of major economies (1970–2018)

continued to today. Upon adopting its Open-Door Policy in 1979, China began to reform its economic system by developing a 'socialist market economy'. The growth and accumulation effect of this economic reform has been evident since the 2000s.

From 2010, construction markets have grown rapidly in Asia and Latin America, particularly in China, India, Indonesia and Brazil, due to rapid economic development. While European construction markets have slowed, China and the US have become the biggest construction markets in the past decade. The past decade has also seen an increase in construction in Africa and the Middle East. The African markets are evenly scattered in the coastal areas (for example, South Africa, Angola, Tanzania, Ethiopia, Egypt, Algeria and Nigeria) around the whole of Africa without a particular hotspot. The Middle East market is boosted by the oil price, but it is often impacted upon by oil price shocks and the move towards renewable sources of energy.

It can be seen from the above analyses that economic power has not shifted from traditionally strong economies (for example, North America, Western Europe, Japan and Australia) to the developing world. In fact, the gap between developed and developing economies, as shown in Figure 21.5, is widening. Nonetheless, developing economies have achieved impressive economic development and construction market growth, creating a multipolar global construction market.

Internationalisation in the Global Market

The global market growth provides construction companies with opportunities to pursue growth (Kim and Reinschmidt, 2011), or to minimise the risk of overreliance on a single market with cyclical and fluctuating demands (Jewell, 2011), or both. To compete, these companies bring with them construction know-how, advanced technologies and strong management capability, and configure resources globally. Data was collected from the *ENR*, which, as mentioned earlier, ranks top 250 global contractors based on 'global' or 'total revenue' comprising domestic revenue and international revenue. Lu (2014) has proven the relatability of the *ENR* data, although it is self-reported, and the data set is particularly useful for understanding global construction at individual firm level or as a whole.

Figure 21.6 illustrates international and domestic revenues logged by the *ENR*'s top 225/250 global contractors over the past 21 years. In general, the international revenue has risen with some fluctuations, particularly during 2007–2009 and 2014–2016. Economic uncertainty matters to the international construction markets. The meltdown of the financial markets in 2008 halted much of the financing needed to launch projects, and was followed by a collapse in oil and metals prices (Reina and Tulacz, 2009). The fluctuation during 2014–2016 was also caused by economic uncertainty, such as falling oil price, China's devaluation of its *yuan*, and growing political turmoil in the Middle East and elsewhere (Reina and Tulacz, 2015, 2016). The domestic revenue has been less affected by the economic uncertainty, and has been continuously increasing. In 2019, total revenue of the *ENR* Top 250 global contractors, comprised of international and domestic revenues, totalled US$487.29 billion, up from US$116.39 billion in 1998. Figure 21.6 also shows that international business growth of top global contractors is consistent with that of domestic business before 2013. In recent years the situation has changed as top global contractors, as a group, have taken a more cautious approach to the international market with its diverse and complex range of projects involving high risk and uncertainty amid a dynamic market. In interviews with chief executive officers of contractors

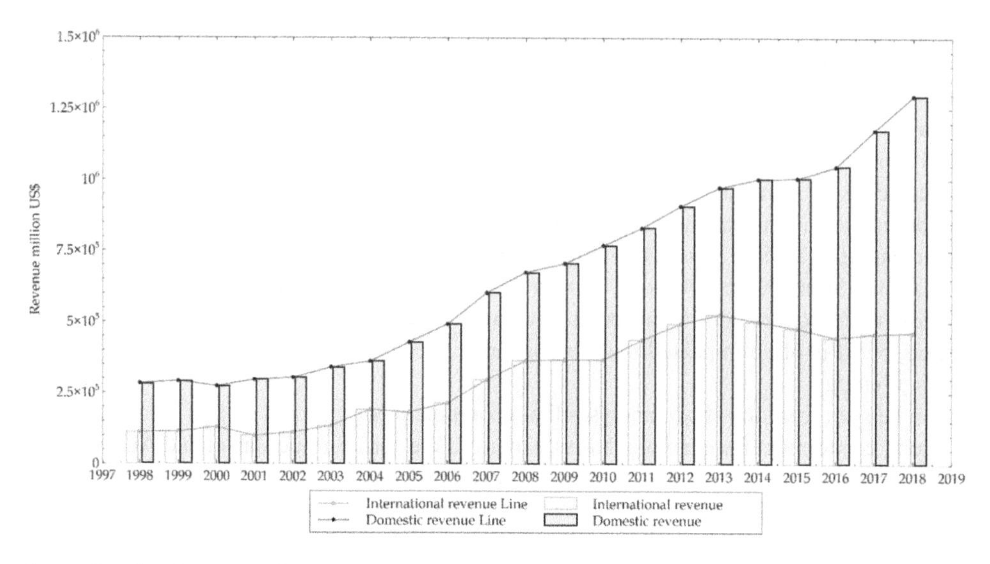

Source: Based on data sources from various *ENR* reports, 1999 to 2019.

Figure 21.6 *The international and domestic revenues for ENR's top 225/250 global contractors, 1998–2018*

published in recent *ENR* annual reports, a widespread depression is easily discerned, except in the case of contractors from the US and China.

Globalisation has changed the nature of competition (Hatzichronoglou, 1996). Researchers have been interested in the mode adopted by firms when entering an international construction market, because the process is onerous and subject to many regulations. These include or relate to controls on land use; building regulations and technical requirements; building permits and inspection; registration of proprietors, contractors and professionals; regulation of fees and remunerations; and environmental regulations (WTO, 1998). Chen and Messner (2009) identified a taxonomy of ten basic entry modes for international construction markets, including strategic alliance, build–operate–transfer, joint venture, representative office, licencing and local agent. Ngowi et al. (2005) describe four ways in which construction firms enter the international markets. These are: during economic booms such as the one resulting from the sale of oil; through bilateral and sometimes multilateral agreements setting up protocols that open markets to each other; by becoming involved in large international projects; and through growing and expanding multinational operations because they have outgrown the domestic market. Ngowi misses some motivations that take companies overseas, such as a client wanting the company to construct a project overseas, and mergers and acquisitions (M&A) undertaken by companies to broaden their exposure to different markets and reduce reliance on their domestic market, which invariably has fluctuations in levels of demand.

Emerging Construction Forces

In Figure 21.7, the top 225/250 global contractors are categorised by nationality/home country and plots are made for the number of firms, and international and total revenues. Two clear trends can be observed. First, global contractors that originated from developed economies have remained strong performers over time. The US has the largest share of companies on the list and share of total revenues, while Europe and Japan are respectable construction forces despite declining numbers of contractors and revenues. Second, companies from several developing or emerging economies, particularly China, South Korea and Turkey, are becoming new forces. Figure 21.7 presents a decreasing trend of international revenue for the US contractors, but an increasing trend of Chinese contractors to expand to overseas markets as well as develop domestic markets. Just five Chinese companies appeared on the *ENR* list of top 225 global contractors in 1994, while since 2013, around 60 Chinese companies have consistently appeared on the top 250 global contractors list, earning around US$900 billion a year from both domestic and international markets. In 2018, Turkey and South Korea had 14 and 12 companies on the list and logged a total revenue of US$27.5 billion and US$88.5 billion, respectively. A caveat for this discussion is that firm nationality counts may not be correct; construction firms can be held in full or in part by capital from another country or international organisations.

According to Low and Jiang (2003), the internationalisation of Chinese construction companies has occurred in three stages: (1) through Chinese government economic and technical aid before 1979; (2) with the emergence of Chinese construction companies from the early 1980s; and (3) with the development of multinational enterprises since the early 1990s. Lu et al. (2009) add a new stage occurring after China entered the World Trade Organization

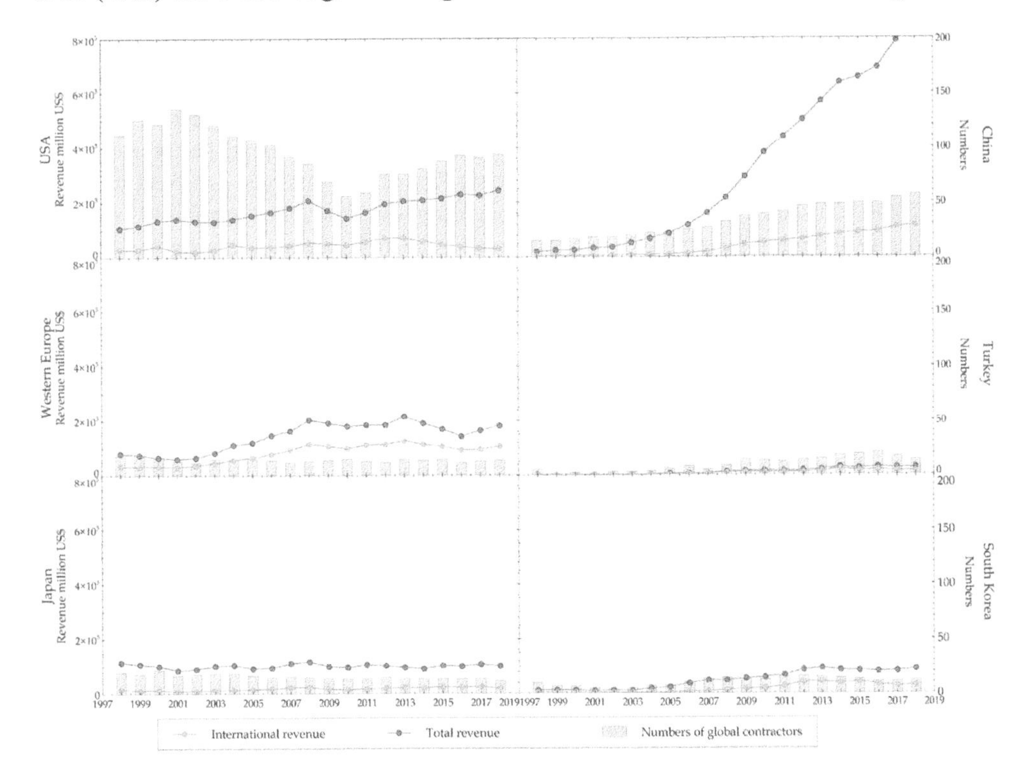

Figure 21.7 *Number of companies listed on the ENR's top 225/250 global contractors and their revenues, 1998–2018*

(WTO) in 2001. At this stage, construction internationalisation took off in line with the Going Global Strategy; an overall strategy formally set forth in the 1990s by the central government to encourage companies to compete in the international arena (Xing, 2002). The Chinese government's Belt and Road Initiative announced in 2013 aims to improve connectivity and cooperation among countries spread across the continents of Asia, Africa and Europe. By making international construction markets more accessible, the Belt and Road Initiative has generated many opportunities for international construction companies (Reina and Tulacz, 2016). It has also been a source of considerable controversy (Tulacz and Reina, 2019), raising concerns about the increasing geopolitical, economic and military influence of China. China's domestic market also plays a role. According to the latest (2019) *ENR* top 250 global contractors list, the top five companies are all Chinese companies, particularly related to railways and highways construction.

To earn precious foreign exchange, the South Korean government has mobilised construction services as one of the major exporting sectors. Lee et al. (2011) divides the historical development of South Korean international construction business into four major phases that correspond to significant events. Starting in the mid-1960s, South Korean international construction companies had successfully participated in more than 6100 contracts in 112 countries, with a total contract value of over US$260 billion (ICAK, 2008). The oil-exporting countries of the Middle East have been the principal markets through which Korean international contractors have survived and thrived (Kim, 1988). In the early stages, Korean international contractors were strong only in labour-intensive works such as general building and civil engineering projects, taking an energetic, capable and hard-working labour force with them to the Middle East (Kim, 1988). In the mid-1990s, Korean firms shifted the type of projects in their overseas construction portfolios from simple building and civil engineering works to more complex industrial plant projects, and through the mid-2000s the plant sector accounted for about 70 per cent of the total amount of overseas Korean international construction firms (ICAK, 2008). This shift benefited from the strong domestic production capability (such as, steel, concrete and other materials) of conglomerates such as Hyundai, Samsung and Pohang Iron & Steel Company.

Following Porter's (1990) diamond framework, Öz (2001) conducted a detailed account of the sources of competitiveness of Turkish international construction companies. These included not only low-cost advantages and geographic and cultural proximity to several promising markets (for example, the Middle East, Central Asia and North Africa), but also accumulated experience at home, pressures from domestic peers, and several related and supporting industries (for example, cement, bricks, ceramics, and other materials). Reina and Tulacz (2014) echo the idea that cultural and religious proximity in these predominantly Muslim regions has made Turkish firms naturally comfortable in these markets. Turkish companies such as ENKA and Renaissance Construction are also positively seeking to enter Western European and US markets, sometimes through the use of aggressive M&A strategies.

Although the rise of the international construction business of China, South Korea and Turkey is attributable to different circumstances and government policies and chance, they appear to possess some advantages in common: (1) catering to the demand conditions from a particular construction market; (2) amenable factor conditions such as abundant labour force; and (3) related and supporting industries for materials, plant, and research and development (R&D). The rise of the international construction business in those three countries was discussed with industrial leaders who conducted business in Africa, China and Southeast Asia.

They expressed the view that the contracting business has long been considered as a low-end activity, which involves massive 'brick and mortar' and labour. It is particularly risky in the international market. Developed countries (except for the US) typically have ageing populations and lack a labour force willing to work overseas in potentially harsh environments. Cost is also a determinant, with the increasing cost of posting staff to reside overseas; with overheads, accommodation and benefits, the cost of an engineer from Europe overseas can exceed US$180 000 per annum. In line with the strategies to shift manufacturing to low-wage economies, developed countries also have less sources of construction materials. With the risks associated with international contracting business also being great, they tend to downstream this business to others while maintaining strength in knowledge-intensive, upstream business such as financing, design and other consultancy services. Nevertheless, in their recent engagement with the industry, the authors noticed that some top international consultancy firms, which were historically only interested in the upstream business, covet the contracting works. After all, the muddy 'brick and mortar' activities make up 70 per cent of the total construction budget.

Kaleidoscopic Diversification

The entry barriers to the construction market are low. The construction market is typically fragmented (see, for example, Egan, 1998) with numerous small players. Interested in measuring the intensity of competition (IoC) in the international market, Ye et al. (2009) synthesised four major concentration methods – concentration ratio, entropy, Gini coefficient and the Herfindahl index – into a new model. They then applied 28 years of international revenue data of the top ENR 225 international contractors in the new model, finding that the magnitude of the market shares occupied by the top four contractors is 0.2735. This reveals that the international construction market is far from concentrated with, for example, domination by a top four or eight big global companies.

Fragmentation in the construction industry is also caused by multiple layers of specialisation and subcontracting, and long supply chains (Hillebrandt, 2000). There is a long production cycle for construction outputs, with several years from concept to project completion, and then outputs normally having a lifespan of over 50 years. Unlike standardised products produced by major firms, design or construction services are provided by a myriad of companies contracted to do a parcel of the job. This can lead to discontinuity problems. Nevertheless, a trend of AEC integration has been emerging. With reference to the ISIC in Figure 21.1, contracting firms are expanding into professional services, such as architectural and engineering design activities, and related technical consultancy (ISIC 7110) and other professional, scientific and technical activities (ISIC 7490), or vice versa. Lu et al. (2013) have noted that, increasingly, firms such as Fluor, Bechtel, Jacobs and McDermott are appearing on both *ENR*'s top international contractors list and top international design firm lists. It is no longer easy to distinguish between a traditional contracting company and a construction professional service firm specialising in design (Lu et al., 2013).

The motivations of AEC integration have been discussed in the literature. Jewell (2011) report that procurement in the international market is changing, increasingly requiring 'one-stop' services including design, build, finance, and even operation. Top companies tend to integrate design and contracting within their business lines to meet this requirement. Lu et al. (2013) report that such integration is also for the purpose of market penetration because the constraints in the risky international market are less onerous in the design and other professional services business. However, the outcomes of such integration are mixed. UK construction firm Balfour Beatty took over the US design giant Parsons Brinckerhoff in 2009. However, in 2014, it completed 'disposal of Parsons Brinckerhoff' with the intention of reverting to 'a simplified and more focused group' (Balfour Beatty, 2020). Similarly, the UK construction, services and property group Kier acquired and then sold another UK firm, Mouchel Consulting, after a short period of operation as a contractor-consultant (Rowland, 2020).

The global construction market has also witnessed kaleidoscopic diversification among top companies into different market segments and/or geographic locations (see Figures 21.2 and 21.3). To measure this diversification, Ye et al. (2017) developed two indexes based on the entropy method: a firm's diversity index in business sectors (*DIb*), and in geographical markets (*DIg*). Using data gleaned from financial reports of companies, *ENR* data, and databases such as Bloomberg and Capital IQ, they calculated diversity indexes of 49 companies and plotted them, as shown in Figure 21.8 (Ye et al., 2017). Except for a few cases of international contractors remaining in a certain business segment or a location (that is, scoring 0.0 in either *DIb* or *DIg*), most had diversified into other segments, locations, or both, as seen in Clusters 3 and 4 of Figure 21.8.

Companies once grew organically. Nowadays, many expand to new markets through M&A. For example, the Spanish company ACS Group adopted a vigorous M&A strategy that pushed it to the No. 1 ranking on *ENR* top 250 international contractors list in 2018. To strengthen its infrastructure construction capability, for example, ACS acquired German construction giant Hochtief in 2010. However, not all construction companies like M&A. East Asian companies

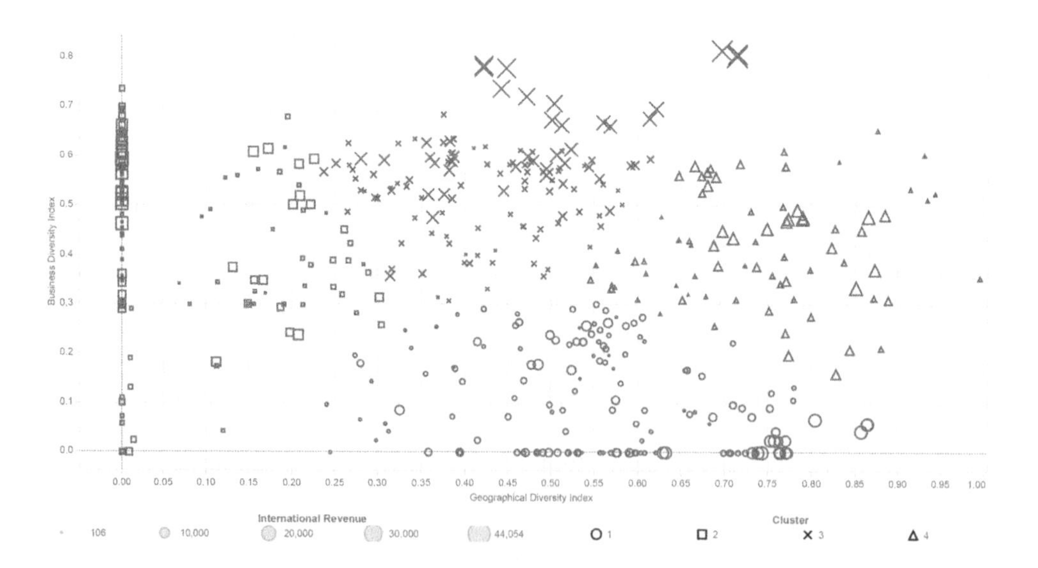

Sources: Data from Bloomberg and Capital IQ.

Figure 21.8 *Construction companies' diversification pattern in different business segments and geographical locations*

are conservative. They prefer to stay close with their construction staff with a similar culture, although theoretically the big concern for delivering a project is to manage the staff.

Construction Globalisation at a Crossroads

Rethinking globalisation has been ongoing even since it began. By looking at the longitudinal development of institutional structures, governance frameworks and technological advancements, the World Economic Forum (WEF, 2019) identified three phases of globalisation: (1) pre-1914; (2) World War II to the late 1990s; and (3) the late 1990s to the present. Its report, *Globalisation 4.0*, describes a fourth phase emerging in the context of Brexit, the Trump administration's policy shifts, China's Belt and Road Initiative, as well as advances in artificial intelligence, robotics, big data, and other technologies. The drawbacks and benefits of globalisation have been discussed, and many have called for it to be reshaped, including Rodrik (1998) who asked, 'Has globalisation gone too far?', as well as Palmisano (2016), Ghemawat (2016), *The Economist* (2017), Reeves and Harnoss (2017) and Wladawsky-Berger (2017).

Since the beginning of 2020, the COVID-19 pandemic has shaken the world. By late July 2020, total confirmed cases of infections had reached 16.7 million and the death toll 660 000 (Johns Hopkins Coronavirus Resource Center, 2020). On a global scale, as of the time of writing in mid-2020, there is no sign that the outbreak has plateaued. In many places, including the US, Spain, Australia and Hong Kong, the epidemic has returned in second or third waves. While the vaccines which have been developed have been tested and approved for emergency use, it will still be necessary to continue to rely on social distancing, mask-wearing and personal hygiene measures to contain its spread. National borders and international airports have been locked down for many months; a situation that was unimaginable a few months ago. Overreliance on Chinese medical equipment and supplies during the period has led to intensification of the criticism of China for its dominance in globalisation, including its Belt and Road Initiative. The COVID-19 pandemic has catalysed a backlash against globalisation which it was already facing from the rise of populism.

What is the future of globalisation? Four interviews with industrial and research leaders involved in international construction were conducted for this study, during February and August 2020. Interviewer 1 is a researcher based in the UK; Interviewers 2, 3 and 4 are industry practitioners based in Beijing and Shanghai, China. Although the interviewees were geographically dispersed, the interviews were all conducted via video-conferencing owing to the pandemic. The views of the interviewees are now shared and analysed.

Interviewee 1 expressed cautious pessimism about economic globalisation. He noted:

> I cannot imagine how the world will go without globalisation. We are locked down at home, but we cannot wait to be unlocked, both literally and figuratively. Globalisation has created huge value, raising many people out of extreme poverty. International construction plays a key role. Indeed, there is some dissatisfaction with globalisation, but it doesn't mean we need a U-turn.

Interviewee 2 shared:

> In addition to the backlash against globalisation, there is an ongoing debate on globalisation excluding China. I believe that this will not affect international construction or Chinese companies too much. In comparison with their Western counterparts, Chinese construction workers are more tolerant of the risks associated with this business. The old-line international construction companies pulled out

> from the Belt and Road Initiative markets, simply because they cannot afford to work at a low price and in a harsh environment.

Interviewee 3, who has a Chinese background, observed:

> I personally feel that the Belt and Road Initiative is a good project. It creates new value for participating countries as well as the Chinese. However, we do hear many criticisms. I don't know why. Perhaps we propagated it too much, as a political or globalisation agenda, or perhaps it was deliberately stigmatised by others.

Finally, Interviewee 4 noted:

> You mentioned the case that some Chinese international construction companies even shipped formwork from China to Africa. We are still too much on the 'efficiency' sphere, assuming that it is more important than anything else. We should share the profit with stakeholders, develop local capacity and train local talent. We should assume more corporate social responsibilities.

Some expressed the view that it is not possible to go back to pre-COVID-19 times, and instead there will be a 'new normal' in many aspects of globalisation, including the global construction market. It is not clear what the new normal will look like, but it is safe to conclude that globalisation is at a crossroads. The increasing trend of internationalisation of construction business may be affected in a distinctly new era where directions towards globalisation are not clear.

SUMMARY AND CONCLUDING REMARKS

Construction materialises the built environment, provides a massive number of jobs, helps to reduce extreme poverty, and improves the quality of people's life. Construction is often used to boost economies, while on the other hand, a prosperous economy demands much construction work. Over the past decades, construction and the economy have enjoyed a virtuous circle, albeit with unavoidable fluctuations. Construction has grown into a massive business sector worth several trillion US dollars per annum. With the globalisation of the world's economy, powered by facilitating technologies and amenable institutional arrangements, the construction business is increasingly undertaken in a globally interdependent marketplace. Marshall McLuhan's (1964) 'global village' in the 1960s is very much a reality of the global construction market now.

While the traditionally strong North American and Western European construction markets have remained so, new markets in China, Turkey, South Korea, India and the Middle East have emerged as hotspots for construction business. While contractors from North American and Western European countries have remained competitive, companies from China, India, South Korea and Turkey are becoming strong construction forces in the global market. The international contracting business is considered as requiring a substantial asset base, and to be dirty and labour-intensive. It often requires dispatching expatriates to remote, relatively harsh physical environments; thus, it is also costly. Therefore, this part of the business has been gradually downstreamed to low-cost economies, on the basis of whoever can afford it. Developed economies, with much stronger know-how, remain dominant in the lucrative upstream construction professional services market. Construction globalisation is not a zero-sum game.

A win–win situation between the demand and supply sides can be achieved by configuring and integrating resources globally. One does not have to survive in the market through head-to-head competition; instead, one can win through professional services and contracting integration, 'co-opetition', or other innovative ways of doing business.

Construction globalisation had been running rapidly along a one-directional carriageway until recently, when a series of geopolitical changes put a question mark over it. These changes grew from increasing populism and have been further underscored by COVID-19. There is increasingly heated discussion on the globalisation backlash, which will certainly impact upon construction. In the aftermath of the pandemic, one cannot expect a quick resumption of business as usual. However, one cannot expect a complete cessation of construction globalisation either. It is hard to imagine that each country would close its doors and redevelop a full set of capabilities with regard to construction expertise, materials, plant and labour of its own. Although one can expect a 'new normal' of global construction, it is not yet possible to know what it will look like. Instead of passively waiting for this new normal to appear, it is opportune for all players, and in particular, construction economics researchers, to consider the vulnerabilities of the old construction globalisation and create a newer, more robust one. It also provides opportunities to examine specific construction activities in the global markets, such as M&A intentions, construction material flows, supply chain management and corporate social responsibility.

REFERENCES

Antràs, P. (2020). De-globalisation? Global value chains in the post-COVID-19 age. Working Paper, Harvard University. https://scholar.harvard.edu/antras/publications/de-globalisation-global-value-chains-post-covid-19-age.

Balfour Beatty (2020). Balfour Beatty completes disposal of Parsons Brinckerhoff. https://bit.ly/2BQdpaO (accessed 15 July 2020).

Chen, C., and Messner, J.I. (2009). Entry mode taxonomy for international construction markets. *Journal of Management in Engineering*, *25*(1), 3–11.

Dainty, A., Moore, D., and Murray, M. (2007). *Communication in Construction: Theory and Practice*. New York: Routledge.

Davis Langdon (2012). *World Construction 2012*. London: Davis Langdon.

The Economist (2017). An economist's bleak view of the future of globalisation. *The Economist*, 25 May.

Egan, J. (1998). *Rethinking Construction: Report of the Construction Task Force on the Scope for Improving the Quality and Efficiency of UK Construction*. https://constructingexcellence.org.uk/wp-content/uploads/2014/10/rethinking_construction_report.pdf.

Elliott, L. (2020). Of course there's a globalisation backlash. *Guardian*. https://www.theguardian.com/commentisfree/2020/feb/13/globalisation-backlash-open-markets-borders-climate-populism-coronavirus (accessed 30 April 2020).

Eurostat (2020). Employment by construction activities. shorturl.at/zFMQV (accessed 3 July 2020).

Ghemawat, P. (2016). *The Laws of Globalisation and Business Applications*. Cambridge: Cambridge University Press.

Global Construction Perspectives and Oxford Economics (2015). Global construction 2030 – a global forecast for the construction industry to 2030. http://shorturl.at/mDV89.

Goodier, C.I., Soetanto, R., Fleming, A., McDermott, P., and Austin, S.A. (2006). The future of construction procurement in the UK: a shift to service provision. In P. McDermott and M.M.A. Khalfan (eds), *Proceedings, CIB W92 Symposium on Sustainability and Value Through Construction Procurement*. University of Salford, pp. 182–193.

Goodman, P. (2020). A global outbreak is fueling the backlash to globalization. *New York Times*, 5 March. https://www.nytimes.com/2020/03/05/business/coronavirus-globalism.html (accessed 7 March 2020).
Hatzichronoglou, T. (1996). Globalisation and competitiveness: relevant indicators. OECD Science, Technology and Industry Working Papers 1996/5, OECD Publishing.
Hillebrandt, P. (1984). *Analysis of the British Construction Industry*. London: Springer.
Hillebrandt, P. (2000). *Economic Theory and the Construction Industry*. London: Macmillan.
Hogan, M.J. (1987). *The Marshall Plan: America, Britain, and the Reconstruction of Western Europe, 1947–1952*. New York: Cambridge University Press.
ICAK (2008). Current status of overseas construction. International Contractors Association of Korea.
Jewell, C. (2011). A typology of construction professional service firms: a consulting engineering perspective. Thesis, University of Reading.
Jewell, C., Flanagan, R., and Lu, W. (2014). The dilemma of scope and scale for construction professional service firms. *Construction Management and Economics*, *32*(5), 1–14.
Johns Hopkins Coronavirus Resource Center (2020). Cumulative cases. https://coronavirus.jhu.edu/data/cumulative-cases (accessed 5 May 2020).
Kim, H.J., and Reinschmidt, K.F. (2011). Diversification by the largest US contractors. *Canadian Journal of Civil Engineering*, *38*(7), 800–810.
Kim, S. (1988). The Korean construction industry as an exporter of services. *World Bank Economic Review*, *2*(2), 225–238.
Lee, S.H., Jeon, R.K., Kim, J.H., and Kim, J.J. (2011). Strategies for developing countries to expand their shares in the global construction market: phase-based SWOT and AAA analyses of Korea. *Journal of Construction Engineering and Management*, *137*(6), 460–470.
Low, S.P. and Jiang, H.B. (2003). Internationalization of Chinese construction enterprises. *Journal of Construction Engineering and Management*, *129*(6), 589–598.
Lu, W. (2014). Reliability of Engineering News-Record international construction data. *Construction Management and Economics*, *32*(10), 1–15.
Lu, W., Li, H., Shen, L., and Huang, T. (2009). Strengths, weaknesses, opportunities, and threats analysis of Chinese construction companies in the global market. *Journal of Management in Engineering*, *25*(4), 166–176.
Lu, W., Ye, K.H., Flanagan, R., and Jewell, C. (2013). Nexus between contracting and construction professional service businesses: empirical evidence from the international market. *Journal of Construction Engineering and Management*, *21*(2), 152–169.
McLuhan, M. (1964). *Understanding Media: The Extensions of Man* (1st edn). New York: McGraw-Hill.
National Bureau of Statistics of China (2019). *China Statistical Yearbook*. http://www.stats.gov.cn/tjsj/ndsj/2019/indexeh.htm.
Ngowi, A., Pienaar, E., Talukhaba, A., and Mbachu, J. (2005). The globalisation of the construction industry – a review. *Building and Environment*, *40*, 135–141.
Öz, Ö. (2001). Sources of competitive advantage of Turkish construction companies in international markets. *Construction Management and Economics*, *19*(2), 135–144.
Palmisano, S. (2016). The global enterprise: where to now? *Foreign Affairs*. https://www.foreignaffairs.com/articles/2016-10-14/global-enterprise.
Pearce, D. (2003). Environment and business: socially responsible but privately profitable? In J. Hirst (ed.), *The Challenge of Change: Fifty Years of Business Economics*. London: Profile Books, pp. 54–65.
Porter, M.E. (1990). The competitive advantage of nations. *Harvard Business Review*, March–April. https://hbr.org/1990/03/the-competitive-advantage-of-nations (accessed May 2019).
Reeves, M., and Harnoss, J. (2017). An agenda for the future of global business. *Harvard Business Review*, 27 February. https://hbr.org/2017/02/an-agenda-for-the-future-of-global-business (accessed May 2020).
Reina, P., and Tulacz, G. (2005). The Top 225 International Contractors. *ENR: Engineering News-Record 255*, 40–54.
Reina, P., and Tulacz, G. (2009). The Top 225 International Contractors. *ENR: Engineering News-Record*: *263*(7), 36–56.

Reina, P., and Tulacz, G. (2014). It's a competitive world after all. *ENR: Engineering News-Record, 273*, 1–20.
Reina, P., and Tulacz, G. (2015). Uncertainty clouds market. *ENR: Engineering News-Record, 275*(5), 33–57.
Reina, P., and Tulacz, G. (2016). Seeking stable markets: political and economic uncertainty in several regions have global firms looking for markets that are reliable and safe. *Engineering News-Record, 277*(6), 31–58.
Rodrik, D. (1998). Globalisation, social conflict and economic growth. *World Economy*, *21*, 143–158. doi:10.1111/1467-9701.00124.
Rowland, M. (2020). The contractor-consultant relationship – destined for divorce. https://bit.ly/315Csza (accessed 16 July 2020).
Sydow, J., Lindkvist, L., and DeFillippi, R.J. (2004). Project-based organizations, embeddedness and repositories of knowledge. *Organization Studies*, *25*(9), 1475–1489.
Tang, F. (2020). Coronavirus: will China opt for massive infrastructure spending spree to save its economy as it did in 2008? *South China Morning Post*. https://bit.ly/31aUrUW.
Tulacz, G., and Reina, P. (2019). Global market is risky business: failures of numerous major international competitors have firms focused more on the bottom line than on the top line. *Engineering News-Record*, 33–54.
Tulacz, G. and Reina, P. (2020). Struggling with COVID-19. *Engineering News-Record*, 33–52.
Turner, J.R. and Müller, R. (2003). On the Nature of the Project as a Temporary Organization. *International Journal of Project Management*, *21*, 1–8.
United Nations (UN) (2008). *International Standard Industrial Classification of All Economic Activities (Revision 4)*. New York: Department of Economic and Social Affairs, United Nations. https://unstats.un.org/unsd/publication/seriesm/seriesm_4rev4e.pdf.
US Bureau of Economic Analysis (2006). FAQ. https://www.bea.gov/help/faq/184 (accessed May 2020).
US Bureau of Labor Statistics (2020). Employment, hours, and earnings from the current Employment Statistics survey (National). https://beta.bls.gov/dataViewer/view/timeseries/CES2000000001.
WEF (2019). *Globalisation 4.0: Shaping a New Global Architecture in the Age of the Fourth Industrial Revolution*.
Wladawsky-Berger, I. (2017). *Globalisation at a Crossroads*. https://blog.irvingwb.com/blog/2017/05/globalisation-at-a-crossroads/comments/.
WTO (1998). *Construction and Related Engineering Services: Background Note by the Secretariat*. S/C/W/38, Geneva.
Xing, H.Y. (2002). *A Brief Guide to Understanding China's Entry into the WTO*. Beijing: RenMinRiBao Publisher (in Chinese).
Ye, M., Lu, W., Flanagan, R., and Ye, K.H. (2017). Diversification in the international construction business. *Construction Management and Economics*, *36*(6), 1–14.
Ye, K., Lu, W., and Jiang, W. (2009). Concentration in the international construction market. *Construction Management and Economics*, *27*(12), 1197–1207.
Zeithaml, A.V., Parasuraman, A., and Berry, L.L. (1985). Problems and strategies in service marketing. *Journal of Marketing*, *49*, 33–46.

22. Relational impacts of corruption on the procurement process: implications for economic growth in developing countries

Albert P.C. Chan and Emmanuel Kingsford Owusu

INTRODUCTION

Corruption in any given context is always frowned upon due to its disastrous nature and the havoc it creates in institutions, states and economies (Chan and Owusu 2017). Contextually, corruption is known to be present not only in the public sector but also in the private sector of both developed and developing economies. However, according to Bai et al. (2013), government corruption tends to be more pervasive in developing economies as compared to the developed. Several studies have investigated the impacts of corruption on economic growth from different contexts as disciplines. For instance, Mo (2001) revealed that a 1 per cent increase in corruption impacts upon economic growth negatively by reducing the overall growth by 0.72 per cent; and that the most critical medium through which corruption impacts upon the economic growth of a given state is through political instability. Political instability, according to Mo (2001), accounts for over 53 per cent of the total 1 per cent impact.

Past studies have examined the impacts and implications of corruption in different contexts and disciplines such as the banking, judicial and political sectors. The assessments have revealed how corruption impacts negatively upon the economic wealth of a context or a nation. Moreover, while the impact of corruption on economic growth is significantly negative in countries such as Korea and Nigeria, other countries may not feel the impact that much (Huang 2016; Egunjobi 2013). Similarly, assessments in other contexts such as finance, marketing and politics, among others (Grabova, 2014; Compton and Giedman 2011; Méndez and Sepúlveda 2006; Mo 2001), demonstrate both 'positive' and negative implications. Other implied implications of corruption, which include the reduction of productivity of productivity, distracting of investors, and the distortion of resource allocation, were identified to be critical indicators that lower economic growth, according to Grabova (2014).

In the context of construction management and economics, a number of studies have examined the impact of corruption on construction projects and the industry at large (see Le et al. 2014a; Kenny 2006; Sohail and Cavill 2008; Bowen et al. 2012). However, studies focused on examining the critical impacts of corruption on the procurement process (including the individual stages and activities) remain lacking in construction management and economics literature. Similar to the analysis of Grabova (2014), it can be inferred that the impact of corruption in public procurement, such as the procurement of construction and other infrastructure-related projects, contributes to the lowering of economic growth. On the flip side, the extirpation of these impacts, in addition to the negative constructs of corruption that render the procurement process vulnerable to corruption, can contribute significantly to economic growth. This chapter, therefore, examines two key constructs or indicators of corruption intended to ascertain the impact of corruption on the procurement process and the relative impact on economic growth: (1) the proneness of the procurement activities to corruption; and (2) the criticalities of the causal factors of corruption and how these two constructs impact upon the overall outlook of the procurement process to ascertain the implied impact on economic growth.

This chapter is intended to contribute to the body of knowledge on corruption-related studies in construction economics by deepening understanding of the subject matter. The work also provides a foundation for further empirical studies on this novel area. The next section introduces the chapter and states the problem, aim and objectives of the study presented here. The following section reviews the relevant literature of the subject matter. The chapter then presents the methods and tools adopted to realize the aim and objectives of the study, before presenting the data analysis and discussion. The chapter concludes by stating the limitations and implications of the study, and recommending possible topics for future research.

THEORETICAL PERSPECTIVES AND DEFINITION OF CORRUPTION

As an omnipresent socio-economic concern, corruption has been debated in various scholarly domains including political science, sociology, business ethics, criminology and management (Breit et al. 2015; Owusu and Chan 2018). This chapter is intended to add to the prevailing discourse on corruption in construction economics. For instance, while economic theory depicts corruption as an opportunistic behaviour that stems from agency theory and rational choice, which ignites an individual's quest to engage in a corrupt act, most of the remaining perspectives attribute or define corruption based on the anomie theory, which connotes the absence of, or the inability to demonstrate, normal social or ethical standards (Martin et al. 2007; Breit et al. 2015; Rose-Ackerman 2007). The underlying driver of all the explications presented is the selfish quest of a person at the expense of the greater good or benefit. While corruption is explicated from a plethora of human behaviour theories from the social sciences, it is also studied in different fields with the theoretical approach of the given field.

While there is no standard definition of corruption within the field of construction economics, this chapter defines corruption drawing inference from the World Bank's working definition. The World Bank (Shah 2006) defines corruption as the abuse of public resources for illicit private gains. Therefore, although corruption may be viewed or defined differently in various contexts, common themes are apparent within all theories of corruption. Recurrent common themes include ethical or moral decadence, misappropriation of resources or funds, and the willingness of parties involved to be lured or forced to engage in a corrupt act. Thus, based on the thematic ideologies underpinning the definition and meaning of corruption, scholars define corruption in projects in a similar manner (Shan et al. 2017). In the field of construction economics, corruption can be defined as the abuse of economic resources intended for the development of construction projects for illicit selfish gains (Le et al. 2014a; Bowen et al. 2012). Projects in this context can be classified as public, private or public–private partnerships, and the various levels where corruption can occur can be identified as the individual, organizational and project levels, as well as the government level.

Figure 22.1 represents the existing constructs of the causal factors of corruption as established under their respective levels of occurrence or influence. Five constructs were established by Owusu et al. (2017): statutory, organizational, regulatory, project-specific and psychosocial-specific constructs. These constructs are presented under three distinctive levels: individual and social influence (level 1); project and organizational influence (level 2); and government or statutory influence (level 3). While the individual constructs were formulated or developed by Owusu et al. (2017), the categorization of the constructs under their respective levels of influence remains one of the contributions offered by this chapter. This chapter follows up on the classification of the constructs by Owusu et al. (2017), and conceptually categorizes the constructs under the specific domains they affect most, even though the individual constructs and their respective underlying variables can occur at any of the established levels or domains. Under the first level (individual and society influence), which is made up of the psychosocial-specific construct, the underlying causal variables captured are psychological and social factors that instigate or worsen corruption within a given context (Owusu et al. 2017; Le et al. 2014a, 2014b; Zhang et al. 2016; Bowen et al. 2012). These factors are intrinsic attributes of either individuals or a group of individuals that make up a community. It is considered to be the first level because it deals with the least underlying unit of measurement to initiate the whole corruption process.

The second level, the project or organizational level, is made up of the causal constructs that affect institutions and related projects. This level consists of three constructs: organization-specific, project-specific and regulatory-specific causal factors. Thus, the level is made up of causal variables that give rise to corrupt practices at the institutional level. While some of these variables can be attributed to human or societal influence, they are mostly predominant at the institutional level. Their impact mostly affects macroeconomic attributes and indicators such as the gross domestic product (GDP) growth rate, consumer price index and commodity prices. Simply put, the overall outcome of a corrupt issue that affects or impacts upon individual organizations in a context collectively affects the broad context (the economy as a whole). Moreover, in terms of the levels of criticalities, the constructs under level 2 come second to level 3 (government or statutory level), as their impacts affect a relatively higher number of persons or projects compared to the construct under level 1. The implications on project-related activities, particularly the procurement process of infrastructure-related projects, which is considered to be one of the most vulnerable processes to corruption, are discussed below.

THE IMPACTS OF CORRUPTION

In determining the economic growth of states and federal governments, the need to pay critical attention to the economic indicators that drive the economy's primary sectors cannot be overemphasized. Each sector of the economy is underpinned by different indicators that are examined to ascertain the overall size and impact of the sector's contribution to the economy. However, while these indicators contribute to the growth of the sectors and the overall economy, corruption tends to be inevitable in these sectors, and hence contributes negatively to the sectors and the economy of a given state or context.

Prior to identifying the negative impacts of corruption on economic growth, it must be noted that several studies, including Acemoglu and Verdier (1998), have argued for the benefits of corruption in several instances, including contribution towards economic growth. According to Bicchieri and Duffy (1997), there are two schools of thought with disparate views on corruption and its associated impacts. The two views can be classified as the moralist and

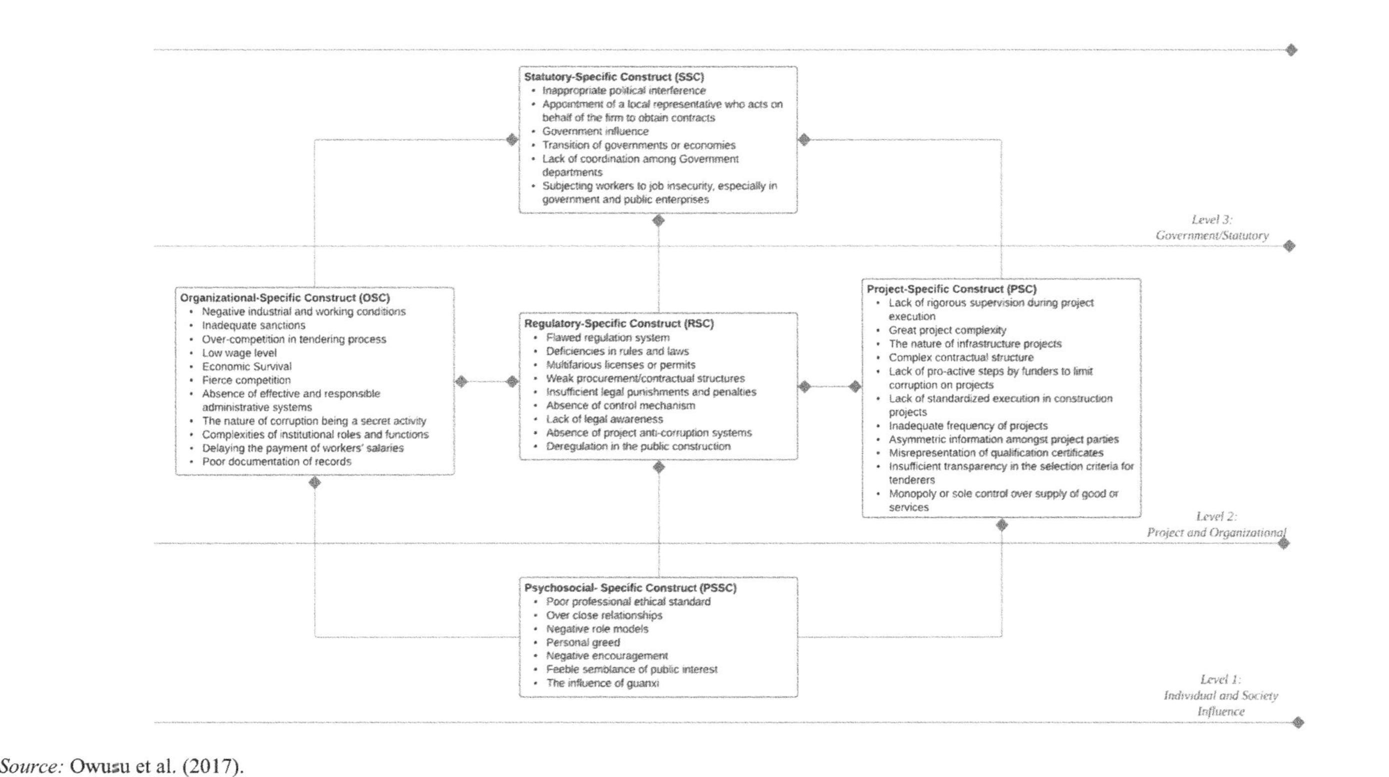

Source: Owusu et al. (2017).

Figure 22.1 Conceptual framework for causes of corruption

revisionist perspectives. Bicchieri and Duffy (1997) note that whereas moralists maintain the view that corruption is harmful and thus hampers economic development while eroding the legitimacy of institutions, the revisionists point instead to the potential rewards of corruption. Revisionists argue that corruption helps to speed up cumbersome procedures, and to purchase political and other access for the excluded (Abueva 1966).

Another typical argument for the benefit of corruption is attributed to the provision of possible ways for investors and entrepreneurs to bypass or evade inefficient stipulations or regulations. Moreover, it is also argued that corruption is one of the natural responses to shortages or access to limited services and facilities such as health care and education in the developing contexts. However, revisionists have mostly centred their works on developing countries experiencing political transformations and developments (Dobel 1978; Benson et al. 1978). This limits the generalizability of their arguments. While there have been counter-assertions against the claimed positive impacts of corruption, each school of thought (for or against corruption) may have its own conclusions for either the positive or negative impacts of corruption, respectively. However, corruption, over the passage of time, has caused more harm than good (Chan and Owusu 2017).

Globally, the economic cost of corruption is estimated to be around US$2.6 trillion, which is more than 5 per cent of the total global GDP (United Nations, 2018). The impact of corruption that stems from construction-related activities not only affects the desirable main objectives of a project (including cost, quality, time, health and safety, and environmental performance) but also reduces the life span of any product of the project, by 50 per cent or more (Kenny 2012). Other more serious negative impacts include the loss and displacement of jobs, properties and lives. Typical examples of such situations are Operation Car Wash in Brazil, which is thought to have involved the misappropriation of over R$6.2 billion (US$2.5 billion), and the collapse of the Rana Plaza building in Dhaka, Bangladesh, which resulted in over 1127 deaths (Rubya 2015; Pattanayak and Verdugo-Yepes 2020). Thus, fighting the prevalence of corruption curbs the negative impacts associated with construction projects and helps to save lives, livelihoods, properties and the quality of services.

IMPACT OF CORRUPTION ON ECONOMIC DEVELOPMENT

The construction industry is seen as the backbone for most economies, not only because it executes infrastructure projects vital for development, but also because of its contribution to the economy, as it is estimated to account for approximately one-third of gross capital formation (Colonnelli and Haas 2016). According to Kenny (2007), the industry is predominantly interlaced with the state; a substantial proportion of public investment goes into construction-related activities. The incidence and proliferation of corruption in the procurement process results in a negative toll on economic outlook. Measuring the impact of corruption on economic development first considers that on macroeconomic indicators, which in turn stem from microeconomic units involving either a single person or group of persons, with greater emphasis on the former as they directly correlate with the performance, efficiency and general outlook of an economy. These indicators are captured as leading and lagging indicators according to the Corporate Finance Institute (2021); the leading indicators include variables such as the performance of the stock market, manufacturing activity levels and retail sales; and some of the lagging indicators are the unemployment rate, GDP growth, corporate profits, interest rates and inflation. These indicators are partly a reflection of government spending, imports, exports and other economic activities that drive the economy.

Studies have found that corruption has a negative impact on most of the mentioned indicators. For instance, using stock market performance, a notable example was the case of the former US energy firm Enron Corporation, a corruption scandal uncovered in 2001 involving 'cooking the books' over a period of time that resulted in the loss of about $11 billion and affected the performance of Enron's stock by reducing the share price from almost US$100 to under US$1 (Flanigan 2002; McLean and Elkind 2013; Tonge et al. 2003; Culpan and Trussel 2005). The adverse effect on the company's workers, creditors and small investors was massive (Gilpin 2001). The snowball effect affected other stocks and resulted in a big blow to the overall market. According to Tran (2002), Wall Street suffered one of its most precipitous plunges, the compounding effect which stemmed from the loss of confidence of investors in the energy sector. A corrupt act involving a small group of persons in one company affected other firms, the overall economy and millions of people.

There are many examples of how corruption starts with individuals and ultimately moves further to have an impact on macroeconomic constructs such as GDP growth, inflation rate, inventory balances, and income and wage growth or decline within a given context. In construction and engineering, the Operation Car Wash scandal in Brazil is typical of cases that commenced at the microeconomic level, then grew to affect macrocosmic indicators that collectively affected the county's economy (Winter 2017; Pattanayak and Verdugo-Yepes 2020).

From the conclusions of Kenny (2007), the most severe impact of corruption in the construction process goes beyond the payment of bribes, to the development of substandard projects with very low economic turnover, along with limited resources for their maintenance. The project's life is threatened by different forms of risks, and the inhabitants using such facilities are exposed to life-threatening conditions. In other instances, it leads to other adverse outcomes, including increased income inequalities, disproportionate distribution of resources and the destabilization of the efficacy of social welfare projects, which in turn results in lower levels of human development. Consequently, many other long-term goals including equality and sustainable development are affected. Thus, finding effective mechanisms for curbing corruption in the procurement process ultimately saves the economy from the repercussions of corruption.

THE IMPACT OF CORRUPTION ON DEVELOPING COUNTRIES

It has been found that corruption has a relatively pernicious impact on the economies of the developing countries (Le et al. 2014a; Søreide 2002; Olken and Pande 2012). It has been observed to be deleterious to economic efficiency in most states; it also distorts the allocation of expenditure among various budgetary functions including infrastructure-related budgets, constricts expenditure efficiency and obstructs budget equilibrium (Dellavalade 2006; Lewis and Hendrawan 2019). Although corruption is reported to reduce state revenues, there are opposing schools of thought regarding its impacts on public spending expenditure (Johnson et al. 1999; Tanzi 1998; Dellavalade 2006). Whereas Tanzi (1998) considers corruption as an inflator of overall public expenditure, Mauro (1997) indicated that there was no significant impact of corruption on the volume of public spending. However, continuous assessment of corruption over time has revealed that the impact of corruption on public projects is costly in either case (Henri 2018).

It has been found that the average productivity of infrastructure investment can be reduced. This is the case when government officials tend to base decisions on infrastructure projects to invest in not because of the intrinsic economic value attached to the development of that item of infrastructure, but because such projects create room for taking rent. This occurs in various forms of corrupt practices, including bribery, solicitation, collusion, kickbacks, money laundering and tax evasion (Tanzi 1998; Chan and Owusu 2017). To create room for such acts, the officials promote capital projects where project complexities and sizes are increased not for any economic benefit but out of selfish interests. Such decisions occur at the project definition stage, where the initial needs and scope of projects are defined. In many developing countries, including Ghana, corruption begins at this point. The various subsequent activity points of the procurement process are also vulnerable; they are also impacted upon by the causal factors. The Organisation for Economic Co-operation and Development (OECD 2016, 2019) reported that about US$150 billion is lost in Africa every year through corruption. Moreover, from the perspective of opportunity cost, the economic impact is that there is no benefit derived; in the context of corruption, the gains move to the corrupt rather than the intended beneficiaries, leaving an adverse impact on the project or the investing agency.

In developing countries, these cases are more severe because of the lack of proactive measures to either detect or prevent corruption. Moreover, the established enforcement measures in most developing contexts, including Ghana, have been found to be less effective; in such contexts the established compliance mechanisms have been found either to have limited effect, or to be ineffective (Owusu et al. 2019). Thus, it would be beneficial for governments to put in place and enforce focused and transparent policies rather than establish numerous complex and poorly implemented regulation; this will have a positive impact on the construction industry and the outlook of the economy.

DETERMINANTS OF CORRUPTION IN A CONTEXT

Corruption indicators determine the degree of the impact of corruption in both developed and developing economies. They include the causes of corruption and associated irregularities, forms of corruption, project stakeholder's involvement, as well as anti-corruption measures, and the barriers that hamper their effectiveness (Le et al. 2014a, 2014b; Tabish and Jha 2011; Zhang et al. 2016; Bowen et al. 2012; Stansbury and Stansbury 2018). Several studies have been undertaken to ascertain the underlying measurement items that drive these constructs, as well as their impacts on construction projects.

Not only is the procurement process identified as the part of the construction project which is most vulnerable to corruption, but also the construction sector is tagged as one of the leading corrupt sectors globally. These two characteristics constitute the primary focal points of ascertaining the impact of corruption on the economy of a given context (Krishnan 2010; Kottasova 2014; Owusu et al. 2019). In other words, the severity of corruption in this context is significantly dependent on the vulnerability of the procurement process to corruption and the prevalence of corruption in the construction sector. The prevalence of corruption in construction stems from consenting project members or stakeholders who are identified in the corruption context as demand-side, supply-side and condoning side (Boyd and Padilla 2009).

An number of studies on the determinants of corruption has been undertaken over the years by construction management and economics scholars (see Le et al. 2014a, 2014b; Tabish and Jha 2011; Bowen et al. 2012; Ameyaw et al. 2017; Zhang et al. 2016). However, this chapter focuses on the criticalities of the causal factors and their relative impacts on the procurement process. The causal factors of corruption are presented in the next section.

CORRUPTION'S CAUSAL FACTORS

The causal factors of corruption can be defined as the indicators that create or provide the grounds for corruption to occur. Among other things, these variables allow for the misappropriation of a project's resources or the distortion of a project's process to obtain illegitimate personal gains, hindering the successful realization of a project or its processes (Le et al. 2014a; Shan et al. 2017). Stansbury and Stansbury (2008) identified a some 47 main examples of corrupt practices throughout the construction process, including price-fixing (identical to the formation of cartels), bid-rigging and suppression, among others.

Owusu et al. (2017) conducted a review of the critical causal factors of corruption in construction and infrastructure-related management. The authors identified over 44 unique factors that were noted to be prevalent in different contexts and projects. Among these 44 variables, the top three most discussed causes were poor professional and ethical standards, over-close relationships, and poor industrial and working conditions. The identified variables were then categorized into five defined constructs using thematic analysis. The five constructs were: psychosocial-specific causes (PSSC), organizational-specific causes (OSC), statutory-specific causes (SSC), regulatory-specific causes (RSC) and project-specific causes (PSC). Each of these constructs contains at least five causal factors (Figure 22.1).

CONSTRUCTION PROCUREMENT AND CORRUPTION

From its etymology *rumpere*, which means 'to break', corruption in any related public procurement, including infrastructure-related works, occurs when there is a loophole that can be exploited for private or selfish gains (Tanzi 1995). Corruption in public procurement, especially in infrastructure-related works, is not a new phenomenon. Classically, some behavioral theorists have linked corruption to negative moral attributes such as greed and the tendency to lie, which are all part of human nature (Weisel and Shalvi 2015). Thus, it is not far from right to claim that corruption is as old as the existence of humanity. However, other than the first documentation of corruption in the form of bribery, corruption has evolved to embody several forms in different disciplines. Chan and Owusu (2017) conducted a review to ascertain over 28 different forms of corruption (under six constructs) prevalent in construction and infrastructure-related works, which are discussed in the next section. These forms are not absent from the activities within the procurement process. As a result, the related thematic structures, including the causal factors, risk indicators, and barriers that hinder the effectiveness of anti-corruption measures within the procurement process and related stages and activities, are worth exploring.

The procurement process is comprised of four main stages: the pre-contract stage, contract stage, contract administration stage and post-contract stage. The susceptibility of these stages and their respective activities to corruption vary in different contexts. Similarly, the impact of corruption on the individual stages and the overall procurement process also vary in contextual terms. Owusu et al. (2019) conducted a detailed empirical study on the proneness of the procurement process to corruption, upon which this chapter builds. Thus, other than ascertaining how vulnerable the procurement process is to corruption, this chapter further explores how corruption's causal factors impact upon the entire process.

RESEARCH DESIGN

A positivist epistemological lens (Fuller 2002) was adopted for this largely quantitative empirical research that contained some elements of the interpretivism paradigm (Schwandt 1994), which involved the interpretation of the literature and secondary information, and the primary qualitative data. From an operational perspective, primary data on the causal factors of corruption were solicited from experts involved in the day-to-day management of procurement and construction management-related work. A questionnaire survey constituted the main data collection instrument and aimed at gathering both quantitative and qualitative views from the experts concerning the constructs. Prior to the development of the questionnaire, a systematic review of the literature was conducted to determine causal factors (that is, variables), which are used as measurement indicators.

As the study's primary goal was to ascertain the criticality indexes of the causal factors stipulated and their impact on the procurement process activities in developing regions, the set of 44 causal factors previously identified by Owusu et al. (2017) was utilized. However, following feedback received from pilot testing the variables, a total of 38 causal factors were considered appropriate for the expert survey and further examination. Eight experts participated in this exercise (the pilot study): three were distinguished professors, two were senior lecturers of the subject under study, and three were leading experts from industry (the World Bank, United Nations and the Global Infrastructure Anti-Corruption Centre, respectively). All the essential comments received were addressed prior to disseminating questionnaires to experts in the main study sample. The questionnaire consisted of both closed-ended and open-ended sections, where the closed-ended section used a five-point Likert grading scale to measure the criticality indexes of the causal measures, and the open-ended section invited respondents to provide their views on any causal factors that were not captured in the closed-ended section. A questionnaire-based approach was adopted because it can offer a valid and reliable source of information in a relatively shorter amount of time and is relatively inexpensive. Moreover, this method enables the anonymity of respondents to be maintained, and the respondents' data to be protected; this is especially important considering the sensitivity of the subject of the study (Hoxley, 2008).

SURVEY PARTICIPANTS

The survey was undertaken in Ghana between 2016 and 2020 (see also Owusu et al. 2018). The respondents included a wide range of construction experts involved in both public and private sector procurement in the developing context. A primary consideration was to secure experts with an appropriate level of knowledge of the different stages of the procurement process and

Table 22.1 *Profile of the respondents*

Construct	Subconstruct	Frequency	Relative frequency	Cumulative frequency
Sector	Public	20	32.26	32.26
	Private	30	48.39	80.65
	Both	12	19.35	100.00
	Total	62	100	
Years of experience	Up to 10 years	45	72.58	72.58
	11–20 years	12	19.35	91.94
	21–40 years	5	8.06	100.00
	Total	62	100	
Professional Background	Engineer	17	27.42	27.42
	Quantity surveyor	31	50.00	77.42
	Contractor	4	6.45	83.87
	Architect	7	11.29	95.16
	Academics	3	4.84	100.00
	Total	62	100	
Involvement in procurement stages	Single stage	7	11.29	11.29
	Multiple stages	28	45.16	56.45
	All stages	27	43.55	100.00
	Total	62	100	
Position in organization	Head of Department	5	8.06	8.06
	Director of Works	8	12.90	20.97
	Senior manager	17	27.42	48.39
	Supervisor	28	45.16	93.55
	Junior staff	4	6.45	100.00
	Total	62	100	

management of construction works, and of the dynamics of corruption within the procurement process. Purposive sampling was therefore adopted. The snowballing technique was also applied; the respondents were requested to help disseminate the questionnaires to their expert colleagues. In total, 62 responses were received, and these were regarded to be valid and suitable for further analysis. Participating experts included professionals from diverse backgrounds such as architects, engineers, quantity surveyors and contractors. Whereas the academics involved were identified through their publications on the subject matter, the industry experts were composed of public and private sector experts who are engaged in the procurement, execution and management of construction and other infrastructure-related works. The profile of the respondents is presented in Table 22.1 (it is pertinent to note that for professionals in the public sector, their administrative positions are indicated instead of their professions).

Table 22.2 Procurement process stages and activities

No	Procurement stage	Activity	Code
1	Pre-contract stage (PCS)	Definition of project's requirements	PCS1
2		Planning of the procurement process and strategy development	PCS2
3		Conducting of pre-tender survey	PCS3
4		Obtaining necessary approvals	PCS4
5		Soliciting of tenders	PCS5
6		Receiving of tenders	PCS6
7	Contract stage (CTS)	Pre-tender meeting to establish evaluation plan and criteria)	CTS1
8		Evaluating tenders to approve or reject bids	CTS2
9		Selecting a suitable contractor	CTS3
10		Awarding of contract	CTS4
11		Preparation and signing of contract	CTS5
12	Contract administration	Issuing of contract revisions	CAS1
13	stage (CAS)	Monitoring progress	CAS2
14		Inspection or following up on delivery	CAS3
15		Administer interim/ progress payments	CAS4
16	Post-contract phase	File final action	PCP1
17	(PCP)	Issue final contract amendment	PCP2
18		Complete financial audits	PCP3
19		Verify delivery/completed reports	PCP4
20		Return performance bonds and close-out contract	PCP5
21		Confirm completeness and accuracy of file documentation	PCP6

Sources: Ruparathna and Hewage (2015) and Owusu et al. (2019).

DATA ANALYSIS AND DISCUSSION

The reliability test tool in SPSSv.23 was adopted to examine the reliability of the data. Following the determination of the reliability index, descriptive and frequency tools (such as the mean index, data dispersion and data distribution) were adopted to estimate the susceptibilities of the procurement process and criticalities of the causal factors. The evaluation of the mean values for each of the activities facilitated the estimation of the criticalities of the constructs. Finally, the network analysis technique was adopted to estimate the relational impacts of the causal factors on the procurement process.

Vulnerabilities of the Procurement Process

This section presents a discussion of the empirical assessment of the proneness of the procurement stages and activities to corruption. From the study of Ruparathna and Hewage (2015), the traditional procurement process consists of four stages, namely the pre-contract stage (PCS), contract stage (CTS), contract administration stage (CAS) and post-contract stage (PCP). Each stage can be further subdivided into at least four different activities. In total, the entire procurement process is made up of 21 different activities. The headings of the stages and their respective activities are presented in Table 22.2.

Studies of the incidence of corruption in construction and other infrastructure-related projects have found the procurement process to be the most vulnerable to corruption globally (Tabish and Jha 2011; Bowen et al. 2012). Owusu et al. (2019) recently explored the vulnerabilities of the activities and the stages of the procurement process. Given that public procurement accounts for more than 30 per cent of the GDP in developing countries, the assessment of the impact of corruption on the procurement process is considered a critical step in mitigating the overall negative impact of corruption on economic growth. Among the 21 activities under the four stages of the procurement process, approximately 50 per cent of both activities and stages were identified as being vulnerable to corruption. At the pre-contract stage, two out of six activities were identified to be vulnerable to corruption, namely: solicitation of tenders with a mean value of 4.02, and obtaining of necessary approvals with a normalized value (NV) of 3.52. Stansbury and Stansbury (2008) pointed out some examples of corrupt acts which occur at this stage of the process. They include facilitation payments to obtain licences or necessary approvals, and the disclosure of the project's information to a bidder who is ready to pay or has already paid a bribe or engaged in some form of corruption to obtain such information. Generically, other than the two activities, the remaining four activities were identified to be neutrally vulnerable. Thus, the overall vulnerability index of the pre-contract stage was realized to be neutral in terms of its susceptibility to corruption.

Unlike the pre-contract stage, the contract stage was revealed by Owusu et al. (2020) to be the most vulnerable stage to corruption among all the stages of the construction project. Three out of five activities were found to be vulnerable to corrupt practices: tender evaluation, contractor selection, and the award of the proposed contract to a corrupt bidder. Their mean values were 4.00, 4.21 and 3.74, respectively. In contrast to the contract stage, the contract administration stage was revealed to be the least vulnerable stage to corruption. Only one activity out of the four (namely, the administration or issuing of progress payments) was identified to be susceptible to corruption, and had a mean value of 3.68. Thus, whereas the overall mean value for the contract stage was interpreted to be vulnerable to corruption, that of the contract administration was found to be neutral.

Lastly, the post-contract phase was revealed to be the only stage after the contract stage that was vulnerable to corruption. This stage encapsulates six different activities with five activities approximated to fall within the vulnerable range. The only activity identified to be neutral was ensuring the accuracy and completeness of file documentation. In all, 11 out of 20 activities and two out of four stages, representing over 50 per cent of the entire procurement process, were identified to be vulnerable to corruption. Following the estimation of the vulnerable stages and activities, the next section presents the assessment and input of the causal factors of corruption in the procurement process. The proneness of procurement activities to corruption is presented in Figure 22.2, and that of the various stages is presented in Figure 22.3.

The results revealed significant disparities regarding the susceptibility to corruption of the activities and stages. Out of the four stages considered, two were identified to be vulnerable to corruption (that is, the contract stage and post-contract stage) while the remaining two (that is, the pre-contract stage and contract administration stage) were noted to be moderately vulnerable. However, despite the susceptibility indexes of the stages, the activities performed under the stages were also found to have different indexes even within the same category, thus indicating that corruption can occur within any of the stages and particular activities. To date, no study has empirically examined the main causal factors that render the procurement process susceptible to corruption, nor identified the extent to which the activities within the process are affected.

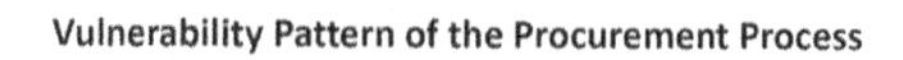

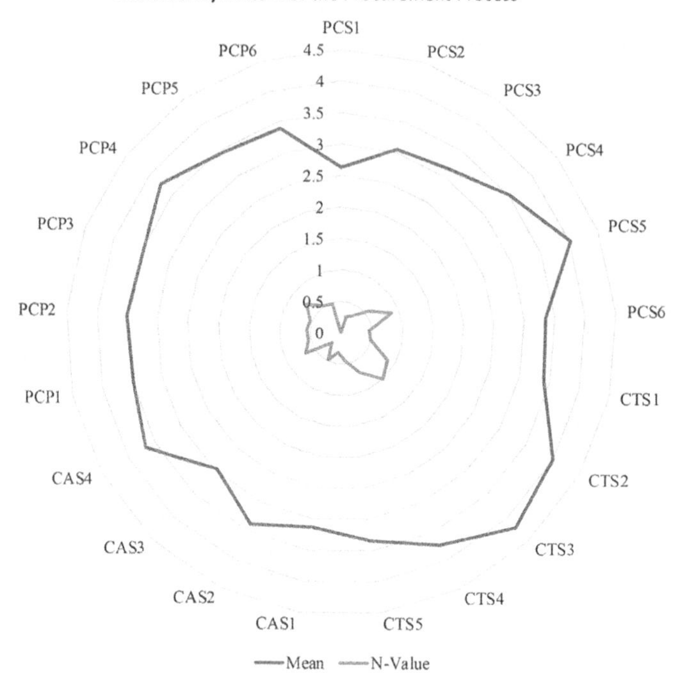

Source: Adapted from Owusu et al. (2019).

Figure 22.2 Vulnerabilities of the procurement activities

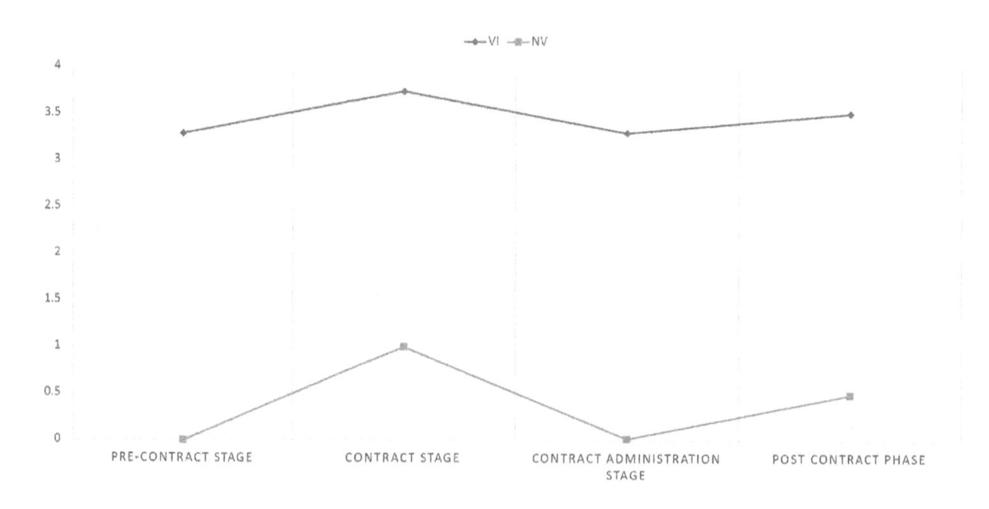

Note: VI = vulnerability index; NV = normalized values.

Figure 22.3 Vulnerabilities of the procurement stages

Criticalities of the Causal Factors

The estimation of the individual criticalities of the causal factors and their respective constructs is reported in Figures 22.4 and 22.5, respectively. The impacts of the causal factors on the individual activities and stages of the procurement process are also presented in Figure 22.6. The 38 variables under their respective constructs are discussed, followed by the Network Analysis results showing the significant relational impacts. The results are presented in descending order of their criticalities beginning from the most highly critical, that is, the psychosocial-specific causes (PSSC) construct, through to the least critical, the organizational-specific causes.

The PSSC in this context refers to the relational interactions of social and psychological factors and the outcome or influence in a given setting (for example, the workplace) or on work execution (Greitzer et al. 2013; Heiser 2001). Owusu et al. (2017) found that variables within this construct are the most discussed in the construction management and economics literature. The results presented in this study concur with this position, and indeed, PSSC was identified as the most critical construct, with a mean index of 3.61. In agreement with the construct criticality, respondents evaluated variables under this construct to be the most critical causative factor of corruption. From the estimation of the experts, four out of the five causal factors captured under this construct were noted to be critical. Personal greed (CC1) was evaluated as the most crucial variable among all the 38 causal factors, with a mean index of 3.92. The remaining critical factors were an over-close relationship (CC6), negative role models (CC9), and poor professional, ethical standards (CC14), with respective mean indexes of 3.71, 3.66 and 3.60.

Statutory-specific causes (SSC) came second to PSSC as the second most critical construct in the developing context, with an overall mean index of 3.58. According to Owusu et al. (2017), SSC constitute government-driven forces that instigate the incidence and proliferation of corruption. Given that the definition of corruption is often attributed to the misappropriation of a state's resources, the government is seen as the system most vulnerable to high levels of corruption (Johnston 2017). While this construct was identified as the second most critical, all the five variables captured under it were identified to be critical (unlike PSSC). With the first two variables obtaining the same mean index of 3.71, the respondents revealed that the inappropriate interferences of political influences in public projects (CC5) and the lack of coordination among government departments responsible for a given public project (CC4) were the most critical factors under this construct. The remaining variables which were subjecting public workers to job insecurity, especially in government and public enterprises (CC18), change of government (CC19), and the appointment of unqualified local representatives who act on behalf of the firm to obtain contracts (CC25), obtained mean indexes of 3.52, 3.52 and 3.45, respectively.

Regulatory or legal-specific causes (RSC) emerged as the third critical construct among the five, with a construct mean index of 3.48. While regulations refer to Acts, directives, norms, or principles for preserving acceptable standards, regulatory-specific causes can be regarded as an objection to stipulated regulatory measures, loopholes in guiding principles, or the partial inclination to the demands of a regulation that result in corruption in the long run. The RSC construct is made up of nine unique variables; five were evaluated to be critical, and the remaining four to be moderately critical. With a flawed regulation system (CC3) emerging as the most critical variable with a mean index of 3.74, the other four critical RSC variables were:

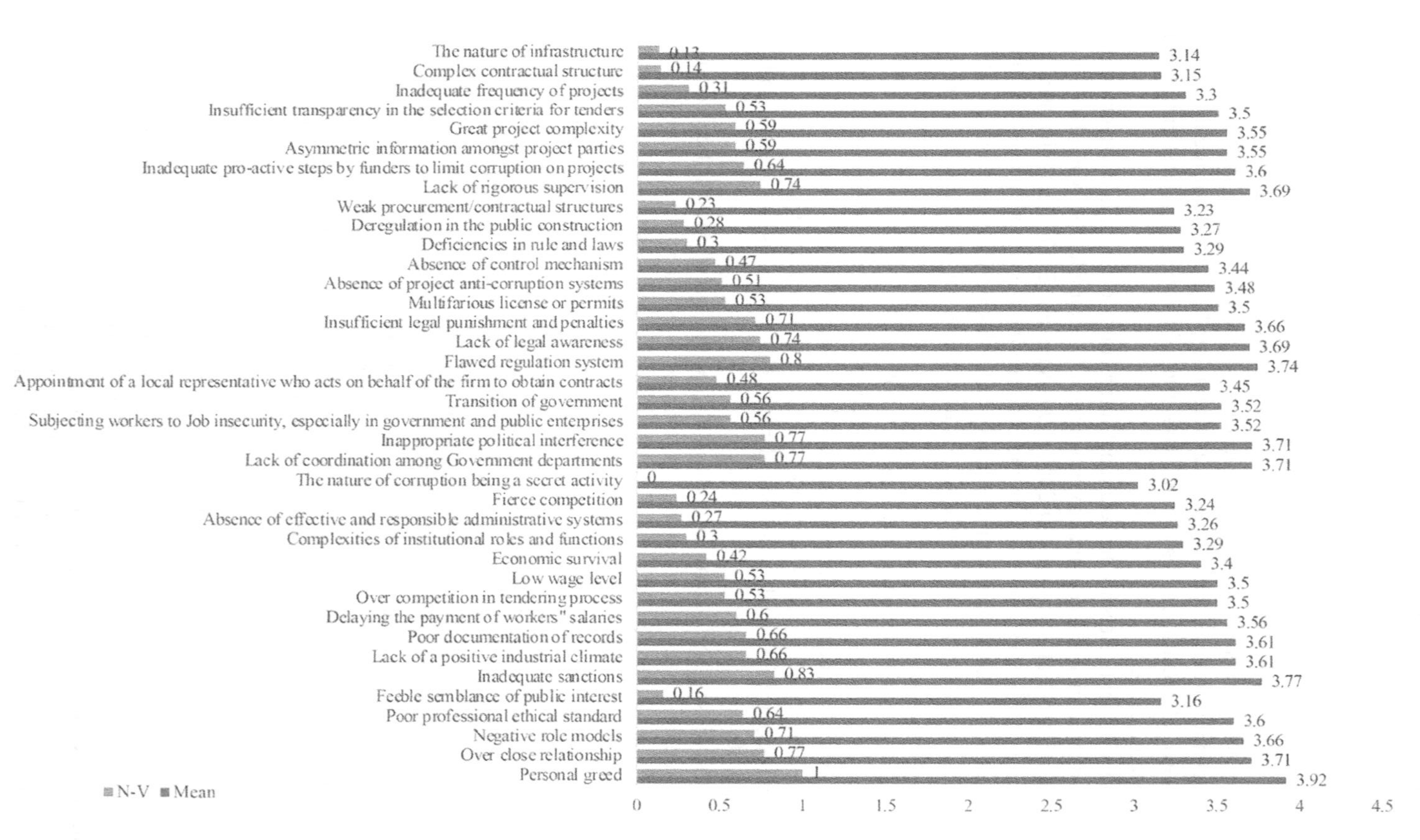

Figure 22.4 *Criticalities of the causal factors*

Figure 22.5 Criticalities of constructs

(1) the lack of legal awareness either in a project setting or a contractual environment (CC7), with a mean index of 3.69; (2) insufficient legal punishment and penalties (CC10), with a mean index of 3.66; (3) multifarious licence or permits (CC21), obtaining a mean index of 3.50; and (4) the absence of project anti-corruption systems (CC24), with a mean index of 3.48.

Even though most projects in the developing world are susceptible to corruption, the project-specific causes (PSC) construct was evaluated or revealed by the respondents to be moderately critical, with a mean index of 3.44. PSC can be defined as the project-oriented loopholes that tend to enable acts of corruption during the planning, procurement and management of construction projects (Le et al. 2014a). With a total of eight causal factors captured under the PSC construct, five were revealed to be critical. They were: (1) the lack of meticulous project supervision (CC8), with a mean index of 3.69; (2) lack of proactive steps by funders to limit corruption on projects (CC13), scoring 3.60; (3) distortion in information flow or symmetric information amongst project parties (CC 16), with a criticality index of 3.55; (4) complexities involved in project and contractual structures (CC17), with a mean index of 3.55; and (5) lack of transparency in the selection criteria for tenders (CC22), with a mean index of 3.50. The project's type and location often influence its degree of vulnerability to corruption. Complex projects and contractual systems in developing countries tend to be more vulnerable to corruption when compared to projects undertaken in a developed region with a more formidable regulatory structure that limits corruption, such as Hong Kong and Singapore (Wai 2016).

The last and least-evaluated construct among the five was the organizational-specific causes (OSC) construct, with a mean index of 3.43. OSC refers to the causal factors of corrupt practices that stem from both the internal and the external structures of a public or private organization or institution (Owusu et al. 2017). Comparatively, this construct had the highest number of variables (that is, 11 unique variables) and the highest number of critical variables (that is, six variables). From the top critical variable under this construct, the practice of inadequate sanctions (CC2), with a mean index of 3.77, was evaluated by the respondents as the most critical, followed by the lack of a positive industrial climate (CC 11), with a mean index of 3.61; and the poor documentation of records (CC 12), scoring 3.61. The remaining three critical variables were: suspending or delaying the payment of workers' salaries (CC15), scoring 3.56; over-competition in the tendering process (CC20), scoring 3.50; and low wage level (CC23), also scoring 3.50. The criticality indexes for the underlying constructs of the individual causal factors are presented in Figure 22.5.

Discussion: Impact of the Causal Factors on Project Procurement Process

This study makes an implied justification following the transitive law in mathematics and logic, which states that if A = B and B = C, then A = C (Lotha 2016). In this context, A represents the causal factors of corruption, B stands for the procurement process, and C represents the economic position. Thus, with reference to the stipulated law, the transitive law implies that high criticalities of the causal factors will result in high levels of corruption, resulting in a high impact on a given context's economic position in a negative manner. This study measures and confirms the first two equations of the logic (that is, A = B and B = C) with the final equation (that is, A = C) established by the law, and further research could examine this supposition empirically. This section examines the criticalities of the individual causal factors of corruption on the procurement process of infrastructure-related works, and a hypothesized implication of the criticalities and impacts on the economy at large. High criticalities of the causal factors are implied to instigate a similar outcome on the procurement process (that is, high level of corruption).

In general, the causal factors were found to have an impact on more than 50 per cent of the activities captured under the procurement process, with the most impacted stages found to be the contract stage and post-contract stage. As presented in Figure 22.6, the procurement process encapsulates four primary stages: pre-contract stage, contract stage, contract administration stage, and the post-contract stage (Ruparathna and Hewage 2015). Each stage is composed of underlying activities that, it is argued, are susceptible to corrupt practices. Therefore, Figure 22.6 presents the overall impact of the causal factors on the procurement process. The results show that the collective criticalities of the causal factors lead to an appreciable level or high impact on the procurement process. Moreover, given the susceptibility of the procurement process to corruption and the significant impact of the causal factors on the procurement process, it is argued that the current situation of the process has a significant negative impact on the economic position of the context of the study.

Taking a closer look at the individual measurement items of the causal factors of corruption and the procurement process, it is evident that variables such as inadequate sanctions on culprits, lack of rigorous supervision, greed, flawed regulation system and lack of legal awareness were the most critical causal factors. Similar revelations were outlined in the studies of Tabish and Jha (2011) and Le et al. (2014a). Moreover, it was found that these factors, in addition to other prominent causal measures such as over-competition in the tendering process, the complexities of institutional roles and functions, and weak procurement and contractual structures (see Figure 22.6), cause a significant impact on the activities of the procurement process. As a result, critical implications on the procurement process could be tracked in activities PCS6, CTS3, CTS4 and CTS5; CAS4; and PCP1, PCP2, PCP3, PCP4 and PCP5).

Explaining the criticalities of the causal measures from extant theories reveals at least three main tested and justified suppositions of the relational attributes of corruption. The first

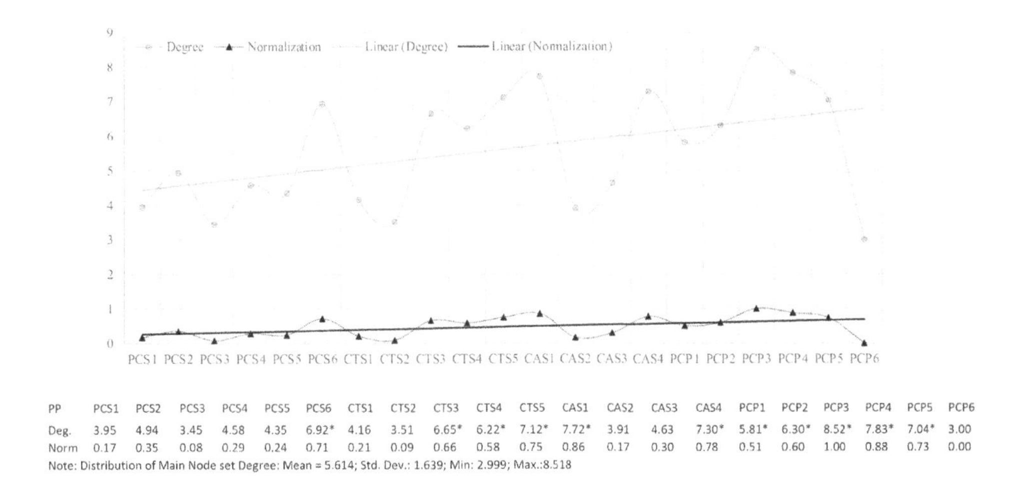

PP	PCS1	PCS2	PCS3	PCS4	PCS5	PCS6	CTS1	CTS2	CTS3	CTS4	CTS5	CAS1	CAS2	CAS3	CAS4	PCP1	PCP2	PCP3	PCP4	PCP5	PCP6
Deg.	3.95	4.94	3.45	4.58	4.35	6.92*	4.16	3.51	6.65*	6.22*	7.12*	7.72*	3.91	4.63	7.30*	5.81*	6.30*	8.52*	7.83*	7.04*	3.00
Norm	0.17	0.35	0.08	0.29	0.24	0.71	0.21	0.09	0.66	0.58	0.75	0.86	0.17	0.30	0.78	0.51	0.60	1.00	0.88	0.73	0.00

Note: Distribution of Main Node set Degree: Mean = 5.614; Std. Dev.: 1.639; Min: 2.999; Max.:8.518

Figure 22.6 *Network analysis of the impacts of the causal factors on the procurement process*

attribute, which relates to the inherent lack of ethical or social norms and standards, can best be explained within the context of behavioural theories such as the anomie theory and cognitive psychology (Dupuy and Neset 2018; Cohen 1965). The second is the rent-seeking theory, and the third is functionalism. While there may be more relevant theories, the discussion is centred around these due to the interconnections of the contextual levels that the study focuses on (that is, the individual, organizational, project and statutory levels).

Starting from the individual unit level and proceeding to the organizational level, corruption is known to spread easily, like a virus (Bertilsson Forsberg and Severinsson 2015). At this level, the most dominant causal construct linked to this unit is the organization-specific construct. Therefore, the organization-specific construct can be termed as the causal factors that instigate the spread of corruption at the organizational level. This is where the chain and the relational attributes of the parties who engage in corruption are best understood. The parties are commonly classified as being on the supply side, the demand side and the condoning side (Boyd and Padilla 2009). The organizational level is where common forms of corrupt acts that involve a common consensus among all the three parties are established. For instance, whereas at the individual level the main corrupt party may coerce other parties to engage in a corrupt act against their will, it is at the organizational level, which exhibits most white-collar crimes and other corrupt activities, where parties agree to come together with agreed-upon intentions to engage in corrupt acts or violate the norms or standards of an organization for an intended mutual benefit.

The common organization-specific causal factors examined in this study include the fight for economic survival, over-competition in the tendering process, poor documentation of organizational records such as financial statements, interim payments for completed works, inadequate sanctions, and the negative industrial or organizational working environment (Le et al. 2014a; Brown and Loosemore 2015; Bowen et al. 2012). However, from the analysis of the experts' responses, the top three most critical OSC variables were identified as inadequate sanctions, poor records of documentation and the lack of a positive industrial climate. Whereas these were found to be the stand-alone critical causal factors, the other OSC factors that had a critical impact on the procurement process included the complexities associated with institutional roles, and both fierce competition and over-competition in the procurement process. The components of this construct manifest and illustrate most of the economic theories of corruption, such as rent-seeking (Majaski 2019; Tullock 1991), for example in situations when construction or consultancy companies or related firms lobby public officers for tariff protection, subsidies and grants. In such situations, the firms are mostly on the lookout to gain additional monetary incentives with no correlative or requisite benefits such as in terms of enhanced productivity (Godwin et al. 2012; Majaski 2019). The persistence of such a situation, especially when it involves several companies that public officials stand to benefit from, not only causes a gradual abuse of public assets, but also the overall impact affects the economic position of a nation.

The discussion now turns to statutory causal factors. Corruption at the statutory level, also known as grand corruption, is known to be the most damaging type of corruption to an economy (Mauro 1998; Owusu et al. 2018). Therefore, the causal factors at this level are noted to be critical, although the levels of criticalities might differ from context to context. The statutory-specific causal construct was the second most critical construct following the psychosocial-specific construct (PSSC), with a criticality index of 3.58. The notable causal factors captured under this construct, in descending order of criticality, included limited coordination among government departments, which could result in the lack of checks regarding the activities of various departments. Other critical SSC factors included inappropriate political interference, government transitioning, and subjecting public servants to job insecurity (Zhang et al. 2016; Le et al. 2014a; Owusu et al. 2017).

Whereas the individual constructs had high criticality mean values, their respective impacts on the procurement process were identified to be neutrally critical as compared to the other causal factors. As highlighted earlier, grand corruption, which takes place at the government level, has a negative impact on the economic growth of a given context. For instance, in a government transition situation, it is common for governments to commence or initiate either feasible or non-feasible projects not for the benefit of the masses, but with the intended purpose of raiding the public coffers. Another case is when an incumbent government launches a project and continues to work on the project until a new government takes up power, and it can choose to either continue or abandon the ongoing project (Owusu et al. 2017; Williams 2017). Notably, in the latter case, projects can be abandoned when there is little opportunity to exploit the resources allocated, or intentionally abandoned in order to initiate new projects where there is sufficient opportunity to corruptly benefit. This case depicts opportunistic behaviour theories. It is the statutory-specific causal construct that some of the other causal constructs, such as the regulatory-specific construct (RSC), stem from.

The RSC construct was the third most critical causal construct of corruption after the SSC construct. The RSC construct included factors such as a flawed regulation system, loopholes or irregularities in stipulated rules and laws, and the lack of legal awareness. However, given that the construct was identified as the third most critical among the six constructs, there were a number of the variables that contributed to the overall criticality of the construct. In descending order of criticality, the top critical factors were: a defective regulation system, lack of legal awareness to prohibit people from engaging in corrupt acts, and multifarious licence aid permits. Moreover, following the determination of the constructs' individual criticalities, the respective impacts of the RSC variables on the procurement process were ascertained. Causal factors such as weak contractual structures and the deficiencies or irregularities in extant rules and law were identified as some of the leading variables that significantly impacted upon the procurement process. Again, similar to the theories discussed in the previous section, the RSC remains a causal construct that perpetrators often deploy to take advantage of the procurement process with the primary intention of exploiting it. The collective impact of the causal factors on the procurement process are presented in Figure 22.6.

The final construct for discussion is the project-specific causal factors, which ranked fourth out of the five causal constructs, with a criticality index of 3.44. Out of the eight causal factors captured under this construct, five were identified to be notably critical, with lack of rigorous supervision emerging as the most critical. The supervision of construction-related activities has always been critical, not just for economic concerns but also for other objectives such as worker safety and health. Without proper supervision and monitoring (of contractual or financial statements), and an effective audit mechanism in place to identify irregularities associated with projects, the project process is vulnerable to misappropriation. This may potentially be why activities such as completing financial audits, administering interim or progress payment, and the verification of completed reports were among the procurement activities that were rated to be vulnerable to corruption. The other critical factors were the inadequate measures by project funders to limit the pervasiveness of corruption in projects, and that they often assume that task to be the responsibility of anti-corruption agencies and the government.

The situation worsens when the funder is the government itself. It is pertinent to note that while the variables under the project-specific causal construct are particularly associated with projects, the other causal constructs, including the psychosocial, organizational, statutory and regulatory-specific causal factors, all have a direct impact on the successful completion of a project.

The results of the vulnerability pattern and the impacted pattern provide useful information regarding the most critical sections of the process that need the most attention. Thus, these stages and activities can be hypothesized as the procurement process sections, where most of the damage from corruption that hinders economic growth occurs. These results show that the allocation of resources for the purpose of the extirpation of corrupt practices in construction must not be generic, but instead it should be prioritized to first cover the most vulnerable and impacted-upon stages and activities of the projects. Thus, the development of anti-corruption measures and frameworks for mitigating corruption within the procurement process must primarily focus on these most vulnerable and impacted activities.

LIMITATIONS

The first limitation is of the study is the non-generalization of the findings. This chapter has focused on the assessment of the procurement process of construction projects and associated impacts in the developing context. The ascertained impacts of corruption on the various stages and activities of the procurement process cannot apply to other contexts. There is, therefore, the need to conduct similar studies in different contexts to ascertain the relevance of the factors considered in this study to the specific conditions of the examined context. Finally, while the empirical analysis of the impact of the causal factors on the procurement process remains justifiable, the implied impacts on economic growth may need further examination with case data to ascertain and possibly measure the real impact on economic growth.

RECOMMENDATION FOR FUTURE STUDIES

As this study has focused on the assessment of the causal factors of corruption in the procurement process of infrastructure-related works, future works can investigate the impact of effective anti-corruption measures on both projects and the economy at large. For instance, an empirical study can examine and estimate the monetary disparities after an anti-corruption framework has been put in place. The assessments can be undertaken in different contexts as well, to measure the differences among different contexts regarding the variations of the estimates.

Moreover, other than examining the relational attributes between corruption and economic growth, future studies can further explore the disparities between growth-reducing and (possibly) growth-enhancing levels of corruption. Finally, as the models for measuring corruption are mostly criticized due to the lack of actual quantification of corruption costs or damages, this study suggests that there is a need for econometric models specifically developed for estimating the monetary misappropriations of construction-related works.

CONCLUSIONS

This chapter sought to examine two key constructs of corruption intended to ascertain the impact of corruption on the procurement process and the relative impact on economic growth. The study therefore examined the proneness of the procurement activities to corruption, and the criticalities of the causal factors of corruption, and how these two constructs impact upon the overall outlook of the procurement process, to ascertain the implied impact on economic growth. An expert survey was conducted with 62 experts selected via purposive and snowball sampling methods. They were invited to evaluate the criticality of the proneness of the procurement process to corruption, the criticalities of the causal factors of corruption, and their impacts on the stages and particular activities of the procurement process. Twenty-one activities were identified under four stages of the procurement process, and 38 measurement items were identified as corruption causal factors captured under five constructs. With at least one activity identified under each stage of the procurement process to be vulnerable to corruption, a total of 11 out of the 21 activities (that is, PSC4 and 5; CTS2, 3 and 4; CAS4; and PCP1–5), and two out of the four stages (that is, CTS and PCP), were identified to be vulnerable.

Regarding the causal factors, PSSC and SSC were identified to be the most critical causal constructs, and the leading three critical causal variables were: personal greed under PSSC, inadequate sanctions under OSC, and flawed regulation system under RSC. The implications of these results were noted to be quite generic until their impacts on the procurement process were examined. Therefore, the results, revealed significant impacts on the procurement process. Typical activities that were identified to suffer significantly from the impact included PCS6, CTS3, CTS4, CAS1 and PCP3; thus revealing the stages in the procurement process and activities where critical attention is needed to strengthen these vulnerable stages. Finally, from the analysis of both indicators (that is, the vulnerable and impacted-upon), it can be inferred that most of the misappropriation that hinders economic growth takes place in the following sections of the procurement process: CTS3 and 4; CAS4; and PCP3, 4 and 5. As a result, resources allocated to fighting corruption within the procurement process can be prioritized based on the most problematic sections of the process mentioned above. Furthermore, new studies can further examine the extent and significance that each of the causal measures directly has on the individual phases of the procurement process, and how the overall impact on the monetary valuation on construction projects towards the economic stance of an economy can be mitigated.

ACKNOWLEDGMENTS

The authors wish to express their sincere gratitude to the Research Grants Council, University Grants Committee of Hong Kong, and to the Hong Kong Polytechnic University for funding this study under the Hong Kong PhD Fellowship Scheme.

REFERENCES

Abueva, J.V. (1966). The contribution of nepotism, spoils and graft to political development. *East-West Center Review*, *3*(1), 45–54.

Acemoglu, D., and Verdier, T. (1998). Property rights, corruption and the allocation of talent: a general equilibrium approach. *Economic Journal*, *108*(450), 1381–1403.

Ameyaw, E.E., Pärn, E., Chan, A.P., Owusu-Manu, D.G., Edwards, D.J., and Darko, A. (2017). Corrupt practices in the construction industry: survey of Ghanaian experience. *Journal of Management in Engineering*, *33*(6), 05017006.

Bai, J., Jayachandran, S., Malesky, E.J., and Olken, B.A. (2013). Does economic growth reduce corruption? Theory and evidence from Vietnam. National Bureau of Economic Research, No. w19483.

Benson, G.C.S., Maaranen, S.A., and Heslop, A. (1978). *Political Corruption in America*. Lexington, MA: Lexington Books.

Bertilsson Forsberg, P. and Severinsson, K. (2015). Exploring the virus metaphor in corruption theory: corruption as a virus? *Ephemera: Theory and Politics in Organization*, *15*(2), 453–463.

Bicchieri, C., and Duffy, J. (1997). Corruption cycles. *Political Studies*, *45*(3), 477–495.

Bowen, P.A., Edwards, P.J., and Cattell, K. (2012). Corruption in the South African construction industry: a thematic analysis of verbatim comments from survey participants. *Construction Management and Economics*, *30*(10), 885–901.

Boyd, J.M., and Padilla, J.D. (2009). FIDIC and integrity: a status report. *Leadership and Management in Engineering*, *9*(3), 125–128.

Breit, E., Lennerfors, T.T., and Olaison, L. (2015). Critiquing corruption: a turn to theory. *Ephemera: Theory and Politics in Organization*, *15*(2),319–336.

Brown, J., and Loosemore, M. (2015). Behavioural factors influencing corrupt action in the Australian construction industry. *Engineering, Construction and Architectural Management*, *22*(4), 372–389. DOI:10.1108/ECAM-03-2015-0034.

Chan, A.P., and Owusu, E.K. (2017). Corruption forms in the construction industry: literature review. *Journal of Construction Engineering and Management*, *143*(8), 04017057.

Cohen, A.K. (1965). The sociology of the deviant act: anomie theory and beyond. *American Sociological Review*, *30*(1), 5–14.

Colonnelli E., and Haas A. (2016). Corruption in construction. International Growth Center. https://www.theigc.org/blog/corruption-in-construction/ (accessed on 13 February 2021).

Compton, R.A., and Giedeman, D.C. (2011). Panel evidence on finance, institutions and economic growth. *Applied Economics*, *43*(25), 3523–3547.

Corporate Finance Institute (CFI) (2021). Economic indicators: a list of measures of the overall state of the macroeconomy. https://corporatefinanceinstitute.com/resources/knowledge/economics/economic-indicators/ (accessed on 9 February 2021).

Culpan, R., and Trussel, J. (2005). Applying the agency and stakeholder theories to the Enron debacle: an ethical perspective. *Business and Society Review*, *110*(1), 59–76.

Delavallade, C. (2006). Corruption and distribution of public spending in developing countries. *Journal of Economics and Finance*, *30*(2), 222–239.
Dobel, J.P. (1978). The corruption of a state. *American Political Science Review*, 958–973. https://www.jstor.org/stable/pdf/1955114.pdf (accessed on 10 July 2020).
Dupuy, K., and Neset, S. (2018). The cognitive psychology of corruption. *Micro-level Explanations for Unethical Behavior, U4* (2018), 2. Available at: https://www.u4.no/publications/the-cognitive-psychology-of-corruption.pdf (accessed on 19 November 2021).
Egunjobi, T.A. (2013). An econometric analysis of the impact of corruption on economic growth in Nigeria. *E3 Journal of Business Management and Economics*, *4*(3), 54–65.
Flanigan, J. (2002). Enron is proving costly to economy. *Los Angeles Times*. https://www.latimes.com/archives/la-xpm-2002-jan-20-mn-23790-story.html (accessed on 9February 2021).
Fuller, S. (2002). *Social Epistemology*. Bloomington, IN: Indiana University Press.
Gilpin K.N. (2001) Enron's collapse: the investors; plenty of pain to go around for small investors, funds, workers and creditors. https://www.nytimes.com/2001/12/04/business/enron-s-collapse-investors-plenty-pain-go-around-for-small-investors-funds.html (accessed on 10 February 2021).
Godwin, K., Ainsworth, S.H., and Godwin, E. (2012). *Lobbying and Policymaking: The Public Pursuit of Private Interests*. Thousand Oaks, CA: CQ Press.
Grabova, P. (2014). Corruption impact on economic growth: an empirical analysis. *Journal of Economic Development, Management, IT, Finance, and Marketing*, *6*(2), 57.
Greitzer, F.L., Kangas, L.J., Noonan, C.F., Brown, C.R., and Ferryman, T. (2013). Psychosocial modeling of insider threat risk based on behavioral and word use analysis. *e-Service Journal: A Journal of Electronic Services in the Public and Private Sectors*, *9*(1), 106–138.
Heiser, W.J. (2001). Corruption: political and public aspects. In: Smelser, N.J., and Baltes, P.B. (eds), *International Encyclopedia of the Social and Behavioral Sciences*, Vol. 11. Amsterdam: Elsevier, pp. 2824–2830.
Henri, N.N. (2018). Impact of corruption on public debt: evidence from Sub-Saharan African countries. *American Journal of Economics*, *8*(1), 14–17.
Hoxley, M. (2008). Questionnaire design and factor analysis. In: Knight, A., and Ruddock, L. (eds), *Advanced Research Methods in the Built Environment*. Chichester: Wiley-Blackwell, pp. 122–134.
Huang, C.J. (2016). Is corruption bad for economic growth? Evidence from Asia-Pacific countries. *North American Journal of Economics and Finance*, 35, 247–256.
Johnston, M. (2017). *Political Corruption: Readings in Comparative Analysis*. New York: Routledge.
Johnson, S., Kaufmann, D., and Zoido-Lobatón, P. (1999). *Corruption, Public Finances and the Unofficial Economy* (Vol. 2169). World Bank Publications. https://documents1.worldbank.org/curated/en/219311468762600809/pdf/multi-page.pdf (accessed on 19 November 2021)
Kenny, C. (2006). Measuring and reducing the impact of corruption in infrastructure. World Bank.
Kenny, C. (2007). Construction, corruption, and developing countries. World Bank Policy Research Working Paper 4271. http://documents1.worldbank.org/curated/en/571281468137721953/pdf/wps4271.pdf (accessed on 9 February 2021).
Kenny, C. (2012). Publishing construction contracts to improve efficiency and governance. *Proceedings of the Institution of Civil Engineers*, *165*(5), 18.
Kottasova, I. (2014). World's most corrupt industries. https://money.cnn.com/2014/12/02/news/bribery-foreign-corruption/index.html (accessed on 19 November 2021).
Krishnan, C. (2010). Tackling corruption in construction. *How corruption is affecting the UK construction industry and the benefits of combatting it.* London: Transparency International. Available at: https://issuu.com/transparencyuk/docs/tackling_corruption_in_construction (accessed on 9 February 2022).
Le, Y., Shan, M., Chan, A.P., and Hu, Y. (2014a). Investigating the causal relationships between causes of and vulnerabilities to corruption in the Chinese public construction sector. *Journal of Construction Engineering and Management*, *140*(9), 05014007.
Le, Y., Shan, M., Chan, A.P., and Hu, Y. (2014b). Overview of corruption research in construction. *Journal of Management in Engineering*, *30*(4), 02514001.
Lewis, B.D., and Hendrawan, A. (2019). The impact of majority coalitions on local government spending, service delivery, and corruption in Indonesia. *European Journal of Political Economy*, *58*, 178–191.

Lotha, G. (2016). The NRICH project – transitivity. Transitive law: logic and mathematics. *Britannica*. https://www.britannica.com/topic/transitive-law (accessed on 1 October 2020).
Majaski C. (2019). Economics – microeconomics: rent seeking. https://www.investopedia.com/terms/r/rentseeking.asp (accessed on 20 September 2020).
Martin, K.D., Cullen, J.B., Johnson, J.L., and Parboteeah, K.P. (2007). Deciding to bribe: a cross-level analysis of firm and home country influences on bribery activity. *Academy of Management Journal*, *50*(6), 1401–1422.
Mauro, P. (1997). The effects of corruption on growth, investment, and government expenditure. *Corruption and the Global Economy (Institute for International Economics, Washington, DC)*, 83–108. https://www.imf.org/en/Publications/WP/Issues/2016/12/30/The-Effects-of-Corruptionon-Growth-Investment-and-Government-Expenditure-2042 (accessed on 19 November 2021).
Mauro, P. (1998). Corruption and the composition of government expenditure. *Journal of Public Economics*, *69*(2), 263–279.
McLean, B., and Elkind, P. (2013). *The Smartest Guys in the Room: The Amazing Rise and Scandalous Fall of Enron*. New York: Penguin.
Méndez, F., and Sepúlveda, F. (2006). Corruption, growth and political regimes: cross country evidence. *European Journal of Political Economy*, *22*(1), 82–98.
Mo, P.H. (2001). Corruption and economic growth. *Journal of Comparative Economics*, *29*(1), 66–79.
Olken, B.A., and Pande, R. (2012). Corruption in developing countries. *Annual Review of Economics*, *4*(1), 479–509.
Organisation for Economic Co-operation and Development (OECD) (2016). *Anti-Bribery Policy and Compliance Guidance for African Companies*. https://www.oecd.org/corruption/anti-bribery/Anti-Bribery-Policy-and-Compliance-Guidance-for-African-Companies-EN.pdf (accessed on 4 March 2021).
Organisation for Economic Co-operation and Development (OECD) (2019). *Policy Note on Africa Infrastructure and Regional Connectivity*. https://www.oecd.org/dev/development-philanthropy/AfricaPolicyNote%20_2019.pdf (accessed on 4 March 2021).
Owusu, E.K., and Chan, A.P. (2018). Barriers affecting effective application of anticorruption measures in infrastructure projects: disparities between developed and developing countries. *Journal of Management in Engineering*, *35*(1), 04018056.
Owusu, E.K., Chan, A.P., and Ameyaw, E. (2019). Toward a cleaner project procurement: evaluation of construction projects' vulnerability to corruption in developing countries. *Journal of Cleaner Production*, *216*, 394–407. https://doi.org/10.1016/j.jclepro.2019.01.124.
Owusu, E.K., Chan, A.P., and Shan, M. (2019). Causal factors of corruption in construction project management: an overview. *Science and Engineering Ethics*, *25*(1), 1–31. https://doi.org/10.1007/s11948-017-0002-4.
Owusu, E.K., Chan, A.P., and Hosseini, M.R. (2020). Impacts of anti-corruption barriers on the efficacy of anti-corruption measures in infrastructure projects: Implications for sustainable development. *Journal of Cleaner Production*, *246*, 119078.
Owusu, E.K., Chan, A.P., and Shan, M. (2017). Causal factors of corruption in construction project management: an overview. *Science and Engineering Ethics*, 1–31.
Owusu, E.K., Chan, A.P., and Shan, M. (2018). Causal factors of corruption in construction project management: an overview. *Science and Engineering Ethics*, *25*(1), 1–31.
Owusu, E.K., Chan, A.P., and Ameyaw, E. (2019). Toward a cleaner project procurement: evaluation of construction projects' vulnerability to corruption in developing countries. *Journal of Cleaner Production*.
Pattanayak, S., and Verdugo-Yepes, C. (2020). Protecting Public Infrastructure from Vulnerabilities to Corruption: A Risk-Based Approach. Well Spent: How Strong Infrastructure Governance Can End Waste in Public Investment. International Monetary Fund (IMF). doi: https://doi.org/10.5089/9781513511818.071.
Rose-Ackerman, S. (ed.) (2007). *International Handbook on the Economics of Corruption*. Cheltenham, UK and Northampton, MA, USA: Edward Elgar Publishing.
Rubya, T. (2015). The ready-made garment industry: an analysis of bangladesh's labor law provisions after the Savar tragedy. *Brooklyn Journal of International Law*, *40*(2), 7.

Ruparathna, R., and Hewage, K. (2015). Review of contemporary construction procurement practices. *Journal of Management in Engineering, 31*(3), 04014038. doi.org/10.1061/(ASCE)ME.1943-5479.0000279.
Schwandt, T.A. (1994). Constructivist, interpretivist approaches to human inquiry. *Handbook of Qualitative Research, 1*. Thousand Oaks, CA: Sage, pp. 221–259.
Shah, A. (2006). Corruption and Decentralized Public Governance. Policy Research Working Paper; No. 3824. Washington, DC: World Bank. https://openknowledge.worldbank.org/handle/10986/8805 License: CC BY 3.0 IGO.
Shan, M., Chan, A.P.C., Le, Y., Hu, Y. and Xia, B. (2017). Understanding collusive practices in Chinese construction projects. ASCE *Journal of Professional Issues in Engineering Education and Practice, 143*(3), 05016012.
Sohail, M., and Cavill, S. (2008). Accountability to prevent corruption in construction projects. *Journal of Construction Engineering and Management, 134*(9), 729–738.
Søreide, T. (2002). Corruption in public procurement. Causes, consequences and cures. Chr. Michelsen Institute.
Stansbury, C., and Stansbury, N. (2008). Examples of corruption in infrastructure. Global Infrastructure Anti-Corruption Centre. http://www.giaccentre.org/ documents/GIACC.CORRUPTIONEXAMPLES.pdf (accessed on 15 June 2018).
Stansbury, C., and Stansbury, N. (2018). Preventing corruption in the construction sector. In: Egbu, C. and Ofori, G. (eds), *Proceedings of the International Conference on Professionalism and Ethics in Construction*, London, 21–22 November, pp. 30–38.
Tabish, S.Z.S., and Jha, K.N. (2011). Analyses and evaluation of irregularities in public procurement in India. *Construction Management and Economics, 29*(3), 261–274.
Tanzi, V. (1995). Corruption, governmental activities, and markets. *Finance and Development, 32*(4). https://www.elibrary.imf.org/view/journals/022/0032/004/article-A007-en.xml (accessed on 20 January 2020).
Tanzi, V. (1998). Corruption around the world: causes, consequences, scope, and cures. *Staff Papers: International Monetary Fund, 45*(4), 559–594. https://www.jstor.org/stable/pdf/3867585.pdf (accessed on 19 November 2021).
Tonge, A., Greer, L., and Lawton, A. (2003). The Enron story: you can fool some of the people some of the time … *Business Ethics: A European Review, 12*(1), 4–22.
Tran, M. (2002). Markets unsettled by the Enron effect. https://www.theguardian.com/business/2002/jan/30/enron.stockmarkets (accessed on 10 February 2021)
Tullock, G. (1991). Rent seeking. In: Eatwell, J., Milgate, M., and Newman, P. (eds), *The World of Economics*. London: Palgrave Macmillan, pp. 604–609.
United Nations (2018). Meetings coverage and press releases. https://www.un.org/press/en/2018/sc13493.doc.htm (accessed on 20 September 2020).
Wai, M.T.K. (2016). Investigation of corruption cases. https://unafei.or.jp/publications/pdf/RS_No79/No79_19VE_Man-wai2.pdf (accessed on 26 June 2019).
Weisel, O., and Shalvi, S. (2015). The collaborative roots of corruption. *Proceedings of the National Academy of Sciences, 112*(34), 10651–10656.
Williams, M.J. (2017). The political economy of unfinished development projects: corruption, clientelism, or collective choice? *American Political Science Review, 111*(4), 705–723.
Winter, B. (2017). Brazil's never-ending corruption crisis: why radical transparency is the only fix. *Foreign Affairs, 96*(3), 87–94.
Zhang, B., Le, Y., Xia, B., and Skitmore, M. (2016). Causes of business-to-government corruption in the tendering process in China. *Journal of Management in Engineering, 33*(2), 05016022.

23. Economic considerations in the procurement and deployment of construction informatics applications

Chimay J. Anumba and Esther A. Obonyo

INTRODUCTION

The construction industry has its roots in the various crafts that are brought together to build a constructed facility. The legacy of this craft-based feature is that many industry practitioners had a sense of pride in their manual skills and there has been a state of inertia to the introduction of new technologies. As such, while other sectors of the economy have embraced automation and realized the associated productivity improvements, the construction industry has not been as receptive (Anumba, 1989) and has lagged behind other sectors in productivity (Teicholz, 2013) and other technology-based benefits such as interoperability (Gallagher et al., 2004).

In the early days of construction, the craftsmen typically worked under the direction of a master builder who had multidisciplinary knowledge and could lead the others in the realization of the end product. With the growth in specializations, the role of the master builder was lost, and the industry suffered from fragmentation in terms of the project team, the stages in the process, the tools used, and project knowledge. This has led to the need for greater integration and collaboration within the construction project delivery process.

Over the last few decades, integration and collaboration have been the foci of many efforts geared towards improving the capacity, capability and performance of the construction industry. While some efforts sought to leverage emerging information and communication technologies, others were more oriented towards changing the culture and business processes of the industry. The technology-based efforts included the adoption of approaches such as computer-aided design (CAD), product modelling, knowledge-based/expert systems, improved wired and wireless communication systems, virtual and mixed reality systems, and building information modelling (BIM). Efforts that were not technology-based per se included partnering or alliancing, collaborative procurement methods – such as design and build, private finance initiative (PFI), public–private partnership (PPP), and integrated project delivery – knowledge management, and other innovative approaches. These efforts were not mutually exclusive, in that the introduction of some technologies facilitated or accelerated the adoption of new collaborative working methods. Conversely, the adoption of new procurement methods made the implementation of several technologies much easier, especially at the project team level.

This chapter focuses on the economic considerations in the procurement and deployment of construction informatics applications in the construction industry. The next section explores, in greater detail, the uptake of technology in the construction industry, including the economic drivers for the changing situation in the industry. This is followed by a review of several current and emerging technologies and how these may impact upon the economics of construction project delivery. A detailed discussion of the economic considerations in technology procurement and deployment is also presented, followed by a summary and some conclusions.

UPTAKE OF DIGITAL TECHNOLOGIES IN THE CONSTRUCTION INDUSTRY

Construction Industry Inertia

Construction was ranked 20th in McKinsey's 2016 Global Institute Industry digitization index, which featured 21 sectors of the economy (Agarwal et al., 2016). This is not surprising, given the unique set of challenges that limit the ability of companies within the sector to deploy digital solutions across multiple projects (Egan, 1998; Adrian, 2004; McKinsey, 2016). Implementing digital solutions along the construction value chain requires significant investment in the coordination of changes across several organizations (McKinsey, 2016). The temporary nature of relationships that exist among organizations that work together to complete most projects, coupled with the high turnover of the construction workforce, make establishing new ways of working and developing new capabilities difficult (Adrian, 2004). The lack of replication across construction projects limits the ease with which the kind of changes required for a digital transformation can be rolled out at scale. The decentralized nature of functions, with projects being executed at sites that are different from the offices hosting the business operation units, usually complicates the deployment of new strategies and tools.

At the macro level, there is a size-related matter that resulted in the construction industry not adopting digital technologies and other innovations at the same rate as other sectors. Small and medium-sized enterprises (SMEs) contribute between 50 and 70 percent of the world's gross domestic product (GDP) (see ILO, 2019). Construction employs 7.6 percent of the global workforce (Lieuw-Kie-Song, 2020). A significant proportion of the construction workforce is employed by SMEs, as shown in Table 23.1. Smaller enterprises hire young people, older workers, less-skilled workers, and other groups with lower chances of securing employment elsewhere. The frequently cited low productivity rate of the construction industry (Morris, 1987; Adrian, 2004; McKinsey, 2016; Green, 2016; Abdel-Hamid and Abdelhaleem, 2020) can be largely attributed to the resulting skills-related labor productivity gaps between SMEs and larger organizations. This gap is compounded by SMEs' limited ability to invest in human capital development that can ensure peak performance (International Labour Organization, 2019).

McKinsey's characterization of the construction industry as one of the "least digitized category" sectors is largely because it is SME-dominated, as shown in Table 23.1. SME employers face unique challenges which limit their ability to address these issues in a substantive manner (International Labour Organization, 2001, 2019). They operate in a landscape riddled with regulatory complexities; this is a disincentive for change. SMEs usually incur higher transactional costs when accessing finance (International Labour Organization, 2019). Because large companies tend to offer higher wages, SMEs experience a shortage of skills at both managerial and workforce levels (McKinsey, 2016; European Union, 2019). The strategic use of digital technologies by SMEs can also enhance management practices, improve market intelligence, and create virtual access to regional and global value chains. Lack of access to

Table 23.1 *US construction company size estimates*

Size of firms in employees	2019	2026
Less than 5	472 794	346 192
5 to 9	119 360	97 953
10 to 19	72 627	70 612
20 to 99	64 162	61 935
100 to 499	8 935	11 431
500 to 999	629	697
1000 to 2499	343	344
2500 to 4999	117	169

Source: Statista (2019).

skills and to financial resources have limited the ability of SMEs to exploit such opportunities (International Labour Organization, 2019).

More recently, the construction industry's inertia towards technology adoption is changing (McKinsey, 2016; Autodesk in association with CIOB, 2018). This is based on several economic and other drivers. First, the cost savings possible from the adoption of digital technologies are now more evident, with numerous examples and case studies. Also, there are quality improvements in the construction product that were hitherto not possible without the precision and improved coordination that construction informatics applications facilitate. The adoption of collaborative procurement methods has accentuated the need for project team members that span different disciplines and geographical areas to deploy digital technologies in working together more effectively and efficiently. Another key driver in the changing posture is government policy and associated mandates, which require that construction projects adopt BIM on public sector projects, or submit documents for planning or building regulations approval in electronic format. These and other changes in the construction industry are driving greater adoption of digital technologies in construction.

Recent Trends

The Fourth Industrial Revolution (4IR) is the biggest catalyst and accelerator for the adoption of cutting-edge digital technologies across all sectors. It is characterized by a fusion of technologies that can seamlessly integrate the physical, digital and biological spheres (Buehler et al., 2018). Within the construction industry, the sophisticated use of digital technologies such as BIM, prefabrication, wireless sensors, three-dimensional (3D) printing, and automated and robotic equipment is proof of the onset of the 4IR era; these will be discussed below. In the 2010s, when several sectors were undergoing waves of transformation triggered by digital technologies, the "infrastructure and urban development industry continued operating as it has for the past 50 years" (World Economic Forum, 2018). However, things are changing. New entrants (start-ups) to construction challenge the status quo through the same entrepreneurial energy that resulted in an unprecedented shake-up of the financial sector, which had remained stable for decades (see Alt et al., 2018). The arrival of new market participants and their start-up companies ushered in the "FinTech" movement. The new start-ups in construction are ushering in a wave of "Construct Tech/ConTech" transformation (Blanco et al., 2018). The impact of ConTech is felt across the entire construction value chain. Companies such as

Table 23.2 *Cost minimization opportunities*

Description		%	
Current cost per housing unit		Total = 100	Cost reductions
Potential areas for cost savings	Unlocking land supply		8 to 23
	Reducing construction costs		12 to 16
	Improved operations and maintenance		2
	Lowering cost of finance		0 to 7
Optimized cost per housing unit		52 to 78	

Source: Percentages extrapolated based on average cost estimates provided by Woetzel et al. (2014).

WeWork are offering end-to-end solutions across a building's life cycle, from identifying, leasing and designing, to building and managing.

Traditional venture capitalist firms have brought Silicon Valley's technology start-up culture into the construction industry (Leonard, 2019). Bricks and Mortar Ventures invests in start-ups developing innovative software and hardware solutions targeting architecture, engineering, construction and facilities management. In 2019, this venture capitalist closed a US$97.2 million institutional venture funding round that was backed by Autodesk, Hilti, Obayashi, CEMEX Ventures, and other major players in the construction industry (Rubenstone, 2019).

Other Macro Drivers of Change

Rapid urbanization in a changing climate

The construction industry can play a pivotal role with respect to helping cities and urban areas in all parts of the world to expand in a more inclusive, sustainable and resilient manner through productivity improvement. For decades, cities and urban areas have been dealing with challenges related to poor housing conditions and an inadequate supply of affordable housing. The existing housing crisis is characterized by an unresponsive housing supply and the scarcity of affordable housing (Kallergis et al., 2018; Wetzstein, 2017; Rohe, 2017; King et al., 2017). Rapid urbanization will further complicate this crisis. According to the UN's *World Population Prospect 2019* (United Nations, Department of Economic and Social Affairs, 2019), the world's population could grow to around 8.5 billion in 2030, 9.7 billion in 2050, and 10.9 billion in 2100. The proportion of people living in urban areas will rise from 55 percent in 2020 to 68 percent in 2050, and 90 percent of such growth will happen in Africa and Asia. As shown in Table 23.2, efficiencies during the construction process can reduce the total cost of housing by between 12 and 16 percent (see Woetzel et al., 2014). The use of digital tools is one of the levers that can be used to significantly reduce construction costs.

Weather-related disruptions are consistently rated as one of the most frequent and harmful causes of construction project delays (see review by Ballesteros-Pérez et al., 2018). Adverse weather conditions are described as either "foreseeable" or "unforeseeable" depending on whether it is reasonable to expect the party performing construction work to plan and account for the disruption in advance. Therefore, it is essential to ensure that contractors and subcontractors have access to reliable weather information. Conventional practice in construction has relied on the use of local weather forecasts and historic trends. This approach does not account for the volatile nature of recent extreme weather events. Contractors need a more robust approach that provides access to daily and extended weather forecasts (Reutter, 2019).

The frequency and severity of extreme events are increasing. The World Economic Forum (2019) *Global Risks Report* ranked extreme weather events as the world's top risk. This has major negative consequences for activities such as concrete pouring; making good short-term decisions will require access to relevant real-time weather information. Automated weather stations can deliver context-specific, real-time information (Reutter, 2019). This may be the only option for accessing weather information for construction projects in remote locations. Big data is being used to advance the trend toward the use of real-time weather information in efforts such as the Climate-I Construction web tool. Its development is being guided by an advisory panel that includes scientists from the United States (US) National Oceanic and Atmospheric Administration.

Demographic trends

At the national level, human capital is both a condition and a consequence of economic growth (Mincer, 1984). Because human input is part of the production function that yields income and other useful outputs over long periods, human capital is a critical component of an organization's capital (see Becker, 2008). Economists regard expenditures on education, training and medical care as investments in human capital. Human capital can be analogized as the "Achilles heel" of the construction industry. This is partly because of some major demographic shifts. The situation in the US exemplifies the challenges that these shifts pose for industrialized countries. According to the US Bureau of Labor Statistics for 2018, 10 million construction workers were over 25 years old. There were just 1 million construction workers under age 25. The median age of a worker in the construction industry was 42.5 years old. It is projected that the construction industry could lose between 14 and 20 percent of specific employee groups, including executives, senior managers, field managers and project managers, by 2023 (see Hardy, n.d.). This will compound the existing skills gap challenge. The construction industry can address this problem by either retaining more of the older workers or attracting a larger number of younger workers. These options are now considered. Also discussed are the shifts in market demand for construction as a result of demographic change.

Retaining aging workforce

The proportion of US workers aged 55 years and older is expected to increase from 11.9 percent in 1994 to 24.8 percent in 2024 (Toosi, 2015). There has been an increase in the average age of workers triggered by the loss of defined-benefit pension plans, loss of retiree healthcare benefits, increasing age eligibility requirements for full Social Security retirement benefits, and the impacts of the 2007–2008 recession (Sokas et al., 2019; McFall, 2011; Tang et al., 2013; Hurd and Rohwedder, 2011; Szinovacz et al., 2015). Because of its sensitivity to business cycles, an aging population has a significant impact on the construction industry (Ringen et al., 2018). Many construction workers intend to work full-time beyond age 65 years; this includes workers with self-reported health challenges (Dong et al., 2015). The average age of construction workers in 2015 was 42.5, and the proportion of those aged 55 and older is expected to increase from 11.9 percent in 1994 to 24.8 percent in 2024 (CPWR, 2018).

The physical demands of construction work have been linked to the growing number of people with work-related disabilities and premature mortality (Ervasti et al., 2019). Fatal and non-fatal traumatic injury rates in the construction industry are three times higher than those in all other sectors (CPWR, 2018; Dong et al., 2015, 2016). The risk factors increase with age (Sokas et al., 2019). A nationally representative longitudinal survey in 1979 established that when participants reached age 40, 38 percent of them will have experienced a days-away-from-work (DAFW) injury, and that these workers are significantly more likely to report diagnosed conditions, musculoskeletal disorders, depression and health problems, limiting their ability to work (Sokas et al., 2019; Dong et al., 2015). A study of US construction workers between the ages of 40 and 59 identified musculoskeletal disorders and other medical conditions as some of the main factors that contribute to their premature exit from the workforce (Welch et al., 2010). Digital technologies such as motion capture devices and wearable sensors are being used to proactively address the underlying risk factors (Obonyo and Zhao, 2018; Zhao and Obonyo, 2018).

Attracting younger workforce: the Generation Z factor

The biggest demographic-related challenges and opportunities for the construction industry revolve around Generation Z (Gen Z). These are individuals born between 1997 and 2012 (Francis and Hoefel, 2018; Dimock, 2019). This generation is increasingly becoming a pivotal force in today's economy. It is estimated that by 2019, Gen Z could account for as much as 30 percent of the global population. Many of them are joining the workforce at a younger age compared to previous generations. Gen Z's relationship with technology has a bearing on their career choices. They have spent their entire lives immersed in a technology-driven world through exposure to on-demand access to information, resources, and people using computers, cellphones and social media. Potential employers must also adapt to the digital culture of Gen Z workers. Trends in the construction industry such as BIM software, drones, wearable devices, virtual reality and three-dimensional (3D) modeling can play a key role in attracting and retaining more Gen Z job seekers to the sector.

Responding to shifts in market demand

The Generation Z factor also has market implications. It puts pressure on the construction industry to deliver more affordable buildings that can fit the budgetary limits of this generation (Hardy, n.d.). Gen Z renters and buyers also actively seek opportunities to minimize their carbon footprint. They want built-in features that support electricity, water and waste reduction. As renters and homebuyers, they prioritize reliable cell reception and high-speed internet over features such as in-unit washers and dryers, parking and pools. Developers are responding to their needs through, for example, installing distributed antenna systems (DAS) to help boost cell signals in large facilities (see Osman et al., 2011). Gen Z renters and buyers have an interest in artificial intelligence-enabled features being incorporated into the controls for heating, ventilation and air conditioning units, and for lighting and security systems. They will expect to interact with the homes through intelligent building capabilities that can extract insights into resource optimization, maintenance, upgrades or planned repairs based on the use of data to learn about the routines and behaviors of building occupants.

CURRENT AND EMERGING TECHNOLOGIES FOR CONSTRUCTION

Numerous information and communications technologies have had an impact on the construction industry over the years. The economic and other benefits of basic applications for design, estimating, scheduling, project management, and other systems is well established and will not be covered here. Instead, this focuses on some of the current and emerging technologies that will influence the future of the construction industry. These are presented with an added perspective on the associated economic implications of their deployment in the construction project delivery process.

Building Information Modeling (BIM)

The emergence of BIM can be traced all the way back to the earliest computer graphics system, Sketchpad, which was developed in 1965 (Sutherland, 1965). However, it was not until the 1980s that the term "BIM" was actually coined. It was based on technical advances made possible by numerous research projects on product modeling at that time, and it has become increasingly popular with further developments in computational modeling and visualization technologies (Sacks et al., 2018).

The growth in the popularity of BIM has resulted in numerous definitions of the concept. For the purpose of this chapter, two of the most authoritative definitions are presented:

- Building information modeling is "a modeling technology and associated set of processes to produce, communicate and analyze building models" (Sacks et al., 2018).
- A building information model is "the digital representation of physical and functional characteristics of a facility. As such, it serves as a shared knowledge resource for information about a facility, forming a reliable basis for decisions during its life cycle from inception onwards" (NIBS, 2015).

The above definitions emphasize several aspects of the technology. First is the focus on the capture and representation of the totality of the information necessary to define a facility; not just in terms of geometrical attributes, but also with regard to its functionality. Another aspect is that there are several processes associated with BIM, making it more than just a modeling technology. It provides a platform for construction project team members to work more collaboratively and efficiently throughout the life cycle of the facility.

The last few years have seen considerable growth in the uptake of BIM in the construction industry. Many countries, such as the United Kingdom, even mandated the use of BIM in public sector projects (HM Government, 2012). This is fueled by the increased recognition of the economic and other benefits that BIM offers. These design, construction and facility management benefits (Sacks et al, 2018; NIBS, 2015) include:

- improved productivity;
- reduced costs;
- greater efficiency;
- less uncertainty and reduced errors and rework;
- improved coordination and constructability;
- greater integration of design and construction;
- improved consistency;
- improved collaboration;
- greater functionality of models due to BIM plug-ins;
- increased sustainability and energy efficiency;
- better interoperability among systems;
- Improved project progress monitoring;

- Improved capacity for pre-assembly and fabrication; and
- more efficient operation and management of facilities.

All the above BIM benefits have economic implications, which are increasingly being realized by those organizations that have adopted BIM. While early adopters did not have case studies to rely on, the industry is now replete with many projects that demonstrate the economic benefits of BIM adoption. There is now little doubt that BIM delivers considerable value to construction sector organizations. However, the most sustainable economic benefits derive from the business process and cultural changes that these organizations adopt, as well as the new capabilities that result from integrating a wide variety of other functionalities with BIM.

Cloud Computing

Cloud computing has grown in importance over the last decade and provides "a model for enabling ubiquitous, convenient, on-demand network access to a shared pool of configurable computing resources (e.g. networks, servers, storage, applications and services) that can be rapidly provisioned and released with minimal management effort or service provider interaction" (Mell and Grance, 2011). It evolved from the concept of "Web services" in the early 2000s and involves end-users remotely accessing data centers and servers on which the computing resources are hosted. These end-users are typically agnostic in terms of the location of the resources, and have no responsibility for directly or actively managing them.

A service-oriented architecture is at the core of cloud computing, with cloud providers typically adopting the service models (Dillon et al., 2010; Mell and Grance, 2011) shown in Figure 23.1 and outlined below:

- Software as a service (SaaS): this involves providing end-users with access to software applications and associated databases running on a cloud infrastructure. These are provided on-demand and obviate the need for end-users to download and install the software on their own computers.
- Platform as a service (PaaS): this enables the consumer "to deploy onto the cloud infrastructure consumer-created or acquired applications created using programming languages, libraries, services, and tools supported by the provider" (Mell and Grance, 2011). PaaS is offered as a service to application developers so that they can run their software on the cloud platform.
- Infrastructure as a service (IaaS): this entails having the consumer "deploy and run arbitrary software, which can include operating systems and applications" on the cloud infrastructure. While application providers may have control over the operating systems, applications and storage, they do not control the underlying cloud infrastructure.

Cloud computing represents a paradigm shift from earlier approaches to the procurement of computing resources, which involved end-users buying each of the resources that they needed, irrespective of the frequency or criticality of use. This was very expensive for many construction sector companies, which are typically SMEs, and lack the technical expertise to leverage and/or maintain the functionality of these systems. With cloud computing, the "pay-as-you-go" model makes these computing resources much more affordable for construction sector SMEs, and obviates the need to own or manage substantial information technology (IT) infrastructure.

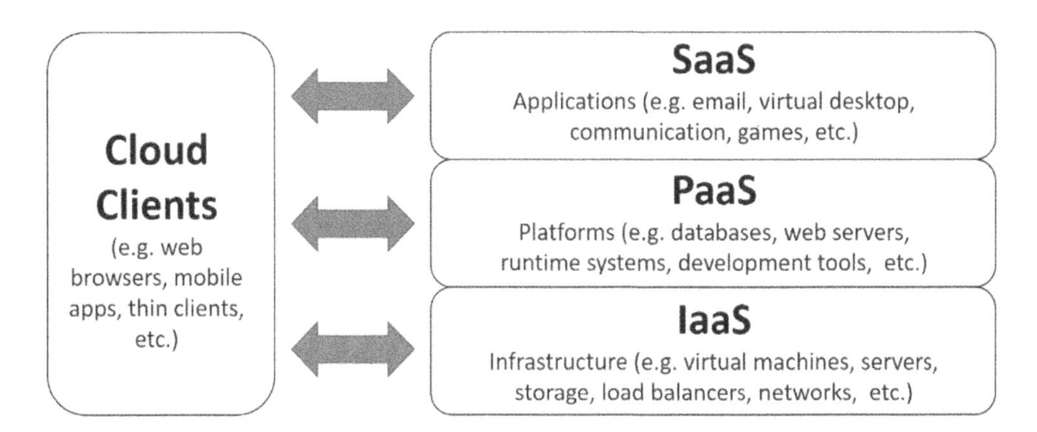

Figure 23.1 *Cloud computing service models*

With regard to deployment, a clouds is classified as private, public or hybrid (Dillon et al., 2010; Marston et al., 2011), depending on whether it is operated for a single organization, open to the general public, or a combination of the two (which allows companies to store proprietary data and information in a secure environment while maintaining access to public cloud infrastructure).

The economic benefits of cloud computing include the following (Marston et al., 2011; AWS, 2020):

- economies of scale are possible based on large numbers of end-users sharing access to computing resources;
- the pay-as-you-go model allows end-users to only pay for those resources needed at a particular point in time without the overhead of making a huge upfront computing infrastructure investment;
- faster access to computer resources;
- lower IT infrastructure management and maintenance costs;
- business agility and flexibility in meeting the variable demand for computing resources;
- reduced requirement for in-house IT personnel; and
- access to novel applications and services not previously available in conventional computing environments.

Sensing

There has been considerable growth in the use of sensing within the construction industry, with a wide variety of sensors now being deployed on job sites and constructed facilities (Akinci and Anumba, 2008). The sensors deployed on construction sites seek to capture real-time data on the status of the facility being constructed; location, flow and levels of material stocks; the stages in the construction process; and the utilization of equipment on the site. The nature of sensing can vary from embedded sensing that assesses some material characteristics of a product, to utilization of spatial sensing and imaging technologies that assess both the quality of products being constructed and the safety of processes being utilized during construction. Through the utilization of a wide array of sensors, both infrastructure components and the processes of managing the construction and operations of those components become intelligent, providing project and infrastructure managers with improved situation awareness and the capacity to make better-informed decisions.

When embedded in constructed facilities such as buildings and civil infrastructure systems, sensors enable a wide range of data to be captured that can facilitate the management of these facilities. For example, in buildings, sensors are routinely used to monitor room occupancy levels, indoor air quality, thermal and visual quality parameters, and the potential danger from gases such as carbon monoxide and methane. In civil infrastructure systems such as dams, bridges and highway structures, sensors are able to detect potential safety hazards and serviceability issues well before they are visible to the naked eye. Sensors are also able to detect missing members in both temporary and permanent structures, thereby enhancing the safety of these structures. Use of digital cameras for visual sensing has grown over the years, and cameras are now regularly used to detect construction defects, alignment issues, and in personnel or equipment tracking.

It is beyond the scope of this chapter to discuss the variety of sensor types. It will suffice to say that the variety and sophistication of sensors and sensor networks is increasing at a rapid pace, thereby facilitating the ubiquitous tracking and monitoring of a wide range of components and processes. The focus here is limited to the various ways in which they are used (described above) and their economic benefits, which can be summarized as follows:

- sensing enables the efficient monitoring of aspects of the construction process;
- the real-time information obtained from sensors enables early detection of potential problems, which may be expensive to address if not found until later;
- sensors save the time and effort hitherto spent on physical inspection of civil infrastructure systems;
- with the transmission of sensing data through wireless communication systems to geographically distributed professionals, there is a reduction in travel time to gather data from job sites and/or constructed facilities;
- quality control is improved when sensing is deployed to track deviations from the design or material specifications; and
- sensing reduces costly on-site errors and accidents as construction personnel have improved situation awareness from the data obtained from sensors.

Cyber-Physical Systems and Digital Twins

Cyber-physical systems (CPSs) enable the synergistic integration of virtual models and the physical environment. They offer additional opportunities for the construction industry to leverage technology to improve the project delivery process. This is because those organizations that have adopted BIM, and virtual design and construction (BIM/VDC) are not maximizing the potential of these technologies. CPS is now increasingly being recognized as vital for improved construction project information management, more efficient project delivery, and enhanced facilities management. A critical aspect of the deployment of CPSs in construction is ensuring bi-directional coordination between the physical components and their virtual representations (Wu et al., 2011) (see Figure 23.2). By providing an effective integration between computational models and associated physical entities, cyber-physical systems offer an appropriate mechanism for construction project teams to bridge the gap between virtual models and

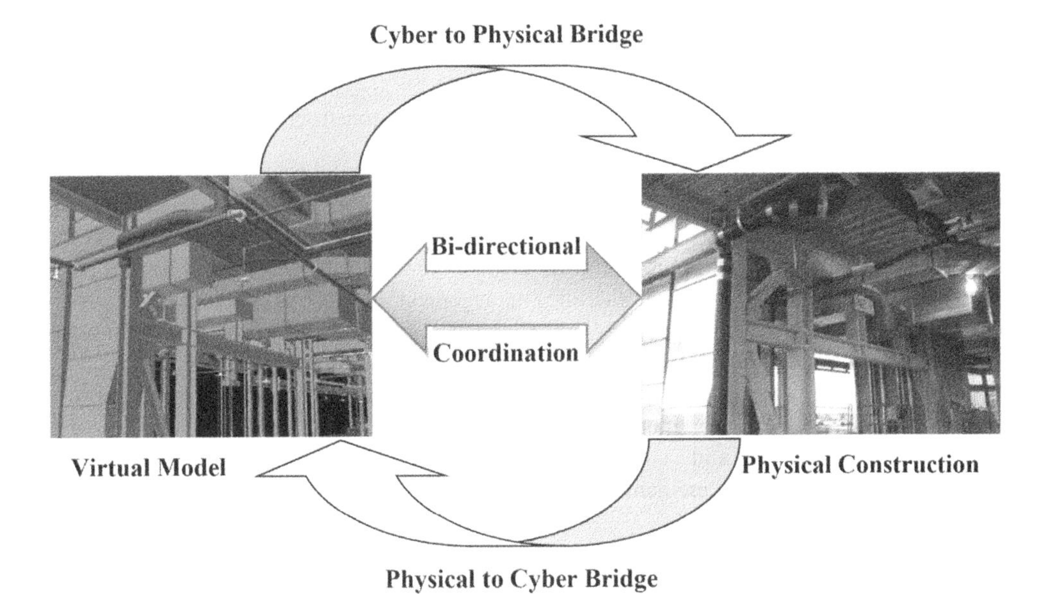

Figure 23.2 Bi-directional Coordination in CPS

the physical construction (Anumba et al., 2020). This will enable a variety of applications, services, and other innovations to be developed and deployed, building on this bi-directional coordination. In particular, it would be much easier to utilize the virtual model as a critical part of the physical construction process.

Cyber-physical systems have been developed for various aspects of the built environment (Anumba and Roofigari-Esfahan, 2020). Those developed for aspects of the construction project delivery process include the application of CPSs to construction component tracking and control (Akanmu and Anumba, 2015), temporary structures monitoring (Yuan et al., 2016), and safety and efficiency of mobile cranes (Kan et al., 2018).

Digital twins (DTs) can be regarded as a specific form of CPSs (Kan and Anumba, 2019) that refers to a near-real-time digital replica of a physical product or process, which includes all information that could be useful throughout all lifecycle phases (Boschert and Rosen, 2016). The origin of the "twin" concept can be traced back to the Apollo program of the US National Aeronautics and Space Administration (NASA) Apollo program. It was used by NASA for the moon exploration mission Apollo 13 and the Mars Rover Curiosity. Two identical space vehicles were built, and the one remaining on Earth during the mission was called the "twin." The twin vehicle in the Apollo program was used for training during the preparation, and while in flight the twin vehicle mirrored the conditions of the vehicle in space (Rosen et al., 2015). The first definition of DT was given by NASA in the Technology Roadmap 2010 as "an integrated multi-physics, multi-scale, probabilistic simulation of an as-built vehicle or system that uses the best available physical models, sensor updates, fleet history, etc., to mirror the life of its corresponding flying twin" (Shafto et al., 2010).

DTs consist of three primary components: a physical product in real space, a virtual product in virtual space, and the connections of data and information that tie the physical and virtual products together (Grieves and Vickers, 2017). Their potential in the construction industry is increasingly being realized and there are emerging applications in construction equipment monitoring, construction progress tracking, site logistics planning, performance of constructed facilities, and facilities in smart cities, to name a few.

The economic benefits of CPS and DT applications in the construction industry include:

- real-time information exchange between the design office and the construction site;
- reduction of construction risks as activities and processes can be more closely monitored and controlled;
- accurate as-built models that are useful for the later phases of the constructed facility's life cycle: commissioning, operation, facilities management, deconstruction, and so on;
- the capacity to actively control key components in the constructed facility beyond the construction stage, thereby delivering efficiency and performance improvements;
- increased capacity for 'What if?' analyses and simulations, which reduce costly errors;
- improved opportunities for more sustainable construction practices, with medium- and long-term cost savings; and
- reduction in costly safety incidents through proactive hazard monitoring.

Internet of Things (IoT)

The Internet of Things (IoT) is witnessing a growth across several industry sectors, and is now being explored in the construction industry (Niu et al., 2019). The concept builds on the existing internet by allowing distributed objects (also referred to as "things" in the IoT) to be sensed and interconnected across communications networks, thereby enabling them to be centrally monitored and controlled (Miorandi et al., 2012). The IoT is formally defined by Sundmaeker et al. (2010) as a dynamic global network infrastructure with self-configuring capabilities based on standard and interoperable communication protocols where physical and virtual "things" have identities, physical attributes and virtual personalities that use intelligent interfaces, and are seamlessly integrated into the information network.

The significance of IoT that surpasses the previous information and communication technology (ICT) systems lies in the view that the IoT itself is beyond the individual application level. Instead, as a critical and integrated infrastructure upon which applications can run, services on the IoT can be scalable, from personalized services (such as digitizing home appliances) to city-wide ones, such as delay-free traffic planning schemes (Stankovic, 2014). Every connected object becomes an integral part of an extensively interconnected and pervasive network through which services are delivered, messages transmitted and devices activated. This is truly powerful, and builds on the other technologies previously described in this chapter: BIM, CPS, cloud computing and sensing.

The IoT offers the potential for considerable economic benefits for construction organizations, as they can leverage the pervasive network and the associated services and applications. Specific benefits include the following:

- access to a wide variety of applications and services;
- reduced upfront costs of establishing an IT infrastructure;
- cost savings in the procurement and management of IT hardware, software and services;
- easy tracking of "things" that are connected to the network;
- efficiency savings in business operations resulting from the ease of access to applications and services;

- ready access to operational information on key equipment and devices;
- reduction in the time for decision-making based on access to real-time information from the connected "things"; and
- cost savings from improved safety management as a result of timely and accurate information.

Artificial Intelligence (AI)

Artificial intelligence (AI) is, simply put, the capacity of computers and computer applications to undertake tasks that require human intelligence or expertise. The term dates back to the 1950s (Simon, 2019) and there is no single universally accepted definition. Over the decades there have been several ups and downs in the evolution of AI (Crevier, 1993). Its applications in the construction industry are more recent and are briefly described here.

Early AI applications in the construction industry were developed in the 1980s and 1990s and included knowledge-based expert systems for a variety of design and construction application areas, artificial neural network systems for decision-making and management applications, fuzzy logic-based systems, and case-based reasoning systems for learning from previous cases (Taffs, 1984; Adeli, 1987; Maher, 1987; Anumba, 1996; Anumba et al., 2001). Many of these systems tackled issues such as estimating, design, defects diagnosis, interpretation of geotechnical data and bidding strategies. While many of these systems solved critical problems, they only had variable success, and widespread adoption was stunted. This also coincided with a decline in the popularity of AI and associated applications.

The last decade has witnessed a strong re-emergence of AI and its integration into mainstream applications often taken for granted by the public. This offers considerable opportunities for the construction industry. The growth in computing power, larger volumes of data, and increased sophistication of analytical tools, algorithms and enhanced visualization capabilities have considerably enhanced the possibilities for AI applications in construction. Emerging AI-based applications now have the capacity to leverage, amongst others, machine learning, deep learning and pattern recognition. Applying AI to existing data from previous construction projects and data harvested from sensing systems will enable more informed and intelligent decision-making. This is also expected to spawn a new range of automated systems in the construction industry; this will include greater use of robotics, unmanned aerial vehicles (UAVs), computer vision systems and environmentally aware equipment on the construction site. This has serious implications for the future management of construction projects, with the potential for project managers to deploy new AI-based tools that are more intelligent and responsive to uncertainties and potential failures.

The economic benefits of new AI-based applications and services in construction are considerable, and include the following:

- more intelligent decision-making;
- reduced processing time and cost for data analysis;
- enhanced knowledge of project and business performance, which enable more timely and economic interventions;
- increased safety of construction operations (with associated cost savings);
- reduction in the cost of human resources;
- increased efficiency;

- enhanced productivity;
- greater automation of mundane operations; and
- more efficient and sustainable use of resources.

All the above technologies proffer considerable economic benefits to the construction industry and will influence the efficiency of the construction project delivery process for many years. While each technology has specific benefits, the greatest benefits and the potential for transformative industry improvements will only be possible from their integrated deployment.

ECONOMIC CONSIDERATIONS IN TECHNOLOGY PROCUREMENT AND DEPLOYMENT

Metrics and Key Performance Indicators

Peter Drucker's "[only] what gets measured, gets managed" mantra applies to the nuances of harnessing a return on investment (ROI) from digital tools and technologies. The measurement of ROI must be tailored to a given company's overall strategy for engaging in a specific industry or sector within a specified geographical area (Hirji and Geddes, 2016). As outlined in Table 23.3, ROI should track digital investments across six strategic focus areas: customers, employees, operations, safety and soundness, infrastructure, and disruption and innovation. Table 23.3 also includes examples of key performance indicators (KPIs) associated with each of these metrics.

Maximizing ROI through Process-Enabled, Technology-Pull Approaches

Organizations that have successfully unlocked value through digitalization have also had to transform their business practices (Hirji and Geddes, 2016) through embarking on a digital transformation journey. Digital transformation is "a process where digital technologies create an impetus for organizations to implement responses to gain or maintain their competitive advantage" (Vial, 2019). The digital transformation journey unlocks new opportunities for innovation through increased avenues for collaboration among distributed networks of diversified actors. While this can improve performance, it leaves organizations vulnerable through loss of control over some of the elements in their operating environment.

Digital technologies must be deployed as part of broader strategies that are implemented to improve business processes and operations (see Tabrizi et al., 2019). The primary objective must be to enhance customer-centricity, new efficiency gains (that is, reducing/eliminating waste) and profitability (that is, value-added creation). Organizations that have realized these outcomes have used a "process-centric" and "technology-pull" approach to drive the adoption of digital technology and tools (Romero et al., 2019). Romero et al.'s (2019) proposed framework for using lean thinking along with other process-centric approaches to support digital transformation initiatives is based on five underlying assumptions: (1) a digital strategy is not simply to digitalize (business) processes but in fact to provide novel ways of working, of doing business, capable of creating new efficiency gains, (digital) capabilities and/or value for the customer; (2) business processes should be standardized/engineered before being digitized to exploit higher effectiveness levels enabled by digital technologies capabilities; (3) technology

Table 23.3 Measuring the return on investment

Specific metrics	KPIs
Customers: creating compelling experiences; meeting and exceeding customer expectations	Net promoter scores or likelihood to recommend Number of social media mentions; social media sentiment Customer reviews and feedback
Employees: enabling and engaging employees	Engagement scores Collaboration Likelihood to recommend Turnover Digital adoption
Operations: digitizing business processes	Manufacturing/production throughput Just-in-time inventory levels Supply chain efficiency Response times of emails/chats/phone calls Number of interactions resolving issue on first contact
Safety and soundness: protecting digital assets and customer data	Number of threats detected and defended Number of privacy breaches Fraud losses
Infrastructure: implementing and running new systems and tools	Speed of new technology implementation Uptime Response time to resolve issues/outages
Disruption and innovation: prototyping, testing and learning; promoting digital culture	Percentage of budget allocated for disruptive technologies and services Proportion of new ideas that reach concept design Number of new customers/segments/sectors from new products and services

Source: Extrapolated from Hirji and Geddes (2016).

cannot magically fix a bad (business) process, and the wrong technology could become an "inhibitor" rather than an "enabler" for the execution of a good (business) process; (4) digital transformation is considered as a sociotechnical phenomenon where all people should be empowered to drive change and use methods and tools to continually improve and digitalize processes; and (5) there is always risk in change, so risk should be managed.

Examples of Potential Savings through Avoided Expenses

Data-driven musculoskeletal disorders (MSDs) prevention

Conventional construction safety management practices tend to focus on the early identification of injury risk factors in safety planning and enhancing workers' safety awareness by safety training. Such approaches are predicated on workers' compliance with safety practices and appropriate modification of behavior to avoid overexposure to risk factors. The prevailing manual observation-based inspection is ineffective for monitoring the emerging risk factors, thus leaving workers developing cumulative injuries through overexposure to unidentified risk factors. This is exemplified in the prevalent musculoskeletal disorders (MSDs), which develop slowly over time and can remain for several months or years. Therefore, there has been a call for monitoring the emerging injury risk in a timely manner, thus allowing preventative response for proactive injury prevention.

Emerging data-sensing technologies, when enhanced with machine learning (ML) techniques, can address this need to monitor injury risk, as shown in Figure 23.3 (see Obonyo and Zhao, 2018; Zhao and Obonyo, 2018). The use of digital technologies based on the approach depicted in Figure 23.3 can help construction companies avoid incurring expenses associated with MSD injuries. The Bureau of Labor and Statistics (BLS) indicates that the US construction industry, a sector which employs approximately 4.5 percent of total laborers, accounts for 18.9 percent of fatal injuries and 7.6 percent of non-fatal injuries across all industries between 2008 and 2018 (see BLS, 2018). Performing construction tasks requires workers to assume awkward postures. The repetition of the postures, coupled with prolonged periods of exposure, can result in MSDs such as a chronic backache and overexertion. It is therefore no surprise that MSD-related problems account for over 37 percent of all injuries among construction trade workers. It is estimated that as much as $11.5 billion is spent annually on injuries to construction workers in the US (Biswas et al., 2017). The situation can be worse, considering the indirect costs of lost workdays, decreased productivity and the impact of chronic pain.

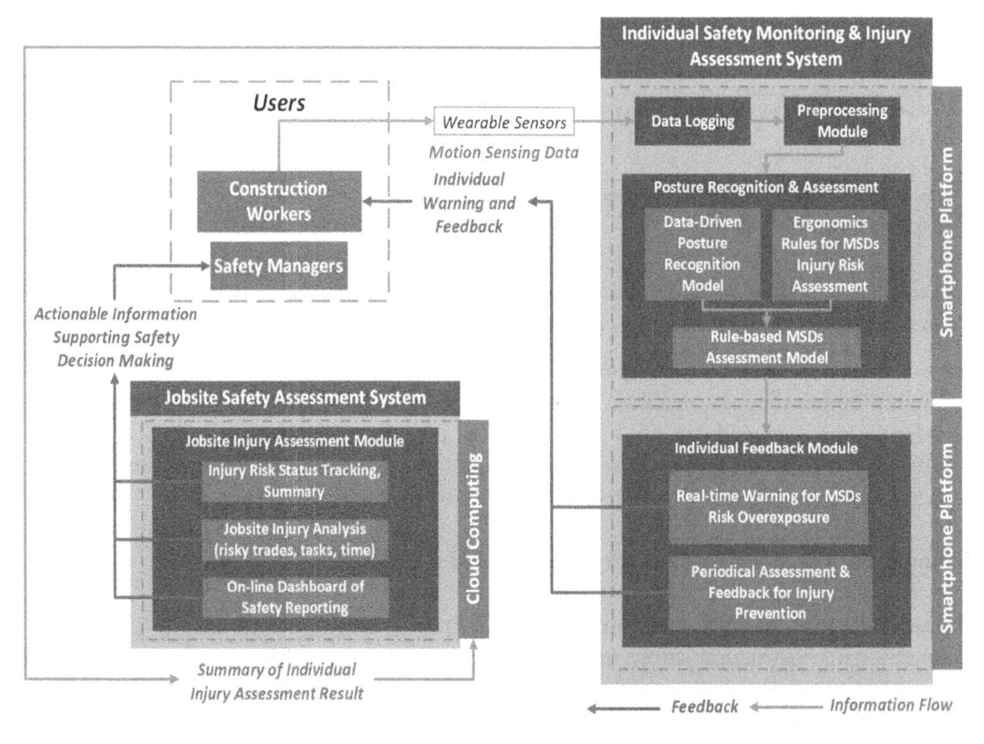

Source: Zhao and Obonyo (2018).

Figure 23.3 Concept for a data-driven MSD prevention decision support system

Designing-out building material waste

Concrete and bricks are some of the main types of construction waste materials. As shown in Table 23.4, the dollar equivalent of bricks wasted during construction in the US peaked close at to $55 million in 2009 (Office of Resource Conservation and Recovery, 2016). The authors

Table 23.4 *US annual construction brick debris generation*

Year	Tons	Number of bricks	Estimated annual costs (USD)
2009	276 945	99 423 255	54 682 790.3
2010	259 572	93 186 348	51 252 491.4
2011	237 394	85 224 446	46 873 445.3
2012	234 836	82 192 600	45 205 930.0
2013	183 865	66 007 535	36 304 144.3
2014	183 597	65 911 323	36 251 227.7

Source: Extrapolated using data from Office of Resource Conservation and Recovery (2016).

estimated these costs using the average of the number of bricks in a ton, which ranges between 333 and 385. One thousand bricks cost $550 on average. Top causes of construction waste include changes in design, leftover materials, packaging waste, errors in design or detailing, and poor weather conditions. The use of digital tools can help to address the underlying issues.

SUMMARY AND CONCLUSIONS

This chapter sought to discuss the economic considerations in the procurement and deployment of construction informatics applications. It introduced the construction industry and its conservative culture in the uptake of new technologies and discussed the economic and other drivers that are leading to a change in the level of technology adoption in the construction industry. Current and emerging technologies such as building information modeling (BIM), cloud computing, sensing, cyber-physical systems (CPSs) and digital twins (DTs), the Internet of Things (IoT), and artificial intelligence were briefly described, as these have considerable potential to influence the economics of construction projects in the future. The economic and other benefits of these technologies were highlighted. The approaches to procuring and deploying technology in the construction industry were also iscussed to show that different approaches have different economic considerations and implications for construction sector firms. The economic considerations in the procurement and deployment of digital technologies in construction were also discussed, including metrics and key performance indicators, maximizing return on investment, and specific examples of how savings were realized through avoiding expenses.

The economic effects of construction informatics include the stimulation of economic growth and creation of new jobs (Kvocho, 2013). Construction informatics has also contributed to economic development through its role as an enabler of innovation and development (Kvocho, 2013). Entrepreneurs see construction challenges such as low productivity in projects, the aging workforce and difficulty of hiring new employees, as an opportunity. According to studies by Agarwal at al. (2016), construction labor productivity has not kept pace with overall economic productivity since the 1990s. They estimate that large construction projects can take 20 percent longer than scheduled to complete, and can exceed the original budget by as much as 80 percent. Digital technologies can be used to unlock additional revenues from the construction industry through eliminating waste and inefficiencies. Some of the underlying issues can be addressed through deploying digital solutions to deliver seamless, real-time experiences across design management, scheduling, materials management, crew tracking, quality control, contract management, performance management and document management (McKinsey, 2016).

Construction informatics has contributed to direct job creation through the growing demand for digitally knowledgeable workers who, in addition to possessing strong computation skills, also understand both construction technology and how the built environment works (Bousquin, 2020). Construction informatics has also triggered the emergence of new start-ups and services (Bartlett et al., 2020; DB Insights, 2020). Digital technologies for construction applications received $25 billion in funding for entrepreneurial activities between 2014 and 2019. The funding levels for 2020 were expected to be significantly higher (DB Insights, 2020). As the use of digital technologies in construction continues to mature, an ecosystem of hubs and platforms is developing, resulting in a significant consolidation of the digital landscape in the industry. Some $17 billion of the $25 billion invested between 2014 and 2019 supported either mergers and acquisitions activity or private equity investment (Bartlett et al., 2020). Construction informatics has also resulted in workforce transformation by enabling freelancing through remote virtual work. Freelancing websites such as Upwork (see URL 1) have been serving as a one-stop work marketplace for micro-work (Kvocho, 2013).

The construction industry stands to reap considerable economic benefits from the adoption of construction informatics applications. This requires that companies make the necessary investments in these technologies. The cost of entry is now much lower than in the early days of these technologies, and the benefits are also much more evident. As such, there are no cogent reasons to remain skeptical. The long-term costs of doing nothing are much more, and in many cases could threaten the very existence of firms that do not adopt digital technologies. Each firm needs to undertake a detailed analysis of its needs relative to the available technologies, investigate the potential return on investment for each technology or deployment combination, make the appropriate business process changes to ensure a smooth transition, and train appropriate personnel in the use of the adopted systems.

REFERENCES

Abdel-Hamid, Mohamed and Abdelhaleem, Hanaa Mohamed (2020). Impact of poor labor productivity on construction project cost. *International Journal of Construction Management*. DOI: 10.1080/15623599.2020.1788757.

Adeli, H. (1987). Knowledge-based systems in structural engineering. In: Topping, B.H.V. (ed.), *The Application of Artificial Intelligence Techniques to Civil and Structural Engineering*, Edinburgh: Civil-Comp Press, pp. 71–78.

Adrian, J.J. (2004). *Construction Productivity: Measurement and Improvement*. Champaign: Stipes Publishing.

Agarwal, R., Sridharm, M. and Chandrasekaran, S. (2016). Imagining construction's digital future. McKinsey. https://www.mckinsey.com/industries/capital-projects-and-infrastructure/our-insights/imagining-constructions-digital-future, accessed August 31, 2018.

Akanmu, A. and Anumba, C.J. (2015). Cyber-physical systems integration of BIM and the physical construction. *Engineering, Construction and Architectural Management: Special Issue on Advanced ICT and Smart Systems for Innovative Engineering, Construction and Architectural Management*, 22(5), 516–535.

Akinci, B. and Anumba, C.J. (2008). Editorial – Sensors in construction and infrastructure management. *ITcon, Special Issue Sensors in Construction and Infrastructure Management*, 13, 69–70.

Alt, R., Beck, R. and Smits, M.T. (2018). FinTech and the transformation of the financial industry. *Electron Markets*, 28, 235–243. https://doi.org/10.1007/s12525-018-0310-9.

Anumba, C.J. (1989). An integrated two and three dimensional data structure for a structural engineering CAD system. PhD thesis, University of Leeds, UK.

Anumba C.J. (1996). *Knowledge-Based Systems in Structural Engineering – Current Developments and Future Trends*. London: Institution of Structural Engineers.

Anumba, C.J., Akanmu, A., Yuan, X. and Kan, C. (2020). Cyber-physical systems development for construction applications. *Frontiers of Engineering Management*. DOI: 10.1007/s42524-020-0130-4.

Anumba, C.J. and Roofigari-Esfahan, N. (eds) (2020). *Cyber-Physical Systems in the Built Environment*. Cham: Springer Publishers.

Anumba, C.J., Ugwu, O.O. and Ren, Z. (eds) (2001). Artificial intelligence in construction and structural engineering. Centre for Innovative Construction Engineering (CICE), Loughborough University, July.

Autodesk and CIOB (2018). Reimagining construction: the vision for digital transformation, and a roadmap for how to get there. Discussion Paper. https://www.autodesk.co.uk/campaigns/ciob-reimagining-future-of-construction/paper/reimagining-construction-paper, accessed December 30, 2020.

AWS (2020). What is cloud computing? Amazon Web Services. https://aws.amazon.com/what-is-cloud-computing/, accessed September 2, 2020.

Ballesteros-Pérez, P., Smith, S.T., Lloyd-Papworth, J.G. and Cooke, P. (2018). Incorporating the effect of weather in construction scheduling and management with sine wave curves: application in the United Kingdom. *Construction Management and Economics*, 36(12), 666–682. DOI: 10.1080/01446193.2018.1478109.

Bartlett, K., Blanco, J.L., Fitzgerald, B., Johnson, J., Mullin, A.L. and Ribeirinho, M.J. (2020). Rise of the platform era: the next chapter in construction technology. https://www.mckinsey.com/industries/private-equity-and-principal-investors/our-insights/rise-of-the-platform-era-the-next-chapter-in-construction-technology, accessed December 31, 2020.

Becker, G.S. (2008). Human capital, Econlib's Basic Concepts collection. Library of Economics and Liberty. https://www.econlib.org/library/Enc/HumanCapital.html, accessed August 15, 2020.

Biswas, G., Bhattacharya, A. and Bhattacharya, R. (2017). Occupational health status of construction workers: a review. *International Journal of Medical Science and Public Health*, 6, 669–675.

Blanco, J.L., Mullin, A., Pandya, K., Parsons, M. and Maria João Ribeirinho, M.J. (2018), Seizing opportunity in today's construction technology ecosystem. McKinsey. https://www.mckinsey.com/business-functions/operations/our-insights/seizing-opportunity-in-todays-construction-technology-ecosystem.

Bureau for Labor and Statistics (BLS) (2018). Injuries, illnesses, and fatalities. https://www.bls.gov/iif/, 2016 (accessed April 22, 2019).

Boschert, S. and Rosen, R. (2016). Digital twin – the simulation aspect. In: Hehenberger, P. and Bradley, D. (eds), *Mechatronic Futures: Challenges and Solutions for Mechatronic Systems and their Designers*. Cham: Springer International Publishing, pp. 59–74.

Bousquin, J. (2020). The most in-demand tech jobs in construction. Construction Dive, Commercial Building Technology. https://www.constructiondive.com/news/the-most-in-demand-tech-jobs-in-construction/584464/, accessed December 31, 2020.

Buehler, M., Buffet, P.P. and Castagnino, S. (2018). The Fourth Industrial Revolution is about to hit the construction industry. Here's how it can thrive. World Economic Forum. https://www.weforum.org/agenda/2018/06/construction-industry-future-scenarios-labour-technology/, accessed October 15, 2019.

CPWR (2018). *The Construction Chart Book: The U.S. Construction Industry and Its Workers*. Silver Spring, MD: CPWR, The Center for Construction Research and Training. https://www.cpwr.com/chart-book-6th-edition-fatal-and-nonfatal-injuries-demographic-trends-fatal-and-nonfatal-injuries, accessed June 31, 2020.

Crevier, D. (1993). *AI: The Tumultuous Search for Artificial Intelligence*. New York: Basic Books.

DB Insights (2020). 24 industries and technologies that will shape the post-virus world. https://www.cbinsights.com/research/construction-tech-funding-trends-2020/, accessed December 31, 2020.

Dillon, T., Wu, C. and Chang, E. (2010). Cloud computing: issues and challenges. *2010 24th IEEE International Conference on Advanced Information Networking and Applications*, Perth, WA, pp. 27–33. doi: 10.1109/AINA.2010.187.

Dimock, M. (2019). Defining generations: where Millennials end and Generation Z begins. Pew Research Center. https://www.pewresearch.org/fact-tank/2019/01/17/where-millennials-end-and-generation-z-begins/, accessed July 1, 2020.
Dong, X.S., Wang, X., Largay, J.A. and Sokas, R. (2015). Long-term health outcomes of work-related injuries among construction workers – findings from the National Longitudinal Survey of Youth. *American Journal of Industrial Medicine*, 58, 308–318. doi: 10.1002/ajim.22415.
Dong, X.S., Wang, X., Largay, J.A. and Sokas, R. (2016). Economic consequences of workplace injuries in the United States: findings from the National Longitudinal Survey of Youth (NLSY79). *American Journal of Industrial Medicine*, 59, 106–118. doi: 10.1002/ajim.22559.
Egan, J. (1998). *Rethinking Construction: Report of the Construction Task Force*. London: HMSO. (Copy of report available on Constructing Excellence website: https://constructingexcellence.org.uk/wp-content/uploads/2014/10/rethinking_construction_report.pdf, accessed: 20 October 2020.)
Ervasti, J., Pietilainen, O., Rahkonen, O., Lahelma, E., Kouvonen, A., et al. (2019). Long-term exposure to heavy physical work, disability pension due to musculoskeletal disorders and all-cause mortality: 20-year follow-up – introducing Helsinki Health Study Job Exposure Matrix. *International Archives of Occupational and Environmental Health*, 92, 337–345. doi: 10.1007/s00420-018-1393-5.
European Union (2019). Supporting digitalisation of the construction sector and SMEs, including building information modelling. https://ec.europa.eu/docsroom/documents/38281, accessed December 30, 2020.
Francis, T. and Hoefel, F. (2018). "True Gen": Generation Z and its implications for companies. McKinsey & Company. https://www.mckinsey.com/industries/consumer-packaged-goods/our-insights/true-gen-generation-z-and-its-implications-for-companies, accessed July 4, 2020.
Gallagher, M.P., O'Connor, A.C., Dettbarn, J.L. and Gilday, L.T. (2004). *Cost Analysis of Inadequate Interoperability in the U.S. Capital Facilities Industry*. NIST Technical Report NIST GCR 04-867.
Green, B. (2016). Productivity in construction: creating a framework for the industry to thrive. Chartered Institute of Building (CIOB). https://www.ciob.org/industry/research/Productivity-Construction-Creating-framework-industry-thrive, accessed December 30, 2020.
Grieves, M. and Vickers, J. (2017). Digital twin: mitigating unpredictable, undesirable emergent behavior in complex systems. *Transdisciplinary Perspectives on Complex Systems*, 85–113. https://doi.org/10.1007/978-3-319-38756-74.
Hardy, B. (n.d.). Understanding Generation Z is the key to multifamily housing success. https://www.constructormagazine.com/understanding-generation-z-is-the-key-to-multifamily-housing-success/, accessed July 4, 2020.
Hirji, N. and Geddes, G. (2016). What's your digital ROI? Realizing the value of digital investments. Strategyand, part of the PwC network. https://www.strategyand.pwc.com/ca/en/whats-your-digital-roi.html (accessed October 20, 2020).
HM Government (2012). *Building Information Modelling*. Crown Copyright, UK.
Hurd, M. and Rohwedder, S. (2011). Trends in labor force participation: how much is due to changes in pensions? *Journal of Population Ageing*, 4, 81–96. doi: 10.1007/s12062-011-9042-8.
International Labour Organization (ILO) (2019). Small matters: global evidence on the contribution to employment by the self-employed, micro-enterprises and SMEs. https://www.ilo.org/wcmsp5/groups/public/---dgreports/---dcomm/---publ/documents/publication/wcms_723282.pdf, accessed July 4, 2020.
International Labour Organization (2001). The construction industry in the twenty first century: its image, employment prospects and skill requirements. TMCIT/2001. https://www.ilo.org/public/english/standards/relm/gb/docs/gb283/pdf/tmcitr.pdf, accessed December 30, 2020.
Kan, C. and Anumba C.J. (2019). Digital twins as the next phase of cyber-physical systems in construction. In: Cho, Y., Leite, F., Behzadan, A. and Wang, C. (eds), *Proceedings ASCE International Conference on Computing in Civil Engineering*, June 17–19, Georgia Institute of Technology, Atlanta, GA, pp 256–264. https://doi.org/10.1061/9780784482438.033.
Kan C., Fang Y., Anumba, C.J. and Messner, J.I. (2018). A cyber-physical system (CPS) for planning and monitoring mobile cranes on construction sites. *Management, Procurement and Law – Proceedings of the Institution of Civil Engineers. Themed Issue: Recent Advances in Construction Technology and Management*. https://doi.org/10.1680/jmapl.17.00042.

Kallergis, A., Angel, S., Liu, Y., Blei, A., Sanchez, N. and Lamson-Hall, P. (2018). Housing affordability in a global perspective. Lincoln Institute of Land Policy, Working Paper WP18AK1. https://www.lincolninst.edu/pt-br/publications/working-papers/housing-affordability-in-global-perspective, accessed February 1, 2020.

King, R., Orloff, M., Virsilas, T. and Pande, T. (2017). Confronting the urban housing crisis in the Global South: adequate, secure, and affordable housing. Working Paper. Washington, DC: World Resources Institute.

Kvochko, E. (2013). Five ways technology can help the economy. World Economic Forum. https://www.weforum.org/agenda/2013/04/five-ways-technology-can-help-the-economy/.

Leonard (2019). Trends: Entrepreneurship and intrapreneurship – entrepreneurs boosting the construction industry's digital transformation. https://leonard.vinci.com/en/entrepreneurs-boosting-the-construction-industrys-digital-transformation/.

Lieuw-Kie-Song, M. (2020). The construction sector can help lead the economic recovery – here's how. https://iloblog.org/2020/05/11/the-construction-sector-can-help-lead-the-economic-recovery-heres-how/, accessed August 30, 2020.

Maher, M.L. (1987). Expert systems for civil engineers: technology and application. American Society of Civil Engineers, ASCE.

Marston, S., Li, Z., Bandyopadhay, S., Zhang, J. and Ghalsasi, A. (2011). Cloud computing – the business perspective. *Decision Support Systems*, 51(1), 176–189.

McFall, B.H. (2011). Crash and wait? The impact of the great recession on retirement planning of older Americans. *American Economic Review*, 101, 40–44. doi: 10.1257/aer.101.3.40.

McKinsey Global Institute (2016). The US economy: an agenda for inclusive growth. Briefing Paper. https://www.mckinsey.com/~/media/mckinsey/featured%20insights/employment%20and%20growth/can%20the%20us%20economy%20return%20to%20dynamic%20and%20inclusive%20growth/mgi-us-economic-agenda-briefing-paper-november-2016.ashx, accessed February 20, 2020.

Mell, P. and Grance, T. (2011). The NIST definition of cloud computing (technical report). National Institute of Standards and Technology, US Department of Commerce, Special Publication 800-145. doi:10.6028/NIST.SP.800-145.

Mincer, J. (1984). Human capital and economic growth. *Economics of Education Review*, 3(3), 195–205. https://doi.org/10.1016/0272-7757(84)90032-3.

Miorandi, D., Sicari, S., De Pellegrini, F. and Chlamtac, I. (2012). Internet of things: vision, applications and research challenges. *Ad hoc Networks*, 10(7), 1497–1516.

Morris, A.P. (1987). A survey of productivity in the construction industry: measurement and causes. https://core.ac.uk/download/pdf/36715765.pdf, accessed December 30, 2020,

NIBS (2015). *National BIM Standard, Version 3*. National Institute for Building Sciences.

Niu, Y., Anumba, C.J. and Lu, W. (2019). Taxonomy and deployment framework for emerging pervasive technologies in construction projects. *ASCE Journal of Construction Engineering and Management*, 145(5). DOI: 10.1061/(ASCE)CO.1943-7862.0001653.

Obonyo, E.A. and Zhao, J. (2018). Towards a real-time data-driven approach for proactive injury prevention in construction. *Advancements in Civil Engineering and Technology*, 2(2). DOI: 10.31031/ACET.2018.02.000533.

Office of Resource Conservation and Recovery (2016). Construction and demolition debris generation in the United States. US Environmental Protection Agency. https://www.epa.gov/sites/production/files/2016-12/documents/construction_and_demolition_debris_generation_2014_11302016_508.pdf, accessed August 15, 2020.

Osman, H., Zhu, H. and Alade, T. (2011). Deployment of distributed antenna systems in high buildings, Vehicular Technology Conference (VTC Spring). *IEEE*, 73, 1–5.

Reutter, K. (2019). Building resilience for a new era of extreme weather events. *FMI Quarterly*, September 2, 2019. https://www.fminet.com/fmi-quarterly/article/2019/09/building-resilience-for-a-new-era-of-extreme-weather-events/, accessed August 15, 2020.

Ringen, K., Dong, X.S., Goldenhar, L.M. and Cain, C.T. (2018). Construction safety and health in the USA: lessons from a decade of turmoil. *Annals of Work Exposures and Health*, 62, S25–S33. doi: 10.1093/annweh/wxy069.

Rohe, M. (2017). Tackling the housing affordability crisis. *Housing Policy Debate*, 27(3), 490494. https://doi.org/10.1080/10511482.2017.1298214, accessed August 15, 2020.

Romero, D., Flores, M., Herrera, M. and Resendez, H. (2019). Five management pillars for digital transformation integrating the lean thinking philosophy. 10.1109/ICE.2019.8792650.

Rosen, R., von Wichert, G., Lo, G. and Bettenhausen, K.D. (2015). About the importance of autonomy and digital twins for the future of manufacturing. *IFAC-PapersOnLine*, 48(3), 567–572.

Rubenstone, J. (2019). Construction VC firm brick and mortar ventures unveils $97m fund. *ENR – Engineering News-Record.* https://www.enr.com/articles/47341-construction-vc-firm-brick-mortar-ventures-announces-97m-fund.

Sacks R., Eastman, C., Lee, G. and Teicholz, P. (2018). *BIM Handbook: A Guide to Building Information Modeling for Owners, Designers, Engineers, Contractors and Facility Managers*, 3rd edition. Hoboken, NJ: Wiley.

Shafto, M., Conroy, M., Doyle, R., Glaessgen, E., Kemp, C., LeMoigne, J. and Wang, L. (2010). Draft modeling, simulation, information technology and processing roadmap. Technology Area 11, 2010.

Simon, H.A. (2019). *Sciences of the Artificial*, 3rd edition. Cambridge, MA: MIT Press.

Sokas, R.K., Dong, X.S. and Cain, C.T. (2019). Building a sustainable construction workforce. *International Journal of Environmental Research and Public Health*, 16(21), 4202. https://doi.org/10.3390/ijerph16214202.

Stankovic, J.A. (2014). Research directions for the internet of things. *IEEE Internet of Things Journal*, 1(1), 3-9.

Statista (2019). Projected number of firms in the construction industry in the United States in 2019 and 2026, by firm size. https://www.statista.com/statistics/684083/forecast-of-firms-in-the-construction-industry-us-by-firm-size/, accessed August 15, 2020.

Sundmaeker, H., Guillemin, P., Friess, P. and Woelfflé, S. (2010). Vision and challenges for realising the Internet of Things. *Cluster of European Research Projects on the Internet of Things, European Commission*, 3(3), 34–36.

Sutherland, I. (1965). SKETCHPAD: a man–machine graphical communication system. MIT Lincoln Lab Tech Report 296.

Szinovacz, M.E., Davey, A. and Martin, L. (2015). Did the Great Recession influence retirement plans? *Research on Aging*, 37, 275–305. doi: 10.1177/0164027514530171.

Tabrizi, B., Lam, E., Girard, K., Gerard, K. and Irvin, V. (2019). *Digital Transformation Is Not About Technology*. Harvard Business Publishing Education. https://www.hbsp.harvard.edu/product/H04TO3-PDF-ENG, accessed Feb 1, 2020.

Taffs, D. (1984). Expert or knowledge based systems – consequences for the industry. *Computer Technology in Construction, ICE Conference Proceedings*. London: Thomas Telford, pp. 297–306.

Tang F., Choi, E. and Goode, R. (2013). Older Americans employment and retirement. *Ageing International*, 38, 82–94. doi: 10.1007/s12126-012-9162-3.

Teicholz, P.M. (2013). Labor productivity declines in the construction industry: causes and remedies (a second look). *AECbytes Viewpoint*. https://www.aecbytes.com/viewpoint/2013/issue_67.html, accessed November 25, 2021.

Toosi, M. (2015). Labor force projections to 2024: the labor force is growing, but slowly. *Monthly Labor Review*. doi: 10.21916/mlr.2015.48.

United Nations, Department of Economic and Social Affairs, Population Division (2019). World Population Prospects 2019: Highlights. ST/ESA/SER.A/423. https://population.un.org/wpp/Publications/Files/WPP2019_Highlights.pdf, accessed August 15, 2020.

Vial, G., (2019). Understanding digital transformation: a review and a research agenda. *Journal of Strategic Information Systems*, 28(2), 118–144. https://doi.org/10.1016/j.jsis.2019.01.003.

Welch, L.S., Haile E., Boden, L.I. and Hunting, K.L. (2010). Impact of musculoskeletal and medical conditions on disability retirement – a longitudinal study among construction roofers. *American Journal of Industrial Medicine*, 53, 552–560. doi: 10.1002/ajim.20794.

Wetzstein, S. (2017). The global urban housing affordability crisis. *Urban Studies*, 54(14), 3159–3177. https://doi.org/10.1177/0042098017711649.

Woetzel, J., Ram, S., Mischke, J, Garemo, N. and Sankhe, S. (2014). Tackling the housing affordability challenge. https://www.mckinsey.com/featured-insights/urbanization/tackling-the-worlds-affordable-housing-challenge, accessed July 15, 2020.

World Economic Forum (2018). Shaping the future of construction future scenarios and implications for the industry. Prepared in collaboration with The Boston Consulting Group. https://exed.annenberg.usc.edu/sites/default/files/Future_Scenarios_Implications_Industry_report_2018.pdf, accessed July 15, 2020.

World Economic Forum (2019). The Global Risks Report 2019. https://www.weforum.org/reports/the-global-risks-report-2019, accessed July 5, 2020.

Wu, F.J., Kao, Y. and Tseng, Y. (2011). From wireless sensor networks towards cyber-physical systems. *Pervasive and Mobile Computing*. doi:10.1016/j.pmcj.2011.03.003.

Yuan, X., Anumba, C.J. and Parfitt, M.K. (2016). Cyber-physical systems for temporary structure monitoring. *Automation in Construction*, 66, 1–14.

Zhao, J. and Obonyo, E.A. (2018). Towards a data-driven approach to injury prevention in construction. *Lecture Notes in Computer Science* (Lecture Notes in Artificial Intelligence series, Vol. 10863 LNCS). Springer Verlag. DOI: https://doi.org/10.1007/978-3-319-91635-4_20.

24. The future: new directions of construction economics research

George Ofori

INTRODUCTION

This final chapter considers the future of construction economics after reviewing the chapters of the book. The objectives of the chapter are to:

- establish and stress the importance of construction economics as a field of knowledge and why it is necessary to study it and continuously improve upon it;
- review the state of construction economics by outlining brief summaries of the chapters in the book;
- consider how construction economics can be developed, and progressed towards becoming a recognised scientific discipline.

WHY SHOULD CONSTRUCTION ECONOMICS BE STUDIED, AND FURTHER DEVELOPED?

Why is construction economics important? This question is now addressed by considering the segments of construction economics: the industry, the company and the project, developing the points made in Chapter 2 into more detail.

Construction Industry Economics

Why study construction industry economics? One of the most often highlighted features of the construction industry is its role in the economy and the contribution it makes to national development. This is highlighted in Chapters 2 and 6. At the global level, WEF and BCG (2016) noted that total annual revenues in construction were $10 trillion and added value was $3.6 trillion; construction accounted for about 6 per cent of global gross domestic product (GDP) (over 8 per cent in developing countries). More than 100 million people were employed in construction worldwide in 2016. The authors observed that other sectors of the economy rely on construction in one way or another, and the lives of all in society are shaped by it. If the construction industry plays a strategic role in the economy, it is essential to analyse it appropriately, and to determine how it can be further developed in order for the results to be applied to enhance the efficiency and effectiveness of construction. As the construction industry is subject to many influencing factors which are themselves in a state of flux, it is necessary for the body of knowledge on it to be continually developed to address the changing factors both individually and in their various combinations.

Another feature of the construction industry, its processes and its products, is how different it is from other sectors of the economy (Hillebrandt, 2000; Killip, 2020). This uniqueness is often pointed out as a reason for its relative poor performance, in comparison with other sectors of the economy, in terms of technology development (Farmer, 2016), productivity (HM Government, 2018) and innovation (Development Bureau and KPMG, 2018), client–contractor relationships (WEF and BCG, 2016), interprofessional relationships (Morrell, 2020) and extent of adherence to regulations (Hackitt, 2018). For example, whereas increase in productivity is seen to lie at the heart of economic growth and development (HM Government, 2017), *The Economist* (2017) reports that McKinsey found that over the last 20 years the global average for value added per hour worked in construction grew by 1 per cent per year, about one-quarter of the growth in manufacturing. The industrialised countries saw poor performance. In Germany and Japan there was no growth; in France and Italy it fell by one-sixth; and in the United States (US) it had declined by half since the late 1960s. Similar observations have been made on construction productivity growth in many countries, including Hong Kong SAR (Development Bureau and KPMG, 2018) and the United Kingdom (UK) (HM Government, 2018).

The UK Office of National Statistics (2021) notes: 'Construction is a naturally volatile industry and is responsive to fluctuations in both consumer and business confidence as well as economic variables, such as interest and exchange rates … it [influences] some of the main economic indicators including inflation and employment' (p. 17). The Chairman of the UK Construction Leadership Council (CLC), Andrew Wostenholme (2016), noted that similar councils in other sectors had realised more success than in construction, and it was seeking to emulate those in adopting 'their proven routes to success'. He also noted that construction faces fundamental issues: the combination of its cyclical workload and low level of client leadership results in a fragmented supply chain. This hinders efforts to maximise value across the facility's life cycle. This drives competition on price, resulting in low profit margins in the industry. That makes it impossible for construction to invest in the skills, technology and innovation necessary to enhance productivity.

The study by WEF and BCG (2016) showed that the structure and approaches of the construction industry stem directly from the risk-averse practices of the clients. Thus, Patricia Hillebrandt has a popular remark that: 'Clients get the industry they deserve'. A study on productivity in construction in Singapore suggested that the stakeholders of the construction industry should agree on setting up the construction industry to foster efficiency, productivity and reasonable margins (Singapore Contractors Association Ltd, 2017). If the construction industry is so important and yet so different, and if its performance is poor, then it is necessary to improve the ways in which it is analysed in order to infer results which form the bases for proposing policies and measures which are applicable to this special segment of the economy.

Construction Company Economics

Why study construction company economics? The reasons lie in two related areas: whether construction companies are different from their counterparts in other sectors of the economy; and the special challenges construction firms face. The construction firm is different from enterprises in most of the other fields of activity because its work comprises projects; thus, the orientation of the parties to the contract and most other stakeholders is short-term. Moreover, the performance requirements are not within the control of the construction firm. In most procurement arrangements, the service provider has no input into the design, or the selection of the materials and some of the methods. For most types of construction, the construction firm has little bargaining power vis-à-vis the client.

The challenges which construction projects pose to firms and practitioners are also different. The projects are location-specific; the service provider has no influence on where the work is based. The bulkiness of items used on a construction project limits the geographical area in which firms can be competitive. This leads to the notion that construction is a 'local industry'. Moreover, at the lower levels of construction by project size and complexity, the barriers to entry are low; this has led to a multiplicity of small, fragile firms. On each project, the firms have to work with many other participants and stakeholders (Hillebrandt, 2000).

Owing to these features, as noted in Chapter 2, construction is characterised by high levels of risk, low margins and high frequency of losses, and high company mortality. The value of construction new work in Great Britain rose in 2019 to reach its highest level on record at £118 977 million (Office for National Statistics, 2021). However, the number of registered construction firms fell sharply in 2019, by 10.9 per cent compared with 2018, to 290 374 (after a period of steady increases, which saw a peak in 2018 of 325 736). There were 3502 insolvencies in construction in Great Britain in 2019, the highest of any sector. Construction-related employment in Great Britain also fell by 6.0 per cent in 2019, which was the first annual decline since 2014 when it fell by 0.1 per cent.

The high level of competition and price focus of clients leads to low margins in construction, which hinders the ability of the firms to invest in technology and in their people, giving construction an unattractive image which makes it difficult for firms to engage and retain good-calibre employees (Wostenholme, 2016). Killip (2020) notes that the underlying challenge with respect to construction skills is one of quality assurance in an industry characterised by fragmentation, poor quality and poor customer service. Keep (2016) considers construction companies to be operating in low-margin, high-volume markets, where costs are kept down by focusing on low quality at low cost. Green (2016) describes construction as operating in the 'low-skills equilibrium': low demand for skills is a stable factor in a market where there are also low wages, little job security and low prestige.

It is suggested that the industry does not always adopt appropriate practices. For example, it is found that few firms capture data on project performance (such as on variation orders, productivity and quality) (Dodge Data and Analytics, 2018); thus, they do not record good practice for feeding forward into future projects. Construction firms tend to adopt the flexible firm model suggested by Atkinson (1984), owing to the fluctuations in demand and activity they face. Construction companies tend to prioritise flexibility of their response to the cyclical variations in the market, rather than investment in long-term human resources development (Dainty et al., 2007). UK construction companies may engage subcontractors and self-employed workers instead of directly employing workers and continually developing their skills. The Singapore construction industry is built on a unique traditional subcontracting system (Debrah and Ofori, 2001). Killip (2020) notes that construction has a problem of false self-employment, which leads to 'degenerative competition' where subcontractors lack the resources to invest in their corporate development and the larger contractors avoid the responsibility to do so as they do not directly employ the workforce. This results in a failure to invest in skills, poor productivity, and a culture which hinders innovation (Behling and Harvey, 2015).

Another commonly highlighted negative feature of subcontracting in construction was noted in a report on Hong Kong SAR (Development Bureau and KPMG, 2018):

> Whilst high levels of subcontracting can be beneficial from the point of view of securing specialist skills for complex tasks, this practice also has the potential to encourage the participation of intermediary organisations seeking to generate commissions without delivering any physical work or services. One unintended consequence of this practice is increased revenue and margin pressure on the overall construction value chain, potentially leading to rushed works, lower quality construction outcomes and increased safety risks (p. 22).

Construction Project Economics

Why is construction project economics important? The project is the way in which the construction industry undertakes all its activities, realises its achievements and registers its performance. The project sets the tone for each of the many companies and individuals participating in it. The features of the construction project have been discussed in the last few paragraphs, and especially in terms of: its location specificity; the many participants involved, each pursuing their own interests; the interests of a large number of stakeholders which differ in perspectives and expectations; and so on. It is suggested that each project is unique; many of the inputs for one project cannot be transferred to another; and it is difficult to effectively learn from one project to feed into another. Finally, as discussed in Chapter 2, the construction process is subject to many regulations which are constantly increasing in terms of stringency. The French National Union of Construction Economists (UNTEC) (2020) notes that projects have very ambitious performance objectives in an increasingly difficult regulatory context. The cross-cutting expertise and capacity for synthesis of the construction economist are therefore sought after.

As discussed in Chapter 2, many studies, at both national and international levels, have found that performance on construction projects generally falls below expectations. For example, *The Economist* (2017) notes that 90 per cent of the infrastructure projects in the world are either late or over budget. Governments and agencies in many countries have highlighted these performance gaps. Some examples are: Hong Kong SAR (Development Bureau and KPMG, 2018), Malaysia (CIDB, 2016) and the UK (HM Government, 2018). Some solutions are proposed at both the broad and specific levels. An example of the former is the UK government's *The Construction Playbook*, 'focused on getting projects and programmes right from the start' (HM Government, 2020b, p. 2) which is a handbook for good practice in project economics. It is expected to 'transform how we assess, procure and manage public works projects and programmes' (p. 2). It has broader, industry-level aims, linking the segments of construction economics together:

> Transformational change will only be achieved by systematically approaching risk, sustainability and innovation across portfolios of projects and programmes. We need to harness the excellence which already exists and learn from this to drive progress and strengthen the health of the sector, including by addressing low levels of productivity and future skills shortages. It is in all of our interests to create a profitable, sustainable and resilient industry with a well-trained workforce for the future (ibid.).

To this end, in the UK, in addition to the construction industry development strategies such as the Construction Sector Deal (HM Government, 2018), the Government Construction

Strategy, 2016–2020 (HM Government, 2016) 'sets out the Government's plan to develop its capability as a construction client and act as an exemplary client across the industry' (p. 2).

Over the past decade, initiatives at the international level to harmonise practice or facilitate comparison in construction economics have borne fruit. The European Council of Construction Economists (CEEC) European Code of Measurement provides a standard basis for the subdivision of costs and for measurement of basic quantities of buildings for budgeting, comparison and analysis across Europe,[1] enabling consistent and reliable comparisons between costs and cost analyses presented in national formats.[2] The International Construction Measurement Standards (ICMS), first published in 2017 by a coalition of mainly quantity surveying and cost engineering professional institutions, provides a single method for reporting, grouping and classifying construction project costs; the second edition (ICMSC, 2019) incorporates life cycle costs, and helps to avoid discrepancies when accounting, comparing and predicting project finances. Similarly, the International Property Measurement Standards Coalition (IPMSC) was formed in 2013 with the aim of bringing about the harmonisation of national property measurement standards to develop and implement international standards for measuring property. It comprises more than 80 professional and not-for-profit organisations.[3] The International Property Measurement Standard (IPMS) ensures that property assets are measured in a consistent way, creating transparency in markets and greater public trust.[4]

THE STATE OF CONSTRUCTION ECONOMICS AND ITS FUTURE

This section presents a brief summary of the chapters of the book, and suggests and discusses other points which are relevant to the topic of the chapter, which will contribute to the development of the field of construction economics.

Chapter 2: History of Construction Economics

In Chapter 2, it is suggested that despite the importance of the subject of the field of knowledge of construction economics, it is not yet an established scientific discipline. These questions are addressed: What is construction economics? How has it developed? What are its principles, theories, concepts and techniques? What is its status? What is its future?

Construction economics has some way to go to be established as a bona fide field of knowledge and become recognised as a distinct segment of mainstream economics. There is a need to establish its theoretical base, and its principles, concepts and tools. The chapters in this book contribute to this endeavour.

Chapter 3: Philosophy of Construction

There is a need for construction economics researchers to be able to answer the question: 'What is the philosophy of construction economics?' The answer will guide the work on developing the elements which the field requires to have in order to be considered as a distinct field of knowledge. The discussion in Chapter 3 considers the application of philosophy to construction economics; identifies the principles of a new construction industry which steers a course that is positive and inspiring; and makes an attempt to address the question above.

There should be a continuing debate on, and further development of, the philosophy of construction economics. The following questions could be considered in this debate, The construction industry is, by its nature, fragmented, with different groups pursuing their interests. In this regard, What is the philosophy of construction economics? How can a unified understanding and shared view of the philosophy be attained? How can the new industry, which encapsulates the six principles outlined in Chapter 2 and demonstrates the philosophy of the subject, be developed?

Chapter 4: Nature of Construction Economics

That the authors of a number of the chapters of this book found it necessary to devote parts of their discussion to consider the answer to the question on the nature and status of the field and many of its main segments, such as 'the construction industry', shows the infancy of the subject. The chapters in this book provide useful insights into the nature of construction economics. In particular, Chapter 2 considers the origins of the subject, its components and the need for it to be developed. Chapter 4 explains the nature of construction economics by considering various definitions of it, drawing out its boundaries and tracing its recent history (since the 1960s) and highlighting the key concepts which were developed during the period.

The further discussion on the nature of construction economics which this chapter generates, especially the consideration of the factors which could contribute to its development (such as the interest of mainstream economists in its topics), and those hindering its development (such as the non-availability of relevant data of good quality), should help in seeking to answer questions including these: What more is required for construction economics to become an established field of knowledge? What can be done to develop construction economics? What can construction economics contribute to the body of knowledge on mainstream economics?

Chapter 5: Construction Economics in Antiquity

What does the past teach us? The review of the current state of research on the economics of the construction industry in ancient Greece and Rome in Chapter 5 shows that construction economics has been in existence for centuries. It has always been important. It is key for scholars in construction economics to know the history of the subject. This should contribute to future efforts to build a complete body of knowledge of the subject. The highlighting of the key policy considerations in public infrastructure programmes of different political regimes in Chapter 5, as well as the cost estimating methods and bases, are relevant, as is the importance given in some regimes to planning and budgeting in order to receive approval, to give an account of progress to the citizenry, and to report on the project's completion. Thus, the application of construction economics can be seen to be going beyond the allocation of resources, to form the basis of professionalism and good practice.

Knowing this ancient origin of construction economics as a systematic area of knowledge makes it even clearer that it is critical to develop the field to enable it to play its civic role. A feature that becomes clear is building transparency into project systems in order to report on the project as it progresses and, ultimately, to prevent corruption and mismanagement. It would be a worthwhile endeavour to explore whether construction economics was evident in a formalised form during the industrial revolution, given the large volume of infrastructure works which were undertaken.

Chapter 6: Construction in the Economy and in National Development

The role of construction in the economy and its direct linkages to other sectors is a key principle in construction industry economics. In Chapter 6, it is noted that the economic impact of construction infrastructure can be transformative, especially in low- to middle-income per capita countries. The close association between physical capital and different measures of the national economy makes infrastructure a powerful engine of economic growth and development. The chapter considers the magnitude of investment and the direction of causality between construction investment and economic growth for sub-Saharan countries.

It is necessary to consider also the possible adverse impact of investment in construction, such as overbuilding, and an import burden if most of the inputs are not made in the country. Both of these were evident in Malaysia, where a confluence of a period of overbuilding and thus excess supply of buildings on one hand, and imports of construction inputs and foreign borrowing (by firms) on the other hand, leading to an overstretch, is considered to have been a factor in the Asian financial crisis in 1996 (see, for example, Lee and Tham, 2007). These phenomena are worth exploring further.

There is also the question: Is there a solution to the issue of the application of longitudinal data in analyses of construction in the economy, which is often criticised (Wells, 1986; Ofori, 1993), as against that based on time-series data for individual countries? Data over more than 30 years are now available for some countries which have made the transition from developing-country status, such as Brazil, Malaysia, Singapore and South Africa. There are also comprehensive data on transition economies such as China (as a nation, as well as its provinces) (Han and Ofori, 2001). Data on European Union members can also be used to examine further the role of construction in the economy. It is necessary to find out the way in which investment in construction leads or lags the increase in economic activity. This will help in making the argument on the role construction plays in economic growth (and whether construction can be 'the balance wheel of the economy') and national development, a subject which has been of interest for many decades (Colean and Newcomb, 1952).

Chapter 7: Construction Project Economics

Construction project economics is the segment of the field of construction economics which is most widely put into practice on a routine basis, as its principles are applied, consciously or unconsciously, to the numerous projects which are undertaken in all parts of every country. It has the greatest potential to influence the performance of the construction. As discussed in Chapter 2, Ruegg and Marshall (2013) ask and answer 'the question of construction project economics' by explaining the aspirations of the players and stakeholders, and outlining how they are met. Chapter 7 gives a comprehensive account of the state of construction project economics and its historical development, showing how new considerations and techniques have been introduced into it. Chapter 5 shows that it has been evident over the centuries.

Construction project economics can benefit most directly from further development of the application of information and communication technology (ICT) to construction; see, for example, the special issue of *Built Environment Project and Asset Management* (Aibinu et al. (2019) on the application of data analytics and big data in construction, especially Madanayake and Egbu (2019). As highlighted in Chapter 2, aspects of project economics such as forms of procurement, bidding and subcontracting have been studied by mainstream economists. An area which offers scope for exploration is project finance; it has so far been left to researchers in the mainstream; Kumar et al. (2021) provide a good review. This interest by economists also shows that this segment of the subject offers the opportunity for construction economists to undertake more rigorous studies, and to contribute new concepts and ideas to the mainstream. Thus, there is scope for setting up a research interest group linking mainstream economists (industrial economists) interested in construction problems such as bidding, with construction economists.

Chapter 8: Construction Industry Development

The broad discussion of construction industry development in Chapter 8 covers the strategies and programmes of many countries, dealing with many of the relevant, interrelated issues on the subject. Chapter 2 discusses how this topic (of industry development) emerged after Turin's (1969) work, and how it has influenced the development of policy over the years. Chapter 8 gives more examples of national industry policies and strategies, in India, Malaysia and Sri Lanka. Some of these efforts have been successful. For example, Malaysia reports that the Construction Industry Transformation Programme, 2016–20 attained 90 per cent of its targets (CIDB, 2021); in particular, productivity on public projects increased by 60 per cent over the five years as a result of an industrialisation programme. The results have been mixed in some countries, such as the UK. Thus, construction industry reforms merit broader and deeper investigations.

Future research could be on: how to effectively use the strategic role of the construction industry as an argument for appropriate policy and action on a sustainable basis; how to share good practice in industry development internationally, building on the work in *Revaluing Construction* (Barrett, 2007) and Improving the Performance of Construction Industries for Developing Countries (Rwelamila and Abdul-Aziz (2021); and how to mainstream the lessons. It would be useful to explore why it has been found necessary in the UK, in particular, to continue to undertake strategic reviews of the construction industry periodically; and the merits of establishing construction industry development agencies, considering the mixed track record of the existing ones. Another issue is determination and monitoring of industry performance at the macro level. Annual data on key performance indicators (KPIs) in construction are being collected and disseminated in the UK (Davis et al., 2018). CIDB (2017) in Malaysia set targets for the outcomes of its initiatives. The Hong Kong SAR government has identified construction KPIs and the performance targets for the country (Development Bureau and KPMG, 2018). It would be useful to argue for more of such national data-gathering programmes, and to develop benchmarks and prediction models for construction performance in all countries.

Chapter 9: Applications of Mainstream Economic Theories to the Construction Industry: Transaction Costs

From the definitions in Chapter 2, construction economics applies theories and concepts from mainstream economics. The application of transaction economics is an example (see Winch, 1989; Li et al., 2014, 2015). The theory and its development are explored in Chapter 9. The consideration of how it has been studied in construction shows that researchers' views differ on its applicability to construction. The chapter also suggests what the researchers can do: not expecting any theory or concept to be completely applicable to the industry, but being ready to explore combinations of related theories and concepts.

When construction economics researchers attempt to apply mainstream theories and concepts, they could consider some of the following questions: Why is the concept applicable, and sometimes applied, to construction? What has been good and bad practice in such applications? After the study, they could consider: Is the concept of any real explanatory and practical value? If relevant, they can consider: What does it need to be most effective in application to construction? Construction researchers should also consider in their work what the application of the theory or concept to construction has to contribute to the development of the concept. Possible areas where construction economics can develop concepts and tools for the benefit of the mainstream include sustainability performance (Dias et al., 2017) and life cycle costing. As many previous authors have noted (Bon, 1989; Chang, 2015) and as discussed below, construction economics will only become a recognised field when it has developed theories and concepts which are of interest to researchers in other branches of economics.

Chapter 10: Construction Economics and SDGs

The importance of the construction industry making a contribution to the attainment of the Sustainable Development Goals (SDGs) is established in Chapter 10. It is noted that policies and regulatory frameworks that drive the adoption of sustainable construction practices will be required; and construction firms will have to incorporate sustainability into their business strategies and their operations, and in this way translate global needs and desires into business solutions.

The database being built up as countries collect and release the information on their annual progress in meeting the specific targets of the various SDGs will be of great value to construction economics researchers. It will provide the opportunity to further establish the importance of construction in the economy and in long-term national development and improvement in the quality of life. Models of the SDG targets with relevant aspects of construction, in total and in the mix of types, should be built to see how construction underlies, and supports, the attainment of particular targets and goals. On a broader front, the design and implementation of global development agendas such as the 2030 Agenda for Sustainable Development (built around the SDGs) will become established in the mainstream economics literature (Sen, 1999; Sachs, 2015). The relevant aspects of it should be similarly developed in construction economics. For this to happen, there should be a major change in the level of interest of construction economics researchers in such global declarations and agendas such as the SDGs, and the New Urban Agenda (UN, 2017).

Chapter 11: Sustainable Construction and Economics

In Chapter 11, sustainability economics is explored, in particular, its focus on the problem associated with economic growth and attendant pollution, and the challenges posed to sustaining economic growth given the demands on environmental resources on the one hand, and on the other, the finite, and non-renewable nature of natural resources. It is suggested that the concept of hard sustainability is most preferable. The argument of self-interest should persuade construction firms to act; as construction has a negative impact on the environment, there is a threat to the survival of the industry itself.

Sustainable construction should be linked to the attainment of the SDGs. Some relevant questions include: What are the rewards for attaining sustainable targets, from the perspectives of construction companies and clients? On the drivers of sustainability in construction, can companies be left to decide to do the right thing, or do they need to be compelled, for example through legislation, or given specific incentives? What is the possible impact of sustainability on construction? What is the future of sustainability in construction? A new concept of construction economics focused on SDG attainment and sustainability in general could be developed.

Chapter 12: International Construction Data

Information on construction is important for effective action by all the players of the construction industry in all its segments. Chapter 12 presents the nature of construction data, difficulties in measuring them, and their relative availability and reliability across countries. It suggests that researchers need to understand their raw material and its possible shortcomings, and take appropriate action in the collection, manipulation and analysis of data, and interpretation of the results.

The following questions could be considered in subsequent discussions: What is the truth behind the notion that construction data is inherently inaccurate? Is there any country where the data are credible? How can the level of quality of construction data be improved? Will the new digital technologies which are being applied in construction make a difference? Owing to the strategic importance of construction, it is necessary to measure relevant aspects of it. There is a need for agreement on common measurement conventions and assumptions, and guiding principles to be disclosed when releasing the data. It is necessary to measure for benchmarking purposes, for example, in order to reveal inefficiency, excessive costs and poor practice where it exists, and to guide policy formulation and development of strategies and programmes, as well as good practice, and also help to combat and curb corruption and mismanagement.

Chapter 13: Measuring and Comparing Construction Costs in Different Locations

Chapter 13 reviews the issues related to measuring and comparing construction costs between countries and different locations within countries, considering measurement at the levels of projects, types of work and industries. It is noted that many factors complicate attempts to compare costs of construction in different locations, and researchers search for reliable methods that alleviate the complexities.

As construction work is ubiquitous, this question is pertinent: What really explains differences in construction cost between one location and another? The answer to this question can help in determining what can be done to reduce costs, what is good practice, and what is worth emulating. For example, what are the differences in the costs of regulation from one location to another? Some of the other issues which can be discussed further include the possibility, merits and possible disadvantages of an international cost database, to enable comparison among countries in order to provide a basis for policy formulation. Another issue is the merits and weaknesses of using cost as a single, or the main determinant of project or industry performance. Are there other, more appropriate, or complementary yardsticks? Indicators relating to sustainability could also be a useful set of yardsticks.

Chapter 14: New Trends in International Construction

There are two chapters on international construction in this book. The first one, Chapter 14, reviews the main theories, concepts and frameworks of international construction and analyses the historical development and the current and future trends. International construction is not new; historical and religious texts dating from antiquity outline instances of trade and exchange in construction materials, expertise and service delivery. As discussed in Chapter 2, Linder (1994) produced a long history of international construction. The thoughts he expressed have not been followed up. Much work has also been done on the development of a framework for analysis (Ofori, 2003; Low and Jiang, 2004). These are based on theories of international business such as Dunning's eclectic paradigm (Dunning, 2000) and Porter's diamond framework (Porter, 1985), and the resource-based view (Peng, 2001).

A pertinent question is: Can international construction economics offer any lessons to mainstream economics or any other field of knowledge? Additional questions include: Is international construction inevitable, or will it come to an end? There are winners and losers from international construction; can anything be done to optimise its benefits for all stakeholders? There is scope for, and significant synergies in, collaboration on the major infrastructure schemes which are being sponsored by China, the UK and US. In addition to realising the intended physical items of infrastructure in a globally systematic and coordinated manner, there can be joint action among the agencies and the international construction companies in these three countries, and others, to: develop and disseminate good practice, such as on health and safety; safeguard the local community; demonstrate how to deliver large, complex infrastructure projects in challenging operating and physical environments; and develop a legacy to influence practice around the world.

Chapter 15: Trust

Trust is key in construction. It has many dimensions: among construction project participants; between the internal and external stakeholders; and between the industry and society. Lack of trust among project participants leads to conflict, disputes, adversarialism and ineffective teams, whereas society's lack of trust in the construction industry affects the industry's social image. Chapter 15 considers trust among construction project participants and how it is influenced by information asymmetry. On projects, trust minimises information asymmetry and reduces the communication risk. Thus, building trust and creating a reputation for trustworthy behaviour is valuable for every project and company.

The level of information asymmetry on construction projects will be reduced by the further application of ICT, and the practice of transparency on projects and at the national and industry levels. Developing trust should be factored into project planning and execution; and as the project progresses, the level of trust on it could be assessed periodically, and the results used to further build trust. Some questions are: How does trust manifest itself in construction? How is trust built in a construction project? Is trust more important at any particular stage(s) than the rest? What should be done at the project level to ensure that there is trust among project participants? There could be further research on the cost and other negative impacts of lack of trust, as opposed to the value of trust.

Chapter 16: Builders of Cities – Synergy between Labour and the Built Environment

Chapter 16 considers the nexus among labour, the construction industry and urban development. It stresses the importance of research and policies on the connection among the construction workers who build cities and towns, the industry they work in, and the settlements they produce, and suggests that they should be the subject of research and policy frameworks. It argues that the quality and quantity of employment in construction have an impact on urban development and urban poverty alleviation; while, in turn, trends in urban development have an impact on the construction industry and its workers.

As discussed in Chapter 2, the role of construction in generating employment is one of its key features (see also McCutcheon, 2001). That the project is location-specific means this work can be created in all parts of the country, and that plans can be made to bring work opportunities where they are particularly needed. Chapter 16 considers key aspects of construction economics and urban economics. In this context, the labour–construction–urbanisation nexus demonstrates the possibility of developing a theoretical framework for studying labour in urbanisation. Topics which can be studied in future include the exploration of the right kind of employment in various situations; and modelling of the cost of generating employment at the project, company and industry levels. Policy considerations of the costs and benefits of such work generation can then be made.

Chapter 17: Economic Principles of Bidding for Construction Projects

As outlined in Chapter 2, the practice of bidding has been studied for many years by both construction economics researchers and mainstream economists. Chapter 17 discusses the principles and considerations influencing the bidding price of a project and how economic factors influence client and contractor behaviour in the bidding process, and explains the considerations of the client and the bidders in order to determine the price of the item to be built.

A number of questions can be explored. The macro-level impact of the bidding approaches adopted by construction firms should be considered. These bidding practices are considered to be a major source of financial risk on construction projects (WEF and BCG, 2015). It is appropriate to continue studies to determine good practice on the part of clients inviting bids, and bidders. How the winning bid is selected differs from one country to another, and often within the same country. For example, the engineer's estimate and how it serves as a benchmark differs from one country to the next, and this influences the selection of the winning bid. How can the true cost of the item be determined? The answer will enable the true performance of the construction industry to be established. Another question is: Does the bidding approach in construction offer any lessons to other sectors? Finally, what form will bidding in construction take in the medium and long term?

Chapter 18: Procurement and Delivery Management

Chapter 18 provides a comprehensive review of good practice in procurement, prerequisites for success, and the risk factors. It is noted that a lack of governance and poor procurement and delivery management practices are major contributory factors to failure on many projects. The chapter covers the development and application of the international procurement standards, and presents a case study of a successful delivery of a project in South Africa which was procured in a unique manner. Opportunities should be taken to share widely the main elements and prerequisites of good practice, such as on Heathrow Terminal 5 and The Shard in the UK, and the two universities in South Africa described in Chapter 18 and in Watermeyer and Philips (2020). Leaving a legacy should be a part of project management, and the economics of such a project objective could be explored.

Topics which could be studied further include ensuring value for money in changing operating environments. The lowest-price-based awarding system which is commonly adopted in procurement is blamed for the competition which results in problems in construction such as financial stress and bankruptcies among companies. There is a trend of movement away from price as the determining factor. An example is Singapore's Price Quality Method (BCA, 2018), which translates the qualitative attributes (productivity – including constructability score, technology adoption, workforce development; and quality – comprising the firm's performance in past or ongoing projects in areas such as timeliness, safety and quality, and current work methods and resources) into quantitative scores which, when combined with the price scores, enables the firm providing the best offer to be selected for award.

Authors from Turin (1969) onward have urged governments to use their strong bargaining position to secure change in the construction industry including corporate development and inclusion of local content, linked to technology transfer (see Asian Development Bank, 2018). Ofori (2016) suggests that the project should be awarded to the company with greatest potential to contribute to the development of the industry. Where innovative methods are adopted, how can they be made effective through the institution of incentives (and penalties), and assessed, and monitored, to realise the desired change? In construction, the procurement process is known to be particularly prone to corruption (Stansbury and Stansbury, 2018), although following an international programme of procurement review, regulatory agencies on public sector procurement have been set up in all countries. This issue of corruption should be continually explored, and the most effective approach to solving it found. Elements of good practice which are worth studying include transparency in the process, and an audit of both cost and other technical performance criteria on completion.

Chapter 19: Economics of Housing Policy and Construction

The shortage of housing and lack of affordability are a reality in countries at all stages of development. Chapter 19 notes that the housing policies of most nations attempt to overcome the problems of severe housing shortages and lack of affordability created by prevailing market conditions: utility-maximising households and profit-maximising firms which create market inefficiencies leading to decreased affordability for a significant proportion of society. The chapter notes that the performance of the construction industry has an influence on the cost and affordability of housing; improving performance on the supply side at the level of construction projects will make housing generally more affordable, but this is not usually considered in policy responses.

Singapore is an example of a country where the policy approach of developing the construction industry to increase housing delivery has been adopted with success. The government's housing programme, launched in 1960, saw the attainment of the 'slum-free milestone' in 1985 (Centre for Liveable Cities, 2015). Currently, 82 per cent of the population live in public housing (which is of relatively high quality); and over 90 per cent of them own their homes (Department of Statistics, 2019). An active contractor and technology development programme, and assistance with acquisition of materials, was a major segment of the housing programme (Wong and Yeh, 1985; Centre for Liveable Studies, 2015).

There is a need for research on many features of housing, and how construction can respond, for example involving the occupiers in design considerations, cost minimisation, post-occupancy evaluation and determination of costs-in-use of built items. Other issues include policy considerations, and industry considerations including delivery and project issues. The approach to housing could embrace the multi-parameter paradigm of cost minimisation, value enhancement, affordability, sustainability, resilience and future-proofing. It is necessary to balance a number of factors, such as taking up land and using materials versus providing housing and sanitation, and enhancing the quality of life. Other questions include: Is there a relationship between housing and economic development? Are there good practices in aspects of housing such as policy formulation, citizen and community participation, pricing and support to end-purchasers which can be shared? Finally, housing economics has emerged as a subfield of mainstream economics. It is pertinent to study how this occurred, and what construction economics can learn from it, after also investigating how the theory has held up in practice, in helping to improve quantity and quality in housing. Finally, what can construction economics contribute to housing economics?

Chapter 20: Stakeholder Management

The concept of the stake and the need to understand it in construction, and to manage the stakeholders in a systematic manner, is explained in Chapter 20. Stakeholder management now forms part of organisational developments, negotiations and dispute resolution. Stakeholder interests can align partially or fully, or completely differ. In the latter case, there is a potential for conflicts in the form of opposition, hostility, sabotage, antagonism, and so on (Awakul and Ogunlana, 2002; Kishore Mahato and Ogunlana, 2011). Stakeholder management is now recognised as a major element of construction projects. Chapter 20 notes that areas of future research on stakeholder management include assessments of the time spent on it, the cost of its implementation, its tangible and intangible benefits, and the role of technology in its application.

It is necessary to develop techniques for predicting, determining, costing and monitoring the involvement of stakeholders in projects. Completing the project to the satisfaction of stakeholders should be a project performance indicator. The possible questions include: Should cost and other considerations on a project be defined more widely, from the perspectives of a range of stakeholders? In particular, should the community's needs and interests be determined and considered in project valuation, assessment and monitoring? Next, how can stakeholder management be effectively applied in general in construction? Can construction teach other sectors anything on stakeholder management? Finally, can the study of stakeholder management in construction contribute to the mainstream?

Chapter 21: The Global Construction Market

Chapter 21 is the second chapter on international construction in this book. It analyses the global construction market. It considers market sectors and market segments; the geographic dispersal of major construction work; the historical development of the market and its key players; and trends including the phenomenon of increasing internationalisation, emerging markets and market forces; and the changes in the service offers including the integration of architecture, engineering and construction services.

Construction economics research should seek to provide more guidance to international construction businesses beyond entry-mode decision-making. The unit of analysis could be a global or regional view of the market transcending national boundaries, to enable more accurate analysis. Local content requirements should be clear, systematic, and beneficial to both local and foreign firms. There have been significant changes in the dominance of international contracting firms from various countries. Chinese players are currently among the leaders, halting the trend of the historical development, with firms from the UK, US and Europe as leading players. Are there any implications of this, and what path will it take in future? How can local companies benefit from the presence of the larger international players? Technology transfer is a knotty, long-standing problem concerning mainly the potential competition from companies developed; and the cost of transferring technology, bringing into play the need for appropriate incentives. Research can be undertaken on finding synergies in the collaboration between local and foreign partners for mutual benefit and industry development.

It was predicted by Bennett (1991) that, by the turn of the millenium, the construction market would be highly concentrated; only a few international firms would be operating as 'systems integrators' throughout the world. How likely is this development? Other broader questions include: Will the current tendency towards deglobalisation become an established phenomenon in construction? What will be the impact on international construction companies, and on domestic ones, and what would be the appropriate policy responses?

Chapter 22: Corruption and its Economic Impact

Studies on the occurrence of corruption and its impact, especially on the poor, are well known. Corruption has been found in countries at all levels of development, but it appears to be more pernicious in developing countries (Kenny, 2007). Construction industry has been identified as the part of the economy which is most vulnerable to corruption (Transparency International, 2005). Chapter 22 considers the impact of corruption in construction on the economic growth of a nation. The chapter discusses how measures such as audits, procedural compliance and transparency have helped to mitigate irregularities.

Corruption is a bane on the construction industry and affects its social image and its desirability as a field of work in which talented citizens wish to pursue their careers. As discussed in Chapter 22, although some positive elements of corruption are highlighted, the consensus is that corruption usually has negative impacts. It has also been shown that it is possible to reduce, if not eradicate, corruption, and good practice is available; introducing transparency programmes is one of them. There is a cost of preventing corruption. In construction, there is a need for a level playing field, removing the asymmetry in information. The economic impact and overall value of the approaches adopted to address corruption, some of which are outlined in Chapter 22 and by Stansbury and Stansbury (2018), could be studied.

Chapter 23: Information and Communications Technology and Construction Economics

Chapter 23 discusses the economic considerations in the procurement and deployment of construction informatics applications. It notes the conservative culture of the construction industry in the uptake of new technologies, and discusses the drivers that are leading to a change in the level of technology adoption. It introduces a range of current and emerging technologies and highlights the potential of each of them to influence the economics of construction projects in the future. Economic considerations in the procurement and deployment of the digital technologies and key performance indicators are also discussed.

The modern digital technologies offer possibilities for construction economists to undertake analyses to ensure efficiency and effectiveness in attaining the performance parameters on projects. They also offer possibilities for post-occupancy evaluation and monitoring using occupants' direct input; for complex analyses such as cumulative assessment of environmental and social impacts of projects; and for input–output analyses. They can help to introduce features of prediction and avoidance in the management of the project towards attaining the performance parameters, instead of controlling the parameters in real time. The information technologies can help to establish a dynamic international database on construction practice, processes and procedures, as discussed above in relation to Chapter 12; this will enable the effective sharing, adaptation and application of good practice from one country to another.

Chapter 24: Future of Construction Economics

This final chapter provides a summary of the book and considers some issues which can be further studied in order to contribute to the development of construction economics. Some possible pathways towards a better future for construction economics are discussed. It is stressed that progress will require a systematic and concerted effort, that certain prerequisites should be in place, and that overall leadership, and action by all researchers in the field will be required.

CONSTRUCTION ECONOMICS IN FUTURE

There has been progress in the efforts to develop construction economics over the years. The extent and pace should be increased. There are fundamental gaps to fill, questions to answer, and debates to continue to reach possible consensus. Some of the issues are now discussed in this section.

Principles of Construction Economics

Mankiw (2015), in his text *Principles of Economics*, presents ten principles of economics. In this book, in Chapter 4, it is suggested that analyses of aspects of the construction industry explicitly incorporate its characteristics. In Chapter 3, these six desirable principles of the construction industry are highlighted: one that is competitive while comprising firms which are conscious of their social responsibility; one that is productive and embraces innovation; one that produces a quality output and seeks to provide optimum value to clients and society; one that is efficient and whose output is efficient; one that employs a workforce that is professional in its attitude, behaviour and skills; and one with an excellent reputation, confidence and pride in itself, and is valued and trusted by society. What would be the ten main principles of construction economics? An attempt is now made to outline some principles, in starting a debate among researchers:

1. Construction has characteristics which distinguish it from other sectors. They include the large size of the industry; the project-based nature of the industry; high expense and long period of gestation of each project; the location-specificity of projects; the large number of players and stakeholders; the role of government as a client; and the extensiveness of regulations covering it. The uniqueness of construction should be a driver of the search for improvement, not an excuse for poor performance or lack of progress.
2. Construction stimulates activities in many sectors of the economy; it builds the foundation for national development; it generates employment; it helps to improve the quality of life.
3. The balance wheel of the economy should be applied with care. The effectiveness of stimulus efforts, where to intervene and when, should be related to possible dangers of excessive investment such as trade imbalance and overbuilding.
4. Construction activity is project-based, discontinuous and risky. 'Clients get the industry they deserve.' Risk-averse clients adopt practices which elicit short-term orientation in firms (prioritising flexibility in the light of workload instability), and lack of investment in human resources, technology and innovation, leading to poor performance.
5. Performance parameters on construction projects are numerous and need to be balanced in trade-offs. It is necessary to broaden the parameters of the construction project and study the necessary trade-offs among them, such as between cost and value; between productivity and health and safety; among these factors and cost and duration; and so on.
6. Construction has an extensive value chain and a complex operating environment with many influencing factors, including a wide range of regulation. Modelling construction is a challenge.
7. Construction has many stakeholders, and they have different, often opposing interests, objectives and aspirations on the same project. Win–win procurement and relationships are viable; it is possible to structure the industry for good performance and productivity.
8. Construction firms have increasingly broadening operational agendas. The changing portfolios of major construction companies, with the importance of ownership, operation and management of assets around the world, provides opportunity to structure the industry anew.
9. Construction has a poor social image in many countries. Action is needed to provide decent jobs, and address corruption and mismanagement in construction which have a negative economic impact at many levels.
10. Construction is a local industry with an international dimension. Construction economists can contribute to the global programmes such as the SDGs and the Urban Agenda by providing the economic arguments for aspects of them.

This tentative set of principles should be further developed and a consensus on a set agreed, in the process of developing construction economics.

Construction Economics and Practice

Construction economics research should aim to contribute to government policy, company strategies, and project programmes and practices. It is pertinent to consider what lies at the foundation of the field, by taking into account the needs and wants of the main stakeholders.

This is now done in outline form; it extends the consideration by Ruegg and Marshall (2013) discussed in Chapter 2:

1. What the government wants: the facility; induced economic growth; long-term national development; improving the quality of life.
2. What the client wants: the item meeting the various project parameters, from a life-cycle perspective, providing scope for periodical post-occupancy evaluation to improve performance and future design.
3. What the industry, its companies and its personnel want: work opportunities; supportive procurement and contract and project administration; conducive operating environment; and progressive developmental possibilities.
4. What the users want: affordable price or rent for the facility; consideration of impact on efficiency of the users' operations.
5. What the people and community want: the facility; job opportunities; induced economic activity in the locality; and opportunity to make a contribution to the planning and design processes.

Extending the Focus of Construction Project Economics

Construction project economics research should move beyond its focus on cost, time and quality, and widen the project performance parameters it considers, even beyond health and safety and sustainability. Some of the additional parameters are now used to illustrate these issues:

1. Consider value. There is much debate on it now, and laws have been passed in some countries, such as the UK, to foster it. However, there are a number of questions: Whose value? Who pays for value? How is the process to realise value to be operationalised? For example, the UK Construction Leadership Council (2018) in its report, *Procure for Value*, stresses life cycle assessment and post-occupancy evaluation. Box 24.1 presents a project on value by the Construction Innovation Hub (2020).
2. Address affordability by clients, end-purchasers and users. Affordability should be a major consideration in construction project economics in research and practice. Affordability will involve analyses of the value of the real estate, and help broaden the scope of construction economics.
3. Job creation and training opportunities from the project should also be subjects of analysis. This will enable construction to live up to the potential ascribed to it in the literature.
4. Input–output analyses of projects to determine the wider impact of each project in the economy should also be a research topic. The application of the technique could move beyond the current macro level to the individual project, with norms being developed over time to aid assessment.
5. Future-proofing of the built item is also a subject of possible research. This goes beyond life-cycle assessment of the item to embrace resilience, and both costs and revenue or benefits.
6. A legacy from the project, to influence future projects by making them better in terms of performance and the resulting product, and to help develop the construction industry.

BOX 24.1 VALUE: ITS DEFINITION AND ATTAINMENT

Construction Innovation Hub (2020), in defining 'value', suggests that 'Value does not just mean cost, nor is it something that exists purely in the construction phase of projects. Value needs to consider a broader range of metrics beyond financial. It must also consider wider social, economic and environmental factors – and consider them across the full investment lifecycle' (p. 1). Delivering this definition of value needs a new approach that better reflects broad, strategic policy objectives, responds to local ambition and meets the needs of users, owners and operators. This approach would also support informed decision-making throughout the product's life cycle and give industry the opportunity to innovate to deliver value on projects, leading to overall improvement in the industry. The Value Toolkit of Construction Innovation Hub (2020) provides this new approach.

The value definition framework in the toolkit comprises four main themes:*

Natural Capital – Air, Climate, Water, Land, Resource Use, and Biodiversity
Human Capital – Employment, Skills and Knowledge, Health, and Experience
Social Capital – Influence and Consultation, Equality and Diversity, Networks, and Connections
Produced Capital – Life Cycle Cost, Return, Production, and Resilience.

The Toolkit contains a suite of tools in four linked modules which will (Constructing Excellence, 2021):

- Support policy-makers, clients and advisors to define the unique value profile for a given project and create value indices to enable informed decisions to be made.
- Help clients and their advisers to select a delivery model and commercial strategy, and industry to develop business models, that best meet the value drivers of the project.
- Enable clients to make procurement decisions based on the value drivers of the project and industry to shape their offers accordingly.
- Continuously forecast and measure value performance throughout delivery and operation, helping clients and industry to maximise value on each project and using performance data to help policy makers to make better decisions on future projects.

Note: * https://constructioninnovationhub.org.uk/wp-content/uploads/2020/12/Value-Toolkit_Value-Definition-Framework_v1.0.pdf.

A Global Perspective for the Field

Construction economics researchers should keep up with developments at the global level. For example, Box 24.2 presents an example from the *New Urban Agenda*. The agenda frames the challenges for built environment policy-makers, administrators, researchers and practitioners into the future. It highlights several possible topics for research in construction economics.

BOX 24.2 A GLOBAL AGENDA FOR CONSTRUCTION ECONOMICS

In the *New Urban Agenda*, leaders have committed their nations to actions including (UN, 2017):

- Provide basic services for all citizens including access to housing, safe drinking water and sanitation, nutritious food, healthcare and family planning, education, culture and communication technologies.
- Ensure that all citizens have access to equal opportunities and face no discrimination, as everyone has the right to benefit from what their cities offer.
- Promote measures that support cleaner cities (tackling air pollution, committing to increased use of renewable energy, better and greener public transport, and sustainably managed natural resources.
- Strengthen resilience in cities to reduce the risk and the impact of disasters (with measures including better urban planning, quality infrastructure and improving local responses.
- Take action to address climate change by reducing their greenhouse gas emissions.
- Improve connectivity and support innovative and green initiatives.
- Promote safe, accessible and green public spaces (through sustainable urban design to ensure the liveability and prosperity of the city).

Aspects of the Nature of Construction Economics

Finally, some aspects of construction economics which are relevant to its development are now considered.

Efforts should be made to deepen construction company economics. For example, the importance of financial management (with main contractors using smaller firms to manage their financing) as a major determinant of success on projects and of companies, rather than the pursuit of technical efficiency and innovation, is also a key subject to study. Efforts, such as the passing of laws, to eliminate the practice of main contractors using lower-tier firms (their subcontractors) as their source of funds by delaying payments to them, have failed (Construction Leadership Council, 2018).

Fragmentation in tasks and in firm structure should also be of greater research interest. The level of concentration in construction, the pyramidal structure of the industry which is evident everywhere and is considered to be a cause of low productivity and innovation, also need to be further studied. For example, in Hong Kong SAR in 2016, of the 24,197 establishments in the construction industry, over 75 per cent had a gross value of construction work activity of less than HK$5 million (Development Bureau and KPMG, 2020). Only 321 of them were classified as main contractors, and they accounted for about half of the total gross value of construction works. This indicates the high degree of concentration in the industry, with market control within a few organisations. Are there particular benefits of this particular feature of construction?

There is a need for effort to compile the full body of works of construction economics. There are some hidden gems. For example, although Turin (1969) noted that national development plans seldom cover construction, Singapore's government had appointed a committee of inquiry into the capacity of the nation's construction industry and its ability to deliver on the newly independent country's ambitious industrial building, housing and school building programme, which produced useful recommendations and proposals for a construction industry development programme (Commission of Inquiry, 1961). Another example of a neglected volume is the report by a team of experts on the construction industry in Tanzania in 1977 (Ministry of Works, 1977). In construction project economics, the work of P.A. Stone, which is outlined in Chapter 2, is also worth a revisit.

The field of construction economics tends to lack tenacity; interest in some topics rises and wanes. Some important concepts such as capacity, capability and productivity have yet to be established; their importance is being stressed now in the infrastructure plans and policies which are being released, and examples are those for Ghana (National Development Planning Commission, 2018), New Zealand (National Infrastructure Unit, 2015) and the UK (HM Government, 2020a). The pioneering work of Hillebrandt (1975) on these subjects should be built upon. Post-occupancy evaluation and life-cycle costing went off the radar for many decades, despite the potential usefulness of the former in providing feedback into future projects. The decline in interest in the latter represents a missed opportunity for construction economics to lead in the considerations of sustainability and the circular economy.

Establishing Construction Economics as a Bona Fide Field: A Process of Development

Construction economics has much potential to contribute to mainstream economics because of the complexity of the industry it deals with and the need for sophisticated tools and techniques to analyse it. However, construction economics is not yet a recognised subdiscipline of economics. Chang (2015) notes that, unlike the growth of many economics subfields such as organisational economics, experimental economics and behavioural economics, which emerged around the same time, the development of construction economics appears slow. Chang (2015) suggests that construction economics should learn from these fields; he believes that they succeeded because: they can expand the boundaries of economics by addressing new issues through the development of an indigenous theory applicable to other economic sectors; the new theory has a solid theoretical foundation connected to mainstream economics; and the new theory is supported by solid empirical evidence.

Who is to develop construction economics? Construction economics needs a strong, dedicated community of researchers if it is to develop. Some of the authors of the chapters in this book were not certain that reference to Working Commission 55 of the International Council for Reseach and Innovation in Building and Construction (CIB W55) was useful as they believed that few people knew about it. There was also some push-back against the reference to the profession of quantity surveying as a player in the development of construction economics. A number of international groupings of construction economists are highlighted in this book. So, who will lead the formation of the community of researchers who will take up the task of developing construction economics?

What path should construction economics take? Bon (1999) suggests that building economics needs to shift its focus from investment decisions to problems of managing building portfolios. Thus, the future of building economics lies in fields such as corporate real estate and facilities management which inform the whole building process, and would enable buildings to be designed and constructed with the whole life of the building process in mind. Also, as mentioned in Chapter 2, various economists have written on such construction-related topics as infrastructure and development, procurement, bidding and subcontracting. Chang (2015) believes the issues of interest to construction economists remain a gold mine which economists have been mining for more evidence in support of existing theories.

The layers of industrial economics fit the nature of construction economics. The *Journal of Industrial Economics*,[5] which was first published in 1952, describes the areas of industrial economics as including: organisation of industry; applied oligopoly theory; product differentiation and technical change; theory of the firm and internal organisation; regulation, monopoly, merger and technology policy. Should construction economics aim to become recognised as a distinct subset of industrial economics? Finally, construction economics has been 'twinned' with construction management. This might be convenient, but it might not be beneficial, as the field needs its own space and path to grow in order to establish its legitimacy. Should construction economics set a path for development on its own, or is there scope for developing construction management and economics as a distinct area of knowledge?

CONCLUSION

All fields concerning the built environment, including construction economics, should be highly regarded because constructed facilities form an important segment of national assets. They are an essential ingredient of modern life, in terms of production, social well-being and enhancement of quality of life. Owing to the peculiar factors which apply in the construction industry, construction economics offers a rich field for the development of new theories and concepts, some of which may become crucial for other fields of knowledge and scientific investigation.

Construction economics is a field in its own right, as knowledge and understanding of the construction industry, process and practice is required. It goes beyond being basic mainstream economics applied to construction problems. Construction economics will eventually be recognised as an important field of economics. Construction economics researchers should contribute to the efforts to realise this.

NOTES

1. https://www.ceecorg.eu/?avada_portfolio=european-code-of-measurement.
2. https://www.ceecorg.eu/?avada_portfolio=european-code-of-measurement.
3. https://ipmsc.org/.
4. https://ipmsc.org/standards/.
5. https://onlinelibrary.wiley.com/page/journal/14676451/homepage/productinformation.html.

REFERENCES

Aibinu, A.A., Koch, F. and Ng, S.T. (2019) Data analytics and big data in construction project and asset management. *Built Environment Project and Asset Management*, 9(4), 474–475. https://doi.org/10.1108/BEPAM-09-2019-139.

Asian Development Bank (2018) Procurement review: guidance note on procurement. Manila.

Atkinson, J. (1984) Manpower strategies for flexible organizations. *Personnel Management*, August, pp. 28–31.

Awakul, P. and Ogunlana, S.O. (2002) The effect of attitudinal differences on interface conflict on large construction projects: the case of the Pak Mun Dam project. *Environmental Impact Assessment Review*, 22(4), 311–335.

Barrett, P. (ed.) (2007) *Revaluing Construction*. Wiley-Blackwell, Chichester.

Behling, F. and Harvey, M. (2015) The evolution of false self-employment in the British construction industry: a neo-Polanyian account of labour market formation. *Work, Employment and Society*, 29(6), 969–998. DOI: https://doi.org/10.1177/0950017014559960.

Bennett, J. (1991) *International Construction Project Management: General Theory and Practice*. Butterworth-Heineman, Oxford.

Bon, R. (1989) *Building as an Economic Process: An Introduction to Building Economics*. Prentice Hall, Englewood Cliffs, NJ.

Bon, R. (1999) *The future of building economics: a note*. http://www.reading.ac.uk/kqFINCH/wkc1/source/bon/Future.pdf.

Building and Construction Authority (BCA) (2018) Price Quality Method Framework – effective for tenders called on and after 31st January 2018. Singapore. https://www1.bca.gov.sg/docs/default-source/docs-corp-procurement/pqm_framework.pdf?sfvrsn=bf44a2a5_4.

Centre for Liveable Cities (2015) *Built by Singapore: from slums to a sustainable built environment*. Singapore.

Chang, C.-Y. (2015) A festschrift for Graham Ive. *Construction Management and Economics*, 33(2), 9 1–105. DOI: 10.1080/01446193.2015.1039044.

CIDB (2017) Construction Industry Transformation Programme. Kuala Lumpur.

CIDB (2021) Media statement – construction productivity increased by 60% under the CITP. https://www.cidb.gov.my/sites/default/files/2021-03/Announcement%20of%20CITP%20Achievements.pdf.

Colean, M.L. and Newcomb, R. (1952) *Stabilizing Construction: The Record and the Potential*. McGraw-Hill, New York.

Commission of Inquiry into Construction Capacity (1961) Final Report. Government of Singapore, Singapore.

Constructing Excellence (2021) Construction Innovation Hub Value Toolkit – shift towards a value-based decision model. https://constructingexcellence.org.uk/construction-innovation-hub-value-toolkit-shift-towards-a-value-based-decision-model/#:~:text=Constructing%20Excellence%20is%20delighted%20to%20be%20working%20with,industry%20to%20respond%20with%20innovative%2C%20high%20value%20solutions.

Construction Innovation Hub (2020) *An introduction to the Value Toolkit*. London.

Construction Leadership Council (2018) *Procure for Value*. London.

Dainty, A., Green, S., and Bagilhole, B. (2007) People and culture in construction: contexts and challenges. In A. Dainty, S. Green and B. Bagilhole (eds), *People and Culture in Construction*. Routledge, Abingdon, pp. 3–25. DOI: https://doi.org/10.4324/9780203640913.

Davis, R., Wilén, A., Bryer, L., Ward, D., Pottier, F., et al. (2018) *UK Industry Performance Report 2018: Based on the UK Construction Industry Key Performance Indicators*. Glenigan, CITB, Constructing Excellence, Department for Business, Innovation and Skills, London.

Debrah, Y.A. and Ofori, G. (2001) Subcontracting, foreign workers and job safety in the Singapore construction industry. Asia Pacific Business Review, 8(1), 145–166. DOI: 10.1080/713999129.

Department of Statistics (2019) *Yearbook of Statistics Singapore, 2019*. Singapore.

Development Bureau of Government of Hong Kong SAR and KPMG (2018) *Construction 2.0: Time to Change*. Development Bureau, Hong Kong SAR. https://www.psgo.gov.hk/assets/pdf/Construction-2-0-en.pdf.

Dias, W.P.S., Chandratilake, S.R. and Ofori, G. (2017) Dependencies among environmental performance indicators for buildings and their implications. *Building and Environment*, 123, 101–108. https://doi.org/10.1016/j.buildenv.2017.06.045.

Dodge Data and Analytics (2018) The key performance indicators of construction. Autodesk. http://abcdblog.fr/wp-content/uploads/2019/12/kpis-of-construction-report.pdf.

Dunning, J.H. (2000) The eclectic paradigm as an envelope for economic and business theories of MNE activity. *International Business Review*, 9, 163–190.

The Economist (2017) Efficiency eludes the construction industry. 17 August, London.

Farmer, M. (2016) *Modernise or Die: Time to Decide the Industry's Future – The Farmer Review of the UK Construction Labour Model*. Construction Leadership Council, London.

Green, A. (2016) Low skill traps in sectors and geographies: underlying factors and means of escape. Institute for Employment Research, University of Warwick. www.gov.uk/government/uploads/system/uploads/attachment_data/file/593923/LowSkillsTraps-_final.pdf.

Hackitt, J. (2018) *Building a Safer Future – Independent Review of Building Regulations and Fire Safety: Final Report*. HM Government, London.

Han, S.S. and Ofori, G. (2001) Construction industry in China's regional economy, 1990 to 1998. *Construction Management and Economics*, 19(2), 189–205. DOI: 10.1080/01446190010010003.

Hillebrandt, P.M. (1975) The capacity of the construction industry. In Turin, D.A. (ed.), *Aspects of the Economics of Construction*. George Godwin, London, pp. 225–257.

Hillebrandt, P.M. (2000) *Economic Theory and the Construction Industry*, 3rd edition. Macmillan, Basingstoke.

HM Government (2016) *Government Construction Strategy 2016–20*. London.

HM Government (2017) *Industrial Strategy: Building a Britain Fit for the Future*. London.

HM Government (2018) *Industrial Strategy: Construction Sector Deal*. London.

HM Government (2020a) *National Infrastructure Strategy: Fairer, Faster, Greener*. London. https://assets.publishing.service.gov.uk/government/uploads/system/uploads/attachment_data/file/938539/NIS_Report_Web_Accessible.pdf.

HM Government (2020b) *The Construction Playbook: Government Guidance on Sourcing and Contracting Public Works Projects and Programmes*. London.

International Construction Measurement Standards Coalition (ICMSC) (2019) *ICMS: Global Consistency in Presenting Construction and Other Life Cycle Costs*, 2nd edition.

Keep, E. (2016). Improving skills utilisation in the UK – some reflections on what, who and how? Research Paper 124, Centre on Skills, Knowledge and Organisational Performance (SKOPE), University of Oxford, August. http://www.skope.ox.ac.uk/?person=improving-skills-utilisation-in-the-uk-some-reflections-on-what-who-and-how.

Kenny, C. (2007) Construction, corruption, and developing countries. Policy Research Working Paper No. 4271, World Bank, Washington, DC, https://openknowledge.worldbank.org/handle/10986/7451.

Killip, G. (2020). A reform agenda for UK construction education and practice. *Buildings and Cities*, 1(1), 525–537. DOI: http://doi.org/10.5334/bc.43.

Kishor Mahato, B. and Ogunlana, S.O. (2011) Conflict dynamics in a dam construction project: a case study. *Built Environment Project and Asset Management*, 1(2), 176–194. https://doi.org/10.1108/20441241111180424.

Kumar, A., Srivastava, V. and Tabash, M.I. (2021) Infrastructure project finance: a systematic literature review and directions for future research. *Qualitative Research in Financial Markets*. https://doi.org/10.1108/QRFM-07-2020-0130.

Lee, P.P. and Tham, S.Y. (2007) Malaysia ten years after the Asian Financial Crisis. *Asian Survey*, 47(6), 915–929. DOI: 10.1525/as.2007.47.6.915.

Li, H., Arditi, D. and Wang, Z. (2014) Transaction costs incurred by construction owners. *Engineering, Construction and Architectural Management*, 21(4), 444–458. https://doi.org/10.1108/ECAM-07-2013-0064.

Li, H., Arditi, D. and Wang, Z. (2015) Determinants of transaction costs in construction projects. *Journal of Civil Engineering and Management*, 21(5), 548–558. DOI: 10.3846/13923730.2014.897973.

Linder, M. (1994) *Projecting Capitalism: A History of the Internationalization of the Construction Industry* (Contributions in Economics and Economic History). Greenwood, Westport, CT.

Low, S.P. and Jiang, H. (2004) Estimation of international construction performance: analysis at the country level. *Construction Management and Economics*, 22(3), 277–289. DOI: 10.1080/0144619032000089607.

Madanayake, U.H. and Egbu, C. (2019) Critical analysis for big data studies in construction: significant gaps in knowledge. *Built Environment Project and Asset Management*, 9(4), 530–547. https://doi.org/10.1108/BEPAM-04-2018-0074.

Mankiw, N.G. (2015) *Principles of Economics,* 7th edition. Cengage Learning, Stamford, CT.
McCutcheon (2001) Employment generation in public works: recent South African experience. *Construction Management and Economics*, 19(3), 275–284. DOI: 10.1080/01446190010020381.
Ministry of Works (1977) *Local Construction Industry Report*. Dar es Salaam.
Morrell, P. (2020) *Collaboration for Change: The Edge Commission Report on the Future of Professionalism*, 2nd edition. Edge Books, London.
National Development Planning Commission (2018) *National Infrastructure Development Plan, 2018–2047 – Outline*. Accra. https://s3-us-west-2.amazonaws.com/new-ndpc-static1/CACHES/PUBLICATIONS/2017/10/24/Presentation1.pdf.
National Infrastructure Unit (2015) *The Thirty Year New Zealand Infrastructure Plan, 2015*. New Zealand Government, Wellington. https://www.treasury.govt.nz/sites/default/files/2018-03/nip-aug15.pdf.
National Union of Construction Economists (UNTEC) (2020) *Construction Economics: Several Skills, One Profession*. Paris.
Office for National Statistics (2021) Construction statistics, Great Britain: 2019. London. file:///C:/Users/user.DESKTOP-13V37T9/Downloads/Construction%20statistics,%20Great%20Britain%202019.pdf.
Ofori, G. (1993) Research on construction industry development at the crossroads. *Construction Management and Economics*, 11(3), 175–185. DOI: 10.1080/01446199300000017.
Ofori, G. (2003) Frameworks for analysing international construction. *Construction Management and Economics*, 21(4), 379–391. DOI: 10.1080/0144619032000049746.
Ofori, G. (2016) Construction in developing countries: current imperatives and potential. In Kähkönen, K. and Keinänen, M. (eds), *Proceedings of the CIB World Building Congress 2016, Tampere University of Technology*, Vol. 1, 39–52. Tampere, Finland, 30 May to 3 June.
Peng, M.W. (2001) The resource-based view and international business. *Journal of Management*, 27, 803–829.
Porter, M.E. (1985) *Competitive Advantage*. Free Press, New York.
Ruegg, R.T. and Marshall, H.E. (2013) *Building Economics: Theory and Practice*. Springer Science, New York.
Rwelamila, P.D. and Abdul-Aziz, A.R. (eds) (2021) Improving the Performance of Construction Industries for Developing Countries*:* Programmes, *Initiatives, Achievements* and *Challenges*. Routledge, Abingdon.
Sachs, J.D. (2015) *The Age of Sustainable Development*. Columbia University Press, New York.
Sen, A. (1999) *Development as Freedom*. Oxford University Press, Oxford.
Singapore Contractors Association (2017) *Construction productivity in Singapore: effective measurement to facilitate improvement*. Singapore.
Stansbury, C. and Stansbury, N. (2018) Preventing corruption in the construction sector. In Egbu, C. and Ofori, G. (eds), *Proceedings of the International Conference on Professionalism and Ethics in Construction*. London, 21–22 November 8, pp. 30–38.
Transparency International (2005) *Global Corruption Report 2005 – Special Focus: Corruption in Construction and Post-conflict Reconstruction*. Berlin.
Turin, D.A. (1969) The construction industry: its economic significance and its role in development. University College Environmental Research Group (UCERG), London.
United Nations (UN) (2017) *The New Urban Agenda*. New York.
Watermeyer, R. and Philips, S. (2020) Public infrastructure delivery and construction sector dynamism in the South African economy. National Planning Commission and GIZ, Pretoria.
Wells, J. (1986) *The Construction Industry in Developing Countries: Alternative Strategies for Development*. Croom Helm, London.
Winch, G. (1989) The construction firm and the construction project: a transaction cost approach. *Construction Management and Economics*, 7(4), 331–345. DOI: 10.1080/01446198900000032.
Wong, A.K. and Yeh, S.H.K. (1985) *Housing a* Nation*: 25 years of Public Housing in* Singapore. Maruzen Asia, Singapore.
World Economic Forum (WEF) and Boston Consulting Group (BCG) (2015) *Shaping the Future of Construction*. Geneva. https://www.weforum.org/global-challenges/projects/future-of-construction.
World Economic Forum (WEF) and Boston Consulting Group (BCG) (2016) *Shaping the Future of Construction: A Breakthrough in Mindset and Technology*. World Economic Forum, Geneva.
Wostenholme, A. (2016) Comment from Andrew Wolstenholme. Construction Leadership Council, 17 October. https://www.constructionleadershipcouncil.co.uk/news/narrative-andrew-wolstenholme/.

Index

Printed and bound by CPI Group (UK) Ltd, Croydon, CR0 4YY
16/04/2025
14650390-0002